CRITICAL ISSUES IN WATER AND WASTEWATER TREATMENT

Proceedings of the 1994 National Conference on Environmental Engineering

Sponsored by the Environmental Engineering Division of the American Society of Civil Engineers

Boulder, Colorado
July 11-13, 1994

Edited by Joseph N. Ryan and Marc Edwards

Published by the
American Society of Civil Engineers
345 East 47th Street
New York, New York 10017-2398

ABSTRACT

This proceedings, *Critical Issues in Water and Wastewater*, contains short versions of most of the 114 papers presented at the 1994 Specialty Conference on Environmental Engineering held in Boulder, Colorado on July 11 to 13, 1994. These papers are organized into 23 distinct sessions that focus primarily on water treatment, water distribution, and wastewater treatment. Some of the topics discussed concern microbes in drinking water, contaminated groundwater remediation, and the effects of floods on hazardous waste sites. To summarize this proceedings provides a practical and timely reference for engineers interested in the current state of water and wastewater concerns.

Library of Congress Cataloging-in-Publication Data

National Conference on Environmental Engineering (1994: Boulder, Colo.)
Critical issues in water and wastewater treatment: pro ceedings of the 1994 National Conference on Environmental Engineering, Boulder, Colorado, July 11-13, 1994/edited by Joseph N. Ryan and Marc Edwards.
p. cm.
Includes index.
ISBN 0-7844-0031-8
1. Water—Purification—Congresses. 2. Sewage—Purification-Congresses. 3. Hazardous waste sites—Environmental aspects-Congresses. I. Ryan, Joseph N. II. Edwards, Marc, 1964- . III. Title.
TD430.N284 1994 94-20431
628.1—dc20 CIP

Library of Congress Catalog Card No: 94-20431
ISBN 0-7844-0031-8
Manufactured in the United States of America.

FOREWORD

"What in water did bloom, waterlover, drawer of water, watercarrier returning to range, admire?"

James Joyce, *Ulysses,* 1922

Joyce's words seem to be addressed to us, the participants in the 1994 National Conference on Environmental Engineering, *"Critical Issues in Water and Wastewater"*, as we come together to commemorate our concern for the quality of water in the United States and to debate our ability to maintain and improve water quality.

The purpose of this conference is to identify and discuss scientific and technological issues currently at the heart of the quality of water in the United States. The conference will address the complete cycle of our water system, from watershed management to water treatment and distribution, wastewater handling, and remediation of contaminated groundwater. In addition, a special session will examine the timely issue of hazardous waste site flooding, a topic brought to the fore by problems associated with the 1993 Mississippi River flooding. The breadth of the conference indicates the current wide-ranging interdisciplinary nature of environmental engineering and water science.

The papers presented in these *Proceedings* were selected from abstracts receiving at least two positive reviews from a nation-wide panel of reviewers. Special thanks goes to the reviewers for their diligent work. Of over 200 abstracts submitted, only 114 were selected for inclusion in the conference. These 114 authors were invited to submit papers for inclusion in these *Proceedings*. All papers contained in these *Proceedings* are eligible for discussion in the *Journal of Environmental Engineering* and are eligible for awards from the American Society of Civil Engineering.

The *Proceedings* also includes five prize-winning student papers that were selected by the Environmental Engineering Division Technical Committees of Air and Radiation Management, Solid and Hazardous Wastes Management, Water Supply, Water Pollution Management, and Risk Assessment and Multi-Media Contamination. At the conference, a special session will be devoted to student presentations of their work.

The 1994 National Conference on Environmental Engineering is sponsored by the Environmental Engineering Division of the American Society of Civil Engineers. Cooperating sponsors include the University of Colorado-Boulder Department of Civil, Environmental, and Architectural Engineering, the American Water Works Association, the Water Environment Federation, the Colorado Section of the American Society of Civil Engineers, and the American Academy of Environmental Engineers. The editors of the *Proceedings* would also like to thank the authors for submitting their work for inclusion in this conference and these *Proceedings*. Additionally, the authors acknowledge the editorial assistance of Shiela Menaker, the American Society of Civil Engineering Manager of Book Production, and Sapna Patel of the University of Colorado at Boulder.

Joseph N. Ryan and Marc Edwards, Assistant Professors
Department of Civil, Environmental, and Architectural Engineering
Univeristy of Colorado
Boulder, Colorado

CONTENTS

SESSION 1

CORROSION CONTROL—THE NEW EVIDENCE

Moderator: Steve Reiber

SESSION 2

MICROBES IN DRINKING WATER/SMALL SYSTEMS

Moderator: Ben Lykins

SESSION 3

MEMBRANES: WASTEWATER AND INNOVATIVE APPLICATION

Moderator: Michael Semmens

---

*Manuscript not available at time of printing.

SESSION 4
SLUDGE REGULATIONS AND REUSE
Moderator: Jeannette Semon and C. Koch

SESSION 5
FILTRATION: PRETREATMENT, PARTICLES, AND ORGANICS
Moderator: A. Amirtharajah

*Manuscript not available at time of printing.

SESSION 6
CLARIFIERS
Moderator: Orris E. Albertson

SESSION 7
HEAVY METALS AND INORGANICS
Moderator: M. Benjamin

SESSION 8
SLUDGE MANAGEMENT TECHNOLOGY
Moderator: J. Smith and C. Koch

*Manuscript not available at time of printing.

SESSION 9
SYNTHETIC ORGANIC CHEMICALS/BIOFILMS
Moderator: H. David Stensel

SESSION 10
MEMBRANES FOR DRINKING WATER
Moderator: Joe Jacangelo

SESSION 11
REMEDIATION OF CONTAMINATED GROUNDWATER
Moderator: William P. Ball

*Manuscript not available at time of printing.

SESSION 12
NATIONAL STUDENT PAPER COMPETITION PRESENTATIONS
Moderator: Balu P. Bhayani and Paul Bowen

SESSION 13
NUTRIENT REMOVAL-1
Moderator: Glenn Daigger

SESSION 14
CONTROL OF DISINFECTION BYPRODUCTS IN DRINKING WATER
Moderator: Jim Lozier

*Manuscript not available at time of printing.

SESSION 15
STORMWATER AND WATERSHED MANAGEMENT
Moderator: Daniel Okun

SESSION 16
VOLATILE ORGANIC CHEMICALS AT WASTEWATER AND HAZARDOUS WASTE FACILITIES
Moderator: Al Pincince

SESSION 17
ADVANCED BIOLOGICAL TREATMENT
Moderator: Mary Ann Taverez

SESSION 18
ACTIVATED CARBON
Moderator: Makram T. Suidan

SESSION 19
INORGANICS AND ARSENIC
Moderator: Clint Smith

SESSION 20
TREATMENT MODELING, DESIGN AND CONTROL
Moderator: Robert W. Okey

*Manuscript not available at time of printing.

SESSION 21
INNOVATIVE WASTEWATER TREATMENT
Moderator: Bill Bellamy

SESSION 22
EFFECTS OF FLOOD ON HAZARDOUS WASTE SITES
Moderator: Carol Whitlock

SESSION 23
FUTURE WATER QUALITY ISSUES
Moderator: John H. Sullivan

Controlling Lead and Copper Corrosion and Sequestering of Iron and Manganese

Jonathan A. Clement[1]
Michael Schock[2]
Darren Lytle[3]

ABSTRACT

With the recently enacted Lead and Copper Rule (LCR), many utilities are faced with the conflict of meeting the requirements of the Rule and controlling aesthetic problems caused by source water iron and manganese. The most common approach for utilities to control "red and black water" is to add a polyphosphate based compound. However, the higher pH required for control of lead and copper solubility reduces the effectiveness of polyphosphate to sequester iron and manganese. There is also the threat that polyphosphate may complex lead and copper and increase their concentration. An alternative treatment approach, sodium silicate addition, was evaluated at medium sized water system with elevated source water iron (0.30 - 2.27 mg/L) and manganese (0.11 - 0.27 mg/L). The goal of the study was to examine the viability of sodium silicate to simultaneously control red water complaints, and reduce lead and copper concentrations. Samples for a wide range of water quality parameters were collected before initiating treatment (5 months) and after treatment to gauge the effectiveness of the approach.

INTRODUCTION

Source water iron and manganese control is a widespread problem for small and medium-sized water systems across the U.S. Surveys indicate that nearly 40% of the water suppliers in the U.S. exceed the secondary MCL for iron (0.30 mg/L) and manganese (0.05 mg/L) (Stiles, 1978). One method that can be employed to control the

1. Process Engineer, Black & Veatch, Cambridge, Massachusetts
2. Research Chemist, U.S.E.P.A., Cincinnati, Ohio
3. Environmental Engineer, U.S.E.P.A. Cincinnati, Ohio

the aesthetic problems associated with these metals is to add chemicals to sequester the metals. This option is often preferred by small and medium sized utilities because it is far less expensive than the construction and operation of iron and manganese filtration systems. The most common chemicals used for sequestering are some form of polyphosphate compounds.

Optimal pHs for sequestering these metals by polyphosphates is usually less than 7 (Robinson et. al., 1990) The ability of polyphosphate based compounds to sequester iron and manganese decreases dramatically above this pH value. However, lead and copper corrosion control strategies require pHs between 7 - 10(Schock, 1986; Schock, 1988 and AWWARF, 1990). Furthermore, there exists the danger that polyphosphates may act as chelating agents potentially increasing lead and copper levels (Holm and Schock, 1991). Another disadvantage associated with polyphosphate based compounds is that the formulations are proprietary in nature, and consequently it is difficult to develop generic information that can be used by a wide range of utilities. These problems will limit the applicability of polyphosphate to many water systems. Therefore, an alternative approach to sequester iron and manganese, while not negatively impacting lead and copper needs to be examined.

Many utilities across southern Canada and the Northeastern United States have used sodium silicate effectively to control red water complaints due to source-water iron or iron from unlined cast-iron water mains (Robinson, 1990 and Clement, 1993). The mechanism by which silicates sequester iron and manganese is poorly understood, but a colloidal dispersion method has been hypothesized (Browman et. al., 1989) Dissolved silica appears to alter the charge on the surface of the oxidized forms of iron and manganese. Because the iron and manganese must be oxidized for sequestering silicates, chlorine must be added nearly simultaneously with the silica.

The chief advantage of using silicates over polyphosphates is that sequestering pHs are in the range of 7-8 which are more conducive to reducing lead and copper than lower pHs. In fact, with low calcium (<10 mg Ca/L) as pH increases the effectiveness of silica to sequester iron increases. However as calcium increases the ability of silicate to sequester iron and manganes decreases. Most research (Robinson et. al., 1990 and Robinson et. al., 1985) strongly suggests that silicates are more effective in sequestering iron than manganese. One reason that was hypothesized for this action is that in the pH ranges studied (7-8) manganese oxidation is in most cases incomplete. As a result, higher pHs (>8) may produce better sequestering of manganese. Another key advantage to the use of sodium silicates for sequestering is that the solution contains sodium oxide Na_2O, which increases the pH of the treated water. This frequently eliminates the need for separate feed systems (e.g. pH adjustment and sequestering compound).

The role of silica to control corrosion of lead and copper is poorly understood. Very little research has been performed to determine if silica is effective on reducing lead and copper corrosion. One study found that silicates were not significantly more effective than pH adjustment alone (AWWARF, 1985) The control of lead and copper by silicates appears to be by a surfical coating mechanism (Schock, 1993). Given the lack data, it is difficult for many utilities to consider using silicates for controlling lead and copper. Until data indicates that silicate are effective for controlling lead and copper, utilities will probably rely on the pH increase that is brought about by the sodium oxide contained within the sodium silicate solution.

PROJECT OBJECTIVES

The project's goal was to find a cost-effective and innovative solution to sequester source-water iron and manganese and control of lead and copper to meet the requirements of the Lead and Copper Rule. To accomplish this goal, a one-year field research study was initiated at a water system with significant source water iron and manganese problems and elevated lead and copper levels. Another goal of the study was to ensure that the solution investigated would not be too site-specific, so that the approach could be widely used by a considerable variety of utilities.

METHODS

The study was conducted at a medium sized water system with elevated source water iron (0.30 - 2.30 mg/L) and manganese (0.11 - 0.27 mg/L) concentrations and elevated 90th percentile values for lead (71 ug/L) and for copper (5.87 mg/L). The source for the system is five gravel pack wells. The utility was feeding a glassy bimetallic phosphate compound to attempt to sequester the iron and manganese at two of its five wells with a pH in the range of 6.0 - 6.3. The phosphate compound, therefore only reached the portion of the distribution surrounding the wells where the compound was being fed.

The study consisted of three primary phases: a design of a monitoring program to characterize and understand the conditions before treatment, design of a set of chemical conditions to mutually satisfy source water iron and manganese problems and lead and copper corrosion, and to conduct post treatment monitoring to gauge and understand the effectiveness of the treatment.

The monitoring program consisted of collecting consecutive volumes of water from 22 residences throughout the distribution system.

Of the 22 sites, 11 were located in the region where the polyphosphate compound was being detected and the remainder were outside this region. This permitted a comparison between the sets of data to determine the effect of the polyphosphate compound. Water samples were collected after the water had been stagnant for a period of 6-12 hours. For sites with lead solder, two consecutive volumes of water were collected. An initial one-liter volume which was analyzed for lead, copper, zinc, iron, and manganese and a second 500 ml sample was analyzed for alkalinity, turbidity, silica, total phosphate, and orthophosphate. Water immediately following the 500 ml volumes was dispensed into a 250 ml bottle through a plastic tube to minimize loss of carbon dioxide and the pH was measured. Sites with lead service lines were sampled in the same manner, except that two consecutive one-liter were included to capture the water contained with the lead service line. Samples were collected from the 22 sites once every month for five months before initiating treatment and for eight months after treatment. Source water was monitored at all wells for pH, alkalinity, turbidity, iron, calcium, manganese, silica total phosphate, and orthophosphate on a weekly basis.

Chemical treatment of the water was to be achieved with sodium silicate. The goal was to determine a common set of chemical conditions that would mutually satisfy sequestering of iron and manganese (pH and silica) and control of lead and copper (pH and DIC). It was decided that the minimum pH necessary to reduce copper levels sufficiently, based on solubility models (Schock, 1988) would be at least 7.0. Control of lead would also require a pH of at least 7.0 and possibly higher. The quantity of silica to be added for control of iron and manganese was targeted at approximately 15 - 20 mg SiO_2/L. To determine the appropriate sodium silicate solution to use treatment curves were developed for silica dose as a function of pH for various grades (SiO_2:Na_2O) of sodium silicate (Figure 1). Based on these results, the lowest ratio (SiO_2:Na_2O) 1.6 commercially available was used.

The treatment conditions initially targeted were approximately pH=7 with a silica dosage of approximately 20 mg SiO_2/L. The sodium silicate was BW-50® manufactured by PQ Corporation (Valley Forge, PA). Sodium hypochlorite was added at the wells at dosages of 1-2 mg CL_2/L depending on the concentration of iron and manganese. The silicate and chlorine were added only at the wells where the polyphosphate compound was being added. This enabled a comparison between a section of the system that was receiving treatment with and without the sodium silicate.

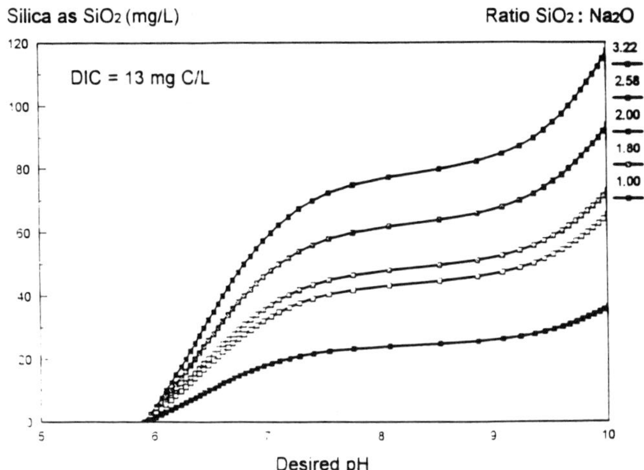

Figure 1. Silica levels as a function of pH for different types of sodium silicate solution.

RESULTS AND DISCUSSION

At the time that this paper was written, only baseline (pretreatment results were available). Therefore, the following presentation and discussion of results is limited to data collected before initiating the silicate treatment. An important aspect of this data is that it permits an examination of the effects of polyphosphate addition on metal levels and water quality.

Water Quality Parameters

Samples were analyzed for pH, alkalinity, turbidity, orthophosphate, total phosphate, iron, manganese and temperature. Calcium was not monitored, because the concentration was below <8mg Ca/L and rarely varied. Table 1 shows average water quality over the 4 month monitoring period. Polyphosphate was estimated by determining the difference between the total phosphate and the orthophosphate. Each section of the distribution system, (treated and non-treated) contained 11 monitoring sites.

The most apparent differences between the water qualities of the two regions were the copper, iron, manganese, orthophosphate, polyphosphate, and turbidity levels. The higher turbidity levels are likely to be primarily a function of oxidized iron which has been witnessed by other researchers (Robinson et. al., 1990). The pH and alkalinity of the untreated section were slightly higher.

Table 1. Water Quality Data for the Polyphosphate treated section and non treated section.

Parameter	Polyphosphate Section	Non-Treated Section
pH	6.08 ± 0.10	6.28 ± 0.04
Alkalinity mg $CaCO_3$/L	17 ± 4	25 ± 9
Turbidity NTU	0.81 ± 0.35	0.44 ± 0.23
Iron mg/L	0.71 ± 0.31	0.03 ± 0.01
Manganese mg/L	0.26 ± 0.05	0.07 ± 01
Polyphosphate mg PO_4/L	1.17 ± 0.17	< 0.15
Orthophosphate mg PO_4/L	0.36 ± 0.17	<0.15
Silica mg SiO_2/L	15 ± 3	12 ± 2
Copper mg/L	4.2 ± 2.36	1.11 ± 0.78
Zinc mg/L	0.56 ± 0.36	0.26 ± 0.08

Copper

Copper levels in the section receiving polyphosphate and orthophosphate were significantly lower than in the section without phosphate (Figure 2). One possible explanation for the lower levels is that orthophosphate may be forming some form of a cuprous or cupric orthophosphate compound. Electrochemical corrosion rate measurements have displayed a reduction in the copper corrosion rate with a orthophosphate dosage of 1 mg PO_4/L at pH of 7 (Reiber, 1990) Theoretical solubility computations indicate that the most likely impact of orthophosphate on copper solubility would be at pHs less than 7. The copper levels measured in this study appear to support the theoretical calculations(Schock et. al., 1994).

Another mechanism which may be producing the lower copper levels is the formation of an iron based surfical coating. Iron levels in the polyphosphate treated section averaged 0.71 mg/L as opposed to 0.03 mg/L in the untreated section. The development of such layer could act as a diffusion barrier limiting the release of copper. X-Ray diffraction will be employed to determine if such a film exists.

The copper results have important implications for water utilities and research. Many times conclusions are drawn from field monitoring data that could indicate that polyphosphate are effective for control corrosion. For example, in this study, without measuring orthophosphate and other important water quality parameters, the conclusion that polyphosphate was effective in reducing copper corrosion, could be erroneously drawn.

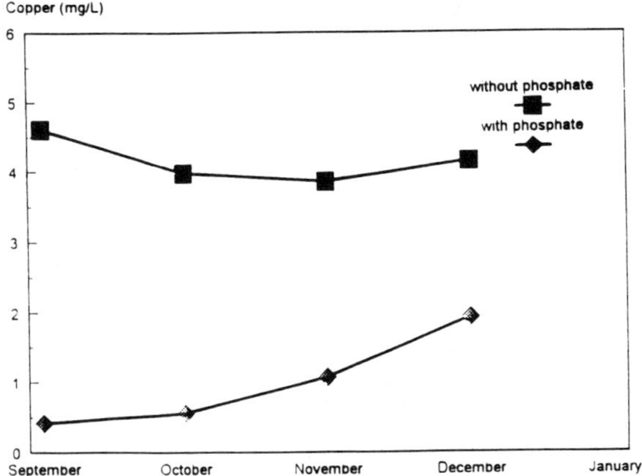

Figure 2. Copper levels in the Phosphate and Non-Phosphate Treated Section.

CONCLUSIONS

At the time this paper was written, only the baseline monitoring was completed. Although the data was limited several important and useful conclusions can be drawn. The final results will indicate whether adjusting silica and the addition of chlorine is successful in simultaneously controlling lead and copper corrosion and source water iron and manganese.

- Adjusting silica, pH, and chlorine appears to be a viable method for piloting by water systems with iron, manganese, lead and copper problems.
- Orthophosphate at low concentrations (0.30 - 0.50 mg PO_4/L) at very low pHs (6.0 - 6.3) appears to be controlling copper levels to 0.54 - 1.0 mg Cu/L, compared to 5 - 10 mg Cu/L without orthophosphate.
- Metal levels at a single water system can vary dramatically due to within system water quality variations, location, and monitoring site characteristics. Therefore, it is impossible to assign a set of water quality parameter to describe the metal levels for a single system. Water quality (eg pH, alkalinity etc.) must be monitored at the same sites where metal concentrations are measured.
- Sodium silicate solution when added to a source-water increases the pH and adds silica. Therefore, the need for separate feed systems (e.g. phosphate addition coupled with pH adjustment) is in most cases eliminated.

REFERENCES

AWWARF. Internal Corrosion of Water Distribution Systems AWWARF/DVGW - Forschungsstelle Cooperative Research Report (1985)

AWWARF, "Lead Control Strategies", AWWARF (1990)

Browman et. al., "Silica Polymerization and Other Factors in Iron Control by Sodium Silicate and Sodium Hypochlorite Additions. Environmental Science Technology 23:5 (1989)

Clement J.A. "Sodium Silicate for Simultaneous Control of Lead, Copper and Iron Based Corrosion." Proceedings on the Conference on the Control of Lead and Copper Drinking Water. USEPA Chicago, IL 1993.

Holm, T. and Schock M.R., "Potential Effects of Polyphosphate Water Treatment Products on Lead Solubility in Plumbing Systems" J. AWWA 83:8 (1991)

Reiber, S. "Copper Plumbing Surfaces: An Electrochemical Study". Journal AWWA 81:7:114 (1989)

Robinson et al., "Iron and Manganese treatment by the Addition of Sodium Silicate and Chlorine". Proceedings of the 1985 AWWA Conference AWWA (1985)

Robinson et. al, "Sequestering Methods of Iron and Manganese Treatment" AWWARF (1990)

Schock, M.R., "Treatment to Attain MCls in Metallic Plumbing Systems" in Plumbing materials and Drinking Water Quality. Noyes Publishing, Park Ridge, NJ (1986)

Schock, M.R., "Understanding Corrosion Control Strategies for Lead," J. AWWA, 81:7 (1988)

Schock M.R., "An Overview of Control Strategies" Proceedings of the Conference on the Control of Lead and Copper in Drinking Water. USEPA Chicago, IL (1993)

Stiles, J.F., "Survey of the Iron and Manganese Problems in the USA". AWWA American Conference, Atlantic City, N.J. June 1978

Schock, M.R., Lytle, D.A., & Clement, J.A. "Effect of pH, DIC, and Orthophosphate on Drinking Water Cuprosolency" Submitted Mans. Journal AWWA (1994).

EFFECT OF ALKALINITY ON COPPER CORROSION

Travis E. Meyer and Marc Edwards
Department of Civil Engineering, Box 428
University of Colorado at Boulder
Boulder, CO 80309-0428

Abstract. The effect of lime/CO_2 treatment on copper corrosion was examined in Boulder tap water using a variety of techniques including electrochemical accelerated aging, conventional aging, electrochemical corrosion rate analysis and copper byproduct release. At pH 7.2 (the unaltered pH of Boulder tap water), a critical alkalinity range of ≥100 mg/l alkalinity as $CaCO_3$ was identified in which corrosion rates and copper byproduct release greatly increased. For higher pH values (≥7.8), corrosion rates and byproduct release were much reduced. An analysis of data from a national survey of large utilities showed nearly identical trends as those found for Boulder tap water. Copper solubility experiments suggested that soluble copper concentrations may be controlled by cupric hydroxide ($Cu(OH)_2$).

Introduction

Corrosion of copper pipe and associated byproduct release have recently received much interest owing to the 1991 U.S. Environmental Protection Agency (U.S.E.P.A.) Lead and Copper Rule. The primary inorganic water quality parameters of importance in copper corrosion have been identified as pH, chloride, sulfate, and bicarbonate (Cruse et al., 1985). Regarding bicarbonate, the AWWA Guide to Internal Corrosion in Distribution Systems states that "*bicarbonate generally tends to reduce the corrosivity ... by an inhibiting action*" (Cruse et al., 1985). A multitude of laboratory experiments as well as practical studies support this hypothesis (Mattsson and Fredriksson, 1968; Shalaby et al., 1990; Drogowska et al., 1992). The objectives of this research were to test the conventional wisdom regarding bicarbonate using both advanced and conventional techniques on a natural water treated with lime/CO_2.

Materials and Methods

Solution preparation. Boulder tap water was used as a base solution to constitute "natural" waters with varying amounts of alkalinity. Boulder tap water had an average pH of 7.2 and an alkalinity of about 15 mg/l as $CaCO_3$ during the experiments (Table 1). This base water was modified to test the effect of lime/CO_2 treatment on copper corrosion in a natural water (Table 2). To maintain pH, an acid solution constituted from tap water bubbled with CO_2 to pH 4.0 was dosed with a pH stat during experiments. Similarly, pH was increased (if necessary) without altering alkalinity by bubbling CO_2-free air through the solution to remove CO_2.

Table 1. Average values for Boulder water quality during testing period.

Parameter	Value	Units
pH	7.0-7.4	pH
Alkalinity	12-18	mg/l $CaCO_3$
Calcium	25	mg/l $CaCO_3$
Chloride	6	mg/l
Copper	<0.01	mg/l
Fluoride	1.1	mg/l
Total Hardness	25	mg/l $CaCO_3$
Iron	1.1	mg/l
Lead	< 0.02	mg/l
Magnesium	7.6	mg/l
Nitrate	0.02	mg/l
Sodium	< 5	mg/l
Dissolved Solids	44	mg/l
Sulfate	6	mg/l
Tot. Org. Carbon	1.28	mg/l

Table 2. Modified levels of pH and alkalinity tested in Boulder tap water.

pH	Total Alkalinity (mg/l $CaCO_3$)
7.2	15, 45, 100, 250
7.4	45
7.8	15, 25, 35, 45, 60, 100, 250
8.5	15, 45, 100, 250

Apparatus. The apparatus used to carry out the electrochemical experiments in this work included a computer-controlled potentiostat, eight corrosion analysis cells, and eight target solution reservoirs and is described elsewhere (Edwards et al., 1994). The core of the system is a Gamry PC3 (Gamry Instruments, Langhorne, PA). potentiostat. A Gamry ECM8 Multiplexer was coupled to the potentiostat, allowing automatic and sequential corrosion analysis of the eight independent cells, or concurrent operation of the eight cells in a potentiostatic mode. Eight Reiber corrosion cells were constructed according to methods described elsewhere (Edwards and Ferguson, 1993). The sole difference is that a 22 gauge 90:10 platinum:iridium wire was used as the counter electrode. The target solution reservoir was a four liter polypropylene container, immersed in a plastic water bath. The water bath was continually purged with cold tap water which removed excess heat produced from each cell loop and maintained the target solution at a constant temperature. The target solution was pumped through each Reiber cell in a closed loop at a flow rate of 0.5 gpm.

Copper coupon preparation. Copper samples used in the experiments were 5/8" diameter nominal copper couplings (internal surface area = 20.0 cm^2 and actual ID of 3/4") purchased from a local plumbing supply shop. Each coupon was washed in

0.1 M NaOH for 2 minutes to remove organic deposits and then rinsed 5 times in Milli-Q water immediately before use in experiments.

Accelerated aging procedure. Potentiostatic methods were used to form characteristic scales on copper surfaces using an accelerated aging technique. The theoretical and practical basis for this technique is described elsewhere elsewhere (Edwards and Ferguson, 1993). The copper surfaces were first subjected to the target solution of interest and allowed to corrode naturally for 30 minutes, at which time corrosion rates were determined electrochemically (see following section). Then, scale was formed by anodic polarization (E = +120 V vs. Ag/AgCl reference) for 72 hours, forcing the copper to corrode at an accelerated rate. At the end of the 72 hour acceleration period, the applied overpotential was removed and the target solutions were replaced with fresh solutions. Each coupon was then exposed to the fresh solutions for 24 hours before corrosion rates were determined electrochemically for the aged copper surfaces.

Electrochemical corrosion rate measurement. For this measurement, E_{corr} was first measured using the Gamry electroanalysis system. The coupon surface was then subjected to a potentiodynamic scan with a standard perturbation of 150 mV (E_{corr}-75 mV to E_{corr}+75 mV), a scan rate of 0.2 mV/sec, and a recording rate of 1 data point per second. These scans were analyzed using the Gamry corrosion analysis software, which calculates the corrosion rate (i_{corr}) according to electrochemical theory (Reiber, 1989).

Byproduct release experiments. Twenty-four inch sections of 3/4" Type M copper pipe were washed in 0.1 M NaOH for 2 minutes to remove any organic deposits and rinsed 5 times with Milli-Q water before initial exposure to the target solutions. Tests were performed at pH 7 and 8 with 15, 45, 100 and 250 mg/l alkalinity as $CaCO_3$ (8 conditions). Each condition was duplicated and the replicate pipes were initially exposed 12 days apart. The pipes were filled with the target solutions 3 times weekly (Monday, Wednesday, Friday) and kept in a horizontal position at all other times with rubber stoppers tightly capping each end. Filtered (0.2 μm cellulose acetate) and unfiltered samples were taken weekly to analyze soluble and total copper concentrations after 72 hours of exposure (Friday to Monday stagnation).

Results

Corrosion rates on aged copper coupons. While each set of water qualities for Boulder tap water was tested using the same water supply, replicate experiments were performed on different days using different water samples. Therefore, replicate experiments were identical for pH and alkalinity values, but the base water quality changed slightly according to changes in tap water over time. Therefore, average values were used to establish overall trends and standard deviations provided information on sensitivity of the copper corrosion process to small changes in water quality.

At pH 7.2, corrosion rates at 100 and 250 mg/l alkalinity were more than 200% higher than those observed at 15 and 45 mg/l alkalinity (Figure 1). Corrosion rates at pH 7.8 and 8.5 were low and not significantly different at all alkalinity values. Thus, it appears that dramatic changes occur in this natural water from pH 7.2 to 7.8 and at alkalinities between 45 and 100 mg/l as $CaCO_3$.

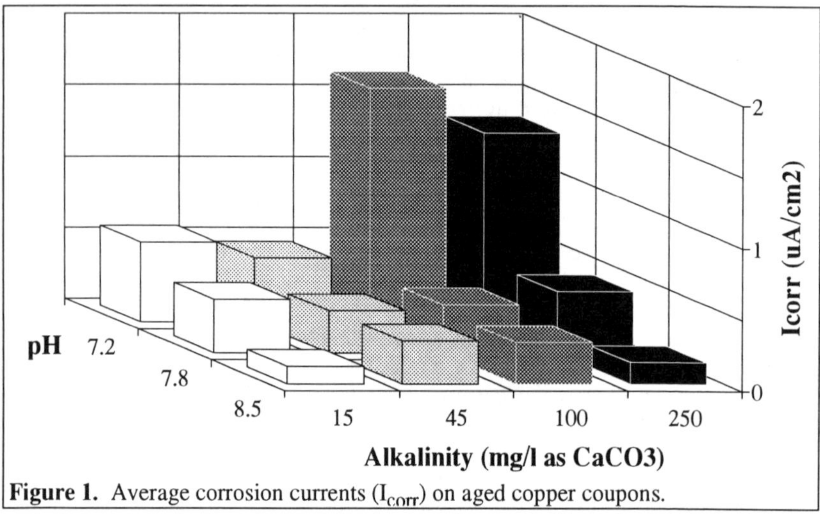

Figure 1. Average corrosion currents (I_{corr}) on aged copper coupons.

Table 3. Standard deviations for I_{corr} in accelerated aging tests.

pH	Alk	Ave. I_{corr}	St. Dev.	# of tests
7.2	15	0.56	0.11	3
	45	0.45	0.09	3
	100	1.64	0.58	2
	250	1.33	0.23	2
7.8	15	0.38	0.10	3
	25	0.39	0.13	2
	35	0.36	0.09	3
	45	0.30	0.07	3
	60	0.32	0.05	3
	100	0.34	0.11	3
	250	0.43	0.15	3
8.5	45	0.30	0.05	3

Alkalinity in mg/l as $CaCO_3$
I_{corr} in $\mu A/cm^2$

Of all the tests that were repeated, the lowest standard deviation between replicate tests (conducted a few days apart) occurred at 45 or 60 mg/l alkalinity (Table 3), indicating that the scale formed in this alkalinity regime was particularly stable and not very susceptible to small changes in water quality. In contrast, while low corrosion rates were occasionally observed at lower alkalinity values (15, 25 and 35 mg/l), conditions were more variable and indicative of greater instability. This instability was also observed at the higher alkalinity values, particulary at pH 7.2 with 100 and 250 mg/l alkalinity.

Byproduct release. Trends in copper byproduct release mirrored the results of the accelerated aging experiments; that is, byproduct release increased with alkalinity and decreased with pH. After about 3 months of exposure, a plot of the last four weeks of data collected in this experiment indicated that byproduct release increased nearly linearly with alkalinity (Figure 2) with r^2 = 0.89 and 0.90 at pH 7 and 8, respectively.

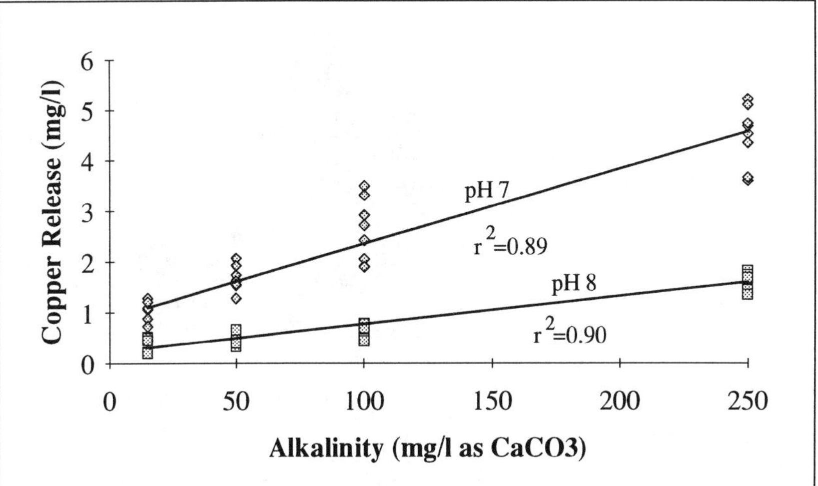

Figure 2. Data over last four weeks of exposure in byproduct release experiments reveal that soluble copper release is nearly linear.

Discussion

Given the controversial nature of these findings and their importance to water treatment practices, still further documentation of the adverse effects of high alkalinity on copper corrosion was sought. Data collected during a national survey of 435 large utilities was analyzed and compared to the observations described in this paper. The analysis of that data proceeded as follows. First, any utilities using corrosion control other than pH and/or alkalinity adjustment were eliminated from the analysis as well as any utility not reporting pH, alkalinity, or 90th percentile copper release. The 151 remaining utilities were then sorted into pH and alkalinity categories which corresponded to the pH and alkalinity ranges used in the accelerated aging experiments. The pH categories included 6.80-7.49, 7.50-8.19 and 8.20-8.99 and the alkalinity categories were 0-29, 30-74, 75-174 and >174 mg/l as $CaCO_3$. The 90th percentile copper release data (collected for the E.P.A. Lead and Copper Rule) for all the utilities in each category were then averaged and plotted.

The results (Figure 3) clearly illustrate the adverse effects of alkalinity on copper corrosion first discovered during the course of this investigation, as well as the sensitivity of this effect to pH. Comparison of the resulting plot to the corrosion rate data (Figure 1) and the byproduct release data (Figure 2) revealed nearly identical trends. In the low pH range (6.80-7.49), bicarbonate had a very adverse effect on corrosion rates, byproduct release, and average 90th percentile copper release. Above pH 7.50, this adverse effect was much reduced but corrosion rates, byproduct release, and 90th percentile copper still generally increased with alkalinity.

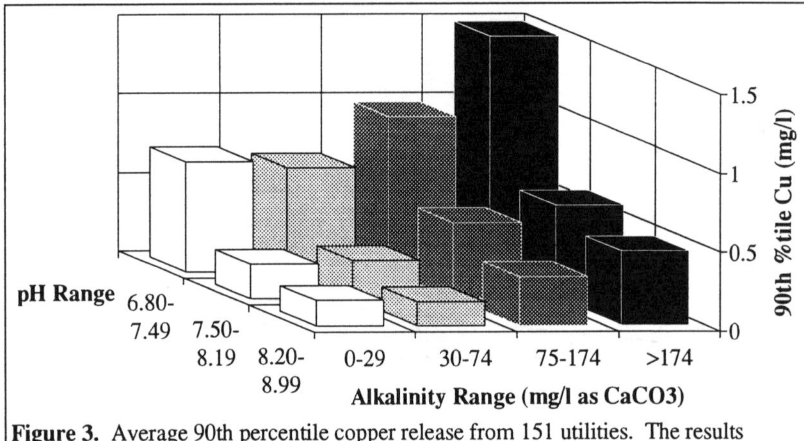

Figure 3. Average 90th percentile copper release from 151 utilities. The results clearly demonstrate the adverse effects of bicarbonate.

In an attempt to explain the adverse effects of high alkalinity, copper solubility in the presence of bicarbonate was studied. One liter solutions of constant ionic strength (0.01 M) were prepared at pH 7.0, 7.5, 8.0 and 8.5. Copper (Cu^{+2}) was slowly added up to 5 mg/l while keeping the solution pH constant with dropwise addition of 0.1 M NaOH. The results (Figure 4) were then compared to theoretical predictions based on thermodynamic data using the computer program Mineql$^+$ (Figure 5). The model was unable to predict copper solubility when malachite or tenorite were considered the predominate solids. However, when cupric hydroxide, $Cu(OH)_2$, was considered the predominant solid, the predictions agreed very well the data.

In this model, the predominant soluble copper species at pH 7.0 to 8.5 include Cu^{+2}, $Cu(OH)_{2(aq)}$, $CuCO_{3(aq)}$, and $CuHCO_3^+$. Using appropriate constants of formation for these copper species, the concentration of soluble copper can be calculated as:

$$\text{Soluble Cu} = [Cu^{+2}] + [Cu(OH)_{2(aq)}] + [CuCO_{3(aq)}] + [CuHCO_3^+]$$
$$= K_{sp}[H^+]^2 + K_1K_{sp} + K_2K_{sp}[H^+]^2[CO_3^{-2}] + K_3K_{sp}[H^+]^2[HCO_3^-]$$

Since K_1, K_2, K_3, K_{sp} and $[H^+]$ are constant at a given pH and $[CO_3^{-2}]$ and $[HCO_3^-]$ are linearly proportional to alkalinity, soluble copper concentrations are therefore predicted to be linearly dependent on alkalinity. This result is in good qualitative agreement with the linear dependence observed in the earlier experiments (Figure 2).

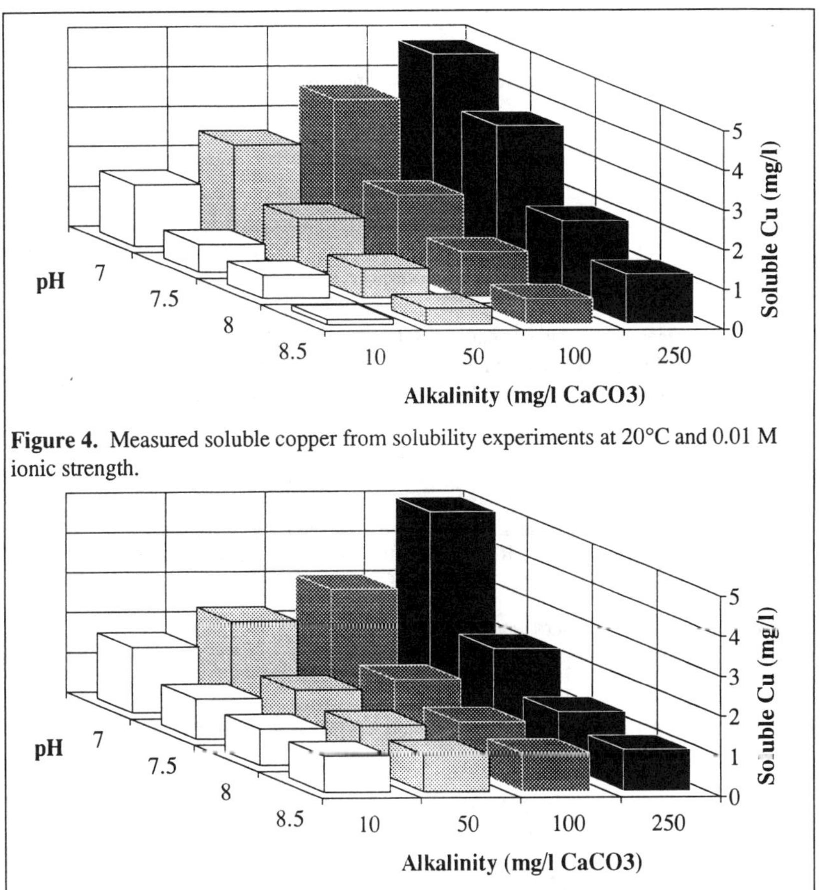

Figure 4. Measured soluble copper from solubility experiments at 20°C and 0.01 M ionic strength.

Figure 5. Predicted soluble copper from Mineql+ at 20°C and 0.01 M ionic strength ($Cu(OH)_2$ assumed to be present).

The similar trends observed for copper solubility and corrosion rates prompted questions as to why this agreement exists. The following analysis forwards a hypothesis that could relate corrosion rate to copper solubility. The basic assumption is that soluble copper-bicarbonate and copper-carbonate complexes become increasingly important at higher alkalinity, tending to reduce the free copper (Cu^{+2}) activity in solution. At a fixed pH and saturated oxygen concentration, the driving force for corrosion increases as free copper activity decreases as evidenced by the overall corrosion reaction:

$$Cu + \frac{1}{2}O_2 + H_2O \rightarrow 2OH^- + Cu^{+2}$$

In turn, the overall corrosion rate may also increase due to the greater corrosion driving force.

Conclusions

(1) Bicarbonate exhibited adverse effects on copper corrosion rates and copper byproduct release in a natural water. An analysis of average 90th percentile copper release data from 151 utilities showed the same trends.

(2) In the presence of bicarbonate, soluble copper concentration appears to be controlled by cupric hydroxide ($Cu(OH)_2$) solubility, causing soluble copper concentration to increase linearly with bicarbonate concentration at a given pH.

References

Cruse, H., O. Von Franque and R.D. Pomeroy. "Corrosion of Copper in Potable Water Systems." in Internal Corrosion of Water Distribution Systems. Cooperative Research Report, AWWARF, USA and DVGW-Forchungsstelle am Engler-Bunte-Institut del Universitat Karlsruhe. Germany (1985)

Drogowska, M., L. Brossard and H. Menard. "Copper Dissolution in $NaHCO_3$ and $NaHCO_3$ + NaCl Aqueous Solutions at pH 8." *J. Electrochem. Soc.* 139:1:39 (1992)

Edwards, M. and J.F. Ferguson. "Accelerated Testing of Copper Corrosion." *JAWWA*. 85:10:105 (1993)

Edwards, M., T. Meyer and J. Rehring. "Effect of Various Anions on Copper Corrosion Rates." *In Press. JAWWA*. (1994)

Mattsson, E. and A.M. Fredriksson. "Pitting Corrosion in Copper Tubes - Cause of Corrosion and Counter-Measures." *Br. Corros. J.* 3:246 (1968)

Reiber, S. "Copper Plumbing Surfaces: An Electrochemical Study." *JAWWA*. 81:7:114 (1989)

Shalaby, H.M., F.M. Al-Kharafi and A.J. Said. "Corrosion Morphology of Copper in Dilute Sulphate, Chloride, and Bicarbonate Solutions." *Br. Corros. J.* 25:4:292 (1990)

Modeling Issues of Copper Solubility in Drinking Water

Michael R. Schock[a]
Darren A. Lytle[b]
Jonathan A. Clement[c]

Historically, uniform copper corrosion has been of little concern. Numerous forms of pitting corrosion and consequent pipe failures have been documented and classified (Cruse and von Franqué, 1985, Edwards, et al., 1994). Recently, because of the monitoring introduced by the Lead and Copper Rule (Federal Register, 1991a, Federal Register, 1991b, Federal Register, 1992), cuprosolvency (copper solubility) has forced many utilities to undertake corrosion control studies and treatment, rather than the replacement of pipes perforated by pitting.

The development of a useful model for cuprosolvency in drinking water is constrained by several areas of uncertainty, many of which were addressed in a recent study (Schock, et al., 1994). These areas are: the nature of "field data;" the selection of appropriate aqueous and solid species for drinking water systems; the uncertainties in selection of the most accurate equilibrium constant data; the role of metastable phases in cuprosolvency control; the effect of temperature; and residual oxidants in the plumbing systems.

The Nature of "Field Data"

Because of the massive collection of data from Lead and Copper Rule monitoring, the use of such data for trying to sort out trends of copper solubility with water quality and to recommend treatment approaches is extremely tempting. Unfortunately, the regulatory monitoring data and random samplings are of limited

---

[a] Research Chemist, Drinking Water Research Division, USEPA, Cincinnati, Ohio, 45268.

[b] Environmental Engineer, DWRD, USEPA, Cincinnati, Ohio 45268.

[c] Process Engineer, Black & Veatch, Cambridge, MA 02140.

use for extracting details of copper chemistry behavior and understanding potential copper passivation strategies for several reasons, such as:

- Water quality parameter data is not required to be reported for utilities meeting the action levels or other exclusionary criteria, so the conditions leading to prior effective "optimization" or the attainment of the action levels by individual utilities are generally not available unless data was tabulated by the utilities on their own and reported.

- For medium-sized water systems, water quality parameter data was frequently, if not almost always, collected at a different time than the samples for lead and copper, making a direct statistical or mechanistic link between background water chemistry conditions and the actual corresponding lead and copper values becomes numerically possible but scientifically meaningless..

- Some important background chemical constituents that might give useful chemistry insights (eg. chloride, sulfate, silica, orthophosphate, etc.) were not necessarily collected, particularly when corrosion inhibitors (phosphates, silicates) are not employed.

- Stagnation times of monitoring samples fell within a ten-hour period (6-16 hours), but copper levels are now known to be very dependent on standing time, more so than with lead.

- Copper dissolution rates depend greatly on the ageing of the passivation films, so interpretation must include adjustment for various plumbing ages to correctly deduce chemical effects.

- Surface films remaining on the interior pipe surface from the manufacturing process can greatly affect the rate of formation and the bulk of passivation films in the plumbing systems (Gilbert, 1966), so copper dissolution will depend greatly on the early cleaning, flushing, flow, and usage pattern of the individual plumbing sites.

- Lead and copper monitoring sites are purposely biased to represent locations tending to contain high levels of lead (Federal Register, 1991a). Consequently, in many cases a substantial portion of the sample volume may have been exposed to lead piping.

Selection of Appropriate Aqueous Species for Cu(I) and Cu(II)

Within a either the Cu(I) or Cu(II) oxidation state, the inclusion of appropriate aqueous complexes in the chemical model is essential to enable accurate prediction of copper solubility. Over a 100-fold negative error in predicted equilibrium copper solubility is introduced by the neglect of the $Cu(OH)_2^\circ$ complex, even if other hydrolysis species are included (Bollinger, et al., 1992). An even greater error is introduced in both the magnitude and trend of solubility when only the free cupric ion (Cu^{2+}) is assumed to be present. In addition to hydroxide complexes, Cu(II) (cupric form) forms several very strong and some weaker aqueous complexes with carbonate species (HCO_3^-, CO_3^{2-}), carbonate complexation tends to become even more important that the hydrolysis reactions as the pH increases, and especially above a pH of about 7.5.

The aqueous chemistry of Cu(I) is dominated by the formation of two particularly significant stabilizing ligands, NH_3(aq) and Cl^-. Cuprous ammine complexes can be formed either directly, or by reduction of cupric ammine complexes. Evidence from various complexation experiments and observations of geochemical mineral assemblages shows that even in the presence of strong oxidants, ammonia can increase the solubility of oxide, hydroxide, oxysulfate, oxycarbonate and oxychloride solids of copper (de Zoubov, et al., 1974, Rickard, 1970a, Rickard, 1970b). Severe attack of copper pipe attributed to the presence of high concentrations of ammonia in the soldering flux has also been noted (Akkaya and Ambrose, 1987).

Cuprous chloride complexes are not as strong as the ammine complexes, but they often may be significant because chloride concentrations in most drinking water are many times those of ammonia. They have been shown to play an important role in retarding the rate of cuprous ion oxidation in natural waters, particularly at neutral to alkaline pH's (Eary and Schramke, 1990, Millero, 1989, Millero, 1990a, Millero, 1990b, Millero, et al., 1987).

The background solubility of cupric copper under oxidizing conditions resulting from the unavoidable hydrolysis reactions is completely defined by the simple total solubility expression ($S_{T,OH}$), which includes the free cupric ion concentration [Cu^{2+}].

$$S_{T,OH} = [Cu^{2+}] + [CuOH^{+}] + [Cu(OH)_2{}^{\circ}] + [Cu(OH)_3^{-}] + [Cu(OH)_4^{2-}] + 2\,[Cu_2(OH)_2^{2+}] + 3\,[Cu_3(OH)_4^{2+}]$$

The selection of the aqueous carbonate species to be incorporated into this solubility model follows the selections of Byrne and Miller (1985) and Stiff (1971), which includes $CuHCO_3^+$, $CuCO_3{}^{\circ}$, $Cu(CO_3)_2^{2-}$, and $CuCO_3OH^-$. Exploratory calculations showed that the $CuCO_3(OH)_2^{2-}$ complex species reported in an uncritical compilation of $\Delta G_f{}^{\circ}$ values (Woods and Garrels, 1987) may have an important impact on copper solubility at pH levels found in many lime-softened or other high-pH water supplies, so it was added.

Considerable experimental uncertainties exist for most of the aqueous hydroxide, carbonate, and mixed complexes, so a wide range of values has been reported in the literature by credible researchers for most species of particular interest, as summarized by Schock, et al. (1994). For example, major inconsistencies exist among most published hydrolysis constants for the complex $Cu(OH)_2°$. Reported values for log β ranging from -13.7 to "less than 17.3" for the reaction:

$$Cu^{2+} + 2H_2O \rightleftharpoons Cu(OH)_2° + 2H^+$$

have been found. To compound the problem, data either conflict or there are very few reports of any determinations of stability constants for the possible complexes $Cu(OH)_3^-$ and $Cu(OH)_4^{2-}$, which would be significant at a pH of approximately 9 and above. Thus, these values must be treated with some skepticism. Baes and Mesmer (1976) could not even select definitive values for the formation constants of $CuOH^+$, $Cu(OH)_2°$, or $Cu(OH)_3^-$. They only chose to report limits on the likely magnitudes of the respective log β constants (eg. "<17.3", "<8"), which unfortunately encompasses almost all of the values reported for these species.

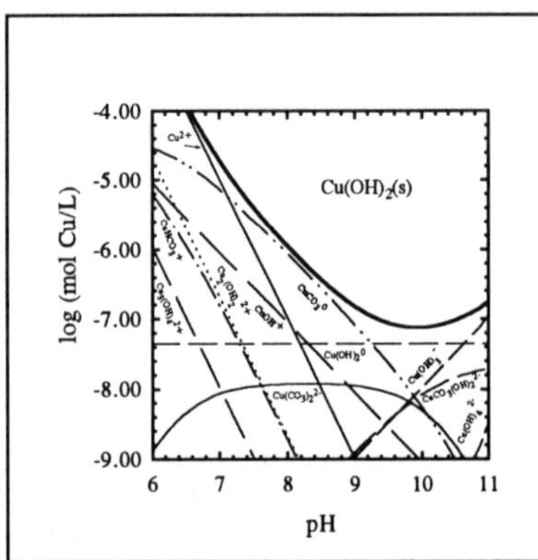

Figure 1. Copper(II) speciation in equilibrium with cupric hydroxide for DIC=4.8 mg C/L (4 x 10^{-4} M), 25°C, I=0.005.

The significance of errors in stability constants of these aqueous complexes are shown in Figures 1 and 2, giving the selected cupric species distribution in equilibrium with cupric hydroxide solid for two different levels of DIC (4.8 mg C/L or 4 x 10^{-4} M; 96 mg C/L or 8 x 10^{-3} M). Because of the importance of complexed forms of Cu(II) relative to Cu^{2+}, formation constant errors will greatly impact the computed species distribution and the copper solubility.

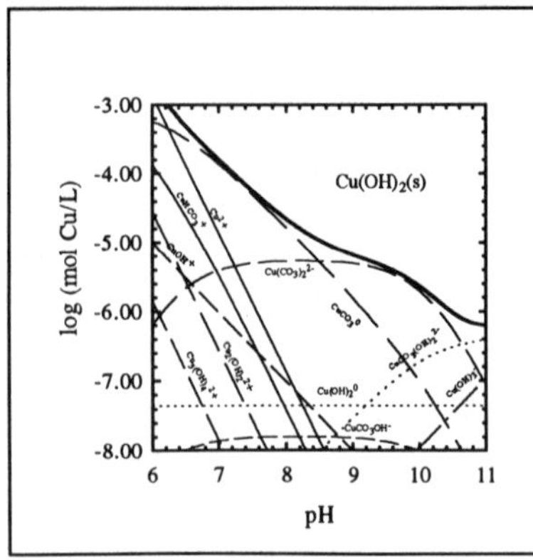

Figure 2. Copper(II) speciation in equilibrium with cupric hydroxide solid for DIC=96 mg C/L (8 x 10^{-3} M), 25°C, I=0.02.

Solid Solubility Constants and Metastable Solids

The differences among several solubility constants for CuO(s) and $Cu(OH)_2(s)$ have been discussed by de Zoubov, *et. al.*(1974), who selected two constants that they felt best agreed with actual copper solubility data. In their critical evaluation of hydrolysis behavior of almost all important environmental metals, Baes and Mesmer (1976) selected somewhat different values, which were an outgrowth of the research of Schindler, *et. al.* on particle size effects and the solubility of oxides and hydroxides (Pankow, 1991, Schindler, et al., 1965, Schindler, 1967). Schindler, et. al. found the log equilibrium constant for the $Cu(OH)_2$(s) solubility reaction (written in Baes and Mesmer format) to vary from 8.92 to 9.13 at I=0.2, as the molar surface increased from 0 to 4570 m^2. Similarly, they found the log equilibrium constant for the tenorite solubility reaction to vary from 7.89 to 8.27 at I=0.2, as the molar surface rose from 0 to 4340 m^2. Particle size effects may therefore have caused much of the scatter in reported experimental equilibrium constants for these solids.

There are considerable differences among reported values for the solubility constant of $Cu_2(OH)_2CO_3$(s). For the reaction written as

$$Cu_2(OH)_2CO_3(s) + 2H^+ \rightleftharpoons 2Cu^{2+} + 2H_2O + CO_3^{2-}$$

the most widely-reported values for log K range from -5.16 to -6.20, with an extreme of -3.99. The most recent careful experimental work by Symes and Kester (1984) leads to a value of log K = -5.48, which has been used for most of the modeling reported here. Compared to the widely-used value of -5.18, there is an uncertainty in the computed copper concentration in equilibrium with malachite is at least about a factor of 2, aside from uncertainties in aqueous speciation.

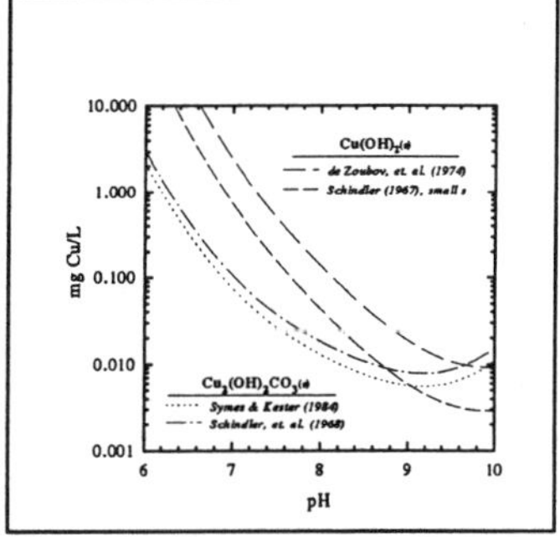

Figure 3. Comparison of effect of different solubility constants on preducted Cu(II) solubility, DIC=4.8 mg C/L (4 x 10^{-4} M), 25°C, I=0.005.

Figures 3 and 4 summarize the impact on predicted copper(II) concentrations produced by different choices for the solubility constants for $Cu(OH)_2$(s) and $Cu_2(OH)_2CO_3$(s). Figure 3 shows that at a fairly low DIC concentration of 4.8 mg C/L (4.0 x 10^{-4} M) the pH boundary between these two solids can shift from approximately 8.8 to 9.9, depending on the combination of values chosen. During recrystallization and ageing, $Cu(OH)_2$(s) will slowly convert to CuO(s). This will cause the pH of transformation from $Cu_2(OH)_2CO_3$(s) stability to that of CuO(s) to occur at lower values than with $Cu(OH)_2$(s).

Temperature Effects on Copper Solubility

Although some researchers have made predictions of copper corrosion behavior at elevated temperature (Adeloju and Hughes, 1986), adequate information is generally not available to characterize equilibrium and solubility constants for many important copper compounds and complexes at temperatures other than 25 °C. Data are particularly lacking for the carbonate complexes (that dominate cupric ion speciation above a pH of approximately 7), hydroxide solids, and phosphate species. Research on $Cu_2(OH)_2CO_3(s)$ solubility in sea water has indicated that the net effect on copper solubility may be a significant decrease at low temperature (Mor and Beccaria, 1975, Paulson and Kester, 1980, Symes and Kester, 1984), but the difference in background chemistry between sea water and most drinking waters affects the speciation to a degree that quantitative extrapolation is unreliable.

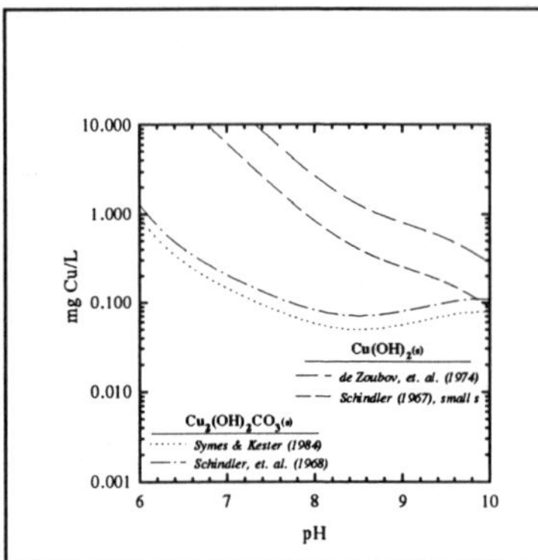

Figure 4. Comparison of effect of different solubility constants on preducted Cu(II) solubility, DIC=96 mg C/L (8 x 10^{-3} M), 25°C, I=0.02.

Residual Oxidants and Redox Potential Effects

In drinking waters, the oxidizing agents (electron acceptors) that will cause the corrosion of metallic copper are predominantly dissolved oxygen and aqueous chlorine species. Several studies have also proven aqueous chlorine species have a significant impact on the copper oxidation and corrosion rates. However, free chlorine species (ie. $HOCl°$, OCl^-, Cl_2) have not been conclusively shown to affect the equilibrium solubility of copper, other than by influencing the valence state of the copper by its presence or absence. Many of the apparent effects of chlorine on copper solubility may merely be the result of observing copper levels under non-equilibrium conditions, such as is shown in Figure 5.

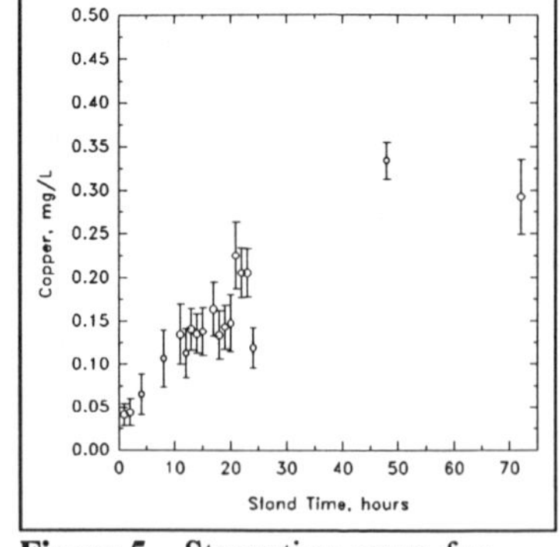

Figure 5. Stagnation curve for copper after 200 days of use, Cincinnati tap water adjusted to pH 7.5.

From these considerations of redox potential conditions of drinking waters, copper plumbing can be exposed to a wide environment of E_H-pH conditions. This environment also evolves, as oxidation and reduction reactions occur through corrosion and deposition processes for different lengths of stagnation times, and in different diameters of pipes. Therefore, all three normal copper valence states (metal, +1, +2) could reasonably occur as either aqueous or solid species under drinking water conditions either geographically or with stagnation or ageing time.

References

Adeloju, S. B. Hughes, H. C. (1986) The Corrosion of Copper Pipes in High Chloride-Low Carbonate Mains Water. *Corros. Sci.* **26**, 851.

Akkaya, M. Ambrose, J. R. (1987) An Electrochemical Technique for the Prediction of Long-Term Corrosion Resistance of Copper Plumbing Systems. *Mat. Perf.* **26**, 9.

Baes, C. F., Jr. Mesmer, R. E. (1976) *The Hydrolysis of Cations.* Wiley-Interscience, New York.

Bollinger, J.-C. et al. (1992) Thermodynamic Study in Aqueous Solutions of Weakly Soluble Ionic Compounds. *Talanta.* **39**, 959.

Byrne, R. H. Miller, W. L. (1985) Copper(II) Carbonate Complexation in Seawater. *Geochim. Cosmochim. Acta.* **49**, 1837.

Cruse, H. von Franqué, O. (1985) Corrosion of Copper in Potable Water Systems Internal Corrosion of Water Distribution Systems. AWWA Research Foundation/DVGW Forschungsstelle, Denver, Colorado.

de Zoubov, N. et al. (1974) Copper Atlas of Electrochemical Equilibria in Aqueous Solutions. National Association of Corrosion Engineers, Houston, TX.

Eary, L. E. Schramke, J. A. (1990) Rates of Inorganic Oxidation Reactions Involving Dissolved Oxygen Chemical Modeling of Aqueous Systems II. American Chemical Society, Washington, D. C.

Edwards, M. et al. (1994) On the Pitting corrosion of Copper. *Jour. AWWA. Submitted, Jour. AWWA.*

Federal Register. (1991a) *Lead and Copper. Final Rule.* 56:110:26460 (June 7).

Federal Register. (1991b) *Lead and Copper. Final Rule Correction.* 56:135:32112 (July 15).

Federal Register. (1992) *Lead and Copper. Final Rule Correction.* 57:125:28785

(June 29).

Gilbert, P. T. (1966) Dissolution by Fresh Waters of Copper from Copper Pipes. *Water Treatment and Examination.* **15**, 165.

Millero, F. J. (1989) Effect of Ionic Interactions on the Oxidation of Fe(II) and Cu(I) in Natural Waters. *Mar. Chem.* **28**, 1.

Millero, F. J. (1990a) Effect of Ionic Interactions on the Oxidation Rates of Metals in Natural Waters Chemical Modeling of Aqueous Systems II. American Chemical Society, Washington, D. C.

Millero, F. J. (1990b) Marine Solution Chemistry and Ionic Interactions. *Mar. Chem.* **30**, 205.

Millero, F. J. et al. (1987) The Effect of Ionic Interaction on the Rates of Oxidation in Natural Waters. *Mar. Chem.* **22**, 179.

Mor, E. D. Beccaria, A. M. (1975) Effect of Temperature on the Corrodibility of Copper and Zinc in Synthetic Sea Water. *NACE Corros.* **31**, 275.

Pankow, J. F. (1991) *Aquatic Chemistry Concepts.* Lewis Publishers, Inc., Chelsea, Michigan.

Paulson, A. J. Kester, D. R. (1980) Copper(II) Ion Hydrolysis in Aqueous Solution. *J. Solution Chem.* **9**, 269.

Rickard, D. T. (1970a) *7. A Note on the Effect of Ammonia on the Solubility of Some Copper Minerals.* Acta Universitatis Stockholmiensis,, Contr. Geology 21:77-87 pp., 7. A Note on the Effect of Ammonia on the Solubility of Some Copper Minerals.

Rickard, D. T. (1970b) *The Chemistry of Copper in Natural Aqueous Solutions.* Acta Universitatis Stockholmiensis,, Contr. Geology 23:1:64 pp., The Chemistry of Copper in Natural Aqueous Solutions.

Schindler, P. et al. (1965) Löslichkeitsprodukte von Zinkoxid, Kupferhydroxid und Kupferoxid in Abhängigkeit von Teilchengrösse und molarer Oberfläche. Ein Beitrag zur Thermodynamik von Grenzflächen fest-flüssig. *Helvetica Chim. Acta.* **48**, 1204.

Schindler, P. W. (1967) Heterogeneous Equilibria Involving Oxides, Hydroxides, Carbonates and Hydroxide Carbonates Equilibrium Concepts in Natural Water Systems. American Chemical Society, Washington, DC.

Schock, M. R. et al. (1994) Effect of pH, DIC, and Orthophosphate on Drinking Water Cuprosolvency. *Submitted, Jour. AWWA*.

Stiff, M. J. (1971) Copper/ Bicarbonate Equilibria in Solutions of Bicarbonate Ion at Concentrations Similar to Those Found in Natural Water. *Wat. Res.* **5**, 171.

Symes, J. L. Kester, D. R. (1984) Thermodynamic Stability Studies of the Basic Copper Carbonate Mineral, Malachite. *Geochim. Cosmochim. Acta*. **48**, 2219.

Woods, T. L. Garrels, R. M. (1987) *Thermodynamic Values at Low Temperature for Natural Inorganic Materials: An Uncritical Summary*. Oxford University Press, New York, New York.

THE EFFECTS OF NOM AND COAGULATION ON COPPER CORROSION

John P. Rehring and Marc Edwards
Department of Civil, Environmental, and Architectural Engineering
University of Colorado, Boulder, CO 80309-0428

Abstract. Copper corrosion was examined in solutions containing natural organic matter (NOM) and in situations where NOM was removed by enhanced coagulation with alum or ferric chloride. Electrochemical methods were used to evaluate the long-term effects of each water quality on copper corrosion. In experiments exploring the role of NOM in copper corrosion, corrosion rates increased with NOM concentration at pH 6, whereas at pH 7.5 and 9 the NOM had less significant effects. Waters treated by enhanced alum coagulation had higher corrosion rates than untreated waters, but enhanced ferric chloride coagulation had the opposite effect. This difference was attributed to the relative effects of added sulfate via alum coagulation versus added chloride via ferric chloride coagulation. Compliance with the EPA Lead and Copper Rule and disinfection byproduct regulations may require that utilities address both regulations simultaneously. That is, water treatment processes, NOM concentration, and copper corrosion behavior are clearly interdependent and should be considered when contemplating changes to meet DBP regulations.

INTRODUCTION

Corrosion has come under increased scrutiny since the 1991 EPA Lead and Copper Rule reduced the acceptable amount of lead and copper in drinking water. Repercussions of the rule will be widespread, and in fact, it has been predicted that as many as 40% of the water utilities in some areas of the United States will fail to meet the new copper "action level" (AWWA, 1993). Failure to meet the rule requires an investigation of alternative corrosion control schemes.

Concurrently, ever-tightening regulation of carcinogenic disinfection by-products (DBPs) has prompted water utilities to remove or alter NOM in water supplies because NOM is a DBP precursor material. A common method of removing NOM is through enhanced coagulation with alum or ferric chloride.

Evidence exists that suggests these two regulatory issues cannot be addressed independently. The fact that NOM can impact copper corrosion has been known for more than 40 years (Campbell, 1954), and its removal for compliance with DBP regulations might be expected to alter copper corrosion. Moreover, changes to the water quality effected by NOM removal measures (such as increased chloride or sulfate concentration due to enhanced coagulation) might also impact corrosion. This work examined the isolated effects of NOM on copper corrosion as well as the effects removing NOM via enhanced coagulation with alum or ferric chloride.

MATERIALS AND METHODS

NOM Concentrate. A concentrate of NOM was prepared for use in dosing into synthetic waters at various concentrations. A series of pre-filters followed by a 200 MW-cutoff nanofiltration membrane (Filmtec, Inc. Model NF70, Mpls., MN) using a recirculation loop was used to concentrate NOM from Silver Lake (Boulder County, CO). The initial dissolved organic carbon (DOC) concentration was about 6 mg/L, and the final concentration of the concentrate solution was 155 mg/L NOM as DOC. After collection, the concentrate was stored in a dark refrigerator at 4 °C. The concentrate was passed twice through a sodium-based cation exchange resin (Rohm & Haas, Ind. 120, Phila., PA) to replace all cations present with sodium. Removal of cations was confirmed by testing total hardness prior to and after cation exchange. The solution was lowered to pH 5.0 with concentrated sulfuric acid to hinder biological growth, and subsequent stripping with nitrogen gas removed carbonate species. As expected, the nanofiltration process concentrated only DOC and not other anions.

Synthetic Solution Preparation. The base synthetic solution used in all experiments contained 2 mM sulfate, 0.2 mM chloride, and 2 mM total carbonate. NOM was dosed at concentrations of 0, 0.2, 2, or 4 mg/L NOM by adding the appropriate amount of NOM concentrate to the base synthetic water. All solutions were oxygen-saturated throughout each test.

For pH control during electrochemical testing, an acid solution containing 2 mM sulfuric acid, 0.2 mM hydrochloric acid, 2 mM sodium hydrogen carbonate, and the correct NOM concentration (treated with the appropriate coagulant, if any) was dosed into the solution reservoirs as needed by pH control pumps. Thus, the anion and NOM concentrations in each solution reservoir were held constant throughout each test and were not changed by acid additions.

Domestic Supply Water Preparation. Two potable supply waters were selected for comparison to the synthetic waters in experiments examining the effects of coagulation on copper corrosion. The supply waters were Silver Lake and Boulder Reservoir (Boulder County, CO). Silver Lake is a pristine high-mountain source with very low anion content and alkalinity, whereas Boulder Reservoir is of relatively high sulfate content and high alkalinity. Typical water quality values for each water are summarized in Table 1. Both raw waters were collected and filtered with a 0.45μm cellulose acetate filter prior to testing. The waters were tested at their naturally-occurring pH values. Acids used for pH control consisted of the supply water (treated with the appropriate coagulant, if applicable) bubbled with CO_2 (g) to pH 4.0. Use of CO_2 as acid maintained constant alkalinity.

Table 1: Typical Water Quality Values for Domestic Supply Waters Tested

Parameter	Silver Lake	Boulder Reservoir
pH	6.4	8.1
NOM (mg/L as DOC)	2.6	3.9
Total Alkalinity (mg/L as $CaCO_3$)	7.8	62
Total Hardness (mg/L as $CaCO_3$)	9.4	120
Chloride (mM)	0.01	0.01
Sulfate (mM)	0.02	0.45

Table 2: Doses Used for Coagulation

Dose	Synthetic Waters	Silver Lake	Boulder Reservoir
Alum (mg/mg DOC)	20	10	20
Ferric Chloride (mg as $FeCl_3$/mg DOC)	10	5	10

Coagulation. Tests were performed to determine the optimal doses of alum and ferric chloride. The optimal dose was defined as the minimum coagulant dose at which maximum NOM removal was realized (Table 2). Alum and ferric doses for a given water were approximately equimolar in terms of Al^{3+} and Fe^{3+}.

The water to be coagulated was first mixed with a magnetic stirrer in a 4 L beaker at approximately 200 rpm, at which time the appropriate dose of alum or ferric chloride was added. Following a 30 second rapid mix period, the rate of mixing was reduced to approximately 30 rpm for an additional 30 minutes to allow flocculation. Thereafter, the water was allowed to settle quiescently for at least 30 minutes prior to filtration with a 0.45 μm cellulose acetate filter. pH was readjusted to the target value using 0.1 M NaOH.

Electrochemical Testing. Potentiostatic techniques were used to accelerate scale formation, simulating several months of conventional exposure to the target water quality. Subsequent potentiodynamic testing provided corrosion rate (i_{corr}) data for the aged samples in accordance with electrochemical theory (Reiber, 1989). Details of the electrochemical testing apparatus and test methods used are provided elsewhere (Edwards et al., 1994; Edwards and Ferguson, 1993). Target solutions were pumped through the electrochemical cells in a closed loop at a flowrate 0.5 gpm and a constant temperature of 15 °C ± 1 °C. Each target water was tested in duplicate or triplicate, and excellent qualitative and quantitative agreement between replicates was observed.

RESULTS AND DISCUSSION

Effect of NOM. Anodic current densities at the conclusion of the potentiostatic tests increased with increasing NOM concentration at each pH. Overpotentials (η = +120 mV - E_{corr}) or "polarization" decreased with increasing NOM concentration at pH 6 but did not show significant trends at higher pHs (Table 3). Drawing an analogy to Ohm's law (V=IR), polarization may be expected to be proportional to current density for a given resistance (as provided by the scale on the pipe surface). Since at pH 6 the overpotential decreased with increasing NOM but the resulting current density increased, the scale formed at higher NOM concentrations was clearly less resistant to corrosion than in the systems with lower NOM concentration. At pH 7.5 and 9, the degree of polarization was fairly constant with increasing NOM content while current density increased, suggesting that less resistant (more aggressive) scales were formed at higher NOM concentration.

Table 3: Overpotentials (η) and log Anodic Current Density (ι) after 72 Hours Polarization

NOM Concentration (mg/L as DOC)	pH 6		pH 7.5		pH 9	
	η (mV)	log ι (A/cm²)	η (mV)	log ι (A/cm²)	η (mV)	log ι (A/cm²)
0.0	118.3	-4.0	119.5	-5.2	146.1	-5.2
0.2	97.6	-3.7	157.5	-4.6	137.0	-5.0
2.0	78.3	-4.4	141.3	-4.2	145.1	-4.5
4.0	88.6	-3.5	123.7	-3.9	135.4	-4.4

Potentiodynamic scans were used to determine the corrosion rates of coupons aged by the potentiostatic methods described above. Corrosion rates were measured in this work as i_{corr}, the unperturbed corrosion current of the pipe sample. In general, for a given NOM content corrosion rates decreased with increasing pH. Addition of NOM to the synthetic water at pH 6 resulted in an increased corrosion rate (Figure 1). Duplicate measurements showed excellent agreement. Although not linear, there is a clear relationship between the amount of NOM added and the resulting i_{corr}. In fact, the rate of corrosion appears to level off at higher NOM concentrations, suggesting the possibility of an asymptote-like value of NOM content above which no additional impact on corrosion rates would be observed.

In contrast, less significant effects were seen at the higher pH values tested (Figure 1). At pH 7.5, the corrosion rate was observed to drop initially with the addition of 0.2 mg/L NOM, but then increased very slowly as the NOM concentration approached 4 mg/L as DOC. At pH 9, no significant trends in i_{corr} were detected as a function of NOM content.

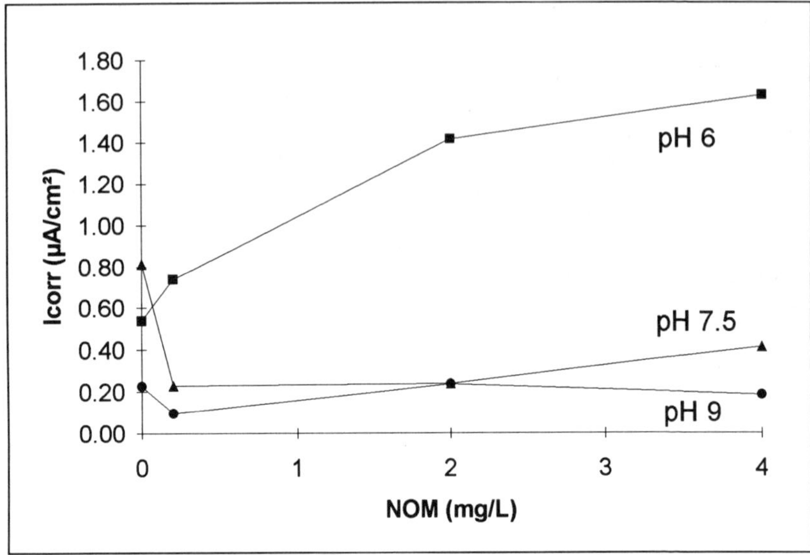

Figure 1: Effect of NOM on corrosion rates for aged coupons in synthetic water.

Alum vs. Ferric Chloride Coagulation. Examination of the anodic current at the end of the potentiostatic tests (72 hours elapsed aging) reveals that in nearly every case, waters treated with alum had a higher current density than those treated with ferric chloride (Table 4). Overpotentials (η = +120 mV - E_{corr}) were generally higher for waters treated with ferric chloride, indicating a higher degree of polarization. Per the Ohm's law analogy discussed earlier, a higher polarization would be expected to yield a higher current density for a given resistance. The fact that ferric chloride waters were more highly polarized than alum but gave *lower* anodic current densities demonstrates the formation of scale that is much more protective in waters treated with ferric chloride than in those treated with alum.

Corrosion rates were measured for each of the coupons aged in the potentiostatic tests 24 hours after their completion using potentiodynamic methods. In synthetic waters without NOM, coagulation with alum caused higher corrosion rates than did coagulation using ferric chloride at each pH tested (Figure 2). Previous research has demonstrated that sulfate alone increases copper corrosion rates, whereas chloride results in passivation (Edwards et al., 1994). The results of the present work are consistent with these reported effects of sulfate and chloride. Alum coagulation, which added a significant amount of sulfate to the water, tended to increase corrosion rates. In contrast, ferric chloride coagulation which added chloride resulted in much lower corrosion rates.

Table 4: Overpotentials (η) and log anodic Current Density (ι) after 72 Hours Polarization

	Alum		Ferric Chloride	
	η (mV)	log ι (A/cm²)	η (mV)	log ι (A/cm²)
pH 6: 0 mg/L NOM	102.1	-4.0	182.8	-4.9
pH 6: 4 mg/L NOM	129.1	-3.8	133.4	-4.6
pH 7.5: 0 mg/L NOM	83.0	-5.1	82.8	-5.8
pH 7.5: 4 mg/L NOM	164.2	-4.1	92.0	-5.5
pH 9: 0 mg/L NOM	110.6	-6.9	163.4	-6.1
pH 9: 4 mg/L NOM	84.4	-4.4	149.4	-6.1
Silver Lake	43.9	-4.1	59.5	-4.5
Boulder Reservoir	75.1	-3.7	131.1	-4.9

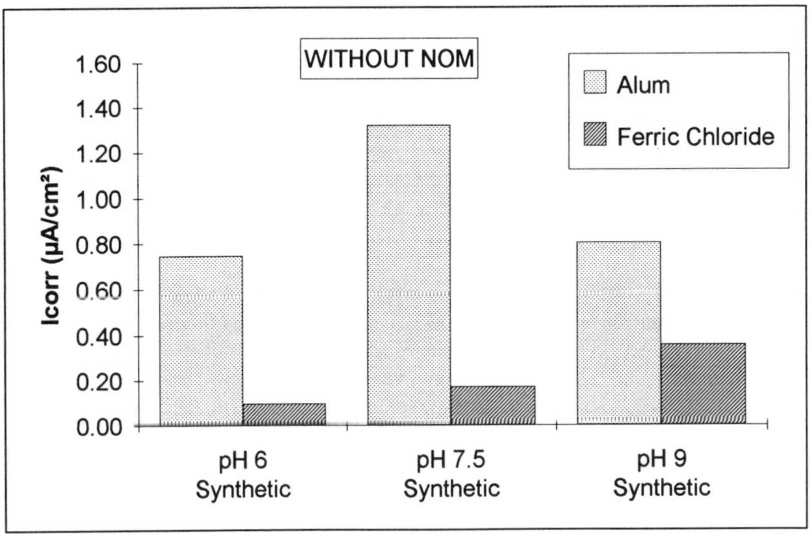

Figure 2: Effects of Alum vs. Ferric Chloride enhanced coagulation on corrosion rates of aged copper samples. Waters shown here contain no NOM.

Effects similar to those seen in waters without NOM were observed in most synthetic and natural waters containing NOM (Figure 3). That is, alum coagulation resulted in higher corrosion rates than did ferric chloride coagulation. The sole exception to this trend was Boulder Reservoir for which ambient sulfate concentrations were initially high, and in that case very little difference in i_{corr} was observed between coagulant chemicals. In the majority of cases alum coagulation yielded higher corrosion rates than untreated water, whereas ferric chloride treatment tended to either reduce or have no effect on the corrosion rate.

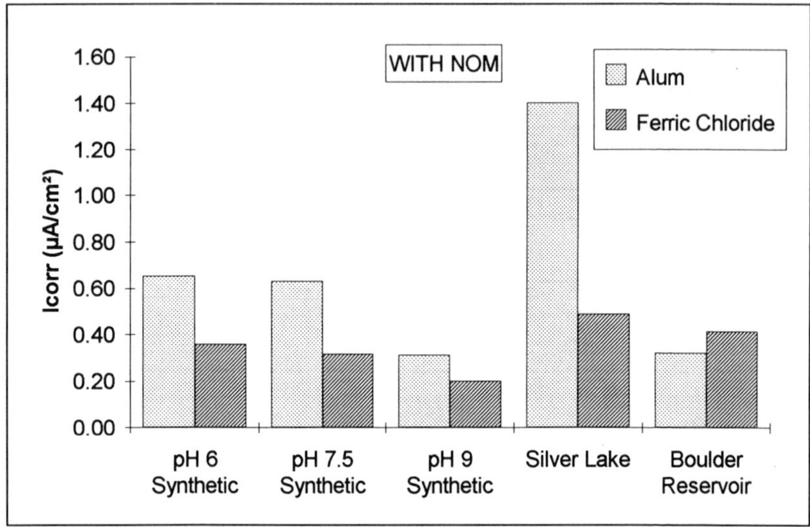

Figure 3: Effects of Alum vs. Ferric Chloride enhanced coagulation on corrosion rates of aged copper samples. Synthetic waters shown here contain 4 mg/L NOM as DOC.

For each water, enhanced alum coagulation removed about the same amount of NOM (within 7%) as ferric chloride. Thus, the significant differences in corrosion rates observed between coagulants cannot be attributed to differences in the amount of NOM removal. Since the qualitative effects of alum versus ferric chloride coagulation were the same regardless of NOM content, the differences may be explained by the contributions of different anions via the coagulant chemicals. That is, with or without NOM present, alum coagulation (adding sulfate) resulted in higher corrosion rates than did ferric chloride coagulation (adding chloride). Interestingly, however, the presence of NOM appeared to reduce the relative impact of the coagulants in synthetic waters. This effect may be due to the formation of NOM-copper complexes, reducing the availability of cuprous or cupric ions for complexation with chloride and sulfate.

In summary, it is clear that copper corrosion is dependent on NOM and can be strongly affected by coagulation processes aimed at removing NOM. Although not typically considered in corrosion control studies, choice of coagulant chemical for removal of DBP precursors has significant effects on the corrosivity of the resulting water. Thus, compliance with the EPA Lead and Copper Rule must be considered simultaneously with steps taken to gain compliance with DBP regulations.

CONCLUSIONS

1) Increasing NOM concentration caused an increase in corrosion rates at low pH for the synthetic water examined in this work; effects of NOM at pH ≥7.5 were minor.

2) Waters treated with alum coagulation resulted in higher corrosion rates than did those coagulated with ferric chloride. This difference was attributed to the addition of chloride vs. sulfate via the different coagulants, with sulfate causing increased corrosion rates and chloride resulting in more passivating scales.

3) The presence of NOM did not alter the qualitative effects of alum and ferric chloride coagulation; however, the magnitude of the differences observed between coagulants was reduced by the presence of NOM.

4) Because of the interactive nature of NOM, water treatment, and copper corrosivity, compliance with the Lead/Copper Rule and DBP regulations must sometimes be considered simultaneously.

REFERENCES

AWWA. *Waterweek*, March 29, 1993, p. 5.

Campbell, H.S. "The Influence of the Composition of Supply Waters, and Especially of Traces of Natural Inhibitor, on Pitting Corrosion of Copper Water Pipes." *Proc. Soc. of Water Treatment and Exam.*, V. 8. 100-116 (1954).

Edwards, M. and J.F. Ferguson. "Accelerated Testing of Copper Corrosion." *J. AWWA*, V. 85, No. 10. 105-113 (1993).

Edwards, M., T. Meyer, and J. Rehring. "Effect of Various Anions on Copper Corrosion Rates." *J. AWWA* (1994, in press).

Reiber, S. "Copper Plumbing Surfaces: An Electrochemical Study." *J. AWWA*, V. 81, No. 7. 114-122 (1989).

Water Chemistry of Lead Corrosion Control

Leland L. Harms[1], Member, Jonathan A. Clement[2], and Michael R. Schock[3]

Abstract

Requirements of the Lead and Copper Rule (U.S. EPA) makes it necessary for regulators, consultants, and water purveyors to understand the fundamental relationships between water quality and lead leaching from lead based plumbing materials. This paper will address the basic concepts of water chemistry as they apply to the control of lead corrosion.

Introduction

National Primary Drinking Water Regulations for lead and copper were promulgated on June 7, 1991. EPA adopted a three-phased approach to controlling these contaminants by specifying treatment, public education, and lead service line replacement. Treatment included source water treatment if the supply had high levels of lead or copper, and the implementation of "optimal" corrosion control to minimize the concentrations of lead and copper at consumer's taps. This paper will address only measures that can be used to reduce lead levels at the tap.

Electrochemical corrosion cells consist of an anode, a cathode, a mechanism of

[1]Senior Water Treatment Engineer, Black & Veatch, 8400 Ward Parkway, Kansas City, MO 64114
[2]Process Engineer, Black & Veatch, 100 Cambridgepark Drive, Boston, MA 02140
[3]Research Chemist, US Environmental Protection Agency, 26 Martin Luther King Dr., Cincinnati, OH 45268

transporting the electrons from the anode to the cathode, and an ionic solution to complete the circuit, allowing corrosion currents to flow. The loss of metal (corrosion) generally is most severe at the anode. Because water contains *oxidizing agents* (electron acceptors) that convert the metal to an ionic form (eg., Cu to Cu^{+} or Cu^{+2}, Pb to Pb^{+2} or Pb^{+4}), corrosion cannot be completely eliminated and all potable waters are corrosive to some degree. This reaction is facilitated by dissolved constituents in the water, primarily *complexing agents*, that stabilize the oxidized forms of the metal in solution, which enhances the corrosion reaction and increases the metal solubility in the water. Common oxidizing agents in drinking water are dissolved oxygen and various chlorine species. Common complexing agents include hydroxide ion, carbonate and bicarbonate ion, ammonia, chloride and others. Corrosion control treatment consists of disrupting the action of the corrosion cell by one or more methods in an effort to limit the corrosion to acceptable physical and aesthetic levels.

Corrosion Control Treatment Options

The Lead and Copper Rule, and thus subsequently the Guidance Manuals (U.S. EPA 1992), identify three available corrosion control strategies:

- Alkalinity and pH Adjustment,
- Calcium Hardness Adjustment, and
- Corrosion Inhibitors.

Identifying these strategies separately can be somewhat misleading, however, in that it implies that one strategy is independent of the other and this generally is not the case. For example, adjusting pH, alkalinity, and calcium levels could all be beneficial for a public water system that practices corrosion control through the formation of a protective calcium carbonate layer. Additionally, it is not unusual to need to adjust a finished water pH in order to effectively gain benefits from a commercially available inhibitor because corrosion inhibitors are only effective in pH ranges appropriate to their chemical action.

Lead levels observed in monitoring programs are a complicated combination of physical and chemical factors, including such non-chemical influences as the presence of natural diffusion barriers protecting the pipe from oxidation, the length of standing time, the inside diameter of the pipe, the flow velocity and turbulence of the water, temperature, the nature of the material (eg., brass, soldered joints, lead pipe), and others (Schock 1989, Schock 1990). Lead levels tend to be lower than predicted by solubility models, because they are controlled by many of these factors, and not equilibrium solubility. Analysis of surficial deposits from many lead pipes has shown a wide variety of coatings (AWWARF 1994).

In some utilities, the coatings predominantly consist of lead corrosion byproduct solids, such as $PbCO_3$ (cerussite), $Pb_3(CO_3)_2(OH)_2$ (hydrocerussite),

$Pb_{10}(CO_3)_6(OH)_6O$ (plumbonacrite), or PbO (litharge). However, elemental analyses and some X-ray diffraction and IR spectroscopy analyses indicate frequent occurrence of many solid phases governed by the source water chemistry or treatment processes, containing such elements as Al, Si, Fe, Ca, S, P, or others. Organic carbon is found on some pipe specimens. From a lead corrosion control strategy standpoint, the state of understanding the nature of these natural corrosion barriers has not yet developed sufficiently to allow utilities to quantitatively manipulate water chemistry to induce the formation of non-lead deposits that will predictably reduce lead dissolution from the various plumbing materials.

Because of these inter-relationships, it seems more accurate to divide the lead control strategies into (1) those that form a protective coating or layer, and (2) those that form a passivating film due to chemical interaction with the plumbing material. The four available lead control strategies can then be defined within this grouping as follows:

1) Surface coating
   - Calcium Carbonate Precipitation, and
   - Silicate Addition.
2) Passivating film
   - Carbonate Passivation, and
   - Phosphate Addition.

Each of the four lead control strategies listed above has its own specific set of water quality parameters which define optimal chemical conditions for reduced lead levels. These conditions will be more completely discussed later in this paper.

Calcium carbonate precipitation is the most commonly described corrosion control technique in water treatment literature. The technique is based on the chemical stability of calcium carbonate (calcite) and the adjustment of the treated water quality to cause some precipitation to occur throughout the distribution system. Thus, the goal is to lay down a uniform, adjacent, thin layer of carbonates on interior piping which will protect the materials. This method is most suitable for waters with calcium levels which exceed 20 mg/L.

Several indices are used to adjust finished water quality to achieve the layer of calcium carbonate within the distribution system. Although these indices are frequently referred to as "corrosion indices", they are in fact indicators of a tendency of water to deposit calcium carbonate and they are not actual indicators of corrosion. The most popular index has been the Langelier Index (LI) which was originally proposed by Langelier in 1936 (Langelier). Traditional water treatment practice has considered a suitable layer to be established when a positive LI is maintained, while a water with a negative index is considered to be aggressive and one that will cause corrosion.

Historically, it has not been unusual to find a water system experiencing "red or brown" water complaints while keeping a positive LI; and vice versa, some systems seemingly never experience these types of corrosion related complaints although their finished water quality would calculate to a negative LI. A balanced approach to attaining a calcium carbonate protective layer is important. It is desired to achieve a continuous coating throughout the distribution system without causing excessive precipitation in some portions of the piping network. Some systems have experienced substantial reductions in hydraulic capacity due to carbonate precipitate buildup. It is difficult to achieve this balance in practice, so some utilities employ small amounts of unstable polyphosphates to impede excessive calcium carbonate crystal growth. Slow reversion of the polyphosphate to the orthophosphate form can allow the film formation to travel further into the distribution system.

The calcium carbonate precipitation potential (CCPP) is a more precise indicator of carbonate equilibrium chemistry than the LI, and its use is recommended. The CCPP has been shown to relate directly to reaction kinetics (Nancollas and Reddy). A desirable range for the CCPP is 4 mg/L to 10 mg/L and the equilibrium conditions needed to achieve the desired CCPP can now easily be determined through the use of computerized spreadsheets or programs.

Passivating film formation control strategies depend on the reaction of the oxidized lead at the surface of the plumbing material with combinations of constituents in the water, such as hydroxide ion, carbonate ion, and orthophosphate ion, to form minutely soluble and adherent films. In contrast to the carbonate precipitation method which attempts to form a physical barrier between the water and the plumbing materials, carbonate passivation is a technique which forms a chemical film with the pipe. This technique is based on the theoretical solubility relationships of basic lead carbonates, $Pb_3(CO_3)_2(OH)_2$ or $Pb_{10}(CO_3)_6(OH)_6O$, or the normal lead carbonate $PbCO_3$ (Schock 1989, Sheiham and Jackson, Schock and Wagner, and AWWARF 1990).

Optimum conditions for the formation of a protective film are at pH levels above 9 and with DIC concentrations of less than 10 mg/L, with a lead-carbonate-hydroxyl film $[Pb_3(CO_3)_2(OH)_2]$ being a common candidate. Thus, the use of carbonate passivation is particularly suited for waters which have a high pH and a low DIC. A low level of DIC, in the 3 to 5 mg/L range, would be considered optimal for film formation. Extremely low DIC conditions should be avoided because of the minimal buffering capacity which would result and the rapid rise in lead solubility. Finished water alkalinities of at least 20 mg/L are needed for normal operations.

Lead can form an orthophosphate solid of low solubility and this compound also can become a means of discontinuing the action from a corrosion cell. Several diverse treatment chemicals have been used for this purpose. The commonly used forms of phosphate inhibitors are (1)orthophosphate, (2) polyphosphate, and (3) ortho/polyphosphate blends. It is the orthophosphate ion which forms the passivating film. The chemical compounds that form with lead usually has been identified as $Pb_5(PO_4)_3OH$ or $Pb_3(PO_4)_2$, likely depending upon the background chemistry of the water.

The lead/phosphate film is theoretically less soluble than a carbonate-based film and forms an excellent protective barrier under certain conditions. The passivation is very pH sensitive and best results are obtained in the pH range of 7.4 to 7.8. Above approximately pH 8 there is evidence that colloidal lead orthophosphate solids form that adhere poorly to the pipe surface, even though the actual lead orthophosphate solubility is lower. Thus, higher pH levels should be avoided for optimal lead control, until further detailed research is conducted to better understand the phenomenon. Again, a low DIC is optimal for passivation with the highest success occurring at DIC levels less than 5 mg/L, but a low DIC must be balanced against the needed alkalinity for a buffered finished water that will maintain the target pH range throughout the distribution system.

The orthophosphate concentration must be maintained for the passivation to be effective as a corrosion control technique. The dosage of orthophosphate is extremely critical to the passivation process. Not only must the orthophosphate residual concentration remain above that required (theoretically) for passivation throughout the distribution system, but it must exceed the minimum necessary by enough of a margin to facilitate rapid and complete coating. Practical demonstration of the effect of orthophosphate dosage on lead levels has been observed (Colling).

Some systems have reported success using polyphosphates, but these compounds are actually sequestering agents which may serve to keep lead and other metals in solution. Polyphosphates will hydrolyze to orthophosphate either in the water or at the pipe surface, thus causing some orthophosphate addition within the distribution system. The rate of reversion to orthophosphate is dependent upon pH, temperature, the presence of certain metals (Fe^{+2}, Ca^{+2}), age, etc. In fact, if sufficient storage time elapses prior to feeding the polyphosphate, a substantial fraction may be in the orthophosphate form and available for lead passivation. Polyphosphates alone are not recommended for general corrosion control purposes, but their use may be necessary for other water quality benefits such as the sequestering of excess calcium and magnesium deposits across filters in softening facilities.

Blends of orthophosphate and polyphosphate in various formulations are also available from many distributors. The use of blends is sometimes advantageous when faced with the dual problems of sequestering and passivation. It is difficult to obtain definitive information for these blends. Details such as the percentage of orthophosphate, types and amounts of other ingredients, etc. are sometimes considered to be proprietary and are not freely disclosed. Thus, it is usually advisable to determine if the potential product has been successfully used elsewhere on similar waters.

The mechanism involved in controlling corrosion with silicates is still unclear. Silicates are manufactured by the fusion of high-quality silica sands to sodium or potassium salts with sodium silicates being most commonly used. Conventional sodium silicates use silica to sodium carbonate molar ratios between 1.5 and 4 to 1. Silicates are considered to be anodic inhibitors which combine with the free metal released at the anode site of corrosion activity to form an insoluble metal-silicate compound. These corrosion byproducts crystallize to form a protective barrier on the surface of pipe walls and other plumbing materials. Microscopic and X-ray examinations have shown more than one film or layer on materials treated with silicates, and whether the formation of the protective film can correctly be termed as a passivation technique is still debateable.

As with phosphate inhibitors, each water treated with silicates will have a silicate demand which must be satisfied before a sufficient residual can be maintained to form the protective film. Calcium and magnesium, as well as other compounds, are known to react with silica. Thus, a sufficient dosage must be maintained to satisfy these demands in addition to providing a silicate residual.

pH Stability

In order to properly implement lead control strategies, a broad range of general inorganic chemical data is important and corrosion control studies should not limit data collection to only some specific analytes such as lead and copper. For example, because all of the control strategies are extremely dependent on pH, it is crucial for water systems to understand the water quality parameters which define or quantify pH stability. Alkalinity is the most widely used measure of pH stability. The classical definition of alkalinity is the acid neutralizing capacity of a water, and it is defined by the following equation:

$$\text{Alkalinity} = [HCO_3^-] + 2[CO_3^{-2}] + [OH^-] - [H^+]$$

The primary mechanism of buffering in water is provided by the carbonate alkalinity system. In other words, the individual carbonate species (CO_2, H_2CO_3, HCO_3, and CO_3) shift from one form to another depending upon the pH of the solution. Alkalinity is closely related to dissolved inorganic carbon, DIC, which is the sum of the carbon dioxide, carbonic acid, bicarbonate, and carbonate species, expressed as mg/L of carbon.

The term buffering intensity is used to denote how much the pH will shift given a unit quantity of acid or base. Most natural waters exhibit minimal buffering intensities near a pH of 8.3, thus only small changes in chemical feeds may cause erratic swings in pH near this range.

Another point to remember is that the water chemistry near the surface of a pipe or fitting may be quite different than that of the bulk solution. Data collected from water samples will reflect the water flowing through the pipe, but they may not be helpful when trying to determine what is happening at the pipe wall. Corrosion control treatment often consists of attempting to form a protective film or layer on the surface of the pipe. Not just any precipitate will form a protective layer; it may simply stay suspended in solution without adhering to the pipe surface.

Summary

The recent promulgation of the Lead and Copper Rule has required every public water system in the United States to examine their corrosion control practices. The basic concepts of water chemistry as related to the control of lead at the tap have been presented in this paper in the hope that increasing the understanding of these fundamentals will result in reduced concentrations of lead in drinking water. The importance of DIC and the carbonate buffering system have been emphasized, and four treatment strategies for lead corrosion control have been identified as carbonate passivation, calcium carbonate precipitation, phosphate addition, and silicate addition.

References

Colling, J. H., *et al* (1992). "Plumbosolvency Effects and Control in Hard Waters." *J. IWEM*, 6(6), 259.

"Internal Corrosion of Water Distribution Systems" (1994). 2nd Ed., AWWA Research Foundation, AWWARF/DVGW Forschungsstelle, in press.

Langelier, W. F. (1936). "The Analytical Control of Anticorrosion Water Treatment." *J. AWWA,* 28(6), 1500-1521.

"Lead Control Strategies" (1990). AWWA Research Foundation and AWWA, Denver, CO.

"Lead and Copper Rule Guidance Manual, Vol. II: Corrosion control Treatment." September 1992. U.S. EPA, Washington, D.C.

Nancollas, G. H., and Reddy, M. M. (1976). "Crystal Growth Kinetics of Minerals Encountered in Water Treatment Processes." In, Aqueous-Environmental

Chemistry of Metals, Ann Arbor Science, Ann Arbor, MI.

"National Primary Drinking Water Regulations for Lead and Copper". June 7, 1991. U.S. EPA, *Federal Register*.

Schock, M. R. (1989). "Understanding Corrosion Control Strategies for Lead". *J. AWWA*, 81(7), 88-100.

Schock, M. R. (1990). "Causes of Temporal Variability of Lead in Domestic Plumbing Systems". *Envir. Monit. & Assess.*, 15, 59.

Schock, M. R. and Jackson, P. J. (1981). "Scientific Basis for Control of Lead in Drinking Water by Water Treatment." *J. Inst. Water Engrs. & Scientists*, 35(6) 491.

CHEMICAL INACTIVATION OF *CRYPTOSPORIDIUM PARVUM* IN DRINKING WATER

Gordon R. Finch and E. Kathleen Black[1]

Abstract

Inactivation of *Cryptosporidium parvum* oocysts by ozone was performed in ozone demand-free 0.05 M phosphate buffer (pH 6.9) in bench-scale batch reactors at 7° and 22°C for contact times ranging from 5 to 15 min. The neonatal CD-1 mouse model was used to measure the viability of *Cryptosporidium parvum*. The inactivation of *C. parvum* by ozone deviated from the simple first-order Chick-Watson kinetic model and was better described by the non-linear Hom kinetic model. The use of the Hom model for predicting inactivation resulted in a family of unique concentration and time values for each inactivation level rather than the simple *CT* product of the Chick-Watson model.

Introduction

Cryptosporidium has become recognized as a frequent cause of water-borne disease in humans (Barer and Wright 1990). Cryptosporidiosis outbreaks from surface water supplies have been documented in the United States and Great Britain (Gallaher et al. 1989; Hayes et al. 1989; Richardson et al. 1991; Rose 1990; Rush et al. 1990).

Chlorine and ultraviolet disinfection of *C. parvum* have been reported to be ineffective (Korich et al. 1990; Lorenzo-Lorenzo et al. 1993; Smith et al. 1988). Chlorine dioxide and ozone have been reported to have the most potential for inactivating *C. parvum* oocysts (Korich et al. 1990; Peeters et al. 1989; Perrine et al. 1990) However, there have been relatively few studies of ozone inactivation of *Cryptosporidium* oocysts to date (Korich et al. 1990; Langlais et al. 1990; Peeters et al. 1989) and none have looked at the kinetics of inactivation.

This paper presents the results of the first part of an American Water Works Association Research Foundation project studying chemical inactivation of *C. parvum*.

---

[1] Environmental Engineering and Science Program, Department of Civil Engineering, University of Alberta, Edmonton, Alberta Canada T6G 2G7

Materials and Methods

The ozone demand-free buffer, ozone residual measurement methods, and disinfection procedures used in this study were the same as used earlier (Finch et al. 1993a). Briefly, ozone demand-free, 0.05 M phosphate buffer (pH 6.9) was used to suspend the test preparation of oocysts. A measured volume of concentrated stock solution of ozone was added to the reactor and stirred. The concentration of ozone in aqueous solution was continuously determined using ultraviolet spectrophotometry at 260 nm. Ozone residual was neutralized by sodium formate at the end of the desired contact time.

Details of the *Cryptosporidium* methods and procedures including production of oocysts, concentration of samples, inoculation into the animal host, and subsequent determination of infection have been described in detail previously (Finch et al. 1993b). The dose-response model used to estimate the infectious dose of oocysts was:

$$\text{Logit} = -6.738 + 3.547 \log (\text{inoculum})$$

The log inactivation was then estimated from:

$$\log \frac{N}{N_o} = \log\left(\frac{n}{n_o}\right)$$

where; n is the estimated infectious dose per animal after ozone treatment and n_o is the initial number of cysts or oocysts in the controls.

Typically chemical disinfection results are expressed in terms of a concentration × time product (CT) for different levels of inactivation (Hoff 1986). The CT product can be derived theoretically from the Chick-Watson, pseudo first-order rate law (Chick 1908; Watson 1908):

$$\log \frac{N}{N_o} = -kC^nT$$

where k is the pseudo first-order rate constant found experimentally; and n is an empirical constant often assumed to be unity. An alternate model has been proposed to account for deviations from the Chick-Watson model encountered in practice (Hom 1972) :

$$\log \frac{N}{N_o} = -kC^nT^m$$

where m is an empirical constant. The assumption that underlies both the Chick-Watson and Hom models is that the disinfectant concentration remains approximately constant during the course of the contact time. In ozone studies, this is an erroneous assumption since ozone is reactive and continuously disappears during the contact time.

Ozone decay in the experiments reported here was reasonably approximated by a first-order ozone consumption model. Consequently, the ozone residual, C, for each experimental trial was estimated from:

$$C = \text{antilog}\left(\frac{\log C_o + \log C_f}{2}\right)$$

where C_o is the initial ozone residual at time zero and C_f is the ozone residual at the end of the contact time, T. This will be called the integrated ozone residual in this paper.

In full-scale systems there are several methods for defining the ozone concentration characteristic, C, depending on the method of adding ozone to the liquid and the type of reactor (Lev and Regli 1992a). In the present study where ozone is added from a concentrated stock solution into a batch reactor, the reactor vessel can be considered as a reactive flow segment with ideal plug flow (Lev and Regli 1992a).

In a continuous flow system, a tracer study would need to be performed to determine the characteristic T for the reactor (Lev and Regli 1992b). When using a batch reactor such as was used in this study, the actual contact time in each experimental trial can be used as the characteristic T. Consequently, when the results of this study are used in a full-scale, dynamic system, the kinetic parameters can be used with whatever mixing regime is determined for the system.

Results and Discussion

Figure 1 summarizes inactivation data determined by infectivity as a function of the integrated ozone residual and contact time product. The method of non-linear least squares (using the Quasi-Newton algorithm available in the Solver module of Microsoft Excel) was used to estimate the parameters, k, n, and m for the Hom kinetic model. The parameter estimates are summarized in Table 1.

Table 1. Summary of parameter estimates using non-linear least squares for Hom model kinetics for *C. parvum* using ozone.

Temperature, °C	k	n	m
7	-0.37	0.61	0.87
22	-1.15	0.22	0.54

Unlike the simple Chick-Watson model where n=1 and the simple CT product is a constant for each level of inactivation, the Hom model has values of n and m that are not unity. Rather the concentration and contact time change in a non-linear fashion according to:

$$C^n T^m = \text{constant}$$

Tables 2 and 3 were developed from the kinetic model based on animal infectivity for 99 per cent and 99.9 per cent inactivation of *C. parvum* using ozone. The ozone dose and contact time become very important decisions for a designer of an ozone disinfection system for controlling *Cryptosporidium*.

Comparison with previous ozone disinfection studies involving *C. parvum* oocysts is somewhat subjective for three reasons: a suitable kinetic model has not been reported in each study; there is a potential disparity in viability measurements; and the ozonation protocol varies from study to study. For discussion purposes, simple CT products will be used as the most expedient means for comparing previous studies. However, recall that the non-linearity of the kinetic model found in this study may be applicable to other studies as well and prevents more rigorous comparison of the literature. Table 4 summarizes the pertinent findings from other studies.

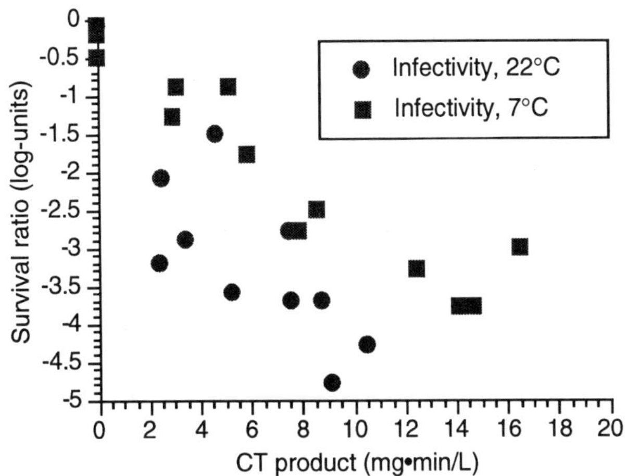

Figure 1. *C. parvum* inactivation determined by animal infectivity at 7 and 22°C as a function of the integrated ozone residual × contact time product.

Table 2. Summary of ozonation design criteria required to achieve 99% inactivation of *Cryptosporidium* oocysts.

Integrated ozone residual, mg/L	Required contact time, min.		Simple *CT*		$C^{0.61}T^{0.87}$	$C^{0.22}T^{0.54}$
	7°C	22°C	7°C	22°C	7°C	22°C
0.25	18.4	4.9	4.6	1.2	5.4	1.7
0.50	11.3	3.7	5.7	1.8	5.4	1.7
0.75	8.5	3.1	6.4	2.4	5.4	1.7
1.00	7.0	2.8	7.0	2.8	5.4	1.7
1.25	5.9	2.5	7.4	3.2	5.4	1.7
1.50	5.2	2.4	7.9	3.5	5.4	1.7
1.75	4.7	2.2	8.2	3.9	5.4	1.7
2.00	4.3	2.1	8.6	4.2	5.4	1.7
2.25	3.9	2.0	8.9	4.5	5.4	1.7
2.50	3.7	1.9	9.1	4.8	5.4	1.7
2.75	3.4	1.8	9.4	5.1	5.4	1.7
3.00	3.2	1.8	9.7	5.3	5.4	1.7

Table 3. Summary of ozonation design criteria required to achieve 99.9% inactivation of *Cryptosporidium* oocysts.

Integrated ozone residual, mg/L	Required contact time, min.		Simple *CT*		$C^{0.61}T^{0.87}$	$C^{0.22}T^{0.54}$
	7°C	22°C	7°C	22°C	7°C	22°C
0.25	29.3	10.4	7.3	2.6	8.1	2.6
0.50	18.0	7.8	9.0	3.9	8.1	2.6
0.75	13.6	6.6	10.2	5.0	8.1	2.6
1.00	11.1	5.9	11.1	5.9	8.1	2.6
1.25	9.5	5.4	11.8	6.7	8.1	2.6
1.50	8.3	5.0	12.5	7.5	8.1	2.6
1.75	7.5	4.7	13.1	8.2	8.1	2.6
2.00	6.8	4.5	13.6	8.9	8.1	2.6
2.25	6.3	4.2	14.1	9.5	8.1	2.6
2.50	5.8	4.1	14.6	10.2	8.1	2.6
2.75	5.5	3.9	15.0	10.8	8.1	2.6
3.00	5.1	3.8	15.4	11.3	8.1	2.6

Table 4. Summary of reported ozonation requirements for inactivation of *Cryptosporidium* sp. oocysts using simple *CT* products compared with published requirements for *Giardia lamblia*.

Species	Ozone protocol	Ozone residual*, mg/L	Contact time, min.	Temp., °C	Simple *CT* for ≥99 percent inactivation*, mg·min/L	Reference
C. parvum	Batch liquid, batch ozone	0.50 0.50	18 7.8	7 22	9.0 3.9	This study
C. parvum	Batch liquid, batch ozone	0.77 0.51	6 8	Room	4.6 4	(Peeters et al. 1989)
C. parvum	Batch liquid, continuous gas	1.0	5 & 10	25	5-10	(Korich et al. 1990)
C. baileyi	Batch liquid, modified batch ozone	0.6 & 0.8	4	25	2.4 - 3.2	(Langlais et al. 1990)
G. lamblia	Batch liquid, batch ozone	0.6	2	22	1.2	(Finch et al. 1993a)
G. lamblia	Batch liquid, continuous gas	0.11 - 0.48 0.03 - 0.15	0.94 - 5 1.06 - 5.5	5 25	0.53 0.17	(Wickramanayake et al. 1984)

* Interpretation depends on protocol used and method for reduction of ozone data by authors.

Korich et al. (1990) used a semi-batch reactor containing 0.01 M phosphate buffer (pH 7) with continuously added ozonated gas to maintain a constant ozone residual of 1 mg/L. Neonatal BALB/c mice were used to determine viability of the oocysts. A constant ozone residual of 1.0 mg/L was reported to effect 99 per cent inactivation after 5 min. of exposure. A potential problem with this study was the use of a semi-batch ozone contactor. The ozone dose and the decay of ozone during the course of the experiments could not be reported, thereby underestimating the actual ozone dose used in the experiments to effect the inactivation.

Langlais et al. (1990) used immune-suppressed male Sprague-Dawley rats to determine the infectivity of *C. baileyi oocysts* after ozonation. They found that a simple *CT* product of 3.2 mg·min/L inactivated between 3 and 4 log-units of oocysts. They also observed that a simple *CT* product of 4.4 mg·min/L inactivated between 4 and 5 log-units of oocysts.

Peeters et al. (1989) followed an experimental protocol very similar to the one used in this study. Neonatal Swiss OF1 mice were used to determine the viability of the oocysts. Using the data from their paper a kinetic model was developed for their data. Assuming first-order decay of ozone residuals in their study and that the dose-response model for the neonatal mice in the present study was similar to the one used in that study, the kinetic model parameters were estimated by non-linear least squares. An estimated 2 log-units of inactivation were obtained for an integrated ozone residual of 0.51 mg/L and contact time of 8 min. The same inactivation was obtained from an integrated ozone residual of 0.77 mg/L and a contact time of 6 min. The resulting C^nT^m values for these data were 1.5 and 1.7 mg·min/L, respectively. For similar integrated ozone residuals, the kinetic model for the present study resulted in a C^nT^m value of 1.7 mg·min/L but at half the contact times of Peeters et al. (1989). This is considered to be very good agreement between the two animal systems and the similar ozonation protocols.

The findings from this study are important to the water industry and regulators. Ozone appears to be one of the best chemical disinfectants for inactivating *Cryptosporidium* in drinking water. Proper ozonation in conjunction with chemical coagulation and filtration will help prevent outbreaks of water-borne disease due to *Cryptosporidium*.

Acknowledgments

The American Water Works Association Research Foundation provided primary funding for this project with additional support from the Natural Sciences and Engineering Research Council of Canada and the University of Alberta. The authors are indebted to Dr. Mike Belosevic and his colleagues in the Faculty of Science, University of Alberta for the use of their animal facilities and laboratories and to Dr. Frank W. Schaefer III of the U.S. Environmental Protection Agency for his help with this research project. The excellent technical support of Qiong Shen was much appreciated.

Cited Literature

Barer, M.R., and A.E. Wright. 1990. A Review: *Cryptosporidium* and Water. *Letters in Applied Microbiology*, 11:271-277.

Chick, H. 1908. An Investigation of the Laws of Disinfection. *Journal of Hygiene (Cambridge)*, 8:92-158.

Finch, G.R., E.K. Black, C.W. Labatiuk, L. Gyürék, and M. Belosevic. 1993a. Comparison of *Giardia lamblia* and *Giardia muris* Cyst Inactivation by Ozone. *Applied and Environmental Microbiology*, 59(11):3674-3680.

Finch, G.R., C.W. Daniels, E.K. Black, F.W. Schaefer III, and M. Belosevic. 1993b. Dose-Response of *Cryptosporidium parvum* in Outbred, Neonatal CD-1 Mice. *Applied and Environmental Microbiology*, 59(11):3661-3665.

Gallaher, M.M., J.L. Herndon, L.J. Nims, C.R. Sterling, D.J. Grabowski, and H.F. Hull. 1989. Cryptosporidiosis and Surface Water. *American Journal of Public Health*, 79(1):39-42.

Hayes, E.B., T.D. Matte, T.R. O'Brien, T.W. McKinley, G.S. Logsdon, J.B. Rose, B.L.P. Ungar, D.M. Word, P.F. Pinsky, M.L. Cummings, M.A. Wilson, E.G. Long, E.S. Hurwitz, and D.D. Juranek. 1989. Large Community Outbreak of Cryptosporidiosis due to Contamination of a Filtered Public Water Supply. *New England Journal of Medicine*, 320(21):1372-1376.

Hoff, J.C. 1986. *Inactivation of Microbial Agents by Chemical Disinfectants.* Report Number EPA/600/2-86/067. Water Engineering Research Laboratory, U. S. Environmental Protection Agency, Cincinnati, OH:

Hom, L.W. 1972. Kinetics of Chlorine Disinfection in an Ecosystem. *Journal of the Sanitary Engineering Division, Proceedings of the American Society of Civil Engineering*, 98(SA1):183-193.

Korich, D.G., J.R. Mead, M.S. Madore, N.A. Sinclair, and C.R. Sterling. 1990. Effects of Ozone, Chlorine Dioxide, Chlorine, and Monochloramine on *Cryptosporidium parvum* Oocyst Viability. *Applied and Environmental Microbiology*, 56(5):1423-1428.

Langlais, B., D. Perrine, J.C. Joret, and J.P. Chenu. 1990. The Ct Value Concept for Evaluation of Disinfection Process Efficiency; Particular Case of Ozonation for Inactivation of Some Protozoan: Free Living *Amoeba* and *Cryptosporidium.* In *New Developments: Ozone in Water and Wastewater Treatment. Proceedings of the Internation Ozone Association Spring Conference, March 27-29 1990, Shreveport, LO.* Norwalk, CT: International Ozone Association, Pan American Committee.

Lev, O., and S. Regli. 1992a. Evaluation of Ozone Disinfection Systems: Characteristic Concentration *C*. *Journal of Environmental Engineering*, 118(4):477-494.

Lev, O., and S. Regli. 1992b. Evaluation of Ozone Disinfection Systems: Characteristic Time *T*. *Journal of Environmental Engineering*, 118(2):268-285.

Lorenzo-Lorenzo, M.J., M.E. Ares-Mazas, I. Villacorta-Martinez de Maturana, and D. Duran-Oreiro. 1993. Effect of Ultraviolet Disinfection of Drinking Water on the Viability of *Cryptosporidium parvum* Oocysts. *Journal of Parasitology*, 79(1):67-70.

Peeters, J.E., E.A. Mazás, W.J. Masschelein, I.V. Martinez de Maturana, and E. Debacker. 1989. Effect of Disinfection of Drinking Water with Ozone or Chlorine Dioxide on Survival of *Cryptosporidium parvum* Oocysts. *Applied and Environmental Microbiology*, 55(6):1519-1522.

Perrine, D., P. Georges, and B. Langlais. 1990. Efficacité de l'Ozonation des Eaux sur l'Inactivation des Oocystes de *Cryptosporidium*. *Bulletin of Academic and National Medicine*, 174(6):845-850.

Richardson, A.J., R.A. Frankenberg, A.C. Buck, J.B. Selkon, J.S. Colbourne, J.W. Parsons, and R.T. Mayon-White. 1991. An Outbreak of Waterborne Cryptosporidiosis in Swindon and Oxfordshire. *Epidemiology and Infections*, 107(3):485-495.

Rose, J.B. 1990. Occurrence and Control of *Cryptosporidium* in Drinking Water. In *Drinking Water Microbiology*. Edited by G. A. McFeters. pp. 294-321. New York, NY: Springer-Verlag.

Rush, B.A., P.A. Chapman, and R.W. Ineson. 1990. A Probable Waterborne Outbreak of Cryptosporidiosis in the Sheffield Area. *Journal of Medical Microbiology*, 32:239-242.

Smith, H.V., A.L. Smith, R.W.A. Girdwood, and E.C. Carrington. 1988. *The Effect of Free Chlorine on the Viability of Cryptosporidium spp. Oocysts*. Report Number WRc PRU 2023-M. Medmenham, Bucks: Water Research Center:

Watson, H.E. 1908. A Note on the Variation of the Rate of Disinfection with Change in the Concentration of the Disinfectant. *Journal of Hygiene (Cambridge)*, 8:536-592.

Wickramanayake, G.B., A.J. Rubin, and O.J. Sproul. 1984. Inactivation of *Giardia lamblia* Cysts with Ozone. *Applied and Environmental Microbiology*, 48(3):671-672.

Cryptosporidium and the Milwaukee Incident

Kim R. Fox and Darren A. Lytle[1]

Introduction

In early 1993, Milwaukee, Wisconsin reported a sharp increase in the number of diarrhea patients and shortages of over the counter drugs for diarrhea control at local pharmacies. The increase in diarrhea was determined to be caused by the organism *Cryptosporidium*. Preliminary investigations conducted by State and City officials suggested that the drinking water may have been partially responsible for distributing the organism around Milwaukee.

On April 6, a doctor ordered a parasitic analysis on a patient's fecal specimen. *Cryptosporidium* was detected in the fecal smear. At that time, the local and state officials were notified of the *Cryptosporidium* detection. A concurrent survey of diarrhea cases in local nursing homes indicated that residents in nursing homes in the southern part of the city were fourteen times more likely to have had diarrhea than those in the northern part of the city. In addition to the nursing home survey, turbidity problems at the southern water treatment plant also implicated drinking water and this plant as being suspect in the cryptosporidiosis outbreak (Figure 1). At that point, the southern (Howard) plant was shut down and all water for Milwaukee was supplied by the northern (Linwood) plant under a boil water order.

The closure of the Howard Water Treatment Plant, the boil water order, the magnitude of the number of reported diarrhea cases, and the media attention, all helped to

[1]Environmental Engineers, U.S. Environmental Protection Agency, Drinking Water Research Division, 26 W. MLK Dr., Cincinnati, Ohio 45268.

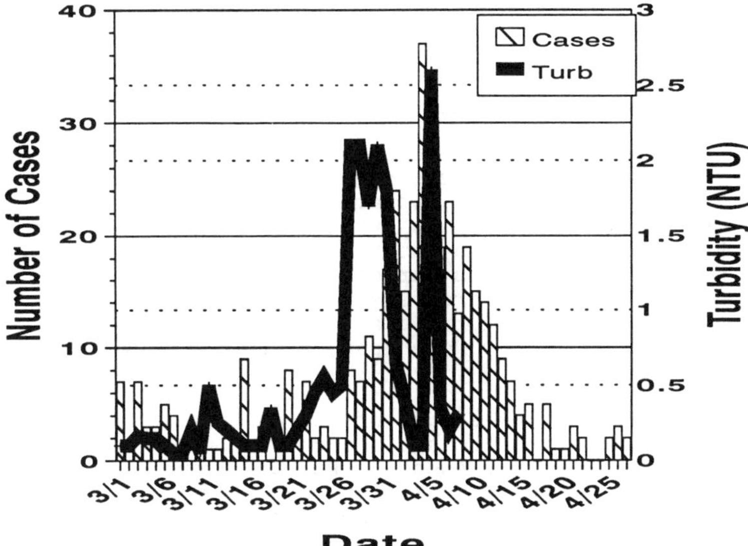

Figure 1. Diarrhea Onset & Max. Filter Turbidity

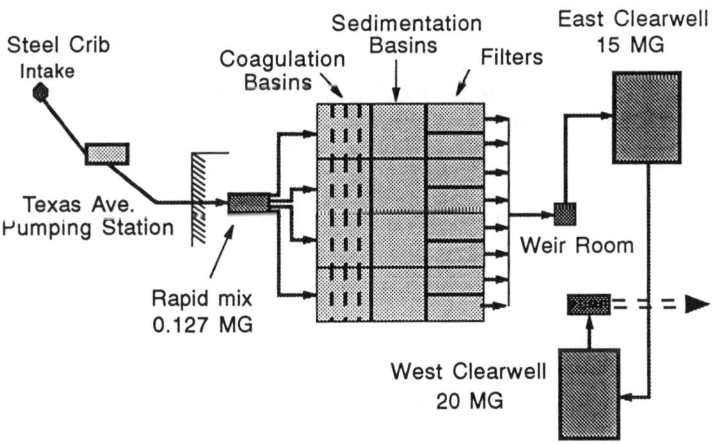

Figure 2 Howard Avenue Water Treatment Plant

focus water utility personnel, engineers, scientists, government officials, and rule-making bodies onto this waterborne outbreak. Follow up surveys have indicated that as many as 403,000 people may have been ill during the Milwaukee incident (Edwards 1993).

Cryptosporidium is a protozoan that if ingested by a healthy adult human may cause some discomfort such as that felt with stomach flu, however, it can be life threatening to others such as infants, AIDS patients and the elderly. Research has shown that most surface waters are contaminated with this parasite (LeChevallier 1991). Cryptosporidium oocysts are very resistant to chlorination and, therefore, their effective removal from surface waters is highly dependent on the operation of a filtration facility.

During the initial stages of the outbreak, the City of Milwaukee requested the assistance of the U.S. Environmental Protection Agency (EPA) to provide technical assistance. A team from the EPA's Drinking Water Research Division (DWRD) familiar with filtration processes and with the removal of Cryptosporidium oocysts by filtration went to Milwaukee. The team concentrated its efforts on evaluating the operational and monitoring data available from the southern Howard Avenue Water Treatment Plant. The goal was to assess how (from an engineering standpoint) Cryptosporidium may have passed the water treatment facility.

As part of the EPA Assistance, the team called upon research previously sponsored or conducted by the DWRD on removing Cryptosporidium from drinking water. This research included both extramural and in-house laboratory and pilot plant filtration studies. Both the in-house studies and the field studies have indicated that good turbidity and particulate removal is necessary for good removal of Cryptosporidium from drinking water.

Plant Inspection

The source water for Milwaukee, Wisconsin, is Lake Michigan (City of Milwaukee 1992). The city's water is treated at two water treatment facilities: the northern situated Linwood treatment plant and the southern Howard treatment plant. Both treatment facilities treat Lake Michigan water by conventional water treatment processes (coagulation, sedimentation, filtration, and disinfection). The Linwood facility has a water treatment capacity of 275 million gallons per day (MGD) and the Howard treatment facility has a filtration capacity of 100 MGD. During typical operation, the

Linwood plant supplies the northern 2/3 of the water district. The Howard plant supplies the remaining 1/3. A large mixing zone exists in the distribution system between the two treatment facilities, but the general flow of water is northern or southern. The initial investigation linked the cryptosporidiosis outbreak to drinking water treated at the Howard treatment facility, resulting in temporary shutdown of the facility. The EPA team focused its efforts on understanding the operation of the Howard treatment plant.

The Howard Plant is a conventional coagulation, sedimentation filtration facility (Figure 2). A complete description can be found in Fox and Lytle (1993b). Alum (aluminum sulfate) was the coagulant used until August of 1992. In August of 1992, the facility switched to polyaluminum chloride (PACL). The switch to PACL was done with the belief that a benefit of higher finished water pH for corrosion control would be met as well as reduced sludge volume and improved coagulation effectiveness in cold raw water conditions. Before switching to PACL, the city consulted with the chemical manufacturer, Wisconsin DNR, and other communities receiving Lake Michigan water and using PACL.

In the time frame immediately preceding March 1993, the Howard Plant (HWTP) was consistently producing a low effluent turbidity water (daily averages of 0.1 NTU or less). During the period of March 18 thru April 8 (plant was shut down April 8), the effluent turbidity from the HWTP was highly variable and ranged from between 0.1 and 1.7 NTU (Figure 1). The team was asked to look at what caused the higher turbidity levels to occur. At all times during this period, effluent water samples were negative for coliforms and met the Wisconsin DNR regulations for turbidity.

The HWTP receives a highly variable quality of influent water (from Lake Michigan) to be processed through the plant. During the time period of March 18 through April 8, the raw water turbidity levels ranged from 1.5 to 44 NTU. Total coliforms ranged from <1 CFU/100 mL to around 3200 CFU/100 mL in the raw waters. All effluent water samples were negative for coliforms. Average raw water turbidity levels for previous months were around 3 or 4 NTU and influent coliform levels <20 CFU/100 mL.

During the time period investigated, plant facility personnel responded to turbidity changes throughout the treatment processes by adjusting coagulant dosages. Coagulation dosage adjustments were also made to compensate for coagulation demands resulting from taste and order controlling treatment. Throughout the period, coagulant dose adjustments were continuously being made to meet the demands of raw water quality.

Although the coagulant dosages were being adjusted, filter effluent turbidities on several occasions exceeded turbidity values that were achieved in previous months. As the coagulant doses approached what might be optimum dosages for the particular raw water conditions, improvements in settled water turbidities and filter effluent turbidities were achieved. This pattern was seen twice. The first time was when polyaluminum chloride was the primary coagulant at the HWTP. The second time was when the primary coagulant was switched back to alum on April 2, 1993. The improvements in filter effluent turbidity demonstrated that the plant is fully capable of producing low turbidity water under optimal chemical coagulation conditions (i.e. dosages applied) even when challenged with high turbidity raw water.

There are several factors that may have increased the time to reach the optimum coagulant dosages for low filter effluent turbidities during this time. One of these factors may include a lack of historical use records for PACL. The previous coagulant had been used for almost thirty years. The new coagulant had only been used for a short time and the historical records were not fully developed. The optimum chemical dosages were sought through laboratory testing and consultation with DNR and chemical supplier to achieve the lowest turbidity possible. The coagulant adjustments were made based on all available data.

Another important factor is the time required to see a result in the treated water quality after chemical adjustment. With a short residence time for the water in the plant and a rapidly changing influent water quality, dosage optimization was difficult.

During the higher effluent turbidity episodes, greater numbers of particulates passed through the HWTP as evidenced by the higher turbidity values exiting the plant clearwell. Although a greater number of particulates passed through the plant, this does not necessarily mean that Cryptosporidium oocysts passed through the plant. This does, however, suggest that if a large number of oocysts were present in the source water at this time, the likelihood of passage would increase. There is no way to know for sure that this scenario resulted in Cryptosporidium oocysts passage. Monitoring for this organism is not a common practice in the drinking water supply industry, nor required in Wisconsin by Wisconsin Administration Code.

Experimental Studies

The Drinking Water Research Division (DWRD) has conducted, and is conducting filtration studies on removing microorganisms from drinking water. Current efforts are focused on removing *Cryptosporidium* oocysts. In addition to the filtration studies, charge/attachment studies are also being conducted to look at the zeta potential of the organisms as they are subjected to the rigors of a water treatment plant.

The DWRD has pilot-scale water treatment facilities located at the Cincinnati laboratory. Various waters can be brought to the treatment facility to challenge the filter units with differing water qualities. Specific organisms of choice are also spiked into the water to achieve levels desired. The organisms are cultivated in-house and are grown in broths, on agar, or in animals.

Experimental Results

In early studies, *Cryptosporidium* parvum oocysts (stored in dichromate) were used in jar test studies. The initial jars were spiked with 10,000 oocysts/liter and the water subjected to optimum coagulation, flocculation and settling conditions. The supernatant from the settling process contained 1,000 oocysts/liter (90% reduction). Tests with fresh (unpreserved) oocysts at the same coagulant dosages only exhibited 50% reduction in oocysts. Although these tests are preliminary, there does appear to be substantial differences between fresh and stored oocysts. One of the pilot plant filter systems was challenged by 100,000 oocysts/liter. The filter effluent was monitored for oocysts and 99% removal of preserved oocysts were routinely seen. The pilot slow sand filter was also challenged with water containing 100,000 oocysts/liter. Effluent concentrations were consistently below 1,000 oocysts/liter and in most cases less than 200. The filter was monitored two weeks after the spiking stopped and no oocysts were found in a fifty gallon sample of the effluent. Particle counting showed a removal of around 96% of particles in the 2.5 to 5 μm range. In all of these tests, the percent total particulate removal measured by particle counters, was less than the percent removal of oocysts.

Slow sand filter tests conducted by Ghosh 1989 showed three log removal of *Cryptosporidium* through the filter during ripening and greater than four log reduction on a fully ripened filter. Ghosh conducted several experiments using various grades of diatomaceous earth (DE) on a precoat filter. Observed oocysts removals exceeded three log removal for all runs, but removals were dependent on DE grade. The larger the grade, the more oocysts found in the effluent.

Zeta potential measurements were made in DWRD labs using clean suspensions of Cryptosporidium oocysts (both muris and parvum) diluted into a river water that had been filtered through 1μm membrane filters. Tests were conducted to evaluate variables such as age of oocysts, storage solution, and pH of suspension fluid on oocysts. Muris oocysts were highly variable in regard to zeta potential as the charge increased from -21 mv to -1.8 mv as the pH was lowered from 10.0 to 4.6 (muris oocysts were not stored in dichromate). Muris oocysts stored in dichromate had a considerable greater charge magnitude than those in clean suspensions (-31.1 versus -20.1 mv, respectively). The parvum oocysts displayed a change from -17.8 mv (in clean suspensions) to -25.2 mv in suspensions stored in dichromate. In one study, the fecal material from a calf was stored in dichromate for twelve hours until the material could be processed the next day. The oocysts were harvested and stored without dichromate. Twenty-four hours after preparation, the measured charge was -6 mv. The suspension was monitored for eight days and the charge gradually changed to -14.1 mv. There were not enough oocysts to continue the study to determine if the charge would recover to -17.8 mv seen with most fresh oocysts. Zeta potential studies by Ghosh showed values 5 mv more negative than the results in the EPA tests with preserved oocysts.

Discussion

The preliminary tests have indicated that most filtration systems are effective in removing greater than 2.5 logs of oocysts under ideal operating conditions. These tests for the most part have been done with oocysts preserved in dichromate or with formalin fixed oocysts. Both the preservation solution and the fixation fluid have shown an effect on the measured zeta potential and may have an effect on the rigidity of the organism. The few studies completed with fresh oocysts have shown less effective removal but more work is planned. Several surveys in the literature (LeChevallier 1991) have indicated that oocysts are passing through some filter plants. The passage of oocysts may indicate that fresh (or untreated) oocysts may be more difficult to remove than those that are pretreated by some manner. The preservation processes may affect the physical characteristics of the oocysts and may make them more susceptible to filtration processes than those in the natural environment.

In order to achieve good reduction in turbidity at all times, stringent controls on coagulant/flocculant dosages are required. This control could be automated or done by operator attention and that determination would be at the discretion of the water utility. Subtle changes in effluent turbidity from a filtration plant may result in large changes in particulates passing through the filters that may or may not be associated with pathogens. The goal therefore, should be to remove these particulates and not have to worry about whether or not they are of concern.

In Milwaukee, the inability to maintain a low filter effluent turbidity may have allowed Cryptosporidium oocysts to pass through the water treatment plant (Fox and Lytle 1993a).

Acknowledgements

The authors wish to acknowledge all the individuals and organizations that worked very hard during the Milwaukee incident. Without the cooperation of everyone, the investigation could not have happened.

References

City of Milwaukee (1992). "1991 Annual Report", Water Engineering Division, .

Edwards, Diane D. (1993). "Troubled Waters in Milwaukee", ASM News, Vol 59, Number 7.

Fox, Kim R. and Lytle, Darren A. (1993a) "Press Conference Presentation". Milwaukee, Wisconsin, April 16.

Fox, Kim R. and Lytle, Darren A. (1993b) "The Milwaukee Cryptosporidiosis Outbreak: Investigation and Recommendations", Proceedings AWWA Water Technology Conference, Miami, Florida.

Ghosh, M.M., et al. (1989). "Field Study of Giardia and Cryptosporidium Removal from Pennsylvania Surface Waters by Slow Sand and Diatomaceous Earth Filtration". Environmental Resources Research Institute.

LeChevallier, M.W., et al. (1991). "Giardia and Cryptosporidium in Water Supplies". AWWA Research Foundation. ISBN 0-89867-569-3.

Control of Microbial Water Quality in the Little Rock, Arkansas Distribution System

Bruno Kirsch, Jr.[1], T. Craig Noble[1], Christopher H. Yu[2], Blaise J. Brazos[3], and John T. O'Connor[4]

Abstract

Beginning in the summer of 1993, seasonal studies were undertaken to evaluate the effectiveness of water treatment processes at Little Rock Municipal Water Works (LRMWW). Two treatment facilities were evaluated for the removal of total bacteria and the growth of bacteria during distribution. In addition to conventional data on coliform and heterotrophic plate count (HPC), total bacterial cell counts were used to evaluate seasonal changes in influent (lake water), treatment plant performance for bacterial removal and microbial populations and growth during distribution. The persistence of disinfectant residuals in finished water and in distributed water was also observed to complement data on microbial populations.

Chlorine demand studies were conducted in chlorine-demand-free glass containers to observe the persistence of chlorine versus chloramine. These studies were replicated under cold (4°C) and warm (24°C) temperature conditions. Each of the two treatment plant finished waters plus water from a location in the distribution system were evaluated in this series.

Non-purgeable organic carbon (NPOC) concentrations were monitored seasonally in the source waters and in both treatment plant effluents. Changes in NPOC during distribution were assessed with respect to the

[1]Little Rock Municipal Water Works, Little Rock, Arkansas.
[2]Burns & McDonnell Engineers-Architects-Consultants, Kansas City, Missouri.
[3]H_2O'C, Limited, Columbia, Missouri.
[4]Chief, Illinois State Water Survey, Champaign, Illinois.

maintenance of disinfectant residuals and the control of bacterial growth during distribution.

Overall, the data collected formed the basis for recommendations on the control of microbial quality in the Little Rock water distribution system. The results also form the basis for a new, comprehensive, year-long pilot plant study to be initiated early in 1994.

Introduction

The LRMWW obtains water from two water supply lakes which have historically been protected from the adverse influences of development. As a result, the influent water to their two water treatment plants is of exceptionally high quality.

Two treatment plants are operated by LRMWW. The Ozark Point Plant, constructed in 1912, is a classical coagulation (alum), sedimentation, sand filtration plant operated at a filtration rate of 81.5 $L/min/m^2$. The Jack Wilson Plant, constructed in 1967 and expanded in 1977 and 1984, is a conventional, present-day facility rated at 4,381 L/s, employing alum coagulation, sedimentation, dual-media filtration of 122.3 $L/min/m^2$. Backwash water is reclaimed at both facilities.

Evaluation of Plant Performance

While both treatment facilities routinely performed well, the older, classical Ozark Point Water Treatment Plant was found to consistently remove a greater percentage of total bacteria, turbidity and total organic carbon than the Jack Wilson Plant. Plant performance on total bacterial cell count removals are shown in Figure 1. Overall, the total bacterial numbers were reduced two orders of magnitude at the Wilson Plant and three orders of magnitude at the Ozark Point Plant. The reasons for the difference are now undergoing study. However, it was observed that the Ozark Point Plant has a lower and more steady-state hydraulic loading. This appears to result in better particles removal during sedimentation. The fine sand filter medium may also contribute to the differences observed. Differences in flocculation energy input and the effect of returning reclaimed backwash water will be evaluated as a part of the next phase pilot plant investigation.

Relationship Between Turbidity and Total Bacteria

Possibly because the source (lake) waters are initially free of clay and inorganic sediments, a strong relationship is apparent between turbidity and total bacteria in the source, settled, filtered and distributed waters. In this

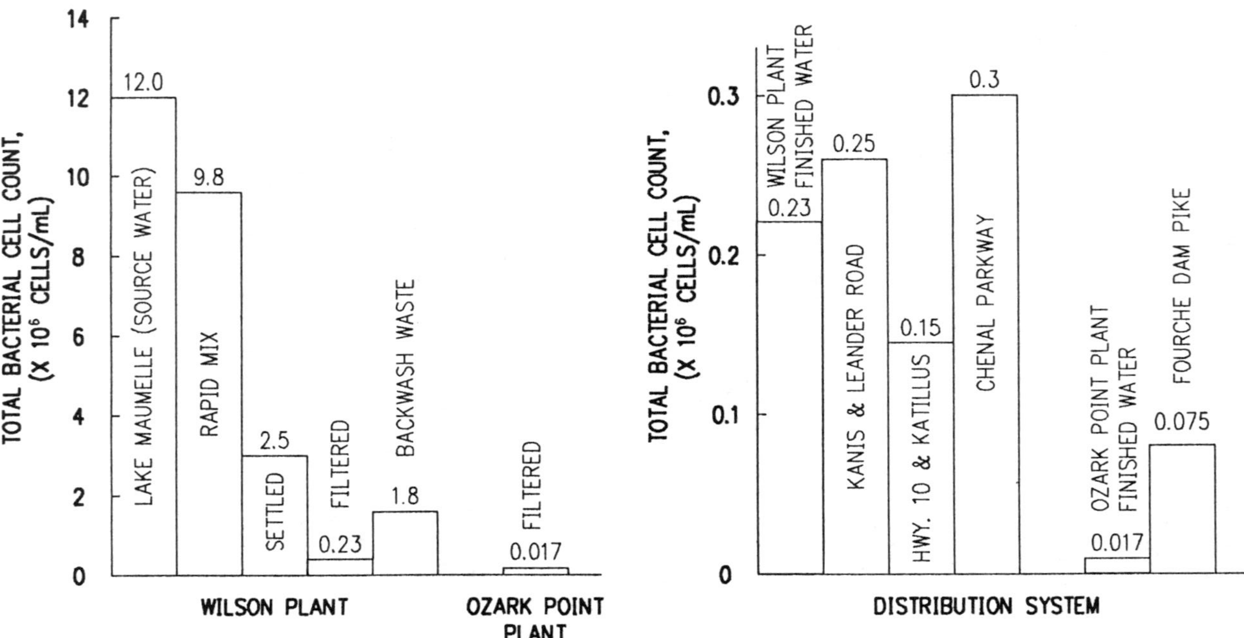

Figure 1: Population of Bacterial Cells at Treatment Plants and in Distribution System

particular instance, turbidity appears to be a good surrogate for microbial population, as shown in Figure 2, although the relationship does not appear to be linear. Turbidities less than 0.2 NTU indicate exceptionally low total bacterial populations. However, an increase in finished water turbidity to 0.5 NTU might indicate a 20 to 30-fold increase in microbial populations.

Total Organic Carbon Removals

During the period of the study, total organic carbon concentrations in the source waters were low and comparable. Again, the Ozark Point Treatment Plant consistently achieved better TOC reductions. Future studies are planned to determine whether the TOC reductions observed bear a significant relationship to bacterial cell removals.

If the TOC reductions presently achieved at LRMWW reflect the 86-99 percent bacterial removals observed, the remaining TOC in the finished water may be due to sub-micrometer colloids and solutes.

Microbial Quality of Finished Water

Actual physical removals of bacterial cells, as measured by direct microscopic count, were generally in the range of 98-99 percent. The low populations of residual cells entering the distribution were at levels typical of high quality groundwaters. From direct removal plus the addition of chlorine, total coliform colonies were consistently absent from the treated waters entering the LRMWW distribution system. HPC were reduced to less than 20 cfu/mL. Table 1 shows the water quality observed at six locations in the distribution system.

Distribution System Evaluation

Little Rock, Arkansas, is a rapidly growing community. Installation of new water mains as part of new construction, however, has been plagued with delays related to failure to pass inspection following the recovery of atypical coliform colonies. As a first step in evaluating the possible cause for these results, chlorine demand studies were conducted using sections of new main.

It was found that cleaning and superchlorination, as currently required by LRMWW, satisfied the chlorine demand imposed by the surface of the new main. Subsequent studies were then conducted to determine whether the disinfectant residuals in finished waters would persist at temperatures representative of winter and summer conditions. To ensure that the studies reflected only the chlorine demand exerted by the reducing agents present in

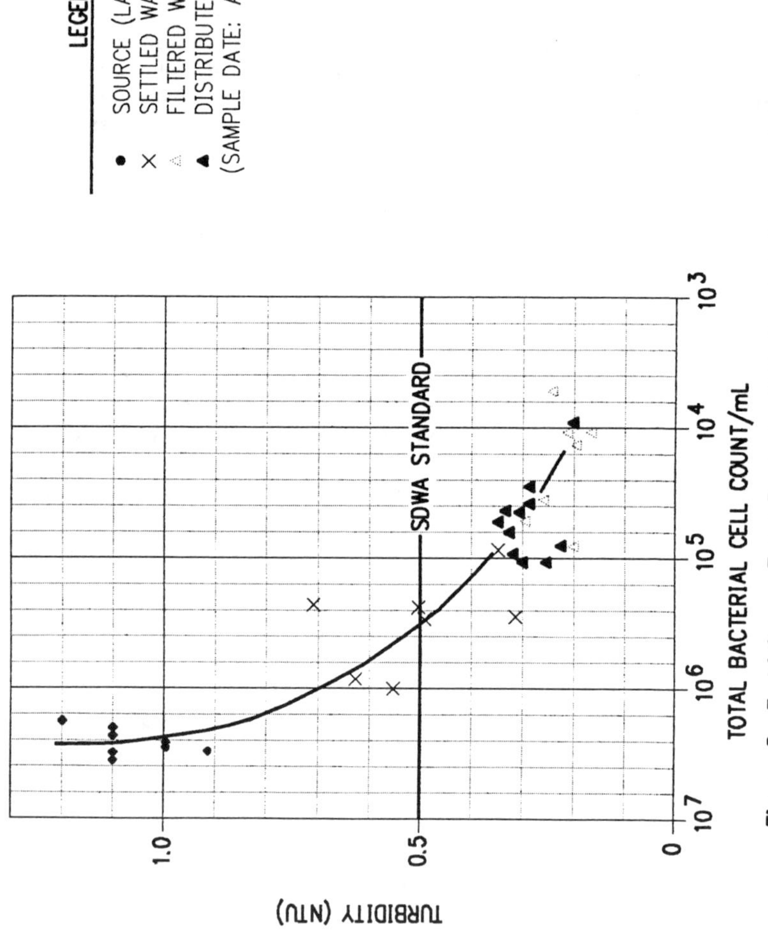

Figure 2: Turbidity vs Total Bacterial Cell Count in Source, Settled and Filtered Waters

Table 1

Water Quality of Distribution System (Sample Date October 11, 1993)

Parameters	Distribution System Sampling Location					
	Denny & Gordon	Fourche Dam Pike	Kanis & Leander	Hwy. 10 & Katillus	Point West (South)	Point West (North)
Temperature (°C)	24	25	23	23	25	24
Total Bacterial Cell Count (10^6/mL)	0.54	0.18	0.23	0.28	0.34	0.23
Description of Bacterial Cells	small, plump	few elongated	small, exogenous	not enlarged	small	very small
Typical/Atypical Total Coliform (CFU/100 mL)	0/394	0/TNTC	0/8	0/277	0/0	0/1
Turbidity, NTU	0.35	0.17	0.23	0.20	0.26	0.22
Free/Total Chlorine Residual (mg Cl/L)	0.0/0.0	0.1/0.2	0.3/0.6	0.4/0.6	0.6/0.7	0.7/0.9
Standard/R2A HPC (CFU/mL)	310/30,200	24/2570	1/113	3/383	0/133	0/3

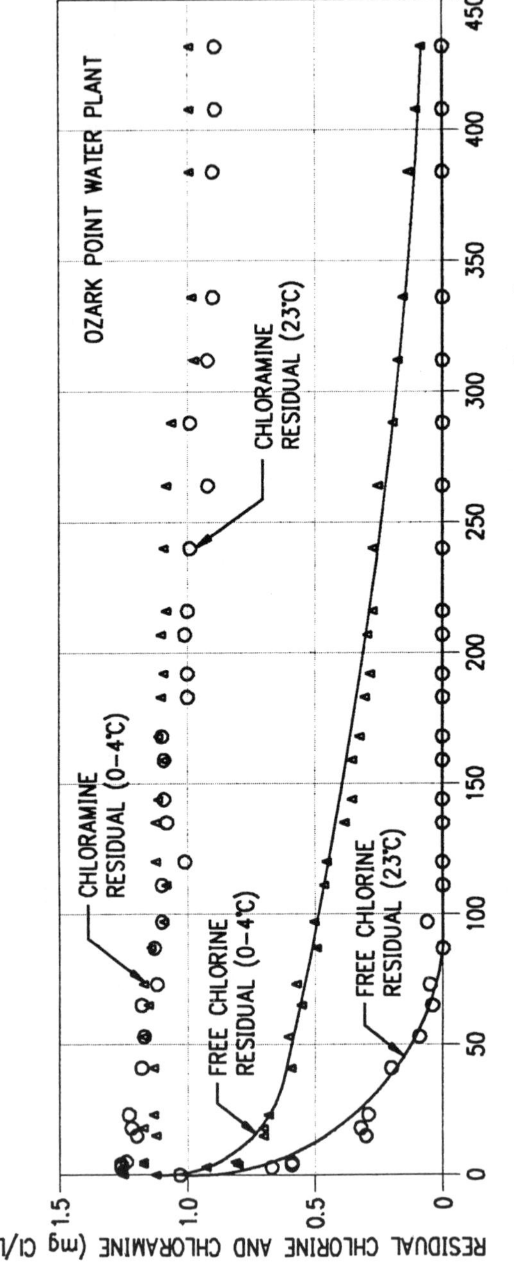

Figure 3: Persistence of Disinfectants in Finished Water under Cold and Warm Temperature Conditions

the plant's finished water, studies were conducted in the laboratory in clean, chlorine-demand-free glass containers.

The results show that chlorine residuals (including the chloramine formed in the finished water as a result of chlorination) persisted at low water temperatures, but were depleted to less than 0.2 mg Cl/L within two days at summer temperatures as shown in Figure 3. The addition of ammonium ion to form chloramine, however, resulted in a highly stable residual which persisted for over 400 hours, under summer or winter temperature conditions.

Conclusions

This study has illustrated the utility of chlorine demand investigation as an adjunct control to further securing microbial water quality during new main installation and finished water distribution. It confirmed that rechlorination may be required in segments of the distribution system on a seasonal basis to maintain a minimum bacteriostatic disinfectant concentration.

THE RELATIONSHIP OF SPECIATION TO IRON AND MANGANESE REMOVAL STRATEGIES

Kenneth Carlson[1], William Knocke[2] , Kevin Gertig[3]

Abstract: Case studies are presented describing how Fe and Mn speciation procedures were used to solve process problems at two water utilities. A series of filtration steps are used to separate these metals into soluble, colloidal and particulate fractions. The species of manganese present in one water source is shown to be dependent on the biogeochemical cycle in the reservoir. The natural oxidation and reduction of Mn was found to be microbially mediated and the application of $KMnO_4$ was adjusted to account for this cycle. A groundwater treatment plant was suffering from a high amount of Fe coming off the filters. The problem was identified as inadequate solids capture and not the oxidant dose allowing a quick solution. Enhanced oxidation of Mn with Cl_2 was postulated to be due to an association with Fe oxide particles.

Introduction

Manganese and iron removal strategies are typically two step processes. The metal is first oxidized from a soluble to an insoluble oxidation state using an oxidant such as potassium permanganate ($KMnO_4$), ozone or chlorine. The insoluble metal oxide is then removed through standard solid-liquid separation techniques. Understanding the concentration of soluble metals is essential for proper targeting of the oxidant dose. For the purposes of water treatment, manganese and iron exist in three physical species; dissolved, colloidal or particulate.

1 Process Engineer, Fort Collins Water Utility, Fort Collins, CO, 80521.

2 Professor, Department of Civil Engineering, Virginia Tech, Blacksburg, VA, 24061-2646.

3 Process and Operations Supervisor, Fort Collins Water Utility, Fort Collins, CO, 80521

In the past, it has been common to base the removal process on the total concentration of iron or manganese measured without regard for what oxidation state or species are actually present. When attempts are made to quantify the soluble metal species the typical method is a 0.45μ membrane filtration step. Knocke (1991) showed that colloidal manganese cannot be captured with a 0.45μ filter and that ultrafiltration (either 30K or 100K molecular weight cut-off) was needed for this level of speciation. Understanding what species of metal exists in the water can help a plant optimize its removal strategy. Case studies are presented here to demonstrate how raw water speciation data shaped the Mn or Fe removal strategy at two utilities.

Materials/Methods

The experimental methods were customized for each case study but several standard procedures were used. Samples were collected in 500 ml., acid-washed bottles and speciated using a combination of 0.45μ filtration (Gelman Sciences, Ann Arbor, MI) and 30K or 100K ultrafiltration (Amicon, Beverly, MA). Manganese species definitions are as follows:

Species	Definition
$Mn(IV)_{particulate}$	$[Mn]_{unfiltered} - [Mn]_{filtrate,\ 0.45\mu}$
$Mn(IV)_{colloidal}$	$[Mn]_{filtrate,\ 0.45\mu} - [Mn]_{filtrate,\ 30K}$
$Mn(II)$	$[Mn]_{filtrate,\ 30K}$

Samples were preserved by acidification with HNO_3 to $pH < 2$. Manganese and iron concentrations were measured using atomic absorption spectrometry.

Several experiments were conducted to determine which ultrafiltration molecular weight cut-off (MWCO) best captured colloidal manganese. A 1 ml. aliquot of $KMnO_4$ was added to 500 ml. of DI water and quenched with a 3:1 excess of standard PAO (0.00564N). Assuming that all of the Mn(VII) had been reduced to Mn(IV), ultrafiltration should be able to remove all of colloidal manganese. Ultrafilter MWCO and cation addition ($CaSO_4$) were varied to optimize removal.

Discussion

Two case studies have been developed to demonstrate how speciation data can be used to optimize Fe and Mn removal.

Case Study 1 : Reservoir With Seasonally High Levels of Mn

In this case, the water source is a reservoir with relatively low manganese levels most of the year. The Mn concentration increases as the reservoir temperature stratifies in the late summer and dissolved oxygen in the hypolimnion decreases to less than 2 mg/L. During these times of the year, the manganese concentration can exceed 0.4 mg/L in the raw water. The treatment strategy in the past has been to begin applying $KMnO_4$ when the total Mn in the raw water increases to about 0.05 mg/L. This approach was generally effective but unacceptable levels of Mn were still detected leaving the filters, particularly during the first month of $KMnO_4$ treatment.

In 1993, raw water manganese was speciated into particulate, colloidal and dissolved fractions using the procedure described. Figure 1 shows a trend of total and dissolved Mn concentrations versus dissolved oxygen for a 3 month period.

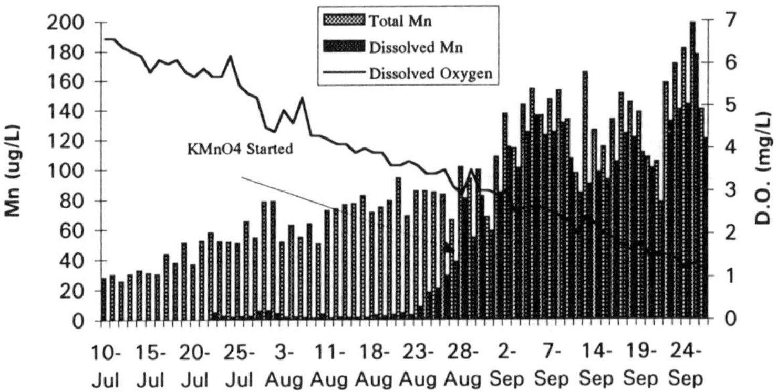

Figure 1 Total and Dissolved Mn Trends for July-Sep, 1993

The total Mn concentration began to increase around the middle of July but the dissolved species concentration was negligible until the end of August. During this time period, the particulate form of manganese was found to be present as an $MnO_{2(s)}$ sheath on the outside of an Mn oxidizing bacteria (*Metallogenium s.p.*). The bacteria and their distinctive brown color were first noticed using microscopic particulate analysis. The presence of manganese was verified using scanning electron microscopy and electron dispersive spectroscopy (see Figure 3). The presence of a microbiologically mediated geochemical manganese cycle has been well documented (e.g. Nealson, 1992). This can be used to explain Figure 1.

It is postulated that both manganese reducing and oxidizing bacteria are present throughout the year. When the reservoir is aerobic, the reducers are present below the sediment surface. Most of the year the concentration of Mn oxidizers is greater than Mn reducers and any Mn that is released (reduced) is quickly oxidized and does not enter the plant's raw water. When the reservoir stratifies and the D.O. decreases, the Mn reducers migrate up to the water/sediment interface facilitating the release of Mn(II) into the water column. At this point the amount of Mn being reduced has surpassed the capacity of the Mn oxidizers and dissolved Mn begins entering the plant (approximately the end of August in Figure 1). A schematic of the Mn cycle is shown in Figure 2.

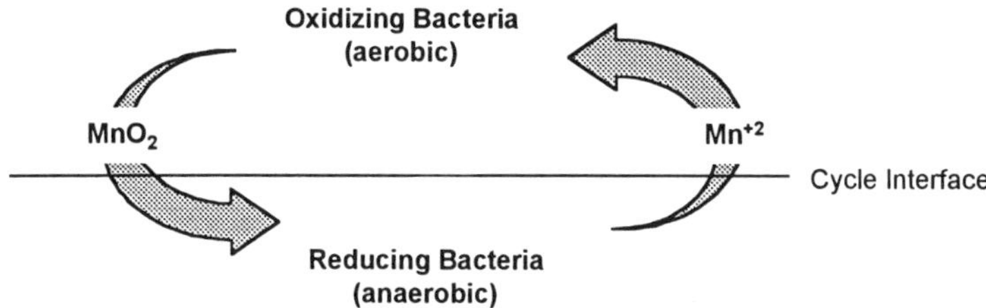

Figure 2 Biogeochemcial Cycling of Manganese

The manganese that was passing through the filters during the first month in past years is thought to be caused by application of $KMnO_4$ to a water without a significant amount of reduced metal. The MnO_2 that is associated with the *Metallogenium s.p.* appears to be disrupted (sloughed off) by the presence of strong oxidants. Figures 3 and 4 show scanning electron micrographs of the MnO_2 encrusted bacteria with and without an excess $KMnO_4$ treatment.

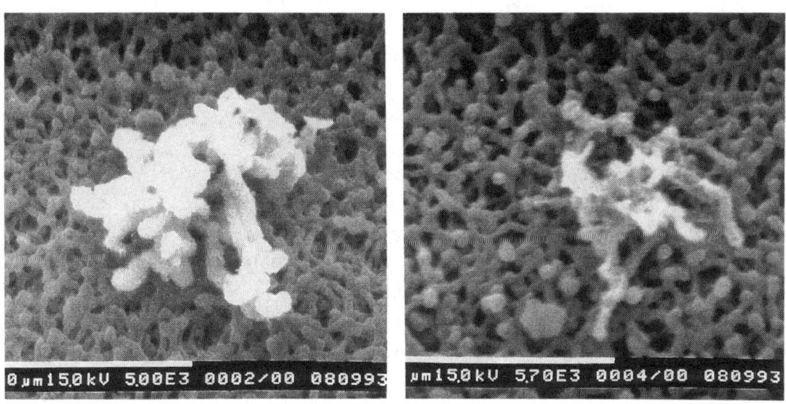

Figure 3 SEM of *Metallogenium s.p.*

Figure 4 SEM of *Metallogenium s.p.* (Treated With $KMnO_4$)

Using electron dispersive spectroscopy, the bacteria in Figure 3 was found to have nine times the mass of manganese as that in Figure 4. The observations presented here are supported by a considerable amount of speciation data that shows the redistribution of Mn from the particulate to the colloidal species solely as a result of $KMnO_4$ treatment. This suggests a negative impact of $KMnO_4$ treatment if the Mn is microbiologically associated and not the dissolved Mn species. Even if the above mentioned phenomenom is not occurring, treating water with $KMnO_4$ that contains little to no dissolved Mn can accomplish nothing positive and potential problems do exist.

The use of speciation allowed this utility to fine tune the application of $KMnO_4$ to optimize Mn removal at the lowest cost. It also provided valuable insight into manganese speciation cycles in the reservoir at different times of the year.

Case Study II: Groundwater Treatment for Fe and Mn Removal

A groundwater treatment facility was monitored for Fe and Mn species to ascertain why periodic problems associated with discoloration of the treated water were occurring. The facility had groundwater sources characterized by high hardness (300-350 mg/L as CaCO3), high alkalinity and elevated Fe (1-2 mg/L) and Mn (0.1 - 0.45 mg/L) concentrations. Water softening was not practiced at the facility; instead, treatment consisted of aeration and chlorination followed by dual-media filtration.

Questions arose as to whether the ineffective Fe and Mn removal was due to the oxidation scheme or the filters. Speciation testing was undertaken as previously described. In addition, certain samples were also filtered through Whatman 40 (rated pore size: 8 μ) or Whatman 42 (rated pore size: 2 μ) filters. Hudson (1981) indicated that passage of water samples through filters of this type could serve as a reasonable surrogate of the anticipated particle removal that would occur in a packed bed of filter media.

The concern regarding metal ion oxidation was prompted by the fact that neither $O_{2\,(aq)}$ or free chlorine is particularly effective for Mn(II) oxidation below pH 8.0 (Knocke, 1991A). Bench scale tests were conducted to determine the speciation of both Fe and Mn after (a) aeration only, and (b) both oxidants followed by thirty minutes of mild flocculation.

Typical results from such speciation testing are shown in Table 1. When aeration alone was practiced, only Fe was oxidized. The oxidation reaction was very effective as evidenced by essentially all of the Fe being retained by filters of size greater than 0.2μ. However, there were significant Fe solids that passed through both the Whatman 40 and 42 filter papers indicating the presence of extremely small Fe particles that may pass through a dual-media filter.

Aeration alone accomplished little soluble Mn oxidation as indicated by the fact that essentially all of the Mn present passed through a 30K ultrafilter. However, when chlorine was added, there was evidence of Mn oxidation. This result is rather interesting since literature sources (e.g. Knocke, 1991A) would not predict this degree of Mn oxidation by free chlorine during a 30 minute contact time in the pH range of this water (7.3-7.5). It was hypothesized that dissolved Mn oxidation by free chlorine was catalyzed by the presence of oxidized Fe particles. As a result, the oxidized Mn was felt to be directly associated with the Fe particles. In comparison, the data show no interaction between Mn and the oxidized Fe particles when aeration alone was practiced, indicating that Mn removal was not due to an adsorption mechanism.

TABLE 1
Speciation of Fe and Mn Following Oxidant Addition

		Speciation (mg/L)				
Oxidant	Metal	Total	<8μm	<2μm	<0.2μm	<30K
Aeration	Fe	1.3	1.1	0.2	<0.1	<0.1
	Mn	0.45	0.45	0.44	0.44	0.43
Aeration &	Fe	1.7	1.0	0.4	<0.1	<0.1
Chlorination	Mn	0.42	0.38	0.33	0.26	0.25

Emphasis was next placed on evaluating the performance of the dual-media filters with respect to particulate Fe capture. The three full-scale filters at the facility were backwashed and then brought on-line at hydraulic loading rates of approximately 3.5 gpm/ft^2. Effluent quality was monitored as total and dissolved Fe. The results, shown in Figure 1, indicate that total Fe in the filter effluent increased rapidly within the first few hours of operation. Since soluble Fe concentrations in these samples were all below 0.03 mg/L, the Fe release into the finished water was due almost completely to particulate Fe solids breakthrough. Further pilot testing demonstrated that this problem could be easily overcome by the addition of small (0.02-0.05 mg/L) quantities of an anionic filter aid polymer.

Removal of Mn species across the filtration step was the remaining area of study. The water typically applied to the filter contained a mixture of both particulate Mn forms and a significant soluble Mn fraction (Table 1; data when chlorine addition practiced). Therefore, filter performance was evaluated with respect to removal of both particulate and soluble Mn species. Typical data for full scale Filter #1 performance are contained in Table 2. During this study the average soluble Mn and Fe concentrations in the filter effluent were less than 0.035 mg/L and 0.03 mg/L, respectively. The data shown in Table 2 indicate a strong correlation between particulate Fe and particulate Mn concentrations in the filter effluent. This result is not surprising since the testing presented in Table 1 had indicated the Mn became associated with the particulate Fe forms following oxidation by free chlorine. The elevated particulate Mn concentrations in the filter effluent were likewise effectively controlled when a filter aid polymer was added.

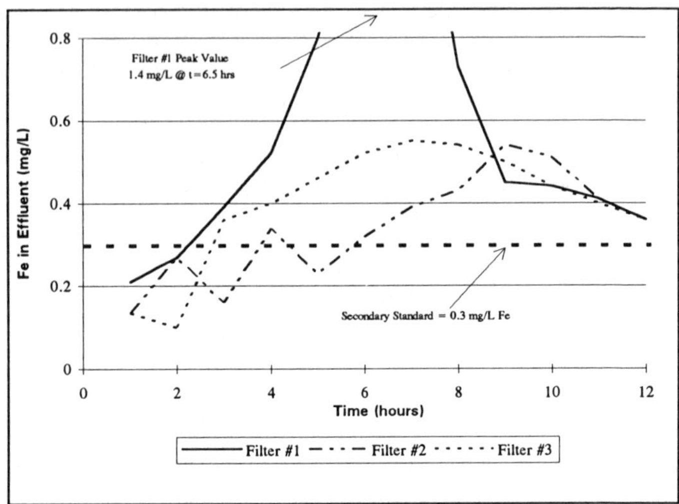

Figure 3 Iron Concentration in Filter Effluent

Soluble Mn removal was accomplished across the dual-media filters as well. Typically, the influent soluble Mn concentration (0.20-0.30 mg/L) was decreased to 0.04 mg/L or less in the filter effluent water. Soluble Mn uptake was accomplished via direct adsorption/surface oxidation of Mn onto the $MnO_{x(s)}$ that coated the filter media. This phenomena has been previously described by Knocke (1991B) and Coffey (1993).

TABLE 2
Typical Finished Water Particulate Iron and Manganese Concentration During a Dual-Media Filtration Cycle (Filter #1)

Filtration Time (hr.)	Particulate Fe Conc. (mg/L)	Particulate Mn Conc. (mg/L)
0.5	0.15	<0.01
2.5	0.27	0.02
4.5	0.56	0.03
6.5	1.40	0.05
8.5	0.42	0.04
12.5	0.31	0.04
16.5	0.35	0.02

In summary, overall plant performance was improved significantly by modifications such as maintaining appropriate free chlorine feed rates and adding a filter aid polymer prior to dual-media filtration. These treatment strategies were able to be conceived once appropriate speciation of the Fe and Mn forms present in the raw and oxidized waters was done. These insights would not have been possible if only total Fe and Mn testing had been undertaken to describe plant performance.

Conclusions

Two case studies have been presented detailing how proper Mn and Fe speciation techniques can be used to understand process problems. In one case, effective oxidation was shown to be occurring even though previous studies using only a 0.45µ speciation step had indicated otherwise. The problem with Mn breakthrough was solved by using $KMnO_4$ only when soluble Mn was present. In addition, the speciation studies led to a better understanding of the microbiological cycles that are present in the source water reservoir. A groundwater source with high levels of both Fe and Mn was characterized using a systematic speciation approach involving both the oxidation and filter processes. Solids capture problems were identified and easy fixes implemented. A more complete understanding of the process was a result of this speciation study also.

Two relatively simple process problems were solved by using the speciation techniques described here. These procedures have been shown to be easy to use and a valuable resource for any water utility with Fe or Mn problems.

Appendix : References

Coffey, B., Gallagher, D., Knocke, W.R., (1993) " Modeling Soluble Manganese Removal by Oxide Coated Filter Media", *Journal of the Environmental Engineering Division - ASCE*, 119(4), 679.

Hudson, H.E., Jr., (1981) *Water Clarification Processes: Practical Design and Evaluation*, Van Nostrand Reinhold, New York.

Knocke, W.R., Van Benschoten, J.E., Kearney, M., Soborski, A., Reckhow, D.A.(1991A). *Alternative Oxidants for the Removal of Soluble Iron and Manganese*, American Water Works Association Research Foundation, Denver.

Knocke, W.R., Occiano, S.C., Hungate, R., (1991B) "Removal of Soluble Manganese by Oxide-coated Filter Media: Sorption Rate and Removal Mechanism Issues, *Jour. AWWA*, 83 (8), 64-69.

Nealson, K.H., Myers, C.R., “Microbial Reduction of Manganese and Iron: New Approaches to Carbon Cycling”, *Appl. Environ. Microbiol.*, 58(2), 439-443.

Comparison of Biological and Chemical/Physical Iron Removal

Charlotte D. Smith[1] and James F. Smith

ABSTRACT

The report compares the efficiency of chemical/ physical and biological iron removal processes during the entire filter cycle. The results of the first three pilot studies performed in the United States indicate that biological iron removal will reduce the iron concentration of groundwater to less then 0.05 mg/L at flow rates of at least 8.0 gpm/sq ft. The high rate filtration, allows for lower capital costs, and less space requirements.

INTRODUCTION

The work described in this report was performed in an effort to help drinking water utilities which have iron or both iron and manganese in groundwater used as the source of supply. The problems associated with incomplete removal of iron during treatment are well known to water purveyors. Iron in finished water can cause discoloration of the water, staining of laundry and plumbing fixtures, and unpleasant tastes. The AWWA/AWWARF Water Industry Data Base contains information on systems serving populations of 10,000 or greater and provides the following information: Five hundred and ninety (54%) of the 1097 utilities in the data base currently use groundwater as the source of supply, and 137 (23.2%) of those systems treat the water for iron or manganese (AWWA, 1991). Numerous texts are available in the United States which describe available treatment options for iron removal (Clark, 1977; O'Conner, 1971). Typically iron removal is described as chemical oxidation followed by filtration. Chemicals used for the oxidation of iron include oxygen, chlorine, chlorine dioxide, potassium permanganate, ozone. If the groundwater has a high organic content, colloidal (organic-iron) compounds can form which are resistant to chemical oxidation alone. Consequently, alum or another chemical coagulant must be used before the filtration (physical removal) step (Knocke, 1994).

---

[1]President, Charlotte Smith Utility Services, Inc.
P.O. Box 490, Collegeville, Pa. 19426-0490.

In Europe, an alternative to chemical/physical removal is commonly practiced, utilizing iron bacteria for the oxidation of iron. The biological process is used either alone, or as a first stage followed by chemical/physical or biological manganese removal in a second stage. The process has been described by Mouchet (Mouchet, 1992). Iron bacteria (which normally exist in the aquifer) are trapped in a granular-media filter. Air is added to the water as it enters the filter to create an environment which will enhance the growth and reproduction of the iron bacteria. As soon as a sufficient number of iron bacteria are resident in the filter, the iron level of the water passing through the filter decreases. Once the iron is removed from the water, low levels of oxidants can be added to create a filterable oxidized manganese compound. The oxidized manganese is filtered through a manganese removal filter.

This report presents the results of the first three pilot projects in the United States which employed the biological process to treat iron for a drinking water supply. The pilot work was performed on the east coast, the west coast and the midwest at three utilities which have iron or iron and manganese in their source of supply. The objective of the pilot work was to provide data to compare biological iron removal with conventional (chemical/ physical) iron removal, so that the utility can choose the best alternative. Comparisons are made on the basis of water quality, water flow rate, filter run length, and water consumed during backwash.

CASE STUDY #1: LINCOLN WATER COMPANY

Lincoln Water Company in Lincoln, Illinois currently operates two water treatment plants to remove iron and manganese from two different well fields. The utility has had persistent red water complaints in sections of the distribution system served by the North Plant. Under the current treatment regime, air, chlorine and potassium permanganate are added to the water before it is filtered through a set of 5 filters connected by a common underdrain. A pilot scale biological filter plant was operated during July 1993, to evaluate the option of installing biological iron removal before the existing filter to remove all of the iron, and therefore improve the treatment of manganese in the existing filter. The existing filter continued to operate during the pilot study. This work marked the first pilot study for biological iron removal for a drinking water utility in the United States.

CASE STUDY #2: TOMS RIVER WATER COMPANY

The Toms River Water Company's Brookside Treatment Plant presently produces 500 to 700 gallons per minute (gpm). The well water treated by the Brookside plant has an iron concentration of 3 milligrams per liter (mg/L); this concentration is 10 times higher than the Safe Drinking Water Act's maximum contamination level (MCL). To reduce the iron level to less than 0.3 mg/L, the utility practices chemical/physical iron removal with aeration, chlorination, detention and filtration. Due to population increases in the Toms River area, the water company must increase their water production for the Brookside area to 2100 gpm. An additional well will be developed at the Brookside site, which will be capable of producing 1400 gpm. Since the new well will be drilled to the same depth, in the same aquifer as the existing well, it is assumed that the well will have the same water quality. Therefore, the utility must triple the capacity of the Brookside Plant. The Brookside Plant is surrounded by wetlands, so a conventional expansion would be difficult. To achieve the treatment goals, studies comparing biological iron removal to chemical/physical iron removal were conducted during August and September 1993.

CASE STUDY #3: OLYMPIC VIEW WATER AND SEWER DISTRICT

In 1990 Olympic View Water and Sewer District developed a well on 228th Street, in Edmonds, Washington with a capacity of 500 gpm. The water from the 228th Street Well contains iron, manganese, and color above the National Secondary Drinking Water Standards of 0.3 mg/L, 0.05 mg/L, and 15 units respectively. Therefore, the District, and various consultants have evaluated traditional chemical/physical processes over the past two years in order to treat the water (Larson, 1993). Due to the high organic content of the well, chemical oxidation alone was found to be ineffective in producing iron, manganese, and color levels below the standards.

Therefore, coagulation/flocculation/sedimentation/ filtration *in addition to* oxidation appeared to be the only alternative available to the District. This process would use potassium permanganate and alum to coagulate organically bound (colloidal) iron. Most of the heavy coagulated particles would settle in a basin before the water passed through granular media filters. The space and capital requirements for this type of treatment are relatively high.

In January 1994, a pilot project was undertaken to evaluate the feasibility of integrating biological iron removal into the treatment regime for the 228th Street Well. Using this approach, the oxidant demand exerted by the colloidal iron is removed during the biological filtration step, allowing for efficient use of the oxidants needed for manganese and color removal in a second stage (dual media) filter. The pilot study is still underway at the writing of this manuscript.

METHODS

The pilot plants were designed and operated based on the source water quality characteristics of the source waters. The pilot plants were 6 or 8 inch pipe filled with 1 meter of sand (1.1 to 1.3 mm effective size), on top of a gravel underdrain. The following methods were used to determine water quality characteristics:

IRON

The samples were collected directly into (or immediately transferred to) optically matched Hach 25 ml (one inch) sampling cells. The cells were thoroughly rinsed with distilled water, between each analysis. Total and soluble iron concentrations were measured with a Hach DR2000 spectrophotometer and Hach Ferrover powder pillows. Soluble iron concentrations were determined by filtering the sample through a 0.45 micron filter before analyzing with Ferrover. The Ferrover (1,10 phenanthroline method) procedure is adapted from Standard Methods for the Examination of Water and Wastewater, (AWWA, 1989)and is EPA approved (USEPA, 1990). This method produces an orange color when iron is present.

MANGANESE

Total manganese was determined using the Hach DR2000 and the Hach PAN method. In this method, ascorbic acid is first added to the sample to reduce oxidized forms of manganese. Next, alkaline-cyanide is added to mask interfering compounds. Then the PAN reagent is added which produces a red-orange color for the spectrophotometric analysis.

DISSOLVED OXYGEN

A YSI Model 50D portable dissolved oxygen (DO) meter and a YSI Model 5739 probe were used to test the DO of the water. The meter was calibrated according to the manufacturers instructions before each DO measurement. To measure the water's DO concentration, Tygon tubing was attached to the sampling tap and the free end was submerged into a 1000 ml beaker. The DO probe was gently stirred under flowing water to achieve an accurate DO measurement.

QUALITY ASSURANCE

In Toms River and Edmonds, samples were sent to a state certified laboratory in order to validate the results of data produced by Charlotte Smith Utility Services, Inc. In Toms River, additional quality assurance techniques were employed as follows:

1)A 1.00 mg/l Hach iron standard was analyzed with the Ferrover method on three separate days and the results each time were 1.04 mg/L. This indicates that the combination of analyst, method, and instrument are within tolerance.

2) A standard addition accuracy test was performed twice. During the first test 0.1 ml of HACH 50 mg/L iron standard was added to 12 ml of distilled water which had a background iron level of 0.02 mg/L. This spiked sample had an iron concentration of 0.22 mg/L and the Ferrover method result was 0.23 mg/L. The addition recovery percentage (AR%)is determined with the following formula:

$$AR\% = \frac{[\text{sample result(mg/L)} - \text{distilled water conc.(mg/L)}]}{\text{standard addition (mg/L)}}$$

$$AR\% = \frac{[0.23 \text{ mg/L} - 0.02 \text{ mg/L}]}{0.2 \text{ mg/L}} = 1.05\%$$

The second sample had 0.02 ml of the 50 mg/L of the standard added to distilled water and the AR% was 95%. Recoveries of one hundred percent plus or minus five percent indicate that the original sample analysis is correct.

RESULTS

WATER QUALITY

The raw water iron levels at the three sites were as follows: Lincoln: 1.8 mg/L; Toms River: 2.9 mg/L; and Edmonds 2.4 mg/L. In all cases, the finished water from the biological iron removal pilot plants produced water with nondetectable concentrations of iron (<0.05 mg/L). In Edmonds, manganese levels in the second stage pilot filter were reduced to "below detection level" when preceded by the biological iron removal filter as a first stage. An analysis of the monthly reports submitted to the States of Illinois and New Jersey, indicates that the average iron concentrations from the Lincoln Water Company's North Plant and Toms River Water Company's Brookside Plant are 0.21 mg/L and 0.26 mg/L respectively. Since biological filtration relies on the retention of a biofilm on the filter media which can be disrupted during backwash, backwash recovery was extensively evaluated. In Toms River, samples were collected 5 minutes after the biological and chemical/physical filters were put back on-line after backwash, and at frequent intervals up to 1 hour into the filter run.

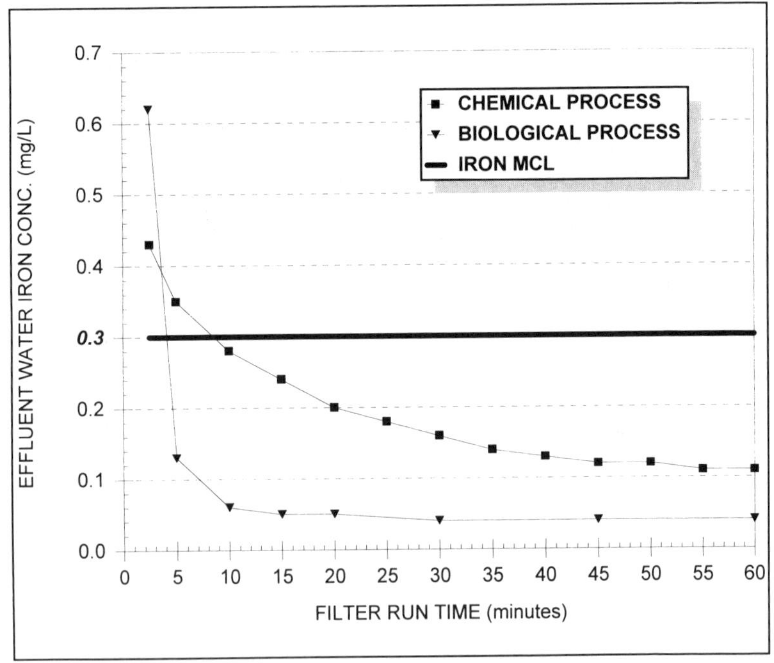

Figure 1: Comparison of the biological iron removal pilot plant and the existing chemical/physical plant at Toms River, New Jersey, immediately after backwash.

Figure 1 shows that the biological pilot plant not only met the 0.3 mg/L standard faster than the existing chemical/physical plant in Toms River after a backwash, but reached a lower finished water iron concentration.

FLOW RATE

Chemical/physical plants typically operate at 3 to 4 gpm/sq ft., with a filter run length of 24 to 48 hours. In Lincoln flow rates of 6, 8, 10 and 12 gpm/sq ft were evaluated. All produced superior water quality when compared to iron concentrations from the existing plant, which operates at 1.3 gpm/sq ft. No differences in water quality were observed at the various flow rates in the pilot plant. However, terminal head loss occurred more quickly at the higher flow rates.

The Brookside Plant currently operates at 2.0 gpm/sq ft., and experiences iron breakthrough above 0.3 mg/L after approximately 36 hours. At the time of the pilot project, the Utility intended to retrofit the existing plant and triple the flow rate utilizing the biological process. Therefore, the pilot plant was operated at 8 gpm/sq ft with the intention of running the full scale plant at 6 gpm/sq ft.

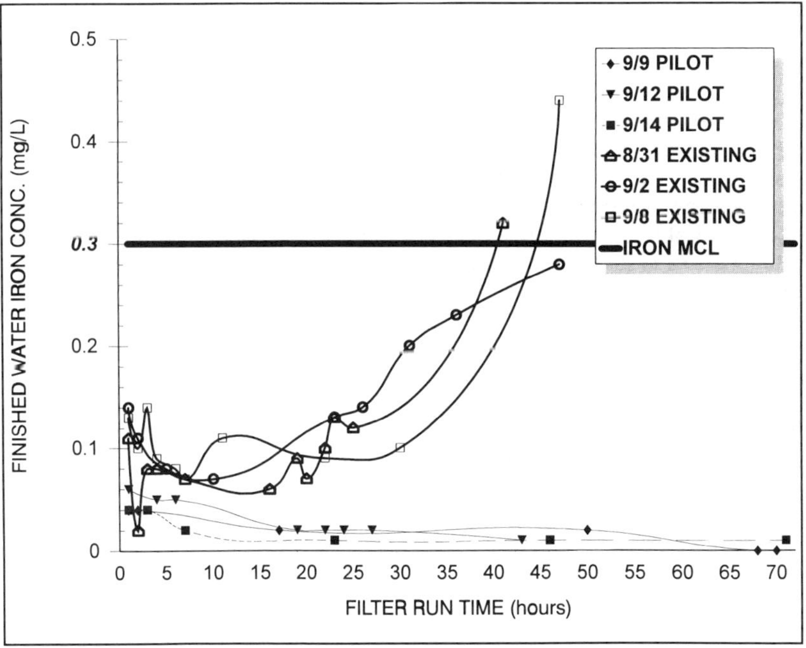

Figure 2: Finished water quality from the biological pilot plant and existing chemical/physical plant, in Toms River. Three filter runs are shown for each plant.

FILTER RUN LENGTH

Figure 2 compares the filter run length and iron concentration throughout the filter run in Toms River. The existing filter is backwashed every 48 hours. However, as seen on the figure, by this time iron levels exceed or nearly exceed the standard. This is due to the fact that most of the removal in the chemical/ physical filter occurs at the top of filter bed and breakthrough is inevitable as the filter run proceeds. Conversely, the bacteria in the biological filter reside throughout the filter bed and continue to oxidize iron throughout the filter run. In fact, no increases were observed during two 72 hour filter runs. Similarly, breakthrough did not occur in Lincoln or Edmonds before backwash of 24 and 48 hours respectively (data not shown). The biological pilot plant experienced air binding which shorted the filter run length. This problem was solved by the time of the Toms River and Edmonds studies, by constructing pressure filters.

CONSERVATION ISSUES

During a 48 hour period the existing Toms River plant produces 2 million gallons. A filter is backwash requires 32,000 gallons or 1.6% of the water produced. The biological pilot plant produced 7,069 gallons during a 48 hour period and required 17 gallons or 0.24% of production volume for the backwash.

CONCLUSION

The report compares the efficiency of chemical/ physical and biological iron removal processes during the entire filter cycle. Performance variables include, flow rates, filter run length, backwash recovery rates and water quality. The results of the studies indicate that biological iron removal will reduce the iron concentration of groundwater to less then 0.05 mg/L at flow rates of at least 8.0 gpm/sq ft. The high rate filtration, allows for lower capital costs, and less space requirements. Since the oxidation occurs within the filter bed, a detention tank is not required. Filter run lengths for the biological pilot filters of 48 hours can be achieved without appreciable head loss. Since the biological process uses untreated water for backwash, and backwashes are performed at lower rates, substantial water savings are realized. Operating costs savings are also obtained, since pre-oxidation with chemicals does not take place. Due to the ease of operation of the biological iron filtration, the process is ideal for small systems. The disadvantage of the process is the delay in initial start-up required as iron bacteria from the aquifer "seed" the biological filter.

At the time this manuscript was submitted for publication, Toms River Water Company was preparing the design for a 3 million gallon per day biological iron removal plant, based on design criteria from the pilot study. Lincoln Water Company continued to run their existing plant under a revised operation plan recommended by the authors, and water quality complaints decreased. The Olympic View Water and Sewer District personnel took over operation of the pilot plant from Charlotte Smith Utility Services, Inc. and continued to collect data to set the design criteria.

REFERENCES

Clark, J.W., Warren, V. & Hammer, M.J. *Water Supply and Pollution Control.* Harper & Row, New York (1977).

Knocke, W.R. et al Examining the Reactions between Soluble Iron, DOC, and Alternative Oxidants During Conventional Treatment. *Jour. AWWA*, 86:117, (Jan. 1992).

Larson, A.L.: Case Pilot Study: Organic Fouling of a Coated Media Pressure Filter. submitted to the *Proceedings of the 1993 AWWA WQTC*, Miami, Florida.

Mouchet, P.: From Conventional to Biological Removal of Iron and manganese in France. *Jour. AWWA*, 84:158, (Apr. 1992).

O'Connor, J.T. Iron and Manganese. *Water Quality and Treatment* (H.B. Crawford and D.N. Fischel, editors). McGraw-Hill Book Co., New York (3rd ed 1971).

Standard Methods for the Examination of Water and Wastewater. AWWA, APHA, WPCF, Washington, D.C. (17 ed., 1989).

USEPA: *Federal Register,* June 27, 45:126:433459 (1980).

Water Industry Data Base: Utility Profiles. AWWA, AWWARF, Denver Colorado. (1991).

ACKNOWLEDGEMENTS

This work was funded by General Waterworks Corporation, and Olympic View Water And Sewer District.

Assistance from the following individuals is gratefully acknowledged: Thomas F. Cleveland, Pierre Mouchet, Phillipe Gislette, Todd Mackey, Pat Radice, Dean Button, John Overstreet, Tracey Liberi, Roger Eberhart, Scott Dunn.

Treatment of Oil in Water Emulsions by Ceramic-Supported Polymeric Membranes

Robert P. Castro[1], Yoram Cohen[2] and Harold G. Monbouquette[3]

Abstract

A novel membrane was developed by growing polymer chains from the surface of a porous ceramic support, resulting in a composite membrane which combines the mechanical properties of the inorganic membrane with the selective interactions of the polymer. The configuration of the grafted polymer brush layer is determined by solvent - polymer interactions, with a hydrophilic polymer being stretched away from the surface by aqueous solutions and collapsed against the surface by organic solvents. This behavior of the grafted chains provides Ceramic-Supported Polymeric (CSP) membranes with unique properties for certain water treatment applications.

One application envisioned for these CSP membranes, in which the selectivity is influenced by interactions between the solvent and the grafted polymer, is the cross-flow filtration of an oil-in-water emulsion. In this case, a hydrophilic grafted polyvinylpyrrolidone (PVP) brush layer expanded into the pore volume due to the affinity of the polymer for water. These extended grafted chains preferentially allow the passage of water over oil, producing a permeate stream with a lower total organic carbon content compared to an unmodified membrane. Another advantage of the CSP membrane is in reducing permeate flux decline believed to be caused by the adsorption of oil onto the membrane surface. For the PVP-modified CSP membrane, the grafted polymer alters the membrane surface character from hydrophobic to hydrophilic, reducing the tendency for oil adsorption. This phenomenon was demonstrated by comparison of permeate flow rate behavior for both unmodified and graft polymerized (CSP) membranes.

---

[1]Graduate Student, Department of Chemical Engineering, University of California, Los Angeles, Los Angeles, CA 90024.

[2]Professor, Department of Chemical Engineering, UCLA.

[3]Associate Professor, Department of Chemical Engineering, UCLA.

Introduction

Ceramic membranes are being used in a wide range of separation applications where polymeric membranes are not applicable due to their superior physical stability and chemical resistance to industrial solvents. However, the advantages ceramic membranes offer are offset by their low selectivity, which makes their use economically unfeasible for many applications. Improvements in selectivity are therefore necessary in order to make ceramic membrane based systems a viable option for applications such as purification and pollution abatement. When microporous ceramic membranes are used in pressure-driven filtration, they separate solutes from solvents predominately by size exclusion and, to a lesser extent, interactions with the membrane surface. Therefore, their selectivity is determined by the pore size of the membrane.

One possible method for increasing the selectivity of these membranes is modification of the membrane surface with a polymer layer (Aptel et al., 1972 and 1976). Surface modification with polymers may be achieved in two ways, either by polymer grafting or graft polymerization. Polymer grafting consists of bonding one end of a live polymer chain to an active site on the surface. This method results in a uniform grafted layer, however surface coverage is limited due to steric hinderance which limits the approach of polymer chains to the pore surface. Graft polymerization is the process of growing polymer chains monomer by monomer from active surface sites by a free radical or other polymerization mechanism. This technique has the advantage of providing a covalently bonded polymer brush layer of high surface coverage since the diffusional limitations and steric hinderance effects evident in polymer grafting are minimized.

The goal in grafting a polymer brush layer on the surface of a microporous ceramic membrane is to develop a new hybrid membrane which couples the selectivity of a polymeric membrane with the mechanical properties of a ceramic membrane. The selectivity of the resultant ceramic-supported polymeric (CSP) membrane is determined by the configuration of the polymer brush layer and the functionality of the grafted polymer. For example, when the solvent-polymer interactions are stronger than the polymer-polymer interactions, the chains extend from the surface, filling the pore volume [Cohen et al., 1992; Osada et al., 1986]. In this situation, the selectivity of the membrane is strongly influenced by solute-polymer interactions. The extent to which the selectivity of the microporous ceramic membrane is altered depends on a number of parameters, such as surface coverage, percentage of pore volume occupied (i.e. length of the grafted chains) and the polymeric material chosen. Because of the composite structure, the polymer layer can be engineered for optimum selectivity with no regard for its macroscopic mechanical properties.

The purpose of this work is to demonstrate the feasibility of altering the separation properties of a ceramic membrane by the graft polymerization of a water soluble polymer, poly(vinylpyrrolidone), onto the surface of silica membranes. Using a graft polymerization method developed for the alteration of silica chromatography resins [Cohen, 1991; Chaimberg and Cohen, 1991], porous silica membranes were modified. The permeability of pure solvents for both the unmodified and modified membranes was determined by conducting flow rate-pressure drop experiments. The separation capability of the membranes was investigated by separating an oil/water emulsion in cross-flow filtration experiments.

Materials

The isotropic porous silica membranes (Microporous Glass, Asahi Glass Inc., Tokyo, Japan)used in this work were 47mm in diameter and had a pore size of 4100 ± 615Å (permeability studies) or 3000 ± 450Å (cross-flow filtration studies). The membranes had surface areas of 2.45 and 3.15 m^2/g, respectively. All the organic solvents used in the flow rate-pressure drop measurements were certified grade (Fisher Scientific, Pittsburgh, PA). For the experiments conducted with water, distilled and deionized water was filtered through a 0.2 μm in-line filter (Whatman Inc., Clifton, NJ) in order to remove any microparticles which may cause membrane fouling (Elmaleh, S. and W. Naceur, 1992). A commercially available cutting and grinding fluid (Castrol 329, Castrol Industrial Inc., Los Angeles, CA) was used to prepare an oil/water emulsion for cross-flow filtration experiments. This proprietary mixture contains a combination of oil and surfactants which allow the formation of a stable emulsion upon mixing with water.

Permeability Studies

The permeability of both an unmodified and then PVP-modified CSP membrane (pore size = 4100Å) for six different solvents were determined by pressure drop-flow rate experiments. The details of these experiments have been published (Castro *et al.*, 1993). The results are summarized in Table 1. The data indicate a significantly greater permeability reduction for water and the aliphatic alcohols (40 - 50%) compared to the hydrophobic solvents cyclohexane and toluene (14 and 24%, respectively).

This solvent permeability behavior can be explained in terms of the configuration of the grafted polymer chains in the various solvents. In a good solvent, where the polymer has a high affinity for the solvent, the grafted polymer brush layer will be swollen with the solvent. This results in chains stretched away from the surface to which they are attached (see Figure 1a) (Milner, 1991). Due to monomer mass transfer limitations during graft polymerization, it is believed that the majority of the grafted chains are on the top surface, with limited penetration into the membrane pores. Therefore, the extended grafted chains hinder flow through the membrane (thus reducing the

Table 1. Solvent Permeability (k) for a 4100Å Pore Size Silica Support and PVP-Modified CSP Membrane

Solvent	Permeability x 10^{12} (cm^2)		Perm. Reduction
	(k_{UNMOD})	($k_{CSP}{}^{a}$)	(%)
Water	29.5	15.4	47.8
Ethanol	24.4	14.5	40.4
1-Propanol	33.1	18.2	45.0
1-Butanol	34.4	16.4	52.3
Toluene	34.9	26.5	24.0
Cyclohexane	36.5	31.4	14.0

a Polyvinylpyrrolidone modified (graft yield = 6.0 mg/cm^2)

permeability) by obstructing or reducing the pore radius. This is believed to be the primary reason for the larger permeability decrease for water and the aliphatic alcohols, since they are the best solvents for PVP.

In a poor solvent, the grafted chains would tend to collapse against the membrane surface or pore walls (see Figure 1b). Thus, for a poor solvent, the grafted chains will have a much smaller effect on the solvent permeability because the chains occupy less of the pore volume compared to chains in a stretched configuration (i.e., in a good solvent). Since both toluene and cyclohexane are poor solvents for PVP, the decrease in permeability is much smaller for these two solvents compared to the other four solvents in which the PVP chains are likely to be in a more stretched (or extended) configuration. It is important to note that regardless of the solvent quality and hence the chain configuration, the polymer chains always remain anchored to the membrane surface by covalent bonds.

Cross-Flow Filtration Studies

The ceramic/polymeric membrane in which the selectivity is influenced by interactions between the solute and the grafted polymer chains was further investigated by performing cross-flow experiments with an oil/water emulsion. In this case, the grafted PVP chains are expected to expand into the pore volume due to the affinity of the polymer to water. However, since the polymer is hydrophilic, it should preferentially allow the passage of water molecules over the oil droplets.

In order to determine the performance of CSP membranes in cross-flow filtration, experiments were conducted using apparatus shown schematically in

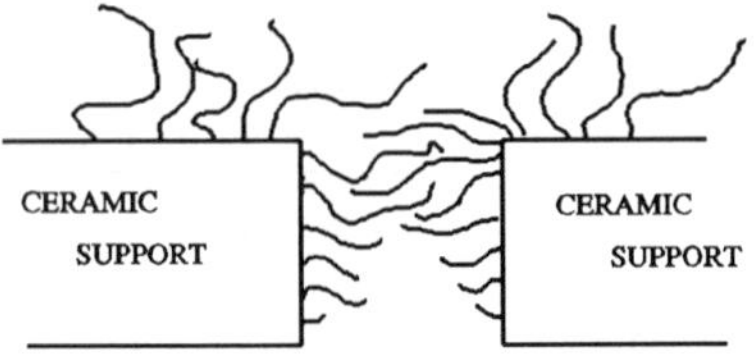

Figure 1a. Configuration of the grafted polymer brush layer when the polymer has a high affinity for the solvent.

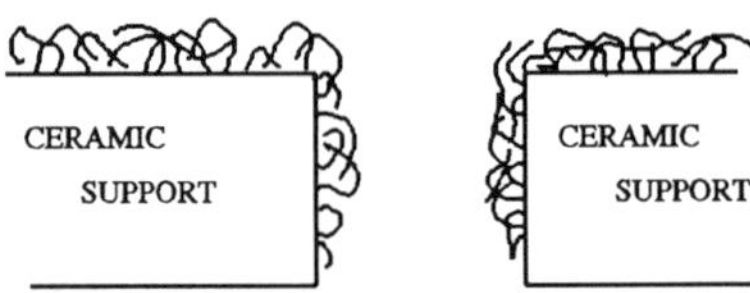

Figure 1b. Configuration of the grafted polymer brush layer when the polymer has a low affinity for the solvent.

Figure 2. For this application, 3100Å silica support membranes were used and the PVP graft yield for the CSP membrane was 1.8 mg/m^2. Both the CSP membrane and an unmodified support membrane were first characterized by operating the system with a water feed solution. In this experiment, the residue and permeate flow rates were measured as a function of time for a set transmembrane pressure (pump setting) with the metering valve in the residue line completely open. The results presented in Figure 3 indicate that the permeate flow rate of the unmodified membrane was greater than that of the CSP membrane due to the expanded configuration of the grafted PVP chains, as expected. The water permeability (k_{H2O}) of the support used for the CSP membrane (9.58×10^{-12} cm^2) was 23.4% less than k_{H2O} of the unmodified membrane (12.5×10^{-12} cm^2), while the permeate flow rate of the CSP membrane was 31.0% less than that of the unmodified membrane. This suggests only a minor resistance to flow provided by the brush layer. One possible explanation is that the grafted polymer brush layer consists of many short chains that do not have a marked effect on the permeate flow. It should be noted that the cross-flow measurement for the unmodified was done at a transmembrane pressure of 2.1 psig while the CSP membrane was measured at 2.3 psig.

Preliminary experiments have been performed with a 4.71 %(w/w) oil/water emulsion using both membranes. The procedure was to measure the system parameters (stream flow rates and concentrations) as a function of time

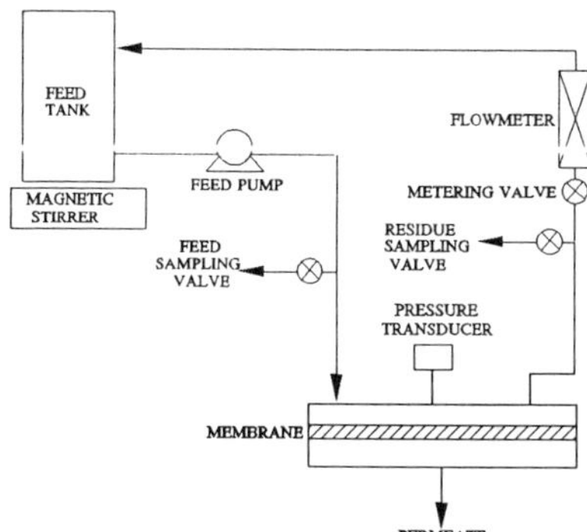

Figure 2. Schematic diagram of the cross-flow filtration apparatus.

for a set transmembrane pressure. Stream compositions were determined by total organic analysis (Model 1270, Ionics, Inc., Watertown, MA). The measured values for the stream flow rates and permeate concentrations for both membranes are presented in Figures 4 and 5, respectively.

Figure 4 shows that both membranes had a comparable permeate flow rate for the emulsion, despite the larger water flow rate for the unmodified membrane (see Figure 3). However, the permeate flow rate for the CSP membrane increased slightly with time, but the permeate flow rate for the unmodified membrane but decreased steadily throughout the experiment. It is believed that this decrease was caused by fouling and/or the immediate formation of an oil gel layer on the surface of the membrane. The fouling of the unmodified membrane expected to be much greater compared to the CSP

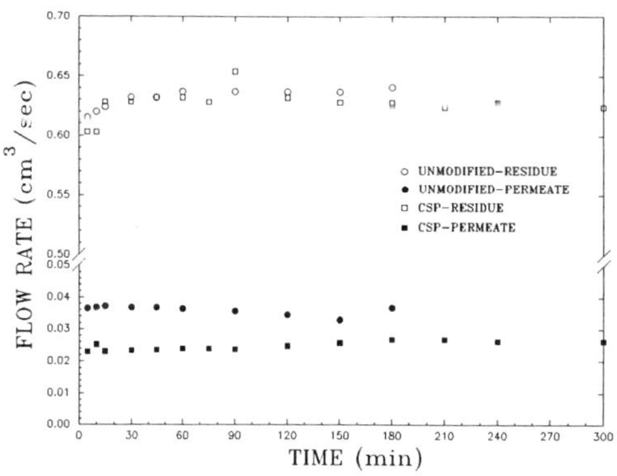

Figure 3. Water flow rate as a function of time for an unmodified and CSP membrane at minimum stage-cut (ΔP = 2.1 and 2.3 psig, respectively).

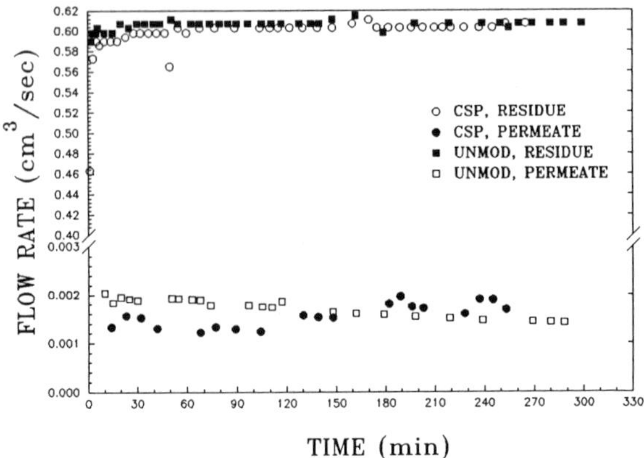

Figure 4. Residue and permeate flow rate as a function of time for an unmodified and CSP membrane (pore size = 3000Å).

membrane because the presence of the PVP brush layer was expected to prevent oil adsorption to the surface of the membrane.

The concentration-time curves of the permeate streams presented in Figure 5 indicate that, for the same feed concentration of 41,800 ppm, the CSP membrane produced a permeate stream that had a lower TOC concentration of approximately 1000 ppm compared to the unmodified membrane. These are preliminary results that demonstrate that the polymer layer has a positive effect. However, further optimization of the membrane rejection will be needed by adjusting the combination of membrane pore size and graft yield.

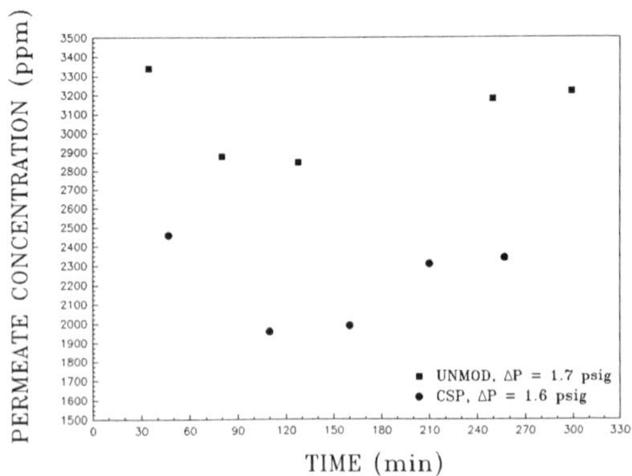

Figure 5. Permeate concentration as a function of time for an unmodified and CSP membrane (pore size = 3000Å).

Conclusion

A new composite membrane combining the mechanical properties of a ceramic membrane and the selectivity of a polymeric membrane is being developed by the attachment of a covalently bonded polyvinylpyrrolidone brush layer to the surface of a porous silica membrane via a graft polymerization process. Permeability studies with these ceramic-supported polymeric (CSP) membranes demonstrate that the configuration of the grafted chains are determined by solvent-polymer interactions, with the hydrophilic polymer chains being stretched away from the surface by aqueous solutions and collapsed against the surface by organic solvents. Preliminary cross-flow filtration experiments with an oil/water emulsion indicate that the modified membrane gives a lower permeate concentration and is more resistant to fouling via oil adsorption compared to an unmodified membrane.

REFERENCES

Aptel, P. et al., Liquid Transport Through Membranes Prepared by Grafting of Polar Monomers onto Poly(tetrafluoroethylene) Films. I. Some Fractionation of Liquid Mixtures by Pervaporation, *J. Applied Polymer Sci.*, 16 (1972) 1061.

Aptel, P. et al., Application of the Pervaporation Process to Separate Azeotropic Mixtures, *J. Membrane Sci.*, 1 (1976) 271.

Castro, R.P., Y. Cohen and H.G. Monbouquette, Permeability Behavior of Polyvinylpyrrolidone-modified Porous Silica Membranes, *J. Memb. Sci.*, 84 (1993) 151.

Chaimberg, M. and Y. Cohen, Free Radical Graft Polymerization of Vinylpyrrolidone onto Silica, *Ind. and Eng. Chem. Research,* 30 (1991) 2534.

Cohen, Y., High Yield Water-Soluble Polymer Silica Separation Resins, U.S. Patent 5,035,803 (1991).

Cohen, Y., P. Eisenberg and M. Chaimberg, Permeability of Graft Polymerized Polyvinylpyrrolidone-Silica Resin in Packed Columns, *J. Colloid and Interface Science*, 148 (1992) 579.

Elmaleh, S. and W. Naceur, Transport of water through an inorganic composite membrane, *J. Membrane Sci.*, 66 (1992) 227.

Milner, S.T., Polymer Brushes, *Science*, 251 (1991) 905.

Osada, Y., K. Honda and M. Ohta, Control of Water Permeability by Mechanochemical Contraction of Poly(Methacrylic Acid)-Grafted Membranes, *J. Membrane Science*, 27 (1986) 327.

Immiscible Organic Liquid Recovery Using Unconfined Membranes

Leonard Jewell[1] and Michael J. Semmens[2]

Abstract

Hydrophobic hollow fiber micropourous membranes have been characterized and applied to the problem of separating and recovering low density hydrocarbons from water surfaces. The merits of hollow fiber technology as applied to this problem are discussed and a model for predicting recovery performance is shown to be accurate. Future work on a recovery device using this technology is outlined and potential applications are identified.

Introduction

Immiscible organic liquids and water mixtures are very common. They can be the result of an unintentional action like a gasoline spill into a body of water or groundwater, or they could be the result of an industrial or manufacturing process like liquid-liquid extraction, parts and equipment cleaning, or cooling in metal cutting and milling operations. Immiscible organic mixtures of these types result in the organic liquid existing either as a dissolved, emulsified, or separate component which floats on top of the water for low density organic liquids. The focus of this investigation was to develop an unconfined microporous hydrophobic hollow fiber membrane apparatus

---

[1] Graduate Student, Department of Civil and Mineral Engineering, University of Minnesota, Minneapolis, MN 55455 (USA)

[2] Professor, Department of Civil and Mineral Engineering, University of Minnesota, Minneapolis, MN 55455 (USA)

that would selectively remove the organic liquid layer from the top of the water and leave the water behind.

While hollow fiber membranes have been used to separate organic liquid and water mixtures before, the previous approaches have used hydrophilic membranes where the water was withdrawn leaving the organic liquid behind. Since the fraction of organic liquid in typical waste streams is often less than 20% by volume, this approach requires the use of large membrane separators. Therefore, in this investigation, the approach was focused on removing the smaller organic liquid fraction and not the larger water fraction by using hydrophobic hollow fiber membranes which resulted in the use of smaller membrane separators. In addition, rather than using in pipe membrane modules, the membranes in this investigation were unconfined and allowed to float directly on the surface thereby saving energy by reducing pumping costs.

Theory

Various hollow fiber membranes manufactured by the Mitsubishi Rayon Company were studied. The inside fiber diameters of these fibers ranged from 270 μm to 280 μm and membrane thicknesses ranged from 54 μm to 56 μm. All membranes were constructed of polyethylene because polyethylene has three properties that make it attractive for this technology. First, the density of polyethylene is less than that of water (i.e. a specific gravity between 0.92 and 0.95) so that the polyethylene fibers will float on top of the water and remain in contact with the low density hydrocarbon spill on the surface of the water thus naturally eliminating the problem of maintaining the recovery device at the interface. Second, the polyethylene polymer has been shown to be resistant to the types of liquid aliphatic and aromatic hydrocarbons likely to be found in typical applications. Third, the polyethylene has a very low surface energy which ensures that the fiber will not be wetted by the water (the contact angle between water and polyethylene is approximately 94°) but, instead it will be wetted by any hydrocarbon present. This preferential wetting allows the fiber to be surrounded by the hydrocarbon even when the thickness of the hydrocarbon layer is less than the outside diameter of the fiber hence allowing for a more

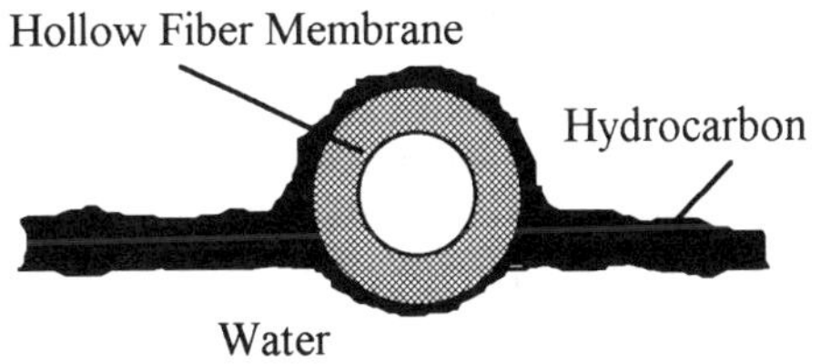

Figure 1: *A preferentially wetted membrane*

complete recovery. Figure 1 illustrates this property.

In addition to the above properties, the membranes studied had the further advantage of being microporous. Microporous membranes are simply membranes where the surface is covered with interconnected pores that are of the order of 0.1 μm in diameter. In particular, for the membranes studied, the pores covered between 60% and 67% of the total membrane surface. The advantage of this microporous structure are: 1) the interconnections promote complete wetting by the hydrocarbons thereby increasing recovery efficiency; 2) the combination of the small pore size openings and the low surface energy of the polyethylene allows for the exclusion of water from the pore, and most importantly; 3) large flux rates through the membrane can be realized because the hydrocarbons can flow through the pores in a laminar flow pattern which is much faster than diffusion through conventional non-porous membrane materials.

While the microporous structure of these membranes does enable larger flux rates through this type of membrane, it also creates the potential for permeate contamination by the non-impregnating fluid. It is important, therefore, that the transmembrane pressure is never large enough to allow hydrocarbon displacement by either water or air. Recent work by Zha *et. al.* (1992) on hollow fiber microporous membranes similar to those used in these experiments showed that the critical pressure necessary for displacement of an impregnating phase by a non-impregnating phase was a function of the pore radius, pore structure, and the interfacial tension between the two phases. From their work, they were able to estimate the critical displacement pressures for many hydrocarbon-air and hydrocarbon-water systems. The results of their studies indicate that the transmembrane pressure necessary to displace nonpolar hydrocarbons from the 0.1 μm pores in polyethylene fibers is between 180 kPa and 250 kPa (26 psi and 36 psi) for water and between 105 kPa and 150 kPa (15 psi and 22 psi) for air.

Therefore, if these hollow fiber microporous membranes were to be deployed onto a hydrocarbon layer floating on top of water, the membranes would naturally locate themselves at the hydrocarbon/water interface, would become preferentially wetted by the hydrocarbon, and if the fibers were manifolded and attached to a vacuum recovery device the hydrocarbon could be continuously recovered. Furthermore, even the maximum transmembrane pressure difference (i.e. 101 kPa or 14.7 psi) would cause hydrocarbon displacement from the pores by either the air on top of the fiber or the water underneath the fiber.

A completed hydrocarbon recovery unit would consist of these fibers woven into a mat and potted into a floating manifold device. This manifold would then be connected to a vacuum regulated suction pump which would draw the hydrocarbon through the fibers and into a waiting recovery tank (see Figure 2). It is important that the vacuum created by the suction pump is large enough to provide for the fastest recovery rate possible but, the lowest pressure in the system must remain greater than the vapor pressure of the hydrocarbon being recovered so that the hydrocarbon is recovered as a liquid and not as a vapor (in order to eliminate the cost of a vapor condenser).

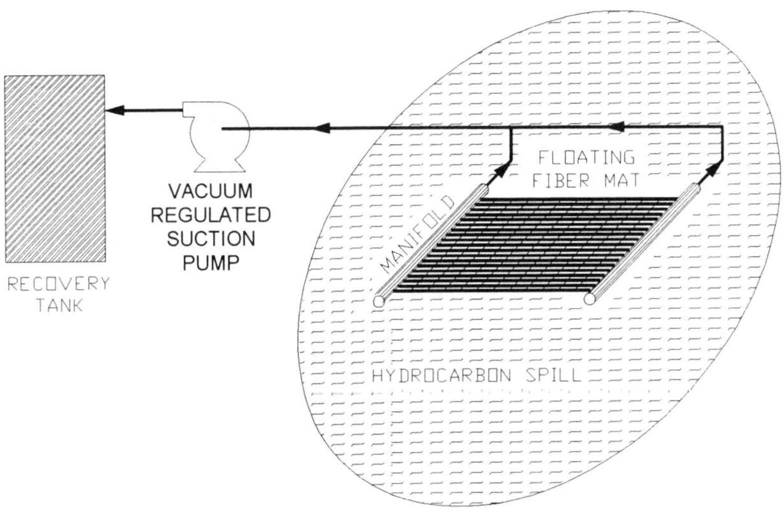

Figure 2: *Proposed free liquid hydrocarbon recovery unit*

Procedure

In order to design any practical hydrocarbon recovery device, the amount of material that can be recovered needs to be accurately characterized. To that end, a model for hydrocarbon recovery was developed that is based upon the following equations describing flux through a membrane:

$$F = \frac{K \cdot \rho \cdot \Delta P}{\mu \cdot \delta} \tag{1}$$

and Poiseuille's Law:

$$\frac{dP}{dz} = \frac{32 \cdot \mu \cdot v}{\cdot d^2} \qquad (2)$$

Where F is the mass flux through the membrane, K is the permeability coefficient for the membrane, ρ is the permeate density, ΔP is the transmembrane pressure, μ is the absolute viscosity of the permeate, δ is the membrane thickness, d is the fiber inner diameter, v is the velocity of the fluid flow in the inside of the fiber, and dP/dz is the pressure drop along the fiber due to friction between the fluid and the membrane wall. While fiber inner diameter and thickness data is readily available from the manufacturers specification sheets supplied with the hollow fibers, the permeability coefficient K is usually not readily available. Therefore, the first phase of this work on hydrocarbon recovery involved the development of a technique to determine the permeability coefficient for a particular membrane (Jewell and Semmens, 1994). In this paper, the permeability coefficients measured experimentally are used to demonstrate the accuracy of the model.

In addition to accurately predicting hydrocarbon recovery rates, the model was also used to identify which design parameters have the largest impact upon recovery performance. This analysis showed that the optimum recovery performance could be achieved by: 1) using fibers with large internal diameter fibers, 2) using membranes with thin walls, 3) using microporous membranes with large porosity, and 4) keeping the membrane potting length to a minimum.

Results

Once a working model to describe the hydrocarbon recovery was developed and the permeability coefficients for the various membranes were determined, experiments were performed to verify the model predictions. The comparisons of the measured and modeled hydrocarbon fluxes for one fiber type and two hydrocarbons are shown in the following figures (Figures 3 and 4). Note that in both cases, the measured hydrocarbon flow rates are within the uncertainty of the model predictions.

In order to verify the resistance of the polyethylene polymer to hydrocarbon mixtures, a fiber was continuously wetted by unleaded gasoline for three days (longer than any gasoline spill is likely to exist) and periodic

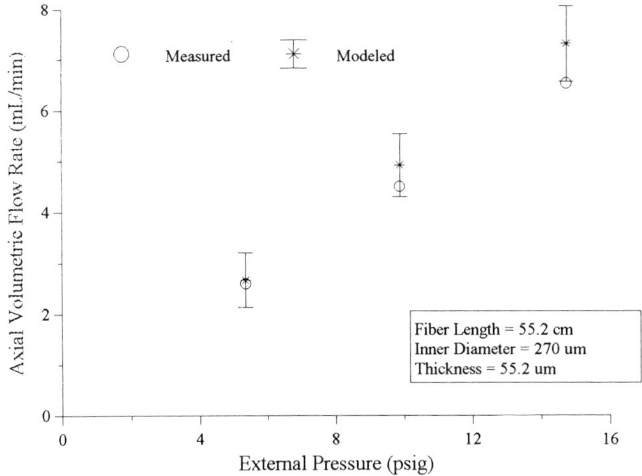

Figure 3: *Gasoline recovery rates from a single fiber*

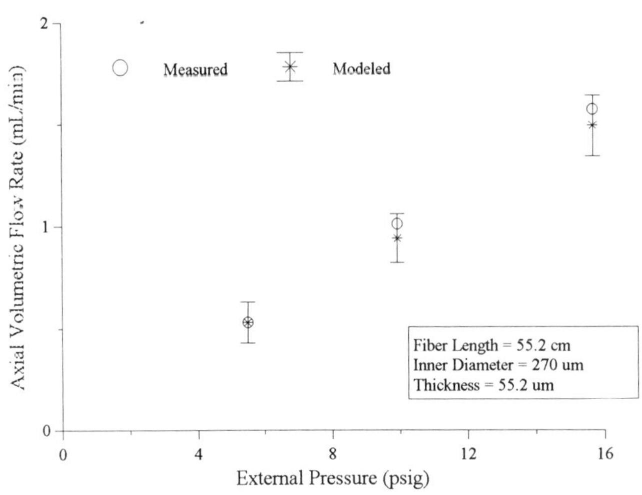

Figure 4: *Diesel fuel recovery rates from a single fiber*

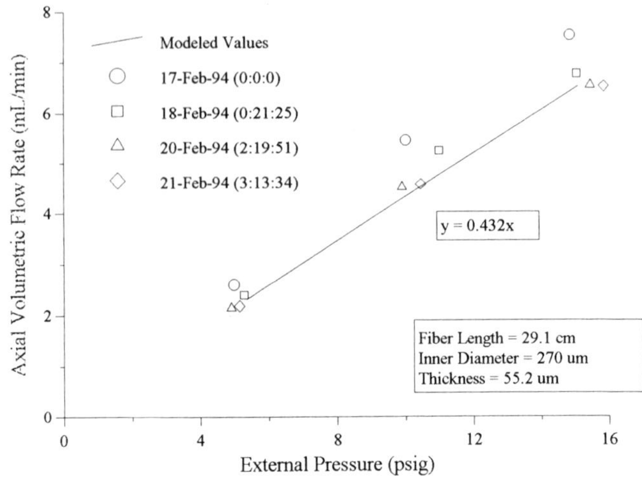

Figure 5: *Effect of prolonged gasoline exposure on membrane flux rate*

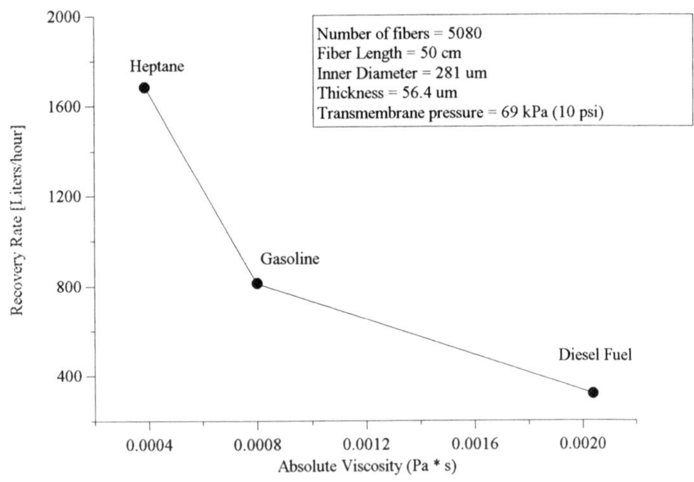

Figure 6: *Expected recovery performance for a 1 square meter fiber mat*

flux rate measurements were taken. As seen in Figure 5, while the flux rate did decline it was due to a reduction of inner diameter in the potted section of the membrane and not due to overall polymer deterioration.

Current research efforts are focused on the development of hollow fiber membrane mats constructed with the fibers that have been characterized using the model and methods described above. Using parameters derived from single fiber tests, expected recovery rates for various hydrocarbons using for a double layer mat are shown in Figure 6.

Conclusion

A new device for the recovery of low density hydrocarbon spills has been demonstrated to be feasible. This new device utilizes properties that are unique to hydrophobic hollow fiber microporous membranes. These properties maintain the recovery device at the hydrocarbon/water interface, enable the recovery of the hydrocarbon while excluding the water, provide very large membrane surface areas per unit volume, resist polymer degradation due to hydrocarbon contact, and provide for large flux rates through the micropores. A method has been developed and verified to accurately predict hydrocarbon recovery rates from such a recovery device. Future work on this project will include the fabrication of hollow fiber membrane mats in order to facilitate deployment and the development of the complete recovery system. A complete system ready for field testing on free solvents in industrial settings, hydrocarbon spills on open bodies of water, or LNAPL (Light Non-Aqueous Phase Liquid) in groundwater contamination should be ready by June of this year.

References

Zha, F.F., Fane, A.G., Fell, C.J.D., Schofield, R.W., "Critical displacement pressure of a supported liquid membrane", *Journal of Membrane Science*, 75, **1992**, 69-80.

Jewell, L.J. and Semmens, M.J., “Hollow fiber membrane permeability model: Development, verification, and parameter effects”, unpublished, **1994**.

Waste Minimization and Recycling by Microfiltration during Nitrocellulose Manufacturing

Lee Clapp[1], Jae K. Park[2], and Byung J. Kim[3]

Abstract

A plant survey was conducted to determine the amount of nitrocellulose (NC) discharged with process water from each of five sequential NC purification processes. Wastewater samples were analyzed for particle size distribution (PSD), total suspended solids concentration (TSS), particle surface charge (zeta potential), and pH. Rough mass balances indicated that approximately 5% of each initial load of NC is lost during the purification process, and that most of this loss occurs during the decanting of various tub supernatants. Tests were performed with a laboratory-scale cross-flow microfiltration (MF) unit to assess the potential of MF for removing and recycling NC fines. Measurements of permeate suspended solids concentrations and flow rates indicated that it may be feasible to remove and recycle NC fines during the various tub decanting procedures using MF. However, high capital and operational costs may result in this technology not being cost effective. Therefore, other methods for recycling NC should also be investigated.

Introduction

The manufacture of nitrocellulose (NC) involves the nitration of cotton linters or wood pulp using a combination of nitric and sulfuric acids. After nitration most of the residual acid is removed from the NC using a counter-current backwash centrifuge. After the nitration phase, it is necessary to remove the residual acid to stabilize the NC. This stabilization is accomplished in several steps: (1) first, the crude NC is repeatedly boiled over several days in the boiling tub house to remove most of the residual acid from the NC; (2) the NC is then cut and beaten in slightly alkaline water in the beater house to reduce the average NC particle size; (3) the NC is next further stabilized in the poacher house by boiling the NC with soda ash; (4) the NC is then screened and vacuum filtered to remove the bulk water and then various NC batches are blended to

[1]Research Assistant, Department of Civil and Environmental Engineer, University of Wisconsin-Madison, WI 53706

[2]Assistant Professor, Department of Civil and Environmental Engineer, University of Wisconsin-Madison, WI 53706

[3]Principal Investigator, U.S. Construction Engineering Research Laboratories, Champaign, IL 61824

achieve the desired final grade; (5) finally, the NC is sent to the wringer house where it is washed again and centrifuged to remove excess water. All of the previous processes use filtered river water as wash and transfer water. After being treated in the wringer house, the NC is transported to another part of the plant for further processing into propellant. Figure 1 shows the generalized flow diagram of NC-manufacturing processes and wastewater generation.

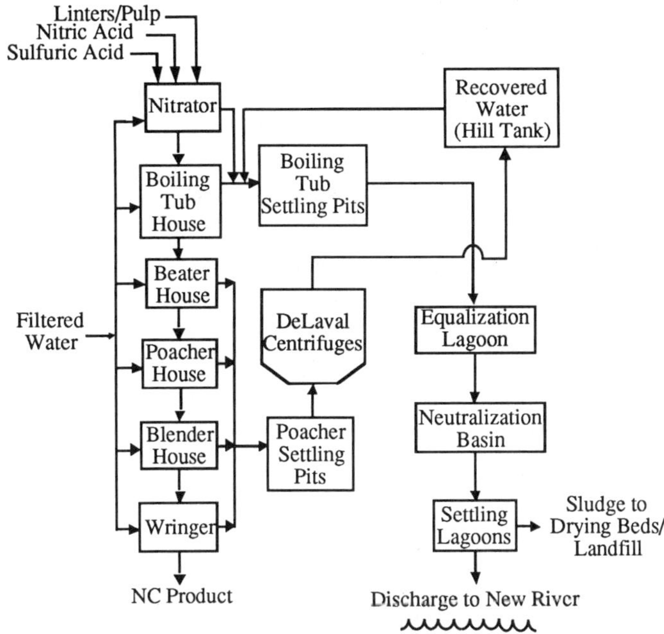

Figure 1. NC-manufacturing and wastewater flow diagram.

All of the processes listed above discharge wastewater containing NC fibers, referred to as NC fines. Low pH wastewater from the nitrator and boiling tub house drains to the boiling tub settling pits. The NC fines in the boiling tub wastewater are larger than for the remaining processes because the NC has not been cut yet. Neutral pH wastewaters from the beater, poacher, blender, and wringer houses all drain to the poacher settling pits. These wastewaters contain a mixture of short fibers and colloidal fines which are generated during the beating operation. The effluent from the poacher settling pits is pumped through DeLaval centrifuges to remove most of the suspended NC fines. The flow rate and TSS concentration for one manufacturing line's poacher settling pit effluent were estimated to be 6435 m^3/day and 143 mg/L, resulting in a total NC mass loading rate to the centrifuges of 920 kg/day (Balasco et al., 1987). The centrifuge backflush water containing the removed NC is discharged back into the poacher pits and the treated centrate is pumped to the recovered water tanks. The NC concentration in the centrate ranges from 30 to 70 mg/L (DeHart, 1993). At one time, this recovered water was used as process water in the boiling tub house; however, the high TSS concentrations gave rise to concerns about product contamination and the recovered water now flows directly to the boiling tub settling pits.

Currently, NC fines in the processing wastewater streams result in the loss of approximately 5% of the each load of NC. Although most of this lost NC is recovered from the settling pits and recycled as "pit cotton" used to produce low-grade propellant, this still presumably constitutes a significant economic loss. In addition, the existing wastewater treatment processes may not be able to meet an Ammunition Procurement and Supply Agency proposed discharge TSS standard of 25 mg/L. Consequently, studies have been performed to assess the potential of using cross-flow microfiltration (MF) for NC recovery and reuse. This technology differs from conventional dead-ended filtration (e.g., reverse osmosis or ultrafiltration) in that the process stream to be concentrated is continually swept past the filtering membrane so that a static layer of solids, that would eventually prevent further filtration, is not formed. A pilot-scale study conducted by Hercules Inc. with a Memtec® (Timonium, MD) hollow fiber type unit (DeHart, 1993) and a separate laboratory-scale study by (Shen et al., 1994) demonstrated that MF could remove fines from the NC manufacturing wastewater, but that large particles (> 50 μm) tended to obstruct the filtrate flow through the membranes. However, the Hercules pilot-scale study showed that the larger NC particles could be removed using a rotary vacuum filter as a prefiltration process, allowing the MF units to operate consistently without clogging.

The objectives of this study were to:

(1) Evaluate the NC particle size distribution, suspended solids concentration, particle surface charge, and pH for each NC-manufacturing process wastewater stream;
(2) Perform a mass balance of NC fine discharge during manufacturing;
(3) Identify potential waste minimization methods for each wastewater stream;
(4) Conduct a laboratory-scale MF study for the determination of NC fine removal efficiency; and
(5) Evaluate the feasibility of using MF for NC recycling.

Experimental Methods

An extensive plant survey was conducted to measure the NC particle size distribution (PSD), TSS, particle surface charge (zeta potential), and pH of the wastewater from each NC-manufacturing process. Samples were taken for both pulp and cotton based NC batches. The PSD tests were performed using a Brinkman Particle Size Analyzer. The PSD measurements were determined based on percentage of the cumulative particle volume assuming all particles to be spherical. The surface chemistry of the NC fines was analyzed by a Pen Kem Inc. System 3000. The TSS measurements adhered to the Hercules Temporary Procedure #L-395 - "Determination of TSS in NC Waste," a standard gravimetric procedure without oven drying.

To assess the potential NC fine removal efficiency attainable using MF, a bench-scale flat-sheet membrane module (Model TM-100, New Brunswick Scientific Co., Inc.) was used. Detailed specifications of the module and feed pumps can be found elsewhere (Shen et al., 1994). The experiments were operated in a feed and bleed mode with effluent from one of the poacher pits. The feed and permeate samples were analyzed for TSS and pH. The flat sheet MF unit had a membrane surface area of 60 cm^2, a channel cross-sectional area of 28.25 mm^2, and a channel height of 0.4 mm.

Wastewater Characteristics

The PSD analyses showed that the NC fines in the various tub supernatant samples were predominantly below 10 μm. However, the PSD results for the vacuum drum tail water, wringer filtrate, and poacher settling pit samples were distributed over a range of 0 to over 100 μm. The larger particle sizes in the vacuum drum tail water and wringer filtrate samples can probably be attributed to these wastewater streams not being settled to remove solids. The larger particle sizes in the poacher settling pit samples can probably be attributed to the poacher pit receiving the vacuum drum tail water and wringer filtrate, and perhaps to the practice of discharging the DeLaval centrifuge backflush water into the poacher pits. No reliable PSD results were obtained for the boiling tub, boiling tub settling pit, DeLaval centrifuge centrate, neutralization basin, or settling lagoon samples due to low TSS concentrations.

The TSS concentrations in the boiling tub drain samples ranged from 0 to 43 mg/L, while the TSS concentrations in the various tub supernatants were generally quite high, ranging from 108 to 601 mg/L. The cotton processing supernatants generally had higher TSS concentrations than the pulp processing supernatants. In addition, the TSS concentrations in the tub supernatants that had been allowed to settle for more than 1 day were significantly lower than in those that had only been allowed to settle approximately only 1 hour. This may be attributed to NC fines being kept in suspension by temperature gradient induced currents while the tubs are cooling; once the tubs are completely cooled they may become more quiescent. The TSS in the poacher pit influent ranged from 53 to 250 mg/L, while the effluent ranged from 107 to 171 mg/L. The TSS concentrations in the DeLaval centrifuge effluent samples were lower than reported in previous studies, ranging from 4 to 17 mg/L. The TSS concentrations in the influent and effluent samples from the boiling tub settling pits and the settling lagoon were generally below 10 mg/L. The TSS for the neutralization basin effluent samples were generally higher than for the influent samples, indicating the formation of calcium sulfate and other precipitates.

In general, the pHs of all the samples were near neutral except for the boiling tub, boiling tub settling pit, and neutralization basin influent samples, which had pHs in the range of 1 to 3. The zeta potentials of the various wastewater samples ranged from -14 to -32 mV, with the poacher, blender, and wringer supernatant samples generally being more highly charged than the remaining samples. Detailed wastewater characteristics are discussed below along with a brief description of each process.

Boiling Tub House. Here most of the residual acid is removed by boiling the NC repeatedly over several days. Although reliable particle size results could not be attained for the boiling tub drain samples due to the low TSS concentrations, the NC particles were clearly large and readily settleable. The TSS concentrations ranged from 0 to 43 mg/L, with the highest TSS concentrations occurring just after opening the drain. This is believed to be due to the formation of a NC mat at the false bottom orifices that reduces the discharge of additional NC.

Beater House. Here the NC is cut and beaten in a slightly alkaline water. The NC particles in the pulp processing tub supernatant samples were predominantly below 10 μm, while the particles in the cotton processing tub supernatant samples were uniformly distributed over a range of 0-100 μm. Although the NC particles in these tubs are generally larger because the NC has not yet been through the fine beaters, it is not known why the larger cotton NC particles apparently did not settle as well. Pulp and cotton supernatant TSS concentrations were measured to be 118 and 108 mg/L, respectively, suggesting that the larger PSD results for the cotton NC may have been anomalous. It should be noted that the observed TSS concentrations were much higher than the 4 mg/L observed in a previous study (DeHart, 1993).

Poacher House. Here the NC is further stabilized by boiling four times, the first time with soda ash. The poacher tubs are decanted after the first three boils. The

NC particles in all the poacher tub supernatant samples were predominantly below 10 µm size (>80%). The TSS concentrations ranged from 139 to 601 mg/L.

Blender House. Here the NC is screened and vacuum filtered to remove bulk water, then various batches of NC are blended to achieve the desired final grade. The NC particle sizes in the blender tub supernatants were predominantly below 10 µm (>93%). The TSS concentrations in the blender tub supernatants that had been settled for approximately 1 hour were 433 and 555 mg/L; however, the TSS in the tub supernatant that had been settled for 7 days was only 155 mg/L. The NC particle sizes in the vacuum drum tail water samples were uniformly distributed from 0 to over 100 µm. Vacuum drum tail water sample TSS concentrations were 290 and 77 mg/L.

Wringer House. Here the NC storage tubs are decanted and the NC is pumped to wringers to be washed and centrifuged. The NC particle sizes in the wringer tub decant samples were predominantly below 10 µm (>86%). The TSS in the tub supernatants that had been settled for approximately 1 hour were 286 and 338 mg/L; however, the TSS in the supernatants that had been settled for over a day were a much lower 112 and 81 mg/L. The NC particles in the various centrifuge filtrate samples ranged from 0 to over 100 µm. The TSS of the filtrate samples collected during a loading of a clean wringer screen were 385 and an extremely high 3660 mg/L for pulp and cotton processing, respectively. However, the TSS of the filtrate sample collected during the actual wringing process, after a layer of NC cake had accumulated on the wringers, was only 120 mg/L. Similarly, the TSS of the filtrate samples collected during a loading of a caked wringer screen were only 40 and 27 mg/L.

Mass Balance

Calculations of the mass of NC discharged to the settling pits from the five processing operations were made using the TSS concentrations measured in the present study and water flows reported previously by Hercules, Inc. (DeHart, 1993). The results are summarized in Table 1. The boiling tub house had by far the lowest estimated NC mass loading to the settling pits, while the poacher tub house had the greatest. The combined NC mass loading to the settling pits was estimated to be 46.3 kg per metric ton of NC being processed, or approximately 4.6%. This is very close to the 4.1% calculated in a previous study (DeHart, 1993). Most of this discharged NC is recovered from the settling pits as "pit cotton" and recycled back into production of low-grade propellant. Rough calculations based on DeLaval centrifuge effluent TSS and flow indicated that approximately 0.3% of the NC discharged to the poacher pits may be ultimately lost to the final settling lagoon, resulting in 4.3% being recovered as pit cotton. These values were close to the calculated ultimate NC loss of 0.6% and recovered pit cotton NC of 3.3% determined in the previous study (DeHart, 1993).

Table 1. Mass balance calculations of the NC discharged to settling pits based on pulp processing wastewater flow estimations (DeHart, 1993).

NC processing operation	Mass NC discharged per tone of NC processed (kg)	Volume of wastewater per tone of NC processed (m^3)
Boiling tub house	0.5	106
Beater house	6.4	53
Poacher house	24.9	72
Blender house	6.8	19
Wringer house	7.7	61
Total	46.3	311

Microfiltration

The results of the bench-scale MF tests are summarized in Table 2. The TSS of the feed NC wastewater was 315 mg/L and the trans-membrane pressure drop was 13.8 kPa (2.0 psi). The feed wastewater overflow rates are two orders of magnitude higher than the permeate flux rates because the feed wastewater is continually swept past the filtering membrane so that a static layer of solids does not accumulate on the surface. The permeate flux rate increased slightly with increasing tangential velocity, but the TSS concentration of the concentrate decreased. The measured permeate flux rates for the bench-scale flat-sheet unit (96, 108, and 114 L/m^2/hr) were very comparable to the reported design permeate flux rate of 113 L/m^2/hr (0.5 gpm/m^2) for the Memtec hollow-fiber MF systems (DeHart, 1993).

Table 2. Summary of bench-scale microfiltration test results.

	Run 1	Run 2	Run 3
Feed overflow rate, L/m^2/hr.	6000	13,000	20,000
Tangential channel velocity, m/sec.	0.35	0.77	1.18
Permeate flux rate, L/m^2/hr.	96	108	114
Concentrate TSS, mg/L	320	317	316
Permeate TSS, mg/L	~0	~0	~0

The permeate flux rate can be used to estimate the number of MF units required to treat a given wastewater stream. For example, the wastewater flow from the poacher house is approximately 6435 m^3/d. If the permeate flux rate for a full-scale MF unit is assumed to be 100 L/m^2/hr as found above, then the membrane surface area required to accommodate the poacher house wastewater flow would be 2685 m^2. Given that the Memtec System model 240M10 has a total membrane surface area of 240 m^2 per unit (DeHart, 1993), a total of 12 MF units would be required. This would of course require the use of surge tanks and scheduled decanting regimes.

An important consideration in the evaluation of MF is whether the NC fines can be concentrated up to the 10 to 25% solids necessary to make recycling feasible. It is also important to assess the MF permeate flow rate at the desired solids concentrations. If the attainable permeate flux decreases significantly as the solids concentration in the waste stream being treated increases, then the calculations above would be inappropriate. However, after conducting preliminary tests with NC wastewater, a vendor (Millipore, Bedford, MA) claimed that MF could be expected to produce a concentrate of greater than 20% solids with average permeate fluxes near 200 L/m^2/hr (Balasco et al, 1987).

Recommended Waste Minimization and Recycling Schemes

Most of the wastewater that flows to the poacher settling pits is generated from the beater, poacher, blender, and wringer tubs. The decanting procedure consists of lowering a pipe into the tubs after the NC slurry has been allowed to settle. The tops of these decant pipes are fitted with perforated heads with approximately 1.5 mm diameter holes. Significant amounts of NC fines escape through the decanter heads to the settling pits during the decanting procedures.

An effective but costly method of removing and recycling NC fines from the various decanted tub supernatants would be to install MF units prior to the discharge to the poacher pit. Previous pilot-scale studies showed that NC fines could be removed from the NC wastewater using MF, but this technology was limited in its application since large NC particles (>50 μm) obstructed the filtrate flow rate through the MF membranes. However, it was proposed that this problem could be circumvented using a rotary vacuum filter (RVF) to remove the larger particles. While implementing RVF and MF would incur high capital costs, these would be offset by the fact that the high quality effluent produced would eliminate the necessity of using the DeLaval centrifuges. It may also be conceivable that the centrifuges could be used for prefiltration in place of RFV. The high quality effluent produced using MF could be recycled as process water. It should also be noted that the high temperatures of the decanted poacher pit supernatants (typically around 77°C) exceed the maximum operating temperature of the tested MF units, and thus heat exchangers would have to be installed if MF were to be implemented. However, the heat exchangers could potentially be used to preheat the water used to wash the NC between boils at the poacher house. A study by Hercules Inc. estimated a potential payback period of 3.9 years if MF were implemented (DeHart, 1993).

A less costly alternative to MF may be to design new decanter heads that would prevent the discharge of the NC fines in the first place. However, the particle size distribution data showed that the NC particle in the various tub supernatants were predominantly below 20 µm. If a finer screen is used to trap these particles, then the decant flow rate would decrease due to clogging. A separate study is being conducted to design a new decanter with a self-cleaning device to overcome these problems.

Another simple procedure that may reduce the amount of NC lost with the tub supernatant would be to wait until the tubs are completely cooled before decanting. This study showed that the suspended solids concentrations in the tub supernatants that had been settled for only 1 hour were significantly higher than in those that had only been settled for over a day (e.g., the TSS concentrations in blender tub supernatant that had been settled 1 hour and 7 days were 555 and 155 mg/L, respectively) . This is believed to be partly attributable to temperature gradient induced currents in the 1 hour settled supernatants that kept the NC fines in suspension. Since the annual NC production rate in upcoming years may be lower than in the past, it may be feasible to accommodate the longer times required for the tubs to completely cool.

Most of the NC lost during the draining of the boiling house tubs occurs during the first few minutes before a NC mat forms at the orifices in the false bottoms. Therefore, a feasible waste minimization technique would be to recycle the drained wastewater for the first few minutes until a mat formed at the orifices.

The total suspended solids concentration in the sample of the wringer filtrate during a clean screen loading with cotton based NC was extremely high (3660 mg/L). Since a reliable value for the rate of wastewater flow discharged to the poacher pits during the wringer screen loading process was not available, it was difficult to evaluate if this constituted a significant fraction of the total NC loading to the poacher pits. The NC solids in the samples collected of the wringer filtrate during the clean screen loading tended to readily settle. This suggests that the NC lost to the poacher pits during the loading of clean wringer screens could be recovered by employing some kind of simple settling unit prior to the poacher pits.

The NC particles in the blender house vacuum drum tail water also tended to be larger and thus settleable. While a reliable value for the vacuum drum tail water flow rate was not available, it is unlikely that this wastewater stream constitutes a significant source of NC loading to the poacher pits given its relatively low TSS concentrations. However, NC recovery could possibly be achieved by installing a decanting device similar to those for the various processing tubs.

Currently the poacher settling pits do not remove the suspended NC fines effectively; poacher pit effluent TSS concentrations were consistently over 100 mg/L.

It may be possible to significantly improve the settling removal efficiency by diverting boiling tub settling pit effluent to the inlet of the poacher pits. A previous study found the iso-electric point (when the particle surface charge is zero) of the poacher pit wastewater was attainable at a pH of 1.9 (Peng et al., 1992). Thus, adding the low pH (~1.5) boiling tub effluent to the poacher pits will reduce the surface charge of the NC fines and promote coagulation and settling.

Implementing the previous recommendation would require discharging the DeLaval centrifuge centrate directly to the neutralization basin. Currently the effluent from the DeLaval centrifuges is pumped to the recovered water tanks, and the effluent from the recovered water tanks is then discharged to the boiling tub settling pits. Apparently the only reason for this practice is to keep an accumulation of several feet of NC at the bottom of the tanks wet. Possible procedures for removing the accumulated NC in the recovered water tanks should therefore be investigated. By improving the settling efficiency of the poacher pits by implementing the recommendation above, it is conceivable that the effluent could be recycled by pumping it to the river water filtration process.

It should be noted that shortly after the sampling survey was conducted, large amounts of NC were observed to be accumulating in the poacher pit, much to the dismay of one of the operators. The source of the excessive NC loading was not known to the operator, but it may have been due to an upset caused by a switch in manufacturing lines. When such upsets occur, it is important that clear communication channels exist between operators and plant engineers.

Conclusions

A plant survey was conducted and production information was obtained for the NC purification processes. It was found that almost 5% of the NC was lost during purification, although most of this is recovered from the settling pits as a low-grade "pit cotton." The NC losses from the boiling tub house were found to be negligible. However, the NC losses from the beater, poacher, blender, and wringer houses were all significant, with the poacher house contributing the greatest NC mass loading to the settling pits. Microfiltration (MF) was evaluated as a means of recycling NC from the beater, poacher, blender, and wringer tub supernatants. It appears that MF could result in almost zero discharge of NC fines during the purification processes. However, high capital and operational costs may result in this technology not being cost effective. Therefore, other potential waste minimization techniques for each waste stream were also identified.

References

Balasco A.A. *et al.,* Arthur D. Little Inc., *Engineering/Cost Evaluation of Options for Removal/Disposal of NC Fines-Final Report,* Sept., 1987.

Clapp, L.W. and Park, J.K., *Nitrocellulose Particle Size Distribution Study,* U.S. Army CERL Draft Technical Report, Mar., 1994.

DeHart, M.G., Hercules Inc. *Evaluation of Reuse Potential of Nitrocellulose and Process Water During Purification Operations,* Apr., 1993

Peng, C.G. *et al, Electrophoretic Properties and Coagulation/Flocculation of Nitrocellulose Fines*, U.S. Army CERL Technical Report, Feb. 1992.

Shen, S., Park, J.K., and Kim, B.J., "Separation of Nitrocellulose Fines by Bench-Scale Flat Sheet Cross-Flow Microfiltration Units", *Separation Science and Technology*, Vol. 29, No. 3, pp. 333-356, Mar., 1994.

The Performance of Unconfined Hollow Fiber Membranes as Pipe Flow and Mixed Flow Aerators

Drew W. Johnson[1] and Michael J. Semmens[2]

Abstract

Unconfined gas permeable membranes situated within mixed reactors are currently being evaluated as a viable technology for gas transfer. The use of sealed hollow fiber membranes that are pressurized with pure oxygen provides rapid and effective gas transfer without bubbles. The ability for gas transfer without bubble formation makes this technology especially well suited for environments where the gas stripping of VOCs (volatile organic compounds) is of concern. The kinetics of gas transfer are fast and well defined since the interfacial area is known and the driving force for transfer is large. In this paper, mass transfer correlations are presented for unconfined membranes which allow the performance of membrane aerators to be compared with other gas transfer technologies.

Introduction

As discussed by Ahmed (1991), the typical method for aeration in waste treatment plants involves the use of various types of bubble diffusers. The efficiency of gas transfer for the process can be characterized by the standard oxygen transfer efficiency (SOTE). The SOTE is the ratio of the oxygen dissolved into the water divided by the oxygen actually delivered to the water under standard conditions. The operating cost for aeration is represented by the standard aeration efficiency (SAE) (Kg Oxygen transferred / kW- hour). As the SAE increases, the operating cost for the diffuser decreases. The SOTE for bubble diffusers is a function of water depth. As the depth of water increases, the gas partial pressures increase and the residence times of the bubbles in contact with the liquid will also increase. This results in a greater proportion of the gas delivered to the system being transferred to the water.

---

[1] Graduate Student, Department of Civil & Mineral Engineering , University of Minnesota, 500 Pillsbury Dr. SE Minneapolis MN 55455

[2] Professor, Department of Civil & Mineral Engineering, University of Minnesota, 500 Pillsbury Dr. SE Minneapolis MN 55455

Unfortunately, the SAE does not necessarily improve with water depth. The benefits of increased transfer efficiencies are offset by increased operating costs at higher pressures

The use of bubble diffusers in waste streams containing VOCs results in VOCs being stripped and released into the atmosphere. The rate and quantity of VOCs stripped is a function of the volumetric gas flow rate and the size of the bubbles formed. The volume of gas pumped through the waste stream can be minimized for some applications by using pure oxygen rather than air. When oxygen is used, it is important that the transfer process is efficient and that SOTE and SAE values are both large. SOTE and SAE values have to be considered when evaluating this solution to the stripping problem.

The release of VOCs into the atmosphere represents a major problem for some treatment plants. As the emissions standards become ever more stringent, the control of VOCs will play an important role in the process design. Aeration using microporous hollow fiber membrane technology can eliminate many of the problems caused by VOCs and odor emissions. Microporous membranes provide a high surface area for mass transfer and they can be used to contain pure oxygen at pressures as high as 6 atm. As a result, the kinetics for gas dissolution are extremely rapid. Perhaps the major benefit of using membranes is that they do no generate bubbles; gas transfer occurs by direct dissolution at the membrane surface. Bubble-less aeration results in SOTE values of 100 percent and in the absence of bubbles, VOCs and odors are not stripped to the atmosphere. To evaluate the cost effectiveness of this technology, SAE values must be considered. With the typical bubble diffusers used in aeration practice, energy requirements arise from compressing air to force it through a diffuser stone at some water depth. The rising bubbles are responsible for transferring oxygen to the water as well as mixing the water. For a membrane system, pure compressed oxygen is used but the bubble free nature means that the water must be mixed to provide a liquid velocity near the membrane/liquid interface. This liquid velocity is necessary to reduce the liquid film resistance to gas transfer as oxygen diffuses into the surrounding waste stream. It is the energy required to provide the liquid flow past the hollow fiber membranes which determines the SAE for a membrane aeration application.

Ahmed and Semmens (1992,a) have shown that the amount of gas delivered to the liquid through the membrane can be calculated using the formula

$$N = kA(C^{*}-C) \quad (1)$$

where:

N - is the gas flux
A - is the membrane surface area
k - is the overall mass transfer coefficient
C - is the dissolved gas concentration in the liquid phase
C*- is the concentration of dissolved oxygen in equilibrium with the

average partial pressure of oxygen inside the hollow fiber membranes and :

$$C^* = \frac{P_{average}}{H_c}$$

where:

H_c is Henry's Constant.

$P_{average}$ is the average pressure along the length of hollow fiber.

According to Ahmed and Semmens (1992,b), the average pressure in a sealed end hollow fiber can be calculated as a function of inlet pressure.

$$P_{average} = 0.26 P_{inlet}^{0.3}$$

Pipe Flow Aerators

The overall mass transfer coefficient (k) is strongly dependent upon the water velocity past the hollow fiber membranes. Ahmed and Semmens(1992,b) developed a mass transfer correlation for large numbers of hollow fiber membranes within a pipe flow configuration. This type of configuration is depicted in figure 1.

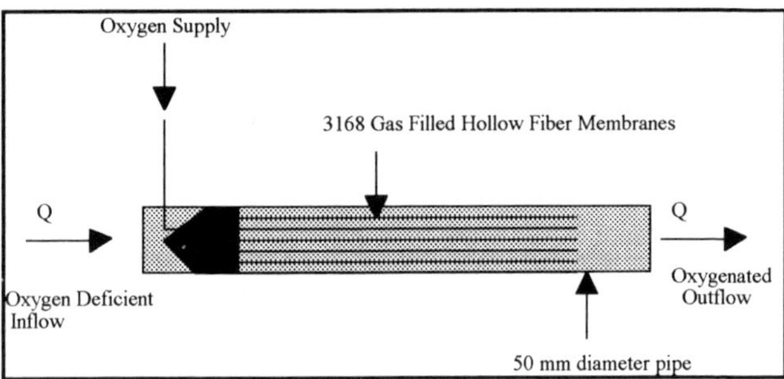

Figure 1 - Membrane Pipe Flow Aerator

The mass transfer correlation for the pipe flow configuration is presented in equation 2.

$$k = 0.018 \frac{D}{d_e} \left(\frac{V_L d_e}{\upsilon} \right)^{0.83} \left(\frac{\upsilon}{D} \right)^{0.33} \qquad (2)$$

Where:

k = mass transfer coefficient

$$d_e = \frac{4(\text{cross - sectional area of flow})}{\text{Wetted Perimeter}}$$

V_L = Liquid Velocity

υ = kinematic Viscosity
D = Gas Diffusivity

For the pipe flow configuration, the gas flux is directly proportional to the liquid velocity to the 0.83 power. However, for turbulent pipe flow, head losses are proportional to the liquid velocity squared and one would expect the aeration efficiency to decrease as the liquid flow rate is increased. Voss (1994) has measured head losses for the pipe flow aerator and calculated aeration efficiency (AE) values for this configuration. Using the mass transfer correlation in equation 2 along with equation 1 and pipe friction relationships, it is also possible to estimate the pipe flow AE. To estimate AE, it is necessary to calculate to headloss due to the membrane module. In this analysis, the turbulent flow correlation of Colebrook and White was used to estimate the pipe/membrane friction factor and a value of 0.5 was used for both the coefficient of contraction and the coefficient of expansion. The estimated AE values are slightly larger than the experimental values reported by Voss. Differences between the measured and estimated values can be linked to differences in the membrane used by Ahmed et al and Voss. Voss tested a composite membrane that has a large membrane resistance. The membrane resistance reduces the rate of transfer by about 30%. If the membrane resistance is incorporated, the experimental data will fit the predicted curve. The measured and estimated AE values as a function of liquid flow rate are shown in figure 2.

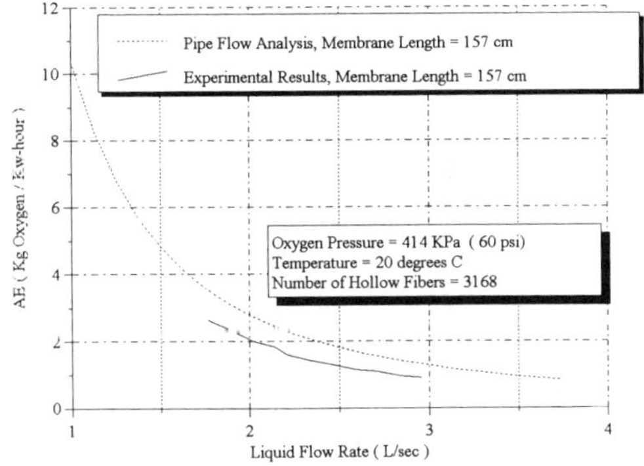

Figure 2 - Aeration Efficiency for a Membrane Module in a 50 mm diameter Pipe

This analysis only considers the energy requirements associated with pumping water through the module and does not incorporate the fixed costs of the pressurized oxygen supply. Never the less, the analysis demonstrates that the aeration efficiency of this module is competitive with typical diffusers. Typical diffusers have aeration efficiencies which vary between 1 and 5 (Kg oxygen / kW - hour) (WPCF, 1988). At high water flow rates through the module the increasing frictional head losses substantially decrease the AE for this configuration. Approximately 70 % of the energy required for this module design is related to frictional losses along the length of the membrane. These losses can be estimated from the Darcy-Weisbach equation as given in equation 3.

$$h_f = f\frac{L}{d_e}\frac{V^2}{2g} \tag{3}$$

This equation illustrates that the head loss through the module can be reduced by decreasing the membrane length (L), However, the gas flux as expressed in equation 1 is a linear function of membrane length and decreasing the membrane length has a small negative effect on aeration efficiency. The losses expressed in equation three can also be minimized by increasing the effective diameter of the pipe, however, if the diameter of pipe is increased, the liquid velocity decreases for a given flow rate and the overall mass transfer coefficient decreases correspondingly.

Mixed Flow Aerators

Johnson and Semmens (1994) developed a mass transfer correlation for a mixed flow reactor with a membrane module submerged and centered within a jet discharge. This configuration is depicted in figure 3

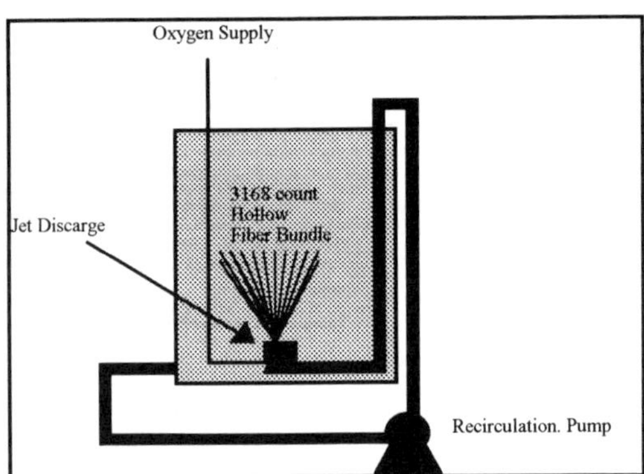

Figure 3 - Mixed Flow Reactor Membrane Aerator

This configuration maintains relatively high liquid velocities while eliminating the losses associated with equation 3. The effective diameter is sufficiently large such that frictional losses become minimal and initially high liquid velocities can be provided for low flow rates at the jet origin. The mass transfer correlation for unconfined fiber in a submerged jet is provided in equation 4.

$$k = 136 \frac{D}{N d_{fiber}} \left(\frac{l_q V_{exit}}{\nu} \right)^{0.39} \left(\frac{l_q}{L} \right)^{0.54} \left(\frac{\nu}{D} \right)^{0.33} \qquad (4)$$

Where:

k = mass transfer coefficient
d_{fiber} = diameter of a single hollow fiber membrane
V_{exit} = liquid velocity at the jet origin
υ = kinematic viscosity
D = gas diffusivity
N = number of hollow fiber membranes
$l_q = \sqrt{A}$ where A is the initial cross sectional area of the jet
L = membrane length

The energy requirements of the module as a function of liquid flow rate were measured and the aeration efficiency for this configuration is presented in figure 4.

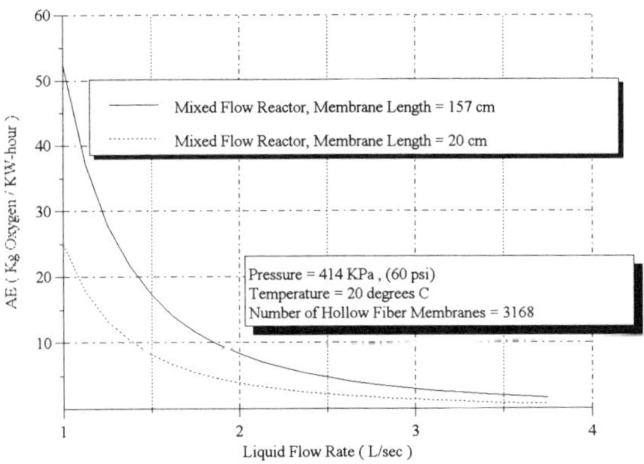

Figure 4 - Aeration Efficiency for a Membrane Module in a Jet Discharge , Jet diameter = 50 mm.

Again, the AE analysis for the unconfined fibers does not incorporate the energy required to provide the pressurized oxygen supply. The AE for the configuration is superior to that observed in the pipe flow configuration (figure 2). At a flow rate of 1 L/sec, the AE for the unconfined fibers is approximately five times that of pipe flow and ten times that reported for conventional bubble aeration. Figure 4 illustrates the fact that AE is highly dependent upon membrane length. With the performance increasing as the length of membrane is increased. In this configuration the principal energy loss is associated with the discharge through the orifice and the length of fibers has little impact on the energy requirements. As the fiber length is increased the mass transfer coefficient decreases (equation 4) but the surface area for gas transfer increases. The result is that AE increases with increasing membrane fiber length.

Operational Concerns

Unconfined hollow fiber membranes in a submerged jet had aeration efficiencies much greater than conventional aeration, and membranes in a pipe flow configuration were shown to be competitive with conventional bubble aeration. Both analyses assumed an operating pressure of 414 KPa (60 psi) within the membrane. The operating pressure for the membrane systems is a variable which offers an operator a high degree of control. The operating pressure regulates both the rate and amount of gas dissolved into the waste stream. Depending on the type of membranes used, the operating pressure can be as high as 689 KPa (100 psi). The AE is a function of the operating pressure and for low dissolved oxygen concentrations in the waste stream, the AE can be expected to increase almost linearly with operating pressure.

The issue of membrane fouling also needs to be addressed. The high shear velocities that can be maintained in a pipe flow application may make this membrane configuration less susceptible to fouling mechanisms (Semmens and Gantzer, 1993). The unconfined fiber configuration, with the membrane module centered within the jet discharge, provides initially high shear velocities at the base of module but as the distance from jet origin is increased, the shear velocities fall dramatically as a result of entrainment effects. The unconfined configuration may require a minimum velocity or minimum flow rate to avoid fouling problems. This is an area that warrants further research.

Summary

The use of membranes for bubble-less oxygen transfer provide gas transfer efficiencies approaching 100 percent. With the use of membranes for bubble-less oxygenation, VOC and odor emissions from conventional aeration tanks can be greatly reduced. Membrane aeration technology therefore provides an effective and inexpensive means of providing VOC emissions control as required by the new Clean Air Act. When operated at an internal pressure of 414 KPa (60 psi), the aeration efficiency of a membrane pipe flow oxygenator is comparable to standard bubble diffusers. When operated at 1 L/sec and 414 KPa (60 psi), the aeration efficiency of a unconfined membrane oxygenator is five times that of the pipe flow membrane oxygenator and ten times that of conventional bubble diffusers. Both pipe flow and mixed flow membrane oxygenators can be operated over a range of gas pressures and this provides the operator with a high degree of control over the amount and rate of gas transfer. In this paper we have shown that membrane oxygenators are superior to conventional aeration technologies in terms of SOTE and SAE. The issue of membrane fouling needs to be addressed and is identified as an area for further research.

Acknowledgments

The financial support of the Center for Clean Industrial and Treatment Technologies is acknowledged. In addition, all membrane modules in this study were supplied by Membran Corporation. 1037 10th Ave SE Minneapolis, MN 55414, USA

References

Ahmed, T, Ph.D. Thesis " Gas Transfer Through Membranes", University of Minnesota, 1991

a Ahmed, T. and Semmens, M.J., " Use of Sealed end hollow fibers for bubbleless membrane aeration : experimental studies", Journal of Membrane Science vol. 69, 1992

b Ahmed, T. and Semmens, M.J. ,"The use of independently sealed microporous hollow fiber membranes for oxygenation of water : model development" Journal of Membrane Science vol. 69, 1992

Johnson , D.W. and Semmens, M.J., " The Development of a Mass Transfer Correlation for Unconfined Membranes in Mixed Flow Reactors" unpublished , 1994

Semmens, M.J. and Gantzer, C., "Gas Transfer Using Hollow Fiber Membranes" Proceedings of a Research Symposium, WEF meeting, Los Angles CA October, 1993

Voss, M. Masters Thesis, Unpublished, University of Minnesota, 1994

Water Pollution Control Federation. Aeration: Manual of Practice, FD-13, 1988

Meeting 503 Regulations for Ten Biosolids Incinerators (And Associated Stories)

Mark Prentice[1], Norman LeBlanc[1], Thomas E. Sadick[2], F. Michael Lewis[3]

Abstract

This paper provides an overview of the Hampton Roads Sanitation District's Part 503 regulations compliance program regarding incineration of biosolids at five of its treatment facilities. Results of performance testing and of the operating techniques and modifications undertaken to improve performance and comply with regulations are presented. Experiences with start-up, maintenance, and reliability of the CEM systems are included. The problem of hydrogen cyanide occurrence in the scrubber water of the multiple-hearth incinerators is addressed, as are the methods and techniques used by the District to identify and deal with the cyanide issue.

Introduction

The Hampton Roads Sanitation District (the District) serves approximately 1.5 million people in the Tidewater region of southeastern Virginia. The District owns and operates nine wastewater treatment plants ranging in size from 0.4 m^3/sec to 1.75 m^3/sec (10 to 40 mgd). Incineration facilities are located at five of the nine plants. Incineration is the primary source of biosolids disposal for these five plants plus another which trucks biosolids to the nearest incineration facility. Each of the five incineration facilities has two multiple-hearth furnaces (MHFs) followed by two-stage wet scrubbing. Prior to the promulgation of the 503 regulations, the District embarked on a program to attempt to ensure compliance with the regulations regarding incineration of biosolids. The major elements of the District's program included:

---

[1]Hampton Roads Sanitation District,P. O. Box 5911, Virginia Beach, VA 23455

[2]CH2M HILL, 11818 Rock Landing Drive, Suite 200, Newport News, VA 23606

[3]F. Michael Lewis, Inc., 319 West Grand Avenue, El Segundo, CA 90245

- Facility inspections and evaluation
- Baseline incinerator performance testing
- Development of 503 permit applications
- Procurement and installation of total hydrocarbon (THC) Continuous Emissions Monitors (CEMs)

In addition, cyanide from the incinerator scrubber water was also found to be causing nitrification inhibition at the District's Virginia Initiative Plant (VIP). Efforts to minimize incinerator production of cyanide along with an evaluation of treatment options for the scrubber water were also undertaken by the District. This paper provides a summary of each major element of the District's program along with conclusions and status of on-going aspects of the project.

Facility Inspections and Evaluations

The initial task in the District's 503 implementation program was an inspection and evaluation of each of the five incineration facilities to determine overall condition of the facilities and to assess operation with respect to achieving performance requirements under 503s. The name, location, size, configuration and average throughput for each of the District's five incineration facilities is summarized in Table 1. Each of the facilities has two multiple hearth incinerators followed by venturi and impingement-tray scrubbers.

Table 1: HRSD Incineration Facilities Summary

Facility Name	Location	Shell Diameter (ft)	# of Hearths	Daily Avg. Throughput (Dry Tons)
Army Base	Norfolk	22.25	6	10.1
Boat Harbor	Newport News	22.25	8	3.6
Chesapeake-Elizabeth	Virginia Beach	18.75	7	14.8
Virginia Initiative	Norfolk	23.17	10	23.6
Williamsburg	Williamsburg	18.75	9	12.5

The District assembled an evaluation team that included staff engineers and operating specialists, along with the engineering firm CH2M HILL and incinerator consultant F. Michael Lewis. Following preparation and review of background material and operating data the team visited and inspected each of the incineration facilities.

The inspection team found all five facilities to be in good overall condition, well operated and maintained. Preliminary recommendations for low cost modifications to improve efficiency and performance with respect to 503 regulated emissions included the following:

- Install burners on upper hearths of all incinerators where missing, using gas burners where possible. This recommendation was provided to enable operation at a higher outlet (breeching) temperature and thus reduce THC emissions.

- Install air flow measurement capability on burners to provide more accurate adjustment to air-to-fuel ratios.

- Modify throat sections of the venturi scrubber to allow maximum head loss and scrubbing efficiency. Use the venturi to control furnace draft instead of the ID fan damper.

- Create a "zero-hearth" afterburner on one of the VIP incinerators by cutting a hole in the top hearth to allow sludge to drop directly on hearth number 2. This incinerator would then be used to determine the potential benefits of operating at higher breeching temperatures to reduce THC and cyanide concentrations.

Baseline Testing

The purpose of the baseline testing program was to operate the District's incinerators under existing and several improved combustion modes to establish performance. An additional objective was to determine the operational parameters needed to achieve compliance at minimum cost, i.e., lowest auxiliary fuel usage.

THC, carbon monoxide, and cyanide (in the scrubber water) were measured. THC is a parameter which will be regulated, and continuously monitored, under the Part 503 regulations for incinerators. Carbon monoxide was measured as a surrogate indicator for goodness of combustion. Both THC and cyanide are products of incomplete combustion, and it was hoped that a correlation between THC, cyanide, and CO could be made. (The correlation would be most useful for cyanide since samples are obtained from the liquid scrubber stream and are not available for real time purposes). A continuous emissions monitor trailer from DEECO, INC. of Cary, North Carolina was set up at each of the plants to measure and record performance data. The portable CEM system monitored THC, carbon monoxide, moisture, and oxygen at the furnace exhaust and stack outlet. The CEM data and information from the plant instrumentation system were fed to a computer for reporting and calculation of corrected THC values.

Baseline testing was conducted at all five of the District's incineration facilities. Each test was conducted over a three- to four-day period. The first day of testing established total hydrocarbon (THC) emission performance under existing operating parameters. On the subsequent test days at each facility, modifications were made to operating techniques (temperature profiles, automatic operation, increased turbulence in burning hearths, etc.) to bring THC emissions into compliance with the 503 regulations (100ppmv).

Although the multiple-hearth furnaces at the different plants have different numbers of hearths, diameters, and burner arrangements, THC compliance was achieved by modification of firing techniques in the furnace in each case. In general, only a modest increase in top hearth temperature was required to bring the furnaces into THC compliance. Typically, compliance was achieved at approximately 1,050°F. However, at the Williamsburg plant, where sludge is conditioned with lime for dewatering, THC compliance was achieved with outlet temperatures in the range of 600 to 700°F. THC compliance was also achieved with autogenous burning at the Army Base facility. Solids concentration of biosolids feed to the incinerator during the autogenous burning period was in the range of 21 to 23%.

Examples of the testing results for the Army Base Plant are shown in Figure 1. Figures 1A, 1B, and 1C all show hourly time series traces of THC (corrected to 7% oxygen) for March 1, 2, and 3, 1993. Also plotted on each graph is the top hearth (hearth No.1) temperature. Figure 1A shows the results of increasing outlet temperature under automatic (oxygen set-point) control. Condition #1 indicates that as temperature is increased by using recycled shaft cooling air, corrected THC values show a corresponding decrease.

Figure 1B, condition #2, shows the results of stable operation in automatic control while feeding the furnace at maximum capacity of 8,000 wet lbs. per hour. During this period totally autogenous conditions prevailed (no fuel usage) and corrected THC values were approximately 50 ppmv or less. Condition #3 in Figure 1B shows the impact of turbulence on THC production. Turbulence induced by providing maximum burner airflow and an increase in temperature caused a dramatic decrease in THC and in carbon monoxide. THC values decreased to approximately 10 ppmv, and CO values dropped from approximately 1400 to 600 ppmv. A lean mixture of gas (approximately 10 cu ft gas per hour) was used on the burners in hearth 2 during this period.

Figure 1C, Condition 4, shows THC values using a relatively wet (16 to 18% solids) foreign sludge from another District plant. A high supplemental fuel usage (70 to 80 cu ft per minute) was used during this period. This test demonstrated that the furnace could be operated to comply with the THC requirements while burning the foreign biosolids. Finally, in Condition 5, outlet temperatures were reduced to determine the threshold temperature for compliance. This work showed that a minimum of approximately 850°F was required to stay below the 100 ppmv limit.

The baseline testing program provided valuable insight into the capabilities of the furnaces and the impact of alternate modes of operation. Information obtained

FIGURE 1

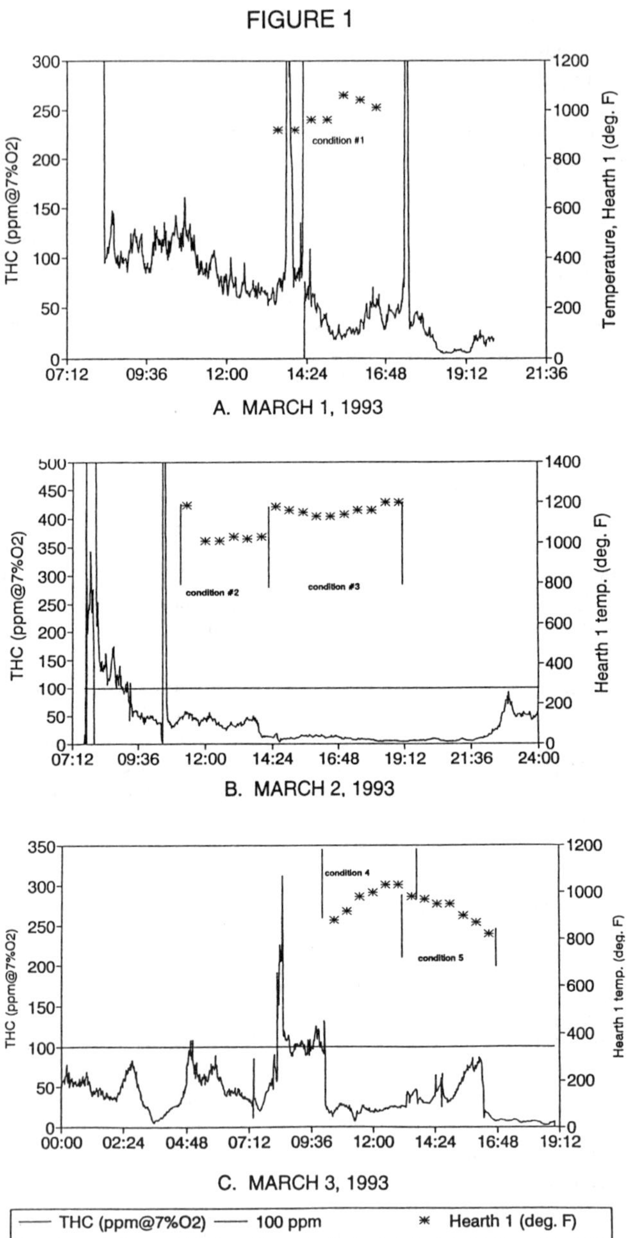

RESULTS OF ARMY
BASE BASELINE TESTING

from the tests was used for planning to meet the 503 requirements and for longer-term facilities planning. For example, data generated during the tests was used to develop and calibrate mass and energy balances. These balances were then used for budgeting purposes to ensure sufficient fuel would be available for operation in higher temperature modes of operation. The balances and baseline testing information were also valuable to the District for biosolids management planning.

Development of Part 503 Permit Applications

The third phase of the compliance program involved preparation of permits for each of the ten incinerators. Guidance was received from EPA in terms of what was expected in the 503 permit application. The general permit requirements for each incinerator included:

- General background, facility description, certification, etc.
- Dispersion modeling
- Pollution equipment control efficiency testing
- Threatened and endangered species evaluation
- Operating information such as biosolids feed rate, temperature profiles, furnace and support system operating parameters

Results of control efficiency testing and determination of maximum allowable biosolids feed rate are probably of greatest interest of all permitting activities. Tests revealed that the biosolids feed and the control efficiencies were acceptable at all the facilities tested. Testing indicated that emissions for beryllium were, in general, approximately an order-of magnitude below the 10 gram per 24 hr National Emission Standard for Hazardous Air Pollutants (NESHAP) standard. For most of the plants, the mercury emissions were approximately two orders-of-magnitude below the 3,200 gram per day limit. Other regulated metals concentrations (arsenic, cadmium, chromium, nickel, and lead) in the feed biosolids and stack were such that all facilities will be permitted to feed biosolids to the incinerator at the maximum capacity of the furnaces. The control efficiencies for metals achieved during the testing at each of the plants is summarized in Table 2. During each of the control efficiency tests biosolids feed rates were set at close to the maximum capacity of each incinerator. THC stack emissions were monitored continuously throughout the tests, and average hourly values were maintained below the 100 ppmv limit.

Table 2: Results of Control Efficiency Testing

Plant Name	Arsenic	Cadmium	Chromium	Nickel	Lead
Army Base	98.2	83.5	99.8	99.3	94.2
Boat Harbor	95.7	94.1	99.7	99.7	98.0

Table 2: Results of Control Efficiency Testing

Plant Name	Arsenic	Cadmium	Chromium	Nickel	Lead
Ches.-Eliz.	96.0	88.0	99.0	99.5	97.4
VIP	94.3	70.3	99.7	99.1	85.9
Williamsburg	96.1	90.6	99.6	99.8	97.1
Average	**96.1**	**85.3**	**99.6**	**99.4**	**94.5**

Procurement and Installation of THC Continuous Emission Monitors

The fourth phase of the program involved the specification, procurement, and installation of continuous emission monitoring (CEM) systems at each of the District's facilities. Bid packages were prepared in August, 1993 for fabrication, delivery, installation, and certification of complete CEM packages at each of the five incineration facilities. Bidders were also required to provide one year of preventive and on-demand maintenance for each system. The performance-based specification also had a January 12, 1994 milestone for the delivery of equipment and a February 11th milestone for certification. The contract was awarded to Spectrum Systems, Inc. of Pensacola, Florida in late September of 1993. As of the date of this writing, all project milestones except certification have been met. The last units were certified on February 18, 1993, within one day of the date required under the 503 regulations. The units are currently operating satisfactorily. An update on the operation and reliability of the CEMs will be provided during the presentation of this paper.

Cyanide Removal Efforts

Cyanide appears to be a product of poor combustion and apparently forms in the scrubber water of many multiple-hearth municipal biosolids incinerators due to localized pyrolysis pathways. The cyanide concentration of scrubber water effluent from the incinerators at the District's high-rate nutrient removal Virginia Initiative Plant (VIP) has averaged 3 to 4 mg/L and is believed to have caused nitrification inhibition at the plant. The inhibition causes problems with start-up of nitrification during the spring and early summer when the coolest water temperatures of the nitrification season are experienced. In addition to nitrification inhibition, other operators with multiple-hearth furnaces and very low water quality-based effluent limitations for cyanide may face difficulties meeting their permits since cyanide removal in secondary systems is not consistent.

The District embarked on a program to remove or reduce the cyanide in the scrubber water recycle stream. Bench-and pilot-scale techniques were used to evaluate a variety of cyanide removal technologies for treatment of the sidestream. The technologies investigated included air stripping, biodegradation, and chemical oxidation using chlorine, hydrogen peroxide, or potassium permanganate. Options to modify the design and/or operation of the incinerator to reduce cyanide concentrations in the scrubber water were also investigated. The results of these investigations indicated that nitrification inhibition at the VIP plant could be eliminated by reducing the cyanide loading on the mainstream biological process from the incinerator scrubber water. On the basis of these studies, a concentration of 1 mg/L CN or less in the recycle stream was established as a target value to meet the desired reduction in CN loading to the mainstream process.

Improving combustion in the incineration process is the preferred option for reducing cyanide since it eliminates the source of the problem. Unfortunately, efforts to date (including experiments to increase turbulence in the burning hearths of the incinerator to improve combustion and operating at higher outlet temperatures) have failed to reliably reduce cyanide in the scrubber effluent to acceptable levels. (Additional experiments with increased combustion turbulence are planned.) Other cyanide removal options, such as chemical oxidation, were determined to be too costly or potentially hazardous to implement. Air stripping of cyanide, while technically feasible and relatively cost effective, was unacceptable to the District because of the other options available. Biological treatment, therefore, appeared to be a favorable alternative.

The District successfully conducted pilot studies of biological treatment of incinerator scrubber water and solids handling recycle stream for a six-month period (December, 1992 to June, 1993). The pilot study verified that cyanide could be reduced to well below the treatment objective of 1 mg/L in a biological system with an average effluent cyanide concentration of 0.47 mg/L. Cyanide removals averaged approximately 85 percent throughout the course of the study. This level of cyanide removal was achieved in the first stage of a 3-stage reactor system with a hydraulic residence time of approximately two hours.

The study also proved that biological degradation of cyanide, not air stripping, was the primary mechanism of CN removal. The biological degradation efficiency of cyanide in the first stage averaged approximately 85 percent as compared to an average of 35 percent in the stripper tank under similar detention times and airflow rates. The pilot data study indicated that a minimum MCRT of four days must be maintained for the biological system to remove CN effectively under the conditions of the test. Increased MCRTs produced lower effluent cyanide levels, but removal was never complete.

The District is currently in the design stage for a 2 mgd sidestream biological treatment plant to treat the solids handling recycle stream which contains the incinerator scrubber water. Two old digester tanks will be retrofitted and configured into a complete mix-activated sludge system. The system should be on line by fall of 1995.

CLEARING THE AIR ABOUT SLUDGE INCINERATOR EMISSIONS

Phillip M. Martin[1], Eugene W. Waltz[2],
and Richard D. Kuchenrither[3], Member ASCE

ABSTRACT

In 1990, a research needs assessment for wastewater treatment agencies conducted by the Water Environment Research Foundation recommended a three-year project to identify and quantify hydrocarbon constituents in emissions from municpal sewage sludge incinerators. The project was designed to evaluate existing emission test data and obtain additional information to more completely characterize hydrocarbon emissions, their associated health risk, and operational factors effecting emissions. This paper presents the results and findings from the first year of the project.

INTRODUCTION

To identify and quantify the typical organic emissions from municipal wastewater treatment plant sludge furnaces, a survey and analysis was made of existing organic emission test data. The emission test data base developed was then compared to the regulated toxic organic lists of the Clean Air Act, the U.S. EPA IRIS list, and those for the states of California and North Carolina. For those organic compounds identified on any of these regulated lists, an analysis was made of their frequency of detection and their relative mass emission rates to determine the most prevalent and predominant compounds. The health risk associated with these most prevalent and predominant compounds was then evaluated. In

---

[1]Environmental Engineer, Black & Veatch, Engineers - Architects, 8400 Ward Parkway, Kansas City Missouri, 64114.

[2]President, Incinerator Rx Corporation, Pac Plaza I Suite B-207, 2346 S. Lynhurst Drive, Indianapolis, Indiana, 46241.

[3]General Partner, Director of Residuals Management, Black & Veatch, Engineers - Architects, 8400 Ward Parkway, Kansas City, Missouri, 64114.

addition to this analysis of the existing emission test data, a series of tests were conducted at three (3) test sites to investigate the thermal decomposition properties of the principal compounds of concern and other related combustion chemistry effects.

BACKGROUND

Over the years, wastewater treatment agencies have conducted numerous incinerator emission tests. Typically, these tests have analyzed particulates and metals. Organics testing had normally been limited to evaluating total hydrocarbons (THCs), as measured by a flame ionization detector, and not individual compounds. Only a relatively small fraction of the emission tests available had identified specific organic compounds.

The survey of existing data included emission test data from 15 wastewater plants, obtained through project team contacts and advertising through various announcements. Some test data were taken from literature published by the Environmental Protection Agency (EPA). The data base included 96 multiple hearth (MH) and 6 fluidized bed (FB) furnace organic tests. The additional testing conducted at 3 test sites primarily explored the effect of temperature on the top hearth or afterburner on total and specific organic emission rates.

RESULTS

Analysis of the existing emissions data base indicated that 326 organic compounds were identified across the 96 MH emission tests. Of these compounds, 49 (15 percent) were associated with potential health risk concerns. Further evaluation of these 49 compounds showed that 17 were detected in 20 percent of the tests, with only 11 of the compounds having reported detection frequencies exceeding 50 percent across all of the tests surveyed. For the six FB furnace tests, only 38 compounds were identified. Of these compounds, 17 (45 percent) were identified as having potential health risk concerns, with only 13 being detected in 20 percent of the tests.

From testing at the three additional MH facility's sludge furnaces, the most predominant organic compounds emitted were found to be benzene, acrylonitrile, and toluene. These compounds accounted for 70 percent of the mass emission rate of the top 17 most frequently detected target organic compounds. The percent mass emission rates for benzene, acrylonitrile, and toluene at these three sites are shown in Figure 1. The prevalence of these same compounds in MH survey data was also quite high with reported detection frequencies of: benzene (97 percent), toluene (90 percent), and acrylonitrile (54 percent).

Since the vast majority of the information used represented MH sludge furnace stack testing, the following results and generalizations were made with respect to MH furnace performance. The THC and target organic compound emission rates encountered in MH testing were found to be predominantly a function of the furnace top hearth or afterburner exit gas temperature when exit temperatures were above 900°F (482°C). The thermal decomposition profiles of THC from two of the

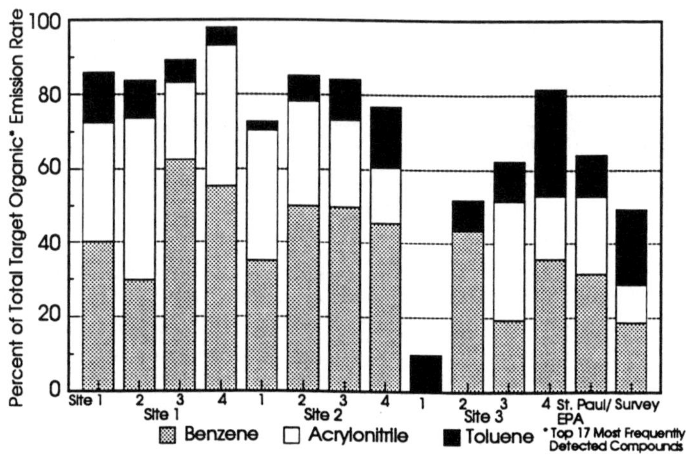

Figure 1. Relative mass emission rates.

additional testing sites are shown in Figure 2. Both THC and target organic compounds (not shown) indicated a common, inverse exponential destruction with temperature over the critical temperature range of 900°F (482°C) to 1200°F (649°C) where combustion chemistry kinetic rate reactions become the dominant mechanism of destruction for most organic species. Below this critical temperature range, it appears that the furnace operating mode and equipment capability factors have more effect upon emission levels.

A comparison of the sum total concentration of the top ten target organic compounds with the average THC levels for each additional testing site, showed THC levels significantly higher than those of the top ten targeted organic compounds. In addition, as temperatures increased, the difference between THC concentration and top ten target organic compounds becomes more pronounced. Overall, thermal destruction

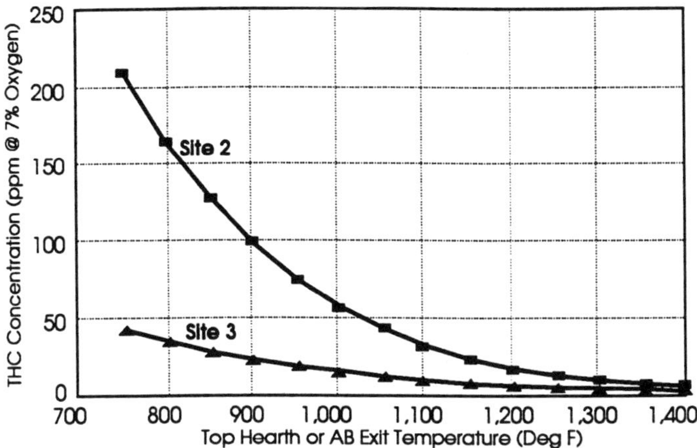

Figure 2. THC thermal decomposition.

efficiencies of the 49 target organic compounds with THC indicated that target organic compounds were destroyed to lower levels than THC over the 800°F (427°C) to 1400°F (760°C) temperature range. Major residual fractions of more thermally stable methane and ethane hydrocarbons (50 to 90 percent) were still present in measured THC emissions at furnace or AB exit temperatures above 1000°F (538°C) as shown in Figures 3 and 4.

These emissions tests results conform with combustion chemistry reaction theory predictions. Most of the organic species which comprise the measured THC emissions have 99 percent destruction tempeatures in the range of 800°F (427°C) to 1000°F (538°C). For methane and ethane, however, the 99 percent destruction level is not achieved until temperatures reach 1600°F (871°C) to 1800°F (982°C), resulting in the THC emission composition at 1000°F (538°C) or more being less potent due to the residual fraction of methane/ethane. As a result, THC emissions were found to be correlated with total and specific target organic compound emission rates as co-variants with temperature, particularly at top hearth or AB exit temperatures below 950°F (510°C). The correlation, however, diminishes with increasing exit temperatures.

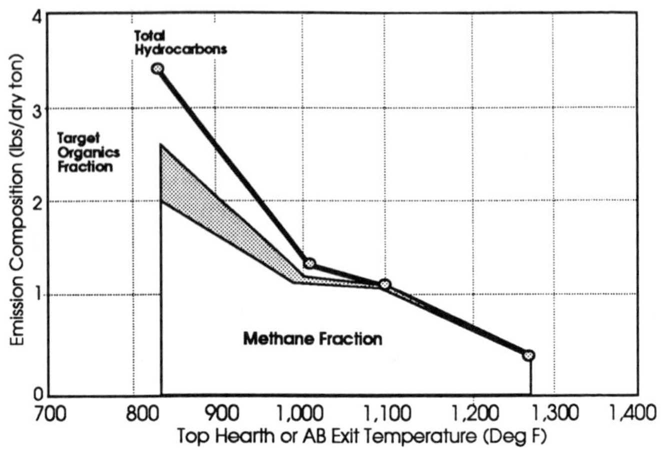

Figure 3. Methane/ethane fraction of THC at site 2.

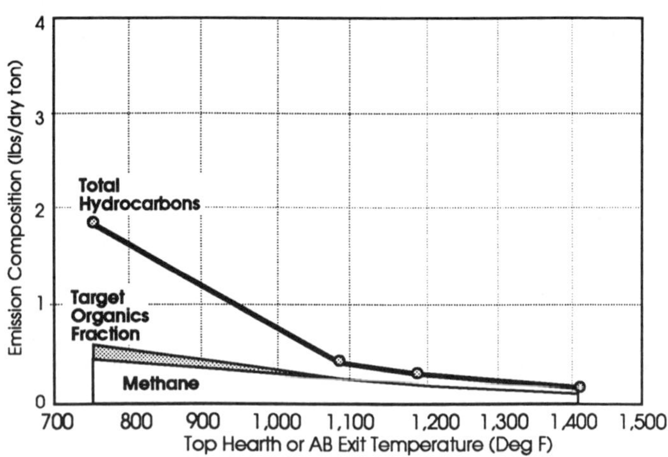

Figure 4. Methane/ethane fraction of THC at site 3.

A comparative analysis of THC emission profiles show existing MH facilities can achieve 100 ppm THC levels in routine operations by operating at a furnace or AB exit gas temperatures in the range of 900°F (482°C) to 1000°F (538°C) on the average. However, THC emission rate profiles with temperature also vary between furnace facilities depending on operating mode, residence times, and other equipment configuration factors. The thermal decompostion shown in Figure 2, differs because of a longer site 3 residence time (twice that of site 2), illustrate the impact of these factors. In addition, elevating top hearth or AB temperatures to reduce THC/organics emission levels were found to increase NO_X emission rates from 1 to 2 lbs per dry ton (0.51 to 1.02 kg/tonne) and fuel consumption rates by 500 to 600 percent as compared to the non-afterburning operating mode.

To better quantify the cancer risk associated with sludge incineration emissions, a cursory risk assessment was conducted at one of the additional test sites for exit temperatures of 982°F (528°C), 1088°F (587°C), and 1308°F (709°C). The results of the assessment indicated that all carcinogenic and noncarcinogenic risks were well below the acceptable levels, since the hazard quotients were less than 0.01 and the total pathway risk ranged from 10^{-8} to 10^{-9} for the evaluated temperatures. Caution must be used, however, when making comparisons of this risk with results from previous incineration risk assessments, mainly because this risk calculation does not describe the health risk associated with all of the organic compounds. While this analysis did include a vast majority of the health related compounds, no attempt was made to account for the small fraction that could be attributable to nondetectable compounds. Other risk assessments have increased the conservatism of their estimate by attempting to account for such compounds.

CONCLUSIONS

The identification and quantification of existing sludge incineration organic emissions found only a few organic compounds with benzene, acrylonitrile, and toluene being the most prevalent and predominant. The percentage of THC has been shown to consist of mostly ethane and methane as temperature increases, with a decreasing fraction of other organic compounds. In addition, the health risk assessment of the identified organic compounds indicated little or no risk and virtually no noncarcinogenic health threat. The relationship of THC and target organic emission rates with furnace or AB exit gas temperature was found to conform with combustion chemistry kinetic rate model predictions in the temperature range of 700°F (371°C) to 1400°F (760°C). The results also showed that MH furnaces can meet the 100 ppm standard adopted in 40 CFR 503 by operating in the 900°F (482°C) to 1000°F (538°C) range.

Operating at temperatures in the range of 1400°F (760°C) to 1500°F (816°C) to control THC/organics emissions, however, will result in negative environmental impacts by increasing NO_x formation and the use of non-renewable oil and gas, which is not commensurate with the emission reduction achieved.

A Study of Alkalinity Leaching and Residual Strength of Alkaline Stabilized Sludge

Andrew P. Kruzic[1], M. ASCE
Thomas M. Petry[2], M. ASCE

Abstract

The addition of sufficient amounts of quicklime or other alkaline earth oxides to dewatered sludge raises both the pH and temperature of the sludge, resulting in microbial inactivation. At typical quicklime doses used for stabilizing sludge a sink of hydrated lime is generated, which maintains the pH in the sludge above 12 and thereby greatly inhibits microbial regrowth. Alkaline compounds also increase the structural strength of dewatered sludge due to drying and pozzolan formation. This paper reports on research to evaluate the effects of rainfall percolation on the residual alkalinity and structural strength of alkaline-amended dewatered sludge, and alkaline-amended dewatered sludge plus soil mixtures.

Introduction

As a result of frequent citizen complaints concerning odors from its sludge landfill at the Trinity River Authority (TRA) Central Wastewater Treatment Plant in Irving, Texas, TRA made a major change in its sludge landfilling operation in 1992. Cement kiln dust (CKD) alone and quicklime alone have been added to dewatered sludge to raise both the pH and temperature resulting in a substantial reduction in the odor producing potential of stockpiled sludge which can not be landfilled during wet weather conditions. Alkaline amendments also produce another desirable effect: they increase the structural strength of the amended sludge, which could result in the elimination of the current practice of adding and mixing soil with the sludge for structural stability and odor control in the landfill. While this change in operation could extend the life of the landfill by up to ten years and greatly reduce or eliminate the potential need for future

---

[1]Assistant Professor and [2]Professor, The University of Texas at Arlington, Department of Civil Engineering, P.O. Box 19308, Arlington, TX 76019

import of soil, it is not clear if the structural strength of alkaline-amended sludge will remain over a long period. Reducing the amount of soil mixed with sludge or completely eliminating soil will result in a larger fraction of the total material in the landfill being organic. It is possible that microbial degradation of the organic material would change the structural properties of the landfilled material to such a degree that a structural failure of the landfill would occur.

Compacted samples of quicklime-amended sludge and a mixture of quicklime-amended sludge plus soil were leached in flexible wall permeameters in an attempt to speed up the expected leaching of hydrated lime, thereby lowering the sample pH and allowing for microbial regrowth.

Materials and Methods

The study was performed utilizing six flexible wall permeameters. Compacted specimens of quicklime-amended sludge and mixtures of quicklime-amended sludge plus soil were leached using distilled-deionized water. Stock materials of natural clay soil and quicklime-amended sludge were provided to the UTA geotechnical laboratory by TRA personnel. The two groups of specimens, one containing only treated sludge and the other containing treated sludge plus soil, were made from different stocks of quicklime-amended sludge generated by TRA. The treated sludge plus soil samples were mixed by hand using a 50% volume ratio of soil to sludge. Cylindrical specimens, of both treated sludge and treated sludge plush soil, were then compacted. Selected specimens were subjected to leaching, followed by shear strength testing. The remainder of the specimens were sealed and placed into a constant temperature/humidity environment. At the conclusion of the leaching program, the held specimens were also tested for shear strength.

Compaction of specimens was accomplished using the equivalent energy levels of standard Proctor (12,375 lb-ft/ft^3) and modified Proctor (56,300 lb-ft/ft^3)tests. Each specimen was 6 inches in diameter and 7 inches high. After compaction specimens had a 6 inch diameter triaxial membrane placed around their circumference and porous stones placed on their top and bottom. This assemblage was then placed into a lucite chamber such that the membrane and chamber formed a space around the perimeter of the specimen which would pressurize the membrane against the sides of the specimen. The rest of the chamber provided inlet and outlet control of fluids which were to permeate the specimen. The confining pressure applied to the membrane surrounding the sides of each specimen was 15 psi and the differential pressure used to move fluid through each specimen was 3 psi.

Following the leaching tests, each specimen was taken out of the chamber and membrane. The dimensions and weights of the resulting cylindrical specimens were determined. Samples of each specimen, leached and held, were taken for water content and further environmental testing. Three approximately 1.0 inch high by 2.5 inch diameter cylinders were cut from the relatively undisturbed part of each specimen. These were used to perform direct shear tests of the unconsolidated-undrained type. The loads chosen for this testing represented overburden pressures of 1000, 2000 and 3000 psf respectively.

A unique aspect of the research was the measurement of caustic alkalinity in the sludge/soil samples to determine the concentration of unreacted hydrated lime. Measurement of pH alone does provide sufficient information to determine the rate of alkaline neutralization. While the obvious advantages of measuring caustic alkalinity for determination of alkaline neutralization rates are clear, a "standard method" for caustic alkalinity measurement in soils or dewatered sludge could not be found. There are several standard test methods associated with the determining the purity of quicklime and the measurement of soil pH, but the most useful model we found for developing a standard test method for determining caustic alkalinity in alkaline-amended dewatered sludge comes from the Texas State DOT for measuring lime purity by acid titration. The procedures used during this study for measuring the residual hydrated lime were:

1. Continuously mix 15 grams of sludge/soil with 150 mL of deionized water in a 400 mL beaker. Carefully titrate the suspension with 1.0 N HCl, never allowing the pH to drop below 10.0 with any single aliquot addition. (pH measurements in the stirred suspension should be determined with an inexpensive electrode because solids in suspension will scratch the glass membrane.) As the titration continues, the pH recovery rate caused by the dissolution of hydrated lime will slow.
2. As the measured pH after recovery approaches 11.0, decrease the volume of HCl added so that the pH does not drop below 10.8 with any single aliquot addition. When the recovery rates slows to the point where the recovery to pH 11.0 from 10.9 requires more than 1 minute, the procedure for measuring pH is changed.
3. Titrate with acid to 10.8 using an inexpensive pH electrode. Cover and stir the sample for 10 minutes and then allow the suspension to settle for 5-10 minutes.
4. Using a high quality combination pH electrode having a concentric ceramic reference junction, carefully lower the electrode into the sludge/soil suspension until the pH sensitive bulb is covered while leaving the reference junction in the supernatant layer. Allow the electrode to reach equilibrium and read pH.
5. Withdraw the electrode, rinse it with deionized water, and check its calibration. Repeat step 4 to assure an accurate pH measurement has been made. If the pH is greater than 11.0 repeat steps 3 and 4.
6. Determine the percent dry weight of the sample and calculate the caustic alkalinity in units of meq/g-dry weight.

Leachate was collected daily during the first weeks of leaching and later every two or three days. Leachate samples were open to the atmosphere during collection but were placed in capped sample storage bottles thereafter. Estimates of caustic and total alkalinity of the leachate were made using pH 11.0 and pH 4.5 as titration endpoints.

Results

Shear strength of most materials is directly related to their dry unit weight and inversely related to their water content. One can see that there is considerable variance in the shear strengths found in Table 1. Some of this variance can be attributed to variance in dry unit weight and moisture content in the specimens.

Table 1. Shear Strength Results

SAMPLE ID	TREATMENT SLUDGE +	COMPACT. ENERGY	GEOTECH W. C. %	DRY UNIT WT. (PCF)	COHESION (PSF)	FRICTION ANGLE
1	10 % QL	STANDARD	155.9	28.2	1548	28.6°
2	10 % QL	STANDARD	165.2	26.1	3361	7.0°
3	10 % QL	STANDARD	156.2	27.6	2394	8.8°
4[1]	10 % QL	STANDARD	187.6	25.3	400	25.1°
5	10 % QL + 1/2 SOIL	STANDARD	91.6	41.6	1296	36.0°
6	10 % QL + 1/2 SOIL	MODIFIED	79.9	42.8	1675	45.5°
7	10 % QL + 1/2 SOIL	MODIFIED	66.4	46.8	1475	43.4°
8[1]	10 % QL + 1/2 SOIL	STANDARD	58.0	58.4	967	39.8°

[1]SAMPLES 4 AND 8 WERE NOT LEACHED

The overall effects on shear strength noted for all the treated sludge specimens during this test sequence include those from differences in dry unit weights, those believed as a result of leaching, and those which result with addition of soil to treated sludge. Values of at least 500 psf cohesion are believed to represent reasonable levels for placing treated TRA sludges in a monofill. All of the specimens tested during this project had cohesions near to or larger than 500 psf. Specimens 2 and 3 had conspicuously higher values of cohesion, but had very low friction angles. A previous study compared treated sludges with a sludge-soil mixture which had provided adequate shear strengths in the field and which had an average cohesion of 500 psf and a friction angle of 28 degrees. It is, therefore, possible to say that specimens 1,5,6,7,and 8 all have shear strengths superior to this mix, while specimen 4 has properties slightly inferior to it. It is clear from the results of the testing discussed here that development of higher dry unit weights provides better friction angles, that addition of soil (50% volume basis) dramatically increases friction angles, and that the levels of leaching done during this study, generally, do not cause reduction in shear strength and may well cause significant increases in the cohesion developed.

A summary of leachate volumes and permeability is provided in Table 2. Specimens 2 and 3 presented problems during the leaching process. Specimen 2 was compressed by the applied horizontal pressure to such an extent that flow was nearly shut off and, eventually, the membrane could not maintain contact along the sides of the specimen. Specimen 3 experienced a membrane failure and its leach test was then terminated. In general, all of the specimens have low permeabilities, ranging from 1×10^{-6} to 1×10^{-9} cm/s. For the test apparatus used, 3 liters of leachate is equivalent to 1 inch of rainfall percolation. The volume of leachate passed through the samples expressed in number of pore volumes varied from 1.3 to 8.1. The variability in the permeabilities is due to the variability in the hydraulic properties of the specimens while the leachate volumes also reflect the varying duration of the leaching tests. The lower permeabilities of specimens 6 and 7 compared to specimen 5 may be attributed to the higher compactive energy used in preparing those specimens. It can also be noted that the permeabilties of the treated sludge containing soil were higher than those of the treated sludge without soil. Comparing initial and final permeability it can be seen that the permeability in all specimens decreased with time.

Table 2. Leaching and Permeability Results

SAMPLE ID	TREATMENT SLUDGE +	TIME LEACHED (Days)	VOLUME LEACHED (Liters)	# PORE VOLUMES	INITIAL PERM. (cm/s)	FINAL PERM. (cm/s)
1	10% QL	200	9.57	4.4	2.5E-7	3.0E-7
2	10% QL	35	0.75	0.3	8.1E-8	1.0E-9
3	10% QL	18	0.67	0.3	6.8E-7	8.5E-7
5	10% QL + 1/2 SOIL	91	15.37	8.1	9.4E-7	1.2E-6
6	10% QL + 1/2 SOIL	111	2.50	1.3	1.4E-7	1.7E-7
7	10% QL + 1/2 SOIL	114	3.09	1.6	1.7E-7	2.1E-7

Measurements of alkalinity in the sludge specimens and the leachate, calibrated to the dry weight solids, are summarized in Table 3. Graphical summaries of the alkalinity results for specimens 1 and 5 are presented in Figures 1 and 2. The caustic alkalinity in the specimens of treated sludge without soil varied between 1.05 and 1.28 meq/g. These values are very low compared with the expected value of approximately 9 meq/g, which have been consistently found in additional studies using sludges of 30% dry weight that are treated with a quicklime dose of 10% on a wet weight basis. Based on this result it appears that the quicklime-amended sludge samples obtained for this study from TRA did not receive the desired dose of 10% quicklime on a wet weight basis. The sludge caustic alkalinity value for specimen 1 is an average of two measurements which were significantly different. The cause of this variability may be in the way the sample was collected for measurement. It should be expected that material at the top of the specimen will be depleted of alkalinity faster than material in the bottom. The amount of caustic and total alkalinity leached from specimen 1 respectively represents 8.7% and 42% of measured caustic alkalinity in the sludge. These values drop to 1% and 5% respectively based on the expected caustic alkalinity value. The caustic alkalinity in the treated sludge plus soil specimens varied between 2.84 and 5.00 meq/g. These values are reasonably close to the expected value of approximately 4.5 meq/g. The amounts of caustic alkalinity leached from the three specimens represent between 0.2% and 2.6% of the expected value of caustic alkalinity in the sludge/soil mixture. The range of total alkalinity leached represents between 3.1% and 12.6% of the expected value of caustic alkalinity in the sludge/soil mixture.

Table 3. Alkalinity Results

SAMPLE ID	TREATMENT SLUDGE +	SLUDGE CAUSTIC ALK. (meq/g-dry wt)	LEACHED CAUSTIC ALK. (meq/g-dry wt)	LEACHED TOTAL ALK. (meq/g-dry wt)
1	10% QL	1.15	0.099	0.482
2	10% QL	1.28	—[2]	—[2]
3	10% QL	1.12	—[2]	—[2]
4	10% QL	1.05	—[1]	—[1]
5	10% QL + 1/2 SOIL	2.86	0.118	0.566
6	10% QL + 1/2 SOIL	5.00	0.016	0.168
7	10% QL + 1/2 SOIL	4.19	0.008	0.138
8	10% QL + 1/2 SOIL	2.84	—[1]	—[1]

[1]SAMPLES 4 AND 8 WERE NOT LEACHED [2]LEACHING OF SAMPLES 2 AND 3 WAS TERMINATED EARLY

From Figures 1 and 2 it can be seen that during the beginning of the leaching tests the pH of the leachate from specimens 1 and 5 was less than 11. The cause for this is unknown although the percolation rates did decrease significantly after the first two weeks. Although there are subtle differences in the rates of caustic and total alkalinity leaching between specimens 1 and 5,

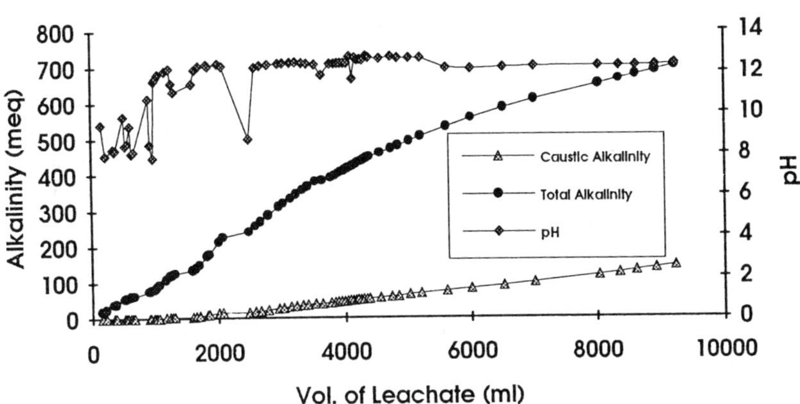

Figure 1. Sample 1 Leachate pH, and Caustic and Total Alkalinity

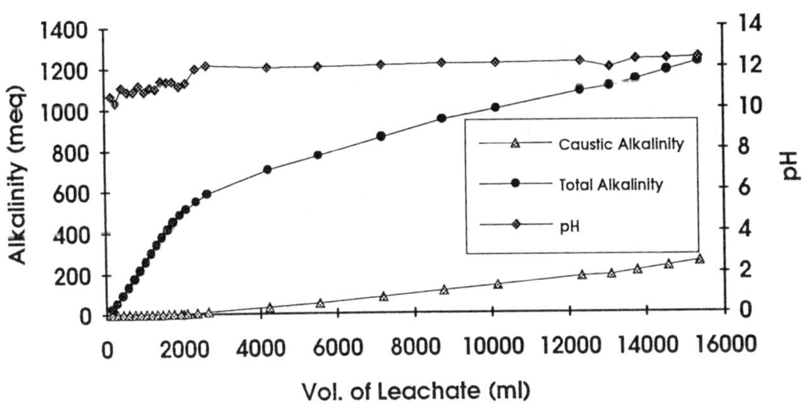

Figure 2. Sample 5 Leachate pH, and Caustic and Total Alkalinity

there are also several common characteristics. In general, the total alkalinity in the leachates is approximately 5 times that of caustic alkalinity. Also, the rate of both total and caustic alkalinity release are fairly linear with time after the passage of 2 liters of leachate.

Using the measurements of leached alkalinity it is possible to estimate the volume of rainfall percolation necessary to dissolve all of the hydrated lime in the sludge. Based on the expected caustic alkalinity in a sludge (30% solids) with a 10% wet weight quicklime dose and the actual rate of caustic alkalinity leaching from the results presented here, 260 inches of rainfall percolation per inch of sludge depth would be required to dissolve the hydrated lime. If it is assumed that the total alkalinity in the leachate comes only from the hydrated lime than the rate of hydrated lime dissolution is 5 times greater and the required rainfall percolation is reduced to 60 inches of water per inch of sludge depth. Using the theoretical value for hydrated lime dissolution and assuming a 10% wet weight quicklime dose to sludge with 30% dry solids, the required rainfall percolation is 120 inches of water per inch of sludge depth.

Conclusions

One goal of this study, to leach significant amounts of alkalinity from quicklime-amended sludge so that microbial regrowth could occur, was not achieved. However, several significant conclusions can be drawn from the results of this study:

1. The shear strengths of quicklime-amended sludge and quicklime-amended sludge plus soil mixtures increases significantly in terms of cohesion as a result of leaching as long as significant levels of caustic alkalinity remain in the mixture.
2. The permeabilities of compacted quicklime-amended sludge and quicklime-amended sludge plus soil (50% volume basis) are fairly low.
3. Sludges treated with 10% quicklime on a wet weight basis will contain significant levels of unreacted hydrated lime that can not be easily leached from the sludge. For a 10% quicklime dose (wet wt) to sludge (30% dry solids) and assuming no soil is added, the required amount of percolation to lower the pH below 11 is between 60 and 260 inches of water per inch of sludge depth.

Although it appears that quicklime leaching by rainfall percolation is low, there are several unknowns to consider. It is possible that microsites exist in the sludge where the hydroxide concentration is considerably lower and microbial regrowth could begin. Poor mixing of sludge with quicklime would improve the likelihood of lower pH microsites. Also, additional studies at UT Arlington have found that gaseous carbon dioxide recarbonation is likely to be the controlling factor in lowering the caustic alkalinity of quicklime-amended sludges.

Beneficial Use Alternatives For Water Treatment Plant Residuals

by
Ann M. Copeland, Associate Member, ASCE[1],
Carel Vandermeyden, P.E.[2], and David A. Cornwell, Ph.D., P.E.[3]

Abstract: The beneficial use alternatives which are currently in practice for water treatment plant residuals and their relative advantages and disadvantages are summarized in this paper. The alternatives of turf farming, top soil blending, land application with biosolids, composting with biosolids, and land application are discussed in terms of technical feasibility, regulatory requirements, and operational methods. Beneficial use programs in Buffalo, New York; Charlottesville, VA; and Marietta, GA are discussed with regards to operational experience and cost impacts.

Introduction: The treatment and disposal of water treatment plant (WTP) residuals is rapidly becoming an integral part of operating WTP facilities as local, state, and federal regulations are requiring more stringent standards for traditionally practiced residuals management programs. The discharge of untreated residuals to most surface waters is severely restricted under the National Pollutant Discharge Elimination System (NPDES) of the Clean Water Act. Discharge of WTP residuals to the sanitary sewer is equally becoming more restrictive through wastewater pretreatment standards, wastewater treatment plant (WWTP) available capacity, digester ability to handle the mostly inorganic WTP residuals, and the WWTP effluent standards. Dwindling availability of space in sanitary landfills has made landfilling of residuals a very costly disposal method. As a result, beneficial use programs for WTP residuals are increasingly being considered by utilities, not only as a cost effective alternative but also as a publicly acceptable management practice.

Beneficial Use Options: The desired residuals management goal for any water utility is to "operate an economically efficient and environmentally attractive residuals management plan by developing one or more long term agreements which

---

[1]Engineer, Environmental Engineering & Technology, Inc., 712 Gum Rock Court, Newport News, VA 23606
[2]Project Manager, Environmental Engineering & Technology, Inc., 712 Gum Rock Court, Newport News, VA 23606
[3]President, Environmental Engineering & Technology, Inc., 712 Gum Rock Court, Newport News, VA 23606

will allow for the proper utilization of residuals in a beneficial application." Though not impossible, this is indeed a difficult goal in light of the relatively low nutrient value of WTP residuals. While residuals from lime softening processes have been beneficially utilized in the past, residuals from coagulation processes are more difficult to market as beneficial products to potential users. At this point, the beneficial use markets that have exhibited the greatest potential for success are the following:

- **Commercial Products**: The use of WTP residuals in manufacturing commercial products has proven successful in such areas as turf farming and top soil blending. In both cases, the WTP residuals used as a substitute for natural soil material and offers economical benefits to the producer. Brick manufacturing is often considered a potential market for WTP residuals and previous experiences by the Santa Clara Valley Water District (Migneault, 1988) and the City of Durham, NC (Rolan, 1976) indicate that this can be a workable program. Brick making, however, has been difficult to maintain over a long period of time.
- **Co-use with Biosolids**: Incorporating WTP residuals in biosolids management programs such as land application and composting can be beneficial. Blended products tend to have lower metal concentrations making the product more marketable. Also, for utilities that operate both water and wastewater facilities, permitting, record keeping, and monitoring requirements are reduced when all the residuals are managed under one program.
- **Land Application**: Land application of WTP residuals to agricultural or forested land is a feasible beneficial use alternative. It is not widely practiced because the need has only been recognized recently and WTP residuals must frequently compete with biosolids.

The following paragraphs will look at these beneficial use alternatives more closely. While it is recognized that innovative alternatives such as yard waste composting, cement manufacturing, and other alternatives are feasible, their use is at this time very limited and often only on a pilot or demonstration scale basis. The focus of this paper is on alternatives that are already practiced by utilities and can be implemented by other utilities.

Commercial Products—Turf Farming: Application of WTP residuals to turf farms provides a residuals management option which can benefit a farmer's harvest and supplements the farmed areas with a new "soil" base. Turf grass has a relatively low nutrient demand but requires significant moisture levels, particularly for initial growth phases. Dewatered WTP residuals applied to a turf farm at the beginning of the seeding process can provide excellent water retention capabilities. The preferred method of application is with a manure-type spreader. The application rate is a function of the type of grass farmed, supplemental fertilization used, and residuals quality available. However, a 2.54 cm application depth of WTP residuals at 20 percent solids concentration is equivalent to a solids loading rate of 38 dry tonnes per hectare or a mass loading rate of 1.7 percent. This beneficial use strategy is currently being considered by the Erie County Water Authority (ECWA) in Buffalo,

New York and by the North Penn/North Wales Water Authorities in Chalfont, Pennsylvania.

ECWA Turf Farming Project: The ECWA operates two water treatment plants, the Sturgeon Point WTP and the Van de Water WTP. Together these plants generate an average of 87 dry tonnes of residuals per month. The solids are currently disposed in a sanitary landfill, but the Authority is working with local turf farmers and the New York State Department of Environmental Conservation to develop a beneficial use program for the residuals. Due to the characteristically low solids levels, dissolved organic content and dissolved inorganic content of Lake Erie and the Niagara River, the required chemical treatment is comparatively minimal at both plants. Residuals produced at both plants are likewise quite "clean" making them ideally suited for beneficial use applications. The residuals are similar in chemical make-up to soils native to the region as shown in Table 1. Full scale demonstration trials with turf farmers are currently underway after successful greenhouse and leaching experiments conducted by the University of Buffalo (Van Benschoten, 1991).

Table 1
Metals Content of ECWA Residuals and Area Soils

	Concentration Range (mg/kg)	
Metal	ECWA Residuals	Buffalo, Tonawanda, and Niagara Falls Soils
Cadmium	0.48 - 1.26	1.0 - 9.0
Chromium	5.56 - 19.20	8.0 - 30.0
Copper	7.12 - 32.26	7.0 - 40.0
Lead	5.54 - 11.90	20.0 - 290.0
Mercury	<0.002	0.08 - 0.28
Nickel	6.42 - 14.48	10.0 - 40.0
Zinc	<0.01 - 37.90	23.0 - 160.0

Residuals mixed with native soils at 25, 50, 75, and 100 percent loadings clearly enhanced turf grass growth in the greenhouse studies. As shown in Figure 1, measured turf grass growth in residuals/soil mixtures was up to three to four times the growth observed in native soils. This improvement in yield translates into a turf farmer being able to bring his fields to harvest sooner and therefore provide a marketable turf product sooner than experienced under current harvesting conditions. Increases in labile phosphorus (as determined in Bray-1 phosphorus tests), indicative of a beneficial increase in the soil of phosphorus available to plants, followed the recorded increases in grass mass. Figure 2 graphically depicts these findings. The residuals apparently contain iron, aluminum, and calcium phosphate (extracted and quantified in the Bray-1 test), in addition to those present in the soil in quantities sufficient to effect increases in grass growth. The presence of a substantial manganese and iron content in the residuals blended with the soil improved conditions for grass growth by enhancing the soils' water holding capabilities. Leaching studies

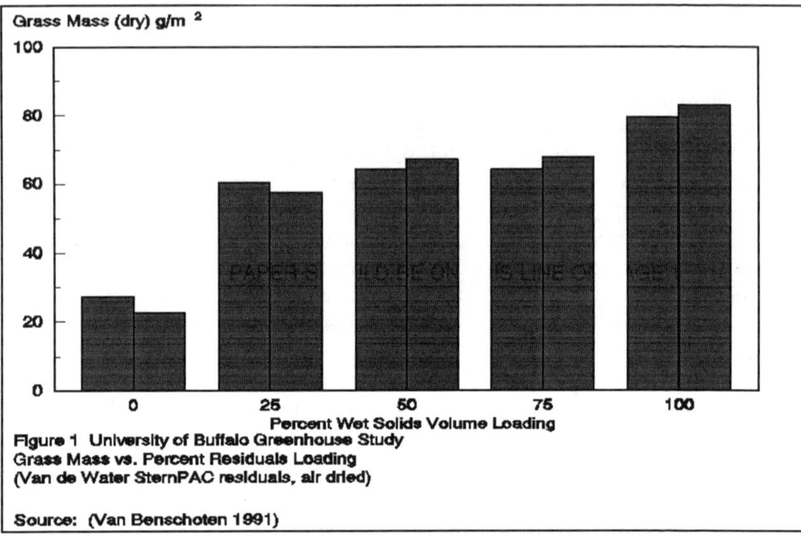

Figure 1 University of Buffalo Greenhouse Study
Grass Mass vs. Percent Residuals Loading
(Van de Water SternPAC residuals, air dried)

Source: (Van Benschoten 1991)

for total dissolved aluminum, total dissolved iron, total dissolved phosphorus, soluble salts, pH, and total organic carbon were conducted for an average yearly rainfall for the Buffalo area over a 77 day period. With the exception of manganese content, measurable leaching of constituents ranged from approximately 0.1 to 3.0 percent of the total concentration present in the residuals. Iron and aluminum leached less than

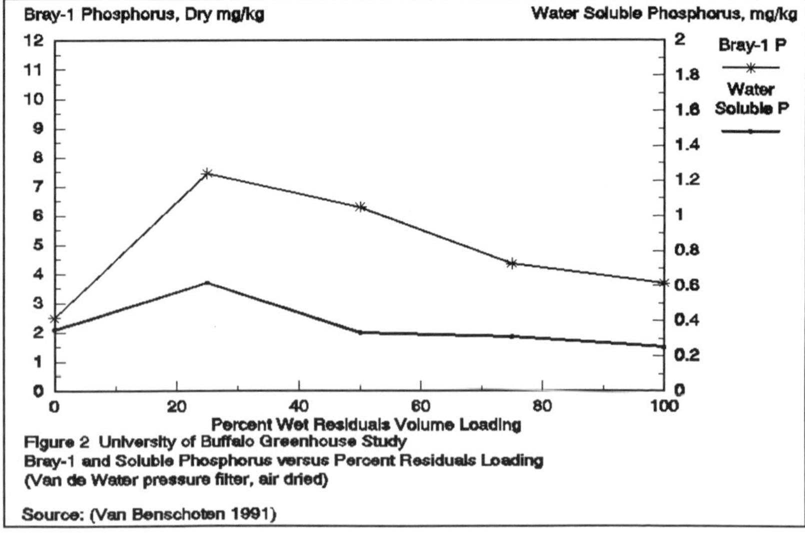

Figure 2 University of Buffalo Greenhouse Study
Bray-1 and Soluble Phosphorus versus Percent Residuals Loading
(Van de Water pressure filter, air dried)

Source: (Van Benschoten 1991)

two percent of the amounts initially present. Aluminum exceeded 100 μg/L after 21 days of testing but subsequently decreased to a mean concentration of 9 μg/L by day 77 of the study. The decrease over time eliminates any concerns over phytotoxicity due to soluble aluminum. Iron was detected in amounts which exceeded the secondary maximum contaminant level (SMCL) of 0.3 mg/L established for drinking water. However, the concentration of iron declined cover time and would not be present in high enough amounts to effect any measurable water quality impact.

The ECWA is currently implementing a full scale beneficial use demonstration project with a local turf farm (Lakeside Sod Supply Company, Inc.) to confirm the viability of this residuals management technique. The beneficial use turf farming program will eliminate approximately 1,300 to 1,700 tonnes of residuals per year from being landfilled, or save 20,000 cubic meters of landfill volume over a 20-year period, if only six months worth of residuals are used in turf farming operations with the remainder being disposed of in the landfill. Should all residuals be used in turf farming, approximately 40,000 cubic meters of landfill volume could be saved over a 20-year period.

Commercial Products—Top Soil Blending: Commercial producers of top soil utilize a variety of raw soil products to develop a marketable product for nurseries, homeowners, professional landscapers, etc. In this process, the raw soils are screened and blended with organic material before being sold as a product. WTP residuals can be blended during the top soil production process to increase the nutrient value and water retention capabilities. To a top soil producer, the WTP residuals is a minimal cost raw material that can increase the profit margin on the final product. The amount of WTP residuals added to the top soil may be 10 percent or less and is a function of consistency, quality, and availability. The acceptable quality of the WTP residuals is determined by individual top soil producers. Earthgro Corporation in Pennsylvania is a commercial top soil producer and has established various classes of quality at which WTP residuals can be accepted as shown in Table 2.

Table 2
Top Soil Blending Metal Content Limits (ppm)

Parameter	Pre-Approved	Requires Review*	Not Accepted
Cadmium	<2.0	2.0-25.0	>25.0
Chromium	<1,000	---	>1,000
Copper	<100	100-1,000	>1,000
Lead	<200	200-400	>400
Mercury	<0.3	0.3-10.0	>10.0
Nickel	<200	---	>200
Zinc	<300	300-1,200	>1,200

*Earthgro must compare analyses with current data and experience.

Blending WTP residuals with top soil can be a mutually beneficial operation for the utility and the producers. Key elements for long term success are reasonably consistent solids quality, reliable delivery of dewatered residuals, and strong contractual arrangement between all parties.

Co-Use With Biosolids: WTP residuals can be mixed with biosolids and subsequently be a co-product in the overall end-use management strategy of the biosolids. For a utility that operates both water and wastewater facilities, this type of WTP residuals management program has certain benefits including (1) avoiding separate permitting and monitoring of the WTP residuals, (2) providing a beneficial and often cost effective end-use avenue for the WTP residuals, and (3) reducing most of the metal concentrations in the biosolids product due to the diluting effect of the WTP residuals. Even in situations where the water and wastewater facilities are owned and operated by separate entities, the WTP residuals can enhance the quality of the biosolids product and provide a source of revenue. WTP residuals can be incorporated with biosolids in a variety of methods, including discharge of liquid WTP residuals to the sanitary sewer, to the WWTP influent, to the WWTP solids handling facility, or blending dewatered WTP with dewatered biosolids.

Regardless of the method employed, the WWTP operations cannot be impacted. Inorganic WTP residuals can significantly impact primary settler overflow quality, digester space, and digester efficiency. WTP residuals should also not degrade the end-use biosolids product quality such as lowering nutrient values or increasing/introducing higher metal concentrations. This is particularly true for land application and composting processes.

Land Application with Biosolids: Land applying a mixture of WTP residuals and biosolids is not a common practice. Aluminum in WTP residuals has the ability to bind with phosphorus, rendering the P available as a nutrient for plant growth. Supplemental fertilization is generally applied to compensate for the loss of P. A proper combination of WTP residuals and biosolids helps the P availability in the soil and provides additional nitrogen for plant growth. The proper mixture of the two solids should be developed on a site specific basis with a focus on the P fixing capability of the mixture. This capability should not exceed the available P content of the mixture. To determine the maximum allowable fraction of WTP residuals that could be mixed with biosolids for land application, the Colorado Department of Health uses the following equation:

$$\text{WTP Residuals Fraction} = \frac{P(DSS) - 1.15\ Al(DSS) - 0.55\ Fe(DSS)}{1.15\ Al(WTPS) - 1.15\ Al(DSS) + 0.55\ Fe(WTPS) - 0.55\ Fe(DSS)}$$

where,

P(DSS)	=	phosphorus content (mg/kg) of the biosolids
Al(DSS)	=	aluminum content (mg/kg) of the biosolids
Fe(DSS)	=	iron content (mg/kg) of the biosolids
Al(WTPS)	=	aluminum content (mg/kg) of the WTP residuals
Fe(WTPS)	=	iron content (mg/kg) of the WTP residuals

If, for example, WTP residuals with 60,000 mg/kg Al and 10,000 mg/kg Fe was mixed with biosolids containing 25,000 mg/kg P, 6,000 mg/kg Al, and 20,000 mg/kg Fe, then the allowable WTP residuals fraction would be 12.5 percent.

Composting with Biosolids: WTP residuals can be composted with biosolids, provided the mixture does not impact the initial temperature of the compost, which must exceed 55°C for 14 days for pathogen destruction. Key to the success of co-

composting is the volatile solids concentration of the biosolids and the WTP residuals, the mix ratio, and the metal concentration in the WTP residuals. The Rivanna Water and Sewer Authority in Charlottesville, Virginia co-composts WTP residuals and biosolids (Potter et al., 1992). Here WTP residuals are dewatered at the WTP and transported to the compost yard where the residuals are mixed with biosolids and wood chips. Demonstration trial conducted early in the process blended 12.5 and 25 percent WTP residuals mixtures with 50 percent wood chips. A temperature profile of the resulting compost pile is shown in Figure 3 and suggest

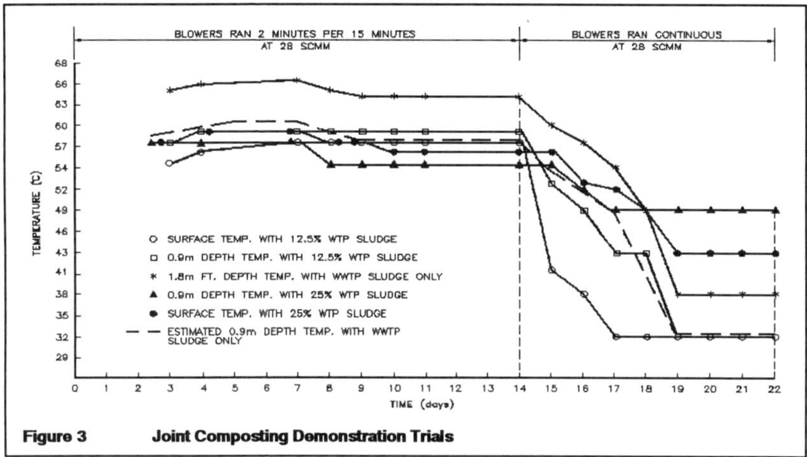

Figure 3 **Joint Composting Demonstration Trials**

that a 25 percent mix ratio is nearly the maximum for maintaining adequate compost temperature. A 12.5 mix ratio performed as well as biosolids alone. No observable differences in color or physical characteristics were noted between the two mix ratios and normal compost. In terms of metal concentrations, the addition of WTP residuals diluted the metal concentrations in the compost and in fact allowed higher application rates for the finished compost. Overall this WTP residuals management strategy is the most cost effective for the Authority and is viewed positively by the public.

Land Application: The land application of WTP residuals, particularly coagulant residuals, can be an environmentally safe and cost effective method for residuals management. Due to the variable characteristics of different raw water sources, allowable residuals application rates must be considered according to the individual WTP residuals composition. The primary concerns associated with land application are the presence of heavy metals and the potential for phosphorus binding. However, with good soil management and crop selection, P depletion can be prevented with proper loading rates and P fertilization. The aluminum concentration of WTP residuals can range from 5 to 1[illegible] percent of the total dry solids mass, which is 50 to 100 percent higher than the concentration of the Al in most soils (Elliott and Dempsey, 1991). Elliott and Dempsey (1991) reported only a 0.3 percent increase in soil Al level when a 22 tonne per hectare residuals loading rate with a residuals Al concentration of 30 percent was land applied. Aluminum phytotoxicity is dependent on Al solubility and is not a problem when the pH range in the soils is maintained between pH 6 to 6.5 (Elliott and Dempsey, 1991).

The addition of alum residuals to soils can also change the soils' physical structure, or bulk density. A soil with a high bulk density is more compact, which is unfavorable to plant growth because root penetration is restricted. A low bulk density soil has more pore space for air and water which is beneficial for plant growth (Tisdale and Nelson, 1975). Rengasamy et al. (1980) found that soil mixed with alum residuals increased soil aggregation and moisture retention and, as a result, increased the dry yield of maize. Bugbee and Fink (1985) demonstrated that improvements in aeration and moisture retention promoted by the addition of alum residuals were able to offset the phosphorus deficiency in lettuce.

The impact of land application on various crops has been investigated by Virginia Polytechnic Institute and State University. Novak et. al (1993) studied the impact of alum and PACl residuals on corn when applied at 1.3 to 2.5 percent by dry weight (29 to 56 tonnes per hectare). Crop yield from the treated plots were not statistically different from the untreated plots. Mutter et. al (1994) studied the impact of PACl residuals on wheat when applied at 2, 4, and 8 percent. No negative effects on wheat grain or biomass yield was observed at these loading rates. Although soil aluminum levels were significantly increased at these loading rates, after two crop rotations the soil Al concentrations were found to be similar to background Al levels. The wheat leaf tissue had a slightly increased Al concentration as compared with the control. This increase in Al concentration, however, was not found to be statistically significant.

A successful land application program is operated by the Cobb County-Marietta Water Authority in Marietta, Georgia (Parsons, 1993). Residuals (approximately 8,000 cubic meters per year) from two plants are dewatered on filter presses to approximately 35 percent total solids. Lime is added in the dewatering process at approximately 10 to 15 percent on a dry weight basis. Local farmers became interested in the land application program because the lime content of the residuals was useful for pH adjustment of their soil. About 88 percent of land under contract for land application is pasture land, the remainder being crop land. The residuals application rate ranges from 34 to 101 wet tonnes per hectare or about 11 to 34 dry tonnes per hectare. The Authority reports that most of the operation and maintenance costs are covered under the application contract, including program management, monitoring, and actual application. The savings of land application over landfilling (at $33/tonne tipping fee) is about $50,000 per year.

Conclusions: There exist beneficial use alternatives for WTP residuals that can be successfully implemented. The development of a beneficial use plan should include a thorough characterization of the residuals for heavy metals, nutrients, calcium carbonate equivalency (CCE), and salts. In addition, the soil and groundwater to which the residuals are applied should be thoroughly characterized. While the up-front development of a beneficial use plan can be time consuming, the long term benefits include financial savings and public acceptance.

References

Bugbee, G.J. and C.R. Frink. 1985. Alum sludge as Soil Amendment: Effects on Soil Properties and Plant Growth. *Connecticut Agricultural Experiment Station*. Bulletin 827.

Elliott, H.A. and B.A. Dempsey. 1991. Agronomic Effects of Land Application of Water Treatment Sludges. *Journal of the American Water Works Association*, 83(4):126-131.

Migneault, W.H. 1988. Potential for Brickmaking. Freshwater Utility Recycles Its Sludge. *BioCycle*, 4:63.

Mutter, R. et al. 1994. *An Assessment of Cropland Application of Alum Sludge*. Master's Thesis, VPI & SU, Blacksburg, Virginia.

Parsons, James M. *A Land Application Program for a Water Treatment Plant Sludge*. 1993 Joint Residuals Conference, Phoenix, Arizona.

Potter, E. and C. Vandermeyden. *Joint Composting of Water and Wastewater Sludge*. AWWA/WPCF Joint Residuals Management Conference, Raleigh, North Carolina (1991).

Rengasamy, P., J.M. Oades, and T.W. Hancock. 1980. Improvement of Soil Structure and Plant Growth by Addition of Alum Sludge. *Communications in Soil Science and Plant Analysis*, 11(6):533-545.

Rolan, A.T. 1976. Evaluation of Water Plant Sludge Disposal Methods. Proc. Annual Meeting. *Jour. North Carolina Section*, 12:1:56.

Tisdale, S.L. and W.L. Nelson. 1975. *Soil Fertility and Fertilizers, 3rd Edition*. New York: Macmillan Publishing Co., Inc., 105-189.

Van Benschoten, J.E., J.N. Jensen, and A.R. Griffin. *Land Application of Water Treatment Plant Sludge*. Prepared for the Erie County Water Authority (1991).

Beneficial Use of Ash:
An Innovative Approach to Sludge Management

Michael E. Manning, Associate Member (1)
John R. Amend, P.E. (2)
Mark K. Ballerstein, P.E. (3)

ABSTRACT

Disposal of ash resulting from the incineration of municipal sewage sludge has historically been a relatively easy, albeit, expensive requirement. The traditional method of disposal is to deposit ash into permitted landfills. However, escalating landfilling costs, new regulatory initiatives and the national focus on recycling and reusing resources is prompting incinerator operators to investigate the viability of beneficial use options. Increasingly, sludge and its various byproducts are being viewed as resources rather than wastes.

This paper summarizes efforts put forth by Monroe County, New York to develop a controlled density fill material comprised primarily of municipal sewage sludge incinerator ash (ash) and Portland cement. Although financial factors did not allow the full development of an acceptable beneficial use product, much of the information learned as a result of this project, should be transferable to other municipalities interested in developing a beneficially usable ash-based product.

INTRODUCTION

Monroe County Division of Pure Waters (Monroe County), Rochester, New York operates three wastewater treatment plants, the Frank E. Van Lare Water Pollution Control Plant (FEV), Gates-Chili-Ogden Wastewater Treatment Plant (GCO) and the Northwest Quadrant Sewage Treatment Plant (NWQ). Since the early 1970's, the sludge produced at these facilities has been incinerated in multiple hearth furnaces. Until 1985, the resulting ash was processed for metals recovery at a low cost to Monroe County. In 1985, the County began landfilling ash when the precious metals market diminished.

Ash produced at FEV and GCO is stockpiled in uncovered mounds at the facilities and periodically landfilled. NWQ ash is loaded directly onto roll-off trailers at the facility and transported to a landfill.

(1) Project Engineer, Malcolm Pirnie, Inc., 1000 Pittsford-Victor Rd., Pittsford, NY 14534
(2) Associate, Malcolm Pirnie, Inc., 1000 Pittsford-Victor Rd., Pittsford, NY 14534
(3) Engineering Operations Manager, Monroe County Department of Engineering, 350 E. Henrietta Rd., Rochester, NY 14620

Because ash typically contains trace amounts of heavy metals, the New York State Department of Conservation (NYSDEC) requires that sludge ash be disposed of in permitted landfills. Landfill tipping fees vary across the Country depending on the availability of space and permitting requirements. Currently, Monroe County's tipping fees range from $56.25 to $62.60/tonne ($62 to $69/ton), which represents an increase of approximately 45% over the last seven years.

Escalating tipping fees combined with recent trends in regulatory philosophy and an increased interest in beneficial use prompted Monroe County to investigate a beneficial use program which is described in this paper.

KEEPING ONE STEP AHEAD

Monroe County began to investigate beneficial use opportunities for its ash as early as 1989. In August of 1989, Cornell University (Cornell) was retained to test the viability of producing a flowable fill from a blend of ash, Portland cement and water. The County's objective for this initial phase of testing was to investigate the technical and practical implications of producing an ash-based product similar to commercially available material, commonly referred to as K-Crete. K-Crete is a low strength controlled density fill material consisting of coal fly ash, Portland cement, water and lime which has been available since the early 1970's. The main focus of this testing effort was to ascertain if a mix design resulting in a 28 day compressive strength in the range of 6,895 kPa (1,000 psi) could be developed using Monroe County ash as the base material. A target of 6,895 kPa (1,000 psi) was established because this compressive strength would allow the Monroe County ash-product to compete favorably with commercially available K-Crete. Flowable fill was selected as the primary beneficial use product because it can be produced from either a wet (stockpiled) or dry ash.

Cornell developed a mix design similar to K-Crete for which the name "Rockcrete" was coined. As part of the study, Cornell performed basic structural as well as a battery of physical tests on cured Rockcrete to ascertain preliminary physical and chemical properties of the product. Several different mix designs were tested to determine a range of compressive strengths relative to the amount of Portland cement and water in the mixture. Identical tests were performed on locally available K-Crete to form a basis of comparison between the two products. It was concluded that a range of compressive strengths (24,130 to 1,380 kPa [3,500 to 200 psi]) were attainable with curing times as short as 14-days. It was also determined that compressive strengths varied considerably with the cement to dry ash ratios utilized in the mix design. Furthermore, the target 28-day compressive strength of 6,895 kPa (1,000 psi) was found to be attainable with a cement to dry ash ratio of 0.5 and a total water to dry solids ratio of 0.64. Compressive strength tests on various mix designs with a cement/dry ash ratio of 0.5 averaged 7,075 kPa (1,026 psi).

A preliminary evaluation of the environmental properties of ash and ash-based products was also conducted as part of this phase of work. Samples of ash, K-Crete and Rockcrete were subjected to the Extraction Procedure Toxicity Test (EP-Toxicity) and the Toxic Characteristic Leaching Procedure (TCLP). Heavy metals concentrations resulting from the EP and TCLP tests performed on Rockcrete, K-Crete and raw ash were within the limits established for each test. Leaching potential appeared to be dependent upon the mix design.

As a result of this initial beneficial use investigation, it was determined that in order to proceed with the development of a beneficial use product, a number of important issues needed to be addressed including:

- The universe of ash beneficial use technologies needed to be investigated and evaluated to determine if flowable fill was the best technology for Monroe County's application.

- Product formulation and production processes needed to be defined.

- Product markets needed to be carefully evaluated.

- Appropriate environmental tests needed to be identified. Leaching procedures needed to be modified in order to simulate product use conditions. Modified leaching tests were needed to determine the potential environmental impact of trace metals contained in the ash as well as the ability of various product admixtures (cement, lime, etc) to reduce leaching potential.

- Leachate metals concentrations needed to be compared against groundwater discharge standards (6 New York Codes, Rules and Regulations [NYCRR] Part 360-1.2(a)(5)) in order to determine whether the proposed product had the potential to cause contamination under a variety of conditions.

- Comprehensive raw ash characterizations (chemical and physical) needed to be performed to determine whether ash variability could be a factor in product development.

REAFFIRMING THE TECHNOLOGY SELECTION

The early product development work conducted by Monroe County was useful, not only because it identified the potential viability of using ash in flowable fill, but also because it clearly identified the technical and regulatory complexity of developing a commercial product from a waste material. Given the potential cost of research and development, an innovative approach for identifying and evaluating potential beneficial use technologies was necessary.

A technical advisory panel (TAP), consisting of experts from academia, government, industry and private consulting was convened to conduct a comprehensive technology search and critically evaluate the myriad of potential beneficial use technologies. The TAP consisted of seven "consultants" and 12 volunteer members. The following beneficial use technologies were evaluated by the TAP members:

- Vitrification.
- Controlled Density Fill (flowable fill).
- Soil Amendment/Fertilizer.
- Metals Recovery.
- Structural Fill (aggregates for road base or soil replacement).
- Asphalt Products (aggregate substitutes for paving systems or roof shingle manufacturing).
- Concrete Products (aggregates for marine or masonry products).
- Landfill Application (daily/intermediate cover or gas venting material).

The TAP met in March 1991 for a one day session on the beneficial use potential of Monroe County ash in light of current regulatory and market trends. The objective of the TAP meeting was to identify two to three beneficial use strategies which held potential for implementation in Monroe County.

The TAP recommended flowable and granular select fill type materials (viz. cement-based products) for further study. The remainder of this paper focuses on the work conducted to develop a cement-based product from Monroe County ash.

FORMALIZING PROJECT DIRECTION

As a result of the TAP meeting, Monroe County elected to pursue a Research Development and Demonstration (RD&D) permit in accordance with 6NYCRR 360-1.3. An RD&D permit is required by the NYSDEC before field-testing of a solid-waste based product can be conducted. In order to obtain an RD&D permit, it is usually necessary to demonstrate that the product under consideration: is viable from a production standpoint, can be marketed (sold or given away) and is not expected to cause adverse environmental impacts. Supporting documentation for the RD&D permit application usually consists of laboratory-scale data and the engineering evaluations necessary to substantiate market availability and product production economics.

Recognizing that the RD&D permit application would be difficult and costly to prepare, Monroe County elected to perform "screening level" laboratory testing to provide technical support for the permit and to develop protocols for "formal" laboratory- and field-scale testing and an ash characterization which would be conducted only after the permit was granted. The theory was to approach the laboratory work in a phased, step-wise manner; only proceeding to the next step if prior results were promising. Screening level testing was to include only basic materials property and geotechnical tests. If the proposed product failed these tests, there would be no need to proceed with more advanced and costly physical (such as freeze-thaw durability) and environmental (leaching) tests.

AGGREGATE, ADMIXTURES AND ASH

Monroe County retained Cornell in 1991 to perform screening level investigations for the development of flowable and select fill. The testing effort concentrated on developing mix designs adhering to guidelines developed by the TAP.

Flowable Fill Screening

The primary goal of the testing effort was to identify one or more mix designs which could be produced on a full-scale basis and successfully compete with commercially available products. The testing effort evaluated several filler materials to ascertain if the substitution of less costly, but functionally similar, materials for Portland Cement could result in significant cost savings. The following mix designs were tested for workability characteristics such as flow (ASTM C230) and setting time (ASTM C191), and compressive strength (ASTM C109):

- Incinerator ash only.
- Incinerator ash + 10, 20, 30% Lime (by weight of dry ash).
- Incinerator ash and 10, 20, 30% Portland cement.
- Incinerator ash and 10, 20, 30% (50/50 blend of Lime and Flyash).
- Incinerator ash and 10, 20, 30% (50/50 blend of cement and Flyash).

Prior to beginning the tests, target values of 3 and 7.5 hours were selected for the workability with time and setting time respectively. Test results indicated that the optimum water content, as percent of total mixture, ranged from 46 to 50%. The workability with time test indicated that only the mixes with 20% and 30% cement and a moisture content of 50%, retained workability characteristics for over 3 hours. The setting time test indicated that only mixes with 20% and 30% cement had setting times between 5 and 10 hours and that all other mixes had setting times greater than 10 hours.

Prior to testing, it was determined that the acceptable compressive strength for the flowable fill would be 1,034 to 1,379 kPa (150-200 psi) at 14 days. It was presumed that at these compressive strengths, which are considerably lower than those for concrete, the material could be excavated at a later date if necessary. Mixes containing fillers had compressive strengths ranging from 75.8 to 634.3 kPa (11 to 92 psi). Ash combined with water exhibited a compressive strength of 68.95 to 82.7 kPa (10 to 12 psi).

The following conclusions were reached as a result of the 1991 Flowable Fill Screening:

- The fluidity or workability of the ash mixtures is highly sensitive to moisture content. Variations of moisture content of only a few percent can make dramatic differences in the product flow behavior.

- Of the cementing agents tested, Portland cement produced the best performance on the basis of the criteria selected for study.

The flowable fill (Rockcrete) could be utilized for certain applications such as trench backfill and filling of voids (i.e., underground structures, abandoned pipelines, etc...) which require a controlled density fill. Rockcrete could not be utilized for structural applications (i.e., concrete foundations and footings, roadbase subgrade, etc...) because of its low compressive strength.

Select Fill Screening

As part of the 1991 Flowable Fill Study, Cornell also experimented with several production methods for mixing, curing and sizing the hardened flowable fill with the objective of producing a granular select fill type material. During this phase of testing, fillers such as lime and fly ash were not investigated because previous efforts determined that although less expensive, they added little value to the final product. Select fill testing focused on limiting the cement content of mixes to no more than 20% since preliminary economic analysis indicated that a mixture with a cement content greater than 20% would be cost prohibitive.

A key objective of the select fill testing was to produce a crushed (sized) material which could be used "commercially" on Monroe County construction projects. Only mixtures resulting in a select fill meeting Monroe County select fill specifications were subjected to advanced physical testing discussed below. The following conclusions were drawn from the select fill screening study:

- FEV ashes (dry and stockpiled) have pozzolanic activities in the range of 172.4 to 259.8 kPa (25-40 psi) with lime. According to ASTM C618, pozzolans acceptable for use in concrete must have an activity index with lime of at least 5,515.8 kPa (800 psi).

- Replacement of Portland cement in the mix with ash greatly reduces the compressive strength of portland cement mortars.

- Mixtures utilizing stockpiled (i.e., wet) ash yielded higher compressive strengths than those with dry ash.

- Test cubes cured in high pressure, high temperature environments (autoclave) resulted in low compressive strengths.

- Wet ash and 20% cement, cured in a moist, high temperature environment at atmospheric pressure produced the highest compressive strengths 1,585.8 kPa (230 psi) of the mixes and curing regimes tested. These results are based on a 24-hr curing time. Prolonged curing at high temperature and moist conditions increased the compressive strength to 4,067.9 kPa (590 psi). Similar compressive strengths were obtained at room temperature and a moist environment by extending the curing time from 24 hours to seven days.

- At the maximum dry unit weight and optimum moisture content, the degree of saturation for the crushed product was S=69.6%. This is based on a previously determined specific gravity of 2.76 for the ash, a specific gravity of 2.16 for Portland cement and a voids ratio at maximum dry density of 1.415.

- The average coefficient of permeability was calculated to be 5.89*10-5 cm/sec (2.32*10-5 in/sec.) which indicates that the product is a semi-impermeable material and may be suitable for use as a landfill cover. The permeability values also indicate that the material would be unsuitable for applications requiring good drainage.

- The unconfined compressive strength (ASTM D2166-85) of the product was calculated to be 213.7 kPa (31 psi) at a compacted dry unit weight of 471.6 kPa (68.4 pcf). This preliminary test data indicates that the material exhibits high strength qualities.

- The CBR (ASTM D1883-87) was determined to be 69 for a crushed sample of the Ash+20% cement mixture at a Modified Proctor maximum dry unit weight and an optimum moisture content of 38%. After soaking for 96 hours, the moisture content of the crushed product increased to 47.5% and the CBR increased slightly to 72. Acceptable CBR values for subgrade material generally range from 25-50.

- Results of the Direct Shear Test (ASTM D3080-72) indicated that the crushed product of Ash+20% cement mixture has a total stress function angle of 10 degrees and an apparent total stress cohesion of 68.95 kPa (10 psi). These values are indicative of a material that would be expected to have a relatively high friction angle and correspondingly high strength.

To emulate full scale production, larger batches of ash+20% cement were produced with a rotary drum mixer and compacted with a vibratory plate compactor. Geotechnical tests were performed on crushed sample and results compared with those obtained with bench top mixing and curing.

Testing of the crushed product, later named Rockagg, showed that it is technically feasible to produce a backfill-like material from Monroe County ash and Portland cement. Several of the mix designs demonstrated promising physical properties. For example, the high unconfined compressive strength, high shear strength, high CBR and low permeability demonstrated by several of the mix designs indicated that the material would be appropriate for use as a landfill liner and as backfill for trenches and berms, but would be unsuitable for structural applications such as subgrades and subbases. Cornell concluded, pending the results of additional physical and environmental testing, that the product appeared to be suitable for application in Monroe County's highest volume market, that is as select fill on pipeline projects.

Although a viable technology, Rockagg is not without its drawbacks. One of the largest drawbacks to the technology is the production sequence involved. Rockagg production consists of three distinct steps: mixing, curing and sizing. In contrast, Rockcrete production is simply a one step process. Other issues such as quality control, product storage and implementation may also be problematic. Extensive testing remains to be performed to determine ideal full-scale production methods for Rockagg.

MARKET STUDY PROVIDES CLEAR DIRECTION

Concurrent with performance of the 1991 Flowable Fill Screening, a study was conducted to determine available markets for a cement-based ash product. It was determined that the market for flowable fill was restricted to the filling of abandoned pipelines and other below grade structures (approximately 764.5 m^3/year [1,000 yd^3/year]). There are primarily two reasons for this conclusion:

- Concrete suppliers were unwilling and unable to use Monroe County ash in the production of a commercial product. Concrete suppliers were primarily concerned with the potential contamination of their facilities and other products with ash, coordination problems, handling costs and storage requirements.

- There were numerous difficulties associated with the Monroe County owning and operating a batch plant for production of flowable fill, the most noteworthy being that as a material supplier on its own projects Monroe County would be liable for delays caused to contractors.

However, a positive outcome of the market study was the conclusion that there was a much larger market for a granular select fill material (approximately 19,880 m^3/year [26,000 yd^3/year]). The primary use for this material would be as a fill material in pipe trenches above the pipe envelope and below the surface treatment.

SUMMARY AND DISCUSSION

Project activities to date have been termed as Phase I activities, those developing supporting documentation for an RD&D permit application. Phase II activities will be those project tasks necessary to prove that Rockcrete and Rockagg should be granted beneficial use status by the NYSDEC and will be performed only if an RD&D permit is issued.

Phase II activities will likely consist of the following:

- Performance of Formal Laboratory-Scale Testing.
- Performance of Ash Characterization.
- Field Testing of Beneficial Use Product(s).
- Petition for Beneficial Use.

It was determined that field testing would only take place upon the successful completion of formal laboratory scale testing programs. It is envisioned that field scale testing might consist of constructing a variety of pipelines utilizing Rockagg and Rockcrete and monitoring the physical and environmental characteristics of the controlled density fill materials and the adjacent natural soils. Upon successful completion of the field scale testing program and the development of stable markets for the beneficial use products, Monroe County could present all of the research data to the NYSDEC and petition for beneficial use status. Full scale production of an ash based product can only take place after approval of beneficial use status.

The project has been suspended indefinitely pending completion of several other sludge management planning activities by Monroe County and improved economic viability. With the opening of a new County owned and operated landfill, Monroe County has available a long-term and economical ash landfilling source. At this time, landfilling appears to be the most economical ash disposal option available to the County.

Monroe County recognizes that sludge incineration will continue at FEV, the largest of the County's three sludge processing facilities, and that ash disposal will be a major part of its residuals management strategy for several years. It is likely that because of the commitment to incineration, Monroe County will revisit beneficial use of ash sometime in the future. The work performed to date will serve as a valuable stepping stone to the final development of an ash based controlled density fill material.

ACKNOWLEDGEMENTS AND CREDITS

The authors wish to extend their gratitude to Cornell University Department of Civil and Environmental Engineering Department; in particular Drs. Kenneth Hover and Kumar Natasaiyer, who supervised and performed research and testing of the ash-cement products. The authors also wish to credit the members of the TAP, in particular Dr. T. Taylor Eighmy of the University of New Hampshire and Dr. Frank Roethel of SUNY Stony Brook who provided valuable insight and direction for the project.

REFERENCES

1. Properties of Monroe County Sludge Incinerator Ash-Cement Mixture, B.R. Bierck, J.J. Bisogni, R.I. Dick, K.C. Hover, K.C. Natasaiyer, School of Civil and Environmental Engineering, Cornell University, October 1989.

2. Flowable Fill Screening Study, J.J. Bisogni, K.C. Hover, K.C. Natasaiyer, Cornell University, September 1991.

3. Geotechnical Screening Analysis of Backfill-Like Material Made From Incinerator Ash and Cement Mixtures (DRAFT), J.J. Bisogni, K.C. Hover, K.C. Natasaiyer, H. Stewart; Cornell University, February 1992.

4. Beneficial Use of Wastewater Treatment Plant Sludge Ash - Executive Summary, Malcolm Pirnie, Inc., January 1993.

Modifying One of the Nation's Most Successful
Biosolids Lime Stabilization Operations

Randy Naef[1], Ronald J. Matheson[2], Mike Guthrie[3], Tracy Cork[3]

Abstract

Since 1978, the Vallejo, California, Sanitation and Flood Control District has managed a tremendously successful biosolids program consisting of lime stabilization and agricultural land application. In-plant biosolids processing consists of lime stabilization of sludges in conjunction with gravity thickening, followed by vacuum filtration dewatering. These processes evolved primarily from facilities from the plant's original physical/chemical wastewater treatment process configuration, as subsequently modified to conventional secondary treatment. Facilitating land application was the District's farsighted purchase in 1983 of a remote 1,850-acre land application site.

The plant biosolids processing facilities are nearing the end of their useful life and have inadequate redundancy. In 1992, the District undertook an evaluation of its solids facilities to identify necessary modifications. Evaluations included methods of lime stabilization, thickening, and dewatering. An extensive pilot-plant testing program was performed to confirm and refine criteria used in the analysis of process alternatives and to quantify secondary impacts (utility requirements, recycling, etc.). This paper reviews the following pilot-plant results: 1) determination of lime requirements for stabilization of raw liquid sludges prior to dewatering and on dewatered raw sludges, 2) a comparison of centrifuge and belt filter press dewatering

---

[1]Project Director, CH2M HILL, 835 N.E. Multnomah, Suite 1300, Portland, OR, 97232

[2]Director of Plant Operations and Maintenance, Vallejo Sanitation and Flood Control District, 450 Ryder Street, Vallejo, CA, 94590

[3]Project Engineers, CH2M HILL, P.O. Box 3016, Corvallis, OR, 97339

of raw and lime-stabilized sludge feedstocks, 3) evaluation of the new 14-pressure roll belt press configuration currently being marketed in the United States, and 4) dewatering recycle characteristics.

Background

Vallejo Sanitation and Flood Control District was formed in 1952 to address the treatment of sewage and flood control in an area which spanned both the City of Vallejo and parts of Solano County. The initial treatment facilities were placed in service in 1959. They consisted of primary treatment and sludge digestion.

In 1973, design was initiated to upgrade the facilities to secondary treatment. The new facilities consisted of physical-chemical treatment processes including lime treatment and filtration. All sludges were captured in the primary sedimentation tanks. Additional lime was added to the sludge to elevate the pH to 12.0 or greater. Limed sludge was stored in the former digester and dewatered using vacuum filters. The physical-chemical process failed to meet discharge standards due to a number of design issues. A new trickling filter/solids contact secondary process was designed in 1986 and put into service in 1988. The treatment facility currently has a dry weather design capacity of 679 L/s, a 1315 L/s wet weather secondary treatment capacity, and a 2630 L/s primary treatment capacity.

Existing Solids Handling System

The solids handling process consists of blending raw primary sludge with waste activated sludge (WAS) in a gravity thickener. Lime slurry is introduced to the sludge stream as it enters the thickener to provide stabilization. The lime stabilized, thickened sludge is then pumped to the vacuum filter building for dewatering. The vacuum filters produce a sludge cake of approximately 24 percent solids which is then transported by belt conveyor to an elevated sludge storage hopper. District vehicles are then loaded from the storage hopper and haul the dewatered sludge 12 miles to a District-owned 1,850-acre farm in Sonoma County called Tubbs Island.

The dewatered sludge is stockpiled on an impervious pad from November to September. When the tenant farmer has harvested the crops and prepared the fields, a contractor spreads the sludge on those fields specified by the District under controlled loading rates. This project has been in operation since 1978 and was awarded the "1990 National First Place Award for an Outstanding Project Involving and Enhancing Beneficial Use of Municipal Wastewater Sludge" for facilities over 219 L/s by the Environmental Protection Agency.

In 1987, a sanitary facilities masterplan, a storm water master plan and a financial master plan were completed that identified projects to be completed through the year 2005. At that time, the existing solids handling system was identified as needing to be replaced by 1995 due to age, lack of redundancy, and various deficiencies.

Predesign and Pilot Testing

A predesign project for solids handling improvements was initiated in late 1992. A major aspect of the predesign work was pilot testing of different thickening and dewatering technologies and lime stabilization process configurations.

The pilot testing program was conducted between January 21 and February 13, 1993. For pilot testing, primary sludge (PS) was thickened to between 5 and 10 percent in one of the plant's primary clarifiers and pumped directly to a temporary blend tank. Waste activated sludge (WAS) was pumped from the WAS piping system to the gravity belt thickener (GBT) test trailer, and thickened WAS (TWAS) was then pumped to the blend tank. The centrifuge and belt filter press pilot dewatering trailers took suction off of the blend tank. The mixed blend tank provided a homogenous and identical feed stock to the dewatering equipment.

Lime Stabilization Test Process Options

The pilot test program was set up to demonstrate lime stabilization of either thickened liquid sludge or dewatered sludge cake ("Pre-dewatering" or "post-dewatering" lime stabilization). For "pre-dewatering" lime stabilization, a lime slurry was added to batches of thickened sludges in the blend tank. For "post-dewatering" lime stabilization, quickline was mixed directly with dewatered cake in an RDP lime/cake mixer followed by a pasteurization vessel.

Laboratory work indicted that the required pH for lime stabilization in the liquid mode could be achieved under ideal conditions at lime doses as low as 8 percent (kg CaO/kgTS dry weight basis). Pilot test dewatering runs were conducted on batches having lime doses ranging from 8 to 27 percent.

The pilot lime/dewatered cake testing was done at lime doses ranging from 17 to 35 percent based on the manufacturer's experience. All test runs achieved the desired pH levels of greater than 12 after 2 hours and 11.5 after 24 hours. All test runs that were sampled for fecal coliform met EPA 503 Class A pathogen levels regardless of whether or not the pasteurization vessel was used.

Dewatering Pilot Test Results

Belt Filter Press – Raw Sludge (Lime for Stabilization Added to Cake in a Mixer after Dewatering Raw Sludge) Nine runs were conducted on PS and TWAS blends ranging from 41/59 percent to 65/35 percent PS/TWAS on a mass basis. Two runs were also conducted on 100 percent primary sludge. The results are presented in Figure 1. The minimum performance requirements desired for raw sludge dewatering were 28

Belt Press Raw Sludge Test Results

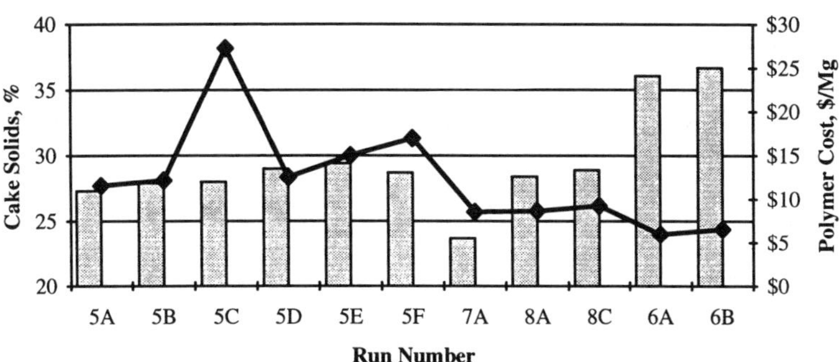

Belt Press Lime Sludge Test Results

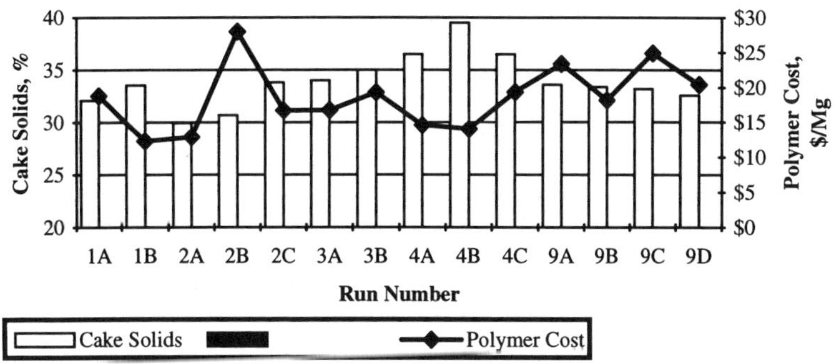

Figure 1. Belt press pilot testing results.

percent cake solids and 95 percent solids capture based on filtrate TSS measurements corrected for addition of belt washwater and dilution water. The belt press exceeded the minimum requirements on six of the nine runs. One of the runs not meeting the requirements (B-7-A) was conducted to determine the maximum throughput of the press. This run had a solids loading of 958 kg/hr/m and met the capture requirements easily; however, cake solids were only 23.7 percent due to the high throughput and low percentage of primary sludge in the blend (41 percent). The two other failing runs (B-5-A and B-5-B) were initial runs that only slightly (less than 1 percent TS) missed the desired 28 percent cake TS concentration.

The six passing runs (B-5C, D, E, and F, and B-8-A and B) represent two different blends of primary and waste activated sludge and met all the performance requirements. For the B-5 runs polymer consumption ranged from 2.7 to 5.8 kg/dry Mg or $12.56 to $27.33 per Mg. The B-8 runs had somewhat lower polymer use, possibly because of the higher percentage of primary sludge in the blend (56 versus 53). Polymer use for these runs was 1.9 and 2.0 kg/dry Mg or $8.69 and $9.27 per dry Mg. All runs had cake solids within 1 percent of the minimum 28 percent. Throughput for the runs ranged from 544 to 753 kg/hr/m at 6 percent raw feed solids. The average polymer use for these six runs is 3.2 kg/dry Mg. Capture for the six runs was 99 percent. The two runs conducted on 100 percent primary sludge achieved excellent performance. Cake solids were over 36 percent, and throughput and solids capture was high. This condition would not be a normal operating mode in the full-scale facility.

Best Filter Press—Lime Sludge (Lime for Stabilization Added to Thickened Sludge prior to Dewatering) Fourteen runs were conducted on PS and TWAS blends ranging from 58/42 percent to 65/35 percent PS/TWAS on a mass basis. Data for the lime belt press runs is presented in Figure 1. Eleven of fourteen runs conducted met or exceeded the desired cake solids and capture requirements of 30 and 95 percent. Run B-2-B had a higher polymer use than any of the other runs but showed no increase in performance. The high polymer dose for this run was the result of an inaccurate estimate of the feed flow to the unit. For successful runs cake solids ranged between 30.0 and 39.5 percent with an average value of 33.9 percent. No correlation was possible between lime dose and cake solids because of limited comparable data. Two runs were conducted using the nip rolls (B-1-A and B-2-C). The average increase in cake solids for these two runs was 0.75 percent.

Five of the runs conducted took samples from the eighth and fourteenth S-rolls to determine the benefit of the additional rolls over a normally configured 8, S-roll Winkle Press. To take these samples the press was stopped and the belts were relaxed enough to grab a sample from the eighth roll. An effort was made to reach as far into the unit as possible to obtain a representative sample. The unit was then retensioned and started. The press was run and the fourteenth roll sample was collected from approximately the same area as the eighth roll sample. Again, the sample was taken away from the edge to obtain a representative sample. The increase in cake solids between the eighth and fourteenth rolls ranged from 0.2 to 1.9 percent. The average increase was approximately 1.0 percent. All of the samples taken from the eighth roll exceeded the minimum 30 percent cake solids required for the test.

Three polymers were used during the testing. Based on jar tests conducted by the manufacturer, testing began using Percol 787, a dry polymer. The manufacturer later switched to Percol 789, also a dry polymer, to try and improve the quality of the filtrate. The average capture for the 787 and 789 runs were 95.4 percent and 95.5 percent, respectively. Additional jar testing by a polymer manufacturer indicated that Excel 100, an emulsion polymer, may perform well and perhaps at lower cost than the dry polymers. This polymer was used on the last day and did show improved capture over the dry polymers.

Centrifuge—Raw Sludge Thirteen runs were conducted on PS and TWAS blends ranging from 41/59 percent to 63/37 percent PS/TWAS on a mass basis. No runs were conducted on 100 percent primary sludge. Data collected for the raw sludge centrifuge runs is presented in Figure 2. The minimum desired performance requirements for raw sludge were 28 percent cake solids and 95 percent solids capture based on TSS measurements. The centrifuge exceeded the minimum requirements on six of the thirteen runs. One of the runs not meeting the requirements (C-2-C) was conducted to determine the maximum throughput of the centrifuge. This run had a solids loading of 635 kg/hr, capture of 85.6 percent and cake solids were 27.5 percent. Some of the earlier (1/23/93) failing runs had difficulty meeting capture using the polymers available at the time. Later runs using different polymers showed better results.

The raw sludge runs considered in developing performance characteristics were C-1-A, B, C, and D and C-4-A and B. Other runs were used to draw conclusions regarding throughput versus cake solids trends. For the C-1 runs, polymer consumption ranged from 4.2 to 6.8 kg/dry Mg or $47.86 to $76.74 per dry Mg using C9490, an emulsion polymer. The C-4 runs had somewhat higher polymer consumption but used a less expensive emulsion polymer (SD2081). All of the C-4 runs achieved excellent capture, but only two (C-4-A and B) achieved cake solids. Polymer use for the two successful runs was 8.2 and 6.1 kg/dry Mg or $58.63 and $43.69 per dry Mg. Cake solids for these runs was 28.1 and 29.9 percent. Several dry polymers were jar tested but screened out in favor of the emulsion polymers above. Based on a comparison between the C-1 and C-4 runs, C9490 was not as effective as SD2081. The centrifuge backdrive was also experiencing problems during the C-1 runs which may have further exacerbated the capture performance. All C-4 runs easily met the capture requirements and were conducted with the centrifuge in good working order.

In comparing the C-4 runs (A-D), the threshold polymer dose to maintain cake solids is illustrated as the polymer dose is lowered from 6.1 to 4.9 kg/dry Mg

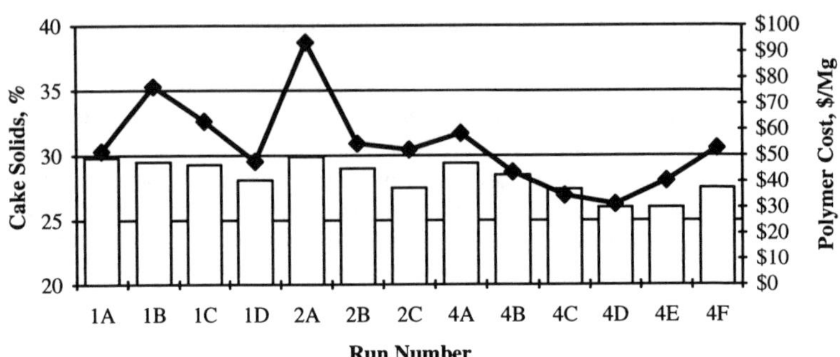

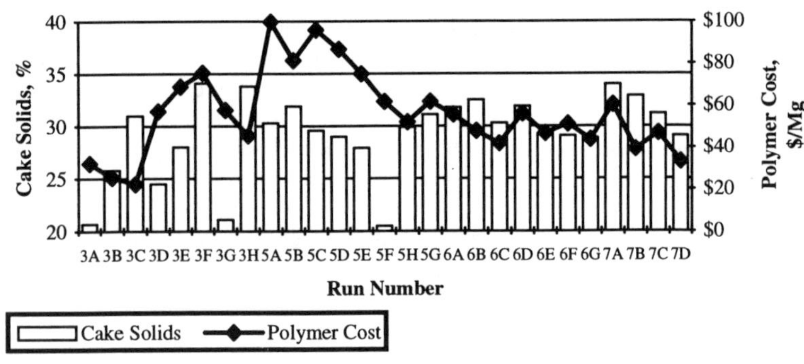

Figure 2. Centrifuge pilot testing results.

($43.69 to $34.69/dry Mg). Interpolation between these points yields 5.6 kg/dry Mg ($40.08/dry Mg) to achieve the minimum 28 percent solids.

Based on the data, an appropriate nominal throughput for the test centrifuge is 408 kg/hr at a minimum of 6 percent raw feed solids. Throughput tests were conducted during the C-4 round up to a solids loading of 603 kg/hr. Capture requirements were easily met; however, cake solids did not meet the minimum 28 percent specified. The cake solids achieved at this throughput were 26.0 and 27.5 percent.

Centrifuge–Lime Sludge Twenty-seven runs at varying lime doses were conducted on PS and TWAS blends ranging from 54/46 percent to 63/37 percent PS/TWAS on a mass basis. Nine of the twenty-seven runs conducted met or exceeded the desired cake solids and capture requirements of 30 and 95 percent. Initially, the polymer injection point was located several meters from the centrifuge inlet to ensure adequate mixing was achieved. In later runs, the injection point was moved to immediately ahead of the feed port and cake solids improved dramatically. Eight of the nine runs that achieved desired performance were conducted using Excel 100 polymer. Data collected for the lime sludge centrifuge runs is shown in Figure 2.

Polymer consumption for the runs identified above that achieved desired performance averaged 6.4 kg/dry Mg and ranged between 4.9 and 7.8 kg/dry Mg. The cost for polymer ranged between $39.20 and $61.82 per dry Mg and averaged $51.25 per dry Mg. During these runs, capture varied between 95 and 99 percent. Cake solids ranged between 30.8 and 34.0 percent with an average value of 32.0 percent. Some correlation was evident between lime dose and cake solids. Throughput for the centrifuge ranged from 401 to 590 kg/hour based on raw sludge. Throughput based on lime conditioned sludge ranged from 456 to 635 kg/hr. In this range, no correlation was observed between cake solids and throughput. In looking at the entire data set, it is also difficult to draw any conclusions regarding cake solids and throughput. A strong relation was observed between percent torque and cake solids.

Recycle Characteristics

Early consideration of pre- versus post-dewatering lime stabilization identified resolubilization of BOD as a concern associated with the pre-dewatering (liquid phase) method of lime stabilization. As a result, filtrate and centrate recycle streams were occasionally monitored for BOD and SBOD in addition to TSS. Limited data indicated BOD:TSS ratios of approximately 1.0 for "pre-dewatering" lime stabilization, and 0.43 for "post-dewatering" lime stabilization.

Flocculation Kinetics Using Fe(III) Coagulant to Coagulate Kaolin Clay in Water: Effects of Temperature

Lim-Seok Kang[1] and John L Cleasby[2]

ABSTRACT

Flocculation kinetics using ferric nitrate as a coagulant to coagulate kaolin clay in water was examined using several experimental factors. Both the particle size distribution data obtained from Automatic Image Analysis (AIA) system and the measurement of the degree of turbidity fluctuation in a flowing suspension by Photometic Dispersion Analyzer (PDA) were used to measure flocculation kinetics. Cold water temperature had a pronounced detrimental effect on flocculation kinetics when using Fe(III) coagulant. For improving flocculation kinetics at low water temperature, maintaining constant pOH was found to be partially effective only in the more acidic pH range studied (pH 6.8).

INTRODUCTION

The study of flocculation kinetics is of fundamental interest in the field of water treatment. However, there has been little work done concerning direct measurement of flocculation kinetics spanning the full range of coagulation domains encountered in public water treatment. This research focuses on the use of Fe(III) salt to make quantitative studies of the rate at which flocculation processes occur, the factors on which these rates depend, and the mechanisms involved. Fe(III) salts are receiving attention as alternative coagulants to alum for several important reasons; (1) concerns about aluminum concentration in treated waters, (2) possible cost savings comparing to other alternative coagulants, and (3) better coagulation efficiency at low water temperature and more efficient removal of color-causing organic materials than alum.

Although the treatment of cold water is encountered in the temperate regions of the world, published information on the impact of cold water temperature on flocculation is both scarce and contradictory. In particular, very little data are available on flocculation kinetics at low temperature covering the adsorption-destabilization (A/D) and sweep floc coagulation domains.

---

[1]Graduate Research Assistant, Ph.D. Candidate.

[2]Anson Marston Distinguish Professor Emeritus,
Dept. of Civil and Construction Engrg., Iowa State University, Ames, IA 50011

The overall objectives of this research are (1) to investigate the kinetics of flocculating kaolin clay in water suspension using ferric nitrate as a coagulant under a number of treatment conditions, (2) to assess the effects of temperature on flocculation kinetics, spanning the full range of coagulation domains including the A/D and sweep floc mechanisms of coagulation.

EXPERIMENTAL METHODS

A summary of experimental conditions is given in Table 1. A stock clay suspension of 800 mg/L was prepared in a 60 L plastic tank equipped with a mixing device and brought to pH 7.0 with 0.1 N NaOH. The stock clay suspension was added to the dilution water to achieve the desired clay concentration. Dilution water was prepared by adding the desired amount of 1 M $NaNO_3$ to distilled water to achieve 0.005 M ionic strength. The ionic strength was selected because there was no sign of coagulation of clay particles by double layer compression and it resulted in consistent measures of zeta potential (ZP). Ferric nitrate was used as a coagulant to allow sulfate concentration to be controlled as an independent variable. The pH of 0.25 M stock coagulant and 10 mg/mL of dosing solution, expressed as $Fe(NO_3)_3.9H_2O$, was checked periodically to ensure consistent speciation in the coagulant.

The flocculation tests were conducted in an 18 L square batch reactor system which was equipped with a two-blade turbine impeller, electric motor and speed controller, and a tachometer. The stirring power input to the reactor was determined experimentally by direct measure of the torque on the impeller shaft at various rotational speeds. At the lower temperature, the impeller speed was maintained constant resulting constant power input but with a lower resulting root mean velocity gradient (G) value.

Table 1. Summary of experimental conditions

Parameter	Description
Primary particle	Kaolin clay (1.8 μm avg. equiv. circular dia.)
Particle concentration	5, 25, 50, and 200 mg/L of kaolin clay
Dilution water	distilled water with 0.005 M $NaNO_3$
Coagulant	dosing solution of 10 mg/mL as ferric nitrate
Temperature	room temperature (23 ± 0.5 °C) and 5 °C
Rapid mixing	250 rpm (G = 455 sec^{-1} at 23 °C) for 1 or 2 min
Slow mixing	30 and 60 rpm (G = 20 and 56 sec^{-1} at 23 °C)
Suspension pH	6.0, 6.5, 6.8, 8.0 selected from experiments

The kinetics of flocculation was monitored by using two-sophisticated systems. For the first method, the flocculation kinetics was quantified by measuring particle size distribution and the rate of disappearance of primary particles with time during the flocculation process utilizing the AIA system. The second method was done by monitoring fluctuations in the intensity of light beam transmitted through a flowing suspension using the PDA. Using the PDA, the ratio value of the root mean square value of the fluctuating signal (Vrms) to the voltage corresponding to the mean transmitted light intensity (V) has been found to increase substantially as particle aggregation occurs, and provided a sensitive measure of the extent of flocculation.

The experimental setup is shown schematically in Fig.1. During the flocculation processes, a sample stream was withdrawn at a rate of 15 mL/min through a glass tube of 2.5 mm internal diameter by a peristaltic pump. The glass tube was passed through a cell of the PDA and the output signal of the PDA was monitored directly by a personal computer based data acquisition system. Also, during the flocculation period, 1 mL of samples were collected at each different mixing time using 1 mL syringe with a needle through a sampling port for the analysis of the particle size distribution using the AIA.

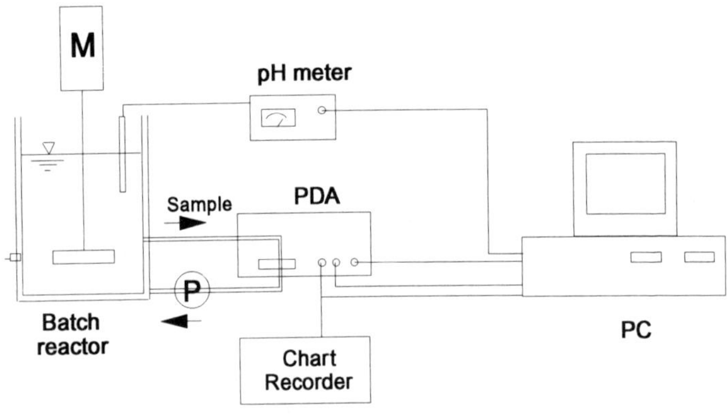

Figure 1. Schematics of Experimental Setup

RESULTS

Figures 2(a) and (b) show the dramatic impact of suspension pH on the rate of flocculation, as evidenced by both the decreases of total particle count fraction and marked increase of Vrms/V (called the flocculation index), respectively. In the x-axis, the beginning of the rapid mix cycle appears as a negative time, and the time zero is the time at which the flocculation was initiated immediately after rapid mixing. Overall, with the same coagulant dose, the curves in Fig.2 show marked increases in the rate of flocculation as the pH increases. The data shown in Fig.2 clearly demonstrates the usefulness of both instruments in monitoring flocculation kinetics, if we define flocculation kinetics as the disappearance of primary particles and the formation of aggregates as a function of mixing time.

a

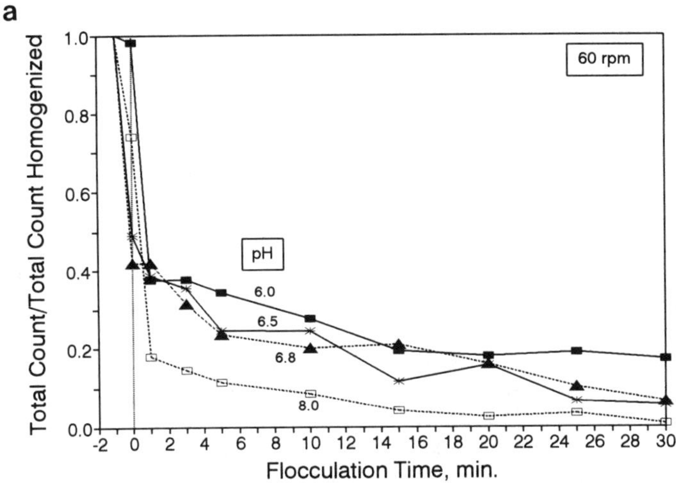

b

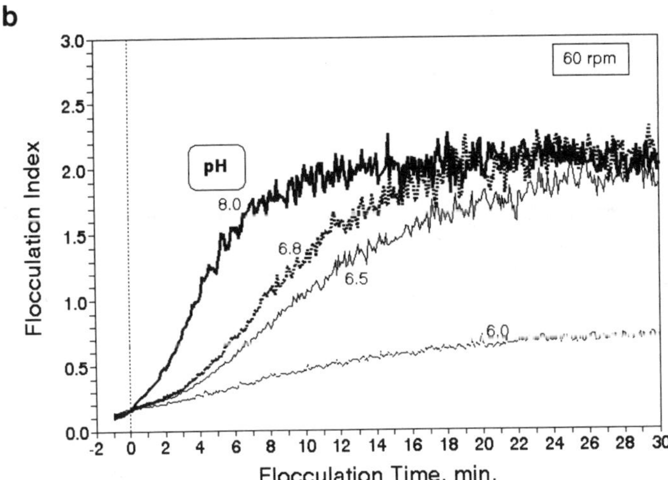

Figure 2. Effect of pH on The Rate of Flocculation (Clay: 25 mg/L, Dose: 5 mg/L as $Fe(NO_3)_3.9H_2O$, Flocculation: 60 rpm, at room temperature). (a) Total Particle Count Fraction vs. Time Using The AIA; (b) Flocculation Index vs. Time Using The PDA

Figures 3 and 4 show the experimental results for flocculation kinetics at 23 ^{o}C versus 5 ^{o}C by presenting the effect of altering system chemistry at 5 ^{o}C by maintaining constant pOH, and the effect of an extended rapid mixing period (Fig. 3b and 4b only). The figures include baseline data at pH 6.8 and 8.0 at 23 ^{o}C, respectively compared with data collected at 5 ^{o}C. Maintaining a constant pOH at 5 ^{o}C results in different pH values at different temperatures due to a change in the ion product of water (pK_w). For example, the pH of 6.8 at 23 ^{o}C results in a pH of 7.5 at 5 ^{o}C to maintain constant pOH at the two temperatures. When comparing flocculation kinetics as evidenced by the rate of disappearance of primary particles at a baseline pH 6.8, as shown in Fig.3(a), the low water temperature did not impair the kinetics of flocculation substantially when holding pOH constant. However, Fig.3(b) indicates a significant detrimental effect on flocculation kinetics at low temperature, even when maintaining constant pOH. At pH 8.0, as shown in Fig.4(a) and (b), it is noted that altering system chemistry was not capable of improving flocculation kinetics at low temperature.

Comparison of the results at pH 8.0 with one at pH 6.8 shows different effect of using constant pOH at low temperature on flocculation kinetics. Thus, holding constant pOH at low temperature instead of using constant pH was favorable for the aggregation of particles when flocculation process was accomplished below a certain pH level, in the more acidic pH range. Also, the detrimental impact of low water temperature was substantially reduced by extending the rapid mixing to 2 minutes. The ZP values for each curve show that Fe(III) coagulant at lower temperature was much more effective at charge neutralization of clay particles than at higher temperature.

The results obtained under the same experimental conditions, except using 50 mg/L of clay, are plotted at pH 6.8 and 8.0 in Figures 5(a) and (b), respectively. As shown in these figures, again, low temperature had a pronounced detrimental effect on flocculation kinetics at both pH 6.8 and 8.0. Also, under this higher particle concentration, it is apparent at the more acidic pH that there was only a slight benefit for holding pOH constant at the lower temperature. Furthermore, at the higher pH, maintaining constant pOH at cold temperature was detrimental.

CONCLUSIONS

1. Both the particle size distribution data obtained from the AIA and the on-line measurement of turbidity fluctuation by the PDA provided reliable and sensitive indications of flocculation kinetics.
2. Low water temperature had the pronounced effects on flocculation kinetics, slowing the rate of flocculation and enhancing charge neutralizing ability of Fe(III) coagulant.
3. The use of constant pOH at 5 ^{o}C was found to be partially effective for reducing the impact of low temperature on flocculation kinetics, but only in the acidic pH range studied (pH 6.8). In addition, the appropriate use of a more extended rapid mixing period was found to offset the detrimental effect of low temperature on flocculation kinetics.

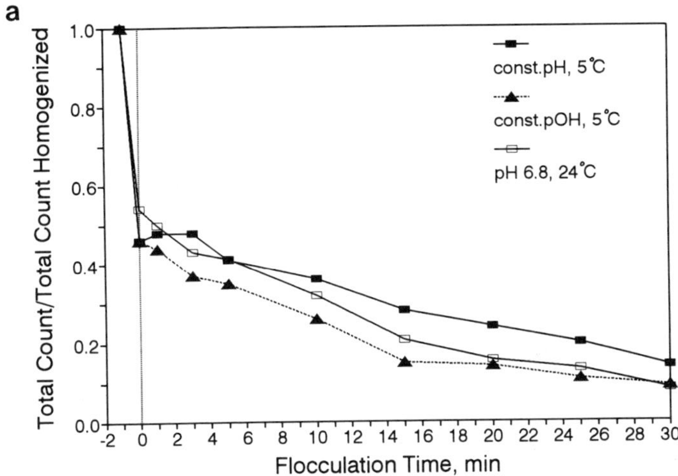

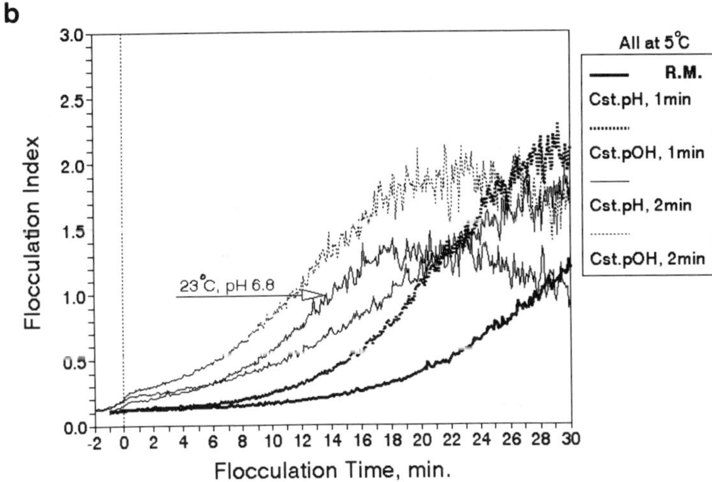

Figure 3. Effect of Temperature and System Chemistry on The Rate of Flocculation (Baseline 23 °C at pH 6.8, Other Curves at 5 °C with Either Const. pH (6.8) or Const. pOH (i.e., pH 7.5), Clay: 25 mg/L, Dose: 5 mg/L as $Fe(NO_3)_3.9H_2O$, Flocculation: 30 rpm). ZP(mV): -8.1 at 23 °C, +16.5 at 5 °C, +14.2 at const. pOH, (a) Total Particle Count Fraction vs. Time; (b) Flocculation Index vs. Time

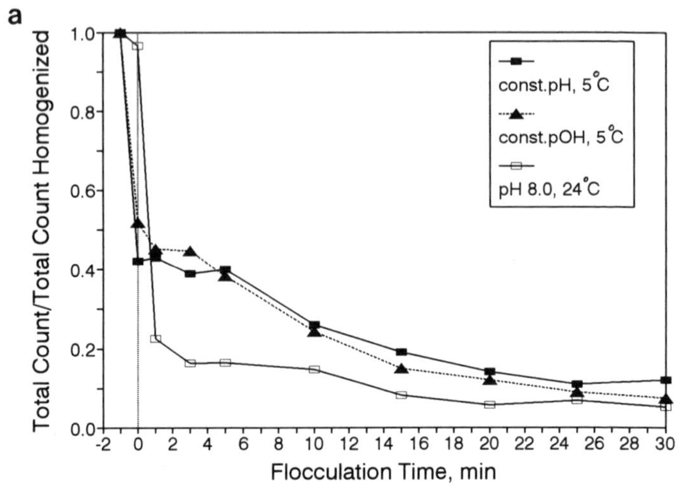

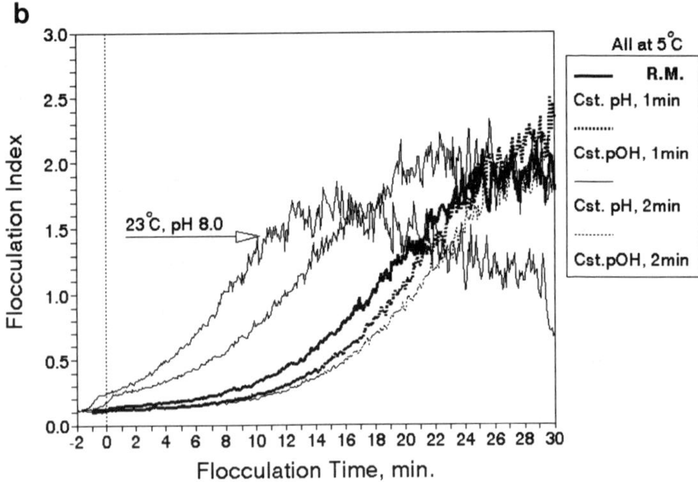

Figure 4. Effect of Temperature and System Chemistry on The Rate of Flocculation (Baseline 23 °C at pH 8.0, Other Curves at 5 °C with Either Const. pH (8.0) or Const. pOH (i.e., pH 8.7), Clay: 25 mg/L, Dose: 5 mg/L as $Fe(NO_3)_3.9H_2O$, Flocculation: 30 rpm). ZP(mV): -11.8 at 23 °C, +11.5 at 5 °C, +9.9 at const. pOH, (a) Total Particle Count Fraction vs. Time; (b) Flocculation Index vs. Time

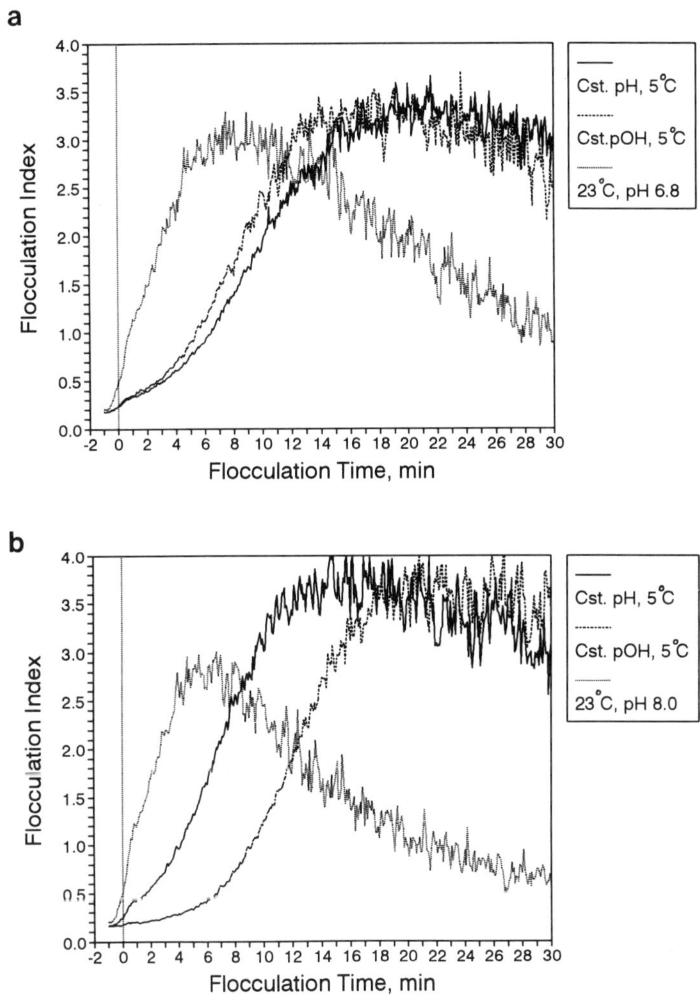

Figure 5. Effect of Temperature and System Chemistry on The Rate of Flocculation, (Clay: 50 mg/L, Dose: 5 mg/L as $Fe(NO_3)_3.9H_2O$, Flocculation: 30 rpm), (a) Baseline 23 °C at pH 6.8, Other Curves at 5 °C with Either Const. pH (6.8) or Const. pOH (i.e., pH 7.5); (b) Baseline 23 °C at pH 8.0, Other Curves at 5 °C with Either Const. pH (8.0) or Const. pOH (i.e., pH 8.7)

Assessing Roughing Filtration Design Variables

Jennifer O. Cole, Catherine Westersund and M. Robin Collins[1]

Introduction

Many small rural communities are affected by the Surface Water Treatment Rule (Federal Register, 1989) and are looking for inexpensive filtration options which are easy to operate and maintain. There are a number of filtration options available to these communities including slow sand, diatomaceous earth, and membrane filtration. Slow sand filtration is especially suited for small communities because the technology is effective, has low operation costs, does not use expensive chemicals or have extensive sludge disposal requirements, and is passive meaning plant operation is self-implementing (AWWA, 1991). However, like membrane filtration and diatomaceous earth, slow sand filters require a high quality raw water source with turbidity levels <10 NTU (Hendricks, 1991) and algae concentrations, measured as chlorophyll-a concentration, <5 mg/m^3 (Cleasby, 1991) to be operated economically. Surface waters are subject to storm runoff and algal blooms suggesting a consistent high raw water quality is difficult to maintain in many source waters. Pretreatment to reduce particle load in the influent stream would increase the application of slow sand filters to many locations previously thought unapplicable to alternative filtration technology.

Roughing filters are a simple, inexpensive pretreatment method which have been successfully used prior to slow sand filters in developing countries (Wegelin, et al, 1986). A roughing filter consists of a filter box divided into compartments (usually three), each filled with progressively smaller gravel media and may be operated horizontally, upflow or downflow. The advantages of using roughing filters for pretreatment over other available methods are their ease of

[1] Department of Civil Engineering, University of New Hampshire, Durham, New Hampshire 03824

operation, minimal maintenance requirements, and lack of chemical dependence. Roughing filters are on a technological level similar to slow sand filtration.

Detailed design information is not available for roughing filters because of their limited exposure, primarily to developing countries. Information about particulate removal by roughing filters and expected treatment performance of slow sand filters following roughing filter pretreatment is also limited.

The main research objective of this presentation was to develop a mathematical model which considered the relative significance of the three major filter design variables (length, hydraulic loading rate and media size) and particle characteristics (as quantified by relative settling velocity) on steady-state treatment performance of gravel roughing filters.

Methods and Materials

Three different raw waters were investigated in the filter element study. The first consisted of deionized water containing 1000 mg/L of kaolinite clay adjusted to a neutral pH and having an ionic strength of 0.003 M in the form of NaCl. The second raw water investigated was deionized water containing *Scenedesmus* algae, which was applied to the filters at an average rate of 1500 μg/L total chlorophyll. The algae was grown at 25°C in 20-liter clear carboys containing a modified B & G media (Jahnke, 1991), for five days prior to being applied to the filters. The third raw water was the same clay solution, but applied to the filters ripened by the application of algae for five days.

Orthogonal filter studies were performed to statistically evaluate the influence of the design variables and raw water quality on particulate removal by downflow roughing filters. The first kaolinite removal study (K1), consisted of an L27 experimental design (Ross, 1988). Table 1 describes the three factors investigated and the number of levels examined for each factor. The algae study (A1) and the kaolinite after algae pretreatment study (K2), consisted of an L18 experimental design (Ross, 1988). Based on the results of the K1 study, flow rate was only examined at two levels for the A1 and the K2 studies as shown in Table 1.

Table 1. Roughing Filter Design Parameters During Kaolinite Removal Study

Factor Levels	Average Gravel Media Size, mm	Hydraulic Loading Rate, m/hr	Filter Depth, cm	Study Evaluated
1	2.68	0.5	30	K1,A1,K2
2	5.50	0.75	60	K1
3	7.94	1.0	90	K1,A1,K2

The filters used were 90 cm high PVC columns. To minimize sidewall effects, 15-cm and 20-cm diameter columns were utilized with three to four rods 0.6-cm in diameter glued to the interior side wall. All filters were loaded by gravity from a constant head tank in a downflow direction. Peristaltic pumps were located on the effluent lines to maintain a constant flow rate throughout the filter run. Sampling ports were located at each 30 cm increment in the filters to measure filter resistance and to sample the filtrate quality as a function of depth. Peristaltic pumps were used in continuous operation at the sampling ports to maintain a constant combined filtrate flow of less than 10% of the overall flow to prevent sedimentation around the sample ports.

The significance of each design variable was determined for steady state treatment conditions for each particle type through statistical ANOVA analysis. The percent contribution to the overall treatment variability, and its polynomial decomposition, was determined for each of the design variables to facilitate model development.

Results and Discussion

Roughing filter performance was assessed in two different ways; steady state treatment performance (Ce/Co) and ultimate filter load (σu). Steady state Ce/Co was determined from the mathematical average of the data points comprising the flat line portion of a plot of filter coefficient (λ) or Ce/Co verses σ for each experimental trial as shown in Figure 1. Ultimate filter load was reached when steady-state Ce/Co for both TSS and turbidity were no longer being achieved. The flat line portion of these curves has been used by other researchers to determine steady state conditions (Wegelin, 1986). Steady state treatment performance (particle removal) was measured by turbidity, total suspended solids (TSS), particle counts and total chlorophyll. The data statistic used to develop models for treatment performance in this discussion was from TSS data.

Kaolinite. A twenty-seven trial orthogonal array was performed to statistically determine the influence of the three principal design variables: media size, hydraulic loading rate and depth (length), each investigated at three levels, on steady state treatment of kaolin clay using clean media. Data trends show increasing treatment performance for decreasing media size, decreasing hydraulic loading rate and increasing depth. These trends were expected and have been reported by other researchers (Westersund, 1993; Pardón, 1989; Wegelin, 1986).

An analysis of variance (ANOVA) was performed on the steady state Ce/Co data generated for each trial run and is summarized in Table 2. The low error term indicates all important factors were considered and interactions between the principal parameters did not contribute significantly to explaining treatment performance.

The principal design variables were all significant factors in explaining experimental variability of roughing filter steady state treatment performance at <p=0.99>. Based on percent contribution to variability, filter length was roughly three times more influential than media size and roughly five times more influential than hydraulic loading rate. All design variables followed a linear rather than a quadratic trend.

Table 2. Summary of Filter Design Variable Contribution to Particle Removals

Particle Type	Gravel Media Condition	Percent Contribution to Steady-State Treatment			
		Gravel Size	Loading Rate	Filter Length	% Error
Kaolin Clay	Clean	21.6	12.7	60.8	4.9
Scenedesmus Algae	Clean	28.7	60.8	3.6	6.9
Kaolin Clay	Algae Ripened	38.4	1.6	46.8	13.2

Algae. The principal design variables were investigated in an eighteen trial orthogonal array for algae steady state treatment performance. Hydraulic loading rate was investigated at two levels and media size and depth were investigated at three levels. Results from an ANOVA analysis showed all three design variables being significant at <p=0.99>. Quantitatively, hydraulic loading rate was twice as influential as media size and nearly 17 times as influential as filter depth in explaining variability in roughing filter performance. The influence of the design variables was opposite of the K1 (kaolinite) study where length was much more important in determining steady state treatment performance than hydraulic loading rate. A breakdown of the parameters into quadratic and linear components showed media size followed a quadratic trend and filter length showed a linear trend. Hydraulic loading rate was evaluated at two levels, so quadratic effects could not be determined.

Kaolinite on Algae-Ripened Media. Kaolinite was applied to the filters after algae had been applied, so the same eighteen trial orthogonal array was performed for the K2 study. Results from an ANOVA analysis showed all principal design variables were significant factors <p=0.99> in treatment performance. Based on percent contribution, length was nearly twice as influential as media size and three times as influential as hydraulic loading rate on treatment performance. All design variables had a linear effect of treatment performance.

It is important to note that the steady state treatment performance for the K2 study was much better than for the K1 study. The percentage of particle removed from suspension in each filter section was higher for the algae-ripened media than the clean media, meaning the algae coated media removed more stable particles than the clean media.

In summary, the relative importance of the individual design variables varied with the characteristics of the influent particulate matter, but also varied with the surface characteristics of the gravel media, as shown in Table 2. All particle transport mechanisms were assumed to be the same between the two kaolinite studies so increased removals can be attributed to favorable attachment mechanisms. Previous studies did not show a significant treatment difference between various types of media, which had differing surface characteristics (Wegelin, et al, 1986); however, the media types investigated where all inorganic in nature.

Modeling. A multi-variable regression analysis which considered the linear and quadratic effects of each design variable was performed on the data statistics for each of the three particle types investigated. The multi-variable regression models for each particle type are shown in Figure 2. The data points are scattered around the line representing perfect agreement between actual and predicted treatment performance. As expected, the deeper filters and the smaller media achieved higher treatment performance.

The multi-variable regression model developed for kaolinite removal on clean media was compared to the multilinear regression model developed by Wegelin, et al. (1986) as shown in Figure 3. Wegelin's model was developed using kaolinite and predicts the removal ratio for a specific particle size. The differences in treatment performance predicted between these two models may be accounted for in a number of ways. Wegelin's model was developed for a horizontal roughing filter and the model described herein was developed for a vertical roughing filter. The average size of kaolinite particle used in the Wegelin study to develop that model was 2.8 μm compared to an average particle size of 1.4 μm for this study. Perikinetic flocculation may have contributed to treatment performance in the smaller media sizes in this study, as evidence by cake development observed on top of the media at the end of a filter run. Also, the Ce/Co values reported for this study represent all particle sizes, not just the average 1.4 μm particles.

The multi-variable regression models adequately predict the treatment performance for the specific particle type they were developed from, but are not applicable to other particle types that have different characteristics. A universal model, which considers the type of particle to be removed, will be able to predict treatment performance for different types of particles. Settling velocity was the method chosen to differentiate between the three particle types used in this study because of the varying size and density of the particles investigated and because sedimentation is considered the primary transport mechanism in gravel roughing filters (Boller, 1993).

The first step in developing this universal model was to normalize the settling velocity of the individual particle to kaolinite. The settling velocity for algae was determined from Stoke's Law, and the settling velocities for kaolinite and a combined kaolinite and algae particle (representing the kaolinite on algae ripened media) were experimentally determined in the laboratory.

Plots were developed for each design variable which related the removal ratio (Ce/Co) to the settling velocity ratio (Vs/Vs kaolinite). Compressing each design variable into one line as shown in Figure 4 yields a graph that can be used to determine the removal ratio for any design variable when the specifics of the design variable (size, depth, hydraulic loading rate) are known. This is accomplished by determining the y-coordinate (Ce/Co)/(Vs/Vs Kaolinite), and multiplying this value by the settling velocity of the particle type being removed to determine the removal ratio for each design variable. The removal ratio (Ce/Co) values for each design variable are added or subtracted together to determine the expected treatment performance for the specific roughing filter configuration being investigated. A comparison of predicted versus actual removals for all particle types using this unifying method is depicted in Figure 5. The general trend for all three design variables was better steady state treatment performance for particles with higher settling velocities. Unfortunately the model predictions were not as close to the actual data as individual particle-based models as shown in Figure 2.

A multi-linear regression analysis using 18 steady state values for each particle type was developed with particle settling velocity considered one of the regression variables. This equation does not consider the quadratic effect of media size on algae removals, but still has a high correlation coefficient ($r = 0.9088$). This model has more data scatter than the individual models developed for each particle type which is expected as more variables are combined into the overall model and the quadratic effects of media size on algae removals were neglected.

Summary and Conclusions

Orthogonally designed pilot filtration studies were successfully used to (i) predict steady state treatment performance and (ii) determine the relative removal contribution of each design variable and their linear and quadratic trends. A universal multi-regression model, incorporating the results from all three particle type experiments, was moderately successful in predicting steady state treatment performance by using settling velocity to differentiate influent particle type.

Acknowledgements

The authors are grateful to the AWWA Research Foundation for funding and specifically to Elizabeth Kawczynski, Dan Dorlack and Kim Hout who served as Project Managers. The authors also acknowledge Joe Roccaro and officials from UNH/Durham, Portsmouth (NH) and Gulf Coast Water Authority (Texas City, TX) for their assistance.

References:

Boller, M. (1993), "Filter mechanisms in roughing filters," *Aqua*, 42(3), 174-185.

Cleasby, J. (1991), "Practical limits of raw water quality for slow sand filters", in Logsdon, G. (ed.) "Slow sand filtration: Manual of praactice", American Society of Civil Engineers, New York.

Cole, J.O. (1993), "Model development and verification of particulate removal by downflow roughing filters", MS thesis, University of New Hampshire, Durham, NH.

Federal Register (1989), "EPA final rule for filtration, disinfection, turbidity, giardia lamblia, viruses legionella and heterotrophic bacteria", 54(124).

Hendricks, D. (ed.) (1991), "Manual of design for slow sand filtration", AWWARF and AWWA, Denver, CO.

Iwasaki, T. (1939) "Some notes on sand filtration", *JAWWA*, **29**:1591.

Jahnke, L. (1991), Personal communication, Department of Botany and Plant Pathology, University of New Hampshire, Durham, NH.

Pardón, M.O. (1991), "Removal efficiency of particulate matter through vertical flow roughing filters", paper presented at the Slow Sand Filtration Workshop, October 27-30, 1991, University of New Hampshire, Durham, NH: AWWA.

Ross, P.J. (1988), "Taguchi techniques for quality engineering", McGraw-Hill, Inc., New York, NY.

Wegelin, M., Boller, M. and Schertenleib, R. (1986), "Particle removal by horizontal-flow roughing filtration", *Aqua*, 3:115-12.

Westersund, K. (1993), "The impact of raw water characteristics on roughing filtration design variables", MS thesis, University of New Hampshire, Durham, NH.

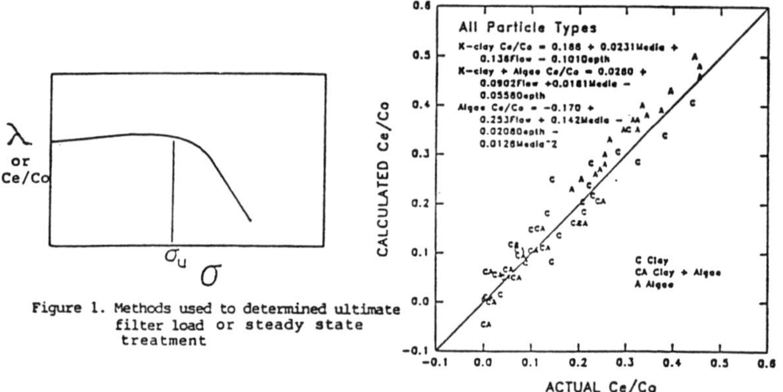

Figure 1. Methods used to determined ultimate filter load or steady state treatment

Figure 2. Performance of multilinear regression models based on particle type

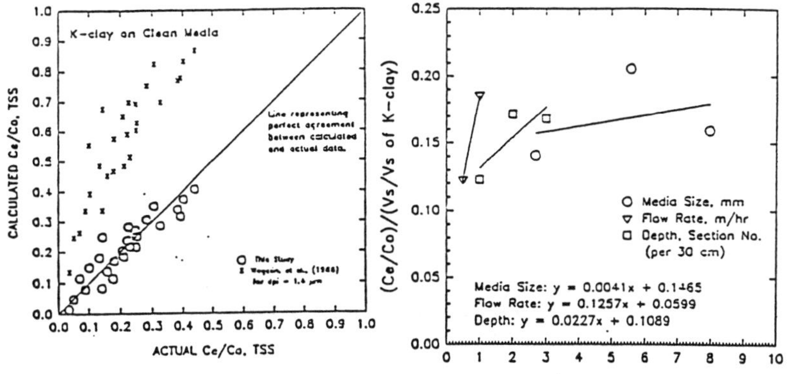

Figure 3. Comparison between roughing filtration models

Figure 4. Removal ratio as a function of settling velocity and design variable

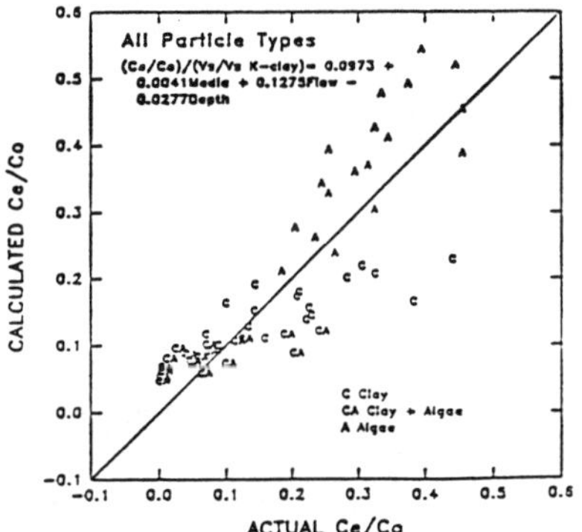

Figure 5. Performance of an universal multi-linear regression model independent of particle type

Filtration and Backwashing Performance of Biologically-Active Filters

A. Amirtharajah[1], Rasheed Ahmad[2],
A. Al-Shawwa[3] and P.M. Huck[4]

Abstract

Before widespread use can be made of biological filters, it is crucial to develop an understanding of the filtration and backwashing performance of these filters. Experimental results indicated that the biological filter impairs microbial quality, reduces filter runs, and produces higher peaks during filter ripening. Anthracite/sand and GAC/sand biological filters were equally good in producing low assimilable carbon (AOC) waters but the latter media generated a higher head loss.

Introduction

Biological filtration may become an essential part of drinking water treatment in the United States during the next several decades. Biological drinking water treatment indirectly reduces disinfection by-products (DBPs) by lowering disinfection requirements. Biological processes can also directly reduce DBPs by removing precursor material, regulated contaminants, preoxidant by-products, certain synthetic organic compounds and taste and odor compounds.

There is now a strong interest in biological processes in North America stemming from recent or impending regulations on both coliforms and disinfection by-products. This interest also results from the recognition that when ozone is

---

[1]Professor and [2]Graduate Research Assistant, School of Civil and Environmental Engineering, Georgia Institute of Technology, Atlanta, GA 30332
[3]Graduate Research Assistant and [4]Professor, Dept. of Civil Engineering, University of Alberta, Edmonton, Alberta T6G2G7, Canada

introduced, biological activity will occur quite often in the distribution system whether it is a specific design objective or not. By operation of filters in the biological mode, it is possible to obtain a water with low assimilable organic carbon (AOC) and hence prevent significant growth of biofilms in the distribution system.

Before there is widespread use of these filters, it is crucial to develop an understanding of the filtration and backwashing performance of biologically active filters. The possible quality degradation during the initial stages of biological filtration in comparison with conventional filters, bacterial detachment mechanisms during filtration and backwashing, backwash strategies for biological filters, the impact of media selection strategies to optimize operation of biological filters in terms of head loss development and filtered water quality are some of the major issues that need to be addressed.

A Fundamental Theoretical Approach

A recent study (Ginn et al., 1992) has indicated that the detachment mechanism needs to be included in any realistic approach towards explaining filtration characteristics using particle size distributions. In addition, only in recent times has the process of backwashing been analyzed in terms of the forces controlling attachment and detachment (Amirtharajah and Raveendran, 1993). Thus, the study of the detachment mechanisms is an essential component for fundamentally understanding filtration and backwashing.

Raveendran (1993) has recently completed a fundamental microscopic study on detachment of particles from collectors which includes forces due to van der Waals' and electrostatic interactions (F_v and F_e), Born repulsion (F_b) and hydrophobic interactions (F_h). Since there is actual contact between interacting surfaces prior to detachment of particles it is necessary to include all four forces ($F_v + F_e + F_b + F_h$) during the initial stages of detachment. For detachment to occur the hydrodynamic forces must exceed the sum of these forces. The equations used to calculate the Born repulsion and hydrophobic forces for a sphere-plane interaction are as follows:

$$F_b = -\frac{dE_b}{dz} = -\frac{Aa\sigma^6}{180z^8} \quad (1)$$

in which E_b = Born repulsion energy; z = separation distance; A = Hamaker constant; a = particle radius; σ = collision diameter.

$$F_h = -\frac{dE_h}{dz} = -2\pi aKh \exp(-z/h) \quad (2)$$

in which E_h = hydrophobic interaction energy; h, K = empirical constants.

Fig. 1 shows calculations with two, three and four forces for 5 μm particles at particular conditions of zeta potential and ionic strength. The hydrodynamic forces may then be compared with these forces for the detachment mechanism. Raveendran (1993) showed that inclusion of all four forces is necessary to consistently explain the results of particle attachment and detachment under changing chemistries of the suspending liquid (pH, ionic strength).

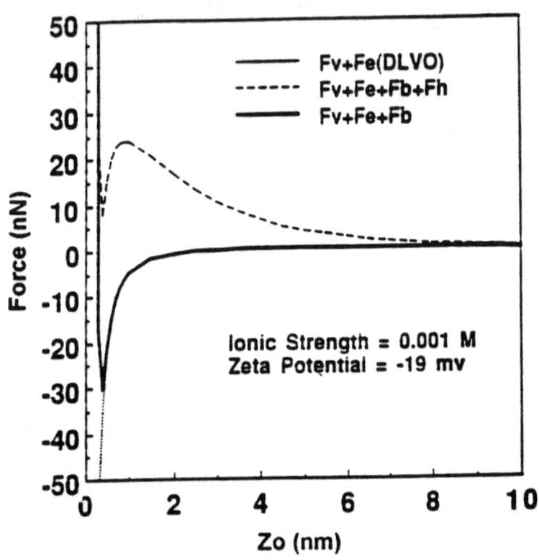

Figure 1. Surface Interactive Forces Between a 5 um Particle and Flat Plate

Bacteria have diverse morphological features in comparison to nonbiological particles and differ markedly in their extent of transport because of differences in cell properties, such as electrophoretic mobility, cell size, and presence of capsules and appendages. Their metabolic activity also can alter the chemical character of the cell surface as well as surrounding media (Martin et al., 1992). The extent to which hydrophobicity influences bacterial adhesion in comparison with other important factors, such as zeta potentials, absence or presence of surface appendages, or production of biosurfactants by adhering cells, is still a subject of research.

The Phenomenon of Filter Ripening

Amirtharajah and Wetstein, (1980) have shown that a large fraction of particles that pass through a well operated conventional filter do so during the initial stages of filtration. Coliform bacteria also tended to pass through the filter during this time period. The major mechanism causing this degradation is the "ripening phenomenon". The initial degradation and improvement phase always occurs in conventional filters. The possible quality degradation during the initial stages of biological filtration in comparison with conventional filters needs to be evaluated. A recent study (Goldgrabe et al., 1993) indicated that biological filtration did not significantly degrade effluent water quality. In recent reviews on biological processes for drinking water treatment (Bablon et al., 1988; Bouwer and Crowe, 1988), the importance of backwashing on long term performance of GAC contactors and GAC-sand filter adsorbers have been emphasized. A sensitive indicator of the relative efficiencies of various backwashing strategies is the headloss development in the following run (DiGiano, 1990).

Objectives

The objectives of the present study are: (1) To experimentally determine particle capture efficiency of traditional filters which are operated to include biological activity, (2) to determine the effluent quality in terms of microbial counts and organic carbon removal, (3) to determine the impact of backwashing on filtration performance, (4) to understand and control particle detachment to optimize the effluent.

Experimental Design

The majority of research for this project was conducted at a pilot plant located at the University of Alberta in Edmonton, Canada. The pilot plant had a common pretreatment section, consisting of screening, presedimentation, coagulation, flocculation, and sedimentation.

The pilot scale experimental program was divided into two phases. For Phase 1, the flow was split following sedimentation into three streams. Approximately half the flow was directed to an ozone contactor which fed three biologically active filters which utilized different flow rates and media. The other half of the flow was directed to two control streams, one of which was chlorinated. The other was not chlorinated and was intended as a control stream representing "natural" biological activity in the absence of ozonation. Most of the filtration and backwashing experiments were done in Phase 2 which had a slightly different configuration of the pilot plant.

The objective of the filtration performance experiments was to evaluate and compare turbidity removal, and consequently water quality for three different streams and two different filter media. HPC, AOC, and TOC were analyzed. Turbidity measurements were made by continuous monitoring of the effluent. Extensive sampling was done in the beginning of a filter run. Backwashing optimized with air scour was at collapse-pulsing rates.

At Georgia Institute of Technology, a laboratory scale biological filter was operated for some of the studies under this research project. The filter influent (nonchlorinated) was obtained from the nearby Hemphill Water Treatment Plant of the City of Atlanta, GA with the addition of a carbon source and other nutrients. A constant filtration rate of 5 m/hr (2 gpm/sq ft) and GAC/sand media were some details of the operation and characteristics of the laboratory scale filter.

Results and Discussion

The experimental results of filter performance suggested that there is not a substantial difference between biological and non-biological streams in terms of turbidity removal except that the pattern and initial peaks during filter ripening are quite different (Fig. 2). The initial peak was highest in the biological filter with GAC/sand media. The nonbiological filter (chlorinated stream - anthracite/ sand) had a dual peak characteristic. The start of breakthrough in biological filters occurred much earlier than in the conventional filter. However, there is no substantial and fast deterioration in effluent turbidity during the breakthrough phase. Nevertheless, filter runs beyond 24 hrs will impair the effluent quality. Another recent study (Goldgrabe et al., 1993) found that filter run times for biological filters were longer than 3 days. One of the major reasons for these long runs was the filtration rate which was 5 m/hr as opposed to 10 m/hr (4 gpm/sq ft) in the current study.

Biological filters with GAC/sand media had considerably higher headloss and headloss gradient in comparison to anthracite/sand filters whether they were biological or nonbiological (Fig. 3). This higher headloss gradient may be attributed to the microbial growth on a GAC bed and the increased removal efficiency of these filters for a short period (~ 6 hr). The effective size of GAC was also smaller than anthracite. Hence, the headloss is partly a result of smaller media size.

Bacterial counts in the effluent of biological filters were found to be about two orders of magnitude higher than that of nonbiological filters. A GAC-sand biological filter produces a poorer effluent in terms of bacterial counts in comparison to an anthracite/sand filter (Fig. 4). However, the effluent quality in terms of turbidity was better for a short period (< 6 hrs) prior to breakthrough.

In biological filters the effluent heterotrophic plate counts (HPC) were considerably higher than the influent HPC. In many cases, the difference of influent and effluent HPC was of two orders magnitude (Table 2).

The TOC removal was marginally better in a GAC-sand biological filter (Fig. 5). However, based on AOC results (Table 1), both GAC-sand and anthracite/ sand biological filters are equally good in producing low assimilable organic carbon. This has been observed by other researchers too. One of the reasons for this may be that the micropores of GAC are smaller than 1 μm. A biological filter operating at a hydraulic loading of 10 m/hr can still produce biostable water.

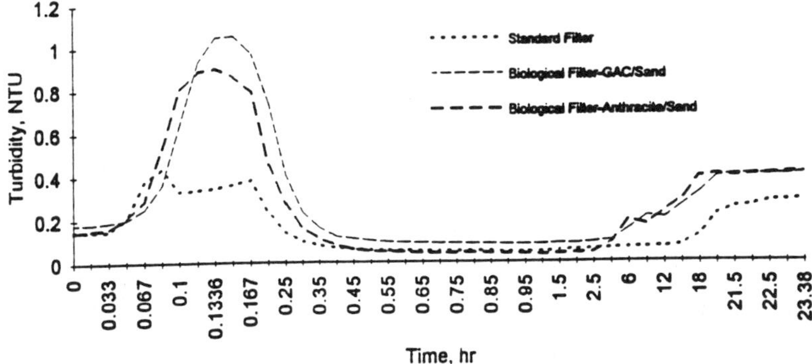

Figure 2. Filtration Performance of Different Streams

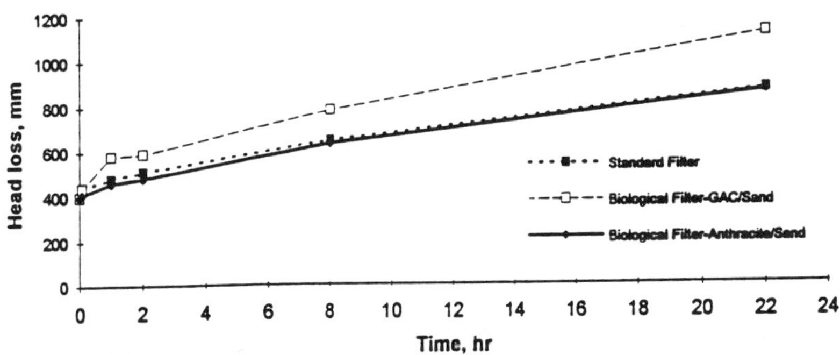

Figure 3. Head Loss in Different Streams During Filtration

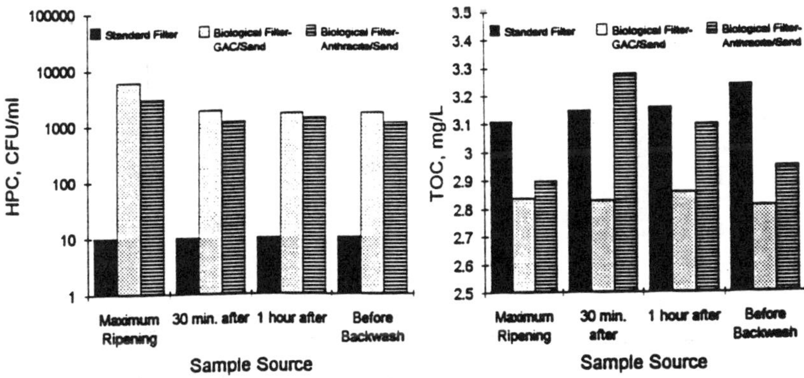

Figure 4. Microbial Counts During Filter Run

Figure 5. TOC During Filter Run

Table 1

AOC data during filter run

	Total AOC (μg C/L as Acetate)		
	Standand Filter	Biological Filter - GAC/Sand	Biological Filter - Anthracite/Sand
Before Backwash	-	9.0	7.0
At Maximum Ripening	-	20.0	6.0
30 min. after Ripening	-	6.0	-
1 hr after Max Ripening	-	-	-

Table 2

Heterotrophic plate count and turbidity results

	HPC* (CFU/mL)		Turbidity, NTU	
Date	Influent	Effluent	Influent	Effluent
11/11/93	80	1.1E+05	1.61	0.28
11/13/93	1.0E+02	1.0E+05	1.31	0.21
11/15/93	60	1.4E+04	1.45	0.30
11/17/93	40	1.2E+04	1.21	0.20
11/19/93	1.5E+02	1.8E+05	1.60	0.32
11/21/93	30	3.0E+04	1.70	0.18

*Heterotrophic Plate Counts (CFU/mL) at 22°C for 7 days - R2A agar - pour plate method.

Acknowledgments

The funding for this research was provided by American Water Works Association Research Foundation (AWWARF). The authors of this paper are thankful to the project team at the University of Alberta, Canada. The efforts of M. J. Mitton and S. Zhang for operation, sample and data collection and analysis are gratefully acknowledged.

References

Amirtharajah, A., and Wetstein (1980). Initial Degradation of Effluent Quality During Filtration. Jour. Am. Wat. Wrks. Assn., 72:9:518-524.

Amirtharajah, A., and P. Raveendran (1993). Detachment of Colloids from Sediments and Sand Grains. Colloids and Surfaces, 73:221-227.

Bablon, G.P., C. Ventesque, and R. Ben Aim (1988). Developing a Sand-GAC Filter to Achieve High Rate Biological Filtration. Jour. AWWA, 80:12:47-53.

Bouwer, E.J. and P.B. Crowe (1988). Biological Processes in Drinking Water. Jour. AWWA, 80:9:82-93.

DiGiano, F.A. (1991). Microbial Activity of Filter Adsorbers. AWWARF Report dated March 1991, pp. 183-210.

Ginn, Jr., T.M., A. Amirtharajah, and P.R. Karr (1992). The Effect of Particle Detachment in Granular Media Filtration. Jour. AWWA 84:2:66-76.

Goldgrabe, J.C., R. S. Summers, and R.J. Miltner (1993). Particle Removal and Head Loss Development in Biological Filters. Jour AWWA 85:12:94-106.

Martin, R.E., E.J. Bouwer, and Linda M. Hanna (1992). Application of Clean-Bed Filtration Theory to Bacterial Deposition in Porous Media. Environ. Sci. Technol. 26:1053-1058.

Raveendran, P. (1993). Mechanisms of Particle Detachment During Filter Backwashing. Ph.D. Dissertation. Library, Georgia Institute of Technology, Atlanta, GA.

Application of Numerical Models in the Design of Clarifiers

by
J.A.McCorquodale[1] and S.P.Zhou[2]

1. Professor, 2. Research Associate, Department of Civil and Environmental Engineering, University of Windsor, Windsor, Canada N9B 3P4

Abstract

In this study, a numerical model of a circular clarifier was applied to reveal the effects of the water depth on the hydraulic and removal efficiency. The velocity and solids concentration fields at different time steps during tank operation were predicted. The model accounts for a number of practical conditions including sludge storage, density variations, settling characteristics, turbulence, ratio of return activated sludge (RAS) withdraw and tank geometry variations at full scale.

Introduction

There are several contentious design issues relating to the geometry of clarifiers which can now be addressed at least, in part, by the latest numerical models. One of these issues is the optimum design depth for a secondary clarifier for a given solids and hydraulic loading regime. The secondary clarifier has three main purposes, i.e., separation of solids and liquid, thickening of the activated sludge and storage of solids. The best tank dimensions will depend on the solids loading, the hydraulic loading, the settling characteristics and the management of the sludge inventory. Each plant is unique and therefore, it is difficult to recommend a general tank geometry.

It is interesting that many clarifiers in Europe are performing satisfactorily at depths of the order of 3 meters while in North America there are many tanks with depths up to 6 m. The results of the authors' modeling research indicates that both schools of design may be correct within the context of their plant operations. This paper discusses the current status of numerical modelling and illustrates how a numerical model can be used as a tool to assist in the selection of appropriate tank dimensions for a specific design condition.

Model Description

The hydrodynamic modelling of clarifiers was initiated by Larsen(1977); these models are meant to replace earlier simplified models based on the work of Camp(1946) and Dobbins(1944). The development of applicable turbulence models such as the k-ϵ model of Rodi(1980), has resulted in a better representation of the mixing processes in clarifiers. Another problem, which has recently been overcome, is the numerical instability encountered with strong density flows in clarifiers (Zhou et al 1992). The bottom boundary of clarifiers has traditionally been treated by an empirical boundary condition with a specified scouring parameter K_r.

In this paper a new "fully mass conservative clarifier model" is applied in which the conventional scouring formula with a specified scouring parameter K_r is replaced by a mass accumulating boundary cell. The relationship (Eq.1) for the solids transport is exactly satisfied in the time integration of the model, i.e.

$$\int [Q_o C_o]\,\Delta t = \int [Q_{RAS} C_{RAS}]\,\Delta t + \int [Q_{eff} C_{eff}]\,\Delta t + \int C_{cell} \Delta V_{cell} \quad (1)$$

where Q_oC_o = mass flux of influent (Q_o=Q_{RAS}+Q_{eff}); $Q_{RAS}C_{RAS}$ = mass flux of RAS withdrawal at bottom; $Q_{eff}C_{eff}$ = mass flux of effluent and $C_{cell}dV_{cell}$ = mass storage in the tank. This fully conservative model can simulate the unsteady sludge blanket development and movement due to solids accumulation.

Hydraulic Model- The governing equations for two-dimensional, unsteady, and density stratified flow in a circular setting clarifier are:

Continuity equation:

$$\frac{\partial ru}{\partial r} + \frac{\partial rv}{\partial y} = 0 \quad (2)$$

r-momentum equation:

$$\frac{\partial u}{\partial t}+u\frac{\partial u}{\partial r}+v\frac{\partial u}{\partial y}=-\frac{1}{\rho}\frac{\partial p}{\partial r}+\frac{1}{r}\frac{\partial}{\partial r}(r\nu_t\frac{\partial u}{\partial r})+\frac{1}{r}\frac{\partial}{\partial y}(r\nu_t\frac{\partial u}{\partial y})+S_u \quad (3)$$

y-momentum equation:

$$\frac{\partial v}{\partial t}+u\frac{\partial v}{\partial r}+v\frac{\partial v}{\partial y}=-\frac{1}{\rho}\frac{\partial p}{\partial y}+\frac{1}{r}\frac{\partial}{\partial r}(r\nu_t\frac{\partial v}{\partial r})+\frac{1}{r}\frac{\partial}{\partial y}(r\nu_t\frac{\partial v}{\partial y})+S_v \quad (4)$$

where

$$S_u=\frac{1}{r}\frac{\partial}{\partial r}(r\nu_t\frac{\partial u}{\partial r})+\frac{1}{r}\frac{\partial}{\partial y}(r\nu_t\frac{\partial v}{\partial r})-2\frac{\nu_t}{r^2}u \quad (5)$$

and

$$S_v=\frac{1}{r}\frac{\partial}{\partial r}(r\nu_t\frac{\partial u}{\partial y})+\frac{1}{r}\frac{\partial}{\partial y}(r\nu_t\frac{\partial v}{\partial y})-g\frac{\rho-\rho_r}{\rho} \quad (6)$$

where u and v = mean velocity components in the r and y directions, respectively; p = the general pressure less the hydrostatic pressure at reference density ρ_r, ρ = the fluid density, g = the component of gravitational acceleration in the vertical direction and ν_t = eddy viscosity.

Mass transport equation:

$$\frac{\partial C}{\partial t}+u\frac{\partial C}{\partial r}+v\frac{\partial C}{\partial y}=\frac{1}{r}\frac{\partial}{\partial r}(r\nu_{sr}\frac{\partial C}{\partial r})+\frac{1}{r}\frac{\partial}{\partial y}(r\nu_{sy}\frac{\partial C}{\partial y}+rV_sC) \quad (7)$$

in which ν_{sr} = eddy diffusivity of suspended solids in the r-direction; ν_{sy} = eddy diffusivity of suspended solids in the y-direction; and V_s is particle settling velocity. By using the Reynolds analogy between mass transport and momentum transport, the sediment diffusion coefficient is closely related to ν_t by the formulae

$$\nu_{sr}=\frac{\nu_t}{\sigma_{sr}};\qquad \nu_{sy}=\frac{\nu_t}{\sigma_{sy}} \quad (8)$$

in which σ_{sr} and σ_{sy} is the Schmidt number in the r-direction and the y-direction, respectively.

Turbulence Model- The eddy viscosity is calculated from the k-ϵ turbulence model (Rodi, 1980) which relates μ_t to the turbulence kinetic energy of k and the turbulence dissipation rate of ϵ.

Model Application and Results

Test cases- The simulation of the tank depth effect is used to illustrate the application of the model in the design of a clarifier. The following ranges of design conditions were considered in the investigation of the effect of tank depth on clarifier performance: Flow from 2000 to 3500 m^3/h, MLSS of 2500 mg/L, 50 % RAS, depth from 3.4 to 5.4 m, radius of 21 m, the tank bottom slope of zero and SVI of 80 L/mg. The remaining geometry parameters of the prototype clarifier are the distance from the reaction baffle to the effluent weir, R_s = 15.8 m; the distance from tank influent to the baffle, R_{in} = 2.8 m; the radius from tank centre to inlet, R_o =2.4 m; the depth of the influent stream opening H_{in} = 1.0 m; height of reaction baffle H_b = 50% of water depth. The hydraulic and solids loading were fed at the centre of the clarifier through a slotted flocculation chamber.

Hydraulic and removal efficiency- Figure 1(a) and 1(b) show typical flow patterns under normal loading conditions (2000 m^3/h) with water depths of 3.4 and 5.4 m, respectively. The flow pattern has the distinct bottom density current and a well-developed recirculation pattern that was first observed in 1946 by Anderson. Figure 2(a) shows the associated solids distribution in a 3.4 m clarifier at the normal design loading. Figure 2(b) shows the results under the identical loading and settling velocity conditions with a 5.4 m deep tank. It is noted that the surface concentrations near the effluent are almost the same in both cases. To illustrate the effect of a deeper tank, the hydraulic loading was increased from 2000 m^3/h to 3500 m^3/h in both the shallow and the deep circular tanks while maintaining the MLSS at 2500 mg/L with the same settling velocities at both flows. Figure 2(c) shows that the shallow tank fails while the deep tank in Fig. 2(d) still gives an acceptable effluent. The main reason for the failure of the shallow tank can be traced to the lack of adequate sludge storage and an adverse hydraulic flow pattern. In particular, the increase in the sludge blanket above the mid depth of the tank appears to cause an increased upward deflection of the influent and consequently, a higher solids concentration in the effluent.

Effluent processes- The total model integration time for each of the four cases was 441 minutes to achieve the equilibrium in tank operation, i.e. $Q_oC_o=Q_{RAS}C_{RAS}+Q_{eff}C_{eff}$. Figure 3 gives the trend of the effluent concentration during the integration period for the four cases. An overshoot of the effluent can be found at the time 30 to 40 minutes which corresponds to the strongest bottom density current which occurs before a thickened sludge blanket builds up.

Figure 3 illustrates the effect of hydraulic detention time and overflow rate. It is noted for $t \leq 50$ min that for a constant overflow rate the longer hydraulic detention time gives lower effluent concentration.
For long operational time($t>100$ min) the model revealed both deep and shallow tanks can give similar removal performance at the final equilibrium stage for certain range of hydraulic detention times or surface loadings. However, if the tank volume does not provide sufficient sludge storage then the tank may fail. Based these finding it is concluded that the tank performance is not a unique function of either overflow rate or detention time.

Conclusions

Recent numerical models are sufficiently well advanced that they can be used as a tool in the selection of critical tank dimensions such as depth, diameter, launder locations, bottom slope and skirt dimensions.

Acknowledgment

This research was supported by a grant from the Natural Sciences and Engineering Research Council of Canada.

References

Camp, T.R. (1946), "Sedimentation and the Design of Settling Tanks", Transaction, ASCE, Vol. 111, pp. 895-936.
Dobbins, W.E. (1944), "Effect of Turbulence on Sedimentation", Transaction, ASCE, Vol. 109, No. 2218, pp. 629-656.
Larsen, P. (1977), "On the Hydraulics of Rectangular Settling Basins, Experimental and Theoretical Studies", Dept. of Water Resources Engineering, Lund Institute of Technology, Lund University, Report No. 1001, Lund Sweden.
Rodi, W. (1980), "Turbulence Models and Their Application in Hydraulics A State-of-the-Art Review", Int. Assoc.

for Hydraulic Research, Delft, Netherlands.
Zhou, Siping and McCorquodale, J.A. (1992),"Influences of Density on Circular Clarifier with Baffles", Journal of Environmental Engineering, ASCE, Vol. 118, No. 6, pp. 829-847.

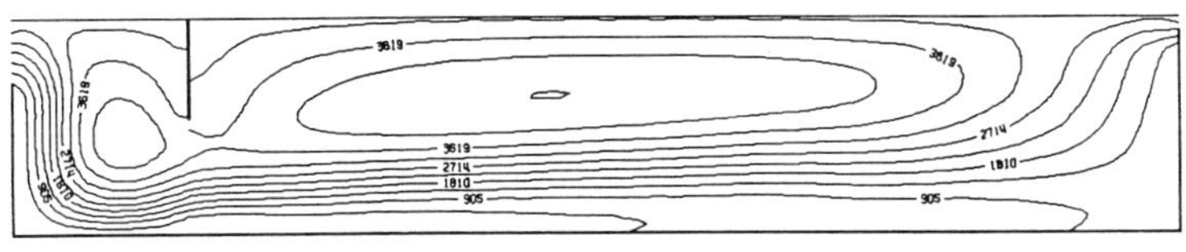

(a)

Inlet Solids Co(mg/L) =2500.0
Ratio of Return Sludge RAS(%) =50.0
Water Depth in Tank H(m) =3.4
Radius Ro(Centre to Inlet m)=2.4
Radius Rs(Baff. to Weir m)=15.8
Effluent Con. Ceff(mg/L)=20.9

Hydr. Loading Qo(m3/h) =2000.0
Stokes Velocity Vo(m/h) =9.0
Height of Rect. Baffle Hb(m)=1.7
Radius Rin(Inlet to Baff. m)=2.8
Simulat. Time T(min) =441.0
Bottom RAS Con. (mg/L)=2641. to 9116.

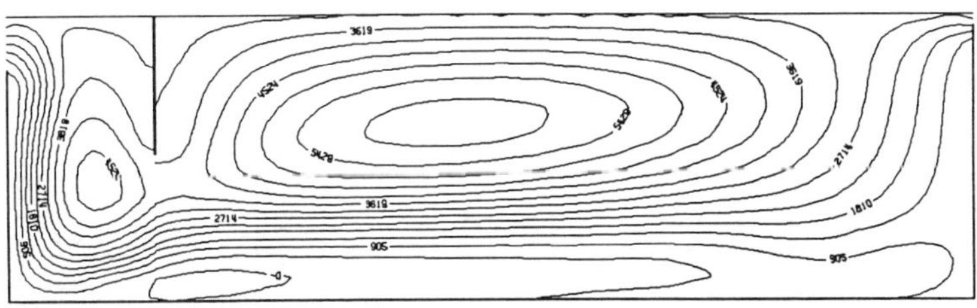

(b)

Inlet Solids Co(mg/L) =2500.0
Ratio of Return Sludge RAS(%) =50.0
Water Depth in Tank H(m) =5.4
Radius Ro(Centre to Inlet m)=2.4
Radius Rs(Baff. to Weir m)=15.8
Effluent Con. Ceff(mg/L)=21.4

Hydr. Loading Qo(m3/h) =2000.0
Stokes Velocity Vo(m/h) =9.0
Height of Rect. Baffle Hb(m)=2.7
Radius Rin(Inlet to Baff. m)=2.8
Simulat. Time T(min) =441.0
Bottom RAS Con. (mg/L)=2319. to 8544.

Fig. 1 Flow pattern in Circular clarifier

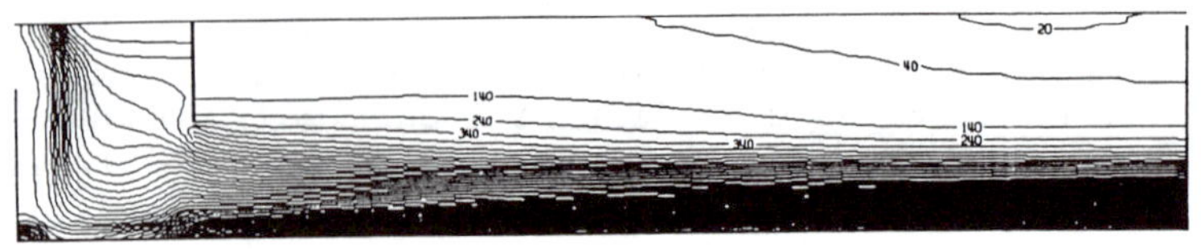

(a)

Inlet Solids Co(mg/L) =2500.0
Ratio of Return Sludge RAS(%) =50.0
Water Depth in Tank H(m) =3.4
Radius Ro(Centre to Inlet m)=2.4
Radius Rs(Baff. to Weir m)=15.8
Effluent Con. Ceff(mg/L)=20.9

Hydr. Loading Qo(m3/h) =2000.0
Stokes Velocity Vo(m/h) =9.0
Height of Rect. Baffle Hb(m)=1.7
Radius Rin(Inlet to Baff. m)=2.8
Simulat. Time T(min) =441.0
Bottom RAS Con. (mg/L)=2641. to 9116.

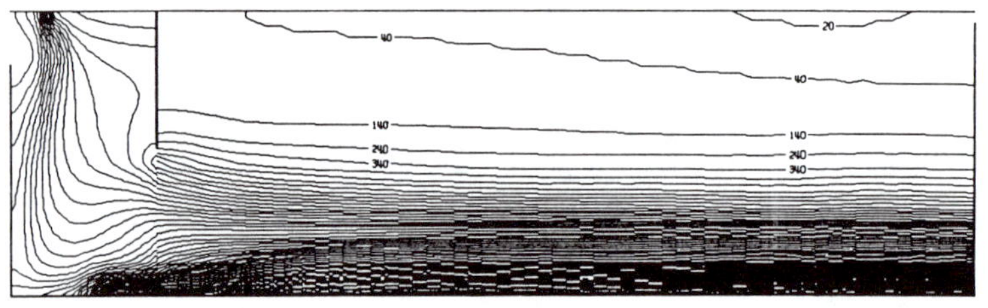

(b)

Inlet Solids Co(mg/L) =2500.0
Ratio of Return Sludge RAS(%) =50.0
Water Depth in Tank H(m) =5.4
Radius Ro(Centre to Inlet m)=2.4
Radius Rs(Baff. to Weir m)=15.8
Effluent Con. Ceff(mg/L)=21.4

Hydr. Loading Qo(m3/h) =2000.0
Stokes Velocity Vo(m/h) =9.0
Height of Rect. Baffle Hb(m)=2.7
Radius Rin(Inlet to Baff. m)=2.8
Simulat. Time T(min) =441.0
Bottom RAS Con. (mg/L)=2319. to 8544.

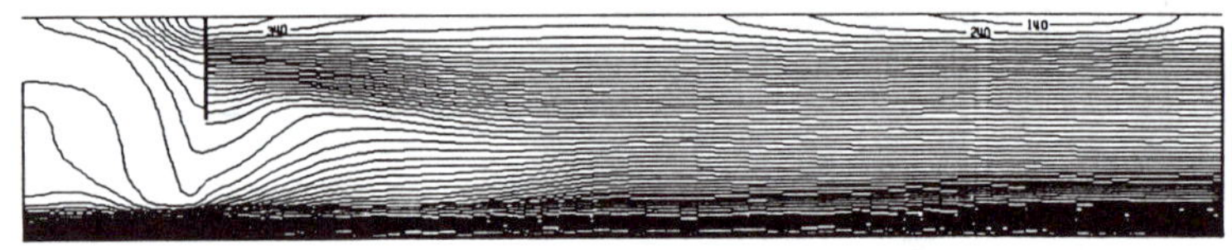

(c)

Inlet Solids Co(mg/L) =2500.0
Ratio of Return Sludge RAS(%) =50.0
Water Depth in Tank H(m) =3.4
Radius Ro(Centre to Inlet m)=2.4
Radius Rs(Baff. to Weir m)=15.8
Effluent Con. Ceff(mg/L)=288.7

Hydr. Loading Qo(m3/h) =3500.0
Stokes Velocity Vo(m/h) =9.0
Height of Rect. Baffle Hb(m)=1.7
Radius Rin(Inlet to Baff. m)=2.8
Simulat. Time T(min) =441.0
Bottom RAS Con. (mg/L)=5053. to 8697.

Fig. 2(a)-(c)

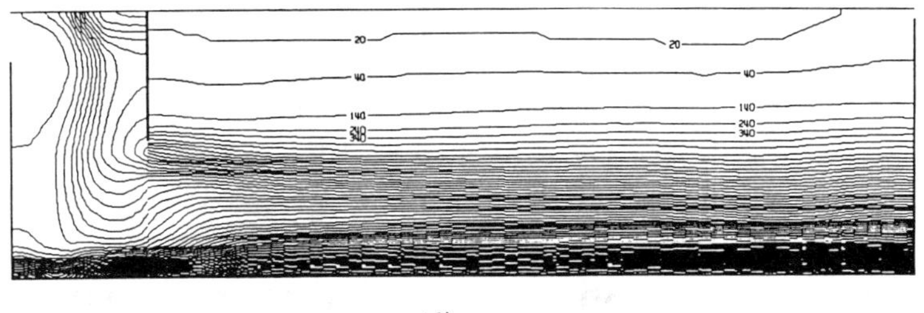

(d)

Inlet Solids Co(mg/L) =2500.0	Hydr. Loading Qo(m3/h) =3500.0
Ratio of Return Sludge RAS(%) =50.0	Stokes Velocity Vo(m/h) =9.0
Water Depth in Tank H(m) =5.4	Height of Rect. Baffle Hb(m)=2.7
Radius Ro(Centre to Inlet m)=2.4	Radius Rin(Inlet to Baff. m)=2.8
Radius Rs(Baff. to Weir m)=15.8	Simulat. Time T(min) =441.0
Effluent Con. Ceff(mg/L)=26.0	Bottom RAS Con. (mg/L)=3717. to 8772.

Fig. 2 Contour plot of predicted solids concentration

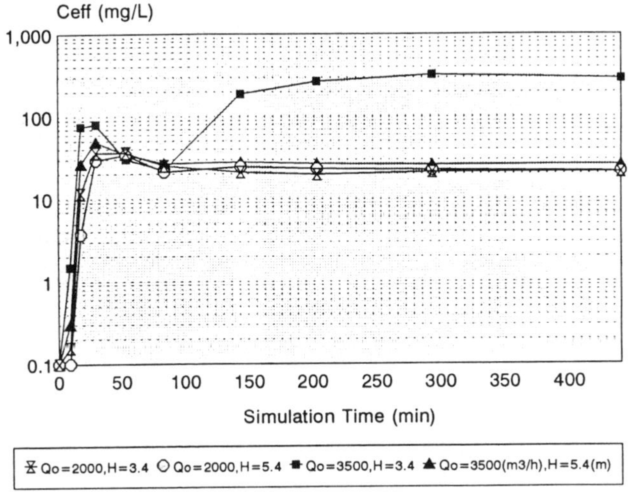

Fig. 3 Effluent concentration process in clarifier

CLARIFIER ENHANCEMENTS YIELD EXCELLENT PERFORMANCE

Cindy L. Wallis-Lage[1]
David B. Hunt[2]

Abstract

The critical role of final clarifiers in meeting effluent permit requirements is indisputable. Poor performance of the final clarification step may adversely affect effluent quality. Experience has shown that effluent quality produced by conventional final clarifiers typically meets the secondary effluent requirements of 30 mg/L BOD and 30 mg/L TSS. However, recent enhancements to final clarifier design have been shown to significantly improve performance over conventional units to the point that filtration may not be required to achieve a high quality effluent. An eight-month full-load demonstration test was conducted at the Jacksonville, Illinois WWTP to establish the capabilities of new final clarifiers to produce an effluent meeting limits of 10 mg/L BOD and 12 mg/L TSS. The test demonstrated that the Black & Veatch clarifier design can produce a high quality effluent which meets stringent effluent limits. As a result, significant cost savings may be realized if effluent filtration can be avoided.

Introduction

Final clarification is often the critical unit process in meeting effluent standards. Poor performance of the final clarifiers may adversely affect effluent quality. Experience has shown that effluent quality produced by conventional final clarifier designs typically meets the secondary effluent

---

[1]Process Engineer, Black & Veatch, 8400 Ward Parkway, Kansas City, Missouri 64114

[2]Partner and Project Manager, Black & Veatch, 8400 Ward Parkway Kansas City, Missouri 64114

requirements of 30 mg/L for both BOD and TSS. However, treatment standards are changing and many wastewater treatment plants (WWTP) are facing more stringent effluent limits. Consequently, many wastewater utilities are confronted with the high cost of filtration to achieve the required effluent quality. Recent enhancements to final clarifier design have been shown to significantly improve performance over conventional units. For some plants this could mean that filtration may not be required and a significant cost savings may be realized.

Black & Veatch final clarifier design practice has evolved over the last several decades through laboratory research and field experience. Several design enhancements have been developed which have improved clarifier performance significantly. These have been described previously by Stukenberg (1983). The key design features which contribute to the superior performance of the Black & Veatch clarifier are discussed below.

- *Large diameter feedwell*. Experience has indicated that operation of clarifiers with large diameter feedwells, improves performance due to the reduction in the downward velocity and better energy dissipation. Mechanical flocculators are sometimes included to encourage reflocculation of the mixed liquor solids.
- *Sidewater depth*. The key element in determining the appropriate sidewater depth (SWD) for a clarifier is to provide sufficient depth to contain the sludge blanket during peak flows. Black & Veatch has concluded that the combined effects of floor slope and clarifier diameter should be considered in this determination. Smaller clarifiers (less than 24.4 meters in diameter) should have 4.6 meters SWD, or greater, for best performance. Larger units may be as shallow as 3.6 meters SWD because of the additional center depth created by the sloping floor.
- *Peripheral baffle*. Studies by Robinson(1974) and Mohart(1978) demonstrated that a circumferential baffle near the effluent weir dissipated some of the energy in the clarifier and resulted in improved settling. This configuration, illustrated on Figure 1, effectively re-directs currents in the clarifier and, thus, reduces solids carryover in the effluent.
- *Conventional sludge removal equipment*. Laboratory studies and full-scale experience have shown that hydraulic sludge removal may induce hydraulic currents in the clarifier which could result in disruption of the sludge blanket, and, subsequently, solids carryover in the effluent. As a result, conventional scraper-type sludge removal equipment is preferred by Black & Veatch over hydraulic sludge removal equipment.
- *Full-surface skimming*. Because solids periodically accumulate on the clarifier surface, Black & Veatch, in cooperation with wastewater equipment manufacturers, developed full-surface skimmers for final clarifiers. The skimmer allows rapid removal of the scum and minimizes solids carryover in the effluent.

It is commonly assumed that properly designed final clarifiers can consistently meet design effluent limits of 30 mg/L for total suspended solids

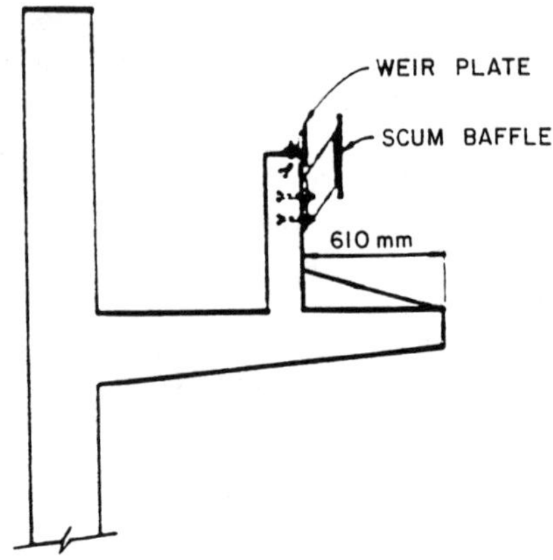

FIGURE I - CLARIFIER BAFFLE AT WEIR TROUGH.

Figure 2. Effluent TSS -vs- SOR

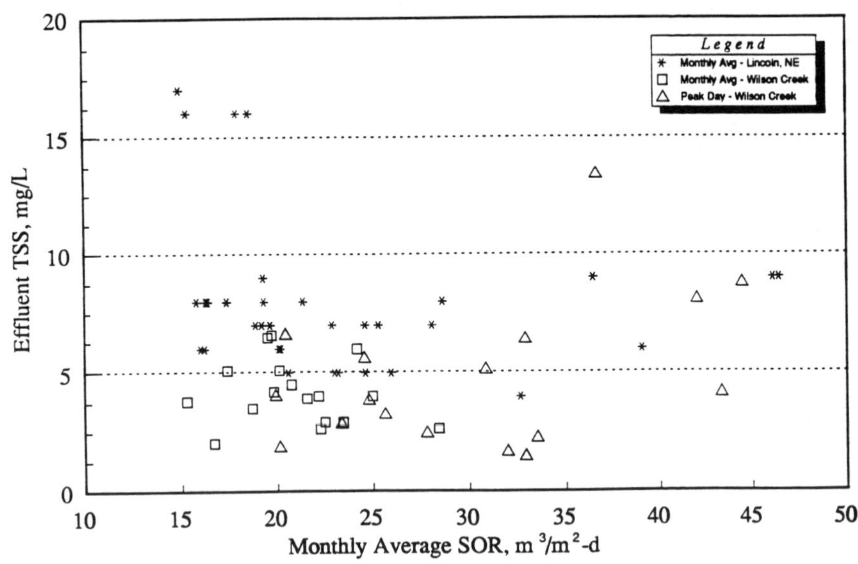

(TSS) and 30 mg/L for biochemical oxygen demand (BOD). However, review of data from several wastewater treatment plants using Black & Veatch clarifier design indicated effluent quality much better than the typical 30/30 design criteria. Data from the Northeast WWTP in Lincoln, Nebraska, and the Wilson Creek Regional WWTP, operated by North Texas Municipal Water District, are presented in Figure 2, to illustrate the clarifier performance at various surface overflow rates (SOR).

This recent experience in final clarifier design was used during the development of wastewater treatment facility improvements for the City of Jacksonville, Illinois. Jacksonville had been issued a new discharge permit with monthly average limits of 10 mg/L BOD and 12 mg/L TSS and peak day limits of 24 mg/L for both BOD and TSS. In addition, new limits on ammonia nitrogen were added to the permit. Although conventional design practice would normally rely on filtration to reliably achieve this level of effluent quality, it was proposed that construction of filters be deferred until performance of the new final clarifiers could be assessed. The regulatory agency agreed to this approach contingent upon the use of conservative design criteria for the new clarifiers and performance of a full-load demonstration program to establish their capabilities under design loading conditions.

Methodology

To determine the required operating conditions to simulate full-load design conditions, plant data for the months of March 1991 through September 1991 were evaluated to determine the monthly average, maximum month, and peak day loading conditions. These data were then used to determine the number of aeration basins and final clarifiers required in service to simulate full-load design conditions with current wastewater characteristics. It was determined that three of the five aeration basins would be required to achieve the anticipated design loading rates based on current average and maximum month organic loads. One final clarifier would be required to simulate the design solids loading and surface overflow rates. The clarifier in operation included the design features discussed above. A comparison of the design conditions and the proposed test loadings is presented in Table I.

Results

The full-load demonstration test was conducted from March through October 1992. During the test period, effluent BOD and TSS data were collected, and sludge blanket depth and SVI's were monitored to evaluate performance. The sludge blanket depth was maintained between 0.45 m and 0.9 m to minimize the impact of a high blanket on clarifier performance. Data from the eight months of testing are summarized in Table II.

Table I. Design and Test Conditions

Parameter	Design Conditions			Test Conditions		
	Avg Month	Max 10-day Avg	Peak Day	Avg Month	Max Month	Peak Day
Flow, m^3/d	28,730	-	56,700	15,120	15,880	29,110
Aeration Basin Volume, m^3	19,110	19,110	19,110	11,470	11,470	11,470
Aeration Hydraulic Retention Time, hr	15.9	-	8.1	18.0	17.3	9.4
Clarifier Surface Area, m^2	1,460	1,460	1,460	730	730	730
Solids Loading Rate, kg/m^2-d		94.1			95.7	
Surface Overflow Rate, m^3/m^2-d	19.7	-	38.9	20.7	21.8	39.9

During the test period, the average surface overflow rate was approximately equal to or greater than the design condition. However, the hydraulic loading required to simulate the peak day design condition was achieved only a few times. As shown by the effluent BOD and TSS data, effluent compliance was maintained throughout the test period with the exception of May 1992 and July 1992. Review of the effluent data indicated that the effluent BOD and TSS concentrations began to increase approximately mid-April and remained above permit limits through mid-May. During the last two weeks in April, the average flow to the plant was 27,220 m^3/d which resulted in an average clarifier surface overflow rate of 37.5 m^3/m^2-d. In addition, sludge wasting was very irregular from mid-April through mid-May. The combination of the hydraulic stress, and the associated high solids load, with the non-steady state operating conditions appeared to be the cause for the effluent BOD and TSS concentration permit violations. Similar operating conditions provoked the peak day excursion in July.

Table II. Test Performance.

Month, 1992	Hydraulic load, m^3/m^2-d	Solids load, kg/m^2-d	Effluent TSS, mg/L	Effluent BOD, mg/L	SVI, ml/g
March					
Avg	33.2	122	9.6	8.4	109
Peak day	58.6	173	23.3	15.9	136
April					
Avg	34.8	107	8.6	9.7	92
Peak day	49.3	137	23.3	21.9	82
May					
Avg	27.0	82.5	12.1	11.6	70
Peak day	36.3	99.1	27	22.2	54
June					
Avg	20.8	62.0	7.9	6.2	86
Peak day	28.6	77.6	13.7	11.4	72
July					
Avg	22.9	67.9	6.6	4.0	94
Peak day	34.8	73.7	25.6	7.8	97
August					
Avg	17.1	60.0	8.6	5.5	69
Peak day	21.3	66.3	23.3	15.0	82
September					
Avg	19.2	76.2	8.0	5.4	86
Peak day	28.6	88.8	16.6	11.7	66
October					
Avg	18.2	62.0	7.8	6.8	82
Peak day	34.3	10.2	17.0	11.4	132

Note: Peak day data for peak hydraulic loading and effluent data may not be for the same day.

Discussion

The variations in surface overflow rate and on effluent quality for monthly average and peak day conditions for the eight-month test period are illustrated on Figures 3 and 4, respectively. As shown by these data, the new clarifiers can accommodate peak day hydraulic loading rates of 57.0 m^3/m^2-d and monthly averages of 32.6 m^3/m^2-d without sacrificing effluent quality under

Figure 3. Surface Overflow Rates

Jacksonville, Illinois

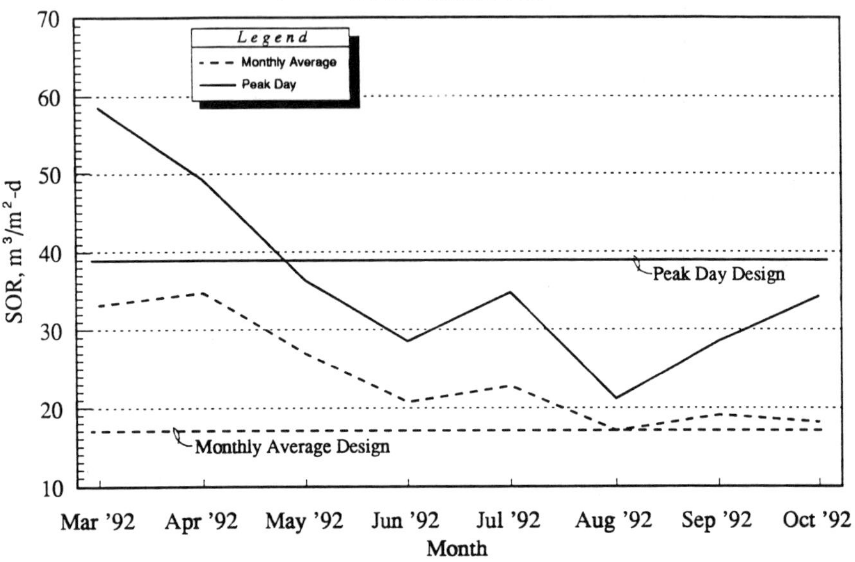

Figure 4. Effluent TSS

Jacksonville, Illinois

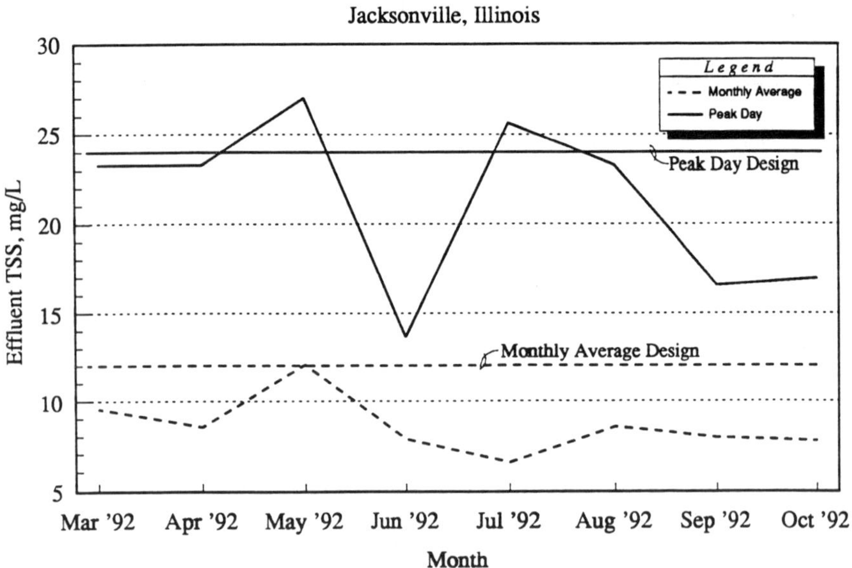

most conditions. The clarifiers also withstood solids loading rates as high as 173 kg/m^2-d for peak day conditions and 122 kg/m^2-d for monthly average conditions without permit violations. The two peak day violations shown on Figure 6 were the result of hydraulic stress, high solids loads, and non-steady state operating conditions due to erratic sludge wasting.

The exceptional performance of the Black & Veatch clarifier is attributed to the enhancements incorporated into the design. The large diameter feedwell and the peripheral baffle improve the hydraulic currents in the clarifier, thus, minimizing the potential for solids overflow in the effluent. The 4.6 meter SWD provided sufficient depth to maintain the sludge blanket, and the full-surface skimmer captured any floating solids.

Conclusions

During the eight-month full-load demonstration test, the monthly average effluent BOD and TSS concentrations were below the permit limits with the exception of May 1992. The test demonstrated that the Black & Veatch clarifier design can produce a high quality effluent which meets stringent effluent limits. As a result, significant cost savings can be realized if effluent filtration can be avoided.

Appendix

Mohart, James L., "Model Dye Tracer Studies of Currents in Final Sedimentation Basins." Special Problem Report, Univ. of Kansas, Lawrence (Nov. 1978).

Robinson, John H., Jr., "A study of Density Currents in Final Sedimentation Basins." M.S. Thesis, Univ. of Kansas, Lawrence (1974).

Stukenberg, John R., Leonard C. Rodman, James E. Touslee, (1983) "Activated Sludge Clarifier Design Improvements." J. Water Pollut. Control Fed., 55, 341.

Optimizing Secondary Clarifier Performance:
A Presentation Outline

John K. Esler[1]

Abstract

Optimizing secondary clarifier performance is an activity that involves a consideration of the design, the maintenance, and the operation of the clarifier. Although the basic shape of the unit is thought to be a major factor, the internal details are the most important design factors in determining a clarifier's capacity. The ability to analyze the impact of these details, along with the operational and maintenance practices, is critical to the evaluation of its performance. The final challenge is how to most effectively modify a clarifier's hydraulic characteristics in order to optimize its performance.

I. **The Clarifier as Part of a System:**

A. Preliminary Treatment ..... impact of rag removal on plugging
B. Pump Units ..... should be variable speed
C. Aeration Units ..... affect floc formation; heat transfer
D. Fixed-film Reactors induce different currents in clarifiers
E. Flow Distribution .... often unbalanced, especially with one unit out of service. Upflow distribution boxes work best. **Always** provide a primary flow measurement device after each clarifier.
F. Return Sludge Flows ..... avoid combined RAS suctions. Use separate RAS lines with individual flow meters. Provide lots of capacity; provide flexibility to throttle down for seasonal low flows. Make adjusting draft tubes easy to do.

----

[1] President, CPE Services, Inc., 19 Elmwood St., Albany, NY 12203

II. Is Shape a Factor in Performance?

Yes ..... and No. Shape alone guarantees nothing. However, the "squircle" is absolutely the poorest shape.
Rectangular units can usually be upgraded more cost-effectively.

III. Is Depth a Factor?

Yes, depth can affect performance. but it's more in relation to the depth of launders. Although increased depth may improve performance, it is probably not the most cost-effective option. Additional depth may be useful for diurnal sludge storage ..... if not nitrifying.

IV. Internal Details Have a Great Effect on Performance.

A. Inlet flow balance ..... always provide a means to measure and control flow.

B. Inlet design ..... avoid jetting; distribute flow horizontally; not as important to distribute flow vertically; provide for flocculation.

C. Circular clarifier sludge withdrawal arms:
1. Should stack tubes horizontally
2. Avoid manifold-type systems
3. Slow them down ?? ..... probably should ..... esp. in squircles.
4. Speed them up ?? ..... if it will minimize denitrification.
5. Avoid arms with pantographic corner sweeps.

D. Hydraulic sludge withdrawal tubes:
1. Submerged gates/slots hardest to control individually.
2. Telescopic valves provide better control, but often rag up.
3. Avoid plugging problems by reducing number of tubes in service (not in design).

E. Plow-type sludge scrapers:
Standard plows OK; may produce better hydraulic conditions than draft tube system; spiral-shaped ..... maybe not ?

F. Circular Centerwells
1. Optimum depth ? no greater than 1/2 SWD
2. Optimum diameter ? Bigger is not better; beware of low flows.
3. Design to enhance flocculation ..... and remove scum
4. Avoid side entry

G. Scum Collection w/ Circular Clarifiers
1. Use full-radius skimmers; avoid perpendicular attachment
2. Use dual skimmers (and multiple scum hoppers?)
3. Rotating scum troughs ..... plan for lots of water w/ scum
4. Scum squeegees are a great retrofit
5. Plan for nocardia
6. Provide safe access to hoppers
7. Provide a simple hopper flushing system (Hinsdale, IL)

H. Rectangular Sludge Collection Mechanisms
   1. MUST PROVIDE for scum removal !!
   2. How fast ? 4+ fpm, especially when nitrifying
   3. How slow ? 1 fpm ..... w/ fixed film ..... if not nitrifying
   4. Traveling Bridges ..... create problems due to flow forced to launders at far end; difficult to correct short-circuiting.

I. Inlets for Rectangular Clarifiers
   1. Effect of large inlet baffle may be counter-productive
   2. Submerged gates w/ head differential OK, but plan for scum removal from distribution channel.
   3. L.A. County San. District has best inlet .... w/ baffled inlet pipes

J. Location of RAS Hoppers in Rectangular Clarifiers
   1. At inlet end ..... typical condition ..... usually short-circuit
   2. At middle (Gould II) ..... promotes short-circuiting in 1st half; reduces internal currents in second half.
   3. At effluent end ?? Works great at L.A. County San. District
   4. Consider V = Q / A vs V = (Q + R) / A

V. Weir Placement is Critical !!

A. In Rectangular Clarifiers (w/o baffles)
   1. Worst Conditions:
      a. At or near end wall ..... or, close together
      b. Launders <u>deeper</u> than necessary to convey flow
      c. Short weirs perpendicular to end wall
      d. Submerged pipes
   2. Better Conditions:
      a. Weirs covering at least 20% of surface
   3. Best Conditions:
      a. Weirs covering at least 30 % of surface
      b. Have ability to measure the effluent flow / head
      c. Adjustable / able to be taken out of service
      d. No deeper than necessary
      e. Parallel to flow (see L.A. Co. San Jose Creek)

B. In Circular Clarifiers (w/o baffles)
   1. Worst Conditions:
      a. Single perimeter weir flush w/ face of wall or cantilevered inward.
      b. Inboard launder deeper than necessary
      c. Single perimeter weir + close inboard launder
   2. Questionable Conditions:
      a. Peripheral feed (downward flow) w/ peripheral effluent weirs (Denver Metro; EBMUD)
   3. Better Conditions:
      a. Inboard launder ..... but, not too deep !

4. Best Conditions:
    a. Spiral Flow (peripheral feed) w/ central launders
    b. Flocculating centerwell w/ radial launders
    c. Flocculating centerwell w/ inboard launder

VI. The Effect of Maintenance on Performance

A. Periodic Maintenance ..... dewater and inspect 2X / year.

B. Remember ..... high ground water can float your clarifier !!

C. Torque overload protection ..... check frequently.

D. Eel seals (on suction manifold type) ..... a problem to check.

E. Centerwell seal ..... needs periodic replacement.

F. Check scraper blades for nominal clearance

G. Weir leveling is very important.

H. Algae ..... must be controlled; design for automatic brushing system; some hypochlorite systems work.

VII. The Effect of Operations on Performance

A. Sludge blanket level control ..... a most critical activity !
    1. Manual core samplers are easy and accurate
    2. Automatic blanket detectors ??? (use "fish finders")

B. SVI Control ..... very important; must learn to <u>identify</u> and control filaments !!

C. MLSS Control:
    1. Your call; use whatever works best for your process.
    2. Generally ..... lower is better.

D. Scum Removal is important to reduce odors ..... and prevent problems in freezing weather.

E. Flow Balancing:
    1. Must CONTROL flow ..... to each clarifier
    2. Must CONTROL RAS ..... from each clarifier.

F. RAS Control:
1. RAS rate can effect hydraulic performance, esp in circulars
2. Use RAS tube selection / control to optimize concentration
3. Consider "solids flux / state point" concept ??

G. Effluent Monitoring:
1. From **each** clarifier
2. Use turbidimeters ..... on-line or with manual grabs.

VIII. Steps in Analyzing Clarifier Performance

A. Determine individual clarifier effluent flows (easier than influent)

B. Determine individual clarifier RAS flows; check all meters.

C. Determine individual clarifier ETSS performance ..... at various overflow rates and blanket conditions.

D. Monitor turbidity.

E. Look for diurnal ETSS variations (esp. time of peak ETSS)

F. Monitor diurnal blanket level changes.

G. Optimize activated sludge quality ..... for your conditions.

H. Determine "Clarifier Hydraulic Characteristics":
1. Actual detention time
2. General flow patterns ..... for different conditions
3. Look for reverse currents
4. Morrill Index ??? (T_{90} / T_{10})
5. Effect of individual launders and weirs
6. Location and intensity of short-circuiting currents
7. Impact of RAS rates; RAS tubes in service

IX. Improving Clarifier Hydraulics

A. The most cost-effective means of improving performance !!

B. Refer to the work of Bob Crosby in circular clarifiers:
1. Crosby sloped peripheral baffle (at Stamford, CT) improved ETSS by >30%; must have at least 45° slope.

2. Crosby mid-radius baffle (at Morganton, NC) improved ETSS by >30%; provides additional flocculation.
3. Crosby "Distributive Centerwell" (at Stamford, CT) didn't improve ETSS !

C. Combination of peripheral baffle and mid-radius baffle ..... may be more effective than one individual baffle (New Haven, CT)

D. Baffles in rectangular clarifiers (Esler-Miller baffles)
1. The right baffle will improve most clarifiers.
2. The wrong baffle(s) can make them worse.
3. Two baffles are better than one (Herkimer, NY)
4. Three baffles are better than two (Branford, CT)
5. Four baffles are even better ? (work now in progress)

X. Some Innovations (Some are good ...... some are not so good !!)

A. Stacked rectangular units in parallel:
1. A very difficult clarifier to maintain (bottom unit may be a permit-required confined space !) and operate (try checking blankets in bottom unit).
2. Should have the same (poor) hydraulic characteristics as most counter-current sludge withdrawal clarifiers.

B. Stacked rectangular units in series (work in progress in Sweden):
1. Should have similar difficulties with operating and maintaining bottom unit as "A".
2. Should have much better hydraulic characteristics than "A".

C. Multiple compartments in series in rectangular clarifiers:
1. Japan's best-performing clarifier ..... verified by our work.
2. A good retro-fit for existing large clarifiers.

D. End-around / Folded-flow configurations:
1. Worked well at Toronto; Hunts Point (NYC)

E. Spiral scrapers
1. Didn't work well at Wilkes-Barre (PA) ..... 190' diameter.
2. Have they worked well at other locations ??

F. Lamella / Tube / Tray ..... not recommended for biological systems.

H. DEEEEP clarifiers ..... are they worth the extra $$$$$?

Closing Thoughts:

- Use a "holistic" / Comprehensive Plant Evaluation approach to evaluating and improving clarifier performance (i.e. identify all the limiting Design / Operation / Maintenance / and Management factors). There are usually several of these factors responsible for limiting the clarifier's performance.

- Making your 20 to 30% efficient clarifiers more efficient is usually much more cost effective than adding more clarifiers. The time spent on optimizing the performance of your existing clarifiers is well worth the effort.

- If you haven't already done it, read (and re-read) the USEPA Technology Transfer publication on "Hydraulic Considerations That Effect Secondary Clarifier Performance" (1980), by Bob Crosby and Jon Bender. It has a lot of good observations for operations people and design engineers.

- Don't forget your **Primary** Clarifiers !! They often have short-circuiting currents that effect their performance. This is the most cost-effective unit for removing BOD !

Evaluating Activated Sludge Secondary Clarifier Performance: A Protocol

E.J. Wahlberg[1], M. Augustus, D.T. Chapman, C. Chen, J.K. Esler, T.M. Keinath, D.S. Parker, R.J. Tekippe, and T.E. Wilson

for the

ASCE Clarifier Research Technical Committee

Abstract

To establish a consistent methodology for evaluating secondary clarifier performance, the ASCE Clarifier Research Technical Committee prepared procedural guidelines in the form of a draft protocol. This protocol was evaluated in a field study during which four clarifier configurations at three full-scale activated sludge facilities were tested. Based on the experience gained in this study, the protocol was revised. An overview of the revised protocol is presented.

Introduction

Despite the pivotal role played by the secondary clarifier in the activated sludge process, optimum design and operation continue to elude the wastewater treatment profession. Recognizing this, the USEPA sponsored a symposium in 1986 aimed at reviewing past research, and identifying and prioritizing future secondary clarifier research needs. Symposium participants presented 14 key questions and a number of recommendations to guide future secondary clarifier research (Tekippe and Bender, 1987). Perhaps most noteworthy among these recommendations was

----

[1] Corresponding author: Project Engineer, County Sanitation Districts of Los Angeles County, P.O. Box 4998, Whittier, California 90607-4998

the need for a standard procedure for evaluating secondary clarifier performance.

As a result of the USEPA-sponsored symposium, the ASCE Clarifier Research Technical Committee (CRTC) was formed. Based on the recommendations from the symposium and the expertise of the committee, the CRTC developed a draft protocol for evaluating secondary clarifier performance. Environment Canada, USEPA, the County Sanitation Districts of Orange County, the County Sanitation Districts of Los Angeles County (LACSD), the Denver Metro Wastewater Reclamation District (MWRD), the New York City Department of Environmental Protection (NYCDEP), and the Soap and Detergent Association contributed financially to the CRTC for purposes of conducting a full-scale study aimed at evaluating the protocol. In this evaluation, the protocol was field-applied at LACSD's San Jose Creek Water Reclamation Plant (Wahlberg *et al.*, 1993a), MWRD's Central Treatment Plant South Complex (Wahlberg *et al.*, 1993b), and NYCDEP's Rockaway Water Pollution Control Plant (Wahlberg *et al.*, 1994).

Objective

The objective of this paper is to present an overview of the CRTC protocol as revised subsequent to the field study.

CRTC Protocol

The CRTC protocol is comprised of five components: physical description of the test facility and secondary clarifier, plant operating summary, pre-test information collection, stress tests, and data analysis and interpretation. Each of these is discussed below.

Physical description of test facility and clarifier. In making meaningful comparisons of secondary clarifier performance from site to site it is imperative that detailed descriptions of the settling tanks in question be known. The participants of the USEPA-sponsored symposium noted a significant lack of adequate descriptions in their review of 50 years of technical literature, which made comparisons difficult or impossible. To alleviate this shortcoming with regards to future clarifier research, the protocol requires a complete description of the test clarifier including: number of clarifiers in the treatment train, type, dimensions, inlet structures, outlet structures (type and length of effluent weir), sludge collection equipment (type, speed, location of sludge hoppers, RAS pumping capacity and distribution), baffling (dimensions, location), amount of freeboard, and skimmers.

In addition, the protocol specifies a thorough description of the test facility. To be included in this description are: facility ownership and geographic location, design flows (average and peak), actual flows during testing, influent pumping, preliminary treatment, primary treatment (type and dimensions), aeration basins (flow regime, process modification, number of tanks, dimensions, type of aeration equipment), mixed liquor conveyance structures (dimensions, aeration equipment), flow splitting, secondary clarifiers (see above), tertiary treatment, chlorination and dechlorination equipment.

Plant operating summary. The purpose of the plant operating summary is to document the normal operation of the test facility for the year prior to the initiation of the evaluation testing. The historical records at the facility are consulted for the necessary data. The minimum, maximum, and mean monthly averages of the following are reported: plant influent flow, secondary effluent flow, primary effluent BOD_5 (or COD) and suspended solids (SS) concentrations, MLSS and MLVSS concentrations, RAS and WAS flows, the return sludge SS (RSSS) concentration, secondary effluent BOD_5 (or COD) concentration, and secondary effluent SS (ESS) concentration. Other data such as influent ammonia concentration, secondary effluent ammonia and nitrate concentrations, SVI, and influent or effluent temperature also are useful for documenting the normal operation of the facility.

Pre-test information collection. The primary objectives of this component in the draft protocol were (CRTC, 1988): (1) to develop methods for measuring and controlling flows to the test clarifier; (2) to obtain a preliminary description of the response of the test clarifier to changes in flow rates; (3) to determine if denitrification in the test clarifier contributes to ESS; (4) to determine the degree of floc breakup and aggregation through the secondary process including the test clarifier; (5) to describe the changes in the sludge blanket levels in the test clarifier in response to a typical diurnal flow pattern; (6) to determine the rates of air flow to the aeration basins and transfer channels upstream of the test clarifier; (7) to carry out a solids flux analysis; and, (8) to perform a hydraulic characterization study of the test clarifier. As the intent of the CRTC was to develop a protocol that was both rigorous and easily applied, some of these have been modified as a result of the evaluation study. Each of these is discussed below.

(1) Measurement and control of the flow through the test clarifier are vital to the application of the protocol. Of the three flows into and out of the test clarifier -- influent, effluent, and sludge withdrawal -- two must be measured accurately; the third then can be calculated. Calculation of flows based on a flow-splitting assumption is not reliable. If direct flow measurement is not available, provision must be made for doing so. During the evaluation study of the protocol, sharp-crested weirs placed in the effluent launders of the test clarifiers proved useful for measuring effluent flow. These weirs were calibrated by determining the relationship between flow rate and head over the weir. Flow measurement devices must be placed and calibrated prior to the initiation of the stress tests.

(2) The emphasis on accurately measuring and controlling the flows to and from the test clarifier was not developed in the draft protocol. Initially, it was suggested that the stress tests be performed over a 24-hr period, in which case the response of the clarifier as a result of the diurnal flow pattern was of concern. With the degree of flow control and measurement required in the revised protocol, this is no longer a concern.

(3) The potential influence of denitrification on the ESS concentration is indirectly assessed by observation of rising solids during the stirred sludge volume index (SSVI), the zone settling velocity (ZSV), the insitu dispersed solids (IDS), and the settling flux tests used to quantify the settling and flocculation characteristics of the mixed liquor entering the test clarifier during the stress tests. If denitrification is not evident in these tests, its influence on the ESS concentration is assumed negligible.

(4) Results obtained from the IDS test allow the sample's state of flocculation to be assessed. This is accomplished by collecting and settling the sample in the same container, thereby sparing the biological flocs in the sample any breakup or aggregation effects resulting from an intermediate transfer step. Although the IDS test can be used at several points within an activated sludge facility to assess floc breakup and flocculation changes occurring in the system (Das *et al.*, 1993), the primary objective of the test in the protocol is to define the flocculation state of the mixed liquor entering the test clarifier during the stress tests. The test was developed by Parker *et al.* (1970).

(5) For the same reason cited above in (2), sludge blanket levels in the test clarifier in response to a typical diurnal flow pattern was of concern in the draft

protocol. With the emphasis on flow control in the revised protocol and the fact that these responses can be assessed from the results of a settling flux analysis, this is not a concern. A settling flux analysis is performed each day a stress test is performed since settling characteristics can change.

(6) The primary reason for determining the air flow rates in the aeration basins and mixed liquor transfer channels in the draft protocol was to quantify energy inputs that may affect floc integrity. The results from the IDS tests, performed repeatedly during the conduct of each stress test, are used instead as a direct measure of the flocculation state of the mixed liquor entering the test clarifier during the stress tests.

(7) Results from a solids flux analysis can be used to assess the possible redistribution of biosolids between the aeration basin and secondary clarifier during a hydraulic forcing of the clarifier (Keinath *et al.*, 1977).

(8) The hydraulic characteristics of the test clarifier play an important role in suspended solids removal. Two tests are performed to assess the test clarifier's hydraulic characteristics: the slug dye and solids distribution/flow pattern tests. These tests originally were used for this purpose by Crosby (1980). Unique to the CRTC protocol is the fact that these tests are performed under steady flow conditions at several surface overflow rates (SORs). Additionally, these tests are performed after steady-state conditions have been established in the test clarifier. Steady state is assumed to exist after steady-flow conditions have been maintained for a period of time equaling three times the theoretical hydraulic residence time (HRT).

<u>Stress tests</u>. The purpose of the stress tests is to obtain data to establish rating curves for the test clarifier. In the CRTC protocol clarifier performance is defined in terms of ESS concentration. Rating curves are plots of performance (i.e., ESS concentration) versus SOR, solids loading, RSSS concentration, RAS flow rate, settling characteristics, dispersed solids concentration, sludge blanket depth, etc.

The stress tests are performed randomly at several targeted SORs. The range of SORs to be considered depends on hydraulic constraints that exist at the facility and the ability of the test clarifier and other on-line clarifiers to withstand, without failure, conditions of thickening overloads. At high SORs, the test clarifier likely will be overloaded with respect to thickening. If sufficient storage volume is available in the test clarifier and other on-line clarifiers to achieve a steady-state sludge blanket depth in the test clarifier, the test can be taken to completion (i.e., three HRTs). In contrast, if the sludge blanket in the test clarifier continues to propogate to the tank's surface, threatening a potential loss of blanket solids in the effluent, the test must be abandoned.

During each stress test, the protocol requires on-line measurement of effluent flow rate (or influent flow rate), return sludge withdrawal flow rate, MLSS concentration, RSSS concentration, and ESS concentration. Signals from these instruments are accessed and stored frequently for the duration of each test. Examples of instrumentation and data acquisition to achieve this are described elsewhere (Wahlberg *et al.*, 1993a, 1993b, 1994).

A stress test is performed by setting the influent and RAS flow rates to achieve a desired effluent flow rate, or SOR. Adjustments then are made only to the influent flow rate to maintain the SOR as constant as possible. These flow conditions are maintained for a period of time equalling at least three times the theoretical HRT. During this period, the signals from the on-line sensors are accessed and stored. In addition, the flocculation and settling characteristics of the mixed liquor entering the test clarifier are assessed by performing the SSVI, ZSV, IDS, and settling flux tests. The SSVI and ZSV tests are performed in accordance with standard methods (Greenberg *et al.*, 1992) except that the settling column described by White (1976) is used. The IDS test is performed using a 4.2-L Kemmerer sampler. A suggested apparatus and procedure for performing the settling flux analysis is described by Wahlberg and Keinath (1988). Although the draft protocol included these tests in the pre-test information component, changing sludge settling and flocculation characteristics require they be conducted during the stress tests. Finally, the sludge blanket depth is measured frequently during the stress tests to ensure clarifier failure (i.e., rising blanket to the tank's surface) is avoided.

As indicated, the slug dye and solids distribution/flow pattern tests are performed at several SORs. Two approaches have been used: (1) the tests are performed at each of the SORs tested during the stress tests, and (2) the tests are performed at SORs corresponding to the low, medium, and high SORs of the range of SORs tested during the stress tests. Although the hydraulic characterization tests originally were part of the pre-test component of the protocol, it is convenient to do these tests in conjunction with the stress tests. This is accomplished by performing either a slug dye test or a solids distribution/flow pattern test after a stress test was completed. This ensured the requisite three HRTs had passed to assume steady-state conditions existed in the test clarifier prior to the initiation of the hydraulic tests. This also streamlines the execution of the protocol.

Data analysis and interpretation. The protocol generates considerable data. The first step in analyzing the data is to plot effluent flow as a function of time. From this plot, the period of most constant flow lasting at least one theoretical HRT is identified visually. For this period the effluent flow, RAS flow, MLSS, RSSS, and ESS concentrations are averaged. As ESS concentration is used to define performance, plots are made of this variable as a function of SOR, RAS flow rate, RSSS concentration, solids loading rate, SSVI, ZSV, dispersed solids concentration, the settling flux parameters (V_0 and K), and sludge blanket. These plots then are used to explain any variation observed in the ESS concentration.

Concluding Remarks

The intent of the CRTC in developing the protocol is to provide the profession with a detailed and rigorous methodology for evaluating and comparing the performance of different clarifier designs and operational strategies. Use of the protocol on an industry-wide basis will afford the profession the data necessary to establish sound, rational design and operational guidelines as to the optimum depth, shape, inlet and outlet configurations, type of sludge withdrawal equipment, SOR, weir loading, solids loading, blanket depth, settling characteristics, RAS withdrawal rate, etc. While the protocol undoubtedly will be revised with further application, it represents a rational beginning from which the profession can gain the data necessary to base secondary clarifier design and operation on proven scientific knowledge.

A detailed copy of the protocol is available from the corresponding author.

References

CRTC (1988). *Field Test Protocol for Evaluating Secondary Clarifier Rating Curves (Outline and Notes: Draft No. 2)*, unpublished manuscript.

Crosby, R.M. (1980). *Hydraulic Characteristics of Acti-*

vated Sludge Secondary Clarifiers, EPA Contract No. 68-03-2782, U.S. EPA, Cincinnati, Ohio.

Das, D., T.M. Keinath, D.S. Parker and E.J. Wahlberg (1993). "Floc breakup in activated sludge plants," *Water Environ. Res.*, **65**, 138.

Greenberg, A.E., L.S. Clesceri and A.D. Eaton, Eds. (1992). *Standard Methods for the Examination of Water and Wastewater*, American Public Health Association, Washington, D.C.

Keinath, T.M., M.D. Ryckman, C.H. Dana and D.A. Hofer (1977). "Activated sludge -- unified system design and operation," *J. Environ. Eng.*, **103**, 829.

Parker, D.S., W.J. Kaufman and D. Jenkins (1970). *Characteristics of Biological Flocs in Turbulent Regimes*, SERL Report No. 70-5, University of California, Berkeley, Calif.

Tekippe, R.J. and J.H. Bender (1987). "Activated sludge clarifiers: design requirements and research priorities," *J. Water Pollut. Control Fed.*, **59**, 865.

Wahlberg, E.J. and T.M. Keinath (1988). "Development of settling flux curves using SVI," *J. Water Pollut. Control Fed.*, **60**, 2095.

Wahlberg, E.J., J.F. Stahl, C. Chen and M. Augustus (1993a). "Field application of the Clarifier Research Committee's protocol for evaluating secondary clarifier performance: Rectangular, co-current sludge removal clarifier," *Proceedings of the WEF 66th Annual Conference & Exposition*, **3**, 1.

Wahlberg, E.J., M.A. Peterson, D.M. Flancher, D. Johnson and C.S. Lynch (1993b). "Field application of the CRTC's protocol for evaluating secondary clarifier performance: A comparison of sludge removal mechanisms in circular clarifiers," presented at the 1993 Annual Meeting of the Rocky Mountain Water Pollution Control Association, Albuquerque, N.M.

Wahlberg, E.J., J. Zhang, L.A. Carrio and G. Pappaceno (1994). "Performance evaluation of the final settling tanks at New York City's Rockaway Water Pollution Control Plant using the CRTC protocol," presented at the 66th Annual Meeting and Exhibition of the New York Water Environment Association, New York, N.Y.

White, M.J.D. (1976). "Design and control of secondary settlement tanks," *Water Pollut. Con.*, **75**, 459.

Sludge Hopper Design for Activated Sludge Clarifiers

O. E. Albertson[1]

Abstract:

Only in the past few years has any attention been allocated to the transport characteristics of sludge collectors. However, no attention has been given to a design rationale for the sludge hopper. The configuration of a sludge hopper effects clarifier performance and restriction in the sludge removal zone causes a response similar to scraper deficiencies. Further, the design practice is obsolete as it was developed at clarifier solids loadings 30-50% of current MLSS designs for advanced wastewater treatment. This paper provides a set of guidelines for hopper design and placement based on the projected sludge flow profiles within the sludge removal zone.

Introduction

At this time there is no rationale procedure for the design of the sludge hopper. Yet, design and placement influence the capacity of the aeration-clarifier system. The underflow is typically 25 to 40% of the clarifier feed and represents a very large point source of liquid removal from the basin. A hopper of insufficient size will cause high localized velocities, encouraging 'rat holing' or short-circuiting of the overlying liquid and dilute the underflow. The result is an increased RAS/Q ratio to maintain the aeration MLSS which extends the problem of clarifier solids transport and removal. Deficiencies in hopper design and placement can create short-circuiting with increased sludge inventory and sludge blanket depths. In setting guidelines for sludge hopper design and placement there is a two-fold objective: open a discussion of operating characteristics of the hopper and rationalize a set of preliminary sizing guidelines using logic and application of engineering judgment to advance methodology.

---

[1]Enviro Enterprises, Inc., PO Box 65312, Salt Lake City, UT 84165 USA

An important step in establishing a foundation for analysis and discussion is to visualize a sludge flow pattern. The sludge flow pattern induced by the scrapers would be a spiral to the hopper (Warden, 1981) as illustrated in Figure 1. The length of this spiral and time of travel (t_t) is dependent on the scraper design (angle of attack, α, and blade depth, d_b), tip speed (V_{SP}) and transport efficiency (Albertson and Okey, 1992).

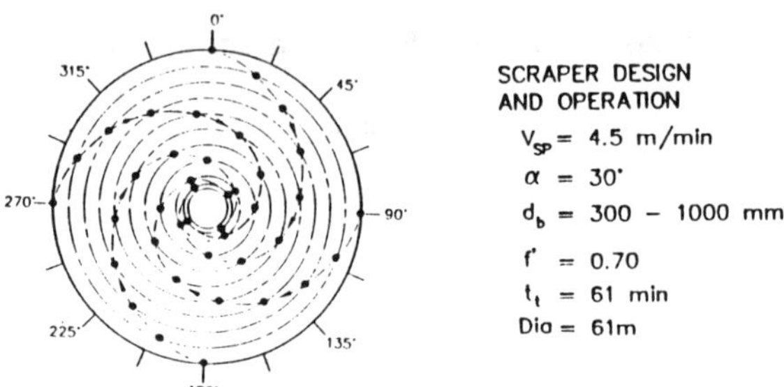

Figure 1. Sludge Path Produced by a Tapered Spiral Collector.

Analysis and Discussion

Albertson and Okey (1992) and the International Association of Water Quality (IAWQ, 1992) recommended specific criteria to encourage the formation and removal of consolidated activated sludge from the circular clarifier. The important considerations were:

1. The sludge transport capacity (V_{SC}) of the scrapers should exceed the rate of return sludge (RAS), i.e., V_{SC} > RAS.
2. Placement of the sludge hopper centerline at about 15% of the tank radius to facilitate the sludge removal.
3. Construction of an adequate floor slope to provide a minimum consolidated sludge depth without the blanket level exceeding the conical zone at average flow conditions.
4. The floor should be flat from the centerline of the hopper to the center column and short outward raking scrapers installed to move sludge toward the hopper.
5. Spiral scrapers, tapered to match sludge depth, have better transport characteristics than straight blades and are preferable to the rsr types.
6. Secondary 1/4-1/2 radius scrapers are often required to enhance transport capacity in larger and heavily loaded clarifiers.

The capacity, or size, and the location of the sludge hopper for activated sludge clarifiers require a different conceptual approach than a primary clarifier where only 0.3-1.0% of the influent flow is removed through the sludge hopper. The sludge hopper retention time may be several minutes while in the secondary clarifier, hopper retention times would generally be a few seconds. Enlarging the hopper has limitations, and there still is a necessity to assure that dense sludge is being moved continuously to the hopper at the rate being withdrawn. The hopper placement must be sufficiently set away from the basin center to reduce restriction of the sludge flow. The IAWQ recommendation (1992) for hopper placement at 0.15 R in an activated sludge clarifier will be revisited in this analysis.

Figure 2 illustrates the hopper and scraper arrangement in the secondary clarifier. Initially a two-dimensional approach helps put the flow modes into perspective. To visualize the flow, imagine that the sludge mass is stationary, and the hopper rotates; a motion similar to the hydraulic flow effects of the rapid sludge removal (rsr) clarifier mechanism. In the case of the rsr, the primary sludge flow is from the front side of the pipe in the apex of the V plow. If the pipe were placed in the middle of the V-plow, flow could enter from all sizes (Albertson, 1991). However, the flow into the leading edge will benefit from the velocity of the mechanism at that point on the radius. As the point of removal approaches the center, this effect will lessen.

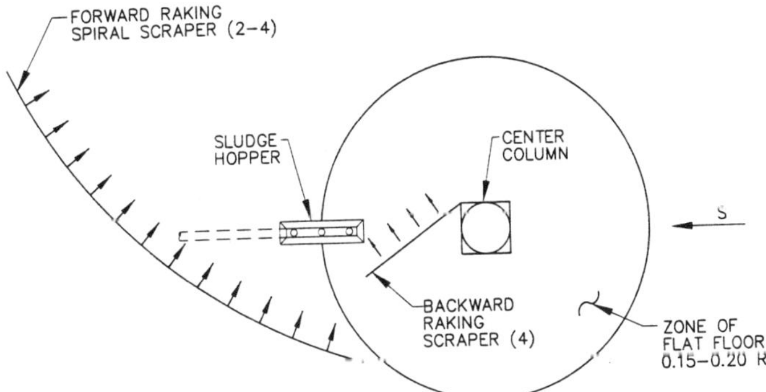

Figure 2. General Arrangement of Scrapers & Hopper with Multiple Inlets (Rake support mechanism not shown)

Figure 3 illustrates the basic objective of removal of the dense phase of sludge without short-circuiting. Some observations and comments are appropriate.

1. The withdrawal of the sludge will result in a density gradient in the area of the hopper inducing flow at some unknown and variable velocity (V_1, V_2, V_3, and V_4).

2. This induced velocity may decrease with increasing solids concentration; i.e., higher viscosity offsets increased density.
3. Sludge will be moved toward and away from the hopper at the velocity of the scraper in this area and be modified by the density gradient formed by the depressed dense sludge blanket over the hopper.
4. If short-circuiting occurs, it will tend to sustain itself by reducing density differences and short-circuiting occurs whenever the sum of the hopper approach velocities through their respective sludge depths is less than the RAS rate.
5. Vortexing must be discouraged since its energy disrupts density driven flow.

Backfilling of dense sludge to help replace the RAS is easier to envision with the hopper moving and the sludge stationary. Sludge flows into the hopper from all sides, but the highest rate would be the leading edge as the density driven component is aided by the velocity of the hopper (scraper). The amount solely attributed to the scraper speed can be estimated which would then define the amount which would need to be driven by density effects to the hopper to prevent short-circuiting.

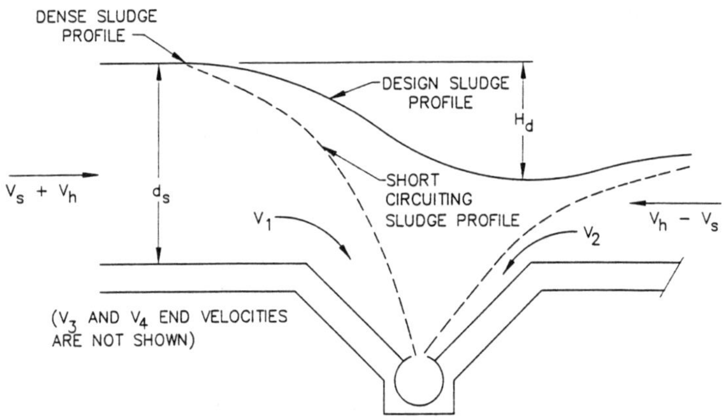

Figure 3. Characterization of Dense Sludge Flow Into Hopper

The logical actions to minimize short-circuiting include:

1. Increase scraper velocity, V_S
2. Increase dense phase sludge depth, d_S'; where d_S-H_d $\geq$ 0.3 m
3. Expand hopper cross-section, reduce the hydraulic velocity, V_h
4. Move the hopper further from center to intercept at increased scraper velocity, V_S

5. Provide two hoppers in larger basins, > 50-55 mØ.

Expanding the hopper cross-section to reduce the required velocity into the hopper seems to be a logical approach. Further, extending the hopper length along the radius would expand the interruption of the spiralling sludge flow and reduce the drawdown (H_d) at the hopper. Lastly, multiple, outlets, with at least 200 mm of head loss across the orifice, should reduce self-sustaining currents (vortices) which support short-circuiting.

These considerations support employment of a long narrow hopper located on a radial line. A longer hopper can employ multiple draw-off ports to further limit drawdown and vortex induced short-circuiting. As the sludge concentration is increased, the higher sludge viscosity should suppress vortexing and enhance density driven transport of sludge.

The circular trough design with a single or dual discharge pipe will still have a constricted flow path to the hopper. The radial hopper with multiple outlets located transverse to the circulating sludge flow will be better positioned to collect and remove the consolidated sludge without short-circuiting.

The sludge hopper located on a radial line would require a length consistent with the volume of sludge intercepted and to be removed. There would be sludge flow from front, ends and backside of the hopper. End and backside flow into the hopper are significant, variable with scraper velocity, and effect the design rational for the hopper. A slower moving scraper partially blocks flow into the hopper when it is in the vicinity of the hopper. The sludge hopper should have multiple RAS inlets as a function of basin size and the RAS flow. The primary consideration may be the length since width may have limited flow over the ends of the hopper aided by the pressure exerted by the forward and backward raking blades and density effects generated by sludge depth and concentration.

The design rational for the hopper employs the following premises:

1. Hydraulic influence is directly related to the sludge velocity.
2. Scraper induced velocity adds or subtracts to hydraulic flow.
3. The minimum dense phase sludge depth over the hopper is 0.3 m; i.e., drawdown (H_d) is d_s - 0.3m.
4. The area of hydraulic influence (drawdown) extends 2.0 m in all directions from the hopper and flow acceleration and drawdown is initiated at that point.
5. Scrapers do not extend over the hopper zone and back flow of sludge into the hopper occurs.

Short-circuiting of this larger than conventional sludge hopper would persist unless a second large hopper is added at about 180°, or a new centrally located hopper design favoring the hydraulic flow pattern is developed. Increasing the level of short-circuiting as the rate of flow which is hydraulically driven cannot increase without first increasing the depth of sludge. A larger sludge inventory is often detrimental to the performance.

Larger hoppers and multiple collection pipes would not completely solve the problem of central sludge removal from basins with nearly flat or shallow floor slopes. Since the concentrated sludge depth would be 0.5-0.75 m, short-circuiting would still occur. Further, constructing the large hoppers in existing basins would be costly, time consuming, and can incur construction problems such as excessive groundwater.

Developing a new centrally located hopper design favoring hydraulic flow patterns was initially conceived to solve a sludge removal problem in retrofit of a 41 mØ header collection unit to tapered spiral scrapers and center sludge draw-off. The floor slope was only 0.01 m/m. The recommended design was a circular sludge collector with multiple ports placed on the floor of the existing basin and connected to the RAS line. The inlet ports would be arranged to ensure a uniform flow into the collector. The design of the central collector is illustrated in Figure 4.

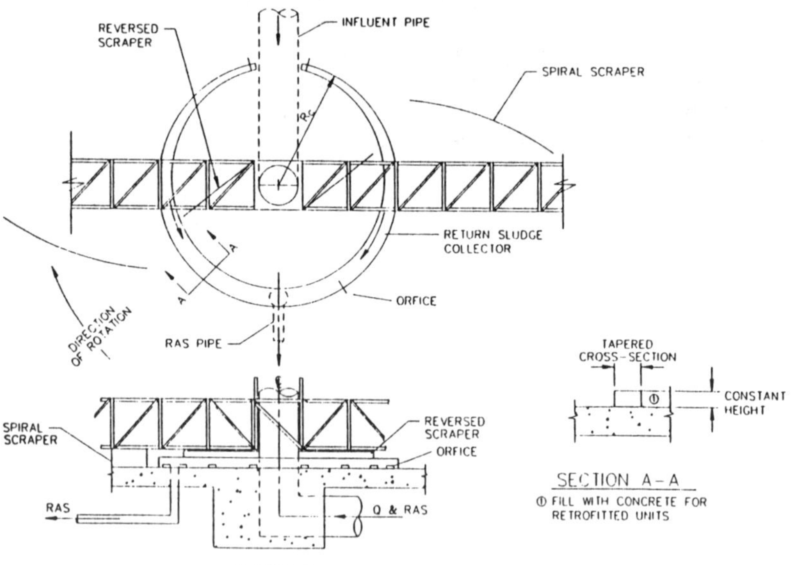

Figure 4. Arrangement of the Central Sludge Collector.

The cross-section of the collector increases with flow to provide a constant velocity allowing use of a uniform spacing and size of inlet. The shallow floor slope and resulting shallow sludge blanket encourages the use of the highest number of openings, but not so small as to cause fouling. The recommended minimum size of opening is 250-300% of the influent bar screen spacing.

The RAS will be withdrawn uniformly around the periphery of the collector enhancing the inward spiral flow produced by the spiral scraper and encouraged by the Coriolis Effect. The removal of sludge at multiple points will control short-circuiting of dilute MLSS to the RAS. The modified spiral blade will present a horizontal barrier to prevent sludge backflow to areas of lower sludge depth until the sludge reaches the zone of collector. The hydraulic effects will extend outward 1-2 m from the collector. Settled sludge in the inner zone will be scraped outward to the collector to prevent deterioration.

Clarifiers with a more conventional slope can employ fewer inlets and the size of the openings can be increased proportionally. This design also reduced the minimum sludge depth required to prevent short-circuiting.

Summary and Conclusions

This work revealed that density motivated and localized hydraulic forces either result in sludge or overlying liquid flowing into the hopper. The analysis focused on conditions which would favor the dense sludge being preferentially removed by the hydraulic forces since even enhanced scraper transport could not provide adequate RAS flow.

The efficient removal of dense sludge is dependent on several factors and are generally summarized as follows:

1. Ensure adequate scraper capacity is available from tank periphery to vicinity of hopper.
2. Employ a floor configuration which provides sufficient hopper coverage (sludge depth) without excessive sludge inventory; use steeper or dual slope designs.
3. Expand scraper capacity in the area of the hopper located at 12-20% (centerline) of the tank radius.
4. Construct the enlarged conventional hopper with a l/w of 3-4 and use multiple sludge withdrawal ports to minimize localized drawdown and vortexing; average vertical flow velocity into the hopper at $\leq$ 3 m/min.
5. Consider an alternative circular collector design which uniformly removes the concentrated sludge at several points on a 15-20% radius. This design will minimize sludge inventory and will be effective to modify clarifiers with minimal floor slopes to central draw-off.

Retrofitting units with shallow floor slopes will favor use of the floor mounted circular collector. Larger clarifiers which have demonstrated problems of sludge hopper short-circuiting may need both scraper modifications to improve outer sludge transport and changes in the hopper configuration. Cost and effectiveness considerations would favor the use of the circular collector. Reducing the sludge blanket depth using the circular collector increases the volume of the clarification zone and can improve effluent quality, flow capacity, or both. The resulting lower RAS/Q ratio also is conducive to improved performance.

This analysis does not supplant the need for field studies using equipment which can determine the flow characteristics in the area of the hopper. This work could be extensive and time-consuming and may not be available for several years. This approach will improve the underflow solids concentrations and thereby reduce the RAS/Q ratio or allow maintenance of the higher MLSS which is more desirable for efficient AWT and BNR processes.

References

Albertson, O.E. 1991. "Improving the Rapid Sludge Removal Collector," presented at the 64th Annual Conference of the Wat. Environ. Fed. Toronto, Ontario, Canada.

Albertson, O.E. and R.W. Okey. 1992. "Evaluating Scraper Designs," Water Environ. Technology. **4**(1): 52-58.

International Association of Water Quality. 1992a. Chapter 5, "Clarifier Design" in Design and Retrofit of Wastewater Treatment Plants for Biological Nutrient Removal, by O. Albertson. Randall, C. *et al.,* editors, Technomic Publ., Lancaster, PA.

International Association of Water Quality. 1992b. Chapter 8, "Control of Bulking and Foaming Organisms" in Design and Retrofit of Wastewater Treatment Plants for Biological Nutrient Removal, by O. Albertson. Randall, C. *et al.,* editors, Techniomic Publ., Lancaster, PA.

Warden, J.H. 1981. "The Design of Rakes for Continuous Thickeners." *Filtration and Separation*, **18**(2):113-114, 116.

Evaluation of Fe oxide-coated GAC for removal and recovery of Cu(II) from water

T.C. Wang, K.P. Chandra, P.R. Anderson[1]

Abstract

A composite solid, prepared by precipitating an Fe oxide onto granular activated carbon (GAC), was evaluated as an adsorbent for the removal and recovery of Cu(II) from water. Relative to adsorption onto uncoated GAC, Cu(II) adsorption capacity increased as the amount of Fe oxide increased, from about 1.4 mg/g for GAC to 5 mg/g for GAC coated with 37 mg Fe oxide/g GAC. Tests in a column process for Cu(II) removal demonstrated that the composite adsorbent could be reused through at least 15 adsorption and desorption cycles. Although a fraction of the adsorbed Cu(II) was retained by the solid, there was no apparent loss in adsorption capacity. Treatment of low concentration solutions was effective; a 100 μg Cu(II)/L solution was reduced to no more than 3 μg/L through 1000 bed volumes processed.

Introduction

A potential treatment technique for the recovery and recycle of metals from water involves their adsorption onto oxide adsorbents. Adsorption processes have long been recognized as playing an important role in the retention of metals by soils (Jenne, 1968) and potential applications of Fe oxides to industrial treatment have been discussed by Benjamin et al. (1982).

The purpose of this research project was to investigate the performance of an Fe oxide-coated carbon to treat metal-bearing waste waters. Previous studies have explored activated carbon by itself as an adsorbent for heavy metals, and observed that it was most effective in the presence of complexing agents (Bhattacharyya and Cheng, 1987; Ku and Peters, 1987). Activated carbon that is partially or fully coated with Fe oxide could be an effective adsorbent for both free and complexed metals.

Methods

Granular activated carbon (GAC) used in this experiment was Calgon Carbon Corporation type TOG, provided courtesy of the manufacturer. According to the product brochure, this GAC has a total N_2 BET surface area of 800-900 m^2/g, an apparent bulk density of 0.54 g/cm^3, a particle density of 0.78 g/cm^3, and a pore volume fraction of 0.82.

----

1. Pritzker Department of Environmental Engineering, Illinois Institute of Technology, 3201 South State Street, Chicago, IL 60616.

Prior to use in this study the GAC was sieved between 425-600 μm, baked in an oven at 200°C for 2 d, washed several times with deionized water, filtered, completely dried in an oven at 110°C, cooled in a desiccator, and stored at room temperature in a covered glass container until further use.

Iron oxide-coated GAC (FeGAC) was prepared by mixing GAC with a known amount of $Fe(NO_3)_3$ solid and adding enough deionized water to cover the solid. The mixture was dried in an oven at 90°C, cooled to room temperature, and washed several times with deionized water to remove detachable and discrete Fe oxide. The resulting FeGAC composite adsorbent was dried in the oven at 105°C and stored at room temperature in a covered glass container until needed. For comparison purposes, a reference Fe oxide was prepared following the same procedures except the GAC was omitted.

Batch adsorption tests were conducted in a 0.01 M $NaNO_3$ background electrolyte, and adsorbate was added from $Cu(NO_3)_2$ stock solutions. Samples were filtered (0.45 μm) and acidified prior to analysis by AAS. Surface chemical characteristics of some of the solids were also evaluated using acid or base in an indifferent electrolyte of constant ionic strength (0.01 M $NaNO_3$) under a nitrogen atmosphere.

In all column adsorption tests, 15 cm^3 of medium was placed in the column (1.5" O.D., 1.0" I.D. polymethylmethacrylate tubes) and the resulting bed depth was about 3 cm. The flow rate was 5 cm^3/min and the theoretical hydraulic residence time was 1 min.

Results and Discussion

Three different composite adsorbents were prepared. These were 37FeGAC, 52FeGAC, and 72FeGAC, with 37 mg, 52 mg, and 72 mg of Fe per g of GAC, respectively. N_2-BET surface areas for the solids were 789, 742, 673, 634, and 92 m^2/g for GAC, 37FeGAC, 52FeGAC, 72FeGAC, and Fe oxide, respectively. Diffractograms for x-ray analyses revealed distinct x-ray peaks for discrete Fe oxide, suggesting it was predominately hematite. The discrete GAC and all three composite solids were x-ray amorphous.

After a 96 h batch adsorption test, the amount of Cu adsorbed on the 52FeGAC and 72FeGAC composite adsorbents was about the same as the amount adsorbed on the Fe oxide (Figure 1). Because GAC by itself had a relatively small Cu(II) removal capacity, most of the removal by the composite adsorbents must have been due to the presence of the Fe oxide. To explore this concept further, the adsorption isotherm data were normalized to the amount of Fe present (Figure 2). Interestingly, all three of the composite adsorbents appear to have the same adsorption capacity for Cu(II). In addition, relative to the discrete Fe oxide, this normalized capacity of the FeGAC adsorbents is at least three times greater.

To assess surface response to potentiometric titration, data from the titration experiments for GAC, discrete Fe oxide, and 37FeGAC were evaluated using FITEQL V.2.0 (Westall, 1982). This evaluation was based on the assumptions that hydroxyl groups were the dominant surface functional groups for each of the solids, and that surface complexation reactions could be described by the diffuse layer model. More detailed descriptions of these types of surface reactions are available from Dzombak and Morel (1990).

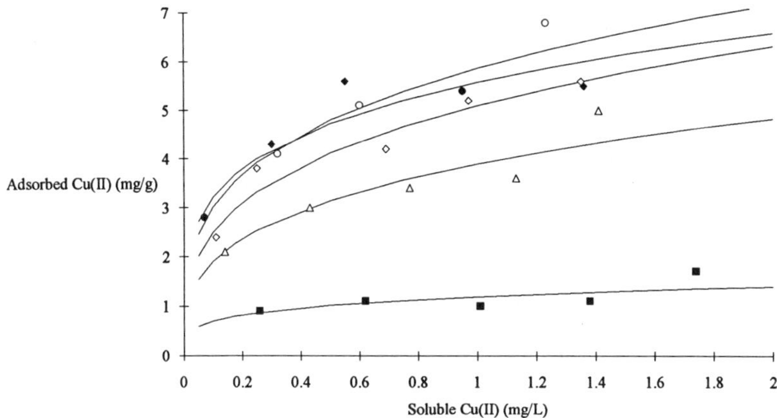

Figure 1. Cu(II) adsorption isotherms for various adsorbents. Solutions were in contact with the solid for 96 h at pH = 6. Solid lines in the figure are Freundlich isotherms. Symbols are 37FeGAC (Δ), 52 FeGAC (◇), 72 FeGAC (○), GAC (■), and Fe Oxide (◆).

Initial input to the FITEQL model included the titration data, adsorbent concentration, and adsorbent specific surface area. However, the model did not converge using these data, so we re-evaluated the titration data using surface area as a fitting parameter. In these tests, the set of model parameters (log K_+, log K_-, site density, surface area) that provided the best fit to the model (lowest values of sum-of-squares/degrees of freedom) was selected as the best description of the surface (Table 1).

Our results for GAC are consistent with the work of Corapcioglu and Huang (1987), who examined the surface chemical characteristics of 14 commercially available GAC's. With the possible exception of the pH_{ZPC}, the surface characteristics for Fe oxide (Table 1) are also in a reasonable range. Measured values of pH_{PZC} for Fe oxides are typically between pH 7 and 9 (Parks, 1964). Except for the surface area estimates, the composite adsorbent (37FeGAC) has surface chemical characteristics that are different from the mass-weighted sum of the two component parts. Surprisingly, the composite solid has a lower pH_{ZPC} value than do either of the components. A comparison of the surface area estimates in Table 1 is especially interesting; the results suggest that FITEQL could be used to provide a meaningful in situ assessment of surface area. For example, our results suggest that Fe oxide on the surface blocked over 50% of the GAC surface when the solid was in solution. After drying, however, the surface Fe oxide blocked less than 10% of the GAC surface.

Removal data were also generated from the column experiments. To assess the capacity of the 37FeGAC, a test was conducted in an attempt to reach complete breakthrough.

However, after about 2000 bed volumes were processed, something changed in the removal process and the effluent Cu(II) concentration appeared to be in a steady state condition (Figure 3). The results may reflect a surface precipitation process in which a solid such as $Cu(OH)_2$ or $CuOHNO_3$ is precipitating at the surface (Farley et al., 1985).

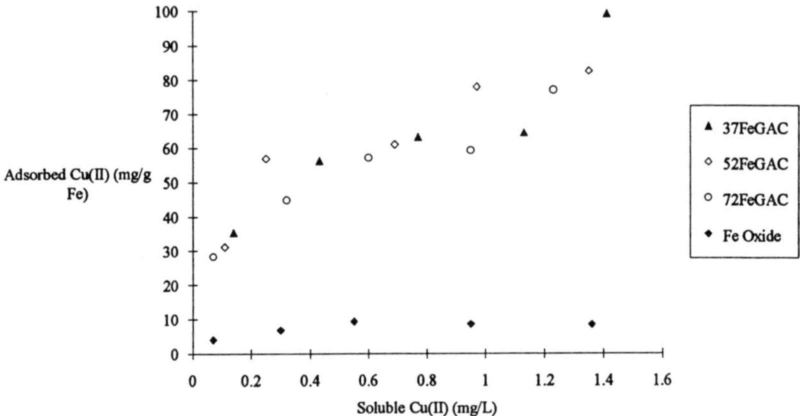

Figure 2. Cu(II) adsorption isotherms normalized to Fe content. Solutions were in contact with solid for 96 h at pH = 6.

Removal capacities estimated from the batch adsorption tests were obtained by using isotherm parameters and extrapolating to calculate the surface loading in equilibrium with a concentration of 2 mg/L. Capacity estimates from column tests were obtained by integrating the area outlined by the breakthrough curves. Because 100% breakthrough was not observed for Cu(II) removal onto 37FeGAC (Figure 3) and because removal after 2000 bed volumes may be due to surface precipitation, the capacity estimate was based on total removal up to 2000 bed volumes. Column and batch capacity estimates for Cu(II) removal by GAC were 1.3 mg/g and 1.4 mg/g, respectively. Column and batch capacity estimates for Cu(II) removal by 37FeGAC were 5.0 mg/g and 4.9 mg/g, respectively.

Table 1. Surface chemical characteristics for GAC, Fe oxide, and 37FeGAC.

Parameter	GAC	37FeGAC	Fe Oxide
$\log K_{+(int)}$	6.30	5.27	8.29
$\log K_{-(int)}$	-7.69	-7.34	-10.60
binding sites (mol/g)	2.0×10^{-4}	3.1×10^{-4}	2.7×10^{-4}
pH_{ZPC}	7.0	6.3	9.5
Model surface area (m^2/g)	800	450	300
N_2 BET surface area (m^2/g)	789	742	92

Additional information about the Cu(II) removal process can be found by examining how the pH of the effluent changes as a function of the number of empty bed volumes processed (Figure 4). Protons released from the solid during adsorption react with surface sites that are not fully protonated. This reaction is equivalent to an acid titration of the medium downstream from the Cu(II) adsorption mass transfer zone (MTZ). In effect, there is a proton MTZ preceding the Cu(II) MTZ through the column. As a result, the column has a pH buffering capacity that modifies the effluent pH curve. The effluent pH curve for FeGAC follows the expected pattern, initially decreasing to an apparent steady state value and gradually increasing toward the influent pH value (Figure 4).

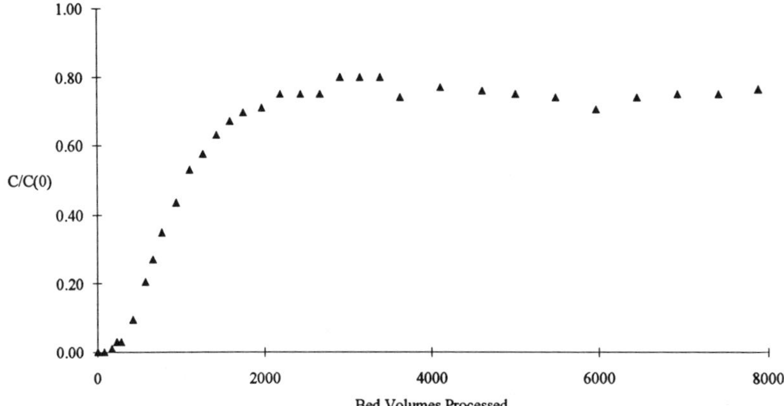

Figure 3. Breakthrough curve for Cu(II) removal by 37FeGAC. Influent Cu(II) = 2 mg/L, loading = 5 mL/min, bed volume = 15 mL, and influent pH = 6.

Retention of adsorbed Cu(II) may influence the ability of the FeGAC adsorbent to be regenerated and reused. To explore this possibility, a column was processed through a series of adsorption and desorption cycles. The flow rate was 5 mL/min and one cycle lasted for 200 bed volumes (10 h). At the end of the cycle, the influent was stopped and the column was regenerated for 2 h by recirculating through the column a 0.01 M $NaNO_3$ solution adjusted to pH 2. The regenerant solution was removed from the column and preserved for analysis, and a new adsorption cycle was started.

A total of 15 cycles was processed (Figure 5). The highest effluent concentration was 0.5 mg/L and that occurred two times, at the end of cycles 2 and 9. The average effluent concentration throughout 3000 bed volumes was 0.06 mg/L, or 97% removal. After 15 adsorption and desorption cycles, the adsorbent showed no signs of decreasing performance.

In this same recycling experiment, the amount of Cu(II) recovered by the regeneration solution was compared to the amount of Cu(II) adsorbed in a cycle (Figure 6). Interestingly, regeneration efficiency increased with increasing cycle number, from only 30% in the first cycle to about 98% in cycle 15. In other words, the fraction of Cu(II) retained by the adsorbent decreased from about 60% in the first cycle to less than 5% in cycle 15. Furthermore, this retained Cu(II) apparently did not interfere with subsequent adsorption cycles. As a result, the accumulated recovery efficiency, also shown in the figure, increased so that the fraction of retained Cu(II) became smaller with increasing number of cycles.

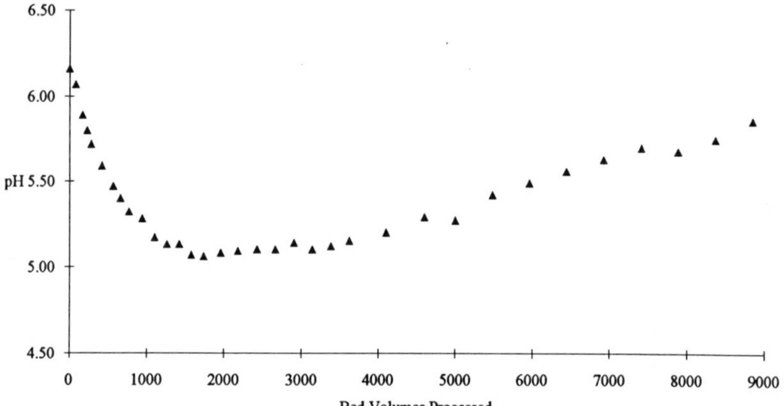

Figure 4. Effluent pH for Cu(II) removal by 37 FeGAC. Influent Cu(II) = 2 mg/L, loading = 5 mL/min, bed volume = 15 mL, influent pH = 6.

One potentially attractive application of oxide-based adsorption processes is as a polishing step, in which metals can be removed to trace levels. To evaluate the FeGAC adsorbent for processing low concentrations of metals, a feed solution containing 100 μg Cu(II)/L adjusted to pH = 6 was pumped through the column. In the first 1,000 bed volumes processed (Figure 7) the effluent Cu(II) concentration was no more than 3 μg/L in all samples.

Conclusions

Fe oxide-coated GAC performed well as a composite adsorbent for the removal of low levels of Cu(II) from solution. Results from these studies indicate that this solid can be used in column adsorption processes to not only remove but also recover adsorbed Cu(II). This technique looks especially promising for treatment to low levels; a column loaded with 100 μg Cu(II)/L produced an effluent with no more than 3 μg Cu(II)/L through 1000 bed volumes processed. Removal capacities of at least 5 mg Cu(II) per g adsorbent were achieved in column tests. The adsorbent was regenerated using a solution adjusted to pH = 2 and the amount of Cu(II) recovered increased from about 30% in the first cycle to over 90% in subsequent cycles. Retention of adsorbed Cu(II) by solid did not interfere with subsequent adsorption steps through 15 cycles.

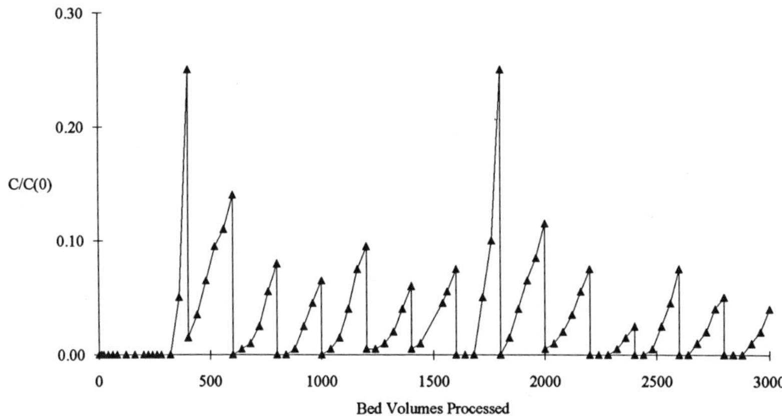

Figure 5. Breakthrough curves for Cu(II) removal onto 37FeGAC through 15 cycles of regeneration. Influent Cu(II) = 2 mg/L, loading = 5 mL/min, bed volume = 15 mL, influent pH = 6.

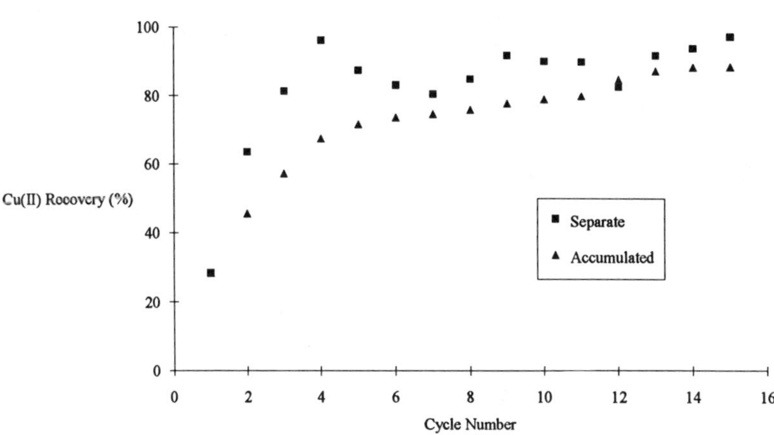

Figure 6. Recovery of adsorbed Cu(II) following removal onto 37FeGAC through 15 cycles. Each cycle was regenerated at pH = 2 for 2 h.

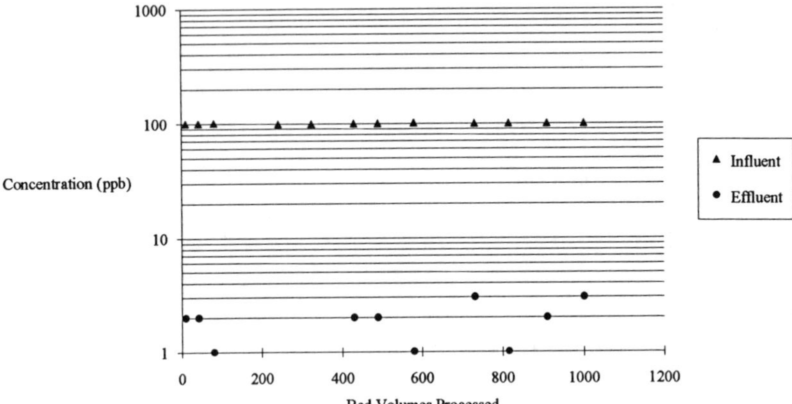

Figure 7. Low level Cu(II) removal onto 37FeGAC. Loading = 5 mL/min, bed volume = 15 mL/min, and influent pH = 6.

References

Benjamin, M.M.; Hayes, K.F.; Leckie, J.O. (1982) "Removal of toxic metals from power generation waste stream by adsorption and coprecipitation." *Journal Water Pollution Control Federation,* 54, 1472.

Bhattacharyya, D.; Cheng, R.C.Y. (1987) *Environmental Progress* Vol. 6, No. 2:110-118.

Corapcioglu, M.O.; Huang, C.P. (1987) "The surface acidity and characterization of some commercial activated carbons" *Carbon* 4:569-578.

Dzombak, D.A.; Morel, F.M.M. (1990) *Surface Complexation Modeling. Hydrous Ferric Oxide*. John Wiley & Sons, New York.

Farley, K.J.; Dzombak, D.A.; Morel, F.M.M. (1985) "A surface precipitation model for the sorption of cations on metal oxides." *Journal Colloid Interface Science* 106, No.1, 226-242.

Jenne, E.A. (1968). "Controls on Mn, Fe, Co, Ni, Cu and Zn concentrations in soils and water: the significant role of hydrous Mn and Fe oxides" in *Trace Inorganics in Water,* Advances in Chemistry Series 73, American Chemical Society, Washington, D.C.

Ku, Y.; Peters, R.W. (1987) *Environmental Progress* Vol. 6, No. 2:119-124.

Parks, G.A. (1964) "The isoelectric points of solid oxides, solid hydroxides, and aqueous hydroxo complex systems." Chemistry Review 65:177.

Westall, J.C. (1982) *FITEQL. A program for the determination of chemical equilibrium constants from experimental data. Version 2.0.* Chemistry Department, Oregon State University, Corvallis, OR.

Sensitized Photooxidation of Dissolved Sulfides in Water

T. F. Brewer, J.G. Curtis, E.A. Marchand, V.D. Adams, and E.J. Middlebrooks
Department of Civil Engineering, University of Nevada, Reno, NV

ABSTRACT

The ability of methylene blue (MB) and riboflavin (RF) to sensitize dissolved sulfides for photooxidation was investigated. Both MB and RF were found to be effective sensitizers for the oxidation of sulfide in water. MB-dosed batch reactors consistently reduced initial sulfide concentrations of 100 mg/l to less than 10-15 mg/l in less than one hour under artificial lighting (91% sunlight corrected fluorescent tubes) at a pH=10 and MB=1mg/l. Preliminary experiments have shown that approximately 80-85% of the removed sulfide is accounted for as accumulated sulfate. RF is also effective at enhancing the removal of sulfide, but experiments similar to those conducted for MB revealed that RF-dosed reactors required approximately 2-3 times longer to achieve sulfide removal comparable to MB (1mg/l), even with an RF concentration of 20 mg/l. The primary product in RF-sensitized photooxidation of dissolved sulfides is also sulfate, with approximately 75-80% of removed sulfide recovered as sulfate. First order plots of experimental data yield reaction rate constants of $k=0.0097\ min^{-1}$ for RF, and $k=0.0273\ min^{-1}$ for MB.

INTRODUCTION

A byproduct of the enhanced recovery of petroleum is flood water that is often contaminated with soluble sulfides (H_2S, HS^-, $S^=$). Underground reservoirs are flooded with water or steam to enhance recovery of petroleum and the flood water becomes contaminated with dissolved sulfides that are mainly produced through the reduction of sulfates by anaerobic sulfate-reducing bacteria (SRB) (Montgomery, et al., 1990).

The toxicity and corrosive properties of sulfides require that strict controls be enforced on their release into the environment and contact with process equipment such as tanks and piping. Sulfide production by SRB is directly or indirectly responsible for extensive damage to production equipment each year due to corrosion (NACE, 1976). In addition, sulfides are toxic to aquatic life, wildlife, and man. The effect is more pronounced in stream dwelling (particularly benthic) organisms however, due to continuous exposure. Consequently, the release of sulfides into the environment must be closely monitored to prevent adverse effects on the surrounding ecosystem. The presence of sulfides in natural waters decreases the concentration of dissolved oxygen which, in turn, heightens the sensitivity of aquatic life to the dissolved sulfides (National Research Council, 1976).

Two methods of control of sulfide contamination have been used or investigated to date. **First**, sulfide production may be prevented by inhibiting the growth of SRB. This can be accomplished (to some degree) by aeration of flood waters used in the secondary production of petroleum, or by addition of a biocide to the flood waters. These methods have had limited success due to attachment of SRB on solid surfaces, and entrapment within a slime layer. Within the slime layer, the SRB are somewhat protected from either the biocide or oxygen in the flood water (Montgomery, et al., 1990, and NACE, 1976).

If *prevention* of sulfide accumulation cannot be sufficiently accomplished by biological methods, then a **second** approach in the form of physicochemical treatment methods has been used to *remove* dissolved sulfides from contaminated waters. Perhaps the most common removal technique is the stripping of sulfide-laden waters under acidic conditions with steam or flue-gas. The dissolved sulfides are recovered as hydrogen sulfide and generally disposed of by incineration. This method serves to create an air pollution problem, however, since the incineration of H_2S produces sulfur dioxide (SO_2), which is a regulated pollutant (Noyes Data Corporation, 1973).

An approach to sulfide removal from contaminated waters that has not been attempted on a widespread scale is photodegradation. Photodegradation occurs by two processes: direct photodegradation and sensitized photooxidation. In the direct photodegradation process, excited molecules lose absorbed light energy by one of several deactivation routes including: dissociation, dehalogenation, and isomerization. Unfortunately, many pollutants absorb light primarily in the ultraviolet (UV, λ <350nm) region of the spectrum and the earth's ozone layer filters nearly all light below 290 nm. The absorption of many pollutants falls below the 290 nm cut-off, resulting in low rates for direct photodegradation and rendering it ineffective as an alternative for treatment of wastes (Watts, 1985).

In the sensitized photooxidation process, sensitizing molecules absorb light in the energy-rich visible region ($\lambda \approx 400$-700nm) of the spectrum. The excited sensitizing molecules may then return to the ground state by transferring the absorbed energy in one of two ways: TYPE I (radical mechanisms), and TYPE II (singlet oxygen mechanism). If molecular oxygen accepts energy from the sensitizer (TYPE II), then singlet oxygen, a highly reactive species of oxygen that will attack a variety of organic and inorganic substrates, is produced (Watts, 1985). TYPE I (sensitizer-substrate) reactions usually involve decomposition of the substrate by a mechanism such as hydrogen abstraction, dissociation, dehalogenation, or isomerization (Heelis, et al., 1981).

A wide variety of sensitizers have been studied for sensitizing capabilities on different substrates. Methylene blue (MB) has been shown to be effective in destruction of coliform and virus populations (Eisenberg, et al., 1987). In addition to sensitized photooxidation studies on biological substrates, a great deal of data has been collected on the photodegradation of organic compounds such as phenols, aldehydes, ketones, nitrogen heterocyclics, and aromatic heterocyclic compounds (Spikes and Straight, 1967). Eisenberg and Middlebrooks (1987), investigated the ability of MB and several other sensitizers such as acetone, malachite green, fluorescein, and rose bengal to photodegrade bromacil, an herbicide, as well as coliform in secondary effluent wastewater. Research has shown that MB sensitized photooxidation often occurs by TYPE II (singlet oxygen) mechanisms, making it an appealing choice for investigation in the photooxidation of sulfides.

RF has also been shown to be effective in degrading a variety of substrates. Watts, et al., studied the RF-sensitized photodegradation of terbacil (an herbicide) and also developed design criteria for sensitizer photooxidation treatment systems (Watts, 1985). A number of studies have shown that RF-sensitized photooxidation occurs by both TYPE I and TYPE II mechanisms. Penzer and Radda (1968) compared the flavin-sensitized mechanisms of amino acids and concluded that singlet oxygen was responsible for the photooxidation of adenosine and guanosine, but the RF-sensitized photooxidation of histidine was found to occur by hydrogen abstraction in addition to the singlet oxygen mechanism (Tomita, et al., 1969).

Both MB and RF are attractive dyes for use in photooxidation treatment of wastes. In addition to their sensitizing capabilities, both dyes are nonhazardous to the environment. MB has a very low toxicity to humans and most aquatic life. It is administered orally and intravenously in humans in the treatment of certain diseases and as an antidote to cyanide exposure (Merck Index, 1976). MB is also used as a treatment for parasitic infections ("ich" or white spot disease) and respiratory illnesses in a wide variety of fish, including trout. Prescribed treatments include immersion of the fish in waters containing 2-5 mg/l MB for periods of up to 7 days (Herwig, 1979, Post, 1983, and van Duijn, 1973). Additionally, MB-sensitized photooxidation often involves the production of singlet oxygen, a highly reactive radical of oxygen, which serves as the pathway for photooxidation. This characteristic makes MB especially appealing for the study of sensitized photooxidation of dissolved sulfides in water. RF is practically nontoxic and is a nutritional requirement (Vitamin B_2) for many species (Machlin, 1984). There have been no known cases of RF toxicity in humans and it has also been observed to be non-toxic in laboratory animals.

The objective of this investigation was to determine the relative ability of (MB) and (RF) to act as sensitizers in the photooxidation of dissolved sulfides. The effect of dye concentration on sulfide removal rates was determined for both RF and MB. Rates of sulfide removal and sulfate accumulation are compared for both MB and RF-sensitized reactions. Rate constants obtained from first order plots of experimental data are also compared.

METHODOLOGY AND MATERIALS

The feasibility of sensitized photooxidation as a treatment method for sulfide-contaminated waters was examined using MB, RF, batch reactors, and artificial sunlight supplied from wavelength corrected tubes (irradiance = 30 w/m^2). The batch reactors used were shallow pans open to the atmosphere (See Figure 1). Analytical methods included titrimetric (Iodometric) determinations of sulfide (*Standard Methods*, 1989) and turbidimetric determinations of sulfate (Adams, 1990). Dissolved oxygen concentration (DO) in reaction solutions was measured with a YSI Model 58 DO meter. pH was monitored with a Fisher Accumet Model 50 pH meter. Light intensity was recorded as irradiance (watts/m^2) with a EG&G Gamma Scientific Photocomp photometer.

Test solutions consisted of DDW adjusted to the desired pH with a buffer solution and spiked with sodium sulfide ($Na_2S \bullet 9H_2O$), reagent grade, obtained from Fisher Scientific. The contents of each reactor were aerated continuously with a Whisper 1000™ aquarium pump equipped with medium bubble diffusing stones. To account for the volatilization of sulfide, all sensitizer experiments were conducted utilizing a control reactor containing sulfide and buffer solution, but no dye. Experiments were conducted using various concentrations of MB and RF under laboratory lights with a mean irradiance of approximately 30 w/m^2. Samples were drawn (50 or 100ml) from each reactor and analyzed immediately by the methods mentioned. DO, pH, and temperature readings were recorded periodically to ensure that experimental conditions were stable in each reactor.

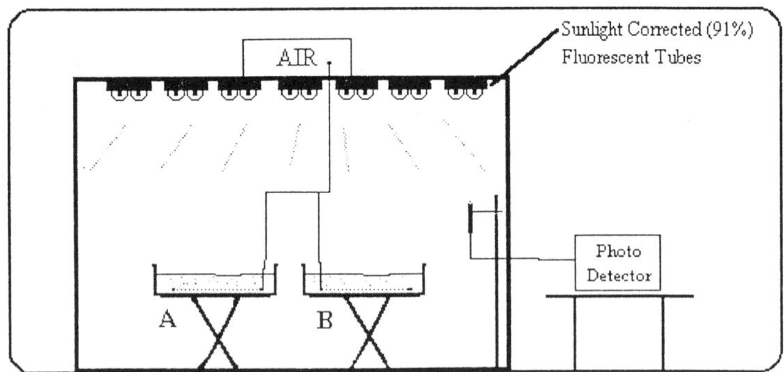

Figure 1: Schematic of experimental set-up for artificial light experiments with batch reactors: A) RF , and B) MB.

RESULTS

Experiments were conducted to determine the effect of varying RF and MB concentration on sulfide removal. The accumulation of sulfate was monitored to determine if the pathway for removal was oxidation of dissolved sulfide to sulfate. Experimental data was used to approximate the rate with a first order equation and obtain reaction rate constants for both MB and RF-sensitized reactions at pH=10.

Figure 2 illustrates the effect of RF concentration on sulfide removal. The data, which is plotted as the ratio of sulfide concentration at a given time (t) to initial sulfide concentration (S/S_o) versus irradiation time, indicates that increased RF concentration results in an increased removal rate. The sulfide data is plotted as (S/S_o) because the initial sulfide concentrations are slightly different for each reactor, although each is ≈ 100 mg/l. The optimum sensitizer concentration of those tested is RF=20mg/l.

The effect of varying MB concentration on sulfide removal rate is shown in **Figure 3.** The data is again plotted as (S/S_o) versus irradiation time for continuity. Although MB-dosed reactors exhibit much faster removal of sulfide than RF-dosed reactors (45-50 minutes for 90% removal versus 210-240 minutes for 90% removal), the variation of MB concentration did not significantly effect the rate of sulfide oxidation. MB concentrations of 1, 2, and 4 mg/l exhibit similar removal rates. A MB concentration of 1 mg/l provides removal as quickly and completely as a concentration of 4 mg/l.

Sulfide removal and sulfate accumulation data for RF=20mg/l and MB=1mg/l at a pH=10 are compared in **Figure 4**. MB obviously sensitizes the photooxidation of sulfide to sulfate much faster than RF under these experimental conditions. Preliminary mass balance data reveal that both RF and MB sensitized reactions of sulfide proceed by oxidation of the dissolved sulfide to sulfate. By accounting for sulfide losses in the control reactors due to volatilization, the amount of sulfide removed due to photooxidation (ΔSp) can be calculated and a theoretical yield for sulfate can be determined from the stoichiometric relationship:

$$(1) \quad 1 \text{ mg/l } (S^{=}) \Rightarrow 3 \text{ mg/l } (SO_4^{=}),$$

so

$$(2) \quad 3 \bullet (\Delta Sp) = \text{Theoretical } (SO_4^{=}) \text{ accumulation},$$

or

$$(3) \quad \% \text{ S closure} = \text{Actual } (SO_4^{=}) / \text{Theoretical } (SO_4^{=}).$$

Using this technique and data from figure 4 to calculate mass balance closures for both RF and MB sensitized photooxidation of sulfide yielded 80-85% closure for both sensitizers. This indicates that the mechanism of removal is through oxidation of dissolved sulfides to sulfate.

Experimental data for both RF and MB were plotted using first-order equations to determine reaction rate constants for the sensitized reactions. The resulting plot is shown in **Figure 5**. Table 1 lists the calculated rate constants for RF and MB and the r-square value for the linear regressions.

TABLE 1: First-Order Rate Constants

Sensitizer	Rate Constant (min^{-1})	R^2 Value
Riboflavin	.0097	.9806
Methylene Blue	.0273	.9416

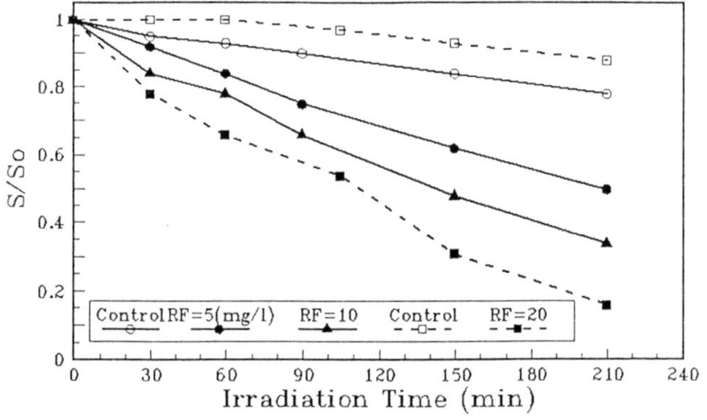

Figure 2: Effect of RF concentration on sulfide removal. Experimental conditions: Artificial lighting, pH=10, continuous aeration, So≈100 mg/l.

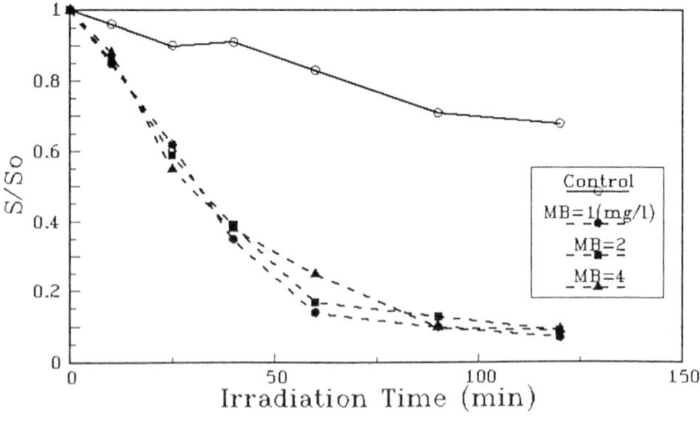

Figure 3: Effect of MB concentration on sulfide removal. Experimental conditions: Artificial lighting, pH=10, continuous aeration, So≈100 mg/l.

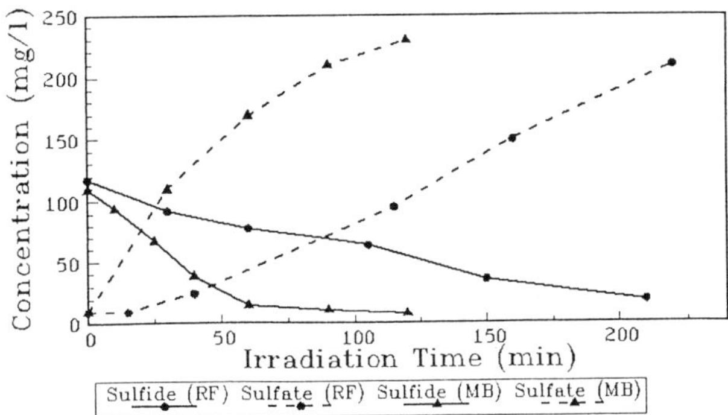

Figure 4: Accumulation of sulfate for MB and RF with respect to sulfide depletion.

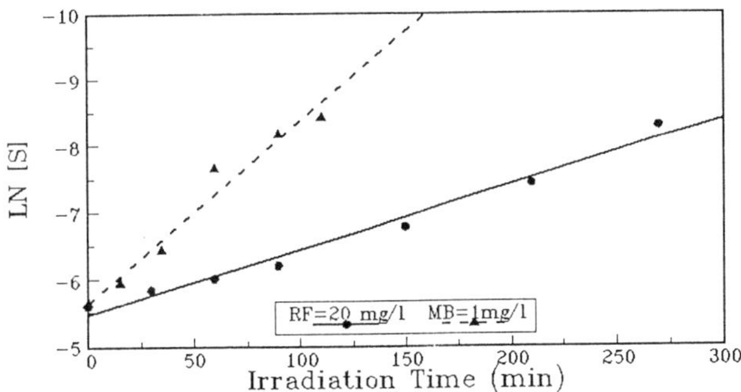

Figure 5: First order plot of RF and MB experimental data. k(RF)=.0097, and k(MB)=.0273.

SUMMARY

The use of MB and RF in sensitizing the photooxidation of sulfide to sulfate warrants further study. Both sensitizers were found to be effective in removing sulfide, although MB removed sulfide 2-3 times faster than RF. The accumulation of sulfate indicates that degradation of sulfide is by oxidation to sulfate. Another interesting observation is the lack of sulfide odor emanation from the sensitizer-dosed reactors. Control reactors consistently gave off the rotten egg odor associated with hydrogen sulfide, but the smell was absent in the dosed reactors. The low cost associated with sensitized photooxidation treatment systems, coupled with the attractive characteristics of both MB and RF (low toxicity to plant and aquatic life) provide additional motives for further investigation of treatment of sulfide wastes with sensitized photooxidation systems.

LITERATURE CITED

Adams, V.D., *Water & Wastewater Examination Manual*, Chelsea, MI: Lewis Publishers, Inc., (1990).

Eisenberg, T.N., Adams, V.D., and Middlebrooks, E.J., "Use of Solar Energy for Wastewater Disinfection for Crop Irrigation," U.S.-Israel Binational Agricultural Research Development (BARD) Research Project #US-613-83 Final Report, (1987).

Eisenberg, T.N., Adams, V.D., and Middlebrooks, E.J., "Sensitized Photooxidation of Bromacil: Pilot, Bench, and Laboratory Scale Studies," Proceedings of the *42nd Purdue Industrial Waste Conference*, Ann Arbor, MI, Lewis Publishers, Inc., (1987).

Heelis, P.F., Parsons, B.J., Phillips, G.O., and McKellar, J.F., "The Flavin-sensitized Photooxidation of Ascorbic Acid," *Photochem. Photobiol.*, **21**, pp. 249-54, (1981).

Herwig, N., *Handbook of Drugs and Chemicals Used in the Treatment of Fish Diseases*, Springfield, IL: Charles Thomas, (1979).

Machlin, L.J., *Handbook of Vitamins:Nutritional, Biochemical, and Clinical Aspects*, New York, NY: Marcel Dekker, INC., (1984).

Merck and Co., *The Merck Index*, Rahway, NJ, (1976).

Montgomery, A.D., McInerney, M.J., and Sublette, K.L., "Microbial Control of the Production of Hydrogen Sulfide by Sulfate-Reducing Bacteria," *Biotechnology and Bioengineering*, **35**, pp. 533-9, (1990).

(NACE) National Association of Corrosion Engineers, *The Role of Bacteria in the Corrosion of Oil Field Equipment*, Houston, TX, (1976).

National Research Council - Subcommittee on Hydrogen Sulfide, *Hydrogen Sulfide*, Baltimore, MD, University Park Press, (1979).

Noyes Data Corporation, *Pollution Control in the Petroleum Industry*, Park Ridge, NJ, (1973).

Penzer, G.R., and Radda, G.K., "Photoreduction of Flanines by Amino Acids," *Biochem. Journ.*, 109, pp.259-68, (1968).

Post, G.W., *Textbook of Fish Health*, Neptune City, NJ: TFH Publications, Inc., (1983).

Spikes, J.D., and Straight, R., "Sensitized Photochemical Processes in Biological Systems," *Ann. Rev. Phys. Chem.*, **18**, pp. 409-440, (1967).

Standard Methods for the Examination of Water and Wastewater, Washington, DC: American Public Health Association, 17th Edition, (1989).

Tomita, M., Irie, M., and Ukita, T., "Sensitized Photooxidation of Histidine and Its Derivatives: Products and Mechanisms of Reaction," *Biochemistry*, **8**, 5149-60, (1969).

Van Duijn, C., *Diseases of Fishes*, London: The Butterworth Group, (1973).

Watts, R.J., "The Development of Design Criteria for Sensitized Photooxidation Lagoons to Treat Toxic and Refractory Organic Wastes," Doctoral Dissertation. Utah State University, (1985).

Recovery of Chromium from Spent Plating Solutions by a Chromyl Chloride Process

Subbarao L. Guddati[1], Thomas M. Holsen[2], and J. Robert Selman[3]

Abstract

A novel chromyl chloride process has been investigated for the recovery of hexavalent chromium from spent plating solutions. In this process chromium is converted to chromyl chloride by reacting it with concentrated hydrochloric acid and then separated as a heavy underlayer, or alternatively, extracted into a solvent as follows:

$$H_2CrO_4 \quad + \quad 2\,HCl \quad \Leftrightarrow \quad CrO_2Cl_2 \quad + \quad 2\,H_2O$$

Purified chromyl chloride is then hydrolyzed and the resulting solution dried and chromium trioxide recovered.

$$H_2CrO_4 \quad [+\ 2\,HCl\,] \quad \Leftrightarrow \quad CrO_3\downarrow \ + \ H_2O\uparrow \quad [+\,2\,HCl\uparrow]$$

In preliminary experiments more than 98% of the chromium has been separated as chromyl chloride (without using any solvent) from an aqueous solution which originally contained 200 g/L chromic acid. Temperature and reactant concentrations were found to greatly affect the stability and the yield of chromyl chloride respectively. Equilibrium conditions have been identified using a geochemical equilibrium speciation model. A statistical analysis of experimental results has been performed to quantify the effects of various parameters on the yield of chromyl chloride.

Introduction

Chromic acid is used in a variety of electroplating and metal finishing processes. Spent plating solutions are generated when the chromic acid baths become contaminated by the accumulation of heavy metals such as copper, iron, nickel, zinc which dissolve from the pieces being plated as well as by the build-up of trivalent chromium formed as result of the reduction of chromic acid. These contaminants have several detrimental effects on the plating operation, including: increased voltages and plating times, increased pitting and noduling, and decreased current efficiency. Spent plating baths from these processes present major waste treatment problems. Currently, most of the dissolved heavy metals from aqueous waste streams are precipitated as hydroxides and stored in long term dump-sites as a sludge (Patterson, 1975). There are several disadvantages to this traditional technique. It has high capital costs, requires extensive instrumentation, and involves long-term

---

[1]Graduate Student, and [2]Associate Professor, Pritzker Department of Environmental Engineering; [3]Professor, Department of Chemical Engineering, Illinois Institute of Technology, IL 60616

liability for the waste. In addition significant quantities of metal are being wasted. To overcome these problems, the metal finishing and electroplating industry is looking for new recovery/recycling techniques. Increasing federal regulations on the disposal of metal sludge and the increasing costs of metals has resulted in an increased emphasis on their reuse.

Most of the previous research on the recovery of chromium from plating solutions has been done for rinse waters (USEPA, 1978a; USEPA, 1978b). For spent plating solutions a number of methods have been proposed including ion-exchange, electrolytic recovery and membrane processes. However, all these treatment methods have serious limitations. Much of the previous reclamation work has centered around using cation-exchange to separate chromium from the spent baths before reclamation. A short coming of this process is the susceptibility of the cation-exchange resins to chemical attack and degradation by strong oxidizing agents such as chromic acid, especially at low pH and high chromic acid concentrations (Wing, 1979). Currently there is no viable recycle and recovery technique for the commercial reprocessing of spent chromium plating solutions. This is in spite of the fact that there are approximately 24,000 electroplating firms in the U.S. producing 6,350 tons of chromium waste (Comptroller General's Report, 1980).

Recovery of Chromium from Spent Plating Solutions

A novel treatment technique, the chromyl chloride process, is being investigated in our laboratories for the recovery of chromium from the spent plating solutions (Guddati *et.al.*, 1994). The process can also be used to recover and recycle chromium from rinse waters. However, concentrating rinse waters prior to the chromyl chloride process will probably be required to make this process economical. This process is expected to be environmentally benign and has some attractive features: the process will be able to recover and recycle all of the acids and solvents back into the recovery process and requires less energy than it does to make pure chromic acid. The process also has several potential advantages over the more conventional treatment processes. For example, this process will be able to:

1. Produce concentrated solutions from spent plating and rinse water solutions that can be recycled back into the plating bath during electroplating;
2. Produce chromium trioxide that can be recycled back into the plating market;
3. Eliminate long term storage of chromium metal wastes from spent plating baths;
4. Decrease energy consumption compared to other recovery processes of chromium;

Process Description

In the proposed reprocessing flow sheet (Fig. 1) chromium is separated from the spent plating solution as chromyl chloride by reacting dissolved chromic acid and concentrated hydrochloric acid:

$$H_2CrO_4 + 2\,HCl \Leftrightarrow CrO_2Cl_2 + 2\,H_2O \quad ...(1)$$

As this reaction shows, in the presence of HCl an equilibrium exists between chromic acid and chromyl chloride. When hydrochloric acid is added to the chromic acid there is essentially quantitative conversion of $HCrO_4^-$ into an intermediate, chlorochromate ion (CrO_3Cl^-):

$$HCrO_4^- + HCl \Leftrightarrow CrO_3Cl^- + H_2O \quad ...(2)$$

This equilibrium can be forced to the right by minimizing the activity of water (e.g. by adding H_2SO_4).

$$CrO_3Cl^- + 2\,H^+ + Cl^- \Leftrightarrow CrO_2Cl_2 + H_2O \quad ...(3)$$

Under the right conditions the chromyl chloride can be separated from the aqueous solution as a heavy underlayer (At room temperature, chromyl chloride is a dark red liquid with a density of 1.92 $g.cm.^{-3}$). Alternatively, chromyl chloride can be extracted into a solvent. The extractant can then be distilled to recover the chromyl chloride and the solvent can be recycled back into the process. Purified chromyl chloride can then be hydrolyzed to produce a solution of dichromic and hydrochloric acids. The resulting solution can be dried and chromium trioxide can be recovered:

$$H_2CrO_4 \quad [+ 2\,HCl] \quad \Leftrightarrow \quad CrO_3\downarrow \; + H_2O\uparrow \quad [+ 2\,HCl\uparrow] \quad ...(4)$$

The hydrochloric acid can be recycled back into the process. The pure chromium product may be recycled back to the plating bath solution or resold. The sulfuric acid rich raffinate solutions can also be recycled back into the recovery system.

Chromyl chloride is quite stable when kept dry in the dark. It decomposes very slowly in daylight forming a series of chromium oxides and chlorine. The decomposition of chromyl chloride is much more rapid in the presence of moisture. Water decomposes chromyl chloride with the development of heat forming a mixture of chromic and hydrochloric acids.

Experimental Methods

The experiments are conducted in a three-neck reaction flask fitted with a dropping funnel, a reflux condenser and a thermometer which is placed in a refrigeration bath to control the temperature of the reaction. The chromic acid solution is placed in the reactor along with a measured quantity of HCl. Concentrated sulfuric acid is then added in small steps. After the reaction is completed, the solution is transferred to a separatory funnel and chromyl chloride is separated as a dense phase. These experiments are carried out at temperatures ranging from 0°C to 20°C and for different concentrations of the feed, hydrochloric acid and sulfuric acid. The lean phase is analyzed by atomic adsorption spectroscopy for chromium and the dense phase is analyzed for water content by Karl-Fisher titration. The purpose of these experiments is to identify and characterize the reaction mechanisms, optimum conditions and product purity. An electrochemical technique using an Ag-AgCl and a PbO_2-$PbSO_4$ electrodes is being currently investigated to determine the equilibrium conditions of the CrO_2Cl_2 reaction.

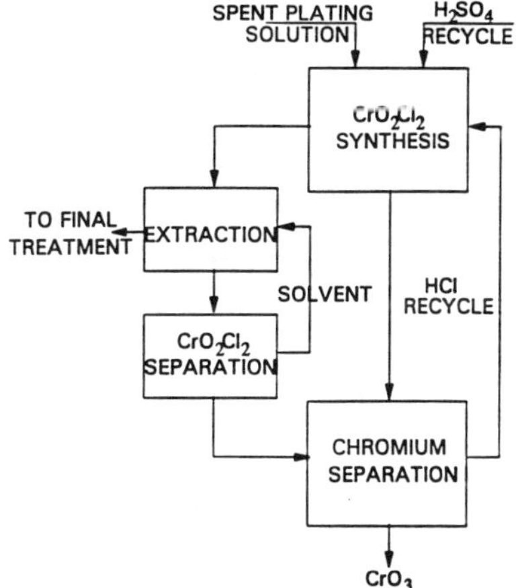

Fig. 1. Chromyl chloride process flowsheet

Results and Discussion

Experiments have been carried out to determine the kinetics and the equilibrium conditions of reaction 1. The experimental results indicate that the yield of chromyl chloride depends strongly on the initial concentration of the chromic in the spent plating solution as well as on the concentrations of hydrochloric and sulfuric acids added to the reaction mixture. The yield of chromyl chloride weakly depends on the temperature of the reaction mixture The decomposition of chromyl chloride strongly depends on the temperature of the system. Stirring provides a better control over temperature by suppressing temperature gradients (thereby the decomposition of chromyl chloride). Centrifugation of reaction mixture (after the completion of reaction) enhances phase separation and thereby increases the chromyl chloride yield.

Effect of HCl

In the presence of lower quantities of sulfuric acid, the yield of chromyl chloride increases sharply with an increase in the amount of HCl added and then decreases. This behavior is due to the fact that the initial addition of hydrochloric acid force the equilibrium of reaction 1 to the right increasing the chromyl chloride yield. It should be noted that an increase in hydrochloric acid addition (which is 62% water) increases the water activity simultaneously. In the presence of higher quantities of hydrochloric acid the activity of water becomes significant and forces the reaction to the left. This effect is much more pronounced in the presence of lower sulfuric acid concentrations. Increasing the amount of sulfuric acid added suppresses the negative effect of water on the chromyl chloride production. The presence of excess HCl leads to a quantitative conversion of chromic acid into chlorochromate acid (reaction 2). In addition, excess HCl leads to chlorine gas formation if the reactor is not kept cool (<20°C). The maximum yield of chromyl chloride is obtained by dissolving chromic acid in more than the equivalent amount of concentrated HCl with an excess of H_2SO_4 acid present. At a given temperature and low sulfuric acid concentrations, excess HCl addition results in a decrease in chromyl chloride yield (Fig. 2). However for high H_2SO_4 concentrations this effect is negligible. The yield

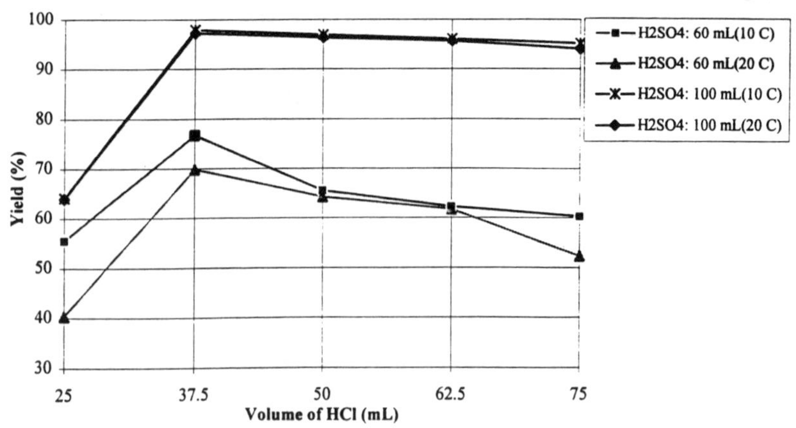

Fig. 2. Effect of HCl on chromyl chloride yield

depends not only on the quantity of HCl acid added but also on its concentration. The yield decreases with decreasing acid concentration until with 20 percent HCl addition, only chlorine is produced.

Effect of H_2SO_4

The yield of chromyl chloride sharply increases with an increase in the quantity of H_2SO_4 added, probably due to a decrease in the water activity. For example, by increasing the quantity of concentrated H_2SO_4 added to the reaction mixture from 60 mL to 100 mL, the yield of chromyl chloride increased from 62 percent to 96 percent (Figs. 2). For increasing quantities of H_2SO_4 addition, the CrO_2Cl_2 yield increases by 10 to 40 percent depending on the concentration and quantity of HCl added initially and the temperature of the system. In the presence of higher concentrations of sulfuric acid, the separation of chromyl chloride phase from aqueous mixture is fast. In this case there is no need to centrifuge the reaction mixture or add solvent to extract chromyl chloride. Though the role of sulfuric acid is to minimize the activity of the water, the chromyl chloride decomposes and forms sulfates in presence of excess concentrated sulfuric acid, oleum ($H_2S_2O_7$).

Effect of Temperature

Statistically temperature has no significant effect on chromyl chloride yield for the range of variables tested here (see next section) but it has a significant effect on the decomposition of chromyl chloride. The yield of chromyl chloride decreases slightly at higher temperatures for low sulfuric acid concentrations but remains constant for high sulfuric acid additions (Fig. 3). Low temperatures suppress the decomposition reaction resulting in higher yields of chromyl chloride. The control of reaction temperature is closely associated with stirring. In the absence of stirring, the reaction mixture is not uniformly cooled and hence large temperature gradients form. These gradients force slower and slower additions of sulfuric acid in order to maintain the low temperatures needed for the reaction. Large temperature gradients allow the decomposition reactions to become significant decreasing the chromyl chloride yield. Stirring suppresses the formation of temperature gradients (thereby decomposition) and allows a better control of the temperature. However, slower mixing rates are better than faster mixing rates for separating of the CrO_2Cl_2 phase.

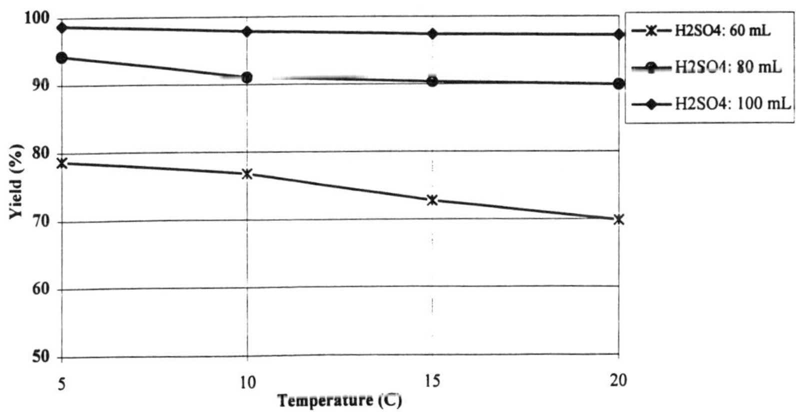

Fig. 3. Effect of temperature on chromyl chloride yield

Equilibrium Data

A geochemical equilibrium speciation model MINTEQA2 (USEPA, 1991) was used for predicting chrome speciation in this system and to compare experimental and theoretical results. The data required to predict the equilibrium composition as well as activities of the components is the total concentrations of the components of interest and other relevant invariant measurements for the system (pH, pE etc.). The chromyl chloride synthesis reaction (reaction 1) and its thermodynamic data (ΔG^o = -2.07 KCal./mol) were added to the MINTEQA2 database since they were not originally present. The simultaneous solution of the nonlinear mass expressions and linear mass balance relationships (equilibrium constant method) approach was used to solve the multi-component chemical equilibrium problem. The model used the Davies equation and/or extended Debye-Huckel relationship for calculating the activities of the species present.

The yields predicted by the MINTEQA2 model is consistent with the experimental results at high H_2SO_4 concentrations (Table 1). At low H_2SO_4 concentrations the theoretical yields are larger than the experimental yields. This difference may be due to poor separation of the phases at low H_2SO_4 concentrations as discussed earlier. Fig. 4 shows the effect of sulfuric acid concentration on the equilibrium activities of the species in the system. With an increase in sulfuric acid concentration, the activity of the CrO_3Cl^- species initially increases and then decreases. This variation is due to the decrease in water activity with an increase in H_2SO_4 concentration which forces both reactions 2 and 3 to the right. Further increasing the H_2SO_4 concentration will decreases the water activity which favors CrO_2Cl_2 synthesis and decreases the CrO_3Cl^- activity. The yield of chromyl chloride is calculated from these equilibrium mass distribution and the results are consistent with the experimental observations. With an increase in sulfuric acid concentration from 1 M to 8 M the chromyl chloride yield increased from 0 to 100 percent. The predicted molalities of water are in agreement with the Karl-Fisher titration results of the chromyl chloride phase. The effect of HCl concentration on the equilibrium composition was also modeled. The theoretical yields of chromyl chloride were again in agreement with the results expected from the experimental results.

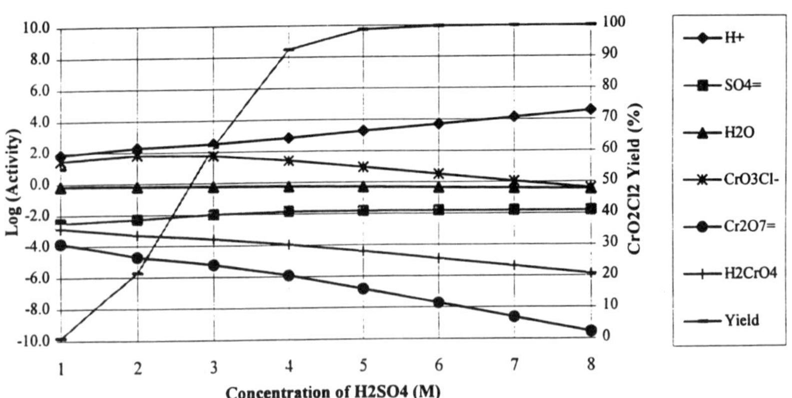

Fig. 4. Predicted activities and yields of chromyl chloride by MINTEQA2

Table 1. Comparison between experimental and theoretical CrO_2Cl_2 yields
Initial CrO_3 concentration = 2 M; HCl concentration = 4 M

No.	Concentration of H_2SO_4 (M)	Yield	
		Experimental	MINTEQA2
1.	2	9.5	21.4
2.	4	76.8	92.4
3.	5	90.4	98.8
4	6	97.8	99.8
5.	7	98.1	100.0

Statistical Analysis

Design of the chromyl chloride process for chromium recovery requires a detailed understanding of the effects of the variables on the process. Also there is a need to interpret the experimental results in order to determine the optimal conditions necessary for the process. A statistical analysis tool was used for this purpose. Table 2 shows a design matrix for a 2^3 factorial experiment in which there are three quantitative variables - HCl, H_2SO_4, and temperature. The response is the yield of chromyl chloride. It lists a set of eight (2^3) possible different experimental conditions and the chromyl chloride yield (response) under these conditions. The data were part of the laboratory-scale investigation of chromyl chloride process and have been simplified somewhat for illustrative purposes. The values of the quantitative variables can also be interpreted in terms of the two levels (high and low). Though the response will depend on the actual change in the levels of the variable, the data in Table 1 can be used for a comparison of the effect of these variables on the response. The statistical analysis of the data in Table 2 is shown in Table 3 .

Table 2. Factorial design of the CrO_2Cl_2 experiments at two levels

Initial chromic acid concentration = 200 g/L; Volume of the sample = 50 mL

Test Condition #	HCl Volume mL	H_2SO_4 Volume mL	Temperature °C	Yield %
1.	37.5	60	10	72.42
2.	62.5	60	10	62.23
3.	37.5	100	10	97.87
4.	62.5	100	10	96.04
5.	37.5	60	20	69.85
6.	62.5	60	20	61.73
7.	37.5	100	20	97.22
8.	62.5	100	20	95.62

From Tables 3, it is clearly evident that the effect of sulfuric acid on the yield of chromyl chloride is predominant and its effect is one order of magnitude higher than other effects. Chromyl chloride yield increases by 30 percentage units when the quantity of sulfuric acid is increased from 60 to 100 mL keeping other variables constant. HCl has a negative effect on the chromyl chloride yield. An increase in the amount of 38% HCl added will lead in a decrease in chromyl chloride yield. The overall effect of HCl addition is estimated to decrease the chromyl chloride yield by about 5 percentage units. By increasing the temperature from 10 to 20°C, the estimated overall effect is a decrease in chromyl chloride yield by about one percentage unit. This decrease in the yield is of the same value of the standard error in carrying out these experiments. Hence it is difficult to interpret the effect of temperature on the chromyl chloride yield for the range of variables tested here.

Table 3. Calculated effects for the 2^3 factorial design of CrO_2Cl_2 experiments

Effect	Estimate ± Standard Error
Average	81.62 ± 0.74
Main effects	
• HCl	-5.45 ± 1.48
• H_2SO_4	30.13 ± 1.48
• Temperature (T)	-1.04 ± 1.48
Two factor effects	
• HCl x T	0.58 ± 1.48
• H_2SO_4 x T	0.05 ± 1.48
• HCl x H_2SO_4	3.72 + 1.48
Three factor effect	
• T x HCl x H_2SO_4	0.46 + 1.48

Summary

The chromyl chloride process offers a very efficient technique for the separation of chromium from spent plating baths. More than 98% of the chromium can be separated as chromyl chloride without using any solvent which corresponds to a reduction in chromic acid concentration from 200 gm/L to 4 gm/L. Higher concentrations of sulfuric acid and lower concentrations of HCl and lower temperatures were found to be the most favorable conditions for chromyl chloride synthesis. The successful completion of this research will result in the development of a novel method for separation and recovery of chromium from spent plating solutions. The proposed method treats the entrained metal at its source without modification of the production scheme, thereby maximizing the amount of material that can be recycled and minimizing the risk and environmental pollution.

References

1. Comptroller General's Report, (1980). "Industrial Wastes: An Unexplored Source of Valuable Minerals". U.S. General Accounting Office. Washington, D.C. EMD-80-45.
2. Guddati *et.al.*, (1994). "The Chromyl Chloride Technique - A Novel method for Chromium Recovery from Spent Plating Solutions". Chromium Colloquium, A.E.S.F. Tech Conf. Orlando, FL.
3. Patterson, J.W., (1975). Wastewater Treatment Technology, Ann Arbor Sci. Pub., Ann Arbor, MI.
4. USEPA, (1978a). "Evaporative Recovery of Chromium Plating Rinse Waters". EPA-600/2-78-127.
5. USEPA, (1978b)."PBI Reverse Osmosis Membrane for Chromium Plating Rinse Waters". EPA-600/2-78-040.
6. USEPA, 1991. MINTEQA2/PRODEFA2, A Geochemical Assessment Model for Environmental Systems. EPA-600/3-91-021.
7. Wing, R.E., (1979). "Case History Reports on Heavy Metal Processes". Proc. 66th Annual A.E.S. Conf. Ind. Finis.

A Coupled Mass Transfer and Surface Complexation Model for Uranium (VI) Removal from Wastewaters

John Lenhart[1], Linda A. Figueroa[1], and Bruce D. Honeyman[1]

Abstract

A remediation technique has been developed for removing uranium (VI) from complex contaminated groundwater using flake chitin as a biosorbent in batch and continuous flow configurations. With this system, U(VI) removal efficiency can be predicted using a model that integrates surface complexation models, mass transport limitations and sorption kinetics. This integration allows the reactor model to predict removal efficiencies for complex groundwaters with variable U(VI) concentrations and other constituents. The system has been validated using laboratory-derived kinetic data in batch and CSTR systems to verify the model predictions of U(VI) uptake from simulated contaminated groundwater.

Introduction

There are at least 25 radiation sites currently proposed or listed on the National Priority List (NPL), including the U.S. Department of Energy's (DOE) Rocky Flats Plant that has reports of elevated levels of uranium in the ground water (U.S. DOE, 1991). The U.S. EPA (1990) has identified the following two areas as critical obstacles to cleanup at sites like Rocky Flats: (1) conduct technological feasibility and treatability studies on removing Ra, Th, and U from water, including field testing of high performance technologies for remediation of Ra and U from contaminated water sites and (2) conduct treatability studies on water contaminated with mixed waste.

We have developed a treatment system for removing U(VI) from waste streams that uses animal chitin as a passive biosorbent. An integral part of the treatment system is a model that allows for the optimization of the treatment system by predicting the removal efficiency based on the influent characteristics of the waste stream. The model accounts for changing influent matrix conditions through the coupling of surface complexation and mass transport models.

---

[1]Environmental Science and Engineering Division, Colorado School of Mines, Golden, CO 80401

Background

A number of conventional treatment processes for metals removal, including conventional coagulation, lime softening, anion exchange, activated alumina, and reverse osmosis, have shown the capability to remove uranium from water (Sorg, 1988). These techniques have been shown to be effective in the laboratory. This laboratory effectiveness has not been translated to a full-scale treatment system due to cost or technological restraints. Iron, alum coagulation and lime softening are not cost effective because they require large doses of coagulating or softening chemical, several sequential unit processes (rapid mix, flocculation, sedimentation, filtration, and pH adjustment), and produce excess sludge to provide an effluent with low U(VI) concentrations (<1 mg/l). Anion exchange is effective in removing uranium from water due to its high affinity for the uranium, a property that makes regenerating the resin very difficult (Sorg, 1988). Activated alumina produces excellent removal ability, but the capacity of the alumina is low (Sorg, 1988). Granular activated carbon was shown to elute increasingly higher concentrations of uranium after breakthrough was reached, suggesting that carbon has a limited affinity for uranium, and will preferentially sorb other ions. Reverse osmosis treatment is nonspecific in its removal ability resulting in an indiscriminate removal of all trace metals, resulting in high treatment cost.

There have been numerous studies involving the adsorption of uranium on various sorbents. The recovery of uranium from aqueous solution has been studied using activated carbon (Saleem *et al.*, 1992), hollow fibers containing amidoxime groups (Takeda *et al.*, 1991), and many biological sorbents (see Benson, 1993). Most biological metal-removal systems in, or approaching, commercialization employ nonliving systems (Brierley *et al.*, 1990; Darnell *et al.*, 1986; Kuyucak and Volesky, 1988; Tsezos, 1988). The use of dead biomass for metal binding circumvents toxicity problems that can occur with living cells and eliminates the economic component of nutrient supply. Subsequent metal and biomass recovery is often easier. Other technological concerns for utilization of natural products for metals recovery/removal include efficiency of removal, biosorbent metal loading capacity, recycle of the biosorbent, contamination of the recovery system leading to septic conditions, and the physical and chemical stability of the biomass.

Chitin has received attention lately as a possible biomass for sorption of heavy metals and has shown great promise in its ability to accumulate chromium, copper, lead, zinc, mercury, and cadmium (Suder and Wightman, 1983). Chitin is a readily available and inexpensive byproduct of the seafood industry. In addition, chitin has good mechanical strength and settling characteristics which eliminate the need for immobilization. Chitin, which is derived from the shells of crustaceans, is not affected by toxic conditions, and does not require nutritional input.

Materials and Methods

Data was collected in three separate reactor configurations (1) a bench scale CSTR system, (2) a bench scale batch reactor system, and (3) a beaker batch study (Benson, 1993). Samples for the bench scale treatment systems were analyzed for uranium using laser phosphoremetry with the Scintrex UA-3 Uranium Analyzer, courtesy of the United States Geological Survey in Golden, CO. This method required the addition of a complexing agent, sodium hexametaphosphate, which forms uranyl-phosphate complexes that fluoresce when excited by the laser. The aqueous uranium was supplied to the bench scale systems using concentrated uranyl nitrate $UO_2(NO_3)_2$ in solution with K^+(0.041 M), Cl^-(0.087 M), Na^+(0.046 M), and atmospheric CO_2. The bench scale reactor system was constructed using a 10 liter polycarbonate carboy as the reactor vessel with two separate feed tanks, one with the concentrated uranium solution and the other containing the mixed chitin solution which were combined after the peristaltic pump to form one feed line into the reactor vessel. In a typical bench scale CSTR adsorption kinetic experiment, the two solutions were fed continuously to the reactor at a total flow rate of 0.2 l/min and a constant pH. The pH in the reactor was monitored and controlled using a VirTis Digital pH Control Module which automatically fed either concentrated acid (~5M HCl) or base (~5M NaOH) to the reactor to maintain the nominal pH of the experiment.

The Predictive Model

The treatment optimization model integrates surface complexation concepts, adsorption rate law expressions, Fick's law and chemical reactor mass balance expressions in a two step process, as schematically represented in Figure 1. Using this model, U(VI) removal efficiency of the treatment system can be estimated for different system conditions.

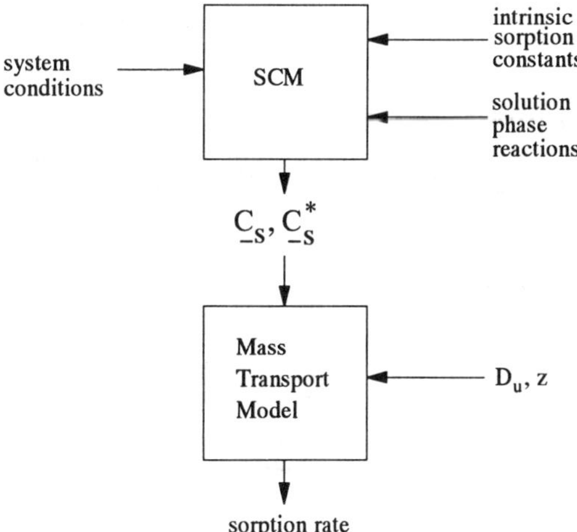

Figure 1. Flow diagram of the model coupling surface chemistry and mass transport of sorbing species to the chitin surface. $\underline{C}_s^*$ and $\underline{C}_s$ represent the equilibrium sorbate concentration immediately adjacent to the chitin surface and sorbed concentrations, respectively.

The model application involves the integration of two distinct modeling steps. The first step involves the use of a surface complexation model, such as the Triple Layer Model (TLM) of Davis and Leckie (1978) as modified by Papelis *et al.* (1988). This model is used to estimate the equilibrium partitioning of U(VI) between the solution phase and the surface of the animal chitin. The surface complexation model provides estimates of these equilibrium concentrations over a range of pH values or general system conditions (ionic strength, dissolved components other than uranium), provided the solution phase reaction and intrinsic sorption constants are known. Reactions representative of those used in the TLM are shown in Table 1.

Table 1. Representative Surface Reactions and Corresponding Mass Action Expressions

Surface Protolysis Reactions	Mass Action Expressions
$SOH_2^+ \overset{K_{a,1}}{=} SOH + H^+$	$K_{a_1}^{\text{int}} = \frac{[SOH][H^+]}{[SOH_2^+]}\exp\left(\frac{-F\Psi_o}{RT}\right)$
$SOH \overset{K_{a,2}}{=} SO^- + H^+$	$K_{a_2}^{\text{int}} = \frac{[SO^-][H^+]}{[SOH]}\exp\left(\frac{-F\Psi_o}{RT}\right)$
Surface Electrolyte Reactions	
$SOH + Na^+ \overset{K_{Na}}{=} SO^- - -Na^+ + H^+$	$K_{Na}^{\text{int}} = \frac{[SO^- - -Na^+][H^+]}{[SOH][Na^+]}\exp\left(\frac{-F(\Psi_o - \Psi_\beta)}{RT}\right)$
Uranyl Surface Reactions	
$SOH + UO_2^{2+} \overset{K_{UO_2^{2+}}}{=} SOUO_2^+ + H^+$	$K_{UO_2^{2+}}^{\text{int}} = \frac{[SOUO_2^+][H^+]}{[SOH][UO_2^{2+}]}\exp\left(\frac{F\Psi_o}{RT}\right)$
$SOH + UO_2^{2+} + 2CO_3^{2-} + H^+ \overset{K_{UO_2(CO_3)_2^{2-}}}{=} SOH_2^+ - -UO_2(CO_3)_2^{2-}$	
	$K_{UO_2CO_3^{2-}}^{\text{int}} = \frac{[SOH_2^+ - -UO_2(CO_3)_2^{2-}]}{[SOH][UO_2^{2+}][CO_3^{2-}]^2[H^+]}\exp\left(\frac{F(\Psi_o - \Psi_\beta)}{RT}\right)$

All reaction stoichiometries and corresponding mass-action expressions and equilibrium binding constants are determined in separate batch sorption experiments.

The results of the surface complexation modeling in Step 1 are used to derive the equilibrium distribution coefficient (K_d), where $K_d = [\underline{C}_s]/[S][\underline{C}_s^*] = k_f/k_r$. The K_d is used in the second step to estimate the desorption rate constant k_r depending upon the magnitude of the k_f. These rate constants and values for molecular diffusion, particle boundary layer thickness, and chemical reactor mass balances, are incorporated into chemical mass balances. The following systems of first order differential equations have been developed for the batch reactor

$$\frac{dC}{dt} = -\frac{k_f S}{1+Da(II)}C + \frac{k_r}{1+Da(II)}C_s \quad ; \quad \frac{dC_s}{dt} = \frac{k_f S}{1+Da(II)}C - \frac{k_r}{1+Da(II)}C_s$$

and for the CSTR.

$$\frac{dC}{dt}=-\left(\frac{k_f S}{1+Da(II)}+\frac{1}{\tau}\right)C+\frac{k_r}{1+Da(II)}C_s \ ;$$

$$\frac{dC_s}{dt}=\frac{k_f S}{1+Da(II)}C-\left(\frac{k_r}{1+Da(II)}+\frac{1}{\tau}\right)C_s$$

These expressions are then simultaneously solved to give an analytical solution describing the mass transport of dissolved U(VI) to chitin surface sites. This submodel provides an estimate of U(VI) removal efficiency (i.e., the fraction of U(VI) sorbed) as a function of time. Models were derived for a nonporous adsorbent, as is chitin, in both batch or continuous flow systems. The analytical solution for a batch reactor configuration is

$$\frac{C}{Co}=\frac{1}{k_f S+k_r}\left\{k_r+k_f S\exp\left[-t\left(\frac{k_f S+k_r}{1+Da(II)}\right)\right]\right.$$

and, for a continuous flow tank reactor (CSTR),

$$\frac{C}{Co}=\frac{k_r+\frac{1}{\tau}(1+Da(II))}{k_f S+k_r+\frac{1}{\tau}(1+Da(II))}-\frac{k_r}{k_f S+k_r}\left\{\exp\left(\frac{-t}{\tau}\right)\right\}$$

$$+\left(\frac{k_f S}{k_f S+k_r+\frac{1}{\tau}(1+Da(II))}-\frac{k_f S}{k_f S+k_r}\right)\left\{\exp\left[-t\left(\frac{1}{\tau}+\frac{k_f S+k_r}{(1+Da(II))}\right)\right]\right\}$$

where, for both reactor configurations, the variables are: D_u (molecular diffusion coefficient, cm^2/sec), S (chitin surface site concentration, mol/cm^3), a_s (area of chitin particle per unit volume, cm^{-1}), k_f (adsorption reaction rate constant, $cm^3/mol\text{-}sec$), k_r (desorption reaction rate constant, sec^{-1}), z (boundary layer thickness, cm), t (time, sec), τ (hydraulic detention time, sec), C (U(VI) concentration at time t), C_o (initial U(VI) concentration), and Da(II) = $\frac{k_f S/a_s}{D_u/z}$ = Damkohler (II) number.

Results and Discussion

Laboratory studies were undertaken to validate the batch and continuous flow models. Results have been obtained for a model U(VI) waste stream containing Cl^-, Na^+, K^+, and ambient carbonate (atmospheric CO_2). Equilibrium distribution coefficients for these solution conditions at pH 4.5 and 6.5 were calculated applying the TLM to equilibrium batch sorption data (Benson, 1993). The fraction of U(VI) removed from the waste stream over time was successfully modeled for the batch configuration in both beaker and bench scale reactors (Figure 2). For the bench-scale continuous flow system, the model initially consistently underpredicted the observed removal efficiency, which we attribute to the interaction of the chitin with the concentrated U(VI) feed solution immediately before being pumped into the reactor. This interaction occurred in a section of tubing approximately 8 inches long where the two separate feed lines containing the concentrated U(VI) and the chitin solution were combined into one feed line into the reactor. To account for this pre-reactor sorption, we modeled (Figure 3) the experimental data using a longer detention time. Future work should be done to eliminate this artifact by providing separate feed lines into the reactor.

Conclusion

The results of this work indicate that the passive biosorption (using flake instead of immobilized chitin) of uranium (VI) is a potentially effective method of cleaning up ground water at facilities such as Rocky Flats. The model also demonstrated effective estimation of the fraction removed over a range of system conditions. This modeling approach is effective due to the use of the surface complexation models. These models provide for estimation of sorption phenomena over a range of pH values or system conditions, which makes them superior to sorption isotherm expressions, which are only good for one set of system conditions. This predictive capability increases the flexibility of the treatment system by allowing us to modify the pH or chitin concentration to obtain optimum removal efficiency.

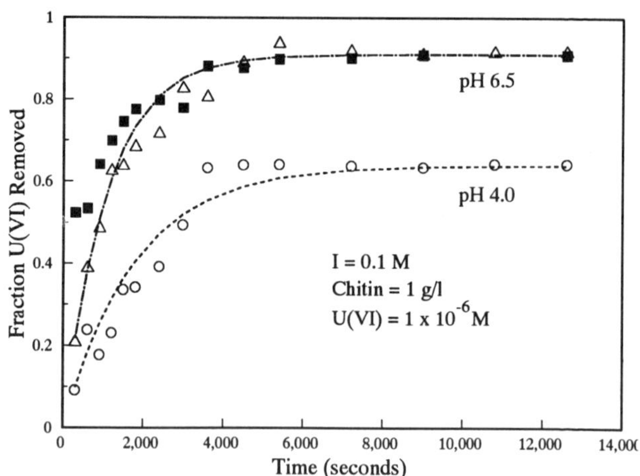

Figure 2. Fractional U(VI) removal as a function of time for a batch system at pH 6.0 and 4.0 (I = 0.1M, U(VI)T = 1 x 10^{-6} M, chitin = 1 g/l). The pH 4 data is from Benson (1993).

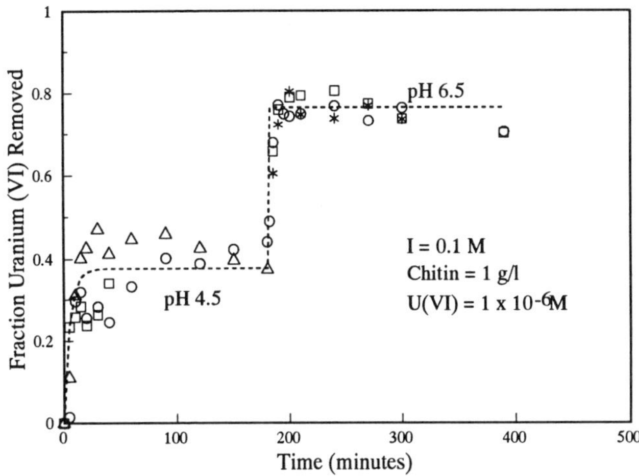

Figure 3. Fraction U(VI) removal as a function of time for a CSTR system. The initial 3 hours were at a pH of 4.5; this was followed by a step change in pH to 6.5. Model calculation are indicated by the dashed line.

References

Benson, L.A., 1993, *Characterization of the Surface Reactions Controlling the Sorption of U(VI) onto Animal Chitin.* MS Thesis, Environmental Science and Engineering, Colorado School of Mines, Golden, CO.

Brierley, C.L., 1990, "Metal Immobilization using Bacteria," In *Microbial Mineral Recovery*, H.L. Erlich and C.L.Brierley(eds.),pp. 303-323. New York: McGraw-Hill.

Darnell, D.W., S. Svec, and M Alvarez, 1986, "AlgaSORB: A New Technology for Removal from Geothermal Waters and Groundwaters," Results for the SITE Emerging Technology Program, U.S. Environmental Protection Agency.

Davis, J.A. and J.O. Leckie, 1978, "Surface Ionization and Complexation at the Oxide/Water Interface: II. Surface Properties of Amorphous Iron Oxyhydroxide and the Adsorption of Metal Ions," *J. Colloid Interf. Sci.*, Vol 67, pp. 90 - 107.

Greene, B., M.T. Henzl, M. Hosea, and D.W. Darnall, 1986, "Elimination of Bicarbonate Interference in the Binding of U(VI) in Mill Waters to Freeze Dried *Chlorella vulgaris*," *Biotechnology and Bioengineering*, Vol. 28, p. 764.

Kuyacak, N., and B. Volesky, 1988a, "Recovery of Cobalt by a New Biosorbent," *The Canadian Mining and Metallurgical Bulletin*, Vol. 81, No. 910.

Kuyacak, N., and B. Volesky, 1988b, "Biosorbents for Recovery of Metals from Industrial Solutions," *Biotechnology Letters*, Vol. 10, No. 2, pp. 137 - 142.

Papelis, C., Hayes, K.F., and J.O. Leckie, 1988, *A Program for the Computation of Chemical Equilibrium Composition of Aqueous Batch Systems Including Surface Complexation Modeling of Ion Adsorption at the Oxide/Solution Interface*, Civil Engineering Technical Report No. 306, Stanford University.

Sorg, T.J., 1988, "Methods for Removing Uranium for Drinking Water," *Journal of the American Water Works Association*, July 1988, pp. 105 -111.

Suder, B.J., and J.P. Wightman, 1983, "Interaction of Heavy Metals with Chitin and Chitosan: Cadmium and Zinc," In *Adsorption for Solution,* R.H. Ottewill, C.H. Rochester, and A.L. Smith (eds.), London: Academic Press.

Takeda, T., K. Saito, K. Uezu, S. Furusaki, T. Sugo, and J. Okamoto, 1991, "Adsorption and Elution in Hollow-Fiber-Packed Bed for Recovery of Uranium from Seawater," *Industrial Engineering and Chemical Research*, Vol. 30, pp. 185 - 190.

Tsezos, M., 1988, "Performance of a New Biological Adsorbent for Metal Recovery: Modeling and Experimental Results," *Biohydrometallurgy,* P.R. Norris and D.P. Kelley (eds.), Science and Technology Letters, Surrey, U.K.

U.S. Department of Energy, 1991, "Proposed Interim Remedial Action: Solar Evaporation Ponds," Rocky Flats Plant, Public Information Document, Golden, Colorado.

U.S. Environmental Protection Agency, 1990, "Assessment of Technologies for the Remediation of Radioactively Contaminated Superfund Sites," EPA/540/2-90/001.

Biological Removal of Heavy Metals by Sulfate Reduction Using a Submerged Packed Tower

George Neserke[1], Linda Figueroa[2] and Nevis Cook[2]

Abstract

The Coors Brewing Co. owns and operates two wastewater treatment plants which handle the combined waste of the City of Golden and the Brewery. The discharge permit for Coors contains very strict limits for metals. Silver and mercury are prohibited from discharge at all and copper and zinc are both at low limits. The copper and zinc limits cannot be achieved with the present plant configuration and several programs are underway to reduce the source concentrations to meet the respective limits. Most of the programs are either very expensive or unlikely to produce the needed results soon enough.

One possible treatment alternative that has been described in the literature is sulfate reduction leading to the generation of hydrogen sulfide. The hydrogen sulfide in turn can precipitate most divalent metals that are available, though there are limits on the precipitation process. The purpose of this research has been to investigate the use of sulfate reduction to remove metals from the effluent of the Coors' Process Waste Treatment Plant (PWTP).

Previous studies have looked at reducing very high levels of sulfate and metals. This research examines much lower levels of both sulfate and metals. Sulfate is less than 80 ppm and metals are in the ppb range. No known studies have attempted to work at these levels. With the low levels of metals, trapping the precipitates is of primary importance. A submerged trickling filter appears to be the best candidate for this.

[1] Manager, Water and Wastewater Environmental Control, Coors Brewing Co. BC 400 Golden, CO 80401

[2] Asst. Professor, Environ. Sci. and Engin. Div., Colo. School of Mines, Golden, CO 80401

Introduction

Biological sulfate reduction is the reduction of sulfate (SO_4^{2-}) to sulfide (S^{2-}) or sulfur (S^o), depending on the active microorganism. In an aqueous environment, sulfide will either bond to metals, be dissolved as sulfide or bisulfide (HS^-) or evolve as hydrogen sulfide gas (H_2S). There have only been infrequent engineering uses of biologically mediated sulfate reduction to date and most significant experiments have looked at the precipitation of metals as the desired result.

The overall stoichiometry for the reduction of sulfate by *Desulfovibrio* strains using ethanol as an electron donor and sulfate as the electron acceptor is shown in equation 1 (Barnes, 1992).

$$2CH_3CH_2OH + SO_4^{2-} \dashrightarrow 2CH_3COO^- + H_2S + 2H_2O \quad \text{(Equation 1)}$$

Free sulfide or bisulfide anions can bind with divalent metals to form precipitates that have very low K_{sp} values. For zinc, the reaction is shown in equation 2:

$$Zn^{2+} + H_2S \dashrightarrow ZnS + 2H^+ \quad \text{(Equation 2)}$$

Once precipitated, it is unlikely that the metal sulfides would again solubilize, but removal of the precipitate may not be a simple matter.

Materials and Methods

The two towers used in this experiment were built by Cook (Cook, 1991) and are illustrated in figure 1. They are made of plexiglass tubes about six inches in diameter and approximately 10 feet tall. The tubes are capped at both ends with a square piece of plexiglass plate sealed using a rubber O-ring. Both towers have an inlet below the bottom of the plate and an outlet near the top of the tube on the side. Each tower has a series of six sample ports along the length of the tower. The two towers were connected by tygon tubing such that the effluent from the first tower became the influent to the second tower. Each tower was equipped with a dump valve so that the contents could be drained as needed. The towers were filled with plastic media, Yaeger, Tripack #2. Black plastic trash bags were placed over both towers and the Tygon tubing connecting the two in order to control photosynthetic growth.

Just before the tubing entered the towers, a three-eighths inch stainless steel tee was placed in the line where ethanol was added to the feed. About two inches after the tee, another tee, this one plastic, was

Figure 1
Tower Configuration

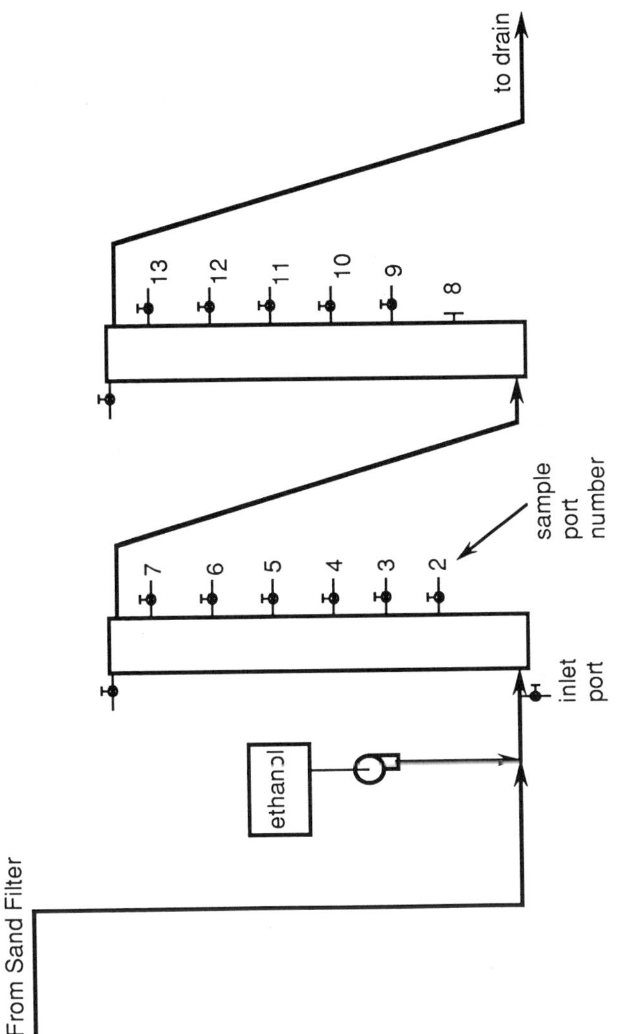

placed in the feed line and a sample valve attached to the tee. Complete mixing was expected by this point. This was where the inlet to the towers was sampled. After the feed flowed through both towers, it was drained to a sump in the basement of the building. Flow through the system was kept at about one liter per minute by restricting the flow at the inlet.

Ethanol was added to the feed stream by using a Laboratory Data Control "miniPUMP" high pressure pump which was originally designed to work with Ion Chromatography equipment. The ethanol feed rate was about 0.5 ml/min. The ethanol solution was approximately one gallon of 200 proof ethanol added to a 10 liter carboy with the remainder filled with tap water. This produced a solution of about 400,000 ppm. The final concentration in the feed was kept close to 200 ppm. No effort was made to keep the influent alcohol precisely consistent; the concentration was added in excess of the carbon requirements of the biomass. No other constituents were added to the feed.

Samples were taken in new 50 ml Nalgene test tubes and were not preserved chemically or filtered except for special tests. All filtered samples were processed through a 28 mm, 0.45 μm Gelman filter using a syringe. Typically, the samples were analyzed for four parameters during each round of experiments: ethanol, sulfate zinc and pH. Dissolved oxygen was measured off line at the towers. Except for dissolved oxygen, all samples were either analyzed that day or refrigerated. Most samples were analyzed within one or two hours of being taken.

Ethanol was measured on a Varian Gas Chromatograph by direct injection, with a capillary column and an FID detector. Sulfate was measured using a Dionex Ion Chromatograph by EPA method 300.0. Zinc was measured on an Atomic Adsorption Spectrophotometer using direct aspiration as outlined in EPA method 289.1. pH was measured on an Orion pH meter-model 611. Dissolved oxygen (D.O.) was measured at various points in the towers using a YSI D.O. meter. Samples were taken using a tygon tube which delivered sample to a large beaker without air entrainment.

Results

Data for determining sulfate kinetics was collected across five sample dates. The data for each test is shown in table 1. A graph of the data is shown in figure 2. Sulfate concentration was not normalized for any of the tests, it was purely a product of the sulfate in the effluent at that time. All runs showed a similar pattern, and not surprisingly, the final concentration was dependent on the initial concentration. Reductions in

Figure 2

Sulfate (ppm) vs. Detention Time

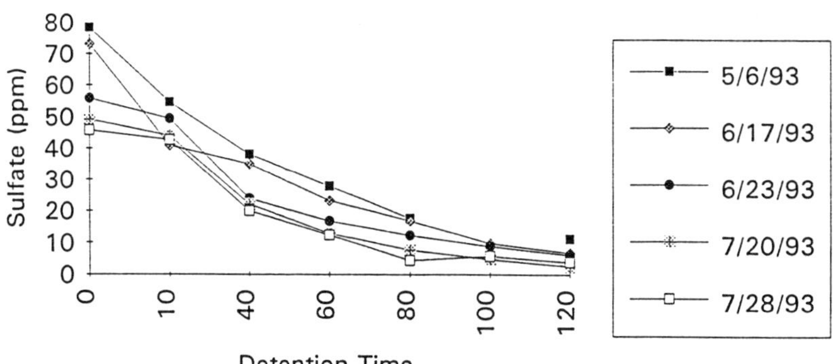

Figure 3

Zinc (ppb) vs. Detention Time

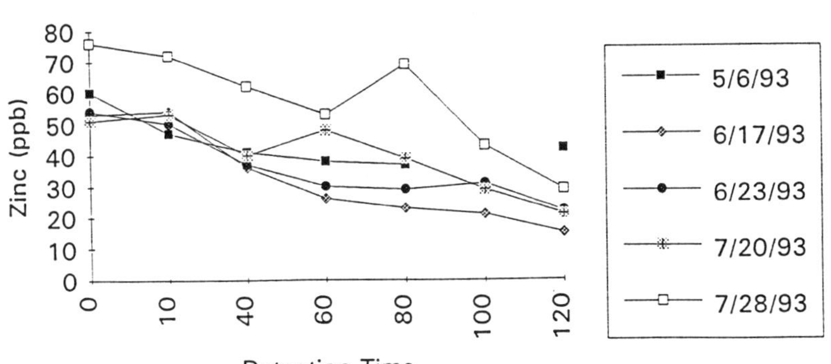

sulfate concentration ranged from 86% to 95%. For all tables to follow, N.S. refers to No Sample taken.

Table 1
Sulfate Concentration (ppm) and Sulfate Reduction (%)

Sample Date	τ= 0	τ= 10	τ= 40	τ= 60	τ= 80	τ= 100	τ= 120	% Change
5/6	78.4	54.5	38.0	27.9	17.6	N.S.	11.2	86
6/17	73.1	40.7	34.9	23.3	16.8	9.8	6.8	91
6/23	55.8	49.2	24.0	16.7	12.2	8.7	6.2	89
7/20	49.1	43.9	22.1	12.9	7.6	4.7	2.5	95
7/28	45.7	42.5	20.0	12.4	4.4	5.7	3.9	92

The kinetic behavior of biotransformations in packed towers usually fits either zero, half or first order models. All three models were evaluated for the data. The first order model had the best regression coefficients for four out of the five data sets. The first order model and the average slope are shown in equation 3.

$$\ln S - \ln S_o = -0.02(\tau) + b \qquad \text{(Equation 3)}$$

where: S = the substrate concentration at time τ
S_o = the initial substrate concentration
τ = time in minutes through the towers

Based on previous studies, it was expected that the pH would increase through the towers due to the creation of alkalinity. However, in these experiments, the pH dropped from 0.5 to 0.9 pH units across the towers.

Zinc removals were not as consistent as the sulfate reduction results. It is possible that this was caused by samples contaminated with suspended solids. It would be expected that solids would carry precipitated or bound zinc. The results were not as spectacular as the sulfate removals. This is a direct contradiction to the published literature. The reason for this result is not clear and will be the subject of further research. Zinc data in ppb for all five runs is shown in table 2 and in figure 3.

Table 2
Zinc Concentration (ppb) and Zinc Removal (%)

Sample Date	T = 0	T = 10	T = 40	T = 60	T = 80	T = 100	T = 120	% Change
5/6	60	47	41	38	37	N.S.	42	30%
6/17	53	54	36	26	23	21	15	72%
6/23	54	50	37	30	29	31	22	60%
7/20	51	53	40	48	39	29	21	58%
7/28	76	72	62	53	69	43	29	62%

The sample from port #13 for the 7/28 run was filtered through a 0.45 μm filter before running. The unfiltered sample result was 36 ppb.

Discussion

The hypothesis of this project was that sulfate reducing bacteria would be able to remove low levels of metals in an upflow anaerobic packed tower. Given information in the literature, it was expected that all metals would be removed to less than detection limits. Though sulfate was greatly reduced, zinc was never removed more than 72% in any of the tests.

Though the goal of this work was not principally sulfate removal, the process worked extremely well. The controlling factor for sulfate removal in this situation was the addition of ethanol. There is no particular need to remove sulfate in most wastewaters, but new regulations concerning sulfate in drinking water could make this aspect of the work very relevant to parts of the country with high ambient sulfate.

Most of the literature cited indicated a higher pH due to the process of sulfate reduction. In many of those cases, dolomite, limestone or outright addition of sodium hydroxide was used to either adjust the pH or to provide buffering during the runs. Several researchers have noted that pH levels lower than 7 were inhibitory to sulfate reducing bacteria. All of the initial pH levels in this project were below pH 7 and all pH values in the effluent from the towers was below pH 6. The lowest pH, 5.26, was accompanied by the highest sulfate reduction for a test: 95%.

The goal of this work was to design a system that removed zinc from effluent. On that point, this experiment was not successful. The literature points to complete removals of metals, while this test only achieved a 72% reduction of zinc. There are no obvious reasons that explain why this failed to happen. However, some theories have emerged that can be easily tested in future work.

Conclusions

This work demonstrated that biological sulfate reduction of a treated industrial effluent was first order with respect to sulfate when ethanol was supplied in excess. Sulfate removal levels of up to 95% were achieved for initial sulfate levels of less than 80 ppm.

In the presence of excess ethanol, pH decreased across the system, possibly due to fermentation processes proceeding concurrently with sulfate reduction and the lack of a viable methanogenic community. Future work is needed to optimize an ecosystem that will keep the pH from declining.

Zinc removals across the system were lower than expected. Though zinc removals reached a high of 72% during one run, the factors that dictate the differences between good runs and poorer runs are not obvious.

References

Barnes, L.J., Sherren, J., Janssen, F.J., Scheeren, P.J.H., Versteegh, J.H., and Koch, R.O. 1992. Simultaneous Microbial Removal of Sulphate and Heavy Metals from Wastewater. *Transactions of the Institution of Mining and Metallurgy. Section C: Mineral Processing and Extractive Metallurgy.* 101: C183.

Cook Jr., Nevis E., Silverstein, JoAnn, Veydovec, Bill, de Mendonca, Maria Marcia, and Sydney, Roger. 1991. Field Demonstration of Biological Denitrification of Polluted Groundwater and Pilot Scale Field Testing of Biological Denitrification with Widely Varied Hydraulic Loading Rates. *Colorado Water Resouces Research Institute, Completion Report No. 162.* Ft. Collins, CO.: Colorado State University.

Key words: Sulfate reduction, Heavy metals removal

Chelating Composite Membrane: Its Application in Selective Heavy-Metal Solid-Phase Separation

Sukalyan Sengupta[1]
Arup K. Sengupta[2]

Abstract

Selective removal of small amounts of heavy metals from the background of large amounts of innocuous sludge or slurry is a challenging separation problem. Conventional fixed-bed sorption processes are unable to handle sludges/slurries with high suspended solids content. This study identifies and investigates a new class of composite sorptive/desorptive ion-exchange membrane made of fine particles of chelating polymers entrapped in thin sheets of porous polytetrafluoroethylene (PTFE). These membranes are not fouled by high concentration of suspended solids but retain the original properties of the chelating exchangers. Various physico-chemical properties of the sorptive/desorptive membranes are discussed in context to decontamination of heavy-metal-laden sludges.

Introduction

Various industrial processes produce an effluent containing heavy metal compounds along with high suspended solids (up to 10% mass/volume). The heavy metals are most likely to be present in small amounts (<5% of the total mass or volume of the solid phase) as precipitates of salts with low solubility products, e.g., as hydroxides, carbonates, sulfides, etc., in the background of bulk amounts of innocuous solids such as clay, sand, calcite, humin, etc. The heavy metals may also be sorbed onto the ion-exchange sites of soil or other solid phase materials. Even though the percentage of heavy metals in the solid phase is very low, the waste sludge is most likely to be categorized as hazardous according to present environmental regulations. An ideal treatment process to decontaminate the sludge/slurry would be to selectively remove the heavy-metal present in the solid phase without affecting the remaining non-toxic materials, thereby

---

[1]Assistant Professor, Civil Engineering Department, University of Massachusetts Dartmouth, North Dartmouth, MA 02747

[2]Associate Professor, Civil Engineering Department, Lehigh University, Fritz Lab # 13, Bethlehem, PA 18015

reducing the volume of hazardous sludge considerably. Solidification/Stabilization of this sludge increases the volume of the waste to be disposed of without guaranteeing non-leachability of the toxic heavy-metals permanently. Some recently developed membrane processes such as surface liquid membrane, Donnan dialysis, and affinity dialysis are effective for removal of dissolved heavy-metals but are not capable of treating sludges with high suspended solids primarily because of membrane fouling. All currently available membrane, fixed-bed and fluidized bed processes require an excessive amount of pretreatment for heavy-metal contaminated sludges. We have identified and investigated a new class of composite sorptive/desorptive ion-exchange membranes in which fine spherical beads (<100 μ in dia) of heavy-metal selective chelating ion-exchangers are physically enmeshed or trapped in thin sheets ($\approx$0.5 mm thick) of highly porous polytetrafluoroethylene (PTFE), as shown in the electron microphotograph (Figure 1). These composite membranes, because of their thin-sheet like physical configuration, can be easily introduced into and withdrawn from any reactor containing sludge/slurry and the target solutes can be adsorbed onto the microbeads. These membranes are not fouled by high concentration of suspended solids but retain the retain the original properties of the chelating exchangers even after use for a number of cycles.

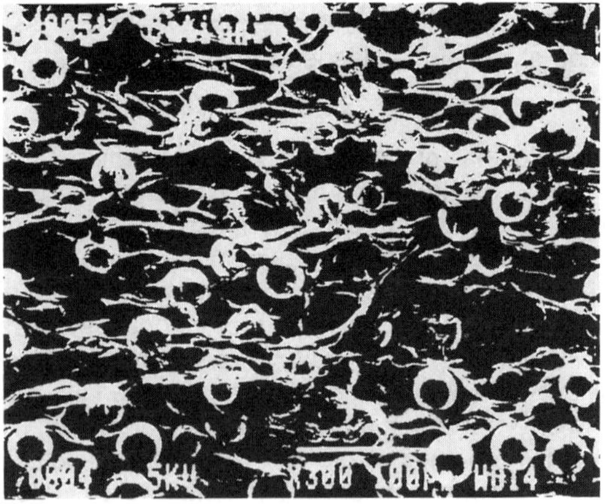

Figure 1. Scanning Electron Microphotograph (300X) of the composite membrane.

The Proposed Decontamination Process

Since this study is aimed at removing heavy-metals, the microbeads chosen were polymeric ion-exchangers containing covalently attached chelating iminodiacetate (IDA) moiety. This functional group was chosen since it has high selectivity towards dissolved heavy-metals over alkali metal or alkaline earth metal cations, which are likely to be present as background innocuous cations in most practical application cases. Table 1 provides the salient properties of the composite IDA membrane, from which it can be seen that the chelating microbeads constitute up to 90% of the composite membrane by mass. Various characterization studies of the composite membrane, e.g., water uptake in different counterion forms, titration, hysteresis, and equilibrium heavy-metal uptake isotherm, etc. proved that the sorption and regeneration behaviors of the composite membranes are essentially the same as the parent chelating microbeads (IDA in this case). Experimental evidence in this regard is provided by Sengupta & Sengupta[1] and Sengupta[2].

Table 1
Properties of the Composite Chelex Membrane

Composition	90% Chelex chelating resin, 10% teflon
Pore size (nominal)	0.4 μm
Nominal capacity	3.2 meq/gm dry membrane
Membrane thickness	0.4-0.6 mm
Ionic form (as supplied)	Sodium
Resin matrix	Styrene-divinylbenzene
Functional group	Iminodiacetate
pH stability	1-14
Temperature operating range	0-75 ^{0}C
Chemical stability	methanol; 1N NaOH; 1N H_2SO_4
Commercial availability	Bio-Rad Inc., California

For a sludge containing small amount (<5%) of heavy-metal, say as $Me(OH)_2(s)$, the composite membrane can be used in a cyclic two-step process as follows:

Sorption Step:

The composite membrane sheet, when introduced into a vigorously stirred sludge reactor after adjusting its pH in the range 3.0-5.0, selectively removes dissolved heavy-metals from the aqueous phase in preference to other non-toxic cations such as Na^+, Ca^{2+}, Mg^{2+}, etc. Consequently, more heavy-metal hydroxide precipitate dissolves to maintain equilibrium aqueous phase concentration and the two reactions occurring in series can be written as follows:

$$\textit{Dissolution--} \quad Me(OH)_2(s) \rightleftharpoons Me^{2+} + 2OH^- \qquad (1)$$

$$\textit{Memb. Uptake--} \quad \overline{RN(CH_2COOH)_2} + Me^{2+} \rightleftharpoons \overline{RN(CH_2COO^-)_2Me^{2+}} + 2H^+ \qquad (2)$$

The overbar represents the membrane phase, in which the polymer matrix R is covalently attached to the chelating iminodiacetate functional group. It may be noted that the overall process involves selective transfer of heavy-metal, Me(II), from one solid phase (sludge) to another (composite membrane) at a moderate pH.

Desorption Step:

The second step of the process entails withdrawing the composite membrane sheet from the sludge and introducing it into a gently stirred tank containing 2-5% v/v H_2SO_4 or any other mineral acid. In this step, the exchanger microbeads in the membrane are efficiently regenerated in hydrogen form (because IDA is a weak acid and thus prefers hydrogen ion the most) according to the following reaction:

$$\textit{Regen.--} \quad \overline{RN(CH_2COO^-)_2Me^{2+}} + H_2SO_4 \rightleftharpoons \overline{RN(CH_2COOH)_2} + Me^{2+} + SO_4^{2-} \qquad (3)$$

The regenerated membrane is subsequently withdrawn and reintroduced into the sludge reactor as a continuation of the cyclic process. Experiments were conducted to determine the kinetics of heavy-metal uptake in the sludge reactor and a batch kinetic model developed which validated the experimental data.

Heavy-Metal Precipitate in a Buffered Sludge at Alkaline pH

Sludges/slurries may contain a heavy-metal precipitate swamped in a solid phase with high buffer capacity (e.g., calcite) and organic substances capable of forming strong complexes with toxic metals. Due to high buffer capacity of the solid phase, pH of the sludge will be alkaline, thereby reducing the aqueous phase free heavy-metal cation concentration, $\{Me^{2+}\}$. This will significantly reduce the uptake capacity of Me^{2+} by the composite membrane, thus making direct application of the above process inefficient. Also, it would be impractical to add acid to the reactor to lower the pH. Our approach was to introduce an aqueous phase ligand into the sludge reactor to selectively increase aqueous phase heavy-metal concentration. However, the chelating functionality of the composite membrane microbead would have stronger preference for the heavy-metal as compared to the aqueous phase ligand and so would split the aqueous phase heavy-metal - ligand bond to attract the heavy-metal cation to its ion-exchange sites. A cyclic recovery process was tried for a sludge phase containing sand, calcite, and copper oxide. Disodium oxalate was added as the aqueous phase ligand and the pH of the sludge reactor maintained at 9.0. It can be seen from Figure 2 that the cyclic process was quite effective for selective separation of copper from a very unfavorable sludge composition. Explanation of oxalate and copper uptake is provided by Sengupta & Sengupta[1] and by Sengupta[2].

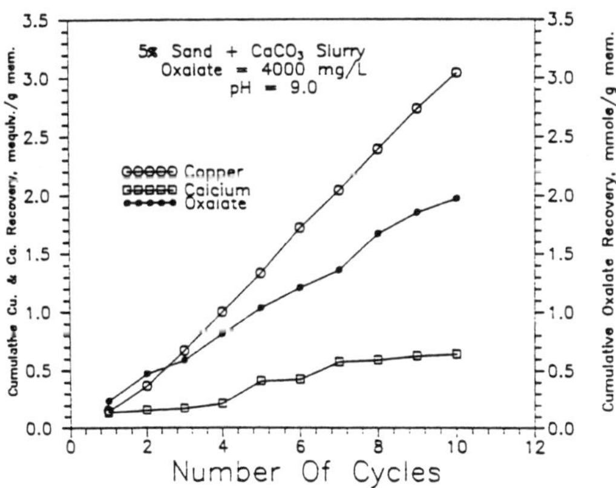

Figure 2. Cumulative Cu(II), Ca(II) and oxalate recovery with cycles at alkaline pH.

Heavy-Metals Extraction from Ion-Exchange Sites of Soil

If the heavy-metal contaminants are bound to the ion-exchange sites of soil, the composite membrane may be used to remove them selectively in a single reactor. In order to confirm the same, analytical grade bentonite (-200 mesh from J. T. Baker Co., Pennsylvania) was chosen as the soil type and was loaded with copper by equilibrating with an aqueous phase containing 400 mg/L copper concentration added as copper nitrate and at pH = 5.5. By mass balance, the total exchange capacity was found to be approximately 50 *mequiv./*100 *g* of dry bentonite (magnesium alumino silicate). Copper-loaded bentonite was then introduced into a plastic container containing 200 *mL* of 500 *mg/L* Na^+ (added as NaCl) solution. Copper-loaded bentonite was the only solid phase present in the slurry and the solids loading was 2.2%. A strip of composite membrane (weighing 0.112 *g*) was introduced into the container and the two-step process (i.e. sorption and desorption) was run, the results of which are presented in Figure 3. It can be seen that >60% of Cu(II) removal was achieved in less than 30 cycles. Copper concentration in the regenerant was used to compute the percentage copper recovery. The mechanism of Cu(II) removal is discussed by the authors in open literature[1,2].

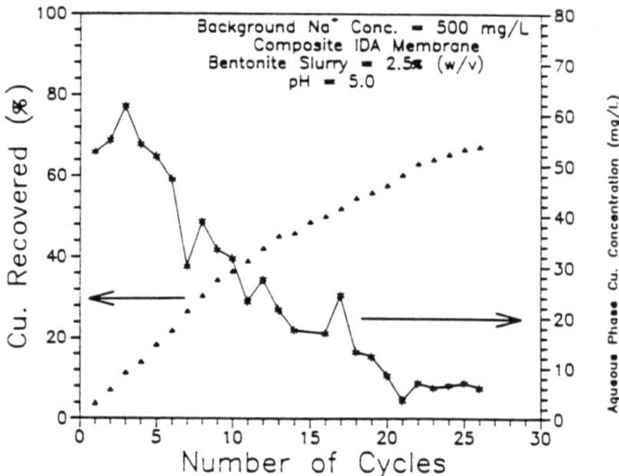

Figure 3. Cu(II) recovery from the ion-exchange sites of bentonite clay during the cyclic process.

Heavy Metals Removal from Humic Acid Sludge

Humic substances are present in all natural aquatic and soil environments; thus it is important to study the fate and transport of heavy metals in the presence of humic substances. It is generally agreed that humic substances contain carboxylic and phenolic groups which act as ligands in the presence of heavy metals to give rise to metal-humic complex. Humic acid extract (HAE), sodium salt [1415-93-6]; supplied by Aldrich Chemical Company was taken to be the reference humic acid sample. Heavy metal capacity of the HAE was determined by the procedure described by Sengupta[2] and the following data was obtained:

At pH = 4.05, pCu = 2.544, νCu = 3.22 ($10^{-3.22}$ moles of Cu/*g* of original HAE)

Experiments were conducted to find out copper removal capacity of a humic acid sludge. Copper-loaded humic acid sludge was taken as the contaminated sample and cyclic extraction test was performed on this sludge at a constant pH of 3.0. A strip of the composite membrane in H^+ form weighing 0.5496 *g* was added to the reactor in each exhaustion cycle, which lasted for 4 hours. At the end of each exhaustion cycle, the membrane was taken out, rinsed with tap water and introduced into the regeneration bottle which had initially 200 mL. of 5% H_2SO_4. Filtered samples from each cycle (exhaustion and regeneration) were collected and analyzed for Cu(II) and Dissolved Organic Carbon (DOC). It can be seen from Figure 4 that after 11 cycles, approximately 44% of Cu(II) was transferred from the humic acid sludge reactor to the regenerant solution. Explanation of the mechanism of Cu(II) transfer is provided by Sengupta[2].

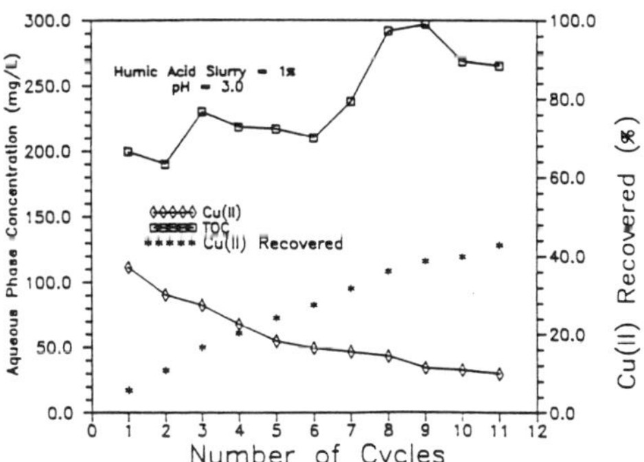

Figure 4. Cu(II) recovery from humic acid-extract sludge in a cyclic process.

Conclusions

The composite sorptive/desorptive ion-exchange membrane shows promise in selective sorption of heavy metals from complex sludges, e.g. from highly buffered alkaline sludges, from soils with cation-exchange capacity, from humic/fulvic acid containing soil, etc. Its tough physical texture allows it to be used in vigorously stirred slurry reactors containing high percentage of suspended solids without any tearing/rupture or membrane fouling by particulate matter. At the same time, the entrapped chelating ion-exchanger microbodies which comprise up to 90% of the material by mass, are uniquely suited to selectively sorb heavy metals from the aqueous phase under severe competition from such common background ions as Ca^{2+} and Na^{+}. The composite membrane can be used in a simple two-step process, sorption and desorption, allowing heavy metals to be selectively recovered and concentrated in an acid solution, which may be suitable for recycle/reuse in any process.

References

1. Sengupta, S., and Sengupta, A. K., "Characterizing a New Class of Sorptive/Desorptive Ion Exchange Membranes for Decontamination of Heavy-Metal-Laden Sludges", *Env. Sci. & Tech.*, Vol. 27, pp. 2133-2140, 1993.

2. Sengupta, S., "A New Separation and Decontamination Technique for Heavy-Metal-Laden Sludges Using Sorptive/Desorptive Ion-Exchange Membranes", *Ph. D. Dissertation*, Lehigh University, Bethlehem, PA, Nov. 1993.

EFFECT OF SLUDGE BLEND ON DEWATERING CHARACTERISTICS

Ali Davarinejad[1], Nikolay Voutchkov[2]

Abstract

This study compares the performance of high solids centrifuges when processing anaerobically digested and undigested mixture of primary sludge and waste activated sludge, in order to evaluate the effect of sludge blend ratio on the dewaterability of the sludge. The operational data (full scale and pilot scale) of six selected wastewater treatment plants were analyzed during the course of this study. Three of the six plants dewater anaerobically digested sludge, two dewater raw sludge, and one dewaters both digested sludge and raw sludge in parallel trains. Several key operational parameters including: primary to waste activated sludge ratio, feed rate, polymer dosage, cake solids concentration, and solids recovery were evaluated in order to provide a basis for comparison of the centrifuge performance. The data indicates that the sludge blend (primary to waste activated sludge ratio) has a significant effect on the dewaterability of both digested and undigested sludges. Under similar operating conditions, cake concentration increases with increasing the primary sludge fraction.

Introduction

The centrifuge technology has been successfully used for solids dewatering at a number of wastewater treatment plants in the U.S. for over a decade. In recent years, the high solids centrifuges, which represent the "second generation" of this technology, have gained significant attention for their ability to reliably and cost-effectively dewater sludge from medium to large municipal wastewater treatment facilities. High solids centrifuges have been mostly utilized for dewatering anaerobically digested sludge. More recently, the high solids centrifuges have also been successfully pilot tested and used in full-scale applications for dewatering undigested solids.

1 Supervising Engineer, Montgomery Watson, 301 North Lake Avenue, Suite 600, Pasadena, CA 91101

2 Senior Engineer, Montgomery Watson, 301 North Lake Avenue, Suite 600, Pasadena, CA 91101

The objective of this study is to compare the performance of high solids centrifuges processing a wide range of blends of anaerobically digested and undigested sludges, in order to examine the effect of sludge blend on the performance of dewatered centrifuges.

Approach

The operational data of a number of high solids centrifuge installations, processing a mixture of primary and activated sludge, was analyzed in order to provide a basis for centrifuge performance comparison at various sludge blends. Subsequently, the data from six selected wastewater treatment plants were analyzed in more detail. Three of the six plants dewater anaerobically digested solids, two dewater undigested solids, and one dewaters both digested and undigested solids in parallel trains. Several key operational parameters including: primary to waste activated sludge blend, sludge feed rate, polymer dosage, cake solids concentration, and solids capture efficiency were compared in order to evaluate the effect of anaerobic blend on the centrifuge performance.

Facilities Description

City of Los Angeles - Hyperion Treatment Plant. The Hyperion Treatment Plant (HTP) provides full primary and partial secondary treatment, with an average design capacity of more than 1,300,000 m^3/d (380 mgd). In addition, HTP provides treatment for solids generated at three upstream water reclamation plants with a combined capacity of approximately 400,000 m^3/d (114 mgd). Ferric chloride is added to the primary clarifiers to enhance the primary effluent quality. The raw primary and thickened waste activated sludges are combined, anaerobically digested, and screened prior to centrifuge dewatering. The plant currently produces more than 275 dry tons per day of digested solids.

City of Los Angeles - Terminal Island Treatment Plant. The Terminal Island Treatment Plant (TITP) provides full secondary treatment with an average design capacity of 100,000 m^3/d (30 mgd). The primary and thickened waste activated sludges are blended, anaerobically digested, and dewatered by centrifuges. Currently, approximately 20 dry tons per day of digested solids are processed at TITP.

Metropolitan Water Reclamation District of Greater Chicago - Stickney Water Reclamation Plant. The Stickney Water Reclamation Plant with an average design capacity of 4,200,000 m^3/d (1,200 mgd) provides full secondary treatment. The primary sludge is screened and blended with the thickened waste activated sludge, anaerobically digested, and dewatered with high solids centrifuges.

Minneapolis Metropolitan Waste Control Commission - Metropolitan Plant. The Metropolitan Plant has an average design capacity of 870,000 m^3/d (250 mgd) and provides full secondary treatment. The primary sludge is gravity thickened and the waste activated sludge is flotation thickened. A portion of the gravity-thickened primary sludge is dewatered by roll presses and the remainder is blended with the thickened waste activated sludge, thermally conditioned, and dewatered by plate and frame presses. The plant is currently producing approximately 250 dry tons of solids per day. Currently, plans are underway for installation of high solids centrifuges to dewater the raw sludge without thermal conditioning.

Prince William County Service Authority, Virginia - H.L. Mooney Wastewater Treatment Plant. The H.L. Mooney Wastewater Treatment Plant is a tertiary treatment facility with a design capacity of 42,000 m^3/d (12 mgd). Chemical coagulation with ferric chloride (at both primary and secondary clarifiers) is provided for phosphorus removal. The waste activated sludge is conveyed to the primary clarifiers for co-settling. The combined undigested solids are gravity thickened prior to centrifuge dewatering. The plant generates approximately 6 dry tons of solids per day.

District of Columbia Water and Sewer Utility - Blue Plains Wastewater Treatment Plant. The Blue Plains Wastewater Treatment Plant in Washington, D.C. with an average capacity of 1,100,000 m^3/d (310 mgd) provides secondary treatment with chemical phosphorus removal. The plant's solids dewatering facilities utilize high solids centrifuges in three separate trains. The first train dewaters a digested mixture of primary and waste activated sludges. The second train dewaters an undigested blend of primary and waste activated sludges, while the third train provides dewatering of thickened waste activated sludge only. The plant handles approximately 450 dry tons of solids per day.

Data Analysis and Evaluation

Table I provides a summary of treatment plant data associated with the above-described facilities. Figures 1, 2, and 3 summarize high solids centrifuge performance at the three plants with anaerobic sludge digestion prior to dewatering: Hyperion Treatment Plant, Terminal Island Treatment Plant, and Stickney Water Reclamation Plant (Harrison et al., 1992; STRR, 1992; Garelli, et al., 1990). Figures 4 and 5 present centrifuge performance data for the two plants where the sludge is not digested prior to dewatering: Minneapolis Metropolitan Plant and H.L. Mooney Wastewater Treatment Plant (Montgomery Watson, 1993; Sharples, Inc., 1989a; Sharples, Inc., 1989b).

Typically, the feed sludge concentrations of the digested sludge ranged from 2 to 4 percent, while the undigested sludge concentration was in the range of 3 to 5 percent. The digested sludge temperatures ranged from 30°C to 35°C and the temperature of the undigested sludge was between 15°C to 20°C. The centrifuge feed rates varied between 20 m^3/h to 60 m^3/h (100 to 300 gpm). Mannich type polymer was used for sludge dewatering at the Hyperion Treatment Plant and all other plants used liquid emulsion polymers. Specific polymer selection was usually determined by trial and error. Dosage rates typically ranged between 5 and 10 kg/ton (10 to 20 lb/ton), and costs ranged from $1.2 to $4.00 per kg of polymer.

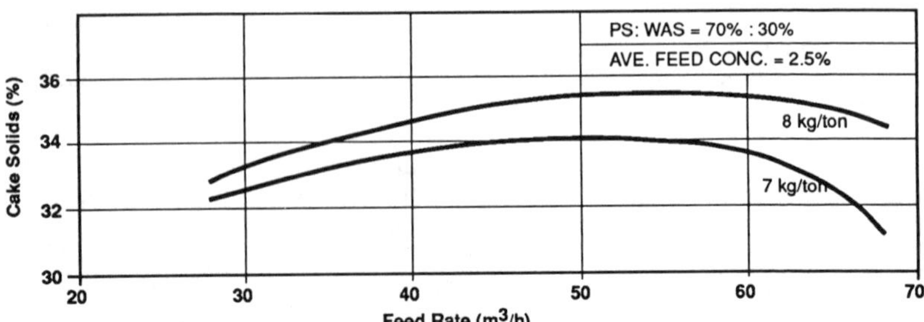

Figure 1. Centrifuge Performance with Digested Solids Hyperion Treatment Plant

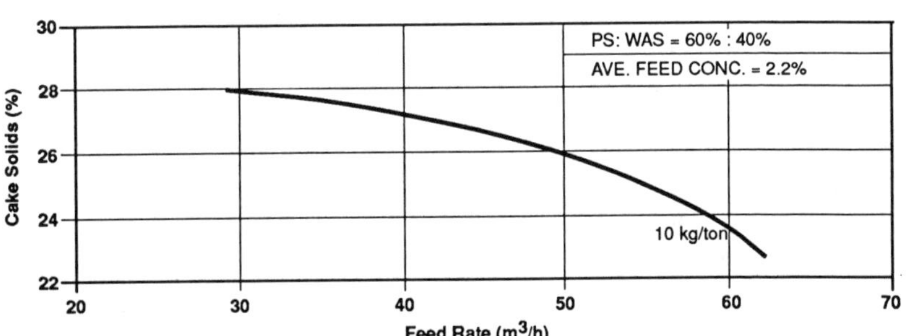

Figure 2. Centrifuge Performance with Digested Solids Terminal Island Treatment Plant

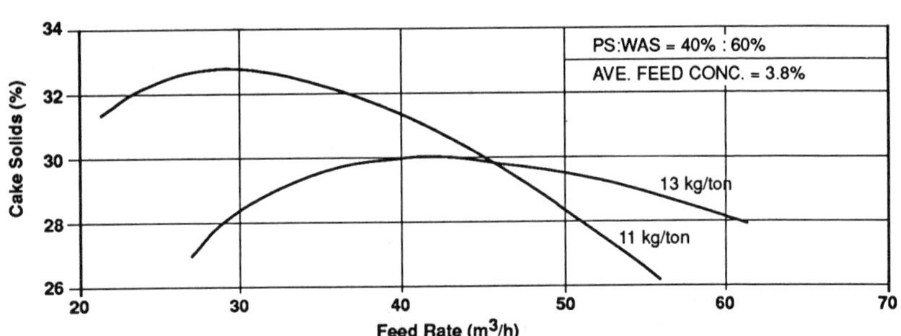

Figure 3. Centrifuge Performance with Digested Solids Stickney Water Reclamation Plant

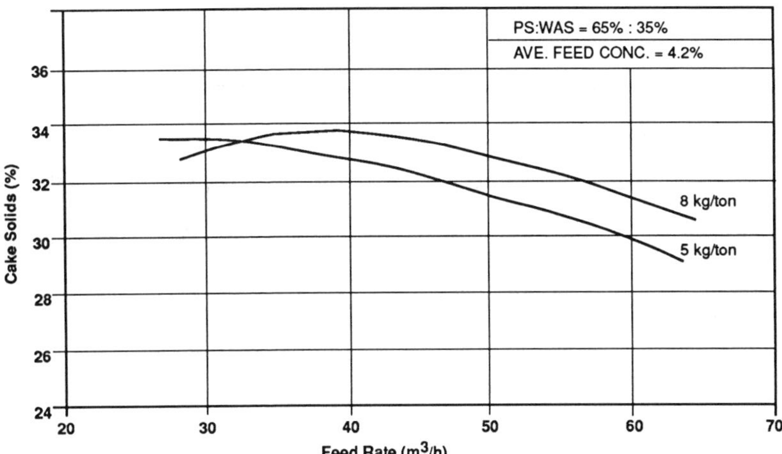

Figure 4. Centrifuge Performance with Undigested Solids Minneapolis Metropolitan Plant

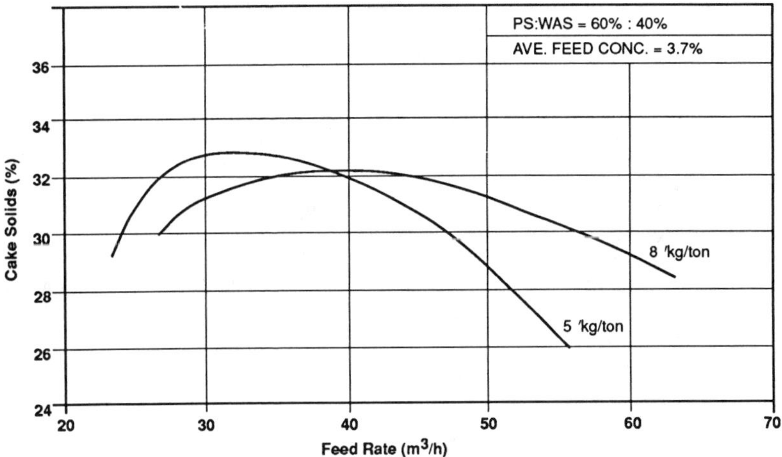

Figure 5. Centrifuge Performance with Undigested Solids H.L. Mooney Wastewater Treatment Plant

TABLE I. SUMMARY OF SELECTED TREATMENT PLANT DATA

Treatment Plant	Plant Capacity, m^3d (mgd)	Primary to WAS, %	Anaerobic * Digestion	Digester SRT, Days
Hyperion	1,300,000 (380)	70:30	Yes	14
Terminal Island	100,000 (300)	65:35	Yes	14
Stickney	4,200,000 (1,200)	40:60	Yes	14
Minneapolis Metropolitan	1,000,000 (300)	65:35	No	---
H.L. Mooney	42,000 (12)	60:40	No	---
Blue Plains	1,100,000 (310)	40:60	Yes/No	28

* Anaerobic Digestion Prior to Dewatering

The results, as presented in the figures, indicate that the high solids centrifuges are able to consistently produce cake with a solids content ranging between 25 to 32 percent, while dewatering digested solids for a wide range of sludge blends. In addition, the data indicates that the high solids centrifuges are capable of consistently producing cake with a solids content ranging between 28 to 32 percent, under similar operating conditions, when dewatering undigested solids.

A solids recovery of 90 to 99 percent can be typically achieved at polymer dosages of 5 to 10 kg/ton (10 to 15 lb/ton) when dewatering both digested and undigested solids with high solids centrifuges. However, solids recovery is slightly higher when dewatering digested sludge as compared to dewatering undigested sludge at similar primary to waste activated sludge (PS:WAS) ratio, polymer dose and feed rate. For both digested and undigested sludge, solids recovery is sensitive to the centrifuge torque and can decrease below 90 percent at flows higher than 60 m^3/h (300 gpm). In addition, solids recovery is enhanced with the increase PS:WAS ratio or polymer dosage.

The sludge blend (PS:WAS ratio) is a significant factor impacting dewatering capabilities of sludge. The dewaterability of both digested and undigested solids is adversely impacted as the ratio of primary sludge to waste activated sludge (on a mass basis) is decreased. Figure 6 presents the impact of sludge blend on cake solids concentration with a solids recovery in the range of 95 percent to 99 percent. For example, at the Minneapolis Metropolitan Plant, at a feed rate of 40 m^3/h (200 gpm) the cake solids concentration increases from approximately 25 percent to 33 percent, when the primary sludge portion increases from 40 percent to 75 percent. In addition, data presented in Figures 1 through 5 indicates that in general, a cake solids concentration higher than 30 percent may be produced only for primary sludge portion of 60 percent or higher. Low PS:WAS ratios can be partially compensated for by increasing the polymer dosage. However, when the primary sludge content is less than 50 percent of the total solids mass, the production of 30 percent cake or higher may not be cost-

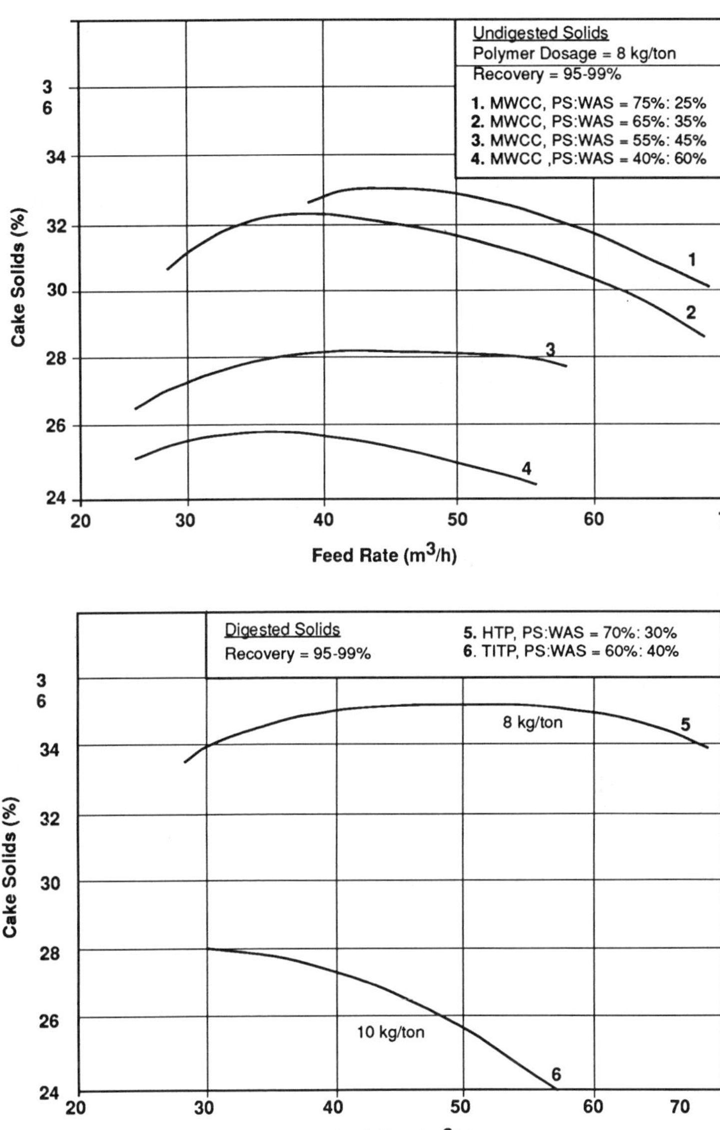

Figure 6. Centrifuge Performance at Various PS:WAS Ratios

effective due to increased polymer cost for both digested and undigested solids. In addition, the sludge recovery rate rapidly decreases with the decrease of primary sludge content.

Summary and Conclusions

The high solids centrifuges can reliably and cost effectively produce cake dryness of 30 percent or higher on both digested and undigested solids with a primary sludge fraction of 50 percent or higher and a polymer dose of 5 to 10 kg/ton (10 to 20 lb/ton), under similar operating conditions. The analysis of data from a number of pilot and full-scale installations indicates that the sludge blend has a significant effect on sludge dewaterability of both digested and undigested sludge. The sludge dewaterability improves with the increase of primary sludge portion. Further, it can be concluded that the anaerobic digestion does not necessarily enhance the sludge dewatering characteristics, when using high solids centrifuges.

Acknowledgement

The authors would like to acknowledge the assistance provided by the District of Columbia, Water and Sewer Utilities Bureau of Sludge Management, Blue Plains Wastewater Treatment Plant.

References

City of Los Angeles, Solids Technology and Resource Recovery Division (1992). Terminal Island Treatment Plant, Sludge Dewatering Test Report.

Garelli, B.A., Kokat, E.L., Schwartz, B.J. (1990). Improved Centrifuge Sludge Dewatering by Automatic Backdrive Torque Control, Water Science Technology, Volume 22, no. 12, pp.303-308.

Harrison, D. S., Kvasnicka, K.W., Berk, J.S., Haug R.T., Zschach, A. (1992). Performance Evaluation of High Torque Sludge Dewatering Centrifuges, 65th Annual Conference Water Environment Federation, New Orleans.

Montgomery Watson (1993). Metropolitan Wastewater Treatment Plant Evaluation Study, Solids Handling Facilities Evaluation, MWCC Project No. 855616.

Sharples, Inc. (1989a). Technical Department Report, Dewatering Mixed Raw Sludge at the H.L. Mooney Wastewater Treatment Plant in Prince William County, Virginia.

Sharples, Inc. (1989b). Technical Department Report, Dewatering Mixed Raw Sludge at the Blue Plains Wastewater Treatment Plant at Washington, D.C.

Reducing Wastewater Sludge Pathogens Using Heat Treatment

Mark Kennedy[1], Associate Member, ASCE
Kondapalli Reddy[2]
Lyle Johnson[3], Member, ASCE

Abstract

The feasibility of using waste heat to reduce pathogen densities in wastewater sludge at a municipal wastewater treatment plant was investigated. Two modifications to an existing two stage mesophilic anaerobic digestion process were studied, 1) pasteurization of primary digester sludge prior to secondary digestion and 2) operation of the second stage digester at thermophilic temperatures. Both options for using waste heat resulted in a sludge which met Class A criteria with respect to fecal coliform and *salmonella* sp. densities. *Ascaris* ova were also inactivated. Destruction of volatile solids was greater than current digester operation and digested sludge did not support the regrowth of bacteria during subsequent storage. Analysis of heat requirements revealed that waste heat generated during summer operation of the anaerobic digesters was more than sufficient to support thermophilic operation of the second stage digester.

---

[1] Associate Professor, Civil and Environmental Engineering, South Dakota State University, Brookings, SD 57007

[2] Graduate Student, Civil and Environmental Engineering, South Dakota State University, Brookings, SD 57007

[3] Water and Wastewater Manager, City of Sioux Falls, 224 West 9th, Sioux Falls, SD, 57102

Introduction

Municipal wastewater sludge disposal is a major concern of environmental regulators and wastewater utility managers. Options for environmentally sound disposal of wastewater sludges are restricted. Landfill space for sludge or incinerated sludge ash is limited. Air pollution and solid waste regulations are making the incineration option less viable both environmentally and economically. Utility managers are increasingly turning to land application of sludge or conversion of sludge into a marketable product as potential options for sludge disposal. Land application of sludge is currently the most widely used method of sludge disposal for small and medium sized treatment plants in the United States (Wastewater 1991). Wastewater sludges contain microorganisms which can cause disease in humans. These microorganisms are referred to as pathogens. Land application of sludge creates a potential for human exposure to pathogens through direct and indirect contact.

To protect human health, the U.S. EPA has promulgated regulations with respect to land application of sludge. Land application of sludge comes under joint authority of the Resource Conservation and Recovery Act and Clean Water Act. Regulations addressing land application are contained in 40 CFR part 503 criteria for classification of solid waste disposal facilities and practices. Land application of sludge is considered a form of solid waste disposal and is subject to these criteria. These regulations protect public health by requiring sludge management practices that eliminate or minimize human contact with sludge contaminants. The EPA promulgated the final 503 regulations on pathogens and vector attraction in Subpart D on February 19, 1993. These regulations detail the classification of sludge and management practices. Sludge is classified into two categories, Class A and Class B sludge, based on pathogen densities in the sludge.

Final sludge regulations state that to achieve s Class A sludge, "either the density of fecal coliform in the sewage sludge shall be less than 1000 Most Probable Number (MPN) per gram of total solids on dry weight basis, or the density of *Salmonella* sp. bacteria in the sewage sludge shall be less than three MPN per four grams of total solids on dry weight basis at the time of disposal" (40 CFR part 503, 1993). One of the alternatives for achieving a Class A sludge is heat treatment. Heat treatment involves raising the sludge temperature to a specific level for a minimum time period. These approximate temperatures and holding periods can be determined using equations outlined in the sludge regulations (40 CFR part 503, 1993). When the solids content is less than seven percent, the temperature of the sludge is more than 50°C and the time period is 30 minutes or longer, the time period and temperature is:

$$D = \frac{50,070,000}{10^{0.1400\,t}}$$

where:

t = temperature degrees Celsius
D = detention time, days

Scope and Objectives

The City of Sioux Falls is presently operating a 5.1×10^4 m^3 per day wastewater treatment plant which produces on average approximately 190 m^3 per day of sludge. The sludge is treated using a two stage anaerobic digestion process. Sludge from a gravity thickener is fed into three primary anaerobic digesters operating in parallel. Primary digester sludge is then fed into a secondary anaerobic digester. Until recently secondary digester sludge had been either dewatered and disposed in a landfill or applied to neighboring farmland. Because of the problems associated with landfills, the City has abandoned this option entirely in favor of farmland application. However, new sludge regulations imposing stringent restrictions on farmland application of sludge has caused uncertainty regarding future farmland application for the City. There is also a growing interest in converting the sludge into a marketable product such as a gardener compost or potting soil.

Currently the City is producing more than 5670 m^3 of methane gas per day from the anaerobic digesters. This gas is being used as an energy source for generating approximately 19% of the electrical needs of the treatment plant. Waste heat from the engines used to drive the generators is recovered for heating the digesters and the digester control buildings. There exists a considerable amount of waste heat which is not used in the summer months. The overall objective of this study is to determine the feasibility of using this waste heat to further treat the digested sludge. The goal of heat treatment is to produce a Class A sludge which can be applied to farmland without any restrictions and at some future time possibly be converted into a marketable product.

Two heat treatment processes were considered in this study. They are pasteurization and thermophilic digestion. These two processes were judged to be the most economically feasible for future addition to the Sioux Falls treatment plant. The primary advantage that pasteurization and thermophilic digestion have over other processes is their compatibility with the existing mesophilic anaerobic digestion and liquid land disposal operations.

Although current practice in Europe is to pasteurize raw thickened sludge prior to feeding the sludge to the primary anaerobic digester, this pre-pasteurization of raw sludge produces unpleasant odors, requiring odor control measures. Alternatively, digested sludge from the primary digester can be post-pasteurized prior to further digestion in the secondary digester. Although not odor free, this modification in the pasteurization process significantly reduces odors. Post-pasteurization was therefore selected for study. As noted earlier, European experience with post-pasteurization indicated problems with bacterial regrowth (Control, 1989). An important objective of this study was to determine whether regrowth would occur in the secondary digester treating pasteurized sludge.

Experimentation with thermophilic digestion in the United States has primarily focused on heating the primary digester to thermophilic levels. Thermophilic digestion has been successfully demonstrated, however, the process is unstable and temperature fluctuations must be kept to a minimum (Garber, 1982). In this study, thermophilic digestion in the secondary digester only was investigated. Most of the volatile solids in sludge are usually destroyed in the primary digester with the secondary digester serving as a settler. With conversion of the secondary digester into a thermophilic reactor, the overall digestion process is not compromised as a result of the inherent instability of thermophilic digestion.

Future plans for the City of Sioux Falls wastewater treatment plant include the construction of sludge storage lagoons. An additional objective of this study was to determine the effects of storage on pathogen survival in both the pasteurized and thermophilically digested sludge.

Experimental Procedures

This study was conducted in two phases. In the first phase, experiments were conducted to evaluate the effectiveness of post-pasteurization of primary digester sludge followed by mesophilic digestion in reducing sludge pathogens. Sludge from the Sioux Falls primary digester was first pasteurized at 70°C for 30 minutes and then fed into a three liter laboratory secondary anaerobic digester. The reactor was initially charged with unpasteurized primary digester sludge. A hydraulic retention time (HRT) and temperature of 15 days and 36°C, respectively, was maintained in the reactor. Concentrations of *Salmonella* sp. and fecal coliform bacteria in the reactor were monitored with time to evaluate pathogen regrowth or destruction in the mesophilic digestion process. Volatile solids destruction and gas production also were monitored to determine the effects of pasteurized sludge on the digester performance. Experiments were repeated using unpasteurized sludge spiked with *Salmonella* sp. and fecal coliform bacteria to charge the reactor.

In the second phase of the study, the laboratory reactor was operated as thermophilic secondary digester with the temperature maintained at 55°C. The reactor was fed sludge from the Sioux Falls primary digester. A fill and draw mode of feeding the reactor was used with 20% of the digesting sludge volume being replaced with feed sludge every three days. This operating scheme allowed a three day contact period while maintaining an average HRT of 15 days. *Salmonella* sp. and fecal coliform bacteria densities were monitored in both the feed and effluent sludges to verify pathogen destruction. Gas production and volatile solids destruction were used as indicators of digestion process stability. A second set of experiments was conducted in which the sludge was spiked with *Salmonella* sp. and fecal coliform bacteria as well as *Ascaris* ova.

At the end of both phases, treated sludges were again spiked with *Salmonella* sp. and fecal coliform bacteria to evaluate the effects of long term storage on pathogen survival. Total solids (TS), volatile solids (VS) and fecal coliform analyses were conducted according to the procedures described in *Standard Methods for the Examination of Water and Wastewater* (1993). *Salmonella* sp. analysis was conducted using procedures developed by Kenner and Clark (1974). Procedures for determining viable *Ascaris* ova are described elsewhere (Parasites, 1981).

Results and Discussion

Sludge pasteurization resulted in the complete destruction of fecal coliform and *Salmonella* sp. bacteria in both raw and primary digester sludges. Initial fecal coliform densities ranged from 8.9×10^4 MPN per gram of TS in primary digester sludge to greater than 2.4×10^8 MPN/g TS in raw thickened sludge. Initial *Salmonella* sp. densities ranged from 5.4 MPN/g TS in raw sludge to 9.2×10^3 MPN/g TS in spiked digester sludge. Viable *Ascaris* ova also were inactivated, although some larvae development occurred before helminths were destroyed during subsequent sludge digestion. Pasteurization of raw sludge produced stringent odors which were not evident with primary digester sludge, therefore, further pasteurization of raw sludge was not investigated. Sludge pasteurization enhanced subsequent mesophilic digestion. The pasteurization process released degradable organics, resulting in an 8% increase in VS destruction compared to current VS reductions being achieved in the secondary digester at the Sioux Falls plant.

Mesophilic digestion of pasteurized primary digester sludge prevented the regrowth of fecal coliform and *Salmonella* sp. bacteria. Initial fecal coliform densities in the mesophilic reactor charged with unspiked digester sludge were 9.2×10^4 MPN/g TS. Fecal coliform densities decreased with time at a rate greater than dilution. Initial *Salmonella* sp. densities in digester sludge were less than 1 MPN/g TS and density changes with time could not be detected. Fecal coliform and *Salmonella* sp. densities in the mesophilic reactor charged with spiked sludge were 5.5×10^6 and 4.8×10^5 MPN/g TS, respectively. Both fecal coliform and *Salmonella* sp. densities decreased with time at a rate greater than dilution, indicating that these types of bacteria could not compete with the methanogenic and fermentative populations indigenous to the mesophilic digestion process. Although fecal coliform and *Salmonella* sp. densities decreased in the anaerobic digestion process, the Class A sludge criterion for fecal coliform densities (1000 MPN/g TS) in both the spiked and unspiked sludge studies was not reached until 40 days after startup of the reactor.

Thermophilic digestion of primary digester sludge was equally effective in reducing fecal coliform and *Salmonella* sp. bacteria. Fecal coliform and *Salmonella* sp. densities in the reactor feed sludge were 2.0 x 10^5 and 96 MPN/g TS, respectively. After two days of thermophilic digestion fecal coliform bacteria were completely eliminated and the density of *Salmonella* sp. was reduced to 0.4 MPN/g TS. *Ascaris* ova were inactivated after a three days of digestion. Volatile solids reductions during thermophilic digestion were 10% higher than those currently achieved in the secondary digester at the Sioux Falls plant.

Sludge from both the pasteurization/mesophilic digestion and thermophilic digestion processes were resistant to fecal coliform and *Salmonella* sp. bacteria regrowth during long term storage. Immediately following digestion, sludge samples were spiked with bacteria. Initial densities for *Salmonella* sp. ranged from 38 to 90 MPN/g TS and fecal coliform ranged from 1.6 x 10^5 to 3.0 x 10^6 MPN/g TS. Fecal coliform and *Salmonella* sp. bacteria in treated sludge decreased with time during storage. *Salmonella* sp. bacteria were not detected after 23 days of storage at both room temperature and 4°C. At room temperature, fecal coliform bacteria were not detected after 34 days of storage. At 4°C, there was a steady drop in fecal coliform density with time, however, after 102 days of storage, the density of fecal coliform bacteria was still 8.3 x 10^2 MPN/g TS. Fecal coliform data suggests that bacterial inactivation in stored sludge was much slower at reduced temperatures. Therefore, if treated sludge were somehow contaminated prior to winter storage there could be some difficulty in meeting Class A sludge criteria when sludge was applied to land in the spring.

Implementation of second stage thermophilic digestion at the Sioux Falls plant appears to be much simpler and more economical than pasteurization of primary digester sludge with subsequent second stage mesophilic digestion. Thermophilic digestion requires only the addition of a water/sludge heat exchanger and additional hot water storage capacity. Pasteurization requires sludge/sludge and sludge/water heat exchangers in addition to a pasteurization reactor.

Modification of the existing continuous flow mesophilic secondary digestion process Sioux Falls to a batch thermophilic seccondary digestion process is illustrated in Figure 1. The proposed operation allows for a minimum contact time of 50 hours during the batch thermophilic digestion process. This contact time meets the requirements specified in the 503 regulations and laboratory results indicate that this contact time should be sufficient to reduce fecal coliform and *Salmonella* sp. bacteria. Heating requirements for maintaining mesophilic digestion in the primary digesters and thermophilic digestion in the secondary digester are summarized in Table 1. Currently 159.3 million kJ of waste heat per three day batch cycle are available during the summer months. Although sufficient waste heat is available in the summer months, the rate of heat transfer required to heat incoming primary digester sludge to thermophilic levels during the ten hour feeding period exceeds the rate of heat recovery from the gas engine/generator. Therefore, the existing 98 m^3 hot water storage tank would need to be enlarged to approximately 681 m^3. The heat exhanger required for proposed modification would cost approximately $70,000.

Table 1. Total heat requirements for second stage thermophilic digestion at the Sioux Falls Wastewater Treatment Plant

Item	Heat Requirement 10^6 kJ per three day cycle
Heating primary digester feed sludge to 36°C	40.8[a]
Maintaining primary digesters at 36°C	9.8[a]
Raising temperature of primary digester sludge from 36°C to 58°C	52.3
Total	103.6

a: (Source: Johnson, 1987)

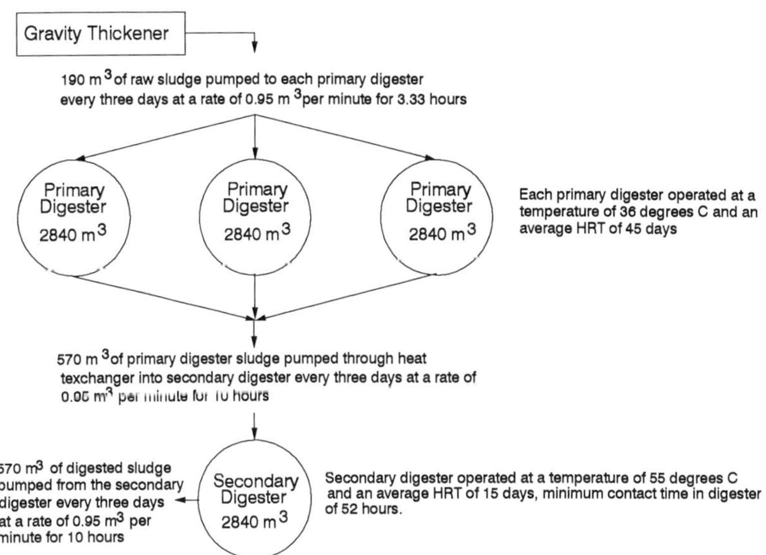

Figure 1. Proposed modifications for thermophilic digestion at the Sioux Falls wastewater treatment plant

Results of this study show that utilization of waste heat to reduce sludge pathogens is feasible with a relatively small capital expenditure. However, this study was conducted using a laboratory scale reactor. Actual digester performance may differ from experimental results because of scale effects. Pilot or full scale testing of the heat treatment option is recommended to verify the results obtained from the laboratory study.

References

Federal Register (1993) part 503, subpart D, February 19.

Control of Pathogens in Municipal Wastewater Sludge, EPA Pub. 625/10-89/006. (1989). EPA Center for Environmental Research Information, Cincinnati, Ohio

Garber, W.F. (1982). "Operating Experience with Thermophilic Anaerobic Digestion." *J. Water Poll. Control Fed.*, 54 (8) 1170 -1175.

Johnson, L.D. and Dostal, G.H. (1987) "Digester Gas Utilization." Presented at the 60th Annual Conference of the Water Poll. Control Fed., Philadelphia, Penna.

Kenner, B.A., and Clark, H.A. (1974) "Determination and Enumeration of *Salmonella* and *Pseudomonas aerugiosa*." *J. Water Poll. Control Fed.*, 46 (9), 2163 - 2171.

Parasites in Southern Sludges and Disinfection by Standard Sludge Treatment, EPA Pub. No. 600/2-81-166. (1981). National Technical Information Service, Springfield, Va.

Standard methods for examination of water and wastewater. (1993). 18th Ed., Am. Public Health Assoc., Washington, D.C.

Wastewater Engineering: Treatment, Disposal, Reuse. (1993). 3rd Ed., Metcalf and Eddy Inc., McGraw Hill Co., New York, N.Y.

Optimization Studies of Anaerobic Expanded Bed Reactor Pretreatment of Domestic Wastewater

Thomas L. Theis[1], Member, ASCE
Bruce J. Alderman[2]
Anthony G. Collins, Member[1], ASCE

Abstract

In recent years the advent of the anaerobic expanded bed reactor (AEBR) has made possible the treatment of wastewaters of comparatively lower strength than have traditionally been treated by anaerobic means. However, regardless of any benefits presented by its use, the process must be shown to be cost effective relative to other options before it will gain acceptance. This study reports on the results of a series of optimization exercises in which the costs of the AEBR as an expansion alternative for organically overloaded treatment plants are compared with costs for more conventional remedies. The optimization model developed includes the unit processes of primary clarification, AEBR, trickling filtration, secondary clarification, activated sludge, gravity thickening, anaerobic digestion, and vacuum filtration. Any of these options may be included or discarded from the overall model in order to represent a given wastewater facility.

Several constraint parameters were varied in order to determine the cases for which the AEBR would be the cost effective option. These were: primary effluent COD, hydraulic loading, temperature, cost of labor, cost of power, cost of sludge disposal, and amortization interest rate. Since the primary motivation for the study was to determine the viability of the AEBR

[1]Professor, Department of Civil & Environmental Engineering, Rowley Laboratories, Clarkson University, Potsdam, NY 13699-5715.

[2]Engineer, Clough Harbour & Assoc., P.O. Box 5269, 3 Winners Circle, Albany, NY 12205.

as a pretreatment method for organically overloaded plants, the primary effluent COD was the main parameter of interest and was varied in multiples of its design value.

Results indicated that the AEBR may be a cost effective alternative when the primary effluent COD is approximately 230 mg/L or above, depending on the type of plant (activated sludge or trickling filter) and other eternal factors.

Introduction

It has generally been recognized that the application of anaerobic processes for the treatment of organic wastes is desirable due to relatively low sludge production in comparison with aerobic processes and possible recovery of methane. Because of the low growth rate of anaerobic bacteria, however, their use has historically been limited to high strength wastes. Advances in anaerobic technology in the past thirty years have resulted in systems with high concentrations of bacteria, creating the possibility of treating more dilute wastes. For example the anaerobic sludge blanket and anaerobic filter have been found capable of treating wastewaters with chemical oxygen demands as low as 750 mg/L (Jewell, 1987). More recently, the development of the anaerobic expanded bed reactor (AEBR) has created the possibility of treating even lower strength wastes, although research has shown that the AEBR alone cannot reliably meet the municipal wastewater discharge limit of 30 mg/L of biochemical oxygen demand without a post-treatment process (Switzenbaum *et al.*, 1984). A pilot scale operation at the Ithaca (NY) Wastewater Treatment Facility has consistently met discharge limits using an AEBR followed by a nutrient film technique (Jewell, 1990).

Regardless of the potential benefits of the AEBR for wastewater treatment, it must prove cost effective if it is to be implemented on a wide basis. The purpose of this study was to establish cost effective guidelines for the construction and operation of AEBR systems at small municipal wastewater treatment plants. The initial point of departure was to use the AEBR as an add-on process, after primary sedimentation, for treatment plants which had become organically overloaded, the Potsdam (NY) activated sludge and Ames (IA) trickling filter facilities being used as base cases (design capacities: Potsdam 3.3 MGD, Ames 2.2 MGD; primary effluent COD: Potsdam 130 mg/L, Ames 240 mg/L). Results from these exercises suggested that specification of an AEBR for newly constructed facilities might also be a viable option.

Approach

Laboratory scale AEBR reactors were constructed and operated in order to determine performance efficiencies when fed domestic wastewater at various temperatures. Results are presented in Figure 1 as fraction of primary effluent COD converted as a function of hydraulic retention time for temperatures varying from 5 to 20°C. Model curves shown assume the reactor behaves as a CSTR with first order biodegradation of COD, and have the mathematical form

$$\text{COD Converted} = \frac{\zeta\,\theta_{an}}{\frac{1}{k_T} + \theta_{an}} \qquad (1)$$

where ζ is the degradable fraction, θ_{an} is the hydraulic retention time, and k_T is the temperature dependent conversion constant. Least square regression of equation (1) to the data resulted in values of ζ and k_T of 0.81 and 1.5 hr^{-1} (20°C), respectively. k_T is corrected using the standard Van't Hoff-Arrhenius relationship. Experimental details can be found in Kilambi (1993).

In order to facilitate design and cost calculations, a wastewater treatment optimization model was assembled. This model used the accepted performance equation for various unit processes (*e.g.*, Metcalf and Eddy, 1991; Reynolds, 1982) and empirical cost equations for each process developed from data collected by Black and Veatch (1971) and published by Craig *et al.* (1978). Cost data were updated using appropriate indices.

The optimal design, in the context of this research, was defined as the treatment alternative (*i.e.*, set of unit processes) which resulted in minimum cost for a given set of external factors. Seven such factors were systematically investigated, as given in Table 1, each with its default value. The model made use of the method of Box (1965), a nonlinear zero order method which has been used previously in wastewater optimization exercises.

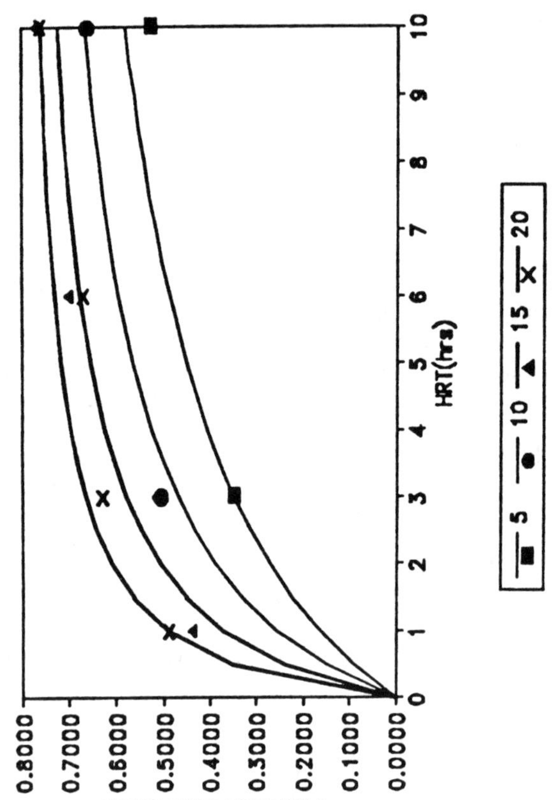

Figure 1. Fraction of COD converted as a function of hydraulic retention time for the laboratory AEBR at various temperatures. Curves are fitted using equation (1).

Table 1 - External Factors

Factor	Default Value
Primary Effluent COD	100% of Plant Design
Temperature	12.5°C
Power Costs	5.7¢/kWhr, 15.6¢/m^3 methane
Sludge Disposal Costs	$25/$m^3$
Labor Costs	$13/hr
Interest Rate	8%
Influent Flow Rate	100% of Plant Design

Results

The principal scenario simulated by the optimal design model was that of an organically overloaded plant undergoing expansion in order to meet existing effluent limitations. This was done by simulating two possibilities: one, expansion using conventional processes (activated sludge, trickling filter, gravity thickener, anaerobic sludge digester, and vacuum filtration), and two, expansion in which the AEBR was allowed to compete with the conventional processes. In all cases, the AEBR was placed following the primary settler, although provision was made for diverting a portion of the flow around the AEBR if technical constraints warranted it (*i.e.*, maintenance of a minimum substrate level for downstream aerobic process needs).

Among the seven factors investigated, the primary effluent COD and hydraulic loading to the plants showed the strongest effects on efficiencies and costs. These effects are illustrated in Figures 2 and 3 for the activated sludge and trickling filter plants, respectively. For each set of results, both factors were varied independently, and are expressed as percentages of the default design values, thus a percent design COD of 200 for the activated sludge plant corresponds to a primary effluent COD of 260 mg/L, while for the trickling filter plant the value would be 480 mg/L. The results are presented in the form of percent cost savings isopleths, that is, the percentage in expansion costs saved in using the AEBR in comparison with conventional means (negative percentages mean that the AEBR option is more expensive). Thus, referring to Figure 2, it can be seen that the AEBR expansion becomes favored when the COD rises to about 1.75 times the design COD (about 230 mg/L). For the trickling filter, this level is lower when given as a percentage (Figure 3), but is about the same when expressed as primary effluent COD. Somewhat more interesting is the more highly favored status of the AEBR as the hydraulic loading of the trickling filter increases. This is due to the assumption that trickling filters have a fixed percent COD removal

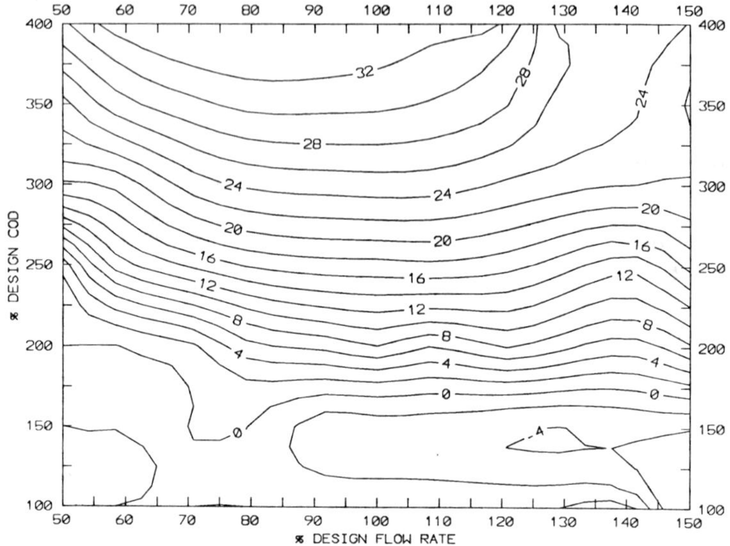

Figure 2. Percent savings in cost for the AEBR in comparison with conventional unit processes -- activated sludge case.

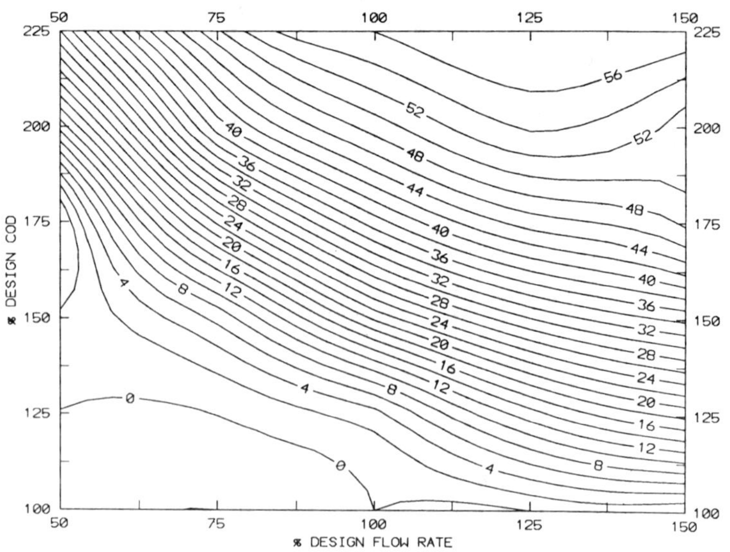

Figure 3. Percent savings in cost for the AEBR in comparison with conventional unit processes -- trickling filter case.

as opposed to fixed effluent concentration for activated sludge systems.

The other external factors of Table 1 were also investigated, with each having comparatively less pronounced effects on the cost patterns shown in Figures 2 and 3. In all cases favored for the AEBR, the optimum design was to have all wastewater anaerobically treated prior to downstream aerobic treatment. The AEBR design HRT most often selected was between five and seven hours, where COD conversion was near the maximum for a given temperature. Further details can be found in Alderman (1994).

Optimization simulations for new facilities, rather than expanded, suggested the AEBR-trickling filter combination without a thickener to be the least cost possibility among all combinations tested.

Conclusions

This study has shown that the AEBR may be a cost efficient alternative for organically overloaded small wastewater treatment plants which must be expanded to meet discharge limitations. The lower limit of primary effluent COD which can be effectively treated is approximately 230 mg/L, depending on the type of plant and other external factors.

Acknowledgments

This research was supported in part by contract 1772-ERER-MW92 of the New York State Energy Research and Development Authority, Mr. Barry Liebowitz, project officer.

Appendix - References

Alderman, B.J. (1994). "Optimization studies for the use of anaerobic pretreatment at small wastewater treatment facilities". Report 94-1, Department of Civil and Environmental Engineering, William J. Rowley Laboratories, Clarkson University, Potsdam, NY.

Box, M.J. (1965). "A new method of constrained optimization and comparison with other methods". *Computer J.*, *8*, 42.

Black and Veatch, Engrs. (1971). "Estimating costs and manpower requirements for conventional wastewater treatment facilities". US Environmental Protection Agency Water Pollution Control Research Series 17090, Washington, DC.

Craig, E.W., Meredith, D.D., and Middleton, A.C. (1978). "Algorithm for optimal activated sludge design". *J. Environ. Engr. Div.*, Proc. Am. Soc. Civ. Engrs., *104*, 1101.

Jewell, W.J. (1987). "Anaerobic sewage treatment". *Environ. Sci. Technol.*, *21*, 14.

Jewell, W.J. (1990). "Energy and biomass recovery from wastewater pilot documentation". Annual Report, New York State Energy Research and Development Authority Contract Number 1365-ERER-ER-90, Albany, NY.

Kilambi, S. (1993). "The treatibility of primary clarified domestic sewage by anaerobic expanded bed reactors". Report 93-14, Department of Civil and Environmental Engineering, William J. Rowley Laboratories, Clarkson University, Potsdam, NY.

Wastewater Engineering. (1991). Metcalf and Eddy, Inc., McGraw-Hill, New York.

Reynolds, T.D. (1982). *Operations and Processes in Environmental Engineering*, PWS-Kent Publishing Company, Boston.

Switzenbaum, M.S., Sheehan, K.C., and Hickey, R.F. (1984). "Anaerobic treatment of primary effluent". *Environ. Technol. Let.*, *5*, 189.

MODELING OF CHEMICAL INHIBITION OF NITRIFICATION

Robert W. Okey, Life Member, ASCE,
and H. David Stensel, Member, ASCE[1]

Abstract

This work contains two quantitative structure-activity relationships (QSARs) modeling chemical inhibition of nitrification, one for heterocyclic compounds and one for acetylenic materials. Both regression models are robust and predict with satisfactory accuracy the behavior of materials of known inhibition characteristics not in the model. The model is also applied to several pesticides, many of which appear able to strongly inhibit nitrification.

Introduction

Agricultural science has produced many organic substances which deliberately inhibit the normal conversion of ammonium to nitrate in soils under cultivation. These compounds are used to retain ammonium in the growth horizon for plant use and reduce the contamination of ground waters by oxidized nitrogen. Some are similar to nitrogen-containing pesticides and many other organics produced by the chemical industry.

The inhibiting substances can be classed in one of three categories: (1) nitrogen-containing heterocyclics, (2) acetylenic compounds, and (3) compounds containing phosphorus and nitrogen. This study is limited to the substances in categories 1 and 2. The N- and P-containing materials will be the subject of future work.

---

[1]Research Associate Professor, University of Utah, Salt Lake City, UT 84112; Professor, University of Washington, Seattle, WA 98195.

The literature from agricultural science, particularly agronomy, is drawn upon heavily for data on the relative effectiveness of these substances. QSAR technology has been employed to develop models describing the molecular features responsible for inhibition.

Background

Agronomists have developed a number of materials designed specifically for the inhibition of nitrification. Early reviews and discussions of the subject of inhibition were carried out by Keeney (1980), and Bundy and Bremner (1973). However, it was not until the mid-80s that definitive studies relating inhibition to molecular structure began to take place. Goring and Laskowski (1982) studied the effects of various types of pesticides on nitrification. McCarty and Bremner (1986) studied and reported on inhibition by acetylenic compounds. Many other reports appeared at approximately the same time including Powell and Prosser (1986) and Slangen and Kerkhoff (1984). The most definitive work was that of McCarty and Bremner (1989) on the inhibition of nitrification in soil by heterocyclic nitrogen compounds. Since that time, there have been many others including a recent study by He and Bishop (1994) on the effect of Acid Orange No. 7 on nitrification.

As the need arose, studies were made and reports began to appear on the control of nitrification during the standard BOD test. Young and Baumann (1972) reported on the ways of controlling nitrification with TCMP which ultimately became known as nitrapyrin. Other work was done by Young (1973) on the use of thiourea and allylthiourea to inhibit nitrification. Callaway and Young (1981) prepared an unpublished paper on the subject. Richterich and Steber (1989) reported on the prevention of nitrogen-caused errors in the BOD test.

Meanwhile, much work was being done on the identification of industrial waste components and other materials which could inhibit nitrification during biological treatment. These studies dealt with selective organic compounds. For example, there are reports on nitrification inhibition by methanol, acetone, formalin and glucose by Oslislo and Lewandowski (1985), and on phenol, nitrophenols, naphthol, and chlorinated naphthol reported by Benmoussa et al. (1986). Studies on cyanide and phenolic nitrification inhibition were reported by Neufeld et al. (1986).

Methods

The statistical and modeling procedures employed are similar to those reported earlier by Okey and Stensel (1993a, 1993b). In this procedure, the software is permitted to identify the relevant or significant system variables as opposed to a pre-selection of variables thought to be significant. The statistical programs utilized were Microstat II and SPSS.

Molecular connectivity and group contribution are used to describe the steric and electronic makeup of the molecule. Two "dummy" variables are employed: (1) "ALIF" meaning the presence of an aliphatic member, and (2) "AROM" meaning the presence of an aromatic member.

The basic data for the model of inhibition by 45 heterocyclics were taken from McCarty and Bremner (1989). The extent of inhibition in three types of Iowa soils @ 0.5 μmol/g is included in the report and it is possible to relate the inhibition of the compounds studied to nitrapyrin (2-chloro-6-(trichloromethyl)-pyridine), one of the most widely-employed commercial compounds. The work also reports on the impact of etridiazole (5-ethoxy-3-trichloromethyl-1,2,4-thiadiazole) and on the effect of 3-methylpyrazole-1-carboxamide (MPC) on nitrification. Having these three basic materials of commercial significance as references provides the basis for including or validating the model developed here with reports by other workers.

Data for the model of inhibition by acetylenic compounds also came from a McCarty and Bremner (1986) study which contained the relative behavior of 21 liquid compounds and four gaseous materials. This paper is important because it compared the effects of acetylenic substances to both nitrapyrin and etridiazole, providing a basis for the general comparison between these two classes of nitrification inhibitors. The non-gaseous materials were studied at 0.5 and 1.0 mmol/kg; the gaseous materials were studied at 0.1, 1.0 and 10.0 Pa.

The Quantitative Structure-Activity Relationship (QSAR) Models

In developing the model describing the inhibiting action of heterocyclics, six outliers were dropped out of the 45 basic data entries. One of these was the acetylenic material, 2-ethynylpyridine, which was not expected to remain in the model as it was assumed that the mechanism for inhibition by acetylenic compounds and by the nitrogen-containing heterocyclics was different. The most striking aspect of the model was the strong influence shown by the presence of heteroatoms. Somewhat surprisingly, the presence of an aliphatic structure (one or more carbons attached to the ring) also was a strong or significant addition to the overall inhibition. The five-member ring structure and molecular complexity both decreased the impact on nitrification. The greatest reduction in inhibition intensity was created by a carboxyl group. Significantly, other than the heteroatom descriptor, no specific group variable remained in the model. Heteroatoms, per se, were more significant than hetero-nitrogen, hetero-oxygen, or hetero-sulfur. Zero chi (^0X), which is a measure of molecular size, was the least significant of the variables but still significant at <0.1. The equation for this model and the significant regression properties are shown below.

$$\text{Log \% Inhibition} = 0.654\ (\text{\# of Heteroatoms}) + 0.541\ (\text{ALIF}) - 2.166\ ({}^{5}X_{ch}) - 0.567\ ({}^{3}X^{v}_{c}) - 1.099\ (\text{COOH}) + 0.104\ ({}^{0}X) - 0.169 \qquad (1)$$

$n = 39$, $r^2 = 0.79$, $F = 20.42$, $Sy = 0.281$

p (Heteroatoms) $<10^{-7}$ $\quad$ $p\ ({}^{3}X^{v}_{c}) <0.05$

p (ALIF) $<10^{-4}$ $\quad$ p (COOH) $<10^{-8}$

$p\ ({}^{5}X_{ch}) <0.05$ $\quad$ $P\ ({}^{0}X) <0.1$

The acetylenic model is based on 21 compounds. The database is small which limits the number of variables that can be used in the regression. At least five cases should be present for each variable employed and each variable should appear in at least five cases. This rule is violated somewhat in that two outliers were dropped to achieve the correlation and regression shown below. A model of 19 cases with 4 variables, in the present case, is satisfactory in that each of the variables appears in more than 5 of the cases. The equation is shown below.

$$\text{Log \% Inhibition} = -0.043\ ({}^{0}X^{v}) + 0.398\ ({}^{3}X^{v}_{c}) - 0.554\ (\text{COOH}) + 0.181\ (\text{AROM}) + 1.985 \qquad (2)$$

$n = 19$, $r^2 = 0.97$, $F = 125.31$, $Sy = 0.055$

$p\ ({}^{0}X^{v}) <0.01$ $\quad$ p (COOH) $<10^{-11}$

$p\ ({}^{3}X^{v}_{c}) <0.05$ $\quad$ p (AROM) $<10^{-3}$

As with the heterocyclic model, the carboxyl strongly reduces the extent of inhibition. Electronic charge also detracts, as measured by ${}^{0}X^{v}$, but complexity (${}^{3}X^{v}_{c}$) increases inhibition strength but decreases it in the heterocyclic model.

Discussion of Results

The model of heterocyclic inhibition has been applied to a number of substances reported in the literature. The model prediction and the comparative rates using nitrapyrin as the base are shown in Figure 1.

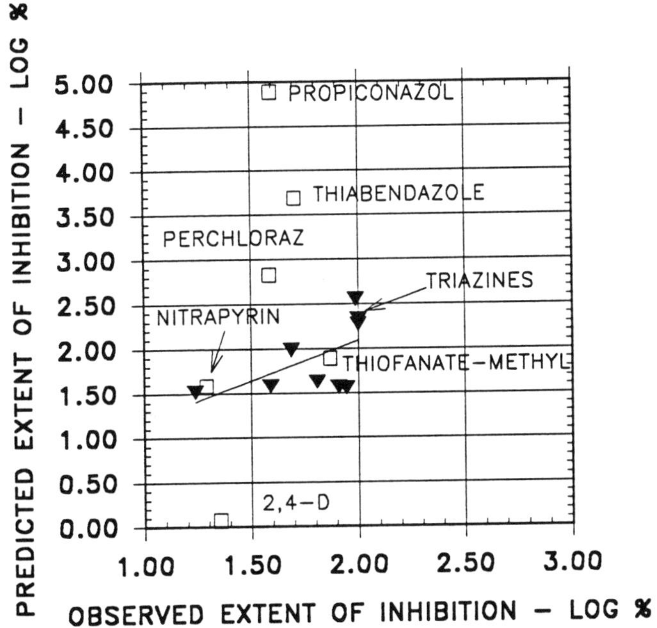

Figure 1. Predicted vs Observed Nitrification Inhibition by Triazines and Certain Herbicides

(Only triazine data used in regression – solid triangles.)
(Triazines: Murakami et al. (1993), 10 ppm in soil;
Herbicides: Hansson et al. (1991), 1–50 μg/g of soil)

The predicted values showing inhibition extent >100% are included in Figure 1 at the actual values to estimate the likely minimum concentration which will produce an effect. Based on these values, concentrations as low as 0.1 mg/L could conceivably produce strong inhibition.

The model of acetylenic inhibition was applied to a limited number of compounds as shown in Figure 2. The limited data show agreement, but, clearly, the model does not reflect the difference between the 1- and 2-butyne. The data were too few to permit a regression.

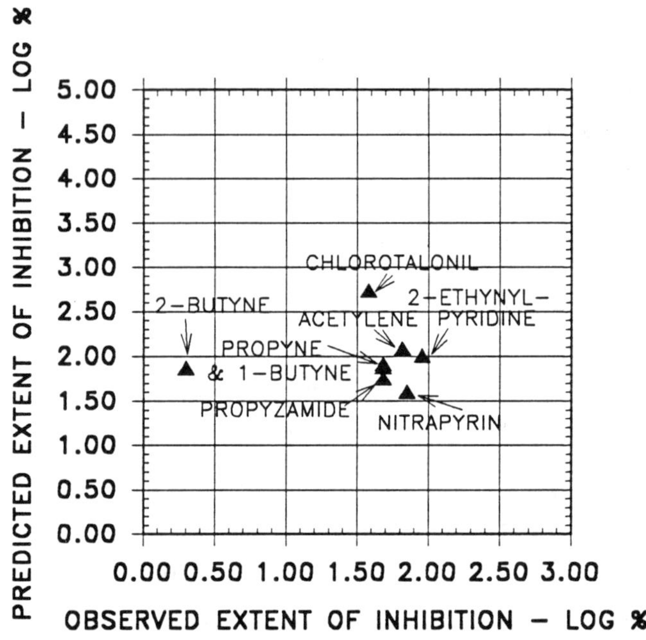

Figure 2. Predicted vs Observed Nitrification Inhibition by Selected Acetylenic Compounds

(Martens & Bremner (1993), 50 mg/kg of soil; Hansson et al. (1991), 1-50 μg/g of soil; McCarty & Bremner (1986), 0.5–1.0 mmol/kg of soil)

Summary and Conclusions

Two QSAR models of chemical inhibition of nitrification, one based on heterocyclic compounds and the other based on acetylenic materials, have been developed and tested for validity. Both models are robust and can predict, with reasonable accuracy, the behavior of materials not in the model.

The work permits the following conclusions:

1. A QSAR modeling nitrification inhibition can be developed utilizing molecular connectivity and group contribution.

2. Many pesticides and pesticide-like substances are strong inhibitors of nitrification and can inhibit at concentrations of 0.1-5.0 mg/L similarly to those materials specifically designed for nitrification control in the BOD test.

APPENDIX. REFERENCES

Benmoussa, H., Martin, G., Tonnard, F., Richard, Y., and Leprince, A. (1986) "Inhibition Study of the Nitrification by Organic Compounds," *Wat. Res.*, **20**(12), 1465-1470.

Bundy, L. G., and Bremner, J. M. (1973) "Inhibition of Nitrification in Soils," *Soil Sci. Soc. Am. Proc.*, **37**, 396-398.

Callaway, O., and Young, J. C. (1981) "Impact of Nitrification in BOD Tests on Treatment Plant Performance," Paper prepared for submission to the *J. Water Pollut. Control Fed.*

Goring, C. A. I., and Laskowski, D. A. (1982) "The Effects of Pesticides on Nitrogen Transformations in Soils." In: Stevenson, F. J. et al., Eds., Nitrogen in Agricultural Soils. Am. Soc. Agron., Madison, WI, *Agron.*, **22**, 689-714.

Hansson, G.-B., Klemedtsson, L., Stenström, J., and Torstensson, L. (1991) "Testing the Influence of Chemicals on Soil Autotrophic Ammonium Oxidation," *Environ. Toxicol. Wat. Qual.*, **6**, 351-360.

He, Y., and Bishop, P. L. (1994) "Effect of Acid Orange 7 on Nitrification Process," *J. Environ. Engn.*, **120**(1), 108-121.

Keeney, D. R. (1980) "Factors Affecting the Persistence and Bioactivity of Nitrification Inhibitors." In: Meisinger, J.J., Randall, G.W., Vitosh, M.L., Eds., Nitrification Inhibitors - Potentials and Limitations. Am. Soc. Agron. Spec. Publ. 38, Madison, WI, 33-46.

Martens, D. A., and Bremner, J. M. (1993) "Influence of Herbicides on Transformations of Urea Nitrogen in Soil," *J. Environ. Sci. Health*, **B28**(4), 377-395.

McCarty, G. W., and Bremner, J. M. (1986) "Inhibition of Nitrification in Soil by Acetylenic Compounds," *Soil Sci. Society of America J.*, **59**(5), 1198-1201.

McCarty, G. W., and Bremner, J. M. (1989) "Inhibition of Nitrification in Soil by Heterocyclic Nitrogen Compounds," *Biol. Fertil. Soils*, **8**, 204-211.

Murakami, M., Tsuji, A., Miyamoto, Y., Yamazaki, C., Ogawa, H., Takeshima, S., and Wakabayaski, K. (1993) "Synthesis of Trichloromethyl-1,3,5-triazines and Their Nitrification-Inhibitory Activity," *J. Pestic. Sci.*, **18**, 147-154.

Neufeld, R., Greenfield, J., and Rieder, B. (1986) "Temperature, Cyanide and Phenolic Nitrification Inhibition," *Wat. Res.*, **20**(5), 633-642.

Okey, R. W., and Stensel, H. D. (1993a) "A Biodegradability Model and a Three-Dimensional View of Biorefractivity," Paper presented at Water Environment Federation Conference, Anaheim, (In press).

Okey, R. W., and Stensel, H. D. (1993b) "A QSBR Development Procedure for Aromatic Xenobiotic Degradation by Unacclimated Bacteria," *Water Environ. Res.*, **65**, 772-780.

Oslislo, A., and Lewandowski, Z. (1985) "Inhibition of Nitrification in the Packed Bed Reactors by Selected Organic Compounds," *Wat. Res.*, **19**(4), 423-426.

Powell, S. J., and Prosser, J. I. (1986) "Inhibition of Ammonium Oxidation by Nitrapyrin in Soil and in Liquid Culture," *Appl. Environ. Microbiol.*, **52**, 782-787.

Richterich, K., and Steber, J. (1989) "Prevention of Nitrification-Caused Erroneous Biodegradability Data in the Closed Bottle Test," *Chemosphere*, **19**(10/11), 1643-1654.

Slangen, J. H. G., and Kerkhoff, P. (1984) "Nitrification Inhibitors in Agriculture and Horticulture: A Literature Review," *Fertil. Res.*, **5**, 1-76.

Young, J. C. (1973) "Chemical Methods for Nitrification Control," *J. Wat. Pollut. Control Fed.*, **45**(4), 637-646.

Young, J. C., and Baumann, E. R. (1972) "Demonstration of the Electrolysis Method for Measuring BOD," U.S. EPA, Final Report, ISU-ERI-AMES-72153.

Analysis of Biomass Distribution and Chemical Interactions in Nitrifying Biofilms

Edna M.C.V. Flora[1], Joseph R.V. Flora[2] (AM.ASCE),
Makram T. Suidan[3] (M.ASCE), Byung J. Kim[4] (M.ASCE)

Abstract

Steady-state models of nitrifying biofilms are developed taking into account the mass transfer of neutral and ionic species, electroneutrality, pH-dependent Monod kinetics, chemical equilibrium, and the presence of a boundary layer. Coupling the base biofilm model with reactor mass balances show significant changes in reactor performance as a function of influent pH and buffer capacity. Within the biofilm model framework, the steady-state biomass distribution in a nitrifying biofilm is sensitive to μ_1, the maximum velocity coefficient of the first step of nitrification. When coupled to a reactor model, the steady-state biomass distribution remains identical over a large range of μ_1.

Introduction

Biofilm processes are important to water quality control and are applied in numerous areas such as in groundwater treatment, in municipal and hazardous wastewater treatment, and in the prediction of the fate and effect of pollutants

---

[1] Grad. Res. Asst., Dept. of Civil and Environmental Eng., U. of Cincinnati, Cincinnati OH 45221-0071

[2] Assistant Professor, Dept. of Civil Eng., U. of South Carolina, Columbia SC 29208

[3] Professor, Dept. of Civil and Environmental Eng., U. of Cincinnati, Cincinnati OH 45221-0071

[4] Environmental Engineer, U.S. Army CERL, 2902 Farber Dr, Champaign IL 61826

in natural stream systems. Traditionally, substrate-utilization kinetics within biofilms has been modeled by coupling Fickian diffusion with Monod reaction kinetics. An inherent assumption in most of the previous models was that the pH remains constant within the biofilm. Experiments have shown differences between the pH in the bulk solution and in the biofilm. Since the rate of substrate utilization in biological systems can be affected severely by changes in pH, recent models have tried to account for the pH variations within the biofilm. However, most of these models neglected ionic interactions during mass transport. Ionic fluxes have to be described using the Nernst-Planck equation.

Nitrogen is one of the principal nutrients of concern in treated wastewater discharges, and nitrification is one of the primary methods of nitrogen control. As reported in the literature, Monod kinetic constants vary as a function of pH (Siegrist and Gujer, 1987). Furthermore, nitrification is a two-step and two-species process. Ammonia is initially converted to nitrite by *Nitrosomonas*, and then nitrite is converted to nitrate by *Nitrobacter*,

$$NH_4^+ + 1.5O_2 \rightarrow NO_2^- + 2H^+ + H_2O \tag{1}$$

$$NO_2^- + 0.5O_2 \rightarrow NO_3^- \tag{2}$$

Both species require oxygen as an electron acceptor for the transformations to occur. Because of the presence of these two species, oxygen and space competition will occur within the nitrifying biofilm.

In this study, a base biofilm model of steady-state nitrifying biofilms is developed taking into account the mass transfer of neutral and ionic species, electroneutrality, pH-dependent Monod kinetics, chemical equilibrium, and the presence of a boundary layer. The base biofilm model is coupled to reactor mass balances, to biomass-distribution-approximation equations, and to both.

Model Development

In the development of the base biofilm model, the species considered are the different forms of nitrogen (NH_3, NH_4^+, NO_2^-, HNO_2, NO_3^-, HNO_3), the dissolved oxygen concentration, H^+, OH^-, Na^+ (to maintain electroneutrality), and inorganic carbon as a buffering compound (CO_3^{-2}, HCO_3^-, CO_2). Following the approach of Flora et al (1993a,b), the model accounts for the mass transfer of neutral and ionic species, electroneutrality, pH-dependent Monod kinetics, chemical equilibrium, and the presence of a boundary layer. The model also neglects activity corrections, electrophoresis, relaxation and hydration effects, surface charges, light inhibition, precipitation of carbonates, and the production of nitrification by-products such as nitrous oxide.

The molar flux of each species is described by Fick's law for neutral species and by the Nernst-Planck equation for ionic species (Cussler, 1984)

$$J_n = -D_n \frac{dC_n}{dx} \tag{3}$$

$$J_i = -D_i \left(\frac{dC_i}{dx} + z_i c_i \frac{\mathcal{F} \nabla_x \psi}{RT} \right) \tag{4}$$

where J is the molar flux, D is the diffusion coefficient in water, C is the molar concentration, x is the distance perpendicular to the surface of the biofilm, z_i is the ion charge, $\mathcal{F}$ is Faraday's constant, $\nabla_x \psi$ is the one-dimensional electrostatic potential gradient, R is the universal gas constant, T is the absolute temperature, and subscripts n and i refer to the neutral and ionic species, respectively.

The biofilm system is analyzed in two discrete sections: in the boundary layer and in the biofilm itself. Neglecting biomass synthesis, respiration and decay, four one-dimensional steady-state mass balance equations are written for oxygen, total ammonium, total nitrite and total nitrate,

$$z \frac{dJ_{O_2}}{dx} = 1.5 r_1 + 0.5 r_2 \tag{5}$$

$$z \left(\frac{dJ_{NH_3}}{dx} + \frac{dJ_{NH_4^+}}{dx} \right) = r_1 \tag{6}$$

$$z \left(\frac{dJ_{NO_2^-}}{dx} + \frac{dJ_{HNO_2}}{dx} \right) = r_2 - r_1 \tag{7}$$

$$z \left(\frac{dJ_{NO_3^-}}{dx} + \frac{dJ_{HNO_3}}{dx} \right) = -r_2 \tag{8}$$

where z is a correction factor for diffusivity, and r_1 and r_2 represent the reaction rates of the first and second steps of nitrification, respectively. In the boundary layer, $z = 1.0$ and $r_1 = r_2 = 0$. In the biofilm, $z = z_{bio}$ (ratio of diffusivity in biofilm to diffusivity in water) and

$$r_1 = \frac{\mu_1 f_1 X_{f_1} C_{t_{NH_4}}}{Y_1 (K_{S_{NH_4}} f_3 + C_{t_{NH_4}})} \frac{C_{O_2}}{(K_{S_{O_2-1}} + C_{O_2})} \tag{9}$$

$$r_2 = \frac{\mu_2 f_2 X_{f_2} C_{t_{NO_2}}}{Y_2 (K_{S_{NO_2}} f_4 + C_{t_{NO_2}})} \frac{C_{O_2}}{(K_{S_{O_2-2}} + C_{O_2})} \tag{10}$$

where μ is the maximum velocity coefficient, X_f is the biomass density, Y is the yield, and K_S is the Monod half-saturation coefficient. The subscripts 1 and 2 represent the first and second steps of nitrification, and the subscript t in C_t indicates total concentration. f_1, f_2, f_3 and f_4 describe the pH dependence of μ and K_S as determined by Siegrist and Gujer (1987). The X_f for each species is estimated from the total biomass density and the yield coefficients,

$$\frac{X_{f_1}}{X_{f_t}} = \frac{Y_1}{(Y_1 + Y_2)} \tag{11}$$

$$\frac{X_{f_2}}{X_{f_t}} = \frac{Y_2}{(Y_1 + Y_2)} \tag{12}$$

Additional mass balance equations for total inorganic carbon and sodium are included in the system of equations. Coupled with expressions for electroneutrality and acid-base equilibria, the resulting system of equations were solved using the techniques outlined by Flora et al (1993a). Values for the parameters used are taken from Cussler (1984), Gee (1990), Lange (1973), Robinson and Stokes (1970), and Siegrist and Gujer (1987), and are listed in Table 1.

A second model is obtained by coupling the base biofilm model with standard, steady-state, completely-stirred tank reactor (CSTR) mass-balance equations. Steady-state effluent conditions are initially estimated and iteratively solved for using the reactor mass balances and fluxes calculated from the base biofilm model.

Equations (11) and (12) are useful as initial approximations of the biomass distribution within the biofilm. However, the amount of biomass in a system should also be proportional to the overall substrate uptake. For a biofilm, this substrate uptake is given by the flux J. A more reasonable approximation for the steady-state distribution of *Nitrosomonas* and *Nitrobacter* in a nitrifying biofilm can be obtained using,

$$\frac{X_{f_1}}{X_{f_t}} = \frac{Y_1 J_1}{Y_1 J_1 + Y_2 J_2} \tag{13}$$

Table 1. Parameters Used In This Study

Equilibrium constants	Diffusion coefficients (cm^2/day)	
NH_4^+ = $5.75x10^{-10}$	O_2 = 1.81	CO_3^{-2} = 0.79
HNO_2 = $5.13x10^{-4}$	NH_3 = 1.42	HCO_3^- = 1.02
HNO_3 = 21.9	NH_4^+ = 1.69	CO_2 = 1.66
HCO_3^{-1} = $5.62x10^{-11}$	NO_2^- = 1.65	Na^+ = 1.15
CO_2 = $4.17x10^{-7}$	HNO_2 = 2.26	OH^- = 4.56
H_2O = $1.01x10^{-14}$	NO_3^- = 1.64	H^+ = 8.04
	HNO_3 = 2.25	
Monod coefficients		
μ_1 = 1.27/day	$K_{S\text{-}NH4(t)}$ = .151 mg NH_4-N(t)/l	
μ_2 = 1.57/day	$K_{S\text{-}NO2(t)}$ = 2.25 mg NO_2-N(t)/l	
Y_1 = 0.28 mg VSS/mg NH_4-N(t)	$K_{S\text{-}O2(1)}$ = 0.3 mg O_2/l	
Y_2 = 0.08 mg VSS/mg NO_2-N(t)	$K_{S\text{-}O2(2)}$ = 0.4 mg O_2/l	

$$\frac{X_{f_2}}{X_{f_t}} = \frac{Y_2 J_2}{Y_1 J_1 + Y_2 J_2} \tag{14}$$

where J_1 and J_2 are the total fluxes of ammonium and nitrate respectively. A third model is obtained by incorporating equations (13) and (14) into the base biofilm model.

A fourth model is obtained by incorporating equations (13) and (14) and steady-state reactor mass-balance equations into the base biofilm model. More realistic bulk conditions result in a more reasonable approximation of the biomass distribution in nitrifying biofilms.

Results and Discussion

Figure 1 shows the profiles of pH and total NH_4, NO_2, and NO_3 within the boundary layer and the biofilm obtained using the base biofilm model. Although changes in nitrogen concentrations are not substantial, a significant drop in the pH across the boundary layer and the biofilm is predicted. This change in pH will have an important impact on the nitrification rates in the deeper portions of the film.

Using the reactor model (second model), the effluent pH and total NH_4, NO_2, and NO_3 concentrations are shown as a function of influent pH and total inorganic carbon in Figures 2a and 2b, respectively. Figure 2a clearly shows that

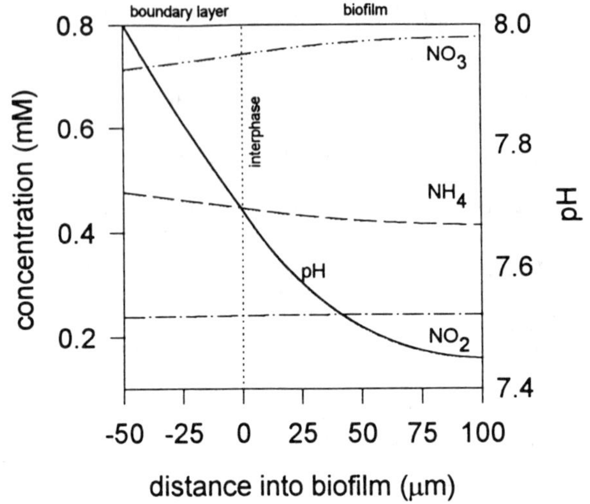

Figure 1. Profiles of pH and total NH_4, NO_2, and NO_3 [bulk : C_{O2}=.25mM, C_{TNH4}=0.48mM, C_{TNO2}=0.24mM, C_{TNO3}=0.71mM, C_{TIC}=1.67mM, pH=8.].

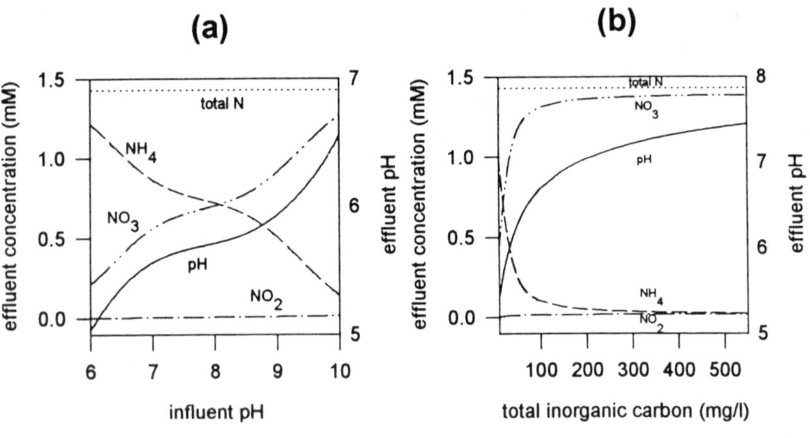

Figure 2. Effluent pH and total NH_4, NO_2, and NO_3 as a function of (a) influent pH (C_{TIC}=1.67mM) and (b) total inorganic carbon (influent pH=8.) [influent : C_{O2}=.25mM, C_{TNH4}=1.43mM, C_{TNO2}=C_{TNO3}=0.].

varying the influent pH can result in a significant change in the performance of the reactor with respect to complete oxidation of ammonia. Figure 2b shows that increasing the buffering capacity of the system also leads to better nitrogen conversions. Both figures illustrate that, under the conditions investigated, ammonia conversion is extremely sensitive to pH, and nitrite does not accumulate in the reactor.

Coupling the biofilm model with biomass distribution calculations (third model), the sensitivity of the fraction of *Nitrosomonas* in the biofilm to the maximum velocity coefficient of the first nitrification step is shown in Figure 3a. As μ_1 increases, the rate of NH_3 utilization increases and a subsequent increase in the *Nitrosomonas* fraction is observed. For values of μ_1 greater than 0.1/day, *Nitrosomonas* is the dominating species in the biofilm. As μ_1 decreases, the fraction steeply drops, reaching near-zero values at $\mu_1 \approx 0.05$/day. Thus, this model predicts that at high and low values of μ_1, there is a tendency for one of the nitrifying species to wash out.

Using the reactor model coupled with biomass distribution calculations (fourth model), Figure 3b shows that the *Nitrosomonas* fraction in the biofilm is not sensitive to wide ranges of values of μ_1 for the conditions investigated. Since nitrite does not accumulate under these conditions, the rate of conversion of ammonia to nitrite is comparable to the rate of conversion of nitrite to nitrate. Hence, the ratio of total ammonium flux to total nitrate flux is nearly constant, resulting in the decreased sensitivity of the *Nitrosomonas* fraction to μ_1.

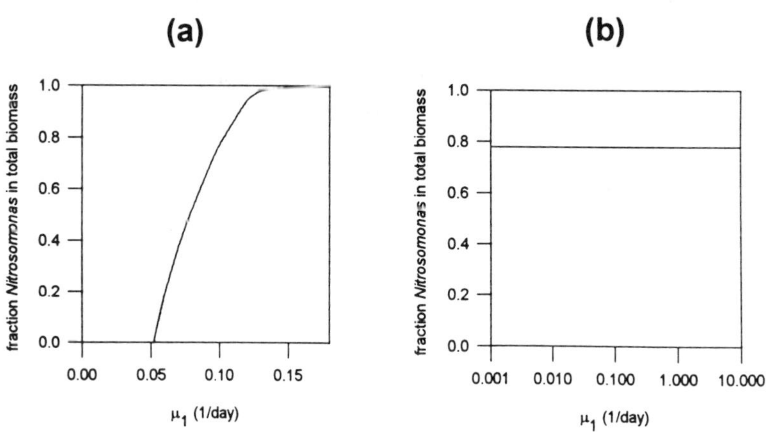

Figure 3. *Nitrosomonas* fraction as a function of maximum velocity coefficient of conversion of ammonia to nitrite, μ_1 : (a) biofilm model and (b) reactor model.

Conclusion

Steady-state models of nitrifying biofilms are developed taking into account the mass transfer of neutral and ionic species, electroneutrality, pH-dependent Monod kinetics, chemical equilibrium, and the presence of a boundary layer. Substantial changes in the pH across the boundary layer and the biofilm are predicted. Varying influent conditions such as pH and buffer capacity can result in significant changes in reactor performance with respect to complete ammonia oxidation. The biomass distribution between *Nitrosomonas* and *Nitrobacter* is approximated. Within a biofilm model framework, one of the nitrifying species washes out at high and low values of the maximum velocity coefficient of the first nitrification step. Within a CSTR model framework, the steady-state biomass distribution is insensitive to wide ranges of values of the maximum velocity coefficient of the first nitrification step.

Acknowledgements

This study was funded by the U.S. Army Construction Engineering Research Laboratories in Champaign, Illinois.

References

Cussler, E.L. (1984) Diffusion : Mass Transfer in Fluid Systems. Cambridge University Press, New York, N. Y.

Flora, J.R.V., et.al. (1993a) Modeling Substrate Transport into Biofilms : Role of Multiple Ions and pH Effects. *ASCE J. Environ. Eng.*, in press.

Flora, J.R.V., et.al. (1993b) Modeling Chemical Interactions in Anaerobic Biofilm Systems. *Proc., 66th Ann. Conf. and Expo. of Wat. Env. Fed.*, Anaheim, California, 307-317.

Gee, C.S., et.al. (1990) *Nitrosomonas* and *Nitrobacter* Interactions in Biological Nitrification. *ASCE J. Environ. Eng.*, **116**, 4-17.

Lange's Handbook of Chemistry, 11th ed. (1973) J.A.Dean, ed., McGraw-Hill Book Co., New York, N. Y.

Robinson, R.A., and Stokes, R.H. (1970) Electrolyte Solutions. Butterworth & Co. Ltd., London.

Siegrist, H., and Gujer, W. (1987) Demonstration of Mass Transfer and pH Effects in a Nitrifying Biofilm. *Wat. Res.*, **21**, 1481-1487.

Assessing the Reliability of Low Pressure Membrane Systems for Microbial Removal

Samer S. Adham[1] and Joseph G. Jacangelo[2]

Introduction and Objectives

Control of microorganisms in drinking water has always been and continues to be a major concern of public health and water treatment practitioners. The current Surface Water Treatment Rule (SWTR) requires all utilities to achieve at least 3 logs (99.9%) removal/inactivation of *Giardia* and 4 logs (99.99%) removal/inactivation of enteric viruses. This includes all utilities using surface water or groundwater under the influence of surface water. Additionally, the water industry will be required to comply with the anticipated Groundwater Disinfection Rule in the near future.

In light of current and anticipated regulations, ultrafiltration (UF) and microfiltration (MF) have received considerable attention as alternative drinking water treatment processes. These membrane processes have been shown to effectively remove measurable particles and selected microorganisms from natural waters. However, one of the most critical aspects of employing membrane technology for microbial removal is assuring that the membrane modules are intact and continue to provide a barrier between the feed and product waters. The objective of this paper is to identify and evaluate at pilot scale existing and potential methods for monitoring membrane integrity and treatment reliability. The study is part of a larger project funded by the American Water Works Association Research Foundation on the use of low pressure membrane technology for compliance with the current Surface Water Treatment Rule and the upcoming Groundwater Treatment Rule.

Methods and Materials

Two pilot plants were used in the study to treat water from Lake Elsman in San Jose, California. This lake water had a TOC concentration range of 2 to 6 mg/L and a turbidity range of 0.5 to 80 NTU. The first pilot plant consisted of a feed pump, a recirculation pump, a backwash pump, and a recirculation loop which

---

1 Professional Engineer, Montgomery Watts, 301 N. Lake Avenue, Suite 600, Pasadena, CA 91101

2 Vice President, Montgomery Watson, 560 Herndon Parkway, Suite 300, Herndon, VA22070

included the membrane module (Figure 1). It was operated in a recycle mode and used the permeate to backwash the membrane module. The second pilot plant consisted of a feed pump and a recirculation loop which included the membrane module. It was operated in a direct filtration mode and used compressed air to backwash the membrane. Typically, each filtration cycle lasted for 30 minutes, afterwhich membrane backwashing was initiated. The sequence of events associated with the backwashing procedure was implemented by a programmable logic controller.

Three commercially available hollow fiber membranes were evaluated in the study. One was a UF membrane and two were MF membranes. The UF membrane was composed of a cellulosic derivative and had a nominal pore size and molecular weight cutoff of 0.01 μm and 100,000 daltons, respectively. The two MF membranes (made of polypropylene and polysulfone derivative) had nominal pore sizes of approximately 0.2 μm.

The study was conducted in two phases. The first was to evaluate the microbial removal capabilities of the various membranes at bench and pilot-scale. The microorganisms selected for the microbial challenge experiments were MS2 bacteriophage, *Giardia muris* and *Cryptosporidium parvum.* The second phase was to evaluate the impact of loss of membrane integrity on microbial and particle passage into the permeate and assess methods to detect the loss of integrity.

Several integrity monitoring techniques were evaluated throughout the course of the study. These techniques included on-line turbidity monitoring, on-line particle monitoring, on-line and batch particle counting, air pressure-hold testing, bubble point testing, and use of sonic sensors. The potential advantages and disadvantages of each method are presented in Table 1.

Results and Discussion

Microbial Rejection

Initial experiments focused on assessing the microbial removal capabilities of all membranes. Table 2 shows representative virus removal data by the three membranes tested. Both synthetic and natural waters were evaluated. The membranes were challenged with feed concentrations of virus ranging from 10^6 to 10^7 plaque forming units per mL. The data show that less than 0.5 logs of virus were removed from the waters when the MF membranes were employed. These results were expected since MS2 virus is an icosahedon particle with a diameter of approximately 0.027 μm. When the UF membrane was employed, more than 7 logs of the virus were removed.

The membranes were also challenged with *Giardia muris* cysts and *Cryptosporidium parvum* oocysts at densities of 10^4 to 10^5 cysts/oocysts per liter. In all cases, none of these organisms were detected in the permeate. These data suggest that if the membrane remains intact, UF should be capable of providing sufficient removal of viruses and protozoa to meet the SWTR requirements; with MF some virus inactivation would need to be achieved by chemical disinfection or

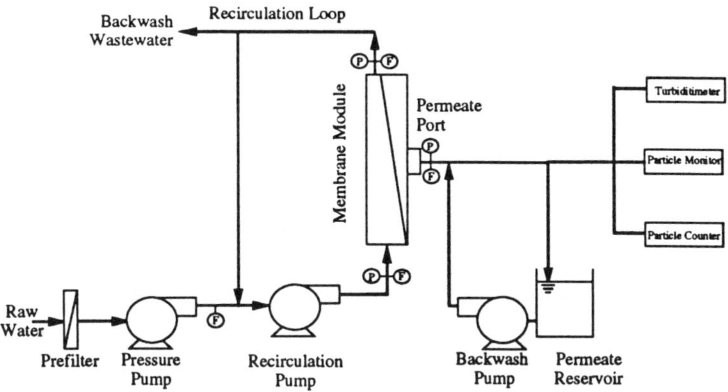

Figure 1. Simplified schematic of the membrane pilot plant used in study

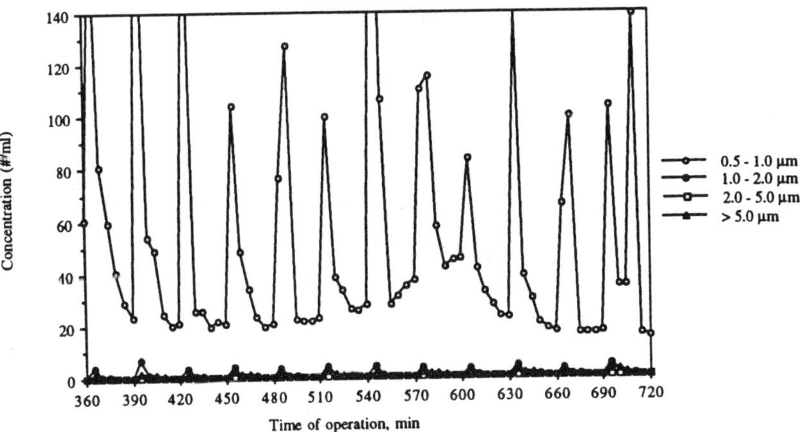

Figure 2: Typical permeate particle concentration as a function of time for Membrane C as determined by the on-line particle counter

Table 1. Various methods to evaluate membrane integrity

Monitoring Method	Advantages	Disadvantages
Particle Counting	• Continuous on-line measurement • Measures several size ranges	• High cost • Indirect measurement of membrane integrity • Requires several sensors for large scale applications
Particle Monitoring	• Continuous on-line measurement • Low cost	• Does not count particle size ranges • Requires several sensors for large scale applications
Turbidity Monitoring	• Extensive water industry experience with its use • Low cost	• Not sensitive to changes in particles at low turbidities • Indirect measurement of membrane integrity
Air Pressure Testing	• Built into membrane system • Direct measurement of membrane integrity	• Not a continuous monitoring method
Bubble Point Testing	• Direct measurement of membrane integrity	• Must take membrane module off line to conduct test • Labor intensive for large plants
Sonic Sensors	• Direct measurement of membrane integrity • Quick and easy to use	• Not a continuous monitoring method (at present) • Minimal experience with its use

Table 2. Removal of MS2 bacteriophage by various membranes from synthetic and natural waters under worst-case conditions (maximum operating pressure, no crossflow)

Membrane	Pore Size or MWCO	Synthetic Water			Natural Water					Mean**		95% CI
		Run #	1	2	12	13	14	15*	16			
Membrane A	0.2 µm		-0.2	0.7	-0.1	0.1	0.1	1.7	0.4	0.2	±	0.3
Membrane B	0.2 µm		0.3	0.2	-0.2	0.4	0.5[1]	2.4	1.4	0.4	±	0.6
Membrane C	100 kD		>6.4	>7.0	>6.9	>7.2	>6.7	>6.0	>7.9	>7.0	±	0.5

Notes:
* MS2 bacteriophage concentration was reduced due to low influent water pH (< 5)
**Includes synthetic and natural waters except for Run 15
MWCO = Molecular weight cutoff
CI = Confidence interval

any other treatment processes. However, viruses are rapidly inactivated by free chlorine, so only short contact times would be necessary.

Membrane Integrity Evaluation

Experiments were also conducted to evaluate the effectiveness of the various membrane integrity monitoring techniques described above at normal operating conditions as well as conditions under which the membranes were compromised.

Particle Monitoring

Continuous particle monitoring of the permeate stream is one of the potential indirect methods for evaluating treatment reliability. Four types of particle detection devices were used in this research to monitor the particles in the permeate stream. These were an on-line turbiditimeter[#], on-line particle monitor[*], on-line particle counter[†] (0.5 - 5 µm), and a batch particle counter[ø] (1 - 120 µm). It should be noted that the particle monitor does not count particle sizes, but rather provide an index of total particle concentration.

Initial data were collected by each particle detection device using new membrane modules. Figure 2 shows a typical on-line particle count data set from the UF membrane permeate. These data show consistent jumps, every 30 minute, followed by a gradual decrease in the particle concentration of the smallest size fractions (i.e. 0.5 - 1.0 µm). This concentration change pattern was due to the disturbance of the permeate stream during each backwashing event. The high permeate particle counts were possibly caused from tubing debris and/or air bubbles created after backwashing. The membrane integrity during this period was not compromised since the MS2, *Giardia,* and *Cryptosporidium* seeding experiments conducted at the same period resulted in complete rejection of all microorganisms by the membrane. The turbidity and particle index measurements of the permeate stream during this period were consistently between 0.02 - 0.03 NTU and 0 - 50, respectively, regardless of the influent concentration. The particle counts obtained from the batch particle counter were below 5 particles/ml, in the 1-120 µm size range.

The sensitivity of the particle monitoring devices to a change in particle concentration was evaluated in two ways. First, the permeate stream was blended with the influent stream at various ratios to simulate compromised membrane modules. Second, the membranes were artificially compromised by cutting one or more fiber(s) from the membrane module. Table 3 show the response of the

---

[#] Hach 1720C, Hach Co., Loveland, CO.

[*] Chemtrac Model PM 3500, Norcross, GA

[†] Met-One Model 211 sensor and Model 9000 counting system , Grants Pass, OR.

[ø] Met-One Model 214 batch sampler with Model 211 sensor and Model 9000 counting system, Grants Pass, OR.

particle monitoring devices to different blending ratios of the feed and permeate streams. As shown in Table 3, the change in permeate turbidity values due to blending was much lower than the change in the particle index and particle count values. These results agree with the previous studies which concluded that turbidity was not a sensitive detection method for membrane systems and that particle counting was much more of a sensitive measure (Jacangelo et al., 1991). More research is currently being conducted to determine if the particle monitor and/or particle counter can still provide reliable results with much lower blending ratios as may be the case when only one membrane fiber is compromised.

Bubble point test

The bubble point test was originally developed to characterize the pore size distribution of MF membranes. It determines the gas pressure needed to overcome the capillary forces holding the liquid (usually water) within the membrane pore. The required gas pressure is inversely proportional to the pore diameter. Thus, the bubble point test can not be applied to UF membranes because the required gas pressure will usually exceed the polymeric membranes tolerance.

The bubble point test can, however, be used to detect major fiber defect in MF or UF membrane modules (Jacangelo et al., 1991). For hollow fiber membrane configurations, the test is usually conducted by immersing the membrane module in a water tank and then gradually applying pressure from the permeate side until gas bubbles start appearing through the compromised fibers or seals. This test has the advantage of being a quick tool to directly measure the membrane integrity. However, it requires interrupting the system operation and taking the module off line which makes it unattractive as a reliable monitoring technique.

The bubble point test was applied in this research before and after each membrane were artificially compromised. This allowed detecting the broken fibers that required plugging for subsequent use of the membrane module. The test was also conducted before the startup of new membrane modules. During this study, one of the new MF membrane modules failed the bubble point test before its installation to the pilot plant. The manufacturer indicated that the membrane fibers and seals were most likely compromised during shipping. Thus, conducting a bubble point test on-site for new membrane modules may be prudent

Air pressure test

The air pressure-hold test is another direct method to evaluate the membrane integrity. The air pressure test is usually conducted by draining the membrane module which has been completely wetted and then applying gas pressure to the membrane from the permeate side and holding the pressure for a period of time (usually 10 minutes). The applied gas pressure should be lower than the bubble point of the membrane. A passed test would result in no filtrate pressure decline over the course of the test while a failed test would result in a gradual or quick decrease of the pressure. The test can be done manually without taking the membrane module off line. One of the pilot plants used in this research had the air pressure test programmed into the system PLC, which allowed for periodic and automated evaluation of membrane integrity.

Table 3: Sensetivity testing of the different particle monitoring devices

	Turbidity (NTU)		Particle index (dimensionless)		Particle count (on-line) (particle/ml)		Particle count (batch) (particle/ml)	
Blending Ratio, %	G. mean	% change	G. mean	% change	G. mean (0.5 - 5 µm)	% change	G. mean (1 - 120 µm)	% change
0 (baseline)	0.03	NA	1.0*	NA	96.7	NA	3.8	NA
0.42	0.03	0	23.4	2242	1101.6	1039	33.7	779
0.93	0.03	0	103.3	10235	2059.9	2030	59.6	1456
1.17	0.04	33	156.8	15580	2993.3	2995	87.6	2186
1.73	0.04	33	222.3	22130	3915.7	3948	138.2	3509
3.34	0.05	67	331.5	33055	5398.9	5481	256.3	6592

NTU = neplometric turbidity unit
G.mean = geometric mean
NA = not applicable
* very low baseline value observed for the experiment; generally between 0 and 50

Sonic sensors

The sonic sensors are an innovative technology for direct monitoring of membrane integrity. Currently, these sensors can be used to detect a compromised membrane module by detecting the sound waves created by a damaged fiber. The sonic sensors sensitivity and reliability are not yet clear. However, they may be further developed in the near future to connect their electronic signals to a computerized data acquisition system. Considering that the cost of this technology may be low, a sonic sensor could probably be attached to each membrane module to provide a continuous monitoring of membrane integrity.

Conclusions

The efficacy and reliability of low pressure membrane processes for microbial rejection was investigated. It was shown that low pressure membrane processes are very effective for microorganisms rejection provided that the membrane nominal pore size is smaller than the rejected microorganism size and that the membrane fibers are intact. Several methods for monitoring the membrane integrity were identified and evaluated. These techniques included the use of particle counters, particle monitors, turbiditimeters, bubble point test, air pressure test, and sonic sensors.

References

Jacangelo, J.G, Laine, J-M, Crans, K.E., Cummings, E.W., and Mallevialle, J. Low-Pressure Membrane Filtration for Removing *Giardia* and Microbial Indicators, Jour. AWWA, 83:9:97, (1991).

Fouling of Nanofiltration Membranes by Natural Organic Matter

Francis A. DiGiano[1], Anne Braghetta[2], James Nilson[2] and Bruce Utne[3]

Abstract

Nanofiltration membranes (with pore sizes on the order of 10^{-9} m) have been shown in pilot plant studies to provide a very effective means of removing natural organic matter (*NOM*). Removal of NOM from product water results in accumulation of rejected species which may foul the membrane. Frequency and intensity of cleaning , as well as the useful life of membranes, are important process optimization issues which cannot be resolved until membrane fouling by NOM is better understood. In this research we report that the concentration and composition of NOM and cross flow velocity are important factors in affecting the rate of fouling; in addition, fouling must be measured not only by permeate flux decline but also by permeate water quality.

Introduction

For smaller treatment plants, membrane separation technology may offer a more effective method of removing *NOM* than enhanced coagulation, the approached recommended by EPA's *Disinfectant and Disinfection By-products Rule*. Pilot studies of membrane filtration, for example, have shown that very high removals are possible with nanofiltration (Laine', et al. 1993). The extent to which this technology will be adopted, however, is not yet clear owing to uncertainties in cost. Two important factors contributing to cost are the frequencies of membrane cleaning and replacement, both of which are affected by *NOM* fouling.

---

[1]Professor, Department of Environmental Science and Engineering, University of North Carolina at Chapel Hill

[2]Graduate Research Assistant, Department of Environmental Science and Engineering, University of North Carolina at Chapel Hill

[3]Environmental Engineer, Malcolm-Pirnie, Newport News, VA

The objective of this paper is to present highlights of our current laboratory experimental program which is directed toward investigation of the fouling of hollow fiber nanofilters by NOM. We vary the feed solution to a bench-scale, nanofiltration membrane module to investigate the effects of the cross-flow velocity, concentration and composition of *NOM*, ionic strength and pH. Fouling is assessed in variety ways including measurements of flux decline, NOM and inorganic ion rejection and *NOM* association with the membrane (by recovery of *NOM* through chemical cleaning). The effects of varying pH and ionic strength on permeate flux decline and solute rejection and related examination of membrane morphology and surface charge are to be detailed elsewhere (Braghetta and DiGiano, 1994). However, some general comments as to their implications will be made at the end.

Methods

The bench-scale, batch recycle nanofiltration system is shown in Figure 1. Zenon Environmental Inc. hollow fiber modules with an inside-out flow configuration (20 fibers with a length of 24 cm) were used. These modules were operated at a pressure of about 6.5×10^5 *Pa (95 psi)* and clean water permeate flux of 68 L/m^2 *-hr* (41 gal/ft^2 *-d*) The membrane is made of polysulfone that has been treated to produce a hydrophilic surface. The molecular weight cut-off (MWCO) is stated by the manufacturer to be approximately 1,000 daltons. NOM was concentrated by reverse osmosis from the Tar River in North Carolina at the water intake for the City of Tarboro. The Tar River is low in turbidity (<10 *ntu*) and alkalinity (<25 mg/l as $CaCO_3$) but high in total organic carbon (*TOC*) concentration (over 7 *mg/L*) thus assuring that *NOM* is the major foulant. Our in-depth studies of ionic strength, pH and surface charge use *NOM* that was concentrated from Suwannee River (Georgia) water, a well characterized source.

The batch recycle system (Figure 1) is designed to return both the permeate and retentate to the feed tank thereby maintaining a nearly constant feed concentration of *NOM*. We assume that the only loss of *NOM* is that due to association with the membrane surface. We have been able to determine the amount associated with the membrane by measurements of *TOC* after cleaning the membrane at the end of each flux decline experiment with a commercial cleaning mix (alkaline solution that contains a detergent, MC-3 provided by Zenon Environmental Inc).

Other important experimental procedures relate to characterization of the composition of *NOM*, both that of the feed to the membrane and that recovered in the membrane cleaning solution. The apparent molecular weight distribution (*AMWD*) of *NOM* was determined by passing sample aliquots through ultrafiltration membranes (Amicon) of different *MWCO* . Studies were also conducted in which the *AMWD* of the feed was varied by contacting of the *NOM* solution with powdered activated carbon (Calgon F400) before applying it to the membrane system. Finally, the *NOM* feed solution was divided into hydrophobic and hydrophilic fractions . The hydrophobic fraction is defined as that adsorbed by an *XAD8* resin at pH 2 (later extracted with 0.1 N *NaOH*) whereas the hydrophilic fraction is that in the solution passing through the resin.

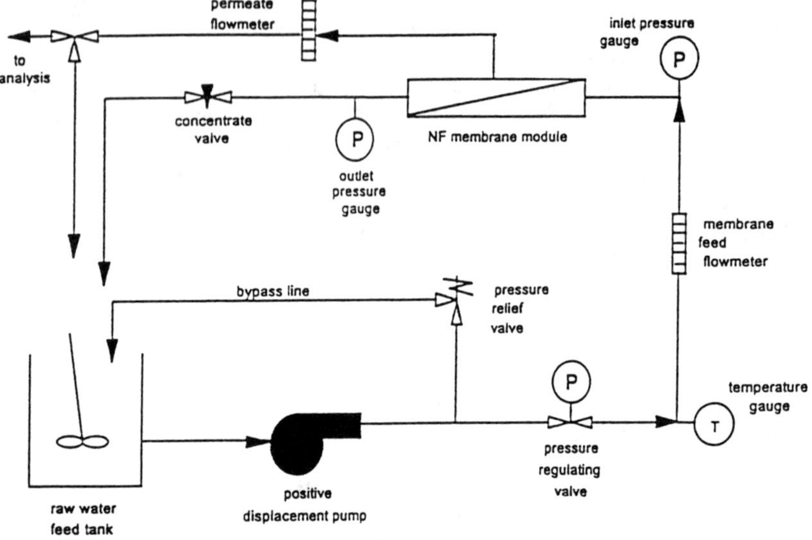

Figure 1. Flow diagram for batch, recycle *NF* system

Results

The effect of cross-flow velocity (*cfv*) on permeate flux decline was examined by using the same feed *TOC* (20 mg/L) concentration in each of three membrane tests. As is shown in Figure 2, the decline in permeate flux is rapid at this high feed *TOC* concentration, thus providing a convenient time-frame for observing the effect of *cfv*. A three-fold increase in *cfv* (which corresponds to an increase in Reynolds number from 760 to 2290) greatly reduced the flux decline. The larger the Reynolds number, the more shear at the membrane surface and the more disruption of the foulant layer at the membrane surface which therefore causes less flux decline. Our earlier studies have shown that the rate of permeate flux decline is also affected by feed *TOC* (DiGiano, et al. 1993). Reducing the feed *TOC* by about a factor of three (from 25 to 8 *mg/L*) at a constant *cfv* produced a similar reduction in the rate of permeate flux decline as did a factor of three increase in *cfv*. A lower feed TOC implies that the rate of accumulation of NOM at the membrane surface will be slower and thus less fouling will occur in a given time period.

The resistance in series model for membrane filtration was used to interpret fouling by *NOM* as the cause for flux decline. This model is:

$$J = \frac{\Delta P}{\mu R}$$

where ΔP is the pressure drop (constant in our system), μ is viscosity, R is the total resistance and J is the flux . The value of R was calculated from each fouling test from the values of ΔP, μ and J after several days of membrane operation. As shown in Figure 3, the resistance to membrane filtration caused by *NOM* fouling (i.e., after subtracting the clean water resistance) was directly related to the mass of *TOC* removed (i.e., recovered) per unit area of membrane surface by the alkaline cleaning process. The results in this figure also support those shown in Figure 2: i.e., the mass of TOC associated with the membrane is greater at at lower *cfv*. By recovery of the *TOC* from membrane surface we have found that most of the *TOC* associated with the membrane is of molecular weight greater than 10,000 (an order of magnitude greater than the *MWCO*).

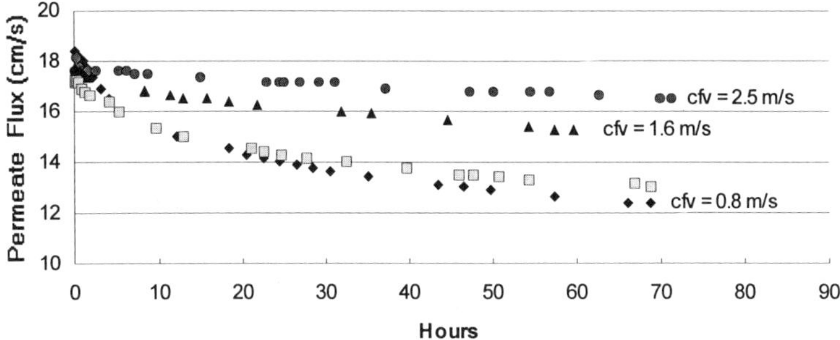

Figure 2. Effect of cross-flow velocity (*cfv*) on permeate flux decline

The permeate *TOC* that corresponds to the membrane resistance measurements (Figure 3) are shown in Figure 4. Permeate *TOC* increased with increased fouling of the membrane surface. It was also found that the mass rate of permeation of *TOC* (*mg/min*) did not remain constant over time of membrane operation but gradually increased as permeate flux declined during the three day test. Taken together, these observations suggest that diffusive transport of *NOM* occurs in addition to convective transport. That is, diffusive transport increases as the local concentration of NOM at the membrane surface increases both due to lower *cfv* or increased time of membrane operation. Although only useful as a rough guide to nanofiltration rejection capabilities, the *AMWD* (from ultrafiltraton experiments) revealed that the feed *NOM* solution contained 3.3 *mg/L* of *TOC* of <1 K molecular weight material fraction. The maximum *TOC* concentration found in the permeate of the nanofilter (Figure 3) was about equal to the <1 K fraction which is also the stated MWCO of the nanofilter.

Adsorption on activated carbon was selected as an empirical method to vary the chemical characteristics of *NOM*. The experimental procedure was as follows. A solution containing a high concentration of *NOM* (25 mg/L as *TOC*) was divided

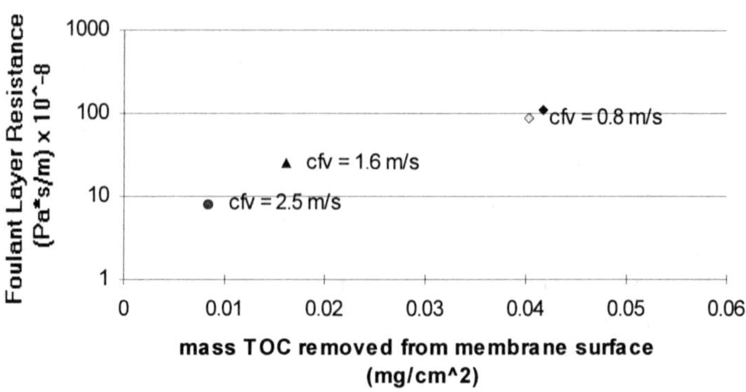

Figure 3. Resistance model to quantify fouling with TOC recovered by membrane surface at end of three-day test

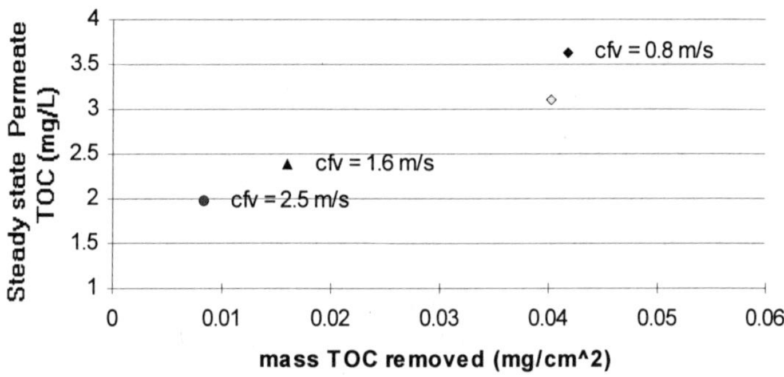

Figure 4. Permeate TOC in relation to TOC associated with membrane at end of three-day test (same experiments as in Figure 3)

into three batches. A dosage of 29 mg/L of powdered activated carbon (*PAC*) was added to each batch and the solutions were contacted for 4 min, 30 min and 120 min. The shorter the contact time, the more selective is removal of faster diffusing and/or more strongly adsorbed *NOM* fractions. Following contact, the *PAC* was removed and each solution was diluted to reach the same *TOC* concentration (about 12 *mg/L*) before conducting the membrane fouling tests. This experimental design eliminated the effect of *NOM* concentration on fouling rate so that the effect of *NOM* composition could be studied independently.

The *AMWD* after different *PAC* contact times (and dilutions to a common TOC concentration) is given in Figure 5. The middle ranges of molecular weight fractions (10-30 *K* and 3-10 *K*) were removed to the greatest extent by adsorption as might be expected. By contrast, the high ($>30\ K$) and low ($<1K$) molecular weight fractions were not adsorbed very effectively. Thus, as contact time increased, the *AMWD* shifted toward a predominance of these less well-adsorbed fractions. It is important to note that *AMWD* is only one indication of differences in *NOM* composition caused by contact with *PAC*. Undoubtedly, chemical composition (e.g., charge density and aromaticity) was also affected.

The permeate flux decline that resulted from each different *NOM* feed solution is presented in Figure 6. The greatest reduction in permeate flux was caused by using the NOM solution that had been contacted for longest time with PAC. The association of NOM with the membrane surface (as measured by recovery of TOC in an alkaline cleaning procedure) after three days of membrane operation using these different NOM feed solutions is shown in Figure 7. The NOM feed solution that had been contacted with PAC for the longest time produced the highest TOC on the membrane surface; this result is consistent with production of the greatest decline in permeate flux by the same NOM feed solution.

The above findings indicate that even though the feed concentration of *NOM* may be the same, the extent of membrane fouling was affected by use of PAC adsorption to change the composition of the NOM. The information provided by shifts in *AMWD* with contact time (Figure 5) suggested that the $>30K$ and/or $<1K$ fraction was responsible for increased fouling with longer *PAC* contact time. The $<1\ K$ fraction is an unlikely contributor to fouling, however, given that the permeate *TOC* was found to be approximately equal to that present in the $<1\ K$ fraction in each of the three solutions (a result similar to that shown in Figure 4). The results suggest, therefore, that the $>30\ K$ fraction was responsible for fouling. This conclusion is also supported by previous studies (DiGiano, et al. 1993) in which the *AMWD* of the *NOM* that was recovered from the membrane surface was found to be dominated by the $>30\ K$ fraction.

Another way of examining the effect of chemical composition on fouling is to sub-divide the *NOM* into its hydrophobic and hydrophilic fractions according to the widely-used operational definition provided by separation by an XAD 8 resin. Using this definition, we found that the *TOC* associated with the hydrophobic and hydrophilic fractions were about equal in this *NOM* sample. It is important to note that the procedures for loading and extracting the XAD 8 resin cause the ionic strength of the hydrophilic fraction to be extremely high. In other work, we have shown that the permeate flux decline increases as the solution ionic strength increases. It was necessary, therefore, to adjust each of the three membrane feed solutions (unfractionated, hydrophobic and hydrophilic) to the same, high ionic strength (equivalent to a conductivity of 3200 $\mu mhos/cm$) to provide a method of isolating the effect of *NOM* composition on flux decline.

The feed *TOC* concentrations of unfractionated *NOM* and the hydrophobic and hydrophilic fractions were set equal (12 mg/L). As shown in Figure 8, the

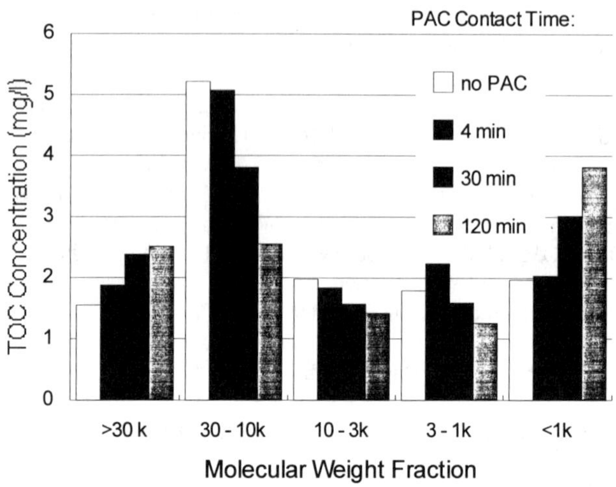

Figure 5. Shifts in AMWD caused by PAC contact time with NOM

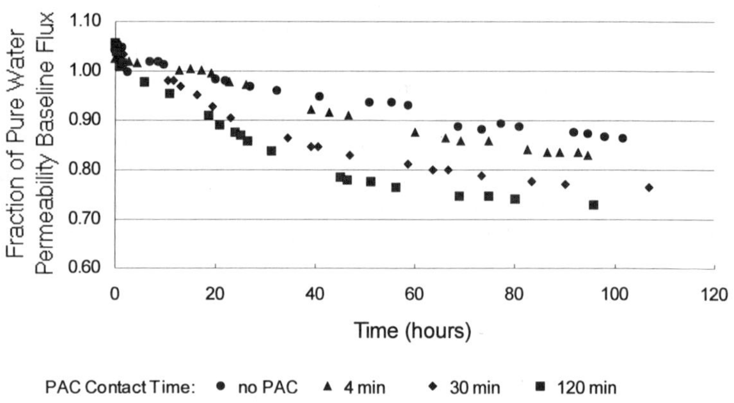

Figure 6. Effect of change in NOM composition on permeate flux decline (all feed solutions at 12.5 *mg/L*

permeate flux decline in control water (organic-free water at high ionic strength) is very substantial . The hydrophilic fraction of *NOM* produces very little additional flux decline compared to this control water. In contrast, the hydrophobic fraction caused a very large flux decline which suggests strong association with the membrane surface. Also of interest is the pattern of flux decline produced by this fraction. It consisted of an initial period (20 hours) of

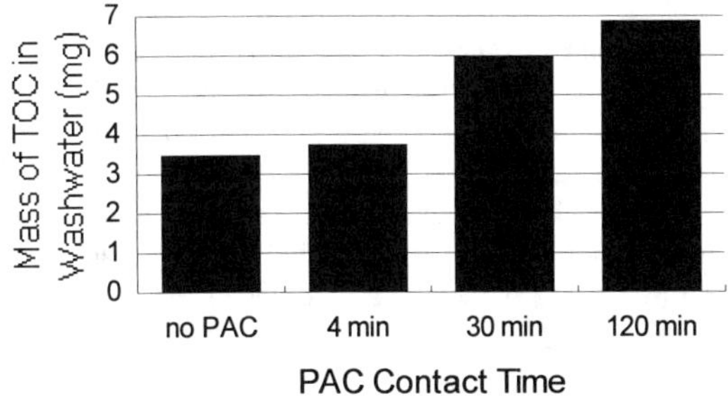

Figure 7. TOC associated with membrane at end of three-day test for each NOM feed composition resulting from different PAC contact times

very rapid decline followed by a rather sudden shift toward a pattern of more gradual decline. This pattern was unlike those observed for unfractionated *NOM* in previous experiments (e.g., Figure 2) and it suggests the possibility of a shift in fouling mechanism as more *NOM* associates with the membrane.

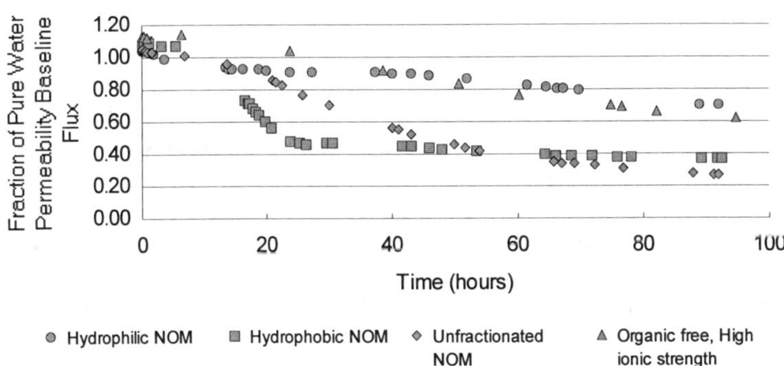

Figure 8. Difference in permeate flux decline produced by hydrophobic and hydrophilic fractions of NOM

Although not shown, the *TOC* and conductivity in the permeate produced by the feed solution containing the hydrophilic fraction were much greater than in that produced by the hydrophobic fraction. Thus the hydrophilic fraction does not associate with the membrane in such a way as to reduce the water flux but its affinity for the water phase causes more transport to occur through the membrane.

The flux decline patterns for feed solutions of the hydrophobic fraction and the unfractionated *NOM* are also interesting to compare. A simple model of fouling would suggest that the rate of fouling is proportional to the mass rate of introduction of *TOC* to the membrane surface. Because the *TOC* of the hydrophobic material in the unfractionated *NOM* feed solution is 6 mg/L, compared to 12 *mg/L* in the hydrophobic fraction, we should expect that the membrane fed with the unfractionated NOM to foul approximately 50% less rapidly than that fed with the hydrophobic fraction. The results in Figure 8 are in qualitative agreement with this hypothesis. However, flux decline caused by the unfractionated feed solution eventually exceeds that of the hydrophobic fraction to a significant extent. This result may be explained by a time-dependent chemical interactions between operationally defined "hydrophobic" and "hydrophilic" fractions and the surface of the membrane.

Experiments are currently underway to probe further into the importance of membrane surface charge. We have found that neutralization of the membrane surface charge of the modified polysulfone membrane used in our experiments both at low pH and high ionic strength causes a significant decline in permeate flux. This effect is probably due to the reduction in intramembrane polymeric charge repulsion. At pH 4, excess positive charge in solution may act to dimish the strength of the negative charge of the hydrophilic membrane surface through charge neutralization. Similarly, compression of the double layer thickness of the charged membrane surfaces at high ionic strength may influence membrane permeability.

Conclusions

Membrane rejection of *NOM* and permeate quality are affected by several important factors including *NOM* composition, *NOM* concentration, pH, ionic strength and cross flow velocity. Our current work is directed toward further elucidation of the mechanisms to explain the effects discussed here. The ultimate goal is a process engineering model to predict the rate of flux decline and the efficiency of cleaning processes as a function of water quality conditions. With greater understanding of the mechanism by NOM associates with the membrane, it may be possible to impact the design of membrane modules and to optimize the pretreatment of *NOM* for control of fouling.

References

Braghetta, A. and F. A. DiGiano. 1994. "Organic Solute Association with Nanofiltration Membrane Surface: Distinction between Adsorption and Gel Layer Accumulation," *Proceedings of the 1994 AWWA Annual Conference*, New York, NY (forthcoming)

DiGiano, F. A., A. Braghetta, B. Utne, and J. Nilson. 1993. Nanofiltration Fouling by Natural Organic Matter and Role of Particles in Flux Enhancement. *Proceedings of the 1993 AWWA Membrane Technology Conference,* Baltimore, MD.

Laine', J. M., J. G. Jacangelo, E. W. Cummings, K. E. Cairns, and J. Mallevialle. 1993. "Influence of Bromide on Low-Pressure Membrane Filtration for Controlling DBPs in Surface Waters, *JAWWA* 85:6:87-99

Colloidal Fouling of Reverse Osmosis Membranes

Menachem Elimelech and Xiaohua Zhu[1]

Abstract

Experiments on the fouling of thin film composite and cellulose acetate reverse osmosis membranes by aluminum oxide colloids are reported. Fouling was significant at high ionic strengths, resulting in reduced water flux and salt rejection. Under the chemical conditions tested, colloidal fouling was found to be reversible, thus providing an indirect evidence that pore blockage is not an important mechanism in colloidal fouling of reverse osmosis membranes. A qualitative model for colloidal fouling of reverse osmosis membranes is proposed.

Introduction

Polymeric reverse osmosis membranes are used extensively in numerous separation processes, including wastewater reclamation and seawater and brackish water desalination. Fouling is a major problem in reverse osmosis operation, resulting in product water flux decline and larger operating costs. Colloidal particles are considered to be the principal cause of membrane fouling. Fouling of reverse osmosis membranes places a large economic restriction on membrane plant operation. A key to this problem lies in a fundamental understanding of the physicochemical mechanisms of colloidal fouling.

---

[1] Department of Civil & Environmental Engineering, University of California, Los Angels, CA 90024-1593

The mechanisms of colloidal fouling are complex and poorly understood. The general objective of the study presented here is to identify the chemical-colloidal factors which control the rate of colloidal fouling of reverse osmosis membranes. More specifically, we will report on: (1) colloidal fouling experiments with model colloidal suspensions and solution chemistries using a laboratory reverse osmosis test unit; (2) identification of chemical-colloidal factors controlling colloidal fouling, and (3) postulating a qualitative model for colloidal fouling.

Experimental

Fouling tests were conducted with cellulose acetate and thin film composite membranes (supplied by *Hydranautics* and *Desalination Systems*). In these fouling experiments, aluminum oxide colloids at concentrations of 10, 30, and 100 mg/L were used. Experiments were conducted with three electrolyte concentrations: 0.001, 0.01, and 0.1 M NaCl (correspond to 58.5, 585, and 5845 ppm NaCl, respectively). In addition, some fouling experiments were conducted in the presence of humic substances. The net driving pressure (*i.e.*, the applied pressure minus the osmotic pressure) was maintained at 400 psi, so that the role of solution chemistry on the extent of colloidal fouling can be investigated without considering pressure effects.

In a typical reverse osmosis fouling test, a particle-free solution with the desired NaCl concentration was circulated through the membrane test unit at the desired flow and pressure. The fluid velocity over the membrane under the hydraulic conditions employed in the tests was 6.7 cm/s, resulting in a Reynolds number close to 590. After about 14 hours of operation at this mode, a concentrated stock suspension of particles was added to the feeding tank to establish a desired particle concentration. This was considered as the initial time of the fouling experiment. The flux just before adding the particles was the reference flux, and all other measured fluxes were referred to this one (*i.e.*, the relative flux is the actual flux at any time divided by this reference flux). Samples to determine the permeate flux and salt rejection were taken at various time intervals. In addition, samples from the feeding tank were taken once or twice a day to measure the particle size of the suspensions by dynamic light scattering.

Results and Discussion

Results of representative fouling tests are presented in Figures 1-3. In this figures, the relative water flux and salt rejection are presented as a function of time. These relative values are obtained by dividing the value of the flux or salt rejection at any time by the corresponding initial value (*i.e.*, just before adding the colloids to the suspension). After 100 hours, the membrane was rinsed with distilled water and the fouling test was resumed. The purpose of this step was to test the reversibility of colloidal fouling. Reversibility of colloidal fouling has direct implications for cleaning of colloid-fouled reverse osmosis membranes.

Figures 1 and 2 are for fouling of composite and cellulose acetate membranes at 0.1 M NaCl. It is demonstrated that after rinsing the membrane with deionized water the flux is restored. Our experiments showed that the thin film composite membranes display a moderate flux decline even in the absence of colloids, which explains why the water flux is not restored to the initial value after rinsing. Similar effects are observed in the fouling experiment with 0.1 NaCl and humic acid (0.6 mg/L TOC), as shown in Figure 3. It should be noted that in the presence of humic acid, the particles and membrane are negatively charged, while in the absence of humic acid the membrane is negatively charged while the particles are positively charged. From the results shown in Figures 1-3, it seems that the fouling behavior is controlled by the high ionic strength, regardless of the electrokinetic properties of the particles and membranes. It is further shown that the presence of a colloidal deposit layer on the membrane surface influences the salt passage through the membrane. A slight decrease in salt rejection is observed. The decrease in salt rejection is attributed to the enhanced concentration polarization caused by the presence of retained particles.

Additional fouling experiments at low and moderate ionic strengths were conducted. The fouling behavior observed in our experiments was related to the solution chemistry and the electrokinetic charge of particles and membranes. Four possible combinations of particle and membrane charges and ionic strengths can be described when particles interact with reverse osmosis membranes. These combinations and their presumed impact on fouling are discussed below.

Low ionic strength; particles and membranes are oppositely charged. In this case, particles deposit favorably onto the membrane surface. However, since the particles are stable at low ionic strengths, suspended particles

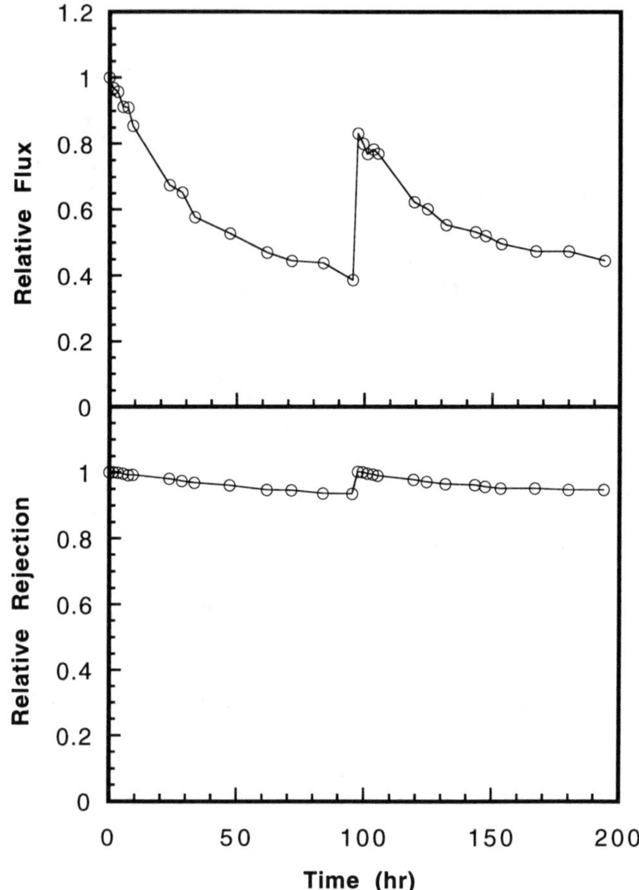

Figure 1. Relative flux and salt rejection of a fouling test with a thin film composite reverse osmosis membrane. Experiment was conducted with a 10 mg/L aluminum oxide colloidal suspension and 0.1 M NaCl solution.

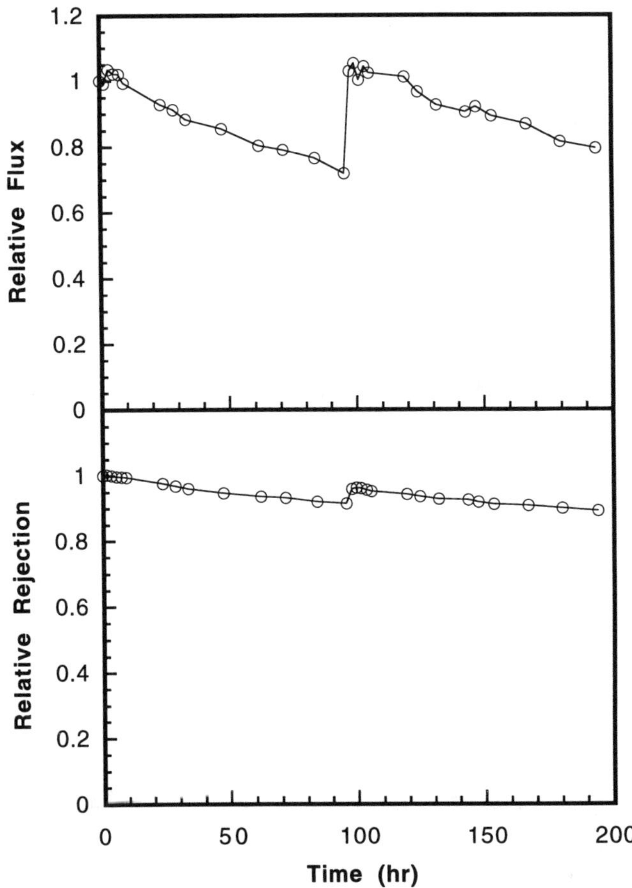

Figure 2. Relative flux and salt rejection of a fouling test with a cellulose acetate reverse osmosis membrane. Experiment was conducted with a 10 mg/L aluminum oxide colloidal suspension and 0.1 M NaCl solution.

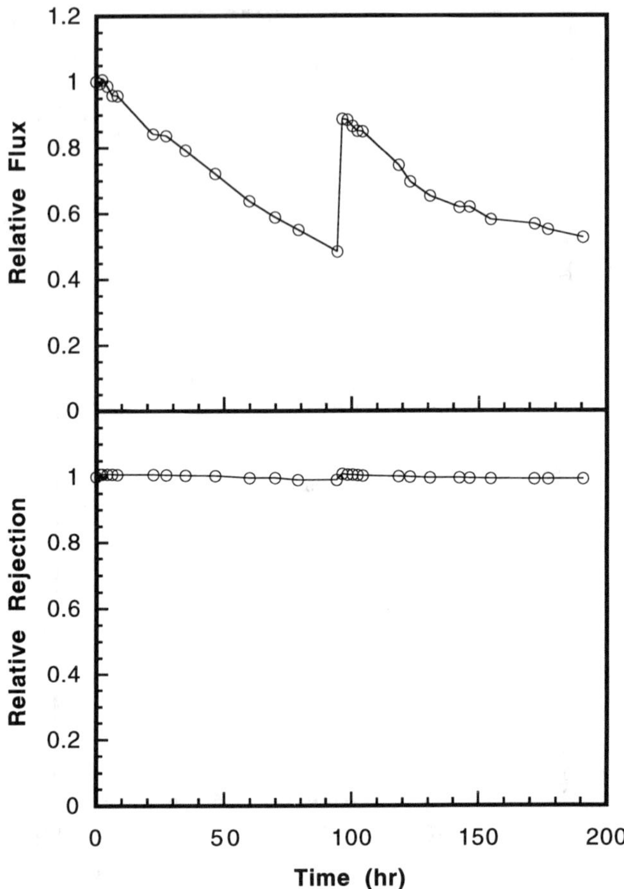

Figure 3. Relative flux and salt rejection of a fouling test with a thin film composite reverse osmosis membrane. Experiment was conducted with a 10 mg/L aluminum oxide colloidal suspension, 0.1 M NaCl, and humic acid (0.6 mg/L TOC).

cannot deposit onto retained particles. Particle deposition studies show that, for this "monolayer deposition", only a small fraction of the surface is covered by particles (Song and Elimelech, 1993). It is hypothesized that in this case there will be no significant water flux decline. Some increase in the salt passage may occur due to the presence of particles on the membrane.

Moderate to high ionic strength; particles and membranes are oppositely charged. As in the previous case, the deposition of particles onto the membrane surface is favorable. Furthermore, since the particles are unstable at high ionic strength, the deposition of particles onto previously retained particles is also favorable. This deposition behavior will probably result in a thick fouling layer, the thickness of which is highly dependent on the hydrodynamic conditions. It is hypothesized that, in this case, there will be a significant water flux decline and an increase in salt passage.

Low ionic strength; particles and membranes are similarly charged. In this case, particle deposition onto the membrane is significantly hindered by repulsive double layer interactions. There will be no buildup of a fouling layer on the membrane surface and it is presumed that no water flux decline will be observed.

High ionic strength; particles and membranes are similarly charged. At high ionic strengths, the repulsive double layer interactions between particles and the membrane surface and between suspended and retained particles are low. As a result, particles can accumulate on the membrane surface and form a thick fouling layer. It is hypothesized that, in this case, there will be a significant water flux decline and an increase in salt passage.

Acknowledgment

The research reported here was funded by the State of California, Department of Water Resources; the National Water Research Institute; and the University of California, Water Resources Center.

References

Song, L., and Elimelech, M. (1993). "Dynamics of Colloid Deposition in Porous Media: Modeling the Role of Retained Particles", *Colloids Surfaces A: Physicochem. Eng. Aspects,* **73**, 49-63.

Pilot Test on Groundwater Organics Removal by Low-Pressure Membranes

Paul Fu[1], Hector Ruiz[1],
Ken Thompson[2], and Carl Spangenberg[2]

Abstract

A 1-year membrane pilot study has been performed to evaluate disinfection-byproduct (DBP) precursors and color removal by low-pressure membranes from groundwater with color values up to 200 color units for the Irvine Ranch Water District (IRWD) Dyer Road Well Field in Southern California. The study results suggest that modified high-flux type nanofiltration (NF) membranes performed better than traditional softening NF membranes for this application. The advantages of the high-flux NF membranes include superior color and DBP precursor removal, reduced inorganic fouling potential, less corrosive permeates, and good performance recovery after membrane cleanings.

Introduction

The membrane pilot test was divided into two sequential phases: membrane screening in Phase I and long-term membrane evaluation in Phase II. The objective for the Phase I testing was to screen eight or more nanofiltration membranes and select two to three for long-term pilot testing. The objectives of Phase II testing were to evaluate long-term membrane performance with respect to constituent removal and fouling characteristics, and to develop criteria for design of a full-scale membrane treatment facility.

Phase I Membrane Screening

For Phase I, a single-membrane-element test unit was used. The test unit has a pressure vessel that contained one spiral-wound membrane element that was 2-1/2 inches in diameter by 40 inches in length.

[1]CH2M HILL, 2510 Red Hill Avenue, Santa Ana, California 92705
[2]Irvine Ranch Water District, 15600 Sand Canyon Avenue, Irvine, California 92718

Phase I tests were performed at low hydraulic recovery rates (approximately 10 percent), where recovery rate is defined as the percentage of feedwater recovered as permeate. A total of eight membranes were screened in Phase I. The characteristics of the membranes are presented in Table 1.

The results of Phase I are documented in a previously published article (Fu, 1993). Based on the Phase I test results, Membranes A, B, and H were recommended for consideration of long-term testing because they produced permeates and concentrates with water quality comparable to or better than that of the other membranes, but with much higher specific productivity.

Table 1
Characteristics of Ultra-Low-Pressure Membranes[a]

Membrane	Manufacturer	Trade Name	Type	Molecular Weight Cutoff (daltons)	Rated Operating Pressure (psig)	Flux at Rated Pressure (gfd)
A	TsiSep	TS 80	NF	280-300	100	33
B	Nitto Denko[b]	NTR 7450	NF/UF	500-1,000[c]	143	55
C	Hydranautics	PVD 1	NF	100-300	143	34
D	Desalination Systems, Inc.	Desal 5	NF	200-300	100	20
E	FilmTec	NF 70	NF	200	70	22
F	Fluid System	TFCS	NF	200-300	80	15
G	Nitto Denko[b]	NTR 7410	NF/UF	Not Available	143	292
H	TriSep	TS 60	NF/UF	500-1,000[c]	100	55

[a] All membranes are of the thin-film composite type.
[b] Represented by Hydranautics in the U.S.
[c] Estimated values

Phase II Long-Term Evaluation

In Phase II testing, a single-membrane, 6-vessel test train was used to operate 18 standard production-type elements at hydraulic recovery rates typical of a full-scale facility. Two membrane types were tested in Phase II over a 10-month period. They were Membrane A, a traditional type NF membrane, and Membrane B, a high-flux type with low inorganic-rejection.

Membrane A Testing

TriSep TS80 membrane was tested for approximately 120 days. Some operational changes were implemented during the run to maximize membrane product water flow and increase membrane fouling potential so that fouling characteristics could be evaluated. These changes included: increasing recovery rate from 75 percent to 90 percent, increasing flux from 15.8 to 20 gfd, and discontinuing antifoulant dosage from Days 79 to 103.

Water Quality. Table 2 summarizes the average water quality of Membrane A's feed, permeate, and concentrate. The permeate and concentrate water qualities are shown for operating conditions at 75 percent and 90 percent recoveries. The data in Table 2 suggested the following findings:

- High rejections were found for inorganic constituents such as calcium, total dissolved solids (TDS), and electrical conductivity (EC). The high concentrations of alkalinity, calcium, and TDS in the membrane concentrate at 90 percent recovery indicated scaling potential by calcium carbonate.

- Membrane permeate has low level of calcium and TDS. Post-treatment may be required for water stability.

- High rejections were found for organic constituents including color, total organic carbon (TOC), and 72-hour total trihalomethanes (TTHM). The permeate's color was consistently less than 5, TOC at approximately 0.3 mg/L, and TTHM was less than 20 μg/L.

System Pressures and normalized permeate flows are shown in Figure 1. The system pressures varied as a function of flux. In general, the pressures were stable if operating conditions remained unchanged. An exception occurred when the membrane was operated without antiscalant from Days 79 through 103. The pressures initially increased gradually and then surged when the flux was raised from 15.8 gfd to 20 gfd on Day 93. The pressure surge was a clear indication of membrane fouling, and a membrane cleaning was performed.

The normalized permeate flow was approximately 15 gpm at 15 gfd, and it decreased to about 13 gpm when the flux was raised to 20 gfd, an indication of membrane fouling.

Membrane Cleaning were performed by recirculating with a 2 percent citric acid solution adjusted to a pH between 3 and 4 and then with a detergent adjusted to a pH of 11. One cleaning cycle was performed for each of the two stages of the membrane system. After this cleaning, the membrane system was turned back on to operation at 90 percent recovery and 20 gfd flux. The system pressures and flows resumed to original normal operating levels.

Table 2
Membrane Long-Term Testing
Summary of Membrane A Water Quality Data

Water Quality Parameter	Membrane Feed	Membrane Permeate		Membrane Concentrate	
		@ 75% Recovery	@ 90% Recovery	@ 75% Recovery	@ 90% Recovery
Alkalinity (mg/L)	215	36	46	770	1,600
Bromide (mg/L)	0.23	0.08	0.13	0.54	0.92
Calcium (mg/L)	5.70	0.30	0.28	15	21
Chloride (mg/L)	18.8	5	9.9	67	123
Color (cu) @ 408 nm	188	<5	<5	--	--
EC (μS/cm)	500	100	130	1,650	3,600
Iron (μg/L)	210	<4	<4	345	450
Magnesium (mg/L)	0.60	0.15	<0.05	1.4	2.7
Manganese (μg/L)	10	<4	<4	10	11
Sodium (mg/L)	128	21	42	405	600
Sulfate (mg/L)	<4	<4	<4	12	--
TDS (mg/L)	328	62	118	1,120	2,040
TOC (mg/L)	11.2	<1	0.3	50	130
TTHM (μg/L), Cl_2	540	6.2	19.4	--	--
Turbidity (NTU)	1.3	0.08	0.13	3.6	4.7
pH	8.8	8.8	9	8.4	8.4

Membrane B Testing

The first 160 days of Membrane B (NTR 7450) test results are presented in this paper. The membranes were operated at 90 percent recovery, 20 gfd flux, and with the addition of antiscalant. Two membrane cleanings were performed on Days 64 and 103, respectively. The following paragraphs discuss the operating conditions and the results of membrane monitoring and performance.

Water Quality. The average TDS of membrane feed, permeate, and concentrate are 290, 239, and 914 mg/L, respectively. The concentrate TDS concentration is less than half of that for Membrane A. Table 3 shows the average concentrations of specific ions that confirm the low inorganic rejection by the membrane. TOC values for membrane feed and permeate are shown to be stable throughout the test period. Membrane feed TOC values averaged 11.0 mg/L and permeate TOC

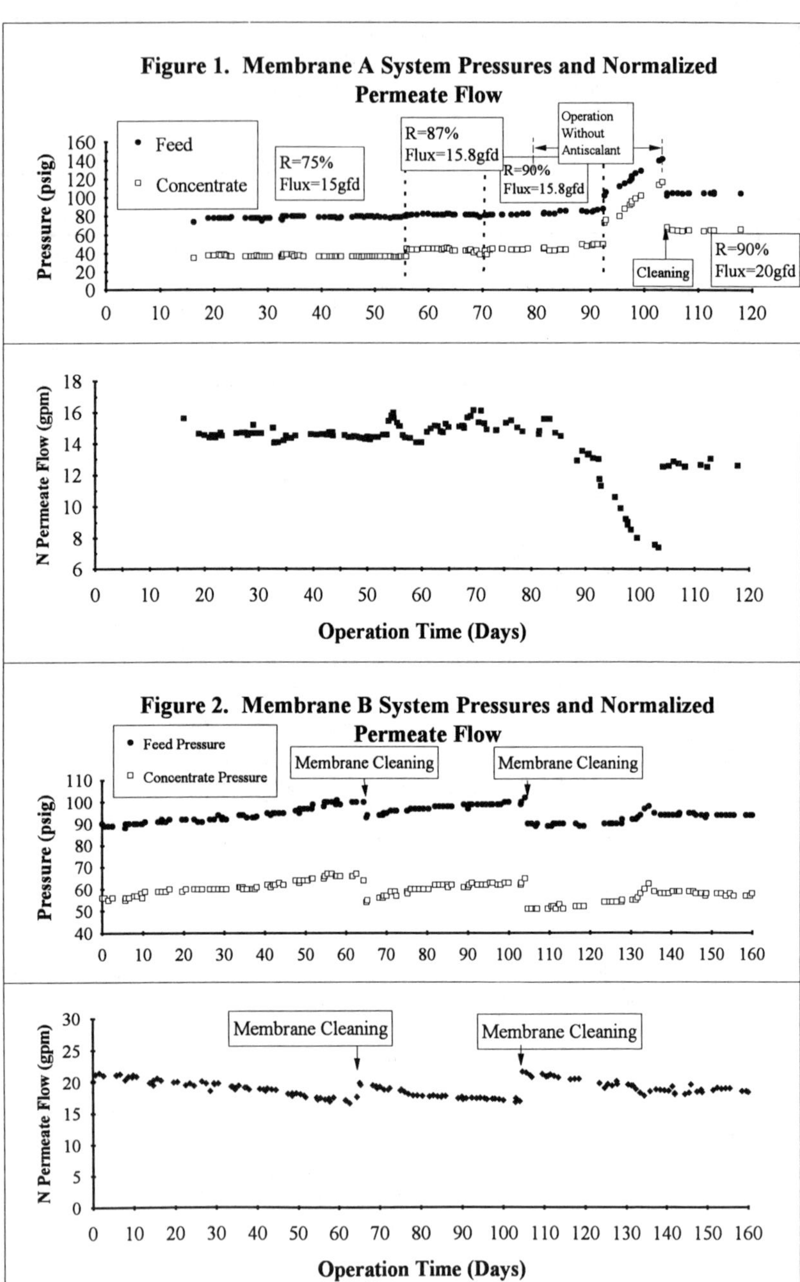

Figure 1. Membrane A System Pressures and Normalized Permeate Flow
Feed
Concentrate
R=75%
Flux=15gfd
R=87%
Flux=15.8gfd
R=90%
Flux=15.8gfd
Operation Without Antiscalant
Cleaning
R=90%
Flux=20gfd
Pressure (psig)
N Permeate Flow (gpm)
Operation Time (Days)
Figure 2. Membrane B System Pressures and Normalized Permeate Flow
Feed Pressure
Concentrate Pressure
Membrane Cleaning
Membrane Cleaning
Pressure (psig)
Membrane Cleaning
Membrane Cleaning
N Permeate Flow (gpm)
Operation Time (Days)

values averaged 0.61 mg/L. Membrane concentrate TOC values averaged about 120 mg/L and ranged from 97 to 146 mg/L. Color concentrations for membrane feed and permeate averaged 170 and <5 color units, respectively. Overall color or TOC removal was greater than 95 percent.

Table 3
Inorganic Water Quality[a]

	Ca (mg/L)	Mg (mg/L)	Mn (mg/L)	Na (mg/L)	Cl (mg/L)	F (mg/L)	SO4 (mg/L)	Fe(T) (mg/L)	Br (mg/L)
Membrane Feed	4.3	0.28	<0.004	104	20	1.16	4	0.073	0.19
Membrane Permeate	2.5	0.13	<0.004	94	17	1.14	<4	0.004	0.16
Membrane Concentrate	18	1.3	0.0147	255	48.6	1.65	6.3	0.61	0.45

[a] Average water quality data based on multiple sets of grab samples

The TTHM of membrane permeate averaged about 49 μg/L. It will meet the Stage 1 (1997) proposed MCL of 80 μg/L for TTHMs; however, the Stage 2 (2002) proposed TTHM MCL of 40 μg/L will not be met using free chlorine for disinfection. Chloramines will be required to meet the Stage 2 TTHM MCL.

System Pressures and Normalized Permeate Flow are shown in Figure 2. Two cleanings were performed over the 160-day test period. Within each run cycle, system pressures show an increasing trend while the permeate flow has a decreasing trend. Those trends are indicators of membrane fouling.

The first cleaning partially recovered the pressure and flow because only sodium hydroxide was used. The second cleaning completely recovered the pressure and flow to the start-up conditions by using detergent with sodium hydroxide.

Comparisons of Membranes A to B

Operating Conditions At 90 percent recovery and 20 gfd flux, the startup feed pressure of Membrane B is 89 psi, or 15 psi lower than that of Membrane A. This implies energy savings with Membrane B.

Water Quality. The average water quality data of the two membranes are compared in Table 4. Both these membranes show comparable water quality with respect to organic constituents except for TTHM. The TTHM, with free chlorine disinfection, for Membrane B permeate is 54 μg/L versus 19.4 μg/L for Membrane A permeate. The future Stage 2 MCL is 40 μg/L.

Table 4
Comparisons of Average Water Quality Data of NTR7450 to TS80[1]

Water Quality Parameter	Average Membrane Feed	Membrane Permeate		Membrane Concentrate	
		NTR7450	TS80	NTR7450	TS80
Alkalinity (mg/L)	210	175	46	450	1,600
Calcium (mg/L)	5.0	2.5	0.28	18	21
Chloride (mg/L)	19.4	17	9.9	49	123
Color (cu)	175	<5	<5	--	--
EC (μS/cm)	480	430	130	1,100	3,600
Magnesium (mg/L)	0.44	0.13	<0.05	1.3	2.7
Sodium (mg/L)	116	94	42	255	600
Sulfate (mg/L)	<4	<4	<4	6.3	--
TDS (mg/L)	310	239	118	914	2,040
TOC (mg/L)	11.1	0.6	0.3	120	130
3-Day TTHM (μg/L)	791/540[2]	54	19.4	--	--
Turbidity (NTU)	1.2	0.2	0.13	4	4.7
pH	8.8	8.8	9	8.6	8.4

[1]Comparisons based on 90 percent recovery at 20 gfd flux operating conditions.
[2]Feed concentrations for NTR7450 and TS80 are 791 and 540 μg/L, respectively.

Overall, the inorganic's concentrations of Membrane B permeate is much higher than those of the Membrane A permeate. On the contrary, the Membrane B concentrate is substantially lower in TDS and in other inorganic concentrations.

Hydraulic Conditions. The specific productivity of the two membranes are presented in Figure 3. At comparable operating conditions (i.e., 90 percent recovery and 20 gfd), the average specific productivity of Membrane B is approximately 1.5 times that of Membrane A. The higher productivity of Membrane B suggests that it is more energy efficient.

Membrane Cleaning Membrane cleanings performed in this study were effective. Membrane A needs separate citric acid cleaning for calcium carbonate scale and alkali detergent cleanings for organic foulants, while Membrane B only needs alkali detergent. However, more than one cleaning cycle is required for an effective cleaning of both membranes.

Conclusions

The following conclusions are derived from this study:

- Both Membranes A and B were successfully operated at 90 percent recovery and an average flux of 20 gfd. Membrane B operated at a 15 percent lower pressure and higher specific productivity than Membrane A.
- Both Membranes A and B produced permeates with acceptable water quality in terms of treatment goals. Potentially, minimal permeate post-treatment for stabilization of the Membrane B permeate is required. Due to the low TDS, alkalinity, and calcium, the Membrane A permeate would require post-treatment to achieve water stability.
- The TDS of Membrane B concentrate is substantially less than the TDS of Membrane A concentrate. The Membrane B concentrate may potentially be discharged to the sewer or used for land applications.

References

Fu, P., D. Bedford, K. Thompson, C. Spangenberg, and S. Malloy. A Membrane Pilot Study on Highly Colored Groundwater in Southern California. American Water Works Association 1993 Annual Conference Proceedings, Water Resources, 23-1, 267 (1993).

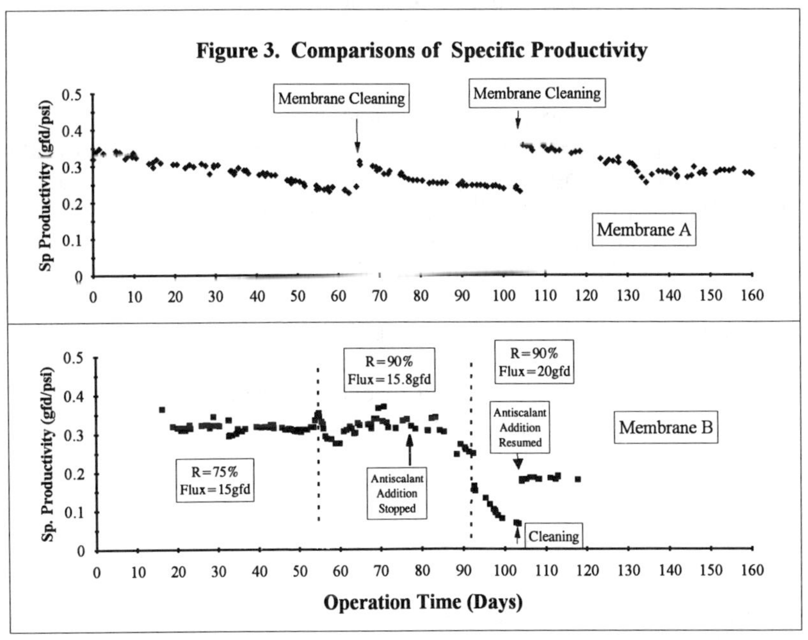

Figure 3. Comparisons of Specific Productivity

DBP and Hardness Control by Membrane Filtration

Charles D. Turner[1] and Srinivasa R. Kadubandi[2]

Abstract

Ultra (UF) and nanofiltration (NF) were evaluated for removal of disinfection by-product (DBP) precursors and hardness from Rio Grande water in New Mexico and Texas. In order to compare hardness and precursor removal at Elephant Butte Reservoir, New Mexico, and El Paso, Texas, pilot plant testing was conducted using membranes with selected molecular weight cutoffs (MWCs). All the UF and NF membranes removed hardness, but only the 600 and 1,000 MWC membranes removed DBP precursors successfully.

Introduction

The current EPA primary drinking water standard for total trihalomethanes (TTHMs), 100 μg/L, is being lowered significantly under the forth coming D/DBP rule. The D/DBP rule and the existing Surface Water Treatment Rule (SWTR) have increased the necessity of finding a workable and cost-effective solution to the DBP problem without increasing the biological health risk. Since El Paso is switching from ground water to surface water sources for its drinking water supplies, these regulations have a significant impact on the El Paso Water Utilities. Increased reliance on the Rio Grande, which is essentially composed of irrigation return flows by the time it reaches El Paso, has led to high DBP formation because of high precursor concentrations.

---

1. Professor and Chairman, Dept. of Civil Engineering, The University of Texas at El Paso, El Paso, TX 79968

2. Graduate student, Dept. of Civil Engineering, The University of Texas at El Paso, El Paso, TX 79968

Research Objectives

The objective of the research was to evaluate the application of low-pressure membrane filtration as an alternative technology for surface water treatment to comply with SWTR and D/DBP rule. Part of the objective entailed the evaluation of the Rio Grande water quality and treatment requirements to meet drinking water standards at El Paso and Elephant Butte Reservoir.

Related Research

The available techniques for precursor removal are i) chemical coagulation, ii) activated carbon adsorption, iii) oxidation by uv-catalyzed ozonation or hydrogen peroxide and iv) membrane filtration. As far as the economics of the above technologies are concerned, the least expensive technology is chemical coagulation. Chemical coagulation, however, is sometimes ineffective in reducing DBPs to a level sufficient to meet the regulations. NF membranes have been shown to produce superior water quality at roughly the same cost as lime softening in many Florida water treatment plants (AWWA report 1989). McGuire et al. (1991) concluded that, though GAC reduced most of the THMFP, it was an expensive technology.

Schnoor et al. (1979) found that most of the organics found in the Iowa River were in the apparent molecular weight (AMW) range of 1,000 and 3,000. Taylor et al. (1987) found that significant rejection of precursors from a ground water source was first observed at a MWC of 10,000 and nanofiltration was as effective as reverse osmosis for precursor removal. Reducing MWC below 400 did not lower THMFP any further. Laine et al. (1990) found that membranes with PAC as pretreatment reduced THMFP by 85%. Kim et al. (1991) found that the AMW fraction of less than 10K was the major precursor for THMs. In a bench scale study, Amy et al. (1990) found that nanofiltration removed DOC successfully. Duranceau et al. (1992) found that nanofiltration reduced THMFP by 95%, TOC by 95% and TOXFP by 93% from ground water.

Pilot Plant Set-up

A pilot plant unit was designed and manufactured in cooperation with Saltech Corporation of El Paso. It had a maximum capacity of 1,500 to 2,000 liters/day, depending upon the membrane flux and surface area. Figure 1 is a flow schematic of the pilot scale membrane

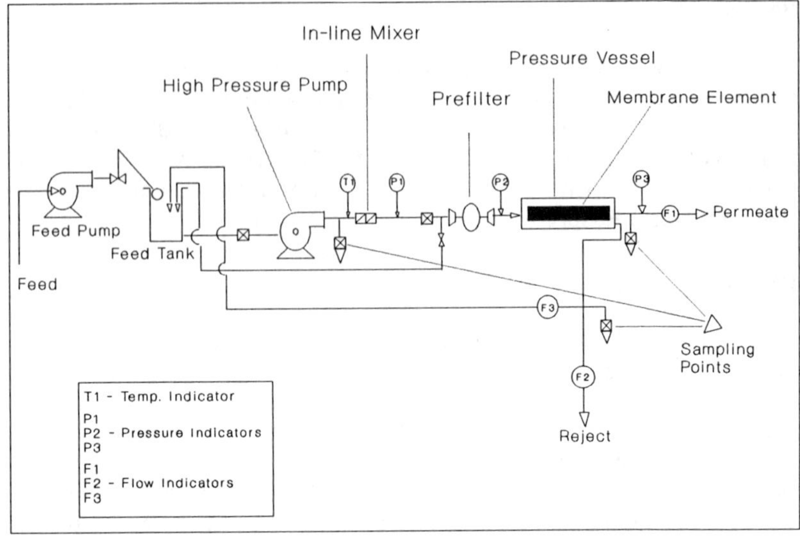

Figure 1. Schematic of the pilot scale membrane unit used to evaluate DBP precursor removal from Rio Grande water at El Paso and Elephant Butte Reservoir

filtration unit. Reject can be returned to the feed line in order to achieve higher recoveries and also simulate the concentration build-up in feed to the last membranes in series operation of prototype plants.

Methodology

The pilot plant was operated at the E. B. Robertson Water Treatment Plant in El Paso and at the Elephant Butte Reservoir dam site. At El Paso, water after the coagulation and clarification processes was supplied as feed water to the unit. Since raw water at Elephant Butte Reservoir had turbidities of approximately 1 NTU, it was fed directly to the unit without pretreatment. Each membrane was tested for 6 to 8 hours a day. Samples of feed, permeate and reject were collected one hour and 6 hours after starting the test. Feed, permeate and reject samples were collected and analyzed for pH, turbidity, conductivity, hardness, TDS, UV-Vis, DOC, chlorine demand and THMFP according to the standard methods.

Results and Discussion

MWD of organics was analyzed at El Paso and Elephant Butte. The MWD did not remain constant during the sampling period. Variations were large in a span of just one week. Nevertheless, most of the organic matter fell in AMW range of 1,000-5,000. Membranes with MWCs less than 10,000 were selected from two different manufacturers: Hydranautics Co. and Desalination Systems Co.. All membranes were spiral-wound, 10 x 100 cm size and made of three different proprietary thin-film composite materials. All membrane elements had a membrane surface area of 6.5 sq.m. Table 1 gives the selected membrane models and their operational characteristics.

Table 1. Selected Membranes and Their Characteristics

No.	Type	Class	MWC	Model	Pressure (KPa)	Manufacturer's Name
1	A	NF	300	DK4040F	700-1050	Desalination Systems
2	B	NF	600	4040LSA PVD1	525-875	Hydranautics Co.
3	C	UF	1000	GE4040F	350-700	Desalination Systems
4	C	UF	2000	GH4040F	350-700	Desalination Systems
5	C	UF	3000	GK4040F	350-700	Desalination Systems

A - Proprietary Material Type-I B - Polyvinyl Alcohol Derivative
C - Proprietary Material Type-II

Rio Grande Water Quality Variation

Rio Grande water at El Paso and at Elephant Butte was analyzed for pH, turbidity, conductivity, TDS, bromides, DOC, UV-Vis, THMFP and hardness. The raw water quality at Elephant Butte was much better than the coagulated, flocculated and settled water at El Paso for all parameters including THMFP. The average feed water THMFP for the test period was 374 μg/L at El Paso and 249 μg/L at Elephant Butte. During a 24 hour period, the THMFP at El Paso varied by more than 100 μg/L. In addition, all other parameters at El Paso varied greatly; whereas, at Elephant Butte, there was little variation.

Membrane Treatment Results

Rio Grande water at both sites was treated using UF and NF membranes. The 300 MWC membrane did not remove TDS or precursors in accordance with its cutoff. Hence, results from this membrane were not used for comparison. The 600 MWC membrane reduced DOC to 0.4 mg/L (96% reduction) at El Paso and 0.2 mg/L (97% reduction) at Elephant Butte. Efficiencies of the other membranes varied from 20% to 84%. The relationship between DOC and THMFP was not linear. However, water with high DOC formed high THMs.

Since UV-Vis is a good indicator of the presence of precursors, membranes which produced water with lower UV-Vis values were expected to form lower levels of THMs. The 600 MWC membrane reduced UV-Vis by 93% at El Paso and by 96% at Elephant Butte. The other membranes reduced UV-Vis by 61% to 81%. The 600 MWC membrane product water had lowest UV-Vis and formed very low THMs (31 μg/L at El Paso and 14 μg/L at Elephant Butte). Figure 2 shows the correlation between UV-Vis and THMFP.

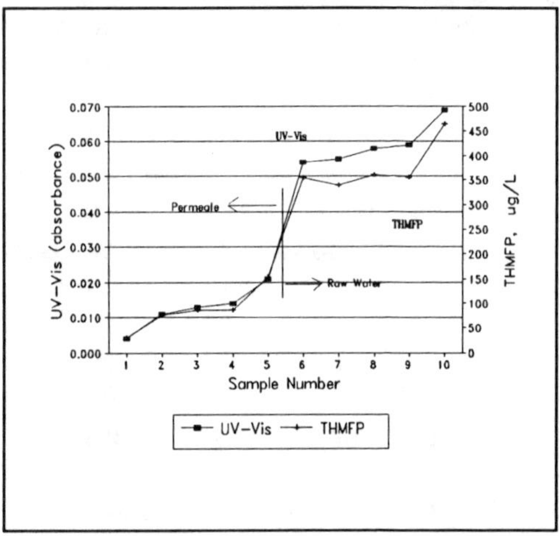

Figure 2. Correlation between UV-Vis and THMFP of Rio Grande water at El Paso

Table 2 gives the THMFP and hardness data for membrane feed and product waters. The average feed water THMFP was 374 μg/L at El Paso and 249 μg/L at Elephant Butte. Figure 3 shows the THMFPs of product waters at both sites.

Table 2. THMFP and hardness data on membrane feed and permeate during testing on Rio Grande water at El Paso and Elephant Butte

MWC	THMFP (μg/L)						Hardness (mg/L) as $CaCO_3$					
	Feed		Permeate		Removal (%)		Feed		Permeate		Removal (%)	
	EP	EB	EP	EB	EP	EB	EP	EB	EP	EB	EP	EB
300	340	245	86	76	75	69	180	178	70	76	61	57
600	361	244	31	14	91	95	188	180	12	10	94	94
1,000	464	258	87	53	81	80	172	184	46	80	73	57
2,000	355	255	76	95	79	63	174	180	72	94	59	48
3,000	354	254	154	100	56	60	184	184	106	120	42	35

EP = El Paso EB = Elephant Butte

Figure 4 shows membrane efficiencies in reducing THMFP and hardness. The 600 MWC membrane produced water with a THMFP of 31 μg/L at El Paso and 14 μg/L at Elephant Butte. This membrane reduced THMFP by 91% and 95% at the two sites, respectively. Efficiencies of the other membranes varied from 56% to 81%. The THMFPs of the product waters of these membranes varied from 76 to 154 μg/L at El Paso and 53 to 100 μg/L at Elephant Butte. The 3,000 MWC membrane reduced THMFP to 100 μg/l at Elephant Butte but at El Paso, the THMFP was 154 μg/l .

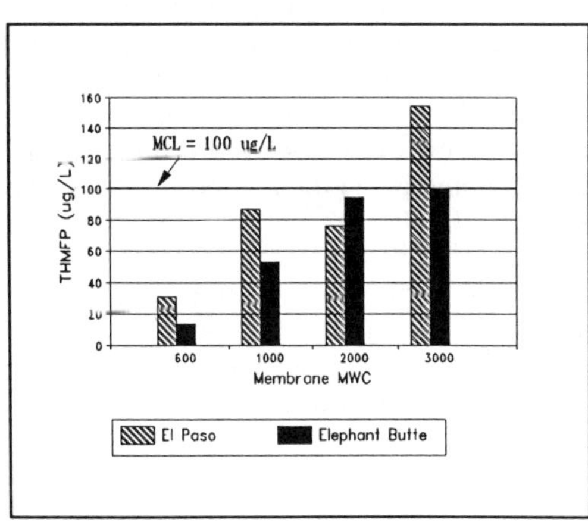

Figure 3. Product water THMFPs at El Paso and Elephant Butte

The average feed water hardness was 180 mg/L, at both sites. At El Paso, the 180 mg/L value represents the hardness of water partially softened prior to use as feed. The 600 MWC membrane produced water with a hardness of 12 and 10 mg/L at the two sites. The reduction achieved by this membrane was 94% at both the sites. Efficiencies of the remaining membranes varied from 35% to 73%. The recommended limit for hardness is 125 mg/L and all membranes produced water below the limit.

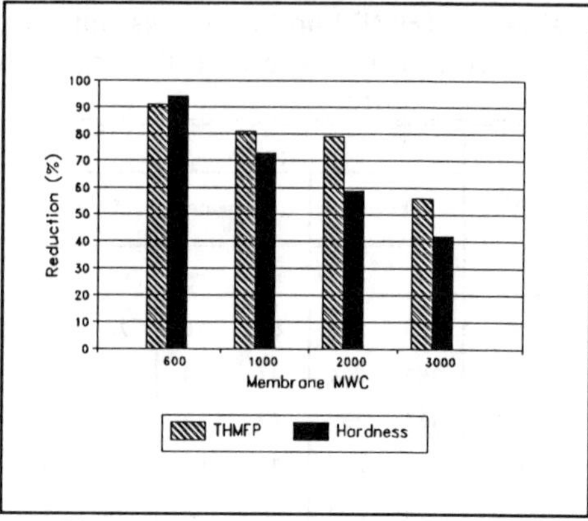

Figure 5. THMFP and hardness reduction by membranes

Conclusions

All membranes, except the 300 MWC, reduced hardness and THMFP in accordance with their MWC. Any of the five membranes tested would provide adequate hardness removal. The membranes with a MWC of 1000 or less provided adequate THMFP precursor reduction for the Rio Grande feed water. None of the membranes could be characterized as ideal because the optimum end points for both THMFP and hardness were not achieved by any single membrane. UV-Vis was shown to be a good surrogate for THMFP.

Membrane treatment of Rio Grande surface water at El Paso would require extensive pretreatment. A membrane treatment system at Elephant Butte Reservoir would require significantly less pretreatment and may be cost competitive with other treatment alternatives.

Appendix I. References

Amy, G.L., Allenman, B.C., and Cluff, C.B. (1990). "Removal of Dissolved Organic Matter by Nanofiltration." *J. Envir. Engrg., ASCE.*, 116(2), p. 200.

AWWA Research Foundation., (Dec 1989). "Assessment of Potable Water Membrane Applications and Research Needs."

Duranceau, J.S., Taylor, J.S., and Mulfold, L.A. (1992) "SOC Removal in a Membrane Softening Process." *J. AWWA*, 84(1), p. 68.

Kim H.S.P., and Symons, J.M. (1991). "Using Anion Exchange Resins to Remove THM Precursors", *J. AWWA*, 83(12), p. 61.

Laine, J.M., Clark, M.M., and Mallevialle, J. (1990). "Ultrafiltration of Lake Water: Effect of Pretreatment on the Partitioning of Organics, THMFP, and Flux." *J. AWWA*, 82(12), p. 82.

McGuire, J.M., Davis, M.K., Tate, C.H., Aieta, E.M., Howe, E.W., and Crittenden, J. C. (1991). "Evaluating GAC for Trihalomethane Control." *J. AWWA*, 83(1), p. 38

Schnoor, J.L., Nitzschke, J.L., Lucas, R.D., and Veenstra J.N. (1979). "Trihalomethane Yields as a Function of Precursor Molecular Weight." *J. EST*, 13(9), p. 1134.

Tan, L., and Amy, G.L. (1991). "Comparing Ozonation and Membrane Separation for Color Removal and Disinfection By-Product Control." *J. AWWA*, 83(5), p. 74.

Taylor, J.S., Thompson, D.M., and Carswell, J.K., (1987). "Applying Membrane Processes to Groundwater Sources for Trihalomethane Precursor Control." *J. AWWA*, 79(8), p. 72.
NF = Nanofiltration

Appendix II. Keywords

AMW = Apparent Molecular Weight
DBP = Disinfection By-Product
MWC = Molecular Weight Cutoff
MWD = Molecular Weight Distribution
NF = Nanofiltration
THMFP = Trihalomethane Formation Potential
UF = Ultrafiltration

Computer simulation of pulsed pumping for remediation under mass transfer limited conditions

Z. Jiang and W. P. Ball
Department of Geography and Environmental Engineering,
The Johns Hopkins University, Baltimore, MD 21218, USA

Groundwater pumping and above-ground treatment (pump-and-treat remediation) is a commonly employed method in the clean–up of aquifers contaminated with synthetic organic pollutants. It has been shown theoretically and experimentally that mass transfer limitations to the elution of sorbed contaminants by this method can have a large impact on the time required for remediation (e.g., Mackey and Cherry, 1989; Goltz and Oxley, 1991). For continuous pumping at a given pumping rate, the contaminant load discharged into extracting wells typically declines rapidly at early time, with high residual concentrations in the aquifer which are only slowly further removed. Such behavior can be attributed to diffusion–limited desorption of contaminant from "immobile" regions of aquifer solids, either at the scale of individual sorbing soil grains or large aggregated regions at low permeability. Thus, following the initial period of rapid aqueous concentration decline, mass transfer limitations to desorption limit the flux of contaminant to the aqueous phase. Under these conditions, the aqueous phase remains at low concentration, and the rate of reduction of the sorbed phase concentration is largely independent of the pumping rate, as shown subsequently. In terms of the amount of water that must be removed in order to meet regulatory standards, continuous pumping at high rate can thus be an extremely inefficient method of remediation.

One method which has been proposed to reduce pumped volume is to use a repetitive process of pumping and interruption, with low or greatly reduced flow rates during the interruption period. This approach is referred to here as "pulsed pumping". During the interruption period, the immobilized and mobile aqueous concentrations of contaminant in the aquifer re-equilibrate such that a relatively large amount of contaminant can be pumped in the early stages of the next cycle. However, a question rises as to proper selection of appropriate time periods for the pumping and interruption. To our knowledge, no systematic approach for answering this question has yet been proposed. In this work, we use computer simulation to calculate a proposed set of criteria for objectively addressing this issue.

In our initial approach to this question, we have idealized the contaminated aquifer as a one-dimensional column. The transport associated with rate-limited sorption during pumping can be described by the following equation (Parker and Van Genuchten, 1984)

$$(\theta_m + f\rho_b K_d)\frac{\partial C_m}{\partial t} + [\theta_{im} + (1-f)\rho_b K_d]\frac{\partial \overline{C_{im}}}{\partial t} = \theta_m D_H \frac{\partial^2 C_m}{\partial x^2} - v_m \theta_m \frac{\partial C_m}{\partial x} \quad (1)$$

and the process of desorption of contaminant from solid particles (or immobilized regions of the aquifer) during flow interruption period can be described by

$$(\theta_m + f\rho_b K_d)\frac{\partial C_m}{\partial t} + [\theta_{im} + (1-f)\rho_b K_d]\frac{\partial \overline{C_{im}}}{\partial t} = \theta_m D_M \frac{\partial^2 C_m}{\partial x^2} \quad (2)$$

where
$\overline{C_{im}}$ = average concentration in the immobile region at position x [M/L^3];
C_m = average (or bulk) mobile region concentration at position x [M/L^3];
θ_m = mobile region porosity [-];
θ_{im} = immobile region porosity [-];
θ = total porosity [-]; $\theta = \theta_m + \theta_{im}$;
f = fraction of immobilized solute in instantaneous equilibrium with the mobile phase [-];
ρ_b = bulk density [M/L^3];
K_d = equilibrium linear partitioning coefficient [M/L^3]
v_m = mobile region velocity [L/T];
D_H = hydrodynamic dispersion coefficient [L^2/T];
D_M = molecular diffusion coefficient [L^2/T];
x = distance measured from the column entrance [L]. In equation (1), the term $\partial C_m/\partial t$ represents the rate of concentration change in the mobile aqueous phase, while the term $\partial \overline{C_{im}}/\partial t$ represents the average rate of concentration change in the immobile region. After integrating equation (1) along the length of the column, we can define the following terms:

$$R_{C_m} = (\theta_m + f\rho_b K_d)\frac{\partial}{\partial t}(\int_0^L C_m dx)/L \quad (3)$$

$$R_{C_{im}} = [\theta_{im} + (1-f)\rho_b K_d]\frac{\partial}{\partial t}(\int_0^L \overline{C_{im}} dx)/L \quad (4)$$

$$R_{total} = \int_0^L (\theta_m D_H \frac{\partial^2 C_m}{\partial x^2} - v_m \theta_m \frac{\partial C_m}{\partial x})dx/L \quad (5)$$

where L is the length of the column. Then, we have

$$R_{C_m} + R_{C_{im}} = R_{total} \quad (6)$$

Physically, the term R_{C_m} represents the rate at which contaminant is being depleted in the aqueous phase; the term $R_{C_{im}}$ represents the rate at which contaminant is being depleted from the solid phase; and R_{total} is the total flux of contaminant out of the system. We define a ratio of the rates from the two phases as

$$RR = R_{C_{im}}/R_{C_m} \quad (7)$$

In the initial stage of pumping (when the column is still close to equilibrium) the contaminant concentration in the aqueous phase is reduced rapidly with time and the term R_{C_m} is initially high. For a column where mass transfer limitations to desorption are significant, the term $R_{C_{im}}$ will be initially low (while the column is still close to equilibrium) and will rapidly rise as the aqueous concentration is depleted and then will decline over time as the remaining sorbed solute is reduced. From the perspective of minimizing total pumped volume, infinitesimally small pumping time is optimal. However, recognizing that this is impractical, we propose RR as a rational means of deciding when to stop pumping during each pumping period. In this context, total pumped water volume is reduced when the rate ratio RR is very small.

After pumping is stopped, the contaminant continues to desorb from the immobilized phase to the normally mobile liquid. Ideally, reequilibration should be achieved before pumping is resumed. Theoretically, however, equilibrium will take an infinite amount of time to be achieved. In this context, we use the theoretical ratio between the concentration of the aqueous phase and the theoretical re-equilibrium concentration as the indicator to determine when pumping should be resumed. The concentration of aqueous phase is chosen to be the average concentration in the aqueous phase along the column, or $\int_0^L C_m dx/L$ and the theoretical equilibrium concentration is defined as the mass remaining in the column divided by retardation factor and total aqueous column volume. This equilibrium ratio, expressed as "percent of equilibrium" and designated as ER, represents how close the column is to re-equilibrium. A larger ER will correspond to longer interruption times.

To quantitatively understand the significance of these two indicators (RR and ER), we have initially examined a case study where an intraparticle diffusion model is assumed to apply (Ball and Young, 1993). For this model, the immobile zone concentration is described by a spherical diffusion equation

$$\frac{\partial C_{im}}{\partial t} = \frac{D_a}{r^2}\frac{\partial}{\partial r}[r^2\frac{\partial C_{im}}{\partial r}] \tag{8}$$

and the average immobile zone concentration is given by

$$\overline{C}_{im} = \frac{3}{a^3}\int_0^a C_{im} r^2 dr \tag{9}$$

where, a is the radius of the particle [L] and D_a is the apparent diffusion coefficient [L^2/T]. D_a is typically much lower than aqueous diffusion and has been interpreted in terms of both retarded pore diffusion and intra-organic matter diffusion (Ball and Roberts, 1991).

Coupling equations (8) and (9) with equations (1) and (2) and initial conditions, we can numerically simulate different pulsed–pumping scenarios. Simulations have been used to calculate the total pumping time and total treatment time (including both pumping time and interruption time to achieve 99.9% clean-up) for a column assumed to be initially at equilibrium. Different combinations of RR and ER have been tested. In our simulations, we use equivalent number of pore volumes fed (at the stated pumping rate) as a dimensionless unit of time. Pore volume P is thus defined as vt/L, where v is the Darcy velocity of the flow in the column at the pumped flow rate and t is the total elapsed time (pumping plus interruption) since initiation of treatment. After pumping is stopped, "pore volumes fed" no longer has physical meaning, but is still used as the unit of time to track the total elapsed time for consistency. In our simulations we also assume that flow in the column under pumped conditions is always steady. The physical–chemical parameters that we have used in our simulations to date (e.g., porosity, bulk density, sorption distribution coefficient, and apparent diffusion coefficient) are based on Borden sand and tetrachloroethene, as reported in the column transport and batch sorption studies of Young and Ball (1993).

Fig. 1 shows how the pumping rate affects the concentration in both the solid phase and aqueous phase. The vertical axis represents the average concentration along the column; and the horizontal axis represents the pumping time. Two sets of simulation data are plotted here. The difference between these two simulation runs is that the pumping rate in the first case is half of that of the second case.

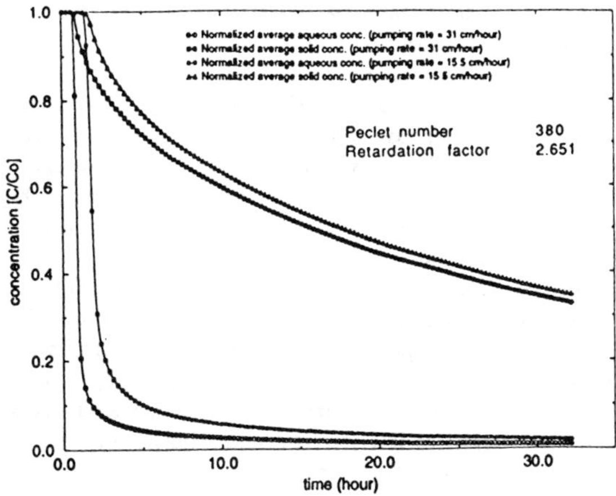

Fig. 1 Concentrations in both solid phase and aqueous phase versus pumping time (hours) under different flow velocities.

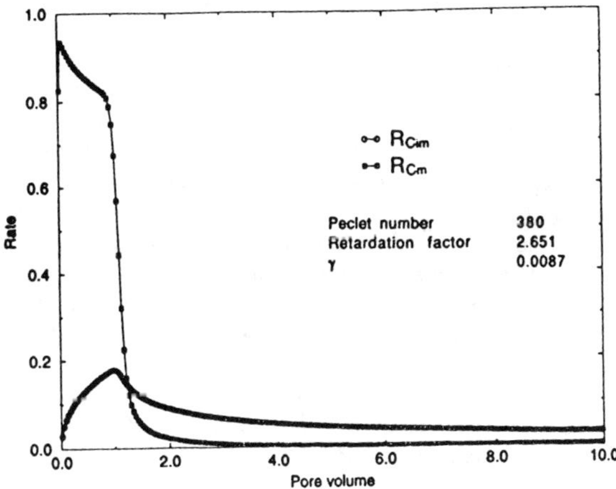

Fig. 2 Rates of concentration change in both solid phase and solid phase versus time (pore volumes).

Concentration in the aqueous phase stays low for both pumping rates after an initially rapid decline. Diffusion-limited desorption results in a longer-term decline of the solid phase concentration and will be a source of extended time to final "clean-up".

For the continuous pumping scenario, Fig. 2 shows the rate of concentration change in both the aqueous and immobile phase as a function of time. Slightly after one pore volume, most of the contaminant in the aqueous phase has been washed out, which is followed by a precipitous decline in average removal rate (squares in Fig. 2). On the other hand, the change in the solid phase concentration is moderate, and the depletion rate also changes less dramatically.

The ratio of these two rates, RR is illustrated in Fig. 3 as a function of time (pore volume pumped) for the continuous pumping case. Also shown is the effect of the parameter γ, where γ is defined as $3D_pL\theta_{im}/a^2v$, where D_p is the effective diffusion coefficient within an aggregate [L^2/T]. Physically, γ represents the time-scale for contaminant diffusion out of the particle as compared to the time-scale for water travel through the column. At low γ values, mass transfer limitations are most significant. As γ increases, local equilibrium is approached, and RR will also approach a constant $[\theta_m + f\rho_b K_d]/[\theta_{im} + (1-f)\rho_b K_d]$. At high RR values, removal becomes inefficient because aqueous concentration is kept at very low value. We subsequently explore the use of RR as a criterion for planning pump duration.

In Fig. 4, a particular scenario of pumping and interruption is shown, in which $ER = 90\%$ and $RR = 10$. Because ER is close to one, the interruption time is much longer than pumping time.

Fig. 5 and Fig. 6 are two contour plots of our simulation results. In Fig. 5, the relationship among total pumping time (measured by pore volume), RR and ER is shown. This plot illustrates that the total pumping time decreases monotonically with increasing percentage of equilibrium ER or decreasing ratio of immobile to mobile flux rate RR. The results in Fig. 5 demonstrate that pumping only while the column is close to equilibrium improves the efficiency of the pulse–pumping treatment scheme. Fig. 6 shows the relationship among total treatment time (including the pumping time and the interruption time, all measured as pore volume), RR and ER. Fig. 6 demonstrates that the total treatment time monotonically increases with the increasing ER. Larger ER is thus associated with longer interruption time and longer total treatment time.

One interesting result of this analysis is that decreases in pumping time can be achieved by reducing RR without severe consequence in terms of total treatment time. On the other hand, both pumping time and total treatment time are highly dependent on the selection of ER, and in opposite ways. The latter result reflects the inherent conflict of these two goals and is expected. Fig. 5 and Fig. 6 thus help to clarify the trade–offs between minimizing total treatment time and minimizing total pumping volume. The proper selection of the interruption period is likely to be determined by some value-laden balance between these conflicting goals. In the analysis above, we have proposed two parameters that could be useful towards rationally selecting such a balance. Other parameters for deciding when to start and stop pumping can also be envisioned.

We note that Harvey et. al have recently published a paper (1993) comparing a pulsed-pumping scheme with continuous pumping. The rate model they used is first order model. They found that continuous pumping at lower flow is always better

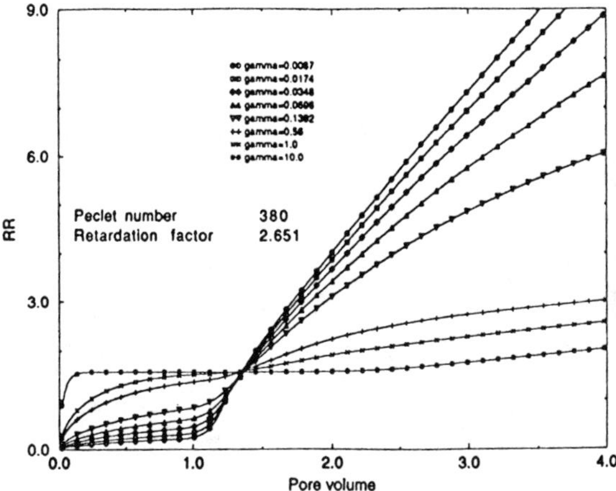

Fig. 3 *RR* versus time under different γ values (see text for detail).

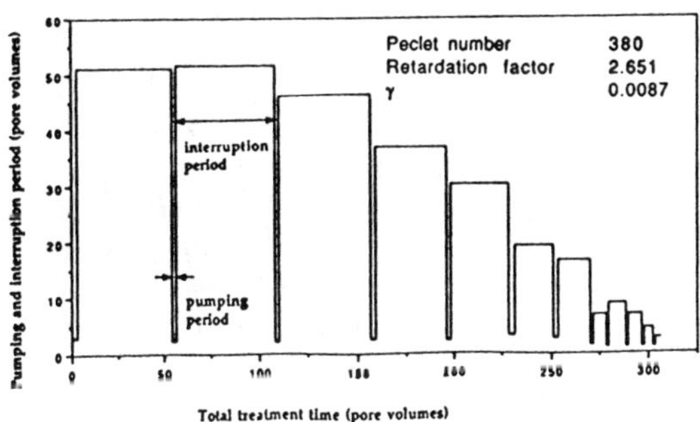

Fig. 4 Time of pumping or interruption versus total (chronological) time. Alternating periods of pumping and interruption are shown, beginning with pumping.

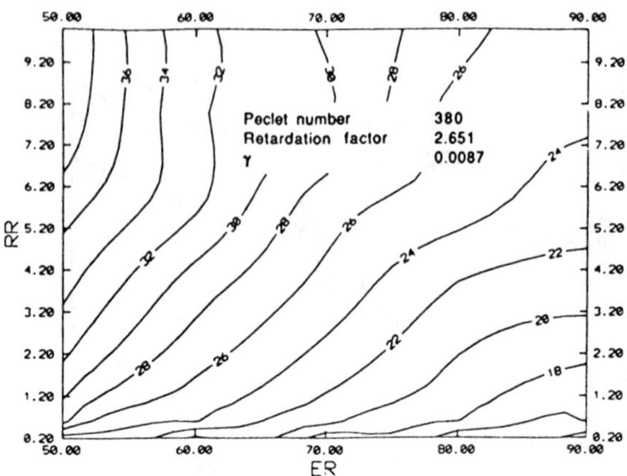

Fig. 5 Pumping time (no. of pore volumes) as a function *ER* and *RR* (see text for detail).

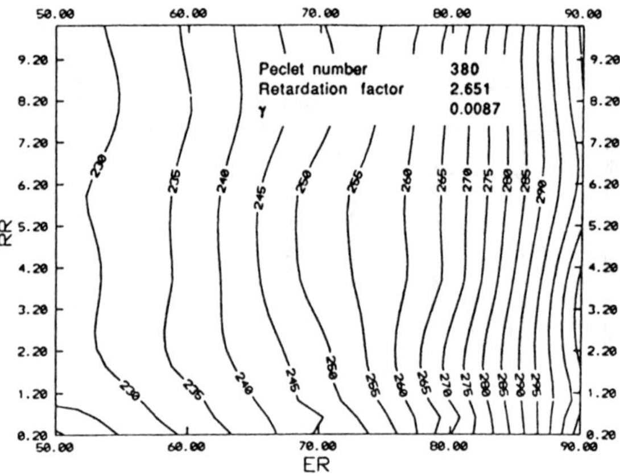

Fig. 6 Total treatment time as a function *ER* and *RR* (see text for detail).

than high flow pulse-pumping to achieve total treatment in less time with equivalent pumped volume. We have independently reached the same conclusion for selected cases by using the spherical diffusion model (Jiang and Ball, 1994). However, despite this recognized advantange to continuous pumping, extremely small pumping rates may not be practical to achieve or maintain. In particular, there will be situations where a minimum practical pumping rate must be established either because of equipment limitations or for purpose of hydraulic control. In these cases, pulsed pumping will offer clear benefits and an appropriate strategy for deciding pump and wait time (as addressed here) will be needed.

REFERENCE:

Ball, W. P. and P. V. Roberts, 1991a, Long–term sorption of halogenated organic chemicals — part 1. equilibrium studies, *Environ. Sci. & Technol.*, 25(7), 1223–1237.

Ball, W. P. and P. V. Roberts, 1991b, Long–term sorption of halogenated organic chemicals — part 2. rate studies, *Environ. Sci. & Technol.*, 25(7), 1237–1249.

Goltz, M. N. and M. E. Oxley, 1991, Analytical modeling of aquifer decontamination by pumping when transport is affected by rate–limited sorption, *Water Resources Research*, 27(4), 547–556.

Harvey, C. F., R. Haggerty and S. M. Gorelick, 1993, Aquifer remediation: a method for estimating mass transfer rate coefficient and an evaluation of pulsed pumping, in press.

Jiang, Z. and W. P. Ball, unpublished data.

Mackay, D. M. and J. A. Cherry, 1989, Groundwater contamination: pump-and-treat remediation, *Environ. Sci. & Technol.*, 23(6), 630–636.

Parker, J. C. and M. Th. Genuchten, 1984, Determining transport parameters from laboratory and field tracer experiments, *Bulletin 84–3*, Virginia Agricultural Experiment Station, Virginia Polytechnic Institute and State University, Blacksberg, VA.

Young, D. F. and W. P. Ball, 1993, A priori simulation of tetrachloroethene transport through aquifer material using an intraparticle diffusion model, *Environmental Progress*, in press.

BIODEGRADATION OF PHENANTHRENE IN SAND COLUMNS IN THE PRESENCE OF NONIONIC SURFACTANTS

David Norris[1] and Tariq Ahmed[2]

Abstract

The effects of three nonionic surfactants on phenanthrene ($C_{14}H_{10}$) removal and mineralization by aerobic bacteria were studied using a bench-scale apparatus. Columns were packed with fine sand coated with a mixture of (9-^{14}C) labeled and unlabeled phenanthrene (0.33 mg/g) and then inoculated by pumping acclimated bacteria. Surfactants at a concentration of 50 mg/L in an oxygenated buffer solution were then pumped through the media for 14 days at average pore velocities of 1 m/d to 3 m/d. Mineralization of phenanthrene was estimated by $^{14}CO_2$ activity in the column effluent and total removal was measured by the change in ^{14}C activity of the sand. Depending on the surfactant, mineralization was either inhibited or enhanced. A two-fold increase in flow rate increased phenanthrene mineralization and total removal greater than the effect of surfactant addition alone. Total removal ranged from 86.4% to 40.3% of the initial phenanthrene present.

Introduction

The polycyclic aromatic hydrocarbons (PAHs) are characterized by low aqueous solubility, high molecular weight and a tendency to sorb strongly to soil organic matter (Verscheuren, 1983) and thus are difficult to remove from contaminated soils with conventional washing or flushing techniques. Nonionic surfactants can be used to increase hydrocarbon solubility in soil cleanup efforts but their effects on PAH biodegradation are not clearly understood.

---

[1]Graduate Assistant, [2]Asst. Prof., Department of Civil Engrg., University of Nevada, Reno, NV 89557

Phenanthrene ($C_{14}H_{10}$), commonly used as a model PAH for research purposes, has been shown to be only used by bacteria in its dissolved state (Wodzinski and Coyle, 1974). This suggests that increasing solubility with a surfactant should increase mineralization and hence bioremediation effeciency. However, Laha and Luthy (1991) found that phenanthrene mineralization in a soil water-system was strongly inhibited with selected surfactants at concentrations above their critical micelle concentration (CMC) and noticed no enhancement at lower concentrations. To investigate surfactant enhancement of PAH mineralization, this experiment simulated saturated flow conditions in a portion of a phenanthrene contaminated, sandy aquifer in such a way that the effects of various surfactants on phenanthrene mineralization and washout could be measured.

Methods and Materials

The experimental apparatus consisted of four glass columns, 7.5 cm in length and 2.5 cm in diameter, filled with pure silica sand (20-40 mesh) which had been lightly coated with a mixture of radiolabelled ($9\text{-}^{14}C$) and unlabelled phenanthrene (0.33 mg/g, activity from 16 nCi/g to 72 nCi/g). A mixed culture of phenanthrene acclimated bacteria was prepared from aerobic seed bacteria obtained from the aeration basin at the local wastewater treatment plant. This was settled and resuspended at a cell mass concentration of approximately 10 mg/L in an oxygen saturated nutrient buffer and then pumped through the columns at flows comparable to aquifer flow rates (specific discharges from 0.3 m/d to 1.0 m/d) for a period of approximately 24 hours. Continuous flow was supplied by a peristaltic pump and columns were oriented vertically and used in an upflow mode to minimize short circuiting (Figure 1).

After bacterial application, three nonionic surfactants, Brij-35 (Sigma Diagnostics, St Louis, MO), Novell-II 1412-60 (Vista Chemical Company, Austin, TX) and Igepal CO-660 (Rhone-Polenc Inc., Cranbury, NJ), were added at a concentration of 50 mg/L to the influent buffer solutions and these were pumped through the columns for an additional 13-14 days.

Phenanthrene mineralization was estimated daily by the $^{14}CO_2$ activity in the column effluent using the apparatus shown in Figure 2. The $^{14}CO_2$ was driven from the sample by the addition of concentrated H_2SO_4. The sample flask was purged under vacuum and the $^{14}CO_2$ recaptured in a series of scintillation vials containing scintillation cocktail and strong alkali. Activity was

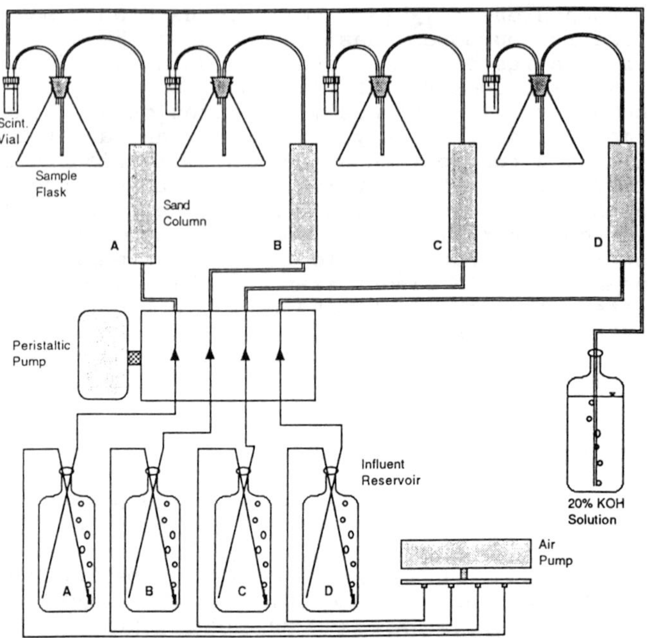

FIGURE-1 Experimental apparatus.

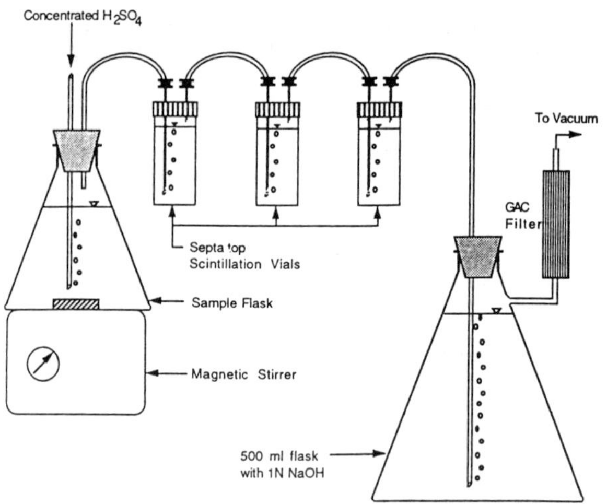

FIGURE-2 $^{14}CO_2$ trap detail.

measured using a 3 minute count on a Beckman (LS 5000 CE) liquid scintillation counter. Recovery of bicarbonate ($NaH^{14}CO_3$) spike activity ranged from 88% to 100% depending on the scintillation cocktail used.

The total amount of phenanthrene removed was measured as a function of column depth by the change in ^{14}C activity on the column sand for the entire 14 day test. Bacterial accumulation as a function of column depth was also estimated using the Lowry protein test (Lowry et al., 1951).

Preliminary tests established the pattern of phenanthrene washout from the columns using the three surfactants in the absence of bacteria and also the pattern of bacterial accumulation after 24 hours of pumping in the absence of phenanthrene (uncoated sand). These results were used as a baseline for comparison against the final phenanthrene removal and protein accumulation tests for the biodegradation experiments.

A total of three column biodegradation experiments were performed. The first compared each of the three surfactants at 50 mg/L in buffer to a plain buffer control at an average pore velocity of 2.93 m/d. The second experiment was identical to the first except that the average pore velocity was reduced to 1.27 m/d. In the third experiment, Igepal CO-660 at 50 mg/L was compared to a plain buffer control at average pore velocities of 1.03 m/d and 2.54 m/d.

Results and Discussion

As shown in Fig.3, mineralization was consistently higher than the control with Igepal CO-660 and lower than the control with Novell-II. The maximum mineralization measured was 26.2% in Expt.#1 using Igepal CO-660 at an average pore velocity of 2.93 m/d and the minimum measured was 2.6% in Expt.#2 using Novell-II at 1.27 m/d. These results are summarized in Table 1.

Measured mineralization as a percent of total removal varied with the surfactant and with the flow. For Igepal CO-660, this averaged 29.3%, with the higher values at the lower flow rates. In contrast, the control averaged 21.9 %, with the higher values at higher flow rates (Table 1).

In comparing the effects of flow in Experiment #3 (Figure 3), the presence of Igepal CO-660 coincided with higher mineralization as compared to the control at both high flow and low flow (average pore velocity of 2.64 m/d and 1.03 m/d respectively). As shown in Table 1,

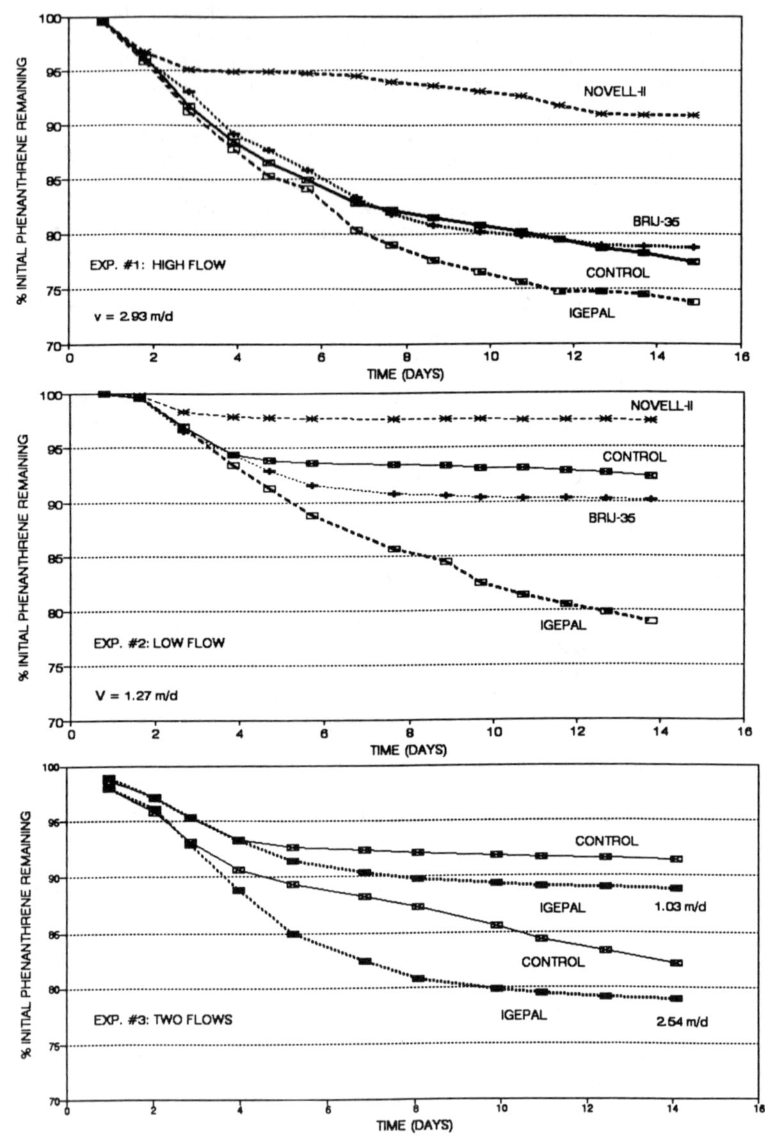

FIGURE-3 Estimated phenanthrene mineralization as measured by the $^{14}CO_2$ activity in the column effluent.

these increases were 3.6% (initial phenanthrene mineralized) at high flow and 3.1% at low flow. However, the difference between controls at high and low flow was 10.5%, indicating that the increase in flow exerted a stronger effect on phenanthrene mineralization than the addition of this surfactant.

An apparent contradiction of this is suggested by comparing the results of Exp.#1 and Exp.#2 in Fig.3. Mineralization in the presence of Igepal CO-660 changed from 26.2% (initial phenanthrene mineralized) to 21% when the flow was reduced by more than 50%, suggesting that a decrease in flow has very little effect on mineralization in the presence of Igepal CO-660. However, precise comparisons between experiments should be considered less reliable than comparisons of results within each experiment. This is because for each experiment there were slight variations in the experimental apparatus and scintillation cocktail used as well as variations in growth conditions such as the initial cell mass, the ^{14}C activity applied to the sand and the ambient room temperature. Hence the results from Exp. #3 alone are considered to be the more reliable indication of the effects of flow.

The overall pattern of phenanthrene removal in the biodegradation experiments, as measured by the change in ^{14}C activity of the sand is shown in Figure 4. This was identical to the pattern observed in previous washout tests which excluded mineralization (results not shown). The pattern of bacterial accumulation on the sand, as indicated by the protein present, can be seen in Figure 5. This was also the same pattern observed in previous cell transport tests using clean silica sand (results not shown). Regardless of flow or the amount of

TABLE 1: Summary of Results

Expt.		Mineralization (% Initial)	Removal (%Initial	Min./Rem (%)	Protein (μg/g)
#1 2.93 m/d	Control	22.5	82.0	27.4	38.5
	Brij	21.3	73.1	29.1	28.9
	Novell	9.2	66.1	13.9	26.0
	Igepal	26.2	83.9	31.2	51.6
#2 1.27 m/d	Control	7.7	45.4	17.0	17.5
	Brij	9.8	55.1	17.8	24.8
	Novell	2.6	40.3	6.50	27.0
	Igepal	21.0	63.3	33.2	27.7
#3 1.03 m/d	Control	8.6	44.4	19.4	27.0
	Igepal	11.2	40.7	27.5	20.5
2.54 m/d	Control	17.8	77.7	22.9	36.7
	Igepal	21.0	86.4	24.3	51.3

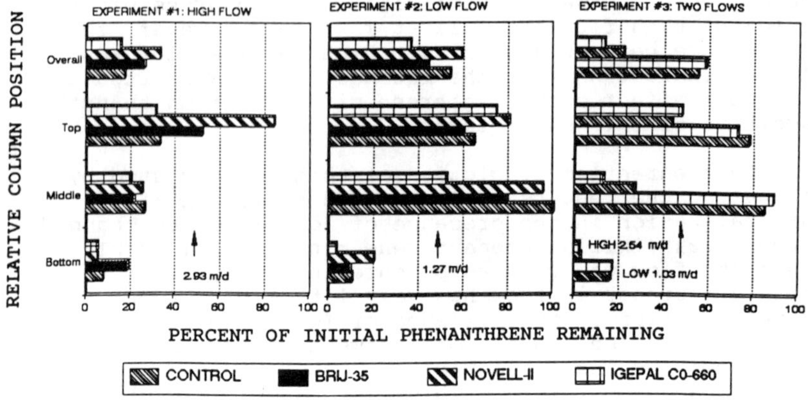

FIGURE-4. Total phenanthrene remaining on column sand after 14-day test as measured by the change in ^{14}C activity.

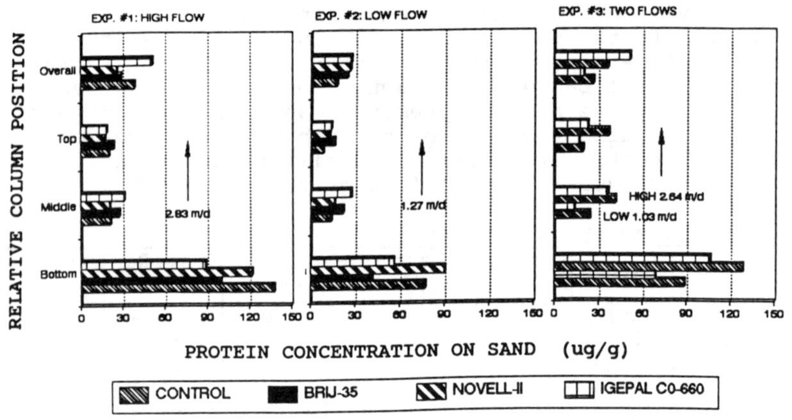

FIGURE-5. Bacterial accumulation on column sand at end of 14 day test as measured by protein present.

mineralization, the point of greatest phenanthrene removal coincided with the point of highest bacterial accumulation. This point was at the entrance to the column (bottom). Though the overall pattern remained more or less constant for all the tests, total phenanthrene removal and total bacterial accumulation did increase with increased flow and with increased mineralization (Table 1).

Conclusions

In phenanthrene contaminated sand (0.3 mg/g) under saturated conditions at pore velocities of 1 m/d to 3 m/d, the presence of non-ionic surfactants at 50 mg/L can either inhibit or enhance phenanthrene mineralization. Novell-II significantly inhibits phenanthrene mineralization where some enhancement is observed with Igepal CO-660, and to a much lesser extent, with Brij-35.

The effect of increasing pore velocity within this range is to generally increase mineralization and total removal. This effect appears greater than the slight increase in mineralization observed by the addition of surfactants.

Bacterial accumulation increases with overall phenanthrene removal, with the zone of highest phenanthrene removal coinciding with the zone of greatest bacterial accumulation. This relationship appears to be due largely to physical parameters such as filtration of the applied bacteria and detachment and solubilization of the phenanthrene and only in part to the biological activity on the sand.

References

Laha, S. and R.G Luthy. 1991. "Inhibition of phenanthrene mineralization by nonionic surfactants in soil-water systems." Environ. Sci. Technol. 25:1920-1930

Lowry, O.H., N.J. Rosebrough, A.L. Farr and R.J. Randall. 1951. "Protein measurements with the Folin phenol reagent." J. Biol. Chem. 193:265-275

Verschueren, K. 1983. Handbook of Environmental Data on Organic Chemicals. 2nd Ed. Van Norstrand. pgs 11,14,91

Wodzinski, R. and J. Coyle. 1974. "Physical state of phenanthrene for utilization by bacteria." Applied Microbiology. 27:1081-1084

Biotransformation of Pesticides in Simulated Groundwater Columns

Gregory G. Wilber*, Associate Member, ASCE, and Stuart R. Garrett*

Abstract

Experiments were performed to investigate the biotransformation of the pesticides alachlor, propachlor, atrazine, and bromacil under aerobic and nitrate-reducing conditions. Glass bead columns with attached, acetate-supported biofilms were used. Degradation of alachlor and propachlor was significant in both columns, and was not sensitive to short-term losses in the primary substrate. Biotransformation of bromacil and atrazine was minimal in these cultures, regardless of the amount of acetate present.

Introduction

Concern regarding the detection of pesticides in both raw ground waters and finished drinking waters has resulted in the addition of many pesticides to the list of drinking water contaminants regulated by the USEPA (Pontius 1992). Among these are four commonly used herbicides: the triazine herbicide atrazine (2-chloro-4-ethylamino-6-isopropylamino-1,3,5 triazine); the acetanilide herbicides alachlor (2-chloro-2',6'-diethyl-N-methoxymethyl acetanilide) and propachlor (2-chloro-N-isopropyl acetanilide); and the nitrogen-heterocyclic herbicide bromacil (5-bromo-3-*sec*-butyl-6-methyluracil). Atrazine, alachlor, and propachlor are commonly used as pre- and post-emergence herbicides on numerous crops including corn and wheat, while bromacil is used primarily in non-cropland applications for the control of annual and perennial weeds and grasses (CPP 1991). All five of these pesticides have been detected in ground water samples in Oklahoma or in numerous other agricultural areas in the U.S. (Ritter 1990, Holden *et al.* 1992). The current standards of the Safe Drinking Water Act include a maximum contaminant level (MCL) of 3 μg/L for atrazine and 2 μg/L for alachlor. Bromacil and propachlor currently have monitoring requirements, and may have specific MCLs established in the near future (Pontius 1992).

---

* Assistant Professor and Graduate Student, respectively, School of Civil and Environmental Engineering, Oklahoma State University, Stillwater, OK 74075

To better assess the risks associated with the introduction of pesticides into the environment, particularly with respect to their transport into drinking water sources, information is needed regarding the fate of these pesticides once they have reached the subsurface. Among the major fate processes which must be considered is biotransformation. What follows is a description of a research project investigating the biotransformation of the four previously mentioned pesticides. Structures of these herbicides appear in Figure 1. Specific objectives of the research were to investigate the effects of two different electron acceptor conditions (aerobic and nitrate-reducing) and the effects of the readily degraded organic compound (acetate) on pesticide transformation. These factors are likely to be important in natural systems as well as in engineered systems aimed at remediating contaminated waters (Lee *et al.* 1988, Cobb & Bouwer 1991).

Biotransformation of these pesticides has been investigated previously, to varying degrees. Atrazine has been studied most extensively, and has been shown to be biodegraded under a variety of conditions (Erickson & Lee 1989). In most cases, degradation has been found to be cometabolic in nature, requiring the presence of a more readily degraded primary substrate. Wilber and Parkin (1991) report the biotransformation of alachlor and, to a lesser degree, atrazine, in continuous-flow, acetate-fed biofilm reactors, under four different electron acceptor conditions. In all cases, the presence of acetate was required for the transformations to continue, though the effect was most pronounced in the sulfate-reducing system. Novick and Alexander (1985) report observing cometabolic biotransformation of alachlor and propachlor in sewage and lake water samples. Propachlor degradation also appeared to require the presence of more readily degraded substrates (Novick & Alexander 1985). However, propachlor has been shown to serve as the sole source of carbon and energy to at least two distinct microbial species (Villarreal *et al.* 1991). Other research which has directly compared the acetanilide herbicides has found in general that those compounds with fewer and smaller substituents (such as propachlor) are more readily degraded than those more heavily substituted (such as alachlor) (Zimdahl & Clark 1982, Alhajjar *et al.* 1990). Less information is available regarding the biotransformation of bromacil. Chaudhry and Cortez (1988) isolated from soil a *Pseudomonas* species capable of aerobically using bromacil as its sole carbon and energy source. More recently, Adrian and Suflita (1990) studied the fate of bromacil in anoxic aquifer slurries. They report minimal transformation occurring in the nitrate- and sulfate-reducing condition, though under methanogenic conditions bromacil was debrominated.

Methods and Materials

These experiments were performed with two identical glass-bead biofilm column reactors. Figure 2 displays the dimensions of these columns, both of which were originally seeded with sediments from Lake Carl

Blackwell, Oklahoma. The columns were operated in the dark at 23°C. A simulated ground water medium, including trace minerals representative of an Oklahoma ground water (USGS 1986), an exogenous carbon and energy source (acetate), as well as one or more of the pesticides of interest, was fed to the columns upflow, via peristaltic pump, in a semi-continuous manner similar to that described by Lanzarone and McCarty (1990). The medium fed to the nitrate-reducing column was stripped of oxygen prior to its introduction into the column. The liquid volume of each column was exchanged on a 12 hour cycle. Each exchange of column liquid took approximately 30 minutes. The subsequent utilization of the electron donor (acetate) and acceptor (oxygen or nitrate) and removal of each pesticide was monitored over time. Abiotic sorption experiments were conducted to confirm that removals due to sorption on any of the column materials were negligible.

Table 1 displays the influent concentrations of the major constituents to each column. Both columns were operated for at least 60 days before inclusion of the pesticides in the liquid media. This allowed the biofilm cultures to become established. Once acetate utilization in the columns became steady, the pesticides were introduced into the column effluents, on 'Day 0' in the figures below.

Table 1. Average influent concentrations of major constituents to each column.

Column	Acetate	Electron Acceptor	Pesticides
Nitrate-Reducing	21 mg/L	54 mg NO_3^-/L	Propachlor (270 μg/L) Atrazine (270 μg/L)
Aerobic	10 mg/L	8 mg O_2/L	Bromacil (250 μg/L) Propachlor (250 μg/L) Alachlor (250 μg/L) Atrazine (250 μg/L)

Aqueous stock solutions of each pesticide were prepared from analytical-grade chemicals. Atrazine was obtained from Supelco (Bellefonte, PA), bromacil from E.I. duPont de Nemours and Co. (Wilmington, DE), and alachlor and propachlor from The Monsanto Company (St. Louis, MO). All other compounds used for the feed solution or for the analyses were analytical grade or better. Pesticide concentrations in the aqueous phase were measured by a C-18 solid-phase extraction method (Thurman *et al.* 1990) followed by GC analysis [Hewlett-Packard 5890-II, with a DB-5 stationary-phase capillary column (J&W Scientific) and electron capture detector]. Acetate was measured by GC [Hewlett-Packard 5890-II, with 60/80 Carbopak/0.3% Carbowax packed column (Supelco)]. Nitrate was measured by ion chromatography (APHA 1989).

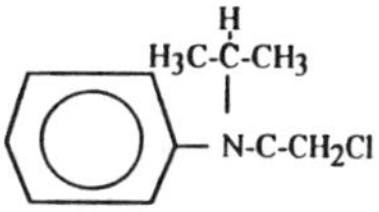

Propachlor
(2-chloro-N-isopropylacetanilide)

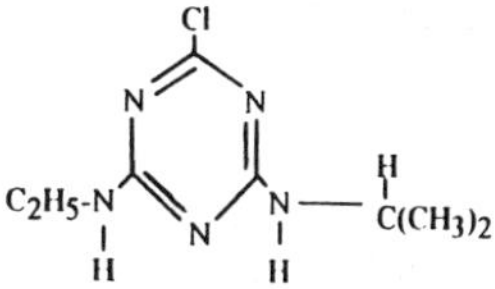

Atrazine
(2-chloro-4-ethylamino-
6-isoproplyamino-*s*-triazine)

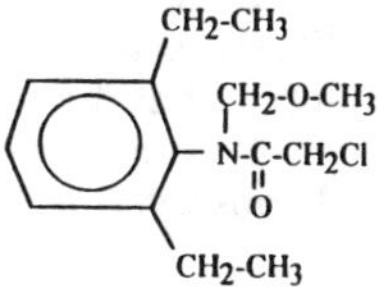

Alachlor
(2-chloro-2',6'-diethyl-
N-methoxymethyl-acetanilide)

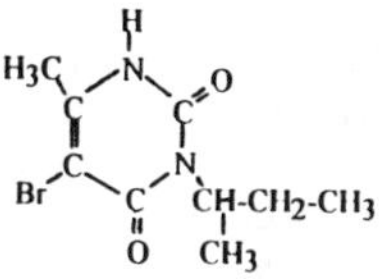

Bromacil
(5-bromo-3-*sec*-butyl-
6-methyluracil)

Figure 1. Structures of the pesticides investigated.

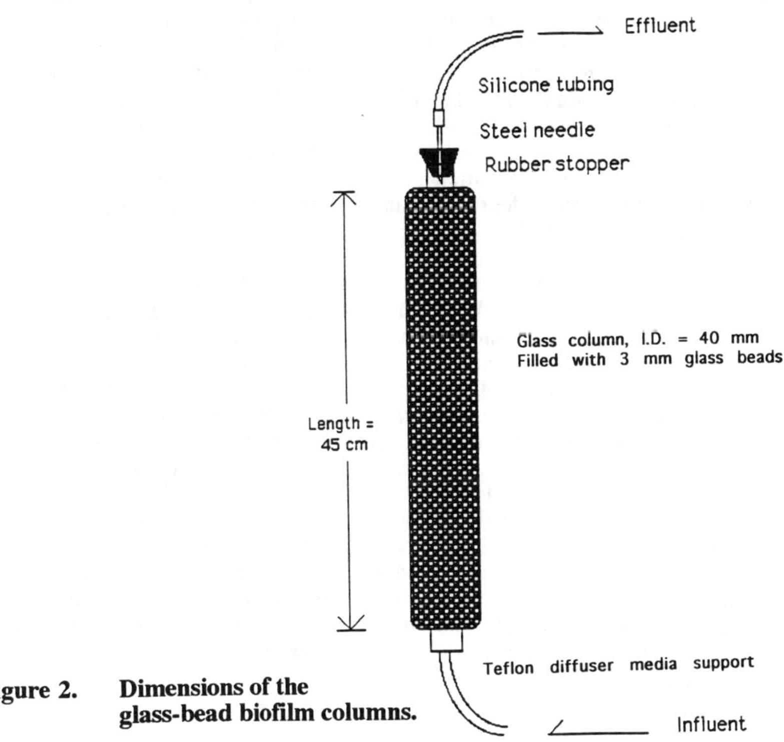

Figure 2. Dimensions of the glass-bead biofilm columns.

Results and Discussion

As mentioned, pesticides were introduced to each column after acetate and electron acceptor concentrations had appeared to reach a steady-state. In the nitrate-reducing column, acetate removal was consistently near 100%, and a stoichiometric amount of nitrate was also utilized. Figure 3 displays the removals of propachlor and atrazine in this column. Several initial observations can be made. Atrazine degradation was minimal throughout this experiment. Propachlor, however, was significantly degraded, down to an average effluent concentration of 185 μg/L (or approximately 30% removal). A stop-flow experiment, in which no exchange of liquid was made for a three day period, was conducted over Days 49-51, resulting in slight dips in the effluent concentrations of both pesticides. That the increase in contact time between the pesticides and cultures resulted in increased removals supports the conclusion that the removals observed were due to biological activity.

Results from the aerobic column are displayed in Figure 4. Recall that this column was maintained under oligotrophic conditions, with an average influent acetate concentration of 10 mg/L. As such, a thick biofilm was not expected or observed. Within a typical twelve hour exchange period, acetate utilization averaged 65%. Effluent oxygen concentrations were usually below 1 mg/L. With respect to the pesticides, from Figure 4, in general, it can be seen that both alachlor and propachlor were significantly degraded, while bromacil appeared to be recalcitrant under these conditions. Atrazine was also not degraded significantly in this column (data not shown). It appears that propachlor degradation was somewhat less than that of alachlor, though the difference between the two can not be considered significant. This contrasts somewhat with earlier reports which found propachlor more rapidly degraded than alachlor (Zimdahl & Clark 1982, Novick & Alexander 1985).

On Day 54, acetate was removed from the influent feed solution, and was not restored until Day 94. During this period, the culture retained its ability to biotransform alachlor and propachlor. There are several possible explanations for this. As mentioned, Villarreal and coworkers (1991) isolated aerobic organisms capable of using propachlor as a sole soucre of carbon and energy. No reports have been found of these or other organisms utilizing alachlor in a like manner. However, given the structural similarities between these two pesticides, this may be occurring. Obviously, further experimentation is required to confirm this. Another hypothesis is that the biotransformation of the pesticides is in fact strictly cometabolic, but this cometabolism can occur, at least for short periods, under conditions of endogenous respiration. Lastly, though sorption to the abiotic components of each reactor was found to be negligible, sorption to the biofilm itself can not be completely ruled out. Subsequent experiments, in which the pesticides were removed from the influent, however, showed no signs of desorption of these compounds.

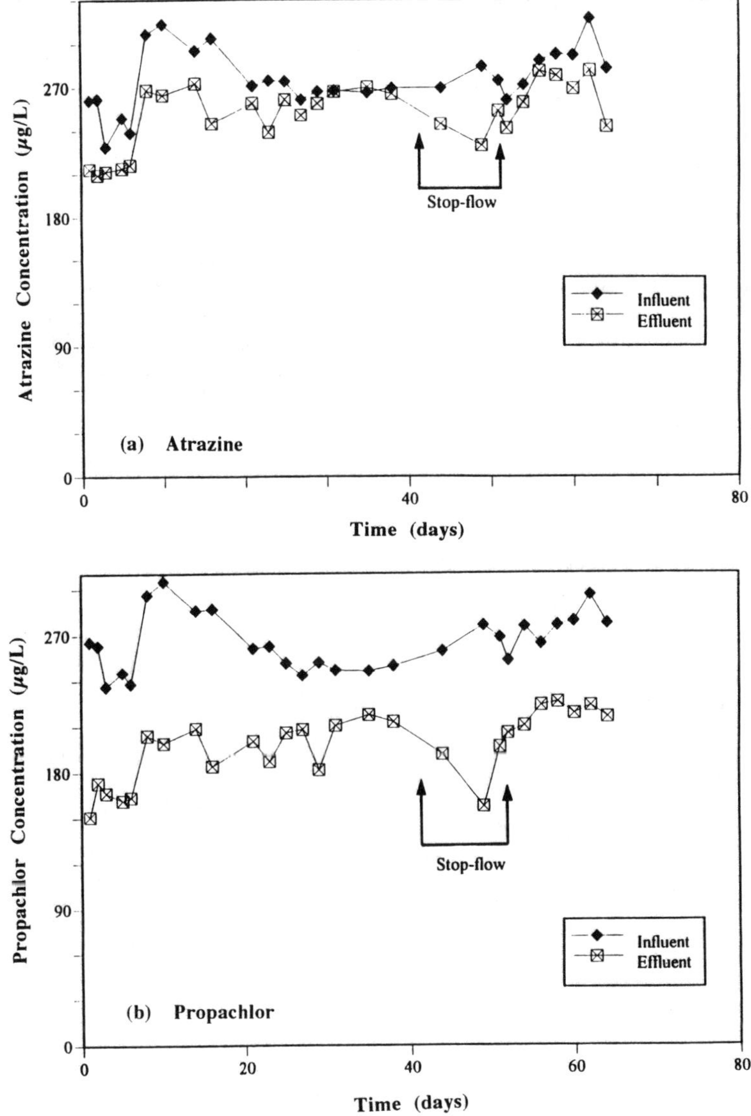

Figure 3. Comparison of influent and effluent concentrations in the Nitrate-Reducing Column for (a) Atrazine and (b) Propachlor.

During this period of primary substrate deprivation, a 3-day stop-flow experiment was conducted, similar to that in the nitrate-reducing column. Again, increased contact time between the pesticides and cultures resulted in increased removals. The effects in this column were somewhat less significant, however, likely due to the fact that during this stop-flow period, oxygen was most likely depleted within the first twelve to twenty four hours. If the culture requires oxygen for the cometabolic transformation of the pesticides, the additional contact time would still not result in increased pesticide removal.

In conclusion, the results of this research indicate that the herbicide propachlor is biotransformed under oligotrophic aerobic and nitrate-reducing conditions. Alachlor is similarly biotransformed under aerobic conditions. Once the cultures were established, biotransformation of these pesticides continued despite the loss of the primary substrate for a 40 day period. Neither of the heterocyclic herbicides, atrazine or bromacil, was degraded to a significant extent in either column.

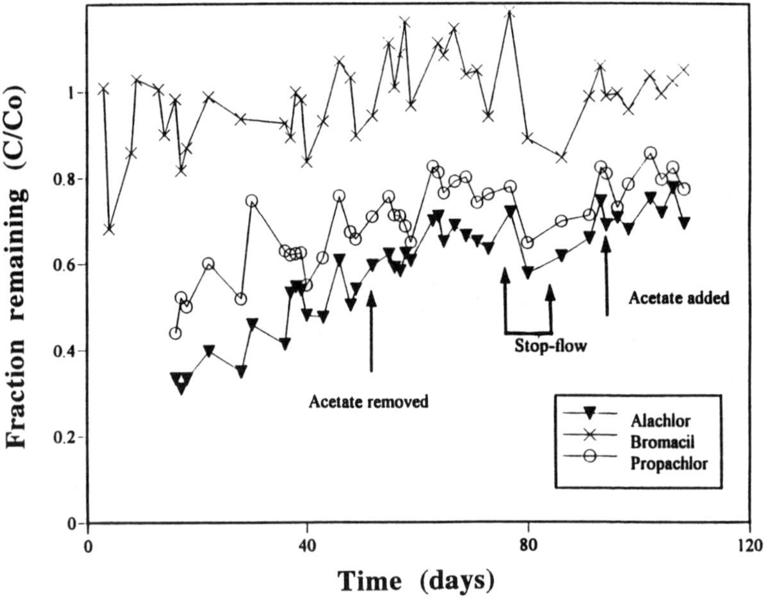

Figure 4. Removals of Alachlor, Propachlor, and Bromacil in the Aerobic Column.

Acknowledgment

This research was supported by a grant from the Oklahoma State University Center for Water Research, Stillwater, OK.

References

Adrian, N. R., and J.M. Suflita. 1990. *Appl. Environ. Microbiol.*, 56(1):292-294.

Alhajjar, N. R. *et al.* 1990. *Water Sci. and Technol.* 22(6):87-94.

American Public Health Association (APHA) 1989. *Standard Methods for the Examination of Water and Wastewater.* APHA, Washington, DC.

Chaudhry, R.G., and L. Cortez. 1988. *Appl. Environ. Microbiol.*, 54(9):2203-2207.

Cobb, G., and E. J. Bouwer. 1991. *Environ. Sci. Technol.*, 25(6):1068-1074.

CPP, 1991. Crop Protection Chemicals Reference. Chemical & Pharmaceutical Press.

Erickson, L.E., and K. Lee. 1989. *CRC Critical Rev. in Environ. Control.* 19:1-14.

Garrett, S.R. 1993. Column Studies on the Biotransformation of Pesticides in Groundwater Under Various Electron Acceptor Conditions. M.S. Thesis, School of Civil and Environmental Engineering, Oklahoma State University.

Holden, L.R. *et al.* 1992. *Environ. Sci. Technol.*, 26(5):935-943.

Lanzarone, N.A. and P.L. McCarty. 1990. *Ground Water*, 28:910-919.

Lee, M. D., *et al.* 1988. *CRC Critical Revi. in Environ.Control.* 18(1):29-89.

Novick, N.J., and M. Alexander. 1985. *Appl. Environ. Microbiol.*, 49(4):737-743.

Pontius, F.W. 1992. *J. Amer. Water Works Ass'n.* 84(3):36-50.

Ritter, W. F. 1990. *J. Env. Sci. Health,* Vol. B, 25:1-29.

Thurman, E. M. *et al.* 1990. *Analytical Chem.*, 62:2043-2048.

U.S. Geological Survey 1986. Oklahoma Ground-Water Quality. In: National Water Summary 1986, US Geological Survey Water-Supply Paper 2325:415-422.

Villarreal, D.T. *et al.* 1991. *Appl. Environ. Microbiol.*, 57(8):2135-2140.

Wilber, G.G., and G.F. Parkin. 1991. Transformation of Pesticides Under Anaerobic Conditions. In: *In Situ and On Site Bioreclamation,* R. L. Hinchee and R.F. Olfenbuttel, eds., Butterworth-Heinemann, 385-402.

Zimdahl, R.L., and S.K. Clark 1982. *Weed Sci.* (30):545-548.

Artificial Recharge of Aquifers

Charles L. Joy[1]
Student Member, ASCE

Abstract

Artificial groundwater recharge is a means of replenishing groundwater supplies with surface water which might otherwise be lost. Artificial groundwater recharge has also been used to protect coastal aquifers against saltwater intrusion; to improve water quality; and to reduce groundwater depletion that causes land subsidence. Three principal methods of artificial groundwater recharge are currently in practice: spreading basins, injection wells, and recharge shafts/pits. The appropriate method for a specific use depends upon the availability of land, the type of aquifer to be recharged, economic constraints, and the purpose of recharge.

Artificial groundwater recharge has created a stir in both the health and legal communities. The quality of source water used for groundwater recharge is of increased concern due to the unknown health effects associated with the consumption of reclaimed wastewater or potentially contaminated surface runoff. Artificial groundwater recharge has forced changes in regional water laws. The ownership rights to treated effluent, the water rights to recharged water, and the potential liability arising from artificial groundwater recharge have only recently been addressed in a few states.

Introduction

Groundwater is a major source of water for municipal, industrial and agricultural use, particularly in arid regions. In the U.S., groundwater from

1 Undergraduate, Department of Civil Engineering, Arizona State University, Tempe, AZ 85287

aquifers is estimated to supply half of the water for municipal use and about one-third of the irrigation water [Bouwer, 1978]. In many areas, the rate at which groundwater is withdrawn exceeds the natural groundwater recharge rate resulting in the depletion of groundwater tables. The depletion and contamination of groundwater reserves are among the most significant challenges in water management. Artificial groundwater recharge is a means of replenishing groundwater supplies with surface water which might otherwise be lost. Recharging aquifers has other benefits including the prevention of salt water intrusion into fresh water aquifers, storage of water, improving water quality, and controlling land subsidence.

Aquifers

An aquifer is a geologic unit capable of transporting and storing usable quantities of groundwater. It consists of a permeable layer through which water may flow and confining layers of less permeable material which restrict the movement of groundwater and form the boundaries of the aquifer. A confining layer of essentially impermeable material is called an aquiclude. A confining layer permeable enough to allow vertical water flow but not permeable enough to transmit water laterally is called an aquitard.

There are two types of aquifers: unconfined and confined. Unconfined aquifers are large subsurface reservoirs of groundwater in porous material. An aquiclude or aquitard underlies the aquifer but there is no overlying confining layer. The permeable layer of soil above the phreatic surface of the groundwater table is called the vadose zone. Surface water percolates through the vadose zone and into the groundwater table. Groundwater from unconfined aquifers must be pumped from the aquifer to the surface for use. Confined aquifers are sandwiched between two confining layers of less permeable strata and hold water under piezometric pressure. The pressure results from the compression of the confining layers. If the pressure is great enough, water will flow freely from a well drilled into the aquifer. The principal source of groundwater in a confined aquifer comes from the infiltration of water at a surface outcropping of the aquifer but water may also seep into the aquifer through overlying aquitards.

Artificial Groundwater Recharge

Artificial groundwater recharge methods may be classified into three groups: *spreading basins, injection wells, and recharge shafts/pits*. Surface spreading basins are commonly used for unconfined aquifers which extend to or near the ground surface. Recharge pits are used for moderately deep aquifers or where the conditions are unfavorable for spreading basins. Injection wells are used to recharge either confined aquifers or unconfined aquifers.

Surface Spreading Basins. Where geologic conditions are suitable and land is available and cheap, surface recharge is the most cost effective approach for recharging groundwater [Bouwer et al., 1990]. Spreading basins may be divided into off-channel systems or in-channel systems.

Off-channel systems consist of large spreading basins in which water is ponded for infiltration into the soil. Fine inorganic solids and organic matter clog pores making periodic drying and cleaning of the basins essential to maintain infiltration rates. Periodic drying reopens soil pores and fissures by allowing the clogging layer to shrink and crack. Drying also increases the rate of organic matter decomposition by increasing the availability of oxygen to microorganisms. Periodic drying also aids in the control of waterborne pests, such as mosquitoes, which breed in stagnant water. Clogging may be reduced by pretreating recharge water in sedimentation basins to remove suspended solids and organics. Plowing or disking the clogging layer will improve infiltration temporarily but fines will continue to accumulate and physical removal of the clogging layer is eventually required to maintain infiltration rates. Groundwater mounding caused by the infiltrating water under spreading basins may also reduce infiltration rates.

In-channel systems spread flowing water through a system of channels, weirs and dams (usually located in a stream bed or flood plain) to increase the infiltration area. In-channel systems experience a reduced rate of clogging due to the continuous flow of water through the system. In addition, the upper channels of the system are often used as sedimentation basins to remove suspended solids and reduce clogging in subsequent channels [Bouwer, 1988]. The continuous flow of water also aids in the control of waterborne pests which might breed in more stagnant water [Bouwer et al., 1990].

Injection Wells. If geologic conditions are not favorable for surface infiltration or sufficient land is not available or too costly, then the groundwater may be augmented by forcing water under pressure into the aquifer through drilled wells. Injection wells reduce the amount of water lost to evaporation during infiltration. The fastest growing type of recharge is the aquifer storage recovery (ASR) well [Bouwer et al., 1990]. ASR wells are able to both store and recover water from the same well.

Injection wells are similar to pumping wells except they are designed for rapid infiltration of water into the aquifer. They are particularly sensitive to clogging at the site of injection into the aquifer and require frequent back flushing to maintain infiltration rates. Extensive pretreatment of recharge water is usually required to remove suspended solids and organics. Soil-water chemical reactions, changes in pH, clay deflocculation, and air binding may also reduce infiltration. Air binding occurs when cold recharge water high in dissolved oxygen warms up and oxygen bubbles out of solution forming air pockets in the aquifer.

Because injection wells intrude directly into the aquifer, the potential for biological contamination and structural damage to the aquifer is greatly increased. Care must be taken to adequately grout well casings, seal aquifer penetration points and limit injection pressures to protect the integrity of the aquifer.

Recharge shafts/pits. Shafts/pits are intermediates between spreading basins and injection wells. They are large well-like openings, usually abandoned gravel pits, filled with water which extend to a point above the groundwater table. Recharge pits are preferable to spreading basins when surface soils overlay more permeable strata. Recharge pits require less land area than spreading basins and reduce the amount of water lost to evaporation by decreasing the exposed surface area. Pits are also considerably less expensive than injection wells and since they do not extend into the aquifer, the potential for structural damage and biological contamination is greatly reduced.

Applications of Artificial Groundwater Recharge

Artificial recharge of aquifers is generally used to store water for future use, but artificial recharge has other applications. In some coastal areas, fresh water aquifers are being protected from salt water intrusion by surface or injection recharge [Atkinson et al., 1986]. Surface or injection recharge is used to create a fresh water ridge along the fresh water/salt water contact zone. The fresh water ridge prevents sea water from flowing into the fresh water portion of the aquifer by raising the piezometric head of the fresh water aquifer above sea level. Groundwater then flows away from the fresh water portion of the aquifer preventing further contamination of the aquifer. This may also be used to protect fresh water aquifers from other sources of contamination.

Artificial groundwater recharge is also used to improve water quality. Generally, the movement of water, particularly wastewater, through the soil renovates the water by decreasing a number of water constituents. [Bouwer, 1978]. Most renovation occurs within the first few feet of movement through the soil. The soil physically filters out large organic and inorganic particles and drastically reduces BOD levels [Bouwer, 1978]. Pathogenic bacteria and viruses find themselves in an hostile environment competing with better adapted native bacteria and thus do not multiply and eventually die. Soils also tend to have a large capacity to adsorb cations on negatively charged clay particles. This cation exchange capacity aids in the removal of metallic ions, salts, ammonium and other positively charged contaminates.

The removal of groundwater from aquifers sometimes results in the collapse of soil layers. In parts of Mexico City, the land has subsided as much as 25 ft. due to excessive groundwater withdrawals [Freeze et al., 1979]. When the buoyant force exerted by water is removed, the full weight of overlying soil layers is applied to underlying layers. The increased overburden pressure may produce excessive settlements in some soil formations. Artificial recharge can control or prevent excessive land subsidence by maintaining historic groundwater levels.

Recharge Water

In the past, natural sources of water, such as surface runoff from rain, generally created little concern when used to recharge potable fresh water aquifers. However, there is increased concern that contaminates present in surface runoff may present potential health problems when used for groundwater recharge. The reuse of municipal wastewater as source water also concerns many health officials [Nellor, 1980]. Soil-aquifer treatment systems (SAT), in which sewage effluent is applied to land and allowed to filter through the soil, have been used to improve water quality and replenish nonpotable aquifers for future use. Such systems have been designed to protect native, high quality drinking water from nonpotable intrusion [Bouwer, 1978].

There is increasing pressure to expand the use of effluent to replenish potable fresh water aquifers. This activity has generally been restricted because of the unknown health effects associated with the consumption of water containing reclaimed water. However, in Arizona, the Department of Environmental Quality has established exemptions for the use of effluent source water for selected recharge projects [Marceau, 1991]. In California, the Orange County Water District has been using wastewater treated to drinking water standards to recharge groundwater since 1962 [Billings, 1990]. Currently, the Orange County Water District and health agencies are conducting an extensive study on the health effects of natural groundwater recharge along the Santa Ana River. For part of the year, 90% of the river flow consists of treated wastewater [Rigby et al., 1990].

Health concerns generally focus on the unknown effects of trace organics, nitrates, minerals, and heavy metals as well as the transmission of enteric viruses and choliforms [Nellor, 1980]. Nellor found that most constituents in reclaimed wastewater are reduced substantially by adsorption as effluent percolates through the vadose zone and moves through the aquifer. Viral adsorption by the soil, however, does not necessarily result in viral inactivation. Some viruses are capable of remaining viable for long periods of time. Adsorption is also readily reversible. Viruses and other charged particles may be easily replaced by other cations in the soil solution and become mobile. Thus, survival times of enteric viruses in groundwater and their mobility must be resolved to determine safe water abstraction distances [Jansons et al., 1989]. Some studies also indicate significant reductions in nitrate concentrations, synthetic organic compounds, and heavy metals in recharged effluent [Bouwer et al., 1990]. Other studies, however, have been less conclusive [Nellor, 1980].

Legal Aspects of Artificial Groundwater Recharge

Artificial recharge of groundwater has created several legal quandaries of which several have only recently been addressed. Groundwater rights differ from state to state so a comprehensive discussion of water rights is not possible, but generally a land owner has at least "reasonable" rights to underlying groundwater based upon the land use and the groundwater supply [Fetter, 1988]. The rights a land owner has on ground water replenished by artificial recharge has recently been addressed in the state of California. In *City of Los Angeles v. City of San Fernando* it was ruled that aquifers may be used for the storage and transport of water in the same manner as surface facilities and that other parties, regardless of overlying property ownership, have no right to this water [Bouwer, 1978].

The ownership of reclaimed wastewater, which may serve as source water for groundwater recharge, has recently been resolved in the state of Arizona. In *Arizona Public Service Co. (APS) v. John F. Long*, a local land developer contested the right of a wastewater treatment facility to sell treated effluent to the APS. The Arizona Supreme Court ruled against Long stating that effluent rights are controlled entirely by the entity which produces it (the treatment facility in this case). The effluent rights remain in control of the producing entity until it is discharged, at which time it becomes subject to the laws governing surface or groundwater [Eden et al., 1992].

The liability aspect of artificial groundwater recharge, especially in the use of reclaimed wastewater, has been only vaguely addressed by Arizona law [Lieuwen, 1990]. The unknown health effects associated with the consumption of water containing reclaimed water is inherently litigious. The potential for litigation may also arise from property damage to crops or equipment caused by the use of reclaimed wastewater. Aquifer contamination or damage is another potential source of litigation.

Summary

The increasing demand for potable water requires society to pursue methods that insure adequate supplies. Artificial recharge is a means to renew, improve, protect, and store fresh water supplies. The use of "natural" water for recharge is generally not questioned and it would be imprudent not to consider it for recharge projects. Increasing populations and the limited availability of potable water has increased pressure to use treated wastewater as a source water to replenish groundwater supplies. The use of effluent should be tempered by concerns over the potential health effects. The future development of artificial recharge projects depend upon the resolution of public health concerns and the legal issues surrounding artificial recharge.

References

Asano, T., Artificial Recharge of Groundwater, Butterworth Publishers, Boston, Mass., 1985.

Billings, C.H., "Water Reclamation by Groundwater Recharge", Public Works, Vol. 121, pg 68-69, Dec. 1990.

Bouwer, H., "Groundwater Recharge with Wastewater: Pre and Post-Treatment", in Proceedings of the Fifth Bieennial Symposium on Artificial Recharge of Groundwater Recharge, May 29-31, 1991, Sponsored by Salt River Project, U.S. Water Conservation Laboratory ARS-USDA, and Water Resources Research Center of the University of Arizona.

Bouwer, H., David, R., Pyne, G., and Goodrich, J.A., "Recharging Groundwater", Civil Engineering (ASCE), pg 63-66, June 1990.

Bouwer, H., Groundwater Hydrology, McGraw-Hill Book Co., U.S.A., 1978.

Chapman, T.G. and Sharma, M.L., "Summary of Symposium on Groundwater Recharge", in Sarma, M.L., ed. Proceedings of the Symposium on Groundwater Recharge Mandurah/6-9 July 1987, A.A. Balkema Publishers, USA, 1989.

Eden, S. and Wallace, M.G., Arizona Water: Information and Issues, Water Resources Research Center, University of Arizona, Issue Paper Number 11, August 1992.

Fetter, C.W., Applied Hydrogeology, Merrill Publishing Co., Columbus, Ohio, 1988.

Freeze, R.A. and Cherry, J.A., Groundwater, Prentice-Hall Inc., Engelwood Cliffs, NJ, 1979.

Huisman L., and Olstoorn T.N., Artificial Groundwater Recharge, Pitman Books Limited, Marshfield Mass., 1983.

Lerner, D.N., Issar, A.S., and Simmers, I., Groundwater Recharge: A Guide to Understanding and Estimating Natural Recharge, Verlag Heinz Heise GmbH & Co., West Germany, Vol. 8, 1990.

References

Lieuwen, A., Effluent Use in the Phoenix and Tucson Metropolitan Areas, Water Resources Research Center, University of Arizona, Issue Paper Number 7, June 1990.

Marceau, W., "Considerations for Aquifer Protection Permits", in Proceedings of the Fifth Bieennial Symposium on Artificial Recharge of Groundwater Recharge, May 29-31, 1991, Sponsored by Salt River Project, U.S. Water Conservation Laboratory ARS-USDA, and Water Resources Research Center of the University of Arizona.

Marron, H., Phinney, D., and Musgrove J., "Artificial Recharge and Well Rehabilitition for Management of Groundwater Nitrates in Peoria, Arizona", in Proceedings of the Fourth Symposium on Artificial Recharge of Groundwater in Arizona May 23 & 24, 1989, Sponsored by Salt River Project, U.S. Water Conservation Laboratory ARS-USDA, and Water Resources Research Center of the University of Arizona

Nellor, M. A. H., Health Effects of Water Reuse By Groundwater Recharge, Technical Report: The Center For Research in Water Resources, Bureau of Engineering Research, University of Texas at Austin, 1980.

Rigby, M. and Mills, B., "Groundwater Recharge with Reclaimed Water: Resolving Regulatory Issues", in Proceedings of the Fifth Bieennial Symposium on Artificial Recharge of Groundwater Recharge, May 29-31, 1991, Sponsored by Salt River Project, U.S. Water Conservation Laboratory ARS-USDA, and Water Resources Research Center of the University of Arizona.

Ozonation as an Alternative to Mandatory Filtration

Dimitrios Katehis[1]

ABSTRACT

The increased levels of required disinfection along with more stringent disinfection by product control as outlined in the Surface Water Treatment Rule and the Disinfectant-Disinfection By Product Rule will require many utilities to consider ozonation as an alternative disinfection process. This is of particular importance to utilities currently not filtering their water, since large costs will be incurred in the building of filtration facilities. It is proposed that using ozone as the primary disinfectant for surface waters of good quality, may simultaneously increase disinfection effectiveness and control disinfection by-products to within the prescribed limits,so that filtration plant costs can be avoided. Since ozone by-product formation is not well understood, pilot plant studies are necessary.

Ozone has been used as a disinfectant for almost a century in Europe and has been recently gaining ground in the United States as well. The 1986 amendments to the Safe Drinking Water Act and the ensuing Surface Water Treatment Rule (54FR 27544) have forced utilities to consider alternative disinfection methods and filtration as possible methods for compliance with the SWTR. Furthermore, the USEPA is expected to regulate more stringently the levels of Disinfection By-Products (DBPs), to include trihalomethanes (THMs), haloacetonitriles (HANs), haloacetic acids (HAAs), chloropicrin, and cyanogen chloride. The SWTR requires a 99.9 percent removal or inactivation of *Giardia* cysts and 99.99 removal or inactivation of enteric viruses Filtration is usually considered necessary to achieve these goals. Depending on the actual quality of the water supply,

[1]Student, Department of Civil Engineering, The City College of New York, 138th Street and Convent Avenue, New York City, NY 10031

different methods have been suggested for compliance with the above stated regulations(3),(13),(15).

A typical water treatment process train consists of prechlorination, coagulant addition, rapid mixing, flocculation, filtration and postchlorination. A large number of variations to this typical treatment train exist, but the main processes are filtration and chlorination. The SWTR will force an estimated 8,200 water utilities using surface water, to modify their treatment to conform to these regulations(16). It is suggested that changes in disinfectant residual concentrations (C), use of alternative disinfectants, or costly plant modifications such as new basins, or baffling of the existing ones will be required to increase the hydraulic detention time(T) so that the CT for the required deactivation can be achieved. Most of the 3,000 utilities that currently do not filter their water will have to do so. The avoidance of filtration must be based among other criteria on low total coliform densities, low turbidity and a protected watershed(4).

Filtration is incorporated into the treatment process to remove particulate matter, turbidity, protozoan cysts and organic carbon (biotreatment)(9) if designed to be biologically active. Taste and odor problems may also be partially alleviated by filtration. Particulate matter and turbidity hamper disinfection by shielding the microorganisms from the disinfectant, but this may not be true for disinfection utilizing ozone(11),(15). When ozone is the primary disinfectant, biotreatment may be necessary to remove aldehydes and ketones. This is especially true for water with high Total Organic Carbon (TOC) levels.
Filtration will not enhance the disinfection process when ozonation is used, unless the water supply has fecal material that will not allow penetration by the disinfectant.

In light of this regulatory action, it is believed that ozone disinfection will become more widespread in the U.S.(8). The strong oxidizing properties of ozone allow it simultaneously solve a wide range of problems associated with a potable water supply. The multiple application aspect of ozone must also be taken into consideration. Ozone will accomplis disinfection, color, taste and odor control, oxidation of organics (pesticides, detergents, phenols), algae oxidation, reduction thru micro-flocculation of turbidity, decomplexing of organically bound manganese, precipitation of iron and manganese, and increase the dissolved oxygen in one contact period.

Ozone is one of the most potent biocides available, having twice the chemical oxidation potential of hypochlorite ion, and in more than three quarters of a century of usage has had no reports of serious negative consequences from its use. It is effective at relatively low dosages and short contact times and it decomposes to O_2 leaving no significant by-products other than aldehydes and in some instances ketones.

There are two general categories of reactions of ozone with the contaminants usually encountered in water treatment. The decomposition of ozone creates free radicals, the most important one being the hydroxyl radical. The reactions involving hydroxyl radicals and constituents of raw water are very fast, second order reactions with rate constants varying by no more than three orders of magnitude. The direct reaction of ozone with contaminated compounds is a highly specific reaction, with second order rates varying by 12 orders of magnitude. Nitrites, hydrogen sulfide, phenols, activated aromatic compounds olefin and simple amines react rapidly with ozone and the reactions are limited only by the mass of ozone transferred into the water. Acetic acid, urea, ammonia, oxalic acid and similar compounds are not easily oxidized even in the presence of large doses of ozone(6). Thus for the oxidation of biorefractory material the rate of ozone mass transfer is secondary to the maintenance of a lower concentration over a longer contact time. Depending on the contaminants present in the specific water source, a large ozone concentration may be employed for a short period of time if disinfection is the primary goal, whereas a low concentration can be used over a longer contact time if taste, odor and color are also to be effectively removed. By employing the latter method appreciable turbidity removal through micro flocculation can be achieved as well.

It has been shown that ozone will produce a 99 percent kill for poliovirus 1 in less than 10 seconds, in contrast with chlorine that requires 100 seconds for similar results at the same concentration(3). *Giardia* cysts, considered among the most resistant to disinfection, were inactivated to within SWTR limits with doses of 1.0 mg/L ozone and a contact time of 6 minutes(15). Of course these values are site specific, (in this case, water originating in the Sacramento-San Joaquin Delta supplied to the Metropolitan Water District of Southern California) but it clearly demonstrates that protozoan cysts are inactivated to required levels within reasonable contact times.

One of the major advantages of filtration is considered to be the removal of turbidity and particulate matter that if allowed to remain will reduce the efficiency of the disinfection process. This is true for chlorine in the range of contact times usually employed in water treatment plants utilizing chlorine ($5<CT<75$)(7) but results of studies of turbidity effects on disinfection on ozone differ. Experimental results have shown that increased levels of turbidity (10 NTU and 50 NTU) had no significant impact on inactivation. It must be noted though, that the turbidity was added artificially (silty sand with a high content of silt fines) and that the results might be different for microorganisms encapsulated or adsorbed in human or animal fecal material, in the range of particle sizes capable of passing through the filters(11). Many of the utilities currently not filtering their water will have to consider filtration in order to meet the disinfection criteria for Giardia cysts and enteric viruses.

To achieve acceptable results in taste, odor and color removal with filtration, it is usually preceded by coagulation, flocculation, and settling. This increases capitol as well as operational costs dramatically. Ozone has been used for taste, odor and color control for more than half a century in Europe. Jiejers has documented the large reduction of color of surface water by ozonation(6). In North America more than 40 plants are using ozone for disinfection, taste and odor control, and color removal (14) with satisfactory results.

A mixture of ozone and hydrogen peroxide (PEROXONE) at a weight ratio of hydrogen peroxide to ozone of 0.2 has produced even better results, with no adverse affects specific to the introduction of hydrogen peroxide. The ozone dosage required to remove a target percentage of 80 to 90 percent of the taste and odor compounds studied (2-methylisoborneol (MIB), and trans-1,10-dimethyl-trans-9-decalol (geosmin))dropped by 36 to 70 percent when PEROXONE was used instead of only ozone(15). This could mean a substantial savings in operating costs for a utility, especially since ozone's main drawback are the increased operating costs.Increases in turbidity, to 10 and 50 NTU, had no impact on the percent removal of geosmin and MIB. The percent removals reported though were the products of a conventional treatment process in which ozone was substituted for chlorine. To evaluate the removals, for a process consisting of only ozonation, pilot plant studies would be required. It is assumed though, that ozonation would be a viable alternative to a conventional treatment process only for utilities that currently only disinfect, and will be required to adopt a conventional treatment process to meet the requirements of the new SWTR.

To conform to the new Disinfectant-Disinfection By Product Rule some utilities have attempted to minimize chlorine contact time with DBP precursors and have thus succeeded in lowering the levels of THMs formed. This method though diminishes the contact time available for disinfection and the inactivation percentages of *Giardia* cysts and of the resistant viruses. Many utilities that were already utilizing a conventional treatment process of their raw water, have attempted to remove DBP-precursors during flocculation. Different methods have been utilized to enhance coagulation, such as the decrease of the pH of the water, which results in higher natural organic matter removal and reduced THM formation. This strategy though results in higher levels of non-THM DBPs, which can be maintained within acceptable limits at elevated pH(12). Other methods used for

DBP control are ammonia addition(chloramination) instead of prechlorination of water and changing the point of the first application of chlorine(16). For the estimated 3,000 utilities that do not currently process their raw water with the conventional manner(16) (prechlorination, coagulant addition, rapid mixing, flocculation, filtration and postchlorination), but simply disinfect and distribute, reverting to one of these methods of control would be extremely costly, since new facilities would be required.

Ozone oxidizes the precursors of a range of DBPs. Substituting ozone in place of chlorine as a preoxidant has consistently resulted in lowering the levels of THM compounds to within acceptable limits(2),(5). The critical variables, with respect to THM formation seem to be the ratios of transferred ozone to dissolved organic carbon (O_3/DOC) and transferred ozone to total organic carbon (O_3/TOC) of the raw water. Subsequent chlorination results in the elevated formation of certain THMs, such as chloropicrin(1) and $CHBr_2Cl$(9). A far more effective THM control strategy can be applied by using ozone as the primary disinfectant and chloramines as the residual disinfectant. A decrease of more than 80 percent in the formation of total organic halide (TOX), HAA, HAN and TTHMs was realized when comparing chlorine to ozone-chloramine treatment. The actual results are site specific and are dependent on the actual constituents of the TOC present in the water supply.

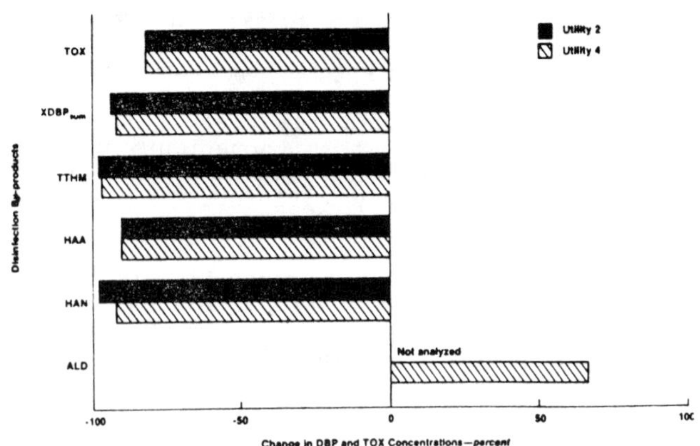

Figure 1. Change in DBP and TOX concentrations after switch from chlorine only to ozone and chloramines treatment(5).

One of the problems encountered when ozone is used as the primary disinfectant or preoxidant is that if bromide is present in the raw water oxidization will result in the formation of free bromine. Ozone converts bromide to hypobromous acid and hypobromite ion. This increases the levels of brominated DBPs in the effluent water. The only effective way of reduction of brominated DBP's would be a system using coagulation/filtration to remove precursors. Caution must be exercised when evaluating ozone as the sole treatment process if bromide is present in even small concentrations in the raw water.

The main by-products of ozonation are aldehydes, ketones and carboxylic acids. The cleavage of the organic carbon bonds lead to the simpler of these organic compounds which are readily biodegradable. Thus formaldehyde, acetaldehyde, glyoxal and methyl glyoxal are the predominant non-halogenated DBPs. The levels of these non-halogenated DBPs increase at least 50 percent after ozonation. The actual values will be a function of the constituents of the raw water TOC. The formation of easily biodegradable organics results in the increase of bacterial nutrients that may lead to increased bacterial growth in the distribution system. One solution to this problem is periodic chlorination to kill any bacteria within the distribution system. No reliable correlation exists between the organic species present in the raw water and the expected organics in the ozonated water. Pilot plant studies are the only reliable method of determining the levels of these organics formed by ozonation. The toxicity of ozonation by-products has not been studied in great detail, and the health risks associated with these compounds are still not well known. Future regulation by the USEPA of the levels of some of these ozonation by products should be anticipated and if the levels are exceedingly high, provisons for some form of biotreatment in the future must be made. The ease of biodegradability of non-halogenated ozone DBPs will allow high removals, while at the same time it will effectively remove the substrate available for bacterial regrowth in the distribution system.

When the surface waters are of good quality with low turbidity levels (less than 1 NTU), such as when water is stored for considerable amounts of time in reservoirs where settling can occur, it is possible that by retrofitting chlorination plants with ozone contacting equipment, disinfection and by-product control can both be achieved. The existence of bromide in the raw water will pose severe problems since brominated DBPs will be created by ozonation. The only effective way of determining if ozonation is a viable solution, are pilot plant studies for the specific surface water since the theory of ozonation by-product formation is not well understood.

In order to avoid mandatory filtration, the switch to ozonation must be a part of an overall plan to protect the watershed area to ensure continuing good quality surface water. It is in the utility's interest to study the feasibility of ozone as a disinfectant, especially if the surface water is of good quality with low turbidity levels.

The savings from not having to upgrade, or in the case of utilities not currently filtering their water supply, building facilities for a conventional treatment train as outlined above, can be substantial.

Bibliography

1. Bader, H. *The Formation of Trichloronitromethane and Chloroform in a Combined Ozonation-Chlorination Treatment of Drinking Water.* Water Resources, 22:3:313 (1988).

2. Glaze, W.H.; Koga M.; Cancilla, D.; Wang K. *Evaluation of Ozonation By-Products from Two California Surface Waters.* Jour. AWWA, 81:8:35(August 1989).

3. Glaze, W.A.; Peyton,G.R.; Huang F.Y.; Burleson,J.L.; & Jones,P.C. *Oxidation of Water Supply Refractory Species by Ozone with Ultraviolet Radiation.* Office of Research and Developement, USEPA, Cinicinnati (1980).

4. *Guidance Manual for Compliance With Filtration and Disinfection Requirements for Public Water Systems Using Surface Water Sources.* Science & Technology Bureau, Office of Water, Cincinnati, Ohio (1989).

5. Jacangelo, J. G.; Patania, N. L.; Reagan, K. M.; Aieta, M. E.; Krasner, Stuart W.; and McGuire, Michael J. *Ozonation: Assessing its Role in the Formation and control of Disinfection By-Products.* Jour. AWWA, 81:8:35(August 1989).

6. Langlais, B.; Reckhow, D.; & Brink, D. *Ozone in Water Treatment: Applications and Engineering.* Lewis Publications, Michigan.

7. Lippy, E.C. *Adequate Disinfection: An Alternative to Mandatory Filtration.* Jour. AWWA. 77:4 (September 1985)

8. McGuire, M. J. *Preparing for the Disinfection By-Products Rule: A Water Industry Status Report.* Jour. AWWA, 81:8:35(August 1989).

9. Miltner,R.J.; Shukairy,H.M.; & Summers,R.S. *Disinfection By-Product Formation and Control by Ozonation and Biotreatment.* Jour. AWWA. 84:53.(November 1992).

10. Rice, R.G.; & Browning, M.E. *Ozone for Industrial Water and Wastewater Treatment.* Office of Research and Developement, Oklahoma (1980).

11. Sproul, O.J.; Buck,C.E.; Emerson,M.A.; Boyce,D.; Walsh,D.; & Howser,D. *Effect of Particulates on Ozone Disinfection of*

Bacteria and Viruses in Water. Office of Research and Developement, Cincinnati (1979).

12. Stevens, A. A.; Moore, L. A.; Miltner, R. J. *Formation and Control of Non-Trihalomethane Disinfection By-Products.* Jour. AWWA, 81:8:35(August 1989).

13. Symons, J.M.; *Ozone, Chlorine Dioxide and Chloramines as Alternatives to Chlorine for Disinfection of Drinking Water.* Office of Research and Developement, USEPA, Cincinnati, (1977).

14. Tate,C.H. *Survey of Ozone Installations in North America.* Jour. AWWA, 83:5:40(May 1991).

15. The Metropolitan Water District of Southern California. *Pilot Scale Evaluation of Ozone and PEROXONE.* AWWA. U.S.A. (1991)

16. 12. Water Quality Division Disinfection Committee. *Survey of Water Utility Disinfection Practices.* Jour. AWWA, (September 1992).

Recovery and Recycle of Copper from Wastewater using Iron Coated GAC

Kotha P. Reddy[1]

Abstract

The purpose of this study was to prepare composite adsorbents that could be packed into a column and have adsorption characteristics similar to Fe oxide. This adsorbent was targeted to remove low levels of copper from water to the adsorbent surface, subsequently desorb material at a different pH, and reuse the adsorbent. GAC (0.5 mm diameter) was coated with various amounts of Fe oxide, from 37 mg Fe/g GAC (37FeGAC) to 72 mg Fe/g GAC (72FeGAC).

Adsorption characteristics were studied for copper removal. As the amount of iron coating on the adsorbent was increased the capacity and rate for copper adsorption increased. Experiments were carried out for short-term and long-term adsorption time periods of 2 and 24 h, respectively. Desorption efficiency (amount of adsorbed material recovered) of copper from GAC was 100% after 2 h adsorption time and dropped to 60% after 24 h adsorption time. Time did not have much effect on Fe oxide desorption efficiency but the FeGAC adsorbents showed the same trend as the GAC with even lower recovery after 24 h. The composite adsorbent can be packed into columns and used for copper removal and reuse.

Introduction

Precipitation has been the traditional treatment process for treating metal contaminated wastewater but it has many disadvantages. Problems include the fact that the minimum solubility of different metals occurs at different pH values and pretreatment may be required to remove complexing agents that may inhibit precipitation (Bhattacharyya and Cheng, 1987). The options for sludge management are becoming limited, the associated disposal costs are increasing, and the long-term liabilities associated with sludge disposal represent an ill-defined, growing concern. An alternative available for these industries is to evaluate recovery and recycle technologies.

---

[1]Graduate Student, Illinois Institute of Technology, 5W 80 Guion Place, New Rochelle, NY 10801.

Adsorption is an alternative process available for removal and recovery of metals from metal contaminated waste streams. Jenne (1968) reported that adsorption processes play an important role in retention of metals by soils. Adsorption processes have a potential role in industrial treatment as discussed by Benjamin et al. (1982). Advantages of adsorption processes over precipitation-based treatment techniques are many. For example, the presence of complexing agents does not necessarily interfere with (Benjamin and Leckie, 1982) and, in some cases, may actually enhance metal adsorption (Davis and Leckie, 1978). The adsorbate-adsorbent ratio decides the metal removal capacity and by increasing the adsorbent concentration, removal to trace levels is possible. An adsorption process is reversible as it is dependent on the pH of the solution; it is possible to recover metals, which could they be recycled back in to an industrial process.

Activated carbon has been used as an adsorbent for removing complexed metal pollutants from wastewater as reported by Sigworth and Smith (1972), Corapcioglu and Huang (1987). Activated carbon adsorption capacity for metals is too small for practical purposes. Metal oxides of iron, manganese, and aluminum are adsorbents that have an edge over activated carbon because of their large adsorption capacity for metals, fast kinetics, and reversibility of the adsorption reaction. Their use has been discussed by many researchers (Davis and Leckie, 1978; Schultz et al., 1987; Anderson and Benjamin, 1990; Edwards and Benjamin, 1989). A characteristic property of the Fe oxide mineral family is its small particle size. These colloidal particles are difficult to remove from aqueous solutions, adding to the cost of treatment and complicating recovery and recycle of metals. Recent research work has recommended the use of iron oxide coated onto other media or combined with a polymer binder to overcome the separation problems associated with colloidal solids (Huang and Vane, 1989; Edwards and Benjamin, 1989; Theis et al., 1992).

The objective of this exercise is to prepare the Fe-oxide coated adsorbents and characterize them. Tests included measuring abrasion resistance and the coating dissolution rate at high, neutral, and low pH. Batch adsorption-desorption and capacity tests were conducted using copper as model adsorbate.

Materials and Methods

This section begins with a description of the methodology and the kind of analyses carried out. It includes the synthesis procedures used to prepare the iron oxide solid and iron oxide-coated GAC, and the GAC cleaning procedure. The prepared FeGAC adsorbent was characterized by durability, batch adsorption-desorption, and capacity tests. Reagent grade chemicals were used throughout this study unless specified, and all the labware was detergent and acid washed and rinsed with deionized water. Standard techniques were used for atomic absorption spectrophotometry (AAS) analysis of iron, and copper with reference to Clesceri et al. (1989). The water used was always deionized water. Solid samples were stored in covered glass bottles.

Adsorbent preparation: Granular activated carbon (GAC) used in this experiment was obtained from Calgon Carbon Corporation(type TOG). The GAC was sieved between 425-600 μm prior to the test. This size fraction was baked in an oven at 200^oC for 2 d, washed several times with deionized water by tumbling it in a roller for a day. This GAC slurry was filtered, completely dried in an oven at 110^oC, cooled in a desiccator, and stored until further use.

The iron oxide-coated GAC (FeGAC) was prepared by mixing GAC with a known amount of $Fe(NO_3)_3$ solid and adding enough deionized water to cover the solid. The contents were mixed by shaking the container, and allowed to sit for a few hours to allow the air trapped in the pores to escape, before the mixture was placed in the oven. The mixture was dried completely in an oven at 90^oC. The resultant composite adsorbent was dried, cooled to room temperature, and washed several times with deionized water to remove detachable and discrete Fe oxide. FeGAC was dried in the oven at 105^oC and stored at room temperature for further usage.

Characterization of the adsorbent: The amount of iron coated on GAC was estimated by extracting the iron from the composite FeGAC by boiling in concentrated HNO_3 for 4 h. The concentrates were analyzed by atomic absorption spectrophotometry. Several experiments were conducted using adsorbents that were representative of the pure end-point adsorbents for comparison purposes. These were uncoated GAC and an Fe oxide prepared following the same procedures outlined above except that no GAC was present.

Each composite adsorbent (FeGAC) was subjected to a durability test. This test was performed by equilibrating each adsorbent at pH values of either 3, 7, or 11 in a mixed reactor. The amount of adsorbent used was 1 g/L, and ionic strength was adjusted to 0.01 M with $NaNO_3$. The resultant solution was filtered through 0.45 μm filter paper. The adsorbent was removed after drying, and the filter paper and filtrate were acidified and digested, concentrated to a small volume, and analyzed for total iron using AAS.

Batch adsorption-desorption and capacity tests: In the batch adsorption experiments adsorbate copper concentration used was 2 mg/L and the adsorbent concentration was 0.5 g/L. Tests were conducted in 0.01 M $NaNO_3$ background electrolyte solutions equilibrated to pH 6. These parameters were used so that the solution would not become saturated with respect to copper hydroxide solid. The contents of each container were mixed and the solution was maintained at pH 6. A sample aliquot was taken at 5 min intervals. Desorption tests followed after a 2 h or 1 d adsorption time. The desorption test was carried out by sequentially decreasing the pH of the solution from 6 to 3; each one unit pH change required 20 min.

Samples taken from the above tests were filtered (0.45 μm) and the filtrate was acidified and analyzed for total Cu by AAS.

In the capacity experiments, separate solutions of the Cu adsorbate and the composite adsorbent (10 mg/L) were prepared in 0.01 M $NaNO_3$ solutions and adjusted to pH 6. Once the pH of the solution had stabilized, the solutions were combined so that the total Cu concentration was 0.4, 0.8, 1.2, 1.6, or 2.0 mg Cu/L. These suspensions were allowed to equilibrate on a rotating rack for 2 h or 4 d. After the equilibration period, solid and solution were separated by gravity settling or filtration and the samples analyzed.

Results and Discussions

This section summarizes the results of the experiments. Whenever possible, adsorbents were evaluated in situ, as was the case for adsorption tests. The evaporative coating process became less effective as the applied $Fe(NO_3)_3$ dosage was increased (Figure 1). For applied Fe dosages of 40, 80, and 120 mg Fe/g adsorbent, the observed coatings were 37, 52, and 72 mg Fe/g adsorbent, respectively. The abrasion test showed that the most of the FeGAC had the greatest Fe loss under acid conditions (Figure 2). The exception was the unusual large loss for 37FeGAC at higher pH. Some kind of hydroxide mineral phase which was not stable pealed off the 37FeGAC; it is not very clear with the data available what could be the reason for such a large loss.

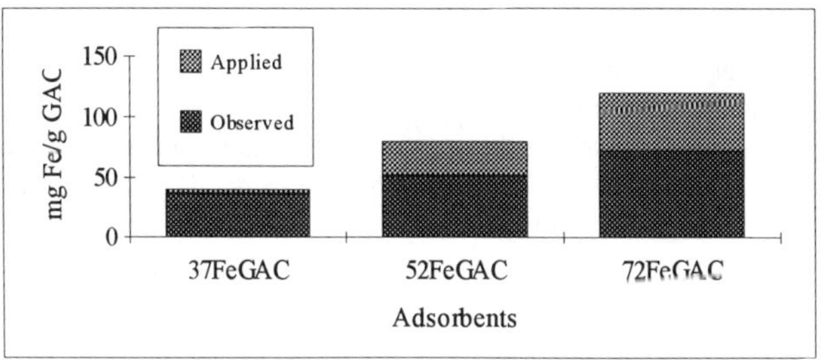

Figure 1. The Coating Efficiency of Fe Oxide on GAC. GAC was combined with $Fe(NO_3)_3$ and evaporated at 90^0C for 24 h.

Cu adsorption: As the Fe coating was increased on the FeGAC adsorbents the amount of copper that could be adsorbed increased (Figure 3). The slopes of the lines connecting these data points suggest that the rate of adsorption was fastest for the Fe oxide, which appears to reach an equilibrium value within about two hours. For FeGAC, the adsorption reaction rate increased as the amount of iron coated on the carbon surface increased. With more iron coated onto the carbon surface, the FeGAC acted more like Fe oxide itself and had a faster surface reaction rate.

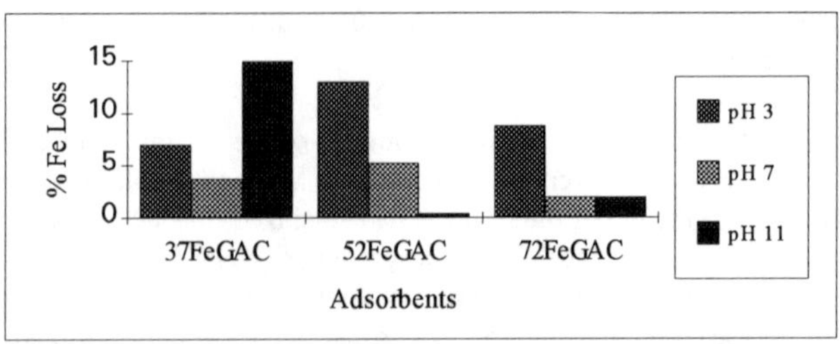

Figure 2. Abrasion Test of FeGAC for 24 h at pH 3, 7, or 11.

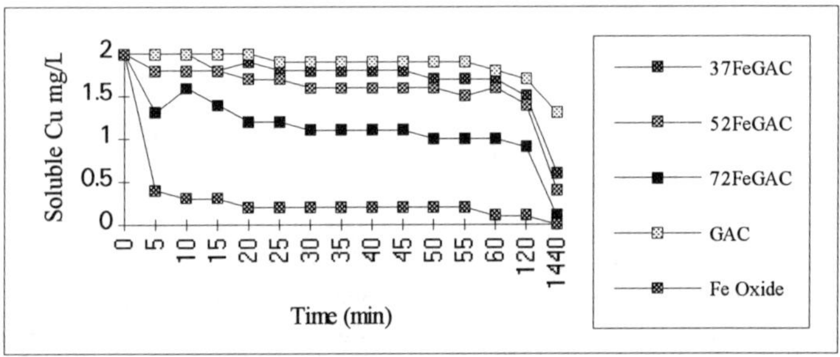

Figure 3. Rate of Cu Adsorption at pH 6. Adsorbent Concentration = 0.5 g/L.

The Cu adsorption capacities of different adsorbents are shown in Figures 4 and 5 for two different equilibration times. The adsorption capacity of the FeGACs clearly increased as the amount of iron on the GAC increased. Capacity expressed as mg Cu / g adsorbent also shows that the approach to equilibrium is much slower for the FeGAC than it is for either GAC or the discrete Fe oxide. It is also interesting to note that after 96 h the capacity of the FeGAC adsorbents approaches the capacity of the discrete Fe oxide.

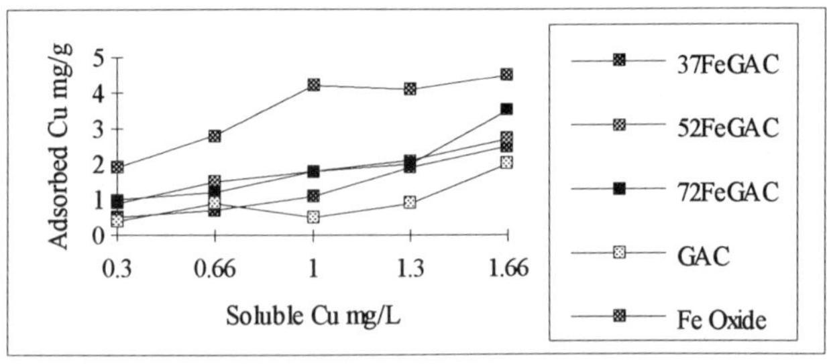

Figure 4. Cu Adsorption Isotherm at pH 6 and 2 h Equilibrium.

Furthermore, because the GAC removal capacity is relatively small, most of the removal capacity appears to be due to the presence of the Fe oxide surface coating. If the Cu adsorption capacity is normalized to the amount of Fe present (Figure 6), all three of the FeGAC solids appear to have about the same adsorption capacity for Cu. Furthermore, relative to the discrete Fe oxide the capacity of the FeGAC adsorbents is at least three times greater. This observation suggests that interactions between Fe oxide and the GAC create additional binding sites at the surface of the composite adsorbent. This also implies that GAC can supply a large surface area for Fe oxide spreading on to it, creating a larger surface area/unit mass ratio. Alternatively the GAC could also change the Fe oxide coating to a different mineral type that has a stronger affinity and larger capacity for Cu adsorption.

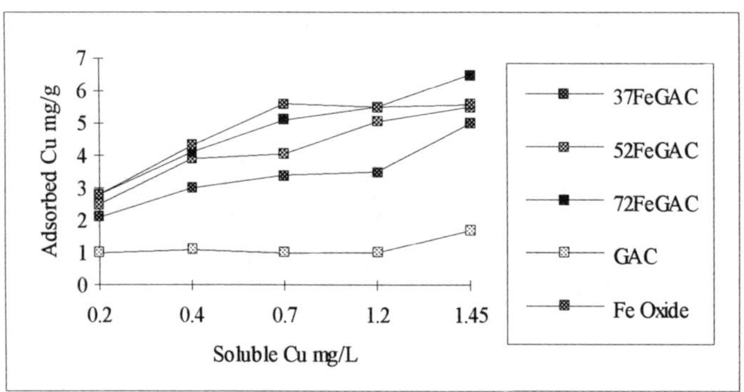

Figure 5. Cu Adsorption Isotherm at pH 6 and 96 h Equilibrium.

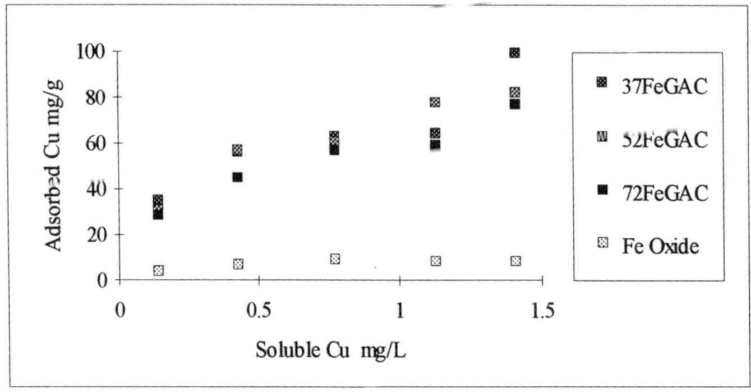

Figure 6. Cu Adsorption Isotherm at pH 6 and 96 h Equilibrium Period Normalized to Fe Content.

The elapsed adsorption time affected the desorption efficiency after the adsorption period (Figure 7). Iron oxide had about the same Cu recovery efficiency after 24 h adsorption time over 2 h adsorption time, but GAC and FeGACs showed lower recovery efficiency after 24 h adsorption. This result implies that mass transfer is an important parameter that affects the Cu desorption efficiency from GAC and FeGACs. Once the copper ion was adsorbed onto the surface site, a slow diffusion process apparently moved the copper ion deeper into the pores of GAC, resulting in lower recovery rates.

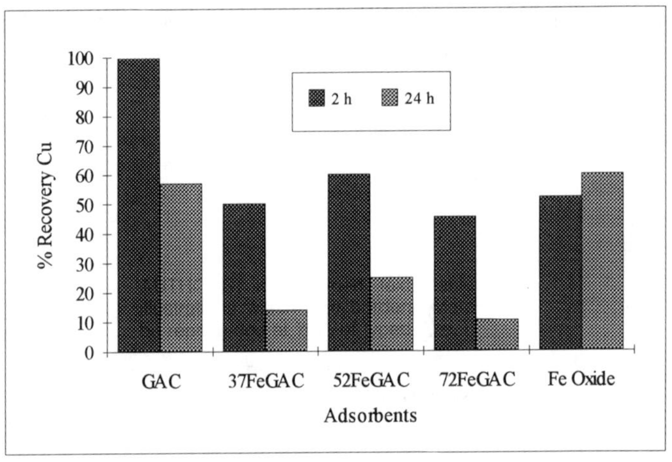

Figure 7. The Desorption Efficiency of Cu from pH 6 to 3.

Conclusions

An evaporative process was used to form an Fe oxide coating on a GAC surface. At relative low applied loading(40 mg Fe/g GAC), the coating efficiency was over 90%, but it dropped to 60% at the higher applied loading (120 mg Fe/g GAC). This Fe oxide coating significantly increased the capacity of GAC for Cu adsorption compared with GAC. When adsorption was compared with discrete Fe oxide and normalized to the amount of Fe present, the FeGAC had a higher capacity. This relatively high adsorption capacity, however, came at the expense of relatively slower adsorption kinetics.

The other useful physical property allows the adsorbents in such a way that it can be packed into a column to treat industrial wastewater on a large scale and can also be regenerated to recover metals for recycle. Once the composite adsorbent adsorption mechanisms are fully understood, it has very good potential in industrial application as it costs less to produce and easy to handle.

References

Anderson, P. R., and Benjamin, M. M., "Surface and Bulk Characteristics of Binary Oxide Suspensions." Environ. Sci. Technol., 24, 692 (1990).

Battacharyya, D., and Cheng, C. Y. R., "Activated Carbon Adsorption of Heavy Metal Chelates From single and Multicomponent Systems." Environ. Prog., 6, 110 (1987).

Benjamin, Mark M., and Leckie, J.(1982) Environ. Sci. Technol. 16, 162.
Corapcioglu, M. O., and Huang, C. P., "The Adsorption of Heavy Metals onto Hydrous Activated Carbon." Water Res., 21, 1031 (1987).

Clesceri. L. S., Greenberg. A. E., Trussell. R. R., "Standards Methods for the examination of water and wastewater 17th edtion" American Public Health Association (1989).

Davis, J. A., and Leckie, J. O., "Effect of Adsorbed Complexing Ligands on Trace Metal Uptake by Hydrous Oxides." Enviro. Sci. Technol., 25, 1309 (1978).
Edwards. M., and Benjamin, M. M., "Regeneration and Reuse of Iron Hydroxide Adsorbents in Treatment of Metal-Bearing Wastes." J. Water Pollut. Control Fed., 61, 481 (1989).

Edwards. M., and Benjamin, M. M., "Adsorptive Filtration Using Coated Sand: a New Approach for Treatment of Metal-Bearing Wastes." J. Water Pollut. Control Fed., 61, 1523 (1989).

Huang, C. P., and Vane, L. M., "Enhancing As^{5+} Removal by a Fe^{2+}Treated Activated Carbon." J. Water Pollut. Control Fed., 61, 1596 (1989).

Jene, E.A.(1968). "Controls on Mn, Fe, Co, Ni, Cu, and Zn concentrations in soils and water: the significant role of hydrous Mn and Fe oxides" in Trace Inorganics in Water, Advances in Chemistry Series 73, American Chemical Society, Washington,

Schultz, M. F.., Benjamin, M. M., and Ferguson, J. F., "Adsorption and Desorption of Metals on Ferrihydrite: Reversibility of the Reaction and Sorption properties of the Regenerated Solid." Environ. Sci. Technol., 21, 863 (1987).

Sigworth, E. A., and Smith, S. B., "Adsorption of Inorganic Compounds by Activated Carbon." J. Am. Water Works Assoc. (1972).

Theis, T. L., Iyer R., and Ellis, S. K., "Evaluating a New Granular Iron Oxide for Removing Lead from Drinking Water." J.Am. Water Works Assoc., 84, 101(1992).

Thermally Enhanced Bioremediation of a Gasoline-Contaminated Aquifer Using Toluene Oxidizing Bacteria

Rula Deeb[1] and Lisa Alvarez-Cohen [2]

ABSTRACT

The combined application of steam injection and vacuum extraction has proved to be very effective for the in situ remediation of a gasoline contaminated aquifer. It is expected that the steam treated zone with its near-sterile nature, increased temperature, and decreased level of contaminant concentration will provide a superior environment for enhanced bioremediation, and will favor the survival of an introduced microbial culture for the destruction of residual gasoline hydrocarbons and especially BTEX compounds (Benzene, Toluene, Ethyl benzene, and Xylene).

A mixed microbial culture seeded from the pre-steamed aquifer material was enriched in a laboratory chemostat on toluene, a major gasoline aromatic. Studies were conducted to determine the optimal conditions for microbial growth and activity. Growth rate studies conducted at different temperatures revealed that cell growth was optimal at 35°C, a temperature at which the aquifer can be maintained using the existing steam injection wells. The enriched culture was shown to degrade all BTEX compounds successfully both individually and in mixtures. Substrate toxicity was observed for some of the gasoline aromatics but at concentration levels well above those found in groundwater. When cells were exposed to mixtures of BTEX compounds, the biodegradation of xylene, the most recalcitrant aromatic among BTEX compounds, was stimulated. When cells were exposed to gasoline, BTEX degradation proceeded with no apparent inhibition by gasoline aliphatics; little aliphatic degradation took place, however, suggesting the absence of monooxygenase enzymes in the mixed culture. In mixtures of both toluene and propane enriched cultures, only dioxygenase activity was observed.

PURPOSE

There is an imperative need for rapid and effective in situ cleanup technologies for implementation at gasoline-contaminated aquifer sites. The

---

[1]Masters Candidate, National Student Essay Competition, First Prize

[2]Graduate Advisor, Assistant Professor, Environmental Engineering, Department of Civil Engineering, University of California, Berkeley, CA 94720

effectiveness of steam injection and vacuum extraction has been demonstrated at both bench and pilot field scales since it increases the mass transfer rate between the steam and semi-volatile contaminants found in gasoline. However, no remediation method will provide complete removal of contaminants in a short time frame, and even steam treatment will leave residues. This leads to the potential of using bioremediation as a post-steam polishing step for the complete destruction of residual contaminants. The distinctive thermodynamic environment remaining at the end of steam treatment, as well as the extreme reduction of concentrations of toxic gasoline hydrocarbons, present an opportunity for enhanced bioremediation.

INTRODUCTION

The release of petroleum hydrocarbons in the environment is a widespread occurrence. One particular concern is the contamination of drinking water sources by the toxic, water soluble, and mobile components of gasoline, especially gasoline aromatics or BTEX compounds: Benzene, Toluene, Ethyl benzene and Xylene (Alvarez et al, 1991).

Several approaches are used for the clean-up of contaminated soil and groundwater aquifers. These include excavation followed by incineration or chemical treatment, in situ vapor-phase stripping of volatiles, the extraction of contaminated water for treatment with either chemical agents or activated carbon. All these processes may be effective if applied properly. However, certain approaches may not result in total decontamination. As a result, there is a growing interest in utilizing microbial degradation or in-situ treatment of contaminated soil and groundwater, either alone, or in collaboration with one of the physiochemical techniques. The aim of microbially-based remediation is to provide optimum environmental conditions so that microbial activity can proceed at the maximum sustainable rate (Morgan et al, 1989).

This study describes the use of steam injection and vacuum extraction as a primary remediation step for the removal of gasoline from a contaminated aquifer followed by bioremediation for the complete destruction of gasoline hydrocarbons and especially gasoline aromatics.

Steam Injection & Vacuum Extraction: The Primary Step

The use of steam injection and vacuum extraction in order to remove both volatile and non-volatile chemicals from a contaminated aquifer is being investigated. It has been shown to be successful as a result of the following mechanisms:

1. Steam displaces mobile fluids from the pore space
2. Steam condensate dilutes immobile water
3. The elevated temperature causes the high vapor pressure non-aqueous phase liquids to boil
4. Evaporation rates of low vapor pressure non-aqueous phase liquids are enhanced
5. Desorption of contaminants from solid surfaces is enhanced due to the increase in temperature (Udell et al, 1989).

The presence of contaminants within the low permeability zones usually limits the effectiveness of many remediation techniques. However, the combination of steam injection and vacuum drying has shown great potential as a method of bypassing this constraint.

In Situ Bioremediation: The Polishing Step

In situ bioremediation, by definition, involves the stimulation of microorganisms within a subsurface aquifer to degrade water contaminants; in other words, it is the management of a subsurface reactor to carry out specific biological degradations (Alvarez-Cohen, 1993). Bioremediation of gasoline contaminated aquifers by the indigenous microorganisms with the addition of inorganic nutrients and oxygen has been successfully implemented in many cases (Morgan et al, 1989; Thomas et al, 1989). In this study, the application of thermally enhanced in situ bioremediation following steam treatment of a gasoline contaminated aquifer is being investigated.

There are several factors that favor post-steam bioremediation; these may be stated as follows:

1. Although no published data exists on the effects of steam injection on the microbial communities in steam-treated soil, it is generally agreed that prolonged exposure of microorganisms to extreme temperatures, as in the case of steam injection, is fatal. This ensures the absence of microbial communities in the steamed zone, and in turn, the absence of predators and competition against the reintroduced enriched culture. This also facilitates quantification of the bioremediation effectiveness of the introduced culture.

2. Steam treatment leaves the site at an elevated temperature. The degradation of the organic contaminants in groundwater has been shown to be significantly enhanced with increased temperatures. For example, studies showed that a tripling of toluene mineralization rates took place due to an increase of temperature from 11°C to 25°C (Armstrong et al, 1991).

3. The steamed zone contains significantly lower concentrations of toxic chemicals that may inhibit microbial activity.

4. Because of the presence of steam injection and extraction wells, the steamed environment can be chemically and thermally altered for the application of thermally enhanced bioremediation.

Thus, a bioremediation strategy designed to take advantage of these conditions would involve the deposition of the appropriate nutrients and oxygen into the steamed zone and the subsequent introduction of an enriched microbial culture seeded from the indigenous subsurface population which is capable of degrading the residual gasoline contaminants.

STUDY FINDINGS & IMPLICATIONS

The Enriched Culture

Both gasoline aliphatics and aromatics can be readily degraded by the indigenous microflora of an aquifer in the presence of appropriate nutrients and an oxidant or electron acceptor. BTEX compounds, the substances of most interest in gasoline because of their documented toxicity, have been shown to be used by microbial communities as primary substrates under both aerobic and anaerobic conditions . However, anaerobic degradation has been shown to proceed at a much slower rate than aerobic degradation (McCarty, 1991). In this study, the aerobic degradation (oxygen being the electron acceptor) of BTEX compounds is investigated.

A microbial culture seeded from the indigenous subsurface population within a gasoline contaminated aquifer site was enriched in a laboratory chemostat at room temperature using toluene, a major gasoline aromatic, as the primary substrate, oxygen as the electron acceptor, and Berkeley Medium as the source of appropriate inorganic nutrients.

Temperature Effects

In order to determine the temperature profile for optimization of microbial growth and degradation activity of BTEX remediating microorganisms, growth rate experiments were conducted. Cell growth of samples incubated at different temperatures was monitored using light spectrophotometry. These studies revealed that cell growth on toluene increased with temperature from 7°C to 35°C, decreased sharply at incubations of 36°C to 40°C, and was inhibited entirely at temperatures above 45°C (cell density = 0.1 g/L, toluene concentration = 80 mg/L) (*Figures 1 & 2*). Substrate utilization was monitored using gas chromatography and was found to be maximal at 35°C. Lag phases were observed at all temperatures and became shorter as the temperature increased from 20°C to 35°C. The cell doubling time on toluene was measured to be 4 - 6 hours. Studies to compare lag times and aromatic degradation rates of two cultures enriched and maintained at room temperature and at 35°C indicate that degradation is enhanced and lag times decreased when the culture is enriched and maintained long-term at the optimal temperature. In conclusion, the results of these temperature studies favor degradation at higher temperatures with an optimum at 35°C. This optimum is obtainable in the steamed aquifer through manipulations of appropriate air and steam injections via the existing wells.

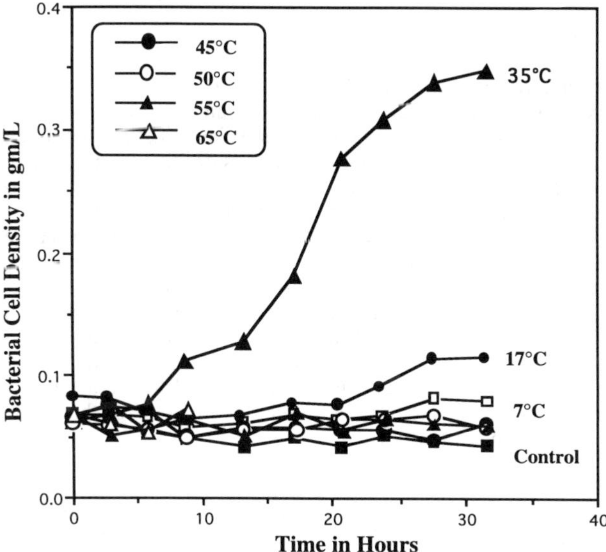

Figure 1. *Determination of Optimal Temperature.* An optimal temperature of 35°C was determined by cell growth and activity studies (cell density = 0.1 g/L, initial toluene concentration = 80 mg/L).

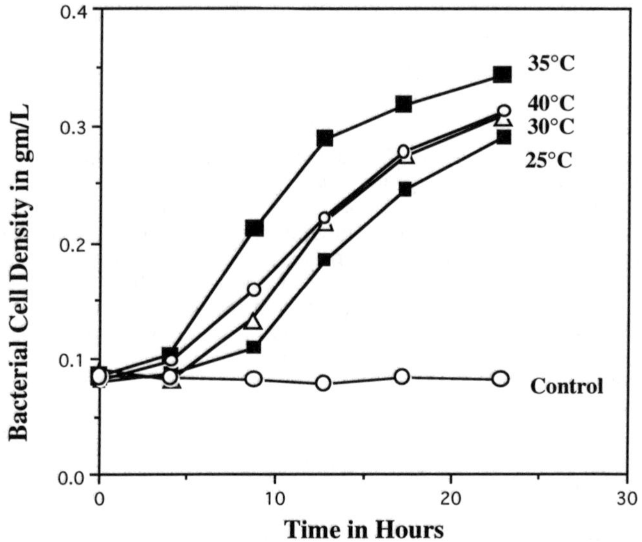

Figure 2. *Determination of Optimal Temperature.* Cell growth decreased at incubations of 36°C to 40° (cell density = 0.1 g/L, initial toluene concentration = 80 mg/L, cell doubling time at 35°C = 4 - 6 hours).

Concentration Effects

Degradation kinetics and toxicity studies with each of the gasoline aromatics were then carried out. The enriched culture successfully degraded all the BTEX aromatics to non-detectable levels. When kinetics were evaluated for the individual aromatics at initial concentrations of 80 mg/L (cell density = 0.1 g/L, time ~ 12 hours), toluene was degraded fastest, followed by benzene and ethyl benzene; no degradation of o-xylene was observed at this concentration *(Figure 3)*. An increase in carbon dioxide and a decrease in oxygen in the headspace of sample vials accompanied degradation of the aromatics. Cell density, monitored using light spectrophotometry, was observed to double in approximately 6 hours. When the initial substrate concentration was halved to 40 mg/L, o-xylene was degraded but at a much slower rate than toluene, benzene and ethyl benzene. At initial concentrations of 20 mg/L, degradation rates increased and a shorter lag time was observed indicating that at higher BTEX concentrations, substrate toxicity may be involved or that the cells may require a longer adaptation time.

Substrate toxicity studies for cells exposed to individual BTEX compounds suggested toluene toxicity at 150 mg/L and o-xylene toxicity at 80 mg/L; no toxicity was observed for benzene at concentrations as high as 200 mg/L. These results indicate that degradation in the steamed aquifer will not be hindered by substrate toxicity since the concentrations of BTEX compounds in the groundwater are well below those found by this study.

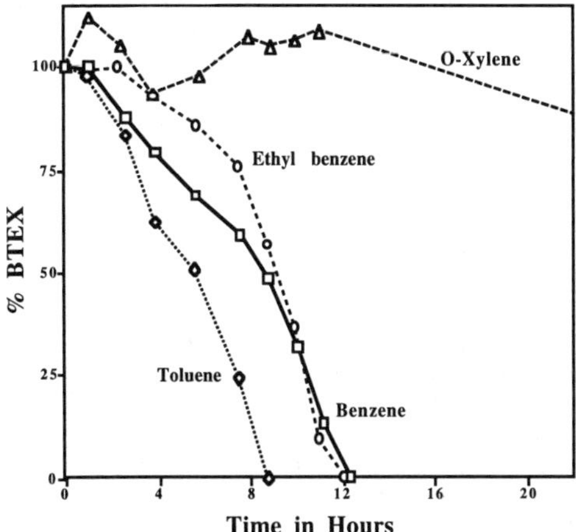

Figure 3. *Degradation of Individual BTEX Compounds.* Toluene oxidizers (cell density = 0.1 g/L) were exposed to individual BTEX compounds at initial concentrations of 80 mg/L).

Mixture Effects

When biodegradation kinetics were studied for the individual aromatics, toluene was degraded fastest, followed by benzene, ethyl benzene and o-xylene *(Figure 3)*. However, when cells were exposed to mixtures of aromatics, degradation rates in the order of ethyl benzene > toluene > benzene > o-xylene were observed *(Figure 4)*.

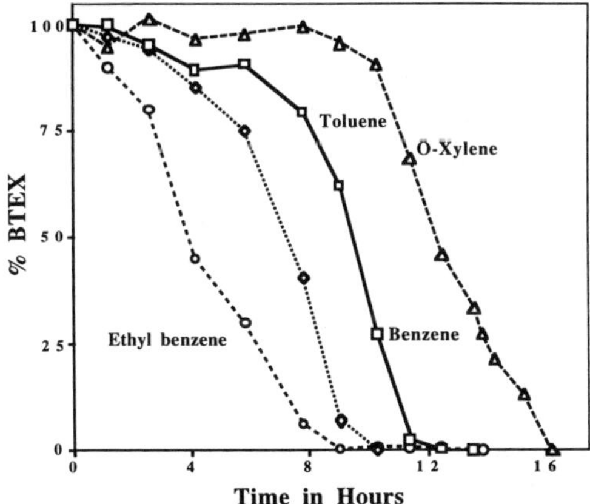

Figure 4. *Degradation of BTEX Mixtures.* Toluene oxidizers (cell density = 0.1 g/L) were exposed to BTEX mixtures at an initial concentration of 80 mg/L).

Degradation of aromatic mixtures clearly followed competitive inhibition kinetics with benzene and o-xylene degradation beginning only after significant disappearance of toluene and ethyl benzene *(Figure 3)*. O-xylene was degraded faster in mixtures with other aromatics than when present alone suggesting that o-xylene degradation was stimulated by the presence of other aromatics compounds *(Figures 3 & 4)*. These results indicate that biodegradation of BTEX mixtures is possible and that in some cases, it will be faster, as with o-xylene, one of the more recalcitrant BTEX compounds.

Gasoline as a Substrate

In order to investigate the degradation of aromatics in complex mixtures, several studies were done using gasoline as a substrate. Toluene oxidizers were successful at degrading each of the BTEX compounds in gasoline as well as other aromatics within several hours (cell density = 0.1 g/L, total liquid volume = 125 mL, gasoline volume = 16 μL). Although all BTEX aromatics were degraded to non-detectable levels, little oxidation of aliphatics took place by the cells suggesting that the mixed culture utilizes dioxygenase rather than monooxygenase enzymes for organic degradation (*Figure 5*). However, the presence of aliphatics in the mixture did not inhibit aromatic degradation. BTEX degradation in complex mixtures such as gasoline, therefore, is successfully carried out by toluene degrading cells and is not inhibited by the presence of other gasoline compounds present in the aquifer.

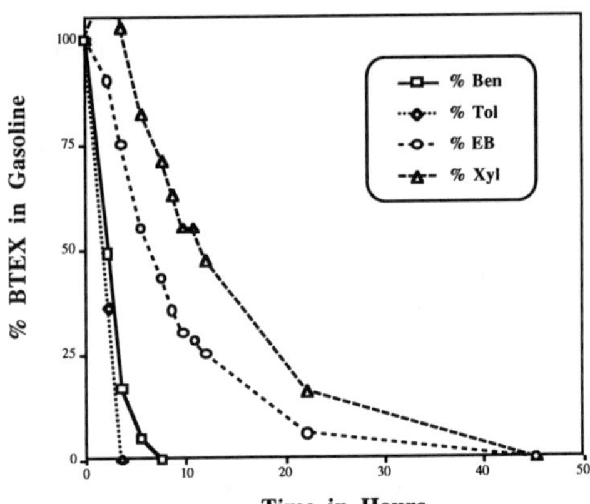

Figure 5. *Degradation of BTEX Compounds in Gasoline.* Toluene oxidizers (cell density = 0.1 mg/L) were exposed to gasoline (concentration =14 mg/L).

Propane Oxidizers

In other studies, propane oxidizers, a microbial population enriched on propane and capable of producing monooxygenase enzymes which facilitate degradation of aliphatics, were used in conjunction with toluene oxidizers. When a combination of propane oxidizers and toluene oxidizers were exposed to gasoline, the aromatics were degraded at a rate comparable to that of toluene oxidizers alone; however, no significant change in aliphatics was observed. When mixtures of hexane and toluene or hexane and octane were exposed to both populations, toluene degradation proceeded but no significant change in hexane or octane took place. Hexane alone, on the other hand, was successfully degraded when exposed to propane oxidizers. These results further support the absence of monooxygenase enzymes in toluene oxidizers and the capability of propane oxidizers to produce monooxygenase enzymes and thus degrade aliphatics, but suggests that the presence of both cultures inhibits the propane oxidizers from functioning. Current experiments are focusing on the effects of mixed microbial populations on degradation kinetics and the concurrent use of several cultures for the complete degradation of gasoline and other complex mixtures of contaminants.

CONCLUSIONS

1. A steam treated, gasoline contaminated aquifer, with its near sterile nature, increased temperature and decreased level of contaminant concentration, provides a superior environment for enhanced bioremediation of residual BTEX compounds.

2. A mixed microbial population was successfully enriched on toluene as the primary substrate with maximal growth and substrate degradation at 35°C (*Figures 1 & 2*).

3. Cells successfully degraded all BTEX compounds with increased degradation rates and decreased lag times at lower substrate concentrations.

4. Toluene exhibited substrate toxicity at 150 mg/L and o-xylene at 80 mg/L whereas no toxicity was exhibited with benzene at concentrations as high as 200 mg/L.

5. Biodegradation kinetics for the **individual** aromatics followed the order of toluene > benzene > ethyl benzene> o-xylene whereas those for **mixtures** of aromatics followed the order of ethyl benzene > toluene > benzene > o-xylene *(Figures 3 & 4)*.

6. Degradation of aromatic mixtures generally followed competitive inhibition kinetics; however, o-xylene degradation was stimulated by the presence of other aromatics *(Figures 3 & 4)*.

7. Toluene oxidizers exhibited only dioxygenase activity whereas a propane enriched culture exhibited monooxygenase activity. A combination of propane and toluene oxidizers exhibited only dioxygenase activity.

REFERENCES

Alvarez, P. J., and T. M. Vogel. 1991. Applied and Environmental Microbiology 57(10):2981-2985.

Alvarez-Cohen, L. 1993. In Situ Bioremediation: When Does it Work? Pp. 136-152. National Academy Press, Washington, D.C.

Armstrong, A. Q., R. E. Hudsen, H. M. Hwang, and D. L. Lewis. 1991. Environmental Toxicology and Chemistry 10:147-158.

McCarty, P. L. 1991. Journal of Hazardous Materials 28:1-11.

Morgan, P., R. J. Watkinson. 1992. Water Research 26(1):73-78.

Thomas, J. M., and C. H. Ward. 1989. Environmental Science and Technology 23(7):760-766.

Udell, K. S. and Stewart, L. D. 1989. UCB-SEEHRL Rept. No. 89-2.

HEAVY METAL REMOVAL USING PEAT/WETLAND TREATMENT

Suzanne Murawski[1]

ABSTRACT

The purpose of this paper is to present an overview of the mechanisms and application of a peat/wetland treatment system for heavy metal removal from wastewater. The mechanisms involved in the removal of heavy metals are complex and difficult to predict, however, peat has been proven to be an effective medium to remove metals. The successful design of a peat/wetland treatment system for acid mine drainage is presented to emphasize the low cost and minimal maintenance involved in this passive metal removal technique.

Introduction

The removal of metals from water has been necessary throughout history. With an increased demand for cost effective and simple methods to remove metals from wastewater, alternative removal methods have been sought. As early as 1962, peat has been suggested as one of the possible mediums to remove metals from wastewater streams (Oreshko et al., 1962). In recent years, peat/wetland treatment systems have been used to remove metals, especially for acid mine drainage applications, landfill leachate, and industrial and municipal wastewaters, with successful results. Application of this low cost and minimal maintenance treatment technology is environmentally attractive in areas where peat or wetlands are locally available.

---

[1]Graduate Student, University of Wisconsin-Milwaukee and Assistant Project Engineer, STS Consultants, Ltd., 11425 West Lake Park Drive, Milwaukee WI 53224

The following sections will present the mechanisms used by the peat/wetland system to remove heavy metals and a general overview of the system application in restoration of acid mine drainage. The chemical, physical and biological mechanisms of metal removal by peat/wetland systems will be discussed first, followed by the general design characteristics of a peat/wetland system.

Peat Characteristics

The most important component of a peat/wetland system is the peat. The physical and chemical properties of the peat depend mainly upon the nature of the plants from which it has originated, the properties of the water in which the plants were growing, and the moisture relations during and following its formation and accumulation. The properties developed in the peat have been shown to be favorable to the removal of metals.

Heavy metal removal efficiency is generally based on the characteristics of the peat such as pH, organic content, fiber characteristics, hydraulic conductivity, flow patterns and sulfate and organic availability (Frostman, 1988). A high hydraulic conductivity and complex flow pattern are favorable to provide a more uniform contact between the organic absorption sites on the peat and the seep waters. Therefore, a longer retention time due to the complex flow pattern is favorable to cause a noticeable increase in the removal efficiency of metal ions. The general trend of decreasing removal efficiency with increasing flow rate is probably a kinetic effect whereby shorter contact times of the metal ions with the peat reduce the occurrence of exchange reactions. The magnitude of this effect on each element would depend on each element's reaction rate with the exchange sites and diffusion rate into the exchanger (Ansted and MacCarthy, 1984).

Peat with a pH of less than 5, generally exhibits poor efficiency in metal removal due to a low buffering capacity. More efficient metal removal has been exhibited with a pH of 6 to 8 for aerobic conditions. Under anaerobic conditions, the pH should remain above 5 to have effective sulfate reduction. The smaller removal efficiencies at low pH's are due to competition of $H+$ ions with the metal ions for the exchange sites. A slight decrease in removal is seen at higher pH's probably due to metal complexation and/or precipitation with $OH-$ (Ansted and MacCarthy, 1984).

A high peat organic content of over 90% is necessary for good heavy metal removal. The organic material is essential for the sulfate reduction reactions. A peat/wetland treatment system with standing vegetation provides additional organics. More vegetative growth on the peat/wetland system will increase the available organics and extend the absorption life of the peat.

Mechanism of Metal Removal

The mechanisms used to remove metals from the wastewater stem from both physical and chemical forces. Physically, metals may be removed by absorption, cation exchange with limited sediment deposition and biosorption. However, most researchers conclude that the bond between metal and peat is chemical in nature in the form of chemisorption, complexation or chelation.

In addition, biological processes also aid in the removal of metals through bioaccumulation of metals by wetland vegetation.

Adsorption is defined as the process by which ions or molecules present in one phase tend to condense and concentrate on the surface of another phase (Sawyer, et al., 1978). There are three general types of adsorption that are involved in metal removal including physical, chemical and exchange adsorption. The physical forces are long range, weak forces that involve the electrostatic attraction between two charges such as negatively charged organic surfaces of peat to positively charged metal ions. Chemical adsorption creates much stronger short range forces that involve the actual chemical bonding of the peat surface to a solute comparable with those leading to the formation of chemical compounds. Exchange adsorption includes ion exchange which involves a weak bond and an exchange of ions on the surface and in solution prior to sorption.

During physical adsorption, the adsorbed molecule is not affixed to another molecule, but is free to move about its surface. In fact, physical adsorption is generally quite reversible. The adsorbed metal ions on peat can easily be released back to the water. Therefore, physical adsorption is likely not one of the main components of metal removal by peat.

Chemical adsorption, on the other hand, plays a bigger role in the removal of metals by peat. The metal ions are more strongly bound to the organic peat surfaces creating a nearly irreversible bond.

Since the peat and the wastewaters are complex in composition, mechanisms other than adsorption are involved in the removal reactions. The precipitation of secondary minerals can occur when wastewaters high in heavy metals are applied to peat. These precipitates can include oxides, oxyhydroxides, hydroxides, and carbonates with phosphate and silicate precipitates less important (Evans, 1989). Also, wastewaters high in sulfate and/or peat with hydrogen sulfide (biological reduction of SO_4^{2-}) may precipitate sulfide minerals under anaerobic conditions (low partial pressures of O_2).

Sulfate reduction occurs by the microbial process of anaerobic microorganisms that are able to decompose simple organic compounds using the sulfate as an electron acceptor. Sulfate reduction within the wetland treatment system is desirable because hydrogen sulfide readily reacts with dissolved metals, precipitating them as sulfide. Additional alkalinity also results to neutralize seep drainage acidity from mine tailing piles. Anaerobic microorganisms reduce sulfate in the presence of simple organic carbon resulting in the formation of sulfides and bicarbonate. When metals are reduced to metal sulfides, the reaction is:

$$M^{2+} + SO_4^{2-} + 2CH_2O > MS\downarrow + 2HCO_3^- + 2H^+$$

The designed peat treatment system must consider optimizing the anaerobic condition by controlling the water/substrate interface and providing a high organic content substrate to enhance growth. The most common limiting factors for sulfate reduction in anaerobic environments are the availability of suitable organic substrate and dissolved sulfates. Sulfate concentrations in the seep waters from acid mine drainage are typically greater than 500 milligrams per liter which is a sufficient concentration for sulfate reduction.

Research has shown that the rate of sulfate reduction through microbiologic activity decreases with lower temperatures. Lower temperatures will decrease the efficiency of the peat treatment system being designed. Therefore, the designed peat treatment system is anticipated to be operated only during the active growing months (May through October). The efficiency of the sulfate reduction in the designed-constructed peat treatment system will be dependent upon maintaining the anaerobic zone in the lower zone of the peat.

In addition, a number of researchers proposed chelation as a dominant reaction mechanism for the binding of metals by organics and further stated that more than one type of chelate is likely involved. The chelating functional groups may be sulfonic groups or groups containing nitrogen or oxygen (Lapakko et al., 1986). The chelating molecules will seize or "sequester" metal ions and hold them to the organic substance (peat) effectively removing them from the waste stream.

Peat/Wetland Treatment System Application

Peat has been shown to effectively remove heavy metals from wastewater streams. One application of a peat/wetland treatment system involves the removal of heavy metals from acid mine drainage below mine tailing piles in northern Minnesota. The waste rock stockpiles are composed of copper-nickel sulfide mineralized rock. The stockpile drainage pH ranges from 5.4 to 7.5, with most stockpile drainages having a pH of greater than 6.5. Nickel is the major trace element in the wastewater with lesser amounts of copper, cobalt and zinc.

Several peat test plots were created under the direction of the Minnesota Pollution Control Authority (MPCA) to design the best configuration for removal of the specific heavy metals. After evaluation of the test plots by the MPCA for more than 3 years, two wetland/peat treatment systems were designed. The designs incorporate two flow regimes due to the varying topographic relief of the site. The first system was constructed with a relatively steep gradient and tends to function as a riverine flow. This system consists of approximately 70,000 square feet of peat and 9 berms. The second system functions as a ponding condition due to its location in a wide flat valley with little relief. The second system was constructed using about 50,000 square feet of peat and 5 berms. The systems, as designed, incorporate both surface water flow and subsurface water flow. The subsurface flow (anaerobic conditions) results in dissimilarity formation of sulfides with react with the metal ions and precipitate as metal sulfides. The surface flow (aerobic conditions) results in the oxidation of metals.

The peat/wetland treatment systems were built across the valley of each system. Figure 1 presents a plan view of a typical peat/wetland treatment system. Construction occurred during the winter months to minimize access and construction problems associated with working in a wetland. The berms were constructed at each 1-foot elevation change to a height that would sustain an approximate 1-foot depth of ponded water upgradient of the berm.

The peat mixture was placed both upstream and downstream of each berm. A typical cross-section of a treatment berm is shown in Figure 2. The peat mixture beds were placed to enhance lateral flow and to increase the contact area and subsurface flow through the peat mixture. The subsurface flow through the peat mixture encourages the development of anaerobic conditions which enhance sulfate reduction activities. Based on the pilot testing, limestone was placed over the top of each berm and extended across the downstream side of the berm to aid in increasing subsurface flow and increasing the pH to optimal metal removal levels.

The peat mixture that was placed was unique in that it consisted of a mixture of reed sedge and process screenings from a peat mining operation. The peat screening materials are considered waste products, therefore incorporation of this material into the peat provided beneficial reuse of a waste product.

The two peat materials were mixed on-site prior to placement. The peat was loosely placed along the berms to minimize a reduction in the hydraulic conductivity of the material due to compaction. Disking of the peat is routinely completed to allow the drainage to contact a greater surface area of peat and, thereby, improve metal removal.

The systems include an anoxic limestone pretreatment system upstream of the peat/wetland system. The drainage enters the limestone pretreatment system prior to release to the peat/wetland system. The anoxic limestone pretreatment system consists of limestone berms with peat beds between the limestone berms. The peat is flooded to simulate anaerobic conditions. The limestone pretreatment increases the pH of the input drainage water which results in increased alkalinity and additional precipitation of metal hydroxides. The increased alkalinity also provides a more compatible water for sulfate reduction processes.

The longevity of the peat, thus the treatment system, is increased by the planting of cattails in the system. The cattails provide for a continued biosorption (metal uptake) and organic source for sulfide precipitation. Once the cattails become saturated with metals, they can be harvested and incinerated. The metals in the ash from the incineration process have been proposed to be covered for reuse. However, no documentation of this process was reviewed.

The use of the passive peat/wetland system requires little or no operational maintenance with the exception of the replacement of spent peat or limestone once it becomes ineffective for metal removal. However, even the replacement of this material is a simple procedure.

Based on one year of data, the peat/wetland systems in Minnesota are removing 95 to 99% of the metals and are meeting the water quality standard/guidelines imposed by the MPCA. For example, for the larger system, nickel concentrations are reduced from 6 to 10 parts per million (ppm) in the influent to 0.1 to 0.2 ppm nickel in the effluent, and in the smaller system, nickel is reduced from 4 ppm in the influent to 0.1 ppm nickel in the effluent. The MPCA standards/guidelines are 213 parts per billion (ppb) (0.213 ppm) for nickel. Based on the results of the pilot testing and two full scale system operations, the mining facility intends to construct 5 additional systems in the future.

Conclusions

The use of the peat/wetland treatment system has shown excellent results for heavy metal removal. The mechanisms that bind metals to peat appear to be chemical in nature, however both physical and biological forces are also involved in the process. These mechanisms and the associated capacities for removal are complex and difficult to predict without treatability and/or pilot testing given the variability between wetland systems, as well as peat. In the case of the acid mine drainage, removal efficiencies of up to 99% have been reported. Based on the first few years of operation, the peat/wetland treatment system is predicted to have sufficient life to treat acid seepage and function effectively, long after mine closure.

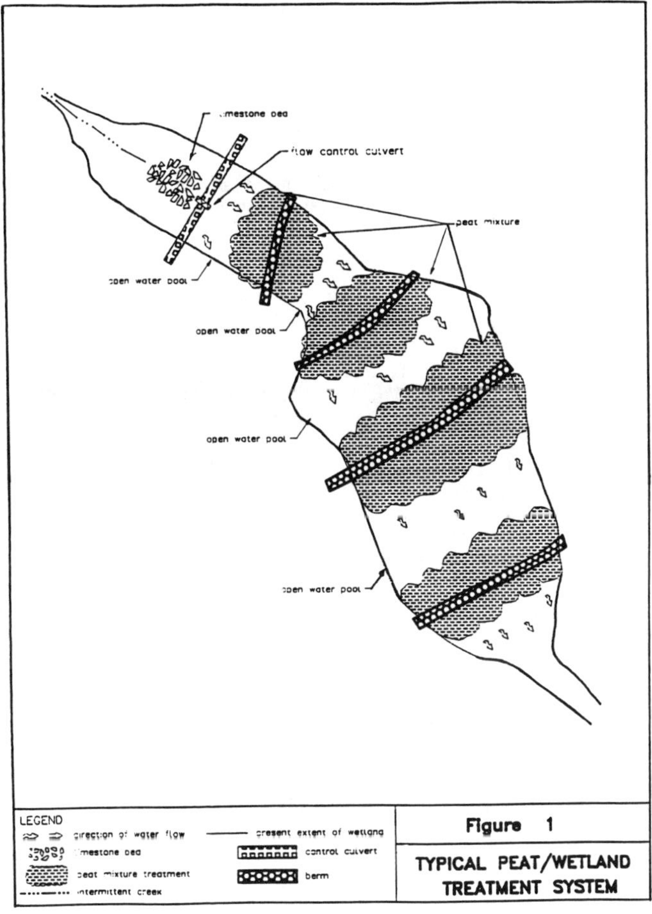

Figure 1

TYPICAL PEAT/WETLAND TREATMENT SYSTEM

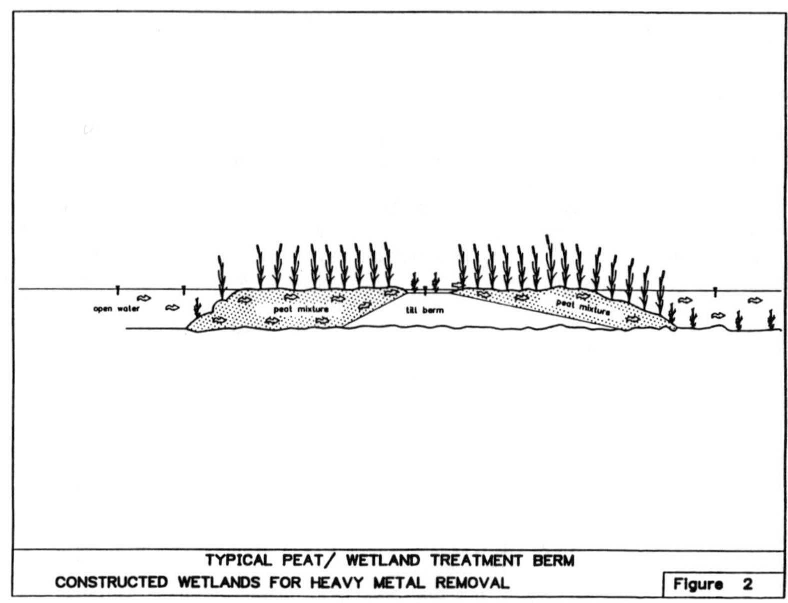

TYPICAL PEAT/ WETLAND TREATMENT BERM
CONSTRUCTED WETLANDS FOR HEAVY METAL REMOVAL Figure 2

REFERENCES

Ansted, J.P. and MacCarthy, P., 1984. Removal of heavy metal ions from solution by chemically modified peat: effects of pH, ionic strength and flow rate. Solvent Extraction and Exchange, 2(7&8): 1105-1122.

Evans, L.J., 1989. Chemistry of Metal Retention By Soils. Environmental Science Technology, 23(9):1046-1056.

Frostman, T.M., 1988. Constructed Wetland for Water Quality Improvement-Heavy Metal Removal. Presented at University of Wisconsin - Green Bay.

Lapakko, K.A. and Eger, P., 1988. Trace metal removal from stockpile drainage by peat. Mine Drainage and Surface Mine Reclamation Vol. 1: Mine Water and Mine Waste, 291-300.

Oreshko, V.F., Berdnikow, A.I. and Prezhbylski, 1962. The sorption of radioactive cobalt on peat. Radiokhimiya, 4:499-502.

Sawyer, Clair N. and McCarty, Perry L., 1978. Chemistry for Environmental Engineering. McGraw-Hill Publishing Company, New York.

Denitrification Inhibition by High Nitrate Wastes

W. Veydovec[1]; J. Silverstein[2]; N.E. Cook, Jr.[3];
L.A. Figueroa[3]; R. Hund[4]; G.D. Lehmkuhl[5]

Abstract

The adaptation of activated sludge and inhibition of denitrification at high nitrate concentrations was studied using pH controlled bench-scale sequencing batch reactors (SBRs), operated with 50% of the SBR volume recycled (recycle volume = influent volume). Denitrification of 1,350 and 2,700 mg/l NO_3^--N was completed after approximately 5 hours and 15 hours, respectively. No denitrification of 5,400 mg/l NO_3^--N was observed. These results suggest that there is a progressive inhibition of denitrification as nitrate concentrations increase from 1,350 to 5,400 mg/l NO_3^--N. In a subsequent series of experiments at an initial reactor nitrate concentration of 1,350 mg/l N, a significant accumulation of nitrite was observed, resulting once in destabilization with loss of denitrification and once in successful adaptation of the activated sludge. At a nitrate concentration of 1,350 mg/l N, the adaptation of activated sludge appears to be unstable, resulting sometimes in stable denitrification and sometimes in biomass washout.

Introduction

The processing of radioactive metal products at nuclear weapons plants and research labs has produced wastewaters containing high concentrations of nitrate, often greater than 50,000 mg/l N (Francis and Hancher, 1980). At the Rocky Flats Site in Golden, Colorado, waste brines consisting primarily of dissolved inorganic salts and small amounts of heavy metals and radionuclides have nitrate levels ranging from 500 to 12,000 mg/l N (Forrey and Lehmkuhl, 1992). The salts from these waste brines

---

1. Graduate student, Department of Civil, Environmental, and Architectural Engineering, University of Colorado, Boulder, CO 80309.
2. Associate Professor, Department of Civil, Environmental, and Architectural Engineering, University of Colorado, Boulder, CO 80309.
3. Assistant Professor, Department of Environmental Science and Engineering, Colorado School of Mines, Golden, Colorado, 80401.
4. American Water Works Association Research Foundation, 6666 W. Quincy Ave., Denver, CO 80235.
5. EG&G-Rocky Flats, P.O. Box 464, Golden, CO 80402.

have been cemented in an attempt to produce a stable wasteform, "saltcrete," destined for final disposal at a radioactive waste storage facility. However, much of the saltcrete does not form a stable concrete block, probably in part due to the solubility of calcium nitrate salts formed during the cementation process. In addition to contributing to the instability of hazardous and radioactive compounds in saltcrete, nitrate itself is a primary drinking water contaminant with an MCL of 10 mg/l N. Thus, removal of nitrate from these waste brines before final disposal may improve the long-term stability of high nitrate wastes.

Biological denitrification of high-nitrate wastewaters produced by nuclear facilities has been investigated at Oak Ridge National Laboratory, where researchers demonstrated denitrification in fluidized bed reactors, continuous-stirred tank reactors (CSTRs), and waste evaporation/percolation ponds (Francis and Malone, 1977; Clark et al., 1975; Napier, 1989). More recently, at the University of Colorado, Boulder, SBRs with the reactor pH controlled between 7.4 and 8.0 were used to denitrify a simulated Rocky Flats Site waste containing 1,350 mg/l NO_3^--N (initial reactor concentration of 675 mg/l NO_3^--N) (Cook et al., 1993). Denitrification was found to follow zero-order kinetics with respect to nitrate concentration and was complete in approximately 2.5 hours.

Other investigators have reported that biological denitrification is inhibited by high nitrate concentrations. In studies performed using CSTRs, Francis and Mankin (1977) reported inhibition of denitrification when the feed nitrate concentration exceeded 1,350 mg/l N. One possible explanation for the observed nitrate toxicity is accumulation of the denitrification intermediate nitrite (NO_2^-). Researchers have reported that nitrite inhibits bacteria at concentrations ranging from 10 to 2,000 mg/l N (Beccari et al., 1983; Chen et al., 1991; Horsley et al., 1982; van Versefeld et al., 1977). Undissociated nitrous acid (HNO_2) has also been implicated in the inhibition of denitrification. Abeling and Seyfried (1992) found that inhibition occurred when concentrations of HNO_2 reached 0.13 mg/l (0.04 mg/l N), and suggested that the concentration of nitrous acid, and not nitrite, controls the inhibition of denitrification. The dissociation constant of HNO_2 is 4.5 (Snoeyink and Jenkins, 1980), which is sufficiently high to allow appreciable amounts of HNO_2 to accumulate at neutral pH values when the nitrite concentration is high. For this reason we have examined both denitrification of high nitrate wastes and nitrite accumulation in SBRs.

Methods

Bench-scale SBRs inoculated with activated sludge taken from a municipal wastewater treatment plant were used to study denitrification of high-nitrate waste brines. The SBRs were operated with a filled volume of 6 liters and 50% recycle. 2.5 liters of feed brine and 0.5 liters of acetate feed were added during the fill period, giving an initial nitrate concentration in the reactor equal to approximately 42% of the feed brine nitrate concentration. A 24 hour SBR treatment cycle was used, consisting of a 0.5 hour fill period, a 20.5 hour react period, a 2 hour settle period, and a 1 hour decant period. The pH of the reactors was maintained between 7.4 and 8.0 using 6 N HCL. Figure 1 shows a schematic of a bench-scale SBR system.

Figure 1: Schematic of bench-scale SBR denitrification system

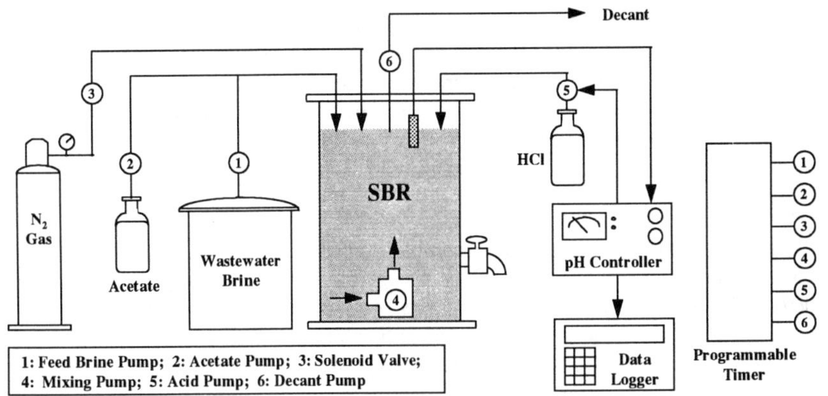

Synthetic salt solutions that simulated high sodium nitrate waste brines reported at the Rocky Flats Site were used as feed solutions. Table 1 shows the components of the feed brine giving an initial reactor concentration of 1,350 mg/l NO_3^--N. The feed brines giving initial reactor nitrate concentrations of 2,700 and 5,400 mg/l N had the same component concentrations as shown in Table 1 with the exception of the NO_3^- and PO_4^{3-} salts. The synthetic brine was made up in tap water with enough phosphate added to maintain a C:P ratio of 30:1. Sodium acetate was fed into the reactor during the fill period as a carbon source for the denitrifying bacteria at a ratio of 1.5 mg acetate organic carbon per mg nitrate-nitrogen (C:N = 1.5).

Table 1: Feed Brine Composition (initial reactor concentration = 1,350 mg/l NO_3^--N)

Component	Concentration	Component	Concentration
$NaNO_3$	12.9 g/l	$NaH_2PO_4*H_2O$	0.25 g/l
KNO_3	8.10 g/l	KH_2PO_4	0.09 g/l
NaCl	0.96 g/l	$NaHCO_3$	0.66 g/l
KCl	0.22 g/l	Matrix Water	Boulder City Tap
Na_2SO_4	0.55 g/l	Ionic Strength	0.28
K_2SO_4	0.17 g/l	pH	8.0

The sum of nitrate + nitrite (NO_x) concentration was analyzed using the UV spectrophotometric method (Standard Methods, 1992) using a scanning double beam spectrophotometer (Varian DMS-100). Nitrite concentration was measured using a chemi-luminescent detector developed at the University of Colorado (Dunham, 1994). The pH of the reactors was monitored and recorded using a submersed pH probe (Orion Model 91-06) connected to a controller (Cole Parmer Model 5656-00) and a data logger (Fluke Model 2285B). Mixed liquor suspended solids (MLSS) and effluent suspended solids (ESS) were determined according to Standard Methods (APHA, AWWA, WPCF, 1992).

Two separate experiments were performed. The first experiment, which lasted two months, tested the ability of activated sludge to adapt to high-nitrate concentrations. Three SBRs were operated in parallel with initial reactor nitrate concentrations of 1,350, 2,700, and 5,400 mg/l N. NO_x profiles were taken to assess denitrification performance. In addition, reactor pH values, which correspond closely to the progression of the denitrification reaction, were continuously recorded. A second experiment of similar duration examined inhibition at high nitrate levels. For this experiment, nitrite profiles were measured in addition to NO_x and pH.

Results and Discussion

Adaptation: A typical NO_x and pH reaction profile for the SBR with an initial nitrate concentration of 1,350 mg/l N is shown in Figure 2. Denitrification was complete (no detectable NO_x) in approximately 5 hours. This corresponds well to the expected doubling of the 2.5 hour zero-order reaction time required for complete denitrification of 675 mg/l NO_3^--N examined in earlier experiments (Cook et al., 1993). An NO_x and pH profile for the reactor with an initial nitrate concentration of 2,700 mg/l N is shown in Figure 3. In this case, complete denitrification took approximately 15 hours, exceeding the expected 10-hour reaction time by 50%. There was no detectable denitrification in the reactor containing 5,400 mg/l NO_3^--N.

Based on these results, there appeared to be increasing inhibition of denitrification as reactor nitrate concentrations increased from 1,350 to 5,400 mg/l N. In the reactor with an initial concentration of 1,350 mg/l NO_3^--N there seemed to be no inhibition, with moderate inhibition in the reactor containing 2,700 mg/l NO_3^--N, and complete inhibition in the reactor with 5,400 mg/l NO_3^--N. Examining the NO_x profile of the moderately inhibited reactor (Figure 3), the denitrification rate is rapid for the first hour of the reaction, and then slows significantly for the remainder of the reaction (indicated in the figure by drawing two separate lines). This suggested inhibition by a denitrification intermediate, with nitrite a likely candidate. Therefore, a second experiment was performed to examine nitrite accumulation during denitrification.

Inhibition: In this experiment, an initial reactor nitrate concentration of 1,350 mg/l N was used. Two separate trials were performed, with the first one lasting two weeks, and the second lasting one month. During the first trial (data not shown), a peak NO_2^--N accumulation corresponding to approximately 100% of the initial NO_3^--N concentration (1,300 mg/l NO_2^--N) was observed in samples taken two weeks after start-up. Shortly thereafter, the activated sludge biomass washed out of the SBR, ending denitrification. The reactor was reseeded with activated sludge from a local wastewater plant and a second trial was started. Figures 4 and 5 show NO_x, NO_3^-, and NO_2^- profiles taken at 2 and 4 weeks after startup, respectively. After two weeks of operation (Figure 4), a maximum nitrite accumulation of 600 mg/l N (approximately 50% of the initial NO_3^--N concentration) was observed, with complete denitrification taking approximately 9 hours. After four weeks of operation (Figure 5), the peak was less than 50 mg/l NO_2^--N, and denitrification was complete

Figure 2: SBR NO_x and pH reaction profile (initial nitrate conc. = 1,350 mg/l N)

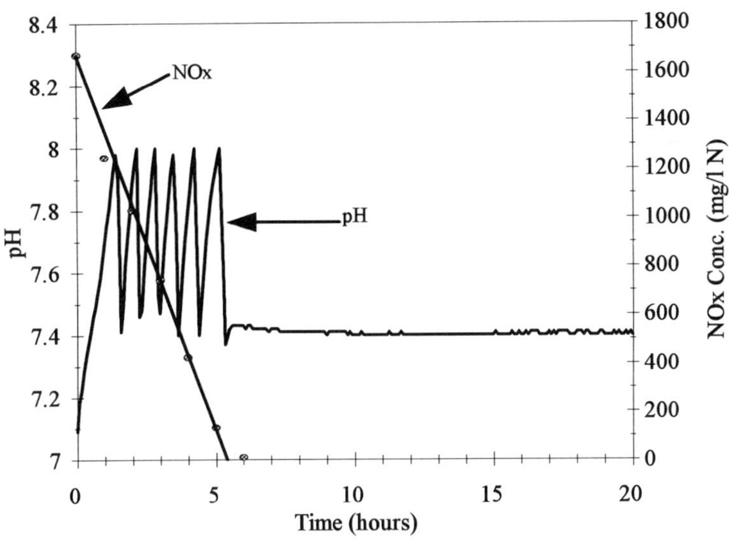

Figure 3: SBR NO_x and pH reaction profile (initial nitrate conc. = 2,700 mg/l N)

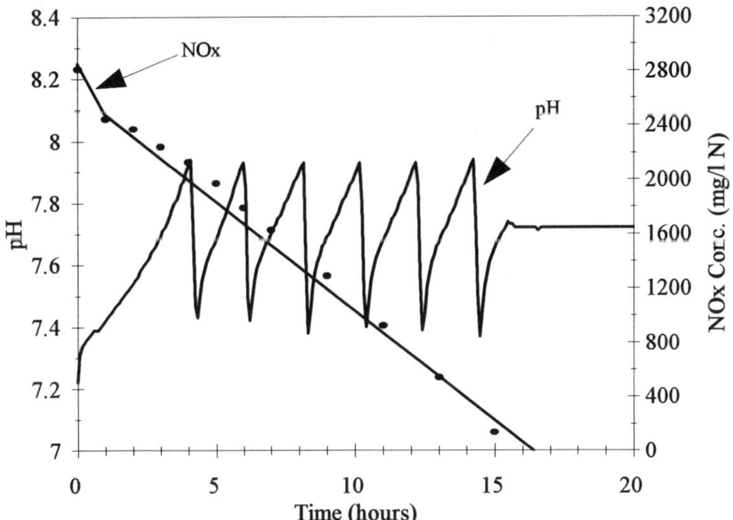

Figure 4: SBR NO_x, NO_3^-, and NO_2^- reaction profile (2 weeks after startup)

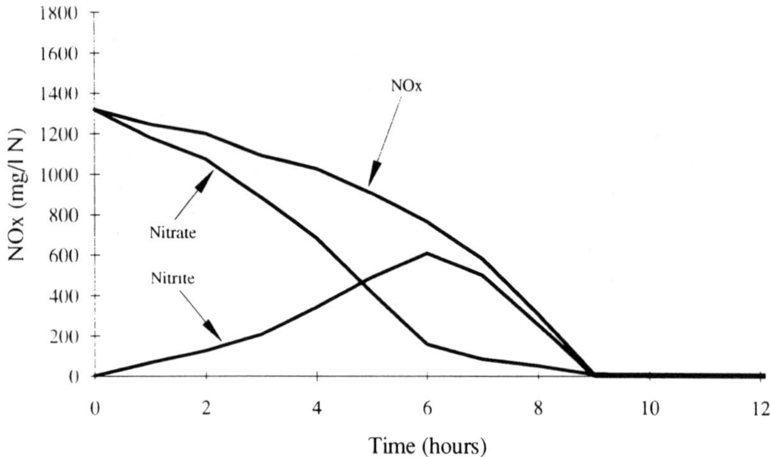

Figure 5: SBR NO_x, NO_3^-, and NO_2^- reaction profile (4 weeks after startup)

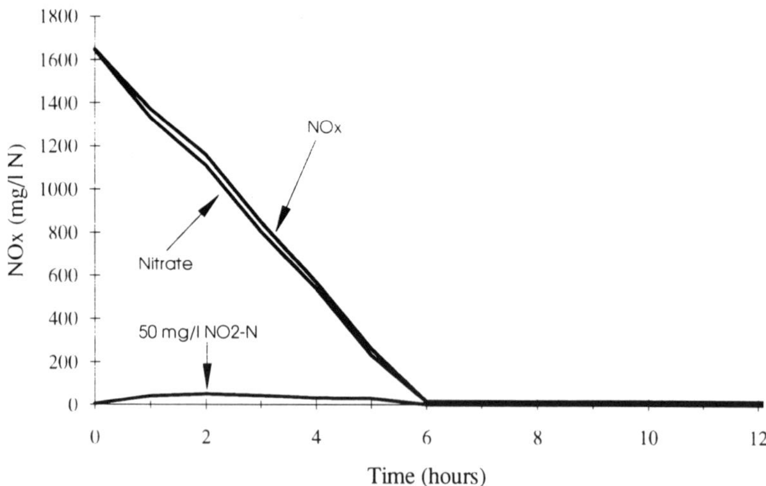

in 6 hours, approaching the high rate of denitrification observed during "steady-state" operation in previous experiments.

One possible explanation for the observed inhibition of denitrification is the accumulation of nitrite. Table 2 shows the peak nitrite values observed during the denitrification of 1,350 mg/l NO_3^--N and the corresponding HNO_2 concentrations calculated at the pH setpoint values of 7.4 and 8.0. The accumulations of nitrite fall well within the range of reported inhibitory nitrite concentrations (10 - 2,000 mg/l NO_2^--N). Calculated HNO_2 values range from 0.02 to 1.63 mg HNO_2-N/l, and, with the exception of the lowest value, are also above the reported toxicity level of 0.04 mg/l HNO_2-N.

Table 2: Peak observed NO_2^- and calculated HNO_2 concentrations

Trial : Operation time	NO_2^- (mg/l N)	HNO_2 (mg/l N) @ pH = 7.4	HNO_2 (mg/l N) @ pH = 8.0
1 : 2 weeks	1300	1.63	0.41
2 : 2 weeks	600	0.75	0.19
2 : 4 weeks	50	0.06	0.02

Further experiments have been planned to examine methods of minimizing HNO_2 accumulation by changing SBR operation parameters (e.g. pH setpoints, fill time, etc.). In this way, SBR denitrification may be extended to treat very high NO_3^- wastes.

Conclusions

- Complete denitrification of 1,350 mg/l NO_3^--N was achieved after approximately 5 hours using pH controlled SBRs. Denitrification of 2,700 mg/l NO_3^--N took approximately 15 hours. No denitrification was observed at an initial nitrate concentration of 5,400 mg/l N.
- There appears to be increasing inhibition of denitrification as reactor nitrate concentrations increase from 1,350 to 5,400 mg/l N. In the reactor with 1,350 mg/l NO_3^--N there was apparently no inhibition. The reactor with 2,700 mg/l NO_3^--N showed moderate inhibition, taking 15 hours for complete denitrification instead of the expected 10 hours. Complete inhibition was observed at an initial reactor concentration of 5,400 mg/l NO_3^--N.
- Significant amounts nitrite accumulated during the denitrification of 1,350 mg/l NO_3^--N. In one trial, a peak nitrite concentration of 1,300 mg/l N was observed. The activated sludge subsequently washed out, ending denitrification. In a second trial, the peak nitrite concentration was reduced to less than 50 mg/l N after one month of operation, forming a stable denitrifying biomass. At an initial reactor nitrate concentration of 1,350 mg/l N, the adaptation of activated sludge appears to be unstable, resulting sometimes in stable denitrification and sometimes in biomass washout.

References

APHA, AWWA, WPCF (1992). *Standard Methods for the Examination of Water and Wastewater*. 18th Edition. APHA. Washington, D.C.

Abeling, U. and Seyfried, C.F. (1992). *Anaerobic-Aerobic Treatment of High-Strength Ammonium Wastewater-Nitrogen Removal Via Nitrite*. Wat. Sci. Tech. 26(5-6), 1007-1015.

Beccari, M., Passion, R., Ramadori, R., Tandoi, V. (1983). *Kinetics of dissimilatory nitrate and nitrite reduction in suspended growth culture*. J. Wat. Pollut. Control Fed., 55, 58-64.

Chen, S.K., Juaw C.K., Cheng, S.S. (1991). *Nitrification and denitrification of high-strength ammonium and nitrite wastewater with biofilm reactors*. Wat. Sci. Tech., 23, 1417-1425.

Clark, F.E., Francke, H.C., and Strohecker, J.W. (1975). *Biological Denitrification of High Nitrate Waste Solutions*. Proc. 30th Ind. Waste Conf., Purdue Univ., Ann Arbor Science, Ann Arbor, MI.

Cook, N.E., Silverstein, J., Figueroa, L.A., Cutter, K., Hund, R., Veydovec, W. (1993). *pH Control of Denitrification of a High Strength Industrial Wastewater*. AC93-039-001, WEF 66th Annual Conf. and Expo., Anaheim, CA.

Dunham, A.J. (1993). *Aqueous Nitrite Analysis by Selective Reduction and Gas Phase Nitric Oxide Chemiluminescence*. Ph D. Thesis, Department of Chemistry and Biochemistry and CIRES, University of Colorado, Boulder, CO.

Forrey, R. and Lehmkuhl, G. (1992). Personal communication, EG&G Rocky Flats Site, Golden, CO.

Francis, C.W. and Hancher, C.W. (1980). *Biological denitrification of high-nitrate wastes generated in the nuclear industry*. Biological Fluidized Bed Treatment of Water and Wastewater, eds. P.F. Cooper and B. Atkinson, Ellis Horwood Ltd., Chichester, England.

Francis, C.W. and Malone, C.D. (1977). *Anaerobic Columnar Denitrification of High Nitrate Wastewater*. Prog. Wat. Tech., 8(4/5), 687.

Francis, C.W. and Mankin, J.B. (1977). *High Nitrate Denitrification in Continuous Flow-Stirred Reactors*. Wat. Res., 11, 289-294.

Horsley, R.W., Roscoe, J.V., Talling, I.B. (1982). *Nitrate reduction by Pseudomonas spp.: antagonism by fermentative bacteria*. J. Appl. Bacteriol., 52, 57-66.

Napier, J.M. (1989). *Recovery and Disposal of Nitrate Wastes*. Rep. No. Y/DZ-457, Oak Ridge Y-12 Plant, Oak Ridge TN.

Snoeyink, V.L., and Jenkins, D. (1980). *Water Chemistry*. John Wiley & Sons, New York, NY.

van Versefeld, H.W., Meijer, E.M., Stouthamer, A.H. (1977). *Energy conservation during nitrate respiration in Paracoccus denitrificans*. Arch Microbiol., 112, 17-23.

Ozonation of Anaerobic Sludge Dewatering Centrate for Ammonia Removal

K. Ramalingam[1], V. Diyamandoglu[2] M.ASCE, and J. Fillos[3]

Abstract

Semi-batch ozonation experiments were carried out on settled anaerobic sludge dewatering centrate. The observed rate of ammonia decay at pH ranging between 7.5 and 8.5 was much slower than the rates observed in clean water solutions. NO_3-N formed was always below 10 percent of the NH_3-N lost and did not follow stoichiometric conversion. Apparent and true color decay was significant. Carbonate alkalinity species distribution and solution pH were not significantly affected at the experimental conditions investigated.

Introduction

Centrate is the liquid stream generated during centrifugation. Anaerobically digested sludges of municipal wastewater treatment plants are frequently dewatered using centrifugation. The centrate produced contains solids that were not captured as cake during centrifugation along with dissolved contaminants that in part are formed during anaerobic digestion. Traditionally, centrate streams are returned to the head of the treatment plant. However, state and local regulations on nitrogen discharge levels often do not allow for such internal recycle procedures without prior removal of nitrogen. In such cases, the nitrogen in the centrate has to be removed prior to recycle.

The composition of centrate is highly dependent on the raw wastewater characteristics, the processes involved in the treatment train, and the polymers used to enhance the dewatering properties of sludges. Centrate produced in the dewatering facilities in New York City is well buffered with alkalinity ranging between 2700 and 3600 mg/L as $CaCO_3$ and has pH ranging between 7.0 and 8.0 (Fillos and Diyamandoglu, 1993).

---

[1] Res. Assoc., [2] Assist. Prof., and [3] Prof., Dept. of Civ. Engrg., The City College, City University of New York, New York, NY 10031.

The NO_2-N and NO_3-N levels were always below 0.5 mg/L. However, the NH_3-N levels were always high averaging at 950 mg/L with an average TKN of 1150 mg/L. The total BOD and COD averaged at 250 and 2050 mg/L and total solids concentration was 3180 mg/L. The heavy metal levels were always below 0.5 mg/L with the exception of vanadium which was observed at 0.6 mg/L. Among the TCL semi-volatile compounds bis(2 ethylhexyl) phthalate and phenol were observed at levels of 320 and 40 μg/L, respectively. Acetone, toluene, chloromethane and 2-butanone were the TCL purgable organics observed at levels below 200 μg/L. Finally, the centrate is produced at about 32°C and was free of volatile organics (Fillos and Diyamandoglu, 1993).

The objective of this study is to investigate the effectiveness of advanced oxidation processes in removing ammonia from settled centrate. This paper presents a summary of the results obtained to date from experiments involving ozonation.

Experimental Methods

The flow diagram of the experimental apparatus is shown in Figure 1. The ozonation experiments were carried out in a glass column reactor (height = 1.5 ft; internal diameter = 0.25 ft) with a centrate volume of 2 liters. The sample was first mixed well in the reactor by bubbling oxygen through it. During this phase gas flow adjustments through the ozone monitors were also carried out. Ozone was introduced into the reactor by turning on the ozone generator (PCI, Model GL-1) at the already operating oxygen gas flow rates. The ozone concentrations in the feed and off-gas were measured using two gas phase ozone monitors (PCI, Model HC-1). The off-gas was bubbled through a potassium iodide solution and passed through an ozone destructor (PCI, Model OD-2) before venting to the fume hood. The initial conditions of the centrate following settling were determined prior to each experiment during the mixing stage with oxygen by analyzing the reactor content for NH_3-N, NO_3-N, color, alkalinity and pH. The ozone concentration and flow rate of both the ozonated oxygen stream (feed gas) and the reactor off-gas were measured at different reaction times.

Analytical Methods

All solutions and working standards were prepared with analytical reagent grade chemicals and distilled-deionized water (Barnstead, NANOpure). Alkalinity and color were determined in accordance with Standard Methods 2320-B and 2120-B, respectively (APHA-AWWA-WEF, 1992). Ozone residual in the liquid phase was measured using the Indigo method (4500-O_3 B). Ammonia was measured using an Orion, model 920 ion analyzer, equipped with an Orion, model 95-12 ammonia electrode and confirmed with the standard automated phenate method (4500-NH_3 H) using a TRAACS 800 analyzer. Nitrate-nitrogen was determined using a single channel high performance ion chromatograph (DIONEX, model 4500i) with in-line gradient eluant mixing. The pH was measured using an Orion model 920 ion analyzer equipped with a dual pH probe.

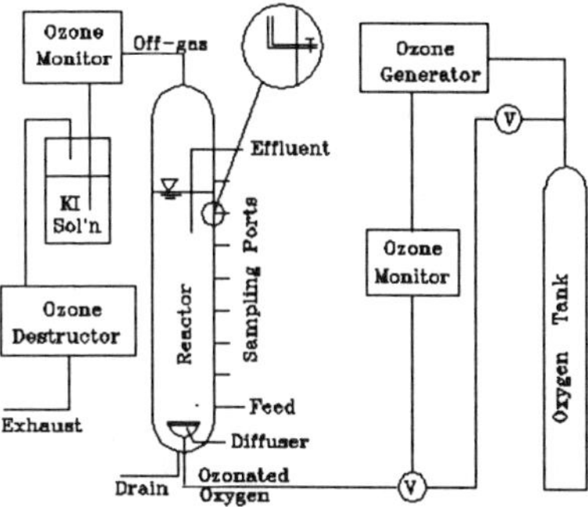

Figure 1. Diagram of the Experimental Apparatus.

Results and Discussion

The results of a typical semibatch kinetic experiment are shown in Table 1. The ozone transfer rate (OTR) in this study ranged between 1 and 4 mg/L.min. OTR was computed as an average of at least 15 sets of observations involving the difference of ozone in the feed gas and the reactor off-gas plus residual ozone. The residual ozone in the aqueous phase was always below 0.5 mg/L. All the experiments were carried out at temperature ranging between 25 and 27°C. Simultaneous ammonia decay and nitrate formation have been observed in all experiments. Ozonation of ammonia had been studies by Singer and Zilli (1975) both in pure ammonia solutions as well as secondary effluent of wastewater treatment plants. These autors determined that the ammonia decay follows first order kinetics and observed that in buffered ammonium chloride solutions and pH<9 the reaction product is mainly nitrate in stoichiometric quantities. Comparison of the results shown in Table 1 with prediction based on Singer and Zilli (1975) are depicted in Figure 2. ΔNH_3-N and ΔNO_3-N are the ammonia lost and nitrate formed during the experiment and the "predicted" ΔNH_3-N was calculated for the experiment shown in Table 1, using the first order rate constant of 1.9×10^{-2} min^{-1} reported by the authors for ammonium chloride solution at pH of 8.4 and $p_{O_3} = 0.055$ atm.

The observed ΔNH_3-N depicted in Figure 2 was consistently higher than ΔNO_3-N. This suggests that a large fraction of NH_3-N is oxidized to end-products different than nitrate and that the stoichiometric conversion of NH_3-N to NO_3-N observed in clean water systems does not hold for the settled centrate. Comparison of the predicted versus observed ΔNH_3-N depicts that the ammonia decay rate in the centrate was much slower than that in pure ammonium chloride solution. Similar observations were also reported by Singer and Zilli (1975) in ozonation experiments carried out with ammonia-enriched secondary effluent. They observed no appreciable ammonia oxidation at pH of 7.4,but significant decay in experiments buffered at pH 9. They also reported stoichiometric conversion of NH_3-N to NO_3-N and significant decay in the dissolved organic matter as measured by COD. The centrate experiments were carried out without any pH adjustment at pH ranging between 8 to 8.5and consistent levels of NH_3-N decay was observed in all of them. However, the conversion of NH_3-N to NO_3-N was not observed to be stoichiometric as reported by Singer and Zilli (1975).

Table 1. Typical Results of an Ozonation Experiment (p_{O_3}= 0.05 atm; OTR = 3.9 mg/L.min). The OTR Values Represent Averages of Several Measurements Between Two Consecutive Detention Times.

Time min	OTR	NH_3-N mg/L	NO_3-N mg/L	$CO_3^{=}$ mg/L as $CaCO_3$	HCO_3^- mg/L as $CaCO_3$	Total Alk. mg/L as $CaCO_3$	Color Pt-Co	pH
0	---	795	0.1	0	2945	2945	1524	7.99
30	4.21	743	1.6	260	2635	2895	940	8.33
60	3.67	720	4.6	550	2335	2885	684	8.46
90	3.63	676	8.5	650	2115	2765	533	8.54
120	3.83	663	15.6	810	1925	2735	489	8.56

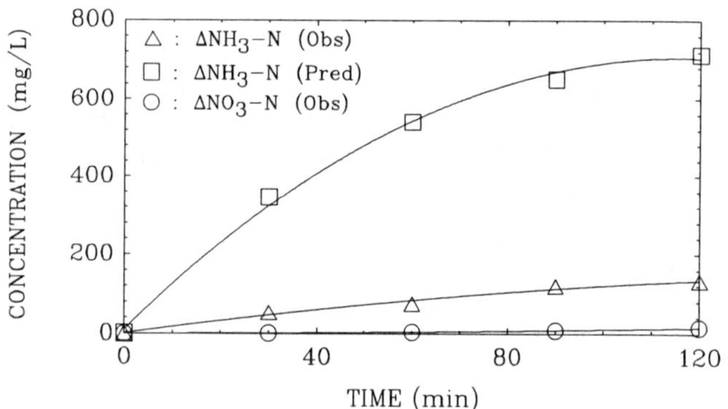

Figure 2. NH_3-N Decay and NO_3-N Formation for the Experiment Shown in Table 1.

Color decay was measured in two experiments with the results shown in Figure (3A) as a pseudo-first order plot. The experimental conditions were similar in both experiments and the results depict a consistent first order decay of apparent centrate color. The decay rate constants were determined to be 3.14×10^{-3} and 3.20×10^{-3} min^{-1} at OTR of 2.7 and 3.9, respectively. These OTR values are in agreement with those reported by Huang *et al* (1993) in ozonation experiments of domestic landfill leachates at OTR ranging between 8 and 66 mg/L.min. The OTR in this study are below those used by Huang *et al* (1993) and the rate constants are in good agreement with the dependency of the color decay rate constant on OTR observed by the aforementioned authors and depicted in Figure (3B). Analysis of the ozonated centrate for various priority pollutants is currently underway. The results are expected to indicate if the process results in undesirable nitrogenous or organic by-products.

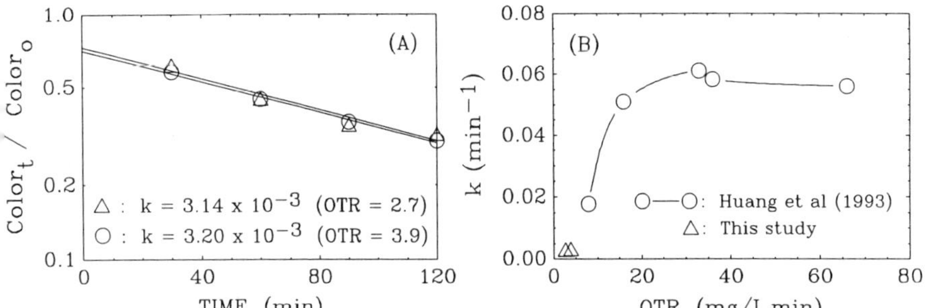

Figure 3. **(A):** Color Decay During Ozonation of Centrate; **(B):** Comparison of Color Decay Rate Constants of Huang *et al* (1993) and This Study.

Despite the very high levels of carbonate alkalinity, the pH of the settled centrate increased, albeit slightly, during all the experiments. This is in contradiction with the results of Singer and Zilli (1975) who observed consistent drop in pH during ozonation of secondary effluent. However, the alkalinity measurements shown in Table 1 substantiate the pH increase as the carbonate species increased with time while total alkalinity decreased. To confirm this observation one experiment was carried out on settled centrate after its alkalinity was titrated down to 740 mg/L as $CaCO_3$ with H_2SO_4. The pH of the solution increased from 6.1 to about 8.1 at the early stages of the experiment and increased further at a slow rate. This causes for the reported pH increase are currently being investigated.

Conclusions

Ozonation of centrate at ambient temperature and pH conditions does result in significant decomposition of ammonia in the anaerobically digested sludge dewatering centrates. Furthermore, the oxidized ammonia is only partially converted to nitrate leaving the possibility for the formation of other reaction end-products that need to be identified. Ozonation of ammonia at elevated pH levels or combination of ozone with other oxidizing agents that promote hydroxyl radical formation through more rapid decay of ozone can be a more effective approach to removal of ammonia from complex water matrices.

References

APHA, AWWA, WEF (1992) "Standard Methods for the Examination of Water and Wastewater" 18th Edition.

Fillos, J. And Diyamandoglu, V. (1993) "Characterization and Treatment of the New York City Sludge Dewatering Centrate: Selection of Treatment Technologies for Bench Scale Studies" Draft Report submitted to New York City Department of Environmental Protection, The City College, Department of Civil Engineering, New York.

Huang, S., Diyamandoglu, V. And Fillos, J. (1993) "Ozonation of Leachates from Aged Domestic Landfills" *Ozone Sci. & Eng.*, **Vol. 15**, pp. 433-455.

Singer, P.C. And Zilli, W.B. (1975) "Ozonation of Ammonia in Wastewater" *Water Research*, **Vol. 9**, pp. 127-134.

Biological Upflow Fluidized Bed Denitrification Reactor Demonstration Project-Stamford, Ct.

Jeannette Semon[1] (M.ASCE), Thomas Sadick[2], Dennis Palumbo[3], Richard Schede[3], Marylee Santoro[3], Marvin Schneider[3], and Steven Lesando[3]

Abstract

Results of a demonstration project using a biological upflow fluidized bed reactor to remove nitrogen from wastewater treatment plant effluent are described. The reactor was operated at a flow rate of 15.14 L/s and a loading of 2371 kg NO_3/1000 m^3/d. Problems were encountered with the mechanical systems during start-up, but they were resolved and the system has performed extremely well throughout the test period.

Performance data obtained from this project are very encouraging. Nitrate removal averaged 97% with effluent total nitrogen concentrations between 1.5 to 2.0 mg/l at 15° C. Approximately 3 mgs of methanol per mg of influent nitrate is used as a carbon source. Reactor empty bed contact time averaged 5.5 minutes. It appears that this type of process can be used successfully to remove nitrogen to meet low discharge concentrations at relatively low temperatures.

Introduction

The discharge of nitrogen in wastewater treatment plant effluent has been identified as a receiving water quality issue throughout the United States and in the European Community. Regulatory agencies are evaluating discharge limits for total nitrogen as low as 2 to 3 mg/l depending on the needs of the receiving water. To attain these limits, treatment plants will need additional tanks and equipment, but in many cases the plants are located in large urban areas with little or no space for expansion.

---

1 Director WPCD, City of Stamford, Harbor View Ave., Stamford, Ct. 06902
2 Senior Environmental Engineer, CH2M Hill, Newport News, Va.23606-4230
3 Operators, Laboratory Staff, WPCD, City of Stamford, Stamford, Ct. 06902

The Long Island Sound region has an especially active program to reduce the amount of nitrogen being discharged to that body of water. However, many of the treatment plants in that region are in densely populated cities, e.g., New York City, and have limited, if any, space for expansion. Technologies which can remove large quantities of nitrogen with relatively small space requirements are, therefore, of particular interest. Upflow fluidized bed reactors appear to meet those criteria.

Biological upflow fluidized bed (BFB) technology has been used since the 1970's for industrial waste treatment and more recently for the denitrification of contaminated groundwater, but there are very few BFB installations being used for denitrification of municipal waste.

In a review of BFB technology presented at the 1990 Water Pollution Control Federation Conference in Washington, D.C., Sutton and Mishra identified three municipalities currently using BFB technology for denitrification of wastewater: Reno, Nevada; Pensacola, Florida; and Riverside, California. The authors suggested that "wider use of the systems was hampered due to factors such as scale-up issues, slow development of economically attractive commercial system embodiments, and proprietary constraints".

Although the BFB technology appears to be well suited to the densely populated, Long Island Sound region, a full-scale demonstration was necessary to yield both design and operating data to confirm the applicability of this technology not only to this area, but also to other densely populated regions where high level nitrogen removal will be required.

The Stamford Water Pollution Control Facility is a 20-MGD, conventional activated sludge treatment facility which discharges directly to Long Island Sound. In recent years, the City of Stamford has received grants to study full-scale nitrification and denitrification using existing plant equipment. These research projects have been very successful with the plant achieving an average of 97% removal of ammonia and 55% removal of total nitrogen. Because the plant achieves such a high removal of nitrogen and because of the staff's experience in the management of research projects, the Environmental Protection Agency selected Stamford as the site for the biological upflow fluidized bed denitrification reactor demonstration project.

Upflow Fluidized Bed Reactor Description

The biological upflow fluidized bed system used for this demonstration project was supplied by ENVIREX, Ltd., Waukesha, Wisconsin, and was designed for a flow rate of 15.14 L/s.. The skid mounted unit is 4.6 m by 6.1 m. The denitrification reactor is 4.6 m high and 1.4 m in diameter. Included on the skid is a control panel and two fluidization pumps. Flow enters the process through two duplex strainers which remove any material which might settle out in the reactor or plug a line or pump. The flow is maintained at a constant rate by a series of flow meters and valves.

Two reactor influent pumps, each capable of pumping 15.8 L/s, were installed in a distribution chamber to supply nitrified, clarified secondary effluent to the reactor. Two methanol feed pumps, each capable of pumping 1.9 L/hr., were installed on the skid to provide carbon substrate to the reactor. Methanol is added prior to the strainers so that it is mixed well with the influent before entering the reactor.

Fine sand (0.5 mm in diameter) is used as the media in the reactor and provides a surface on which the denitrification organisms attach and remain in place even with water flowing through the reactor. The force of the water pumped into the reactor causes the bed to fluidize and expand. The large amount of surface area on which the organisms can attach, provides a very high biomass concentration within the reactor, which enables the denitrification process to proceed faster than in suspended growth systems and in much less space. The reactor bed height is controlled to between 2.9 m and 3.4 m. Wasting of excess biomass is done intermittently with the frequency dependent on the growth rate of the organisms and the operating temperature.

Project Description and Operation

The specific objectives of the project were:

- Determination of the operational and maintenance demands of the system.
- Determination of the complexities and practical aspects of procuring, storing, feeding and handling methanol.
- Validation of full-scale performance and reliability, particularly during cold weather operations.
- Evaluation of process control techniques such as methanol feed and solids/liquid separation.
- Identification of unexpected problems or benefits that emerge only in full-scale operations.
- Development of design criteria of the process and mechanical systems and more accurate cost estimations for construction and operation of BFB systems.

To accomplish these goals, a three phase program was developed; Phase 1-Startup, Phase 2-Baseline Testing, and Phase 3-Experimental.

Phase 1 began in August, 1993. After installation of the skid on a concrete platform, all piping and electrical systems were connected. The reactor was filled with sand to a level of 1.8 m. Water was then added to the unit to wash out the fines from the sand. The final settle bed height was 1.3 m and the initial expanded bed height was 1.8 m. Concurrently, representatives from ENVIREX began training the Stamford operators and laboratory personnel on the operation of the system and process control.

Filtered activated sludge was added to the flooded sand bed as a seed source of microorganisms. The mixture was then recycled for about two days. On August 18, forward flow was started at a rate of about 3.15 L/s and increased to 6.31 L/s on August 19. Full forward flow, 15.14 L/s, was established on September 15.

Beginning on August 18, methanol was fed to the reactor at a dosage determined by the following formula:

$$\text{Methanol dose, mg/l} = 2.47(NO_3) + 1.53(NO_2) + 0.87(DO)$$

Where:
NO_3 = Concentration of nitrate, mg/l
NO_2 = Concentration of nitrite, mg/l
DO = Influent dissolved oxygen concentration, mg/l

By early September, there was a fairly good growth of microorganisms on some of the sand particles and a layer 1 to 2 cm thick above the bed. The bed height had expanded to approximately 1.83 m and remained at that height until October 2. From October 2 to October 7, there was a rapid growth in the bed with the height at 3.5 m on October 7. The sand particles had a thick, reddish-brown growth of denitrifying organisms covering the entire surface.

Problems

During the start-up phase, August 18 to October 14, problems were encountered which required modifications to the reactor and both the influent and methanol pumping systems.

When attempts were made to increase forward flow to the design flow of 15.4 L/s, a problem with the hydraulics of the reactor became evident. As flow increased to about 9.5 L/s, the reactor level rose to the high level shut-off. Two additional attempts were made with the same result. An analysis of the problem showed that the effluent discharge pipe from the reactor (15.2 cm in diameter) was too small and had to be enlarged to 20.3 cm.

Once that modification was made, full forward flow to the reactor was increased, but the influent pumps could not deliver more than 12.9 L/s. The pumps were inspected and found to be wired incorrectly.

Throughout the entire start-up period, there were problems with the methanol feed pumping system. The system was plagued with air leaks which prevented the continuous flow of methanol into the reactor. The diaphragm housings and the four-way valves on the pumps were very fragile and while trying to stop the air leaks, several of them were broken and had to be replaced. In addition, the one-way valves supplied by the pump manufacturer did not work well, allowing air to enter the system. Although another brand one-way valve improved the situation, the final solution was to convert from the suction lift application to a flooded suction application.

Methanol addition is key to the performance of this process. When methanol could not be fed, nitrate removal efficiency decreased dramatically. During this start-up phase, influent nitrate concentrations ranged from 6.8 to 8.3 mg/l and effluent nitrate concentrations ranged from 0.0 mg/l to 7.0 mg/l, which show the effect of the intermittent methanol addition.

Sampling Plan and Analytical Program

Composite samples were taken three days per week and analyzed for TBOD, TCOD, TSS, TKN, NO_2, NO_3, TAN, Alkalinity, Orthophosphate and Total phosphate concentrations. On one composite sample per week, VSS, soluble BOD, and soluble COD concentrations were also determined. Daily grab samples were taken for evaluating the following parameters: pH, temperature, dissolved oxygen, nitrate, and nitrite.

In addition, the operators recorded the following measurements: bed height, process flowrate and methanol consumption.

Discussion and Results

Data collected during November and December represent the steady-state operation of the reactor. The system was operated with full forward flow and no recycle throughout this period. The nitrate loading during this period averaged 2371 kg $NO_3/1000\ m^3/d$ which was within the design loading of 1600 to 3200 kg $NO_3/1000\ m^3/d$. The design loading rate is dependent on temperature and can vary from one manufacturer's design to another. The hydraulic retention time (HRT) of the reactor was 5.5 minutes and the flux rate was 884 m^3/m^2-d.

One of the elements key to this project was the determination of the effect of temperature on the efficiency of the process. Figure 1 shows a temperature profile obtained throughout the entire test period. As can be seen, temperatures varied from a high of 29° C to a low of 13° C. The effluent nitrate concentration averaged 0.3 mg/l during November (average temperature 20.5° C) and 0.2 mg/l during December (average temperature 15° C). Therefore, it is concluded that temperature had no significant effect on nitrate removal efficiency in this temperature range.

Table 1 summarizes the performance data for the reactor during November and

December. Since methanol is added to the influent prior to the sampling point, there is an increase in the influent BOD and COD concentration due to the BOD and COD of the methanol itself. These data indicate that there is very good removal of the total and soluble BOD and COD within the reactor and very little excess methanol is being added to the system.

Table 1. Biological Fluidized Bed Performance, November and December, 1993

Parameter	Influent Conc., mg/l	Effluent Conc., mg/l	Percent Change
Total Nitrogen	10.8	2.2	-80
TKN	1.7	1.9	+12
Ammonia	0.1	0.0	N/A
Nitrite	0.0	0.1	N/A
Nitrate	9.1	0.2	-97.8
Total BOD_5	31.5	13.1	-58
Soluble BOD_5	19.8	3.5	-82
Total COD	82.9	53.6	-35
Soluble COD	64.8	37.3	-42
TSS	7.1	11.1	+56
VSS	7.5	10.7	-43
Alkalinity	55.6	82.1	+48

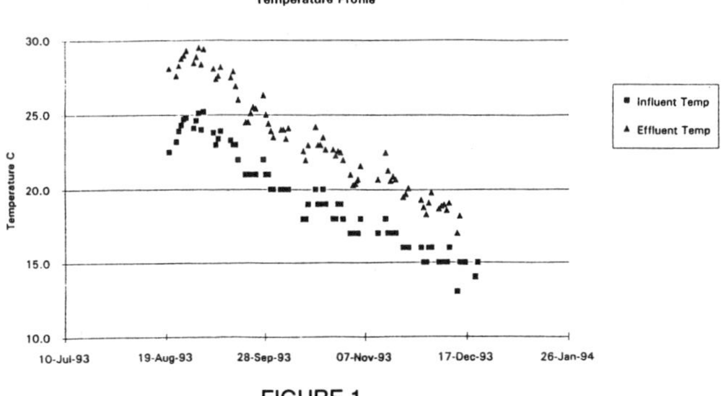

FIGURE 1

Both the effluent nitrate and TKN concentrations were lower than expected and showed that this process can be used to meet very low effluent total nitrogen limits. Figure 2 shows a graph of the influent and effluent total nitrogen concentrations

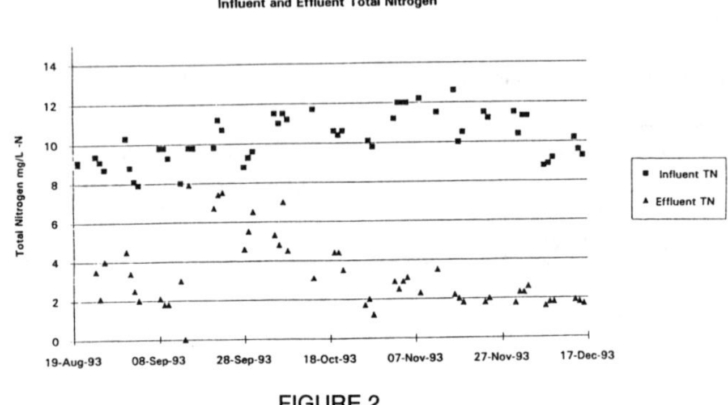

FIGURE 2

throughout the entire test period. Effluent total nitrogen concentrations averaged 2.2 mg/l and were well below the level of three to five mg/l proposed by the regulatory agencies within the Long Island Sound region.

During previous research projects at the Stamford WPCF, testing was done to determine the non-biodegradable, soluble TKN concentration. Non-biodegradable TKN is that combination of organic and ammonia nitrogen that cannot be further reduced in a biological system. It varies from plant to plant, but is usually in the range of 1.5 to 1.0 mg/l. It is especially important when regulatory agencies are establishing effluent nitrogen limits since it represents the practical minimum limit that can be met. At the Stamford WPCF, the non-biodegradable, soluble TKN concentration ranged from 1.5 mg/l to 2.5 mg/l. The average influent TKN concentration to the reactor was 1.7 mg/l and the average effluent concentration was 1.9 mg/l. The average effluent total nitrogen concentration was 2.2 mg/l for this same period. It can be concluded, therefore, that most of the biodegradable nitrogen is removed in this process.

The theoretical stoichiometric production of alkalinity is 3.57 mg as $CaCO_3$ per mg of nitrate reduced to nitrogen gas. Data obtained during this period show an alkalinity production of 3.0 mg $CaCO_3$ per mg of nitrate reduced which is less than the stoichiometric amount. This may be due to incomplete denitrification and/or another reaction occurring which consumes alkalinity.

It has been suggested that the upflow fluidized bed process could result in dramatic increases in effluent suspended solids requiring the use of sand filters to meet effluent standards. However, operational data indicate that a permit suspended solids concentration of 20 mg/l could easily be met using this system (Figure 3), since effluent concentrations of less than 12 mg/l were typically obtained. Further optimization should reduce levels even farther.

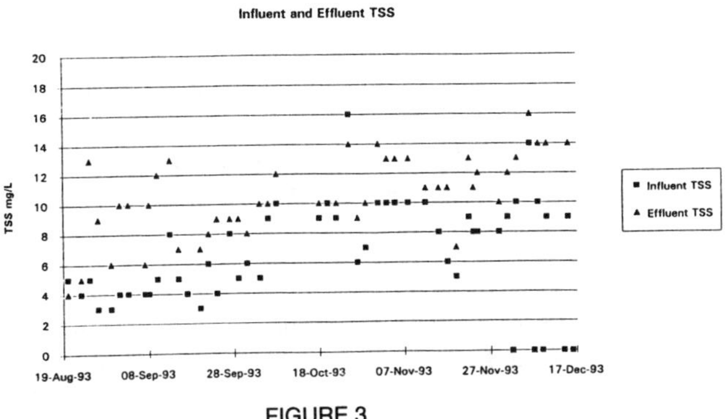

FIGURE 3

Conclusions and Future Studies

Data collected during this project indicate that upflow fluidized bed technology is a viable system for the removal of nitrogen. As expected, the high biomass concentration inherent to this technology creates a highly efficient system requiring only a small reactor size. This is extremely important for plants that are required to remove large amounts of nitrogen, but have only limited space.

During the next several months, cold weather testing will continue and experiments will evaluate the effects of high nitrate loadings (6296 kg NO_3/1000 m^3/d) and alternate carbon sources which may reduce costs. All of this information will then be used to develop the design criteria of the process and mechanical systems and determine more accurate cost estimations for construction and operation of BFB systems.

Acknowledgments

The authors want to acknowledge the contribution to this project by the entire ENVIREX staff, especially Mr. Robert Wenta; by Dr. John Jeris, Professor, Manhattan College; and by Mr. Roger Owens, Ecolotrol.

ASSESSMENT OF BNR SYSTEMS IN FLORIDA

Richard O. Mines, Jr.[1] Affiliate Member, ASCE
and C. William Woods[2]

Abstract

Operating data from eight, full-scale, single-sludge, biological nutrient removal (BNR) wastewater treatment facilities in Florida were collected and evaluated. Three types of BNR facilities were evaluated: A^2/O Process, 5-stage Bardenpho™ Process, and oxidation ditches preceded with anaerobic/anoxic selectors. One year of operating data were examined for each facility. All facilities met stringent effluent requirements of 5, 5, and 3 mg/L for biochemical oxygen demand (BOD_5), total suspended solids (TSS), and total nitrogen (TN). Facilities which had an effluent total phosphorus (TP) limit of 1 mg/L also consistently met the required level of removal. Oxidation ditches preceded with anaerobic/anoxic selectors were found to be as reliable and effective as the conventional BNR systems.

Introduction

In Florida, domestic wastewater treatment plants (WWTPs) that discharge to surface waters must meet effluent limits of 5, 5, 3, and 1 mg/L for five day biochemical oxygen demand (BOD_5), total suspended solids (TSS), total nitrogen (TN), and total phosphorus (TP), respectively (Thabaraj 1993). To meet these effluent requirements, advanced wastewater technologies must be used for removing nitrogen and phosphorus. Biological nutrient removal (BNR) technologies are the most promising since less sludge is generated than from facilities using chemicals for phosphorus removal and biological sludges can be land applied for ultimate disposal.

---

[1]Assistant Professor, University of South Florida, Department of Civil Engineering and Mechanics, 4202 E. Fowler Avenue, Tampa, Florida 33620-5350.

[2]Project Engineer, Gaylor Engineering, 400 Douglas Avenue, Suite C, Dunedin, Florida 34698.

This investigation assessed the performance of eight, full-scale, single-sludge, BNR systems operating in Florida. The approach consisted of collecting and evaluating 12 months of operating data for each facility. Influent and effluent wastewater characteristics evaluated included BOD_5, TSS, TN, and TP. Statistical analyses were performed to determine if there were significant differences in the process performance between the three BNR configurations evaluated. Another objective was to collect intermediate data on process performance at various stages throughout the treatment train at each facility. Unfortunately, this data did not exist or was not available. The results of this study are most useful to design engineers and planners faced with upgrading existing facilities, especially municipalities that operate oxidation ditches.

Description of Treatment Facilities

Table 1 presents a list of the facilities evaluated, design capacity, type of process utilized, chemicals added and dosages used, and if effluent filtration was being utilized. The design flow capacity of the facilities ranged from 11,400 m^3/day (5.0 mgd) to 56,800 m^3/day (15 mgd). The three types of BNR processes examined were: A^2/O, 5-stage Bardenpho™, and oxidation ditch preceded with anaerobic/anoxic selectors.

The A^2/O process was developed by Air Products, Inc. of Allentown, Pennsylvania and is patented by Kruger Process. This process consists of sequential anaerobic/anoxic/aerobic treatment of the wastewater. Return activated sludge (RAS) is recycled from the secondary clarifier back to the head of the anaerobic zone and mixed liquor recycle (MLR) is recycled from the aerobic or oxic zone to the anoxic zone at about twice the influent flowrate (2Q). The anaerobic zone is required for fermentation of the incoming substrate to volatile fatty acids and to enhance the proliferation of phosphorus removing microorganisms such as Acinetobacter. The anoxic zone is required for growing denitrifying microorganisms which remove nitrogen from the wastewater. TN concentrations in the effluent will range from 6 to 12 mg/L, whereas, TP concentrations will vary from 1.2 to 1.7 mg/L for the A^2/O process (The Soap and Detergent Association 1988, Air Products and Chemicals, Inc. 1990). The City of Largo and City of Dunedin utilize the A^2/O configuration at their wastewater treatment facilities. Both facilities use deep-bed denitrification (Denite®) filters with methanol addition for reducing nitrate nitrogen concentrations to < 3 mg/L TN. Methanol doses of 30 mg/L and 40 mg/L are used at the Dunedin and Largo plants, respectively. Alum is added to the mixed liquor entering the secondary clarifier at the Dunedin facility to meet the effluent TP requirement of 1 mg/L. The Largo WWTP is exempt from the phosphorus requirement.

The 5-stage Bardenpho™ process consists of sequential treatment in the following zones: anaerobic, first stage anoxic, oxic, second stage anoxic, and reaeration. EIMCO Process Equipment Company of Salt Lake City, Utah is owner

Table 1. Summary of Treatment Facilities

Location (1)	Design Capacity (m^3/d) (2)	Biological Process (3)	Supplemental Treatment (4)	Effluent Limits BOD_5, TSS, TN, TP (mg/L) (5)
Clearwater, FL (East)	18,900	5-Stage Bardenpho™	Sand Filters	5,5,3,-
Clearwater, FL (Northeast)	51,100	5-Stage Bardenpho™	Sand Filters	5,5,3,-
Clearwater, FL (Marshall St.)	37,900	5-Stage Bardenpho™	Sand Filters 60 mg/L alum added to secondary influent	5,5,3,1
Hillsborough Cty, FL (Dale Mabry)	22,700	Oxidation Ditch with Selectors	Denite® Filters, no methanol added, 60 mg/L sodium aluminate added to secondary influent	5,5,3,1
Hillsborough Cty, FL (Falkenburg)	22,700	Oxidation Ditch with Selectors	Denite® Filters, no methanol added, no alum added	5,5,3,1
Hillsborough Cty, FL (Valrico)	11,400	Oxidation Ditch with Selectors	Denite® Filters, no methanol added, no alum added	5,5,3,1
Largo, FL (Largo)	56,800	A^2/O	Denite® Filters, 40 mg/L methanol added, alum feed equipment provided but not utilized	5,5,3,-
Dunedin, FL (Dunedin)	22,700	A^2/O	Denite® Filters, 30 mg/L methanol added, 3 mg/L alum added to secondary influent	5,5,3,1

of the patented 5-stage Bardenpho™ Process. RAS is recycled from the secondary clarifier to the head of the anaerobic zone and MLR at approximately four times the influent flowrate (4Q) is recycled from the oxic zone to the first stage anoxic zone. Since this process has two anoxic zones, effluent TN concentrations of < 3 mg/L can consistently be maintained and effluent TP concentrations of < 1 mg/L can be realized (The Soap and Detergent Association 1988, Bowker and Stensel 1987). Three 5-stage Bardenpho™ facilities evaluated included: Clearwater East, Clearwater Northeast, and Marshall Street WWTPs. The East and Northeast plants do not have effluent TP limits.

Oxidation ditches preceded with anaerobic/anoxic selectors have been widely used in Hillsborough County, Florida, for meeting surface water discharge requirements. The three oxidation systems preceded with anaerobic/anoxic selectors that were evaluated included: Valrico, Falkenburg, and Dale Mabry. Following preliminary treatment, the wastewater flows through anaerobic/anoxic selectors and then into the oxidation ditches. Following secondary clarification, the effluent passes through Denite^R filters for polishing although methanol is not currently added for enhancing denitrification. RAS is returned to the headworks. MLR pumps are installed at Valrico and Falkenburg for recycling nitrates from the oxic zone to the anaerobic/anoxic selector although they are no longer used at either facility. There are no MLR pumps provided at the Dale Mabry facility. The anaerobic/anoxic selector provides the proper environment for the growth of non-filamentous microorganisms which retards filamentous growth in the ditches and conditions the sludge for phosphorus removal in the oxic zone. Both nitrification and denitrification are accomplished in the oxidation ditches by operating the mechanical aerators and/or brush rotors such that portions of the ditch remain aerobic and other portions remain anoxic. Average monthly effluent TN and TP concentrations from these plants range from 0.83 to 2.98 mg/L (standard error = ±0.56 mg/L) and from 0.12 to 0.56 mg/L (standard error = ±0.11 mg/L), respectively. Alum is not currently used for precipitating phosphorus at the Falkenburg or Valrico WWTPs, whereas, 28 mg/L of sodium aluminate is added to the mixed liquor entering the secondary clarifier at the Dale Mabry WWTP.

Results and Discussion

Table 2 lists the annual average influent wastewater characteristics for each facility. Each facility treats medium strength domestic wastewater with minor contributions from industry. The BOD_5:TP ratio of the raw wastewater ranged from 21 to 45 indicating that each of the facilities were candidates for BNR technologies. Table 3 presents the annual average effluent wastewater characteristics from the 8 facilities. All facilities were able to meet their effluent requirements on a consistent basis.

Figure 1 shows the effluent TN concentration (mg/L) versus percent of observations ≤ stated value. All facilities achieved effluent TN concentrations of 3

Table 2. Annual Average Influent Wastewater Characteristics

Facility	Flow (m^3/day) (1)	BOD_5 (mg/L) (2)	TSS (mg/L) (3)	TN (mg/L) (4)	TP (mg/L) (5)	BOD_5:TP Ratio (6)	Period (7)
Northeast	21,800	244	190	30[a]	6.1	40:1	8/92-7/93
East	11,200	186	131	26[a]	4.8	39:1	8/92-7/93
Marshall St.	27,300	229	181	29	5.1	45:1	8/92-7/93
Dunedin	16,100	162	194	---	5.1	32:1	8/92-7/93
Largo	45,400	140	208	33	---	21:1[b]	10/92-9/93
Falkenburg	11,300	229	229	35	6.1	38:1	9/92-8/93
Valrico	11,400	187	193	40[a]	5.7	33:1	6/92-6/93
Dale Mabry	9,400	210	203	34[a]	5.1	42:1	9/92-8/93

a = TKN value b = After (Morales et al. 1991)

Table 3. Annual Average Effluent Characteristics

Facility (1)	BOD_5 (mg/L) (2)	TSS (mg/L) (3)	TN (mg/L) (4)	TP (mg/L) (5)
Northeast	2.9	1.3	1.9	2.1
East Plant	2.8	1.3	1.7	1.9
Marshall St.	1.2	1.2	2.1	0.25
Dunedin	2.3	1.2	1.3	0.14
Largo	2.7	1.4	1.5	--
Falkenburg	1.3	0.42	1.2	0.25
Valrico	1.0	0.21	1.0	0.33
Dale Mabry	1.6	0.53	2.2	0.38

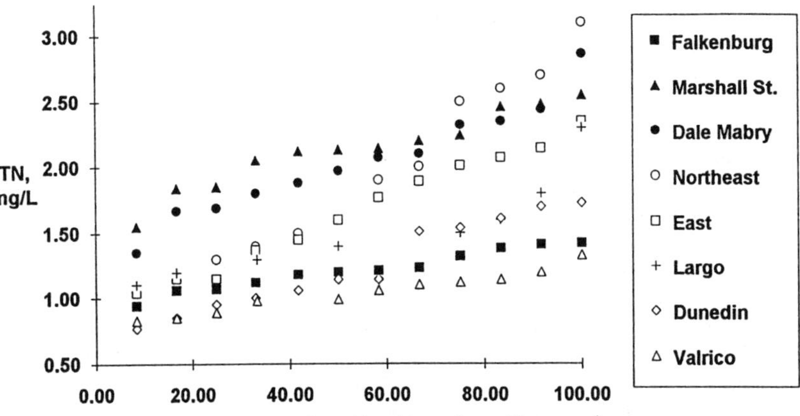

Figure 1. Effluent TN Concentrations

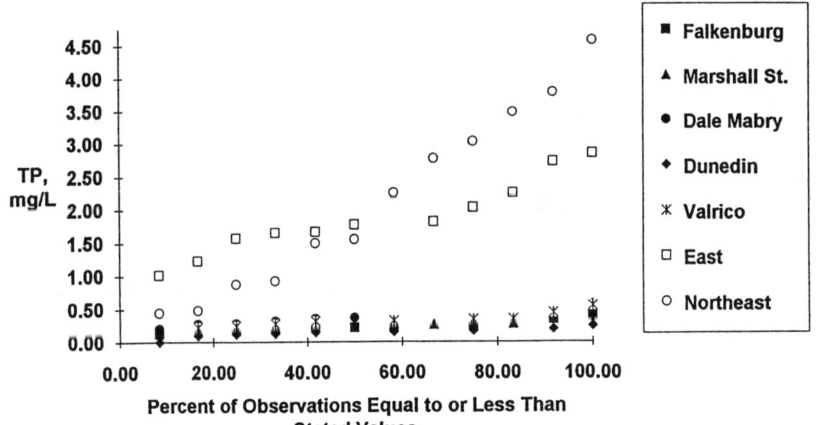

Figure 2. Effluent TP Concentrations

mg/L 95 percent of the time. The Valrico WWTP (oxidation ditches preceded by anaerobic/anoxic selectors) achieved the lowest effluent TN levels consistently. Figure 2 presents the effluent TP concentration (mg/L) versus percent of observations ≤ stated value. This figure indicates that facilities with an effluent TP limit of 1 mg/L were capable of achieving an effluent TP concentration of 0.55 mg/L 100 percent of the time. The Dunedin WWTP (A^2/O process) consistently achieved the lowest effluent TP concentrations. The two facilities with no effluent TP limit (East and Northeast WWTPs) could, most likely, be fined tuned to meet the 1 mg/L TP limit by adding a coagulant. Evaluation of the data also indicated that the oxidation ditches preceded with anaerobic/anoxic selectors consistently removed nitrogen and phosphorus. Efficient removal was thought to have been accomplished in part because two of the three facilities evaluated were not at design flow capacity which provided longer detention times in the selectors and oxidation ditches, thereby enhancing nutrient removal. However, the Valrico plant was operating essentially at design capacity which nullifies this hypothesis.

Statistical analyses were performed on the data to determine if a significant difference in process performance existed between the three types of BNR systems evaluated. A one-way analysis of variance (ANOVA) with equal sample size was conducted on effluent TN and TP concentrations. All facilities with exception to the Dunedin WWTP (influent TN samples are not collected), were used in the ANOVA test on effluent TN concentrations. Facilities examined for the ANOVA test on effluent TP concentrations included all facilities with exception to the Largo WWTP for which influent and effluent TP concentrations are not determined. At a 95 percent confidence level, 6 degrees of freedom for variance between the samples, and 77 degrees of freedom for variance of error, the null hypothesis was rejected for both effluent TN and TP concentrations indicating there was a significant difference in process performance for each of the treatment facilities evaluated.

Two-tailed Student "t" tests were performed on paired samples; tests were conducted on the means of the seven facilities for percent TN removed and on the means of the seven facilities for percent TP removed. These analyses were performed at a confidence level of 95 percent with eleven degrees of freedom. In regard to nitrogen removal, the performance of the Valrico WWTP was found to be significantly different when compared to the other six plants. This agrees with the previous data interpretation that the Valrico facility achieved the lowest effluent TN concentrations. The performance of the Dunedin WWTP was significantly different, at a 95 percent confidence level, in regard to TP removal when compared to the six other WWTPs. This agrees with the previous evaluation of the data showing the lowest effluent TP concentrations were achieved with the A^2/O process.

Conclusions

Evaluation of the A^2/O Process, 5-stage Bardenpho™ Process, and oxidation ditch technology preceded with anaerobic/anoxic selectors indicated that each of the technologies was capable of meeting effluent TN and TP concentrations of 3 mg/L and 1 mg/L, respectively. The A^2/O technology requires a secondary denitrification process such as deep-bed denitrification filters to achieve the 3 mg/L effluent TN limit. The Dunedin plant (A^2/O) achieved the lowest effluent TP concentrations of the 7 facilities evaluated. Effluent TP concentrations of < 0.24 mg/L were obtained with a minimal alum dose of 3 mg/L. The three Bardenpho™ facilities evaluated met their effluent requirements consistently. An alum dose of 60 mg/L was required at the Marshall Street plant to meet the 1 mg/L TP effluent limit. The three oxidation ditches preceded with anaerobic/anoxic selectors were found to be a viable and effective technology for removing nitrogen and phosphorus from wastewater. Annual monthly effluent concentrations for TN and TP ranged from 0.83 to 2.98 mg/L (standard error = ±0.56 mg/L) and 0.12 to 0.56 mg/L (standard error = ±0.11 mg/L), respectively.

Acknowledgements

This work was supported, in part, by the University of South Florida Research and Creative Scholarship Grant Program under Grant No. 21-04-938-RO.

We would like to thank Jeff Elick (City of Largo); Gilbert Gardner, Mark Caranza, Ken Stanczykowski, and Chuck Yawn (Hillsborough County); Jan Magdziasz (City of Dunedin); and Joseph Reckenwald (City of Clearwater) for providing us with the plant operating data.

References

Air Products and Chemicals, Inc., A^2/O Operating Data (1990).

Bowker, R. P. G. and Stensel, H. D., Design Manual for Phosphorus Removal. EPA-625/1-87/001, U.S. Environmental Protection Agency, Washington, D.C. (1987).

Morales, L. M., et. al., "Capability Assessment of Biological Nutrient Removal Facilities." Research Journal WPCF, 63,900-909. (1991).

Thabaraj, G. J., "AWT Processes in Florida". Florida Department of Environmental Regulation, Tampa, FL (1993).

The Soap and Detergent Association. "Principles and Practice of Nutrient Removal for Municipal Wastewater." New York, NY. (1988).

Automated ORP Control for Aerobic/Anoxic Cycling in Oxidation Ditch Treatment

Thomas E. Coleman, Member[1], H. David Stensel, Member[2], Darrel Fleischman[3] and W. Brent Denham, Member[1]

Abstract

The 1140 m^3/d (0.3 mgd) Grand Coulee, WA. oxidation ditch was modified to incorporate a computer-control anoxic/aerobic cycle operation to investigate SVI control and nitrogen removal. A directional mixer placed in the ditch is turned on and the aerators are turned off at selected times to create the anoxic cycle. The control system requires continuous ORP measurements in the ditch, with computer logging, data synthesis and decision making. An automated, single anoxic/aerobic control cycle was studied for a 5-month period. Diluted SVI values were maintained in a range of 70-110 mL/g for a 3-month period after chlorination was used to breakup a filamentous floc. Nitrogen removal averaged between 76 and 80 percent.

Introduction

Hundreds of wastewater treatment facilities in the United States and abroad have been designed for secondary treatment using the oxidation ditch activated sludge process (WEF/ASCE MOP 8, 1992). The distinguishing features of the oxidation ditch are the "race track" configuration of the aeration tank channel and the use of long aeration and solids retention times (SRTs); in the order of 24 hours and 30 days, respectively. Traditional advantages for these systems has been the simple and cost effective treatment scheme, simple and easy operation, and relatively high quality effluent in terms of BOD, suspended solids, and oxidized nitrogen. Improvements in the treatment capability of oxidation ditches should be expected, in view of recent improvements in conventional activated sludge treatment designs that result in better and more reliable sludge settling characteristics and high levels of nitrogen removal.

[1]Senior Engineer and Project Engineer, Gray & Osborne, Inc. Consulting Engineers, P.O. Box 2069, Yakima, WA. 98907

[2]Professor, Department of Civil Engineering, FX-10, University of Washington, Seattle, WA. 98195

[3]Plant Superintendent, Grand Coulee, WA. 98901

Oxidation ditch systems generally develop poor settling sludge as indicated by high sludge volume index (SVI) values in the range of 200 - 250 mL/g. The cause of the poor SVIs in oxidation ditches appears to be related to filamentous bacteria growth due to their low BOD loading rates or food to mass (F/M) ratios, as described by Jenkins et al. (1984). The basic problem is that there are no selective pressures in the system to favor the growth of good floc-forming, non-filamentous bacteria.

Albertson (1987) described the historical development of various modifications to the activated sludge process that resulted in selective pressures to favor the growth and predominance of non-filamentous bacteria, leading to low operating SVIs. The modifications include the installation of anaerobic, anoxic or high F/M ratio contact zones ahead of the activated sludge aeration tank, where a large portion of the influent BOD is removed by non-filamentous bacteria, before filamentous bacteria can compete for it in the aeration tank. When an anoxic contact tank is used, oxygen is limited and non-filamentous bacteria use nitrate to consume the influent BOD. A recycle stream from the aeration tank supplies the nitrate and the anoxic tanks may have detention times ranging from 1 to 3 hours (Stensel, 1981). This method was successfully demonstrated by Heide and Pasveer (1974) for an oxidation ditch and SVIs of 70 mL/g were achieved, compared to SVIs as high as 500 mL/g without the pre-anoxic contact basin.

Besides SVI control, the anoxic/aerobic system can reliably provide high levels of nitrogen removal due to nitrate reduction in the pre-anoxic tank. Nitrogen removals ranging from 80-85 percent are possible, using aeration tank recycle ratios from 3.0 to 4.0, respectively. Nitrogen removal is also possible in oxidation ditch systems without pre-anoxic tanks, by the creation of anoxic zones in the ditch channels downstream from the aerator locations, as biological respiration depletes the dissolved oxygen (DO). Removal levels range from 50-90 percent, but are not as reliable and consistent as the pre-anoxic designs. Consistent nitrogen removal requires careful control of the aeration power draw by varying impeller submergence or aerator speed as the plant load varies.

A disadvantage of using the anoxic/aerobic treatment system is the cost and space associated with providing an additional basin and mixed liquor recycle system. A less expensive and simpler alternative is to use an anoxic/aerobic cycling operation within the oxidation ditch. This method has been applied at the Grand Coulee, WA. wastewater treatment plant for nitrogen removal and SVI control, using an automated, computer-operated control system with oxidation-reduction probe measurements. A directional mixer is placed in the oxidation ditch channel and at selected times the mixer is turned on and the ditch aeration is turned off. During the aerator OFF period, nitrate is used instead of oxygen to consume influent BOD, to result in a non-filamentous growth selector operation and nitrogen removal. Various numbers of OFF/ON cycles can be used daily depending on the treatment objectives (Coleman et al., 1993). An important consideration for successful application of this anoxic/aerobic cycling process is knowing when to turn off the mixer and to turn on the aerators. If the OFF period continues past the time of nitrate depletion, the overall treatment performance is impaired due to the lack of an electron acceptor. If the OFF period is terminated with nitrate present, maximum nitrogen removal and selection pressure is not accomplished. This paper reports on the results of the development and application of an automated computer-operated control system at Grand Coulee, WA. that uses continuous readings from an oxidation-reduction potential (ORP) probe placed in the oxidation ditch to control the length of the aerator OFF time (anoxic time).

ORP is a measurement of the ratio of oxidants to reductants in a system. The control system doesn't depend on the absolute value of ORP, but its changes as oxygen and then nitrate is depleted in the oxidation ditch mixed liquor. Peddie et al. (1990) showed that a dramatic change occurs in the slope of an ORP versus time curve when nitrate is depleted. Wareham et al. (1993) used this information to control anoxic/aerobic cycling in bench-scale aerobic digester testing.

A microprocessor-based control system, installed at the Grand Coulee plant, logged and interpreted ORP measurements every minute. Based on changes in the slope of the ORP versus time curve, the computer determined when to initiate aeration after the start of an anoxic period. The control system was operated using a 486 interface computer with programmable controller handling the analog signals from a Broadley James Corporation ORP probe, Exidyne Instrumentation Technologies DO probe and the discrete outputs for aeration control. The Grand Coulee oxidation ditch system has a design capacity of 1140 m^3/d (0.3 mgd) and the average daily flow during the study was about 760 m^3/d (0.2 mgd). The ditch volume and depth are 1140 m^3 and 3.05 m (10 ft.), respectively and it is aerated by two 11.2 Kw (15 Hp) surface brush aerators. One of two 10.7-m (35 ft.) diameter clarifiers was in operation during the study. A submersible mixer installed in the channel was able to maintain a minimum channel velocity of 30 cm/sec (1.0 ft/sec) with a power draw of 1.8 Kw (2.4 Hp), with the aerators turned off.

Control System Installation and Testing Program

The automated control system was placed on-line on September 20, 1993. One anoxic cycle per day was selected for the initial operation, with the aerator OFF period initiated at 8:00 AM. The length of the anoxic period ranged from 4-8 hours, with most of the longer off periods occurring during the colder temperature months. Prior to automated operation, the system had been operating manually with two anoxic/aerobic cycles per day to observe overall performance and ammonia oxidation and nitrate reduction rates during aerator on and off times, respectively.

A number of plant performance operating parameters were monitored routinely during the anoxic/aerobic cycle testing including aeration basin mixed liquor suspended solids (MLSS) and volatile suspended solids (MLVSS) concentrations, SVI, SRT, and temperature. Microscopic observations on the activated sludge floc characteristics and filamentous growth characteristics were done weekly. Flow proportioned influent and effluent composite samples were obtained daily to monitor $TBOD_5$, TSS, NH_4-N, NO_3-N plus NO_2-N and soluble Chemical Oxygen Demand (SCOD) concentrations. BOD and solids analyses were done in accordance with Standard Methods for the Examination of Water and Wastewater, 18th Ed. (1989). SCOD analyses were done with the Hach COD closed reflux method. NH_4 and oxidized nitrogen analyses were done using the Hach Company reagents and a Hach DR 3000 spectrophotometer.

The SVI did not decrease when the anoxic/aerobic cycling operation was started and remained quite high, in the 200 mL/g range, and a significant level of filamentous bacteria was observed in the floc. There was concern that the selector pressure expected from anoxic cycling was not strong enough to overcome the large filamentous population in a reasonable time period, so that chlorination of the mixed liquor recycle was carried out from November 7 to November 20, 1993. The chlorination dose ranged from (2 to 6 Kg/1000 Kg MLSS-d) and only at the higher dose were the filamentous bacteria broken up. The SVI and treatment performance continued to be monitored up to February 15, 1994.

Results and Discussion

Figure 1 shows a typical curve of the variation in ORP, and DO, NH_4-N, and NO_3-N concentrations during anoxic/aerobic cycling. When the mixer is turned on and the aerators turned off, the DO, NO_3, and ORP values decline and the NH_4-N concentration increases. The ORP curve shows two significant inflection points; the first is where the DO concentration reaches zero and the second is where the nitrate is depleted.

FIGURE 1

CHANGES IN ORP, NH4-N, AND NO3-N DURING ANOXIC/AEROBIC CYCLING

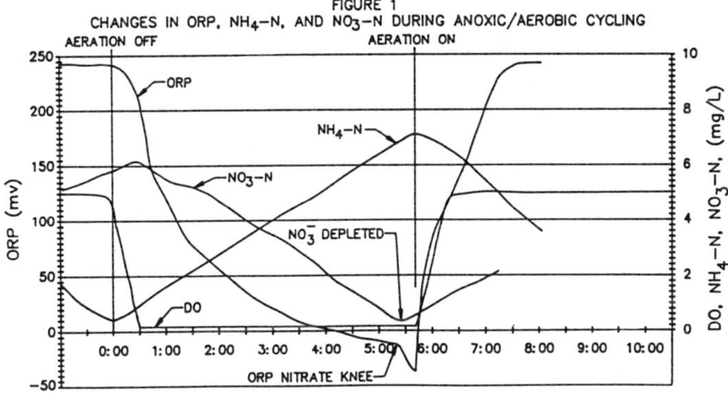

After the latter the aeration is started and the ORP, DO and NO_3 values increase.

Plant performance in terms of sludge settling characteristics and nitrogen removal, are presented here for the anoxic/aerobic cycling operation from September 1, 1993 to February 15, 1994. During this period the MLSS concentration was normally around 3500 mg/L, the SRT from 25 to 35 days, and the F/M ratio around 0.05 g BOD/g MLSS-d. After the chlorination period the MLSS was lower for about a month, at around 3000 mg/L. The average plant flow during this testing period was 893 m^3/d (0.235 mgd). DO profiles around the aeration basin channels were determined at various times and DO concentrations were always found to be 1.0 mg/L or more.

Outside of the chlorination period the plant effluent TSS and BOD concentrations were consistently below 10 mg/L and for many months averaged less than 5 mg/L. During the chlorination period in November and after it during December, these concentrations averaged less than 20 mg/L.

Figure 2 summarizes the plant performance during the anoxic/aerobic cycling. At approximately day 20 the automated operation was started with one cycle per day, and the mixed liquor temperature (2a) was near 20^0C. During November, which is day 60 to day 90, the temperature steadily dropped from about 15^0C to 11^0C. During January (after day 120) it was usually at 10^0C, and during February, (after day 150) it ranged from 7 to 9^0C.

Figure 2b shows that the diluted SVI varied around 200 mL/g until chlorination was used after the first week of November. The chlorination effort lasted almost 2 weeks and after that the SVI value dropped to about 110. The SVI dropped further during December and averaged 70 mL/g during that month. It increased during January and February, but still averaged what are considered good SVI values of 94 and 108 mL/g, respectively. The increase in SVI during February may be related to the decline in the mixed liquor temperature. Microscopic observations showed no significant increase in filamentous growth extending from the floc or bridging of floc particles, as observed previously with high SVI conditions. Fairly dense, zoogleal-like floc particles were observed. In summary low SVIs were observed for almost 3 months after chlorination. Based on historical SVIs of 200 mL/g or above for the plant operation, these results suggest that the anoxic/cycling operation is providing selective pressure to maintain low SVI levels. Longer term results are being obtained at the same operating condition, with automated ORP control, for confirmation.

Figures 2c and 2d summarize effluent nitrogen concentrations and percent nitrogen removal during the test period. The chlorination affected the nitrogen removal efficiency between days 70 and 100, because of some destruction of the nitrifying bacteria population. By day 100, which is December 10th, the nitrification population had recovered. For the anoxic/aerobic periods before and after the chlorination inhibition, the nitrogen removal efficiency weekly average values were between 76 and 80 percent.

Nitrogen removal efficiencies were calculated using daily composite influent ammonia and effluent ammonia and nitrate plus nitrite concentrations, and previous data that showed that the influent ammonia-nitrogen concentration was 85% of the influent Total Kjeldhal Nitrogen (TKN) concentration. The average effluent ammonia-nitrogen and nitrate + nitrite-nitrogen concentrations were 1.1 to 1.3 mg/L and 3.0 to 4.3 mg/L, respectively. During chlorination these increased to 5.3 and 4.2 mg/L, respectively.

FIGURE 2

ANOXIC/AEROBIC CYCLING TEST RESULTS

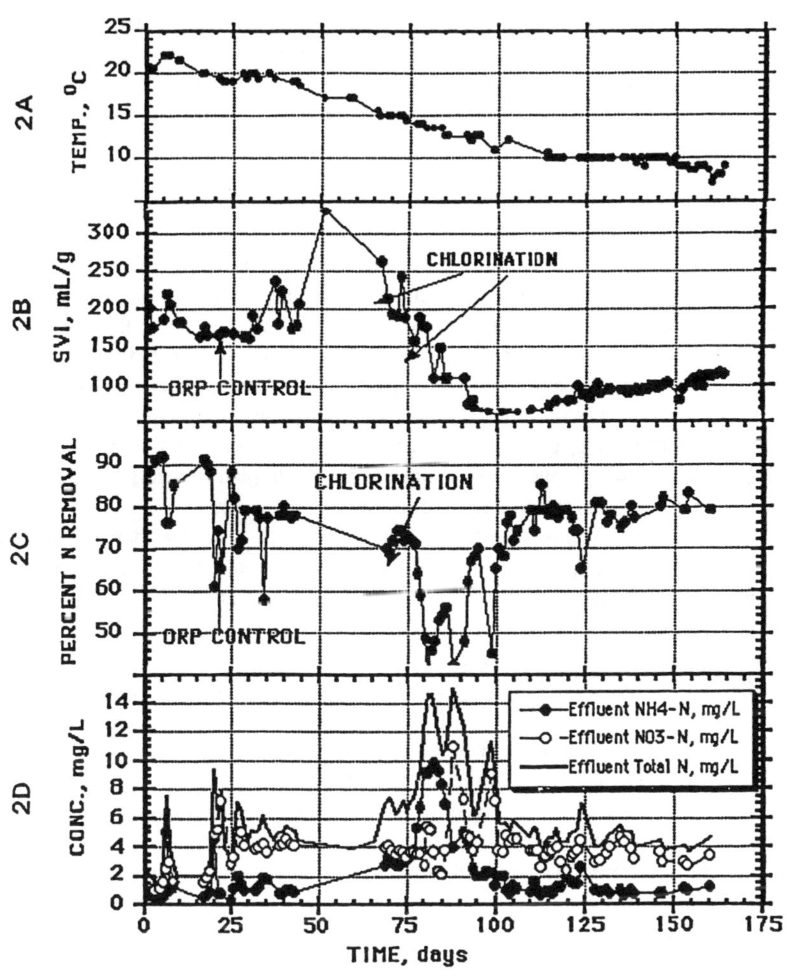

Slightly better nitrogen removal was observed for the two cycle per day operation that was used between days 0 and 20, prior to the one cycle per day operation after September 20th. During that period the percent nitrogen removal averaged 84% and the effluent ammonia and nitrate plus nitrite concentrations were 1.4 and 2.3 mg/L, respectively. Previous work (Coleman et al., 1993) suggested that more frequent cycles would improve nitrogen removal efficiency, but perhaps result in higher SVI values due to more frequent exposure to low DO conditions. This needs further long term investigation.

The utilization of nitrate during the anoxic periods represents a significant potential for energy savings as compared to a continuously aerated system. Total power consumption records for the Grand Coulee wastewater treatment plant were reviewed in relation to the influent BOD loading for the years 1987 through 1993. Since the submersible mixer was installed in 1990 the average kwh use per pound of BOD treated has been 1.95 verses 2.26 kwh/lb prior to 1990. This represents a 13.75 percent reduction in total power use.

Summary and Conclusions

An automated, computer-controlled anoxic/aerobic cycling operation, based on oxidation-reduction potential measurements, was incorporated into an existing oxidation ditch system to investigate its ability to control SVI and provide reliable and efficient nitrogen removal. The control system required the use of a microprocessor and the installation of a directional mixer and ORP and DO sensors in the ditch. The control system was continuously operated for a 5-month period, up to the preparation of this paper, and continues to be used. The following conclusions from the full-scale plant investigation using a one cycle per day operation are as follows:

1. SVI control was maintained for a 3 month period under automated anoxic/aerobic cycling, after chlorination was applied to minimize the filamentous population level. Diluted SVI values ranged from 70 to 110 mL/g during this period.
2. For the single anoxic cycle used, a 4-8 hour anoxic period was needed to deplete the nitrate nitrogen each day, with longer times required during colder temperatures.
3 For the single anoxic cycle used:
    - Nitrogen removal averaged from 76-80 percent.
    - Effluent ammonia-nitrogen averaged 1.1-1.3 mg/L.
    - Effluent nitrate plus nitrite-nitrogen averaged 3.0-4.3 mg/L.
4. Higher nitrogen removal levels are possible with more frequent anoxic/aerobic cycles.
5. Total energy use at the Grand Coulee Wastewater Treatment Plant has been reduced by 13.5 percent based on kwh per pound of influent BOD loading since the installation of the submersible mixer.

References

Albertson, O.E. (1987) "The Control of Bulking Sludges: From the Early Innovators to Current Practice", *Jour. of the Wat. Pollut. Control Fed,* **59**, 172-182.

Coleman, T.E., Stensel, H.D., Fleischman, D. and W.B. Denham. (1993) "Oxidation Ditch Treatment with Anoxic/Aerobic Cycling", *Proceedings of the CSCE/ASCE National Conference on Environ. Engr.*, July 12, 1993, Montreal.

Design of Municipal Wastewater Treatment Plants, Manual of Practice 8. (1992), Water Environment Federation and Amer. Soc. of Civil Engrs., 927.

Heide, B.A. and A. Pasveer. (1974) "Oxidation Ditch: Prevention and Control of Filamentous Sludge", H_2O, **7**, 373.

Jenkins, D., Richards, M.G., and G.T. Daigger. (1984) Manual on the Causes and Control of Activated Sludge Bulking and Foaming. Water Research Commission, Pretoria, Republic of South Africa.

Peddie, C.C., Mavinic, D.S., and C.J. Jenkins. (1990) "Use of ORP for Monitoring and Control of Aerobic Sludge Digestion", *Jour. of Environ. Engr., Amer. Soc. of Civil Engrs.*, **116**, (3), 461

Randall, C.W., Barnard, J.L. and H.D. Stensel. (1992) Design and Retrofit of Wastewater Treatment Plants for Biological Nutrient Removal. Vol. 5, Chapter 4, Technomic Publishing Co., Lancaster, PA.

Standard Methods for the Examination of Water and Wastewater. (1989). 18th Ed. American Public Health Association, Washington D.C.

Stensel, H.D., (1981) "Biological Nitrogen Removal System Design", Water, American Institute of Chemical Engineers, 237.

Wareham, D.G., Hall, K.J., and D.S. Mavinic, (1993) "Real-time Control of Aerobic-Anoxic Sludge Digestion Using ORP. *Jour. of Environ. Engr., Amer. Soc. of Civil Engrs.*, **131**, 120

Modeling of Biofiltration of Natural Organic Matter
in Drinking Water Treatment

Jack Z. Wang[1] and R. Scott Summers, Member of ASCE[2]

Abstract: A multi-substrate model of the biofiltration of natural organic matter in drinking water was developed and successfully applied. This model assumes that biofiltration is a two-step process in which external mass transfer is followed by surface bioreaction. The Monod-type bioreaction rate is used and mass transfer resistance due to the microbial biomass is neglected in this model. The natural organic matter is fractionated into fast-, slow- and non-biodegradable components. The sensitivity of this model to each of the individual parameters indicates that the biodegradation of natural organic matter in drinking water biofilters is controlled by the surface bioreaction rate. Comparison of the model predictions to experimental results indicates that this model works well over a wide range of hydraulic loading rates.

INTRODUCTION

Biofiltration is a promising technique for controlling organic matter, disinfection by-products and microbial regrowth in drinking water distribution systems (Rittmann and Huck, 1989), and as such, much research has been performed on drinking water biofiltration in recent years (Prévost et. al., 1989; Servais et. al, 1991; Miltner and Summers, 1992; LeChevallier et. al., 1992;). A few of these studies have focused on biofiltration modeling using a biofilm model (Zhang and Huck, 1993; Lu and Huck, 1993). However, studies have shown that the amount of biomass present in drinking water biofilters is very low because of low substrate concentrations and that no complete or continuous biofilm has been found on the surface of the filter media (Wang and Summers, 1993). Thus, a biofilm model which includes biofilm thickness and the biofilm density may not be needed for simulating drinking water biofiltration. The objectives of this paper are to develop a more representative model and to validate it under practical conditions.

---

[1] Ph. D. Candidate; [2] Associate Professor
Department of Civil and Environmental Engineering, University of Cincinnati, Cincinnati, OH 45221-0071.

MODEL DEVELOPMENT

Based on the existing understanding of drinking water biofiltration, the following assumptions are made in the derivation of this model: 1) biomass is homogeneously distributed on the surface of filter media; 2) substrate diffuses through a liquid layer to the surface of filter media and then is utilized by the cells attached to the surface; 3) the diffusion of substrates into a bulk biomass can be neglected because of low amount of biomass present; and 4) because the amount of biomass in drinking water filter changes slowly with operation time (Wang and Summers, 1993), an isothermal pseudo steady-state operation is assumed.

A Monod-type bioreaction rate is used to model substrate utilization on the surface of the filter media. Based on the above assumptions, for a single substrate the model can be mathematically expressed as follows :

$$V_0 \frac{\partial C(z)}{\partial z} - \varepsilon E_L \frac{\partial^2 C(z)}{\partial z^2} + \frac{3}{R}(1-\varepsilon))K_L(C(z) - C_R(z)) = 0 \quad (1)$$

$$\frac{3}{R}(1-\varepsilon)K_L(C(z) - C_R(z)) = \frac{V_{max} X C_R(z)}{K_s + C_R(z)} \quad (2)$$

$$\text{At } z = 0, \quad C(z) = C_0 \quad (3)$$

$$\text{At } z = H, \quad C(z) = C_e \quad \text{or} \quad \frac{\partial C(z)}{\partial z} = 0 \quad (4)$$

where V_0 is filter velocity (m/s); z is filter depth (m), C(z) is substrate concentration in the biofilter (mg/L); $C_R(z)$ is substrate concentration at the surface of filter media (mg/L); ε is bed porosity; R is the radius of filter media (m); E_L is axial dispersion coefficient (m^2/s); K_L is external mass transfer coefficient (m/s); V_{max} is maximum substrate utilization rate (mg/mg cells/s); X is biomass concentration in biofilter (mg cells/L) and K_s is half maximum rate concentration (mg/L). Under the hydrodynamic conditions of drinking water filtration, axial dispersion can be neglected. Therefore, Equation 1 can be further simplified to:

$$V_0 \frac{\partial C(z)}{\partial z} + \frac{3}{R}(1-\varepsilon))K_L(C(z) - C_R(z)) = 0 \quad (5)$$

and Equation 4 is no longer needed in this model. By solving Equations 2, 3, and 5 simultaneously, this model can predict the change of a single substrate concentration in drinking water biofilters.

However, natural organic matter (NOM) is comprised of a wide range of organic compounds, especially for ozonated water (Langlais et. al., 1991). The biodegradation kinetics of these components can also be quite different. Prévost et. al. (1990) have found that the biodegradation of the assimilable organic carbon (AOC) fraction was much faster than that measured as biodegradable dissolved organic carbon (BDOC). For a two minute empty bed contact time (EBCT), 62% to 90% of AOC was biodegraded. To achieve the same amount of removal for BDOC, 10 to 20 minutes of EBCT was necessary. Therefore, a single substrate approach is not representative. Based on additional previous studies (Morgen, 1990; Carlson and

DiGiano, 1992) and the results reported herein, a three component approach is proposed. They are a fast biodegradable component which may be represented by aldehydes, ketoacids or AOC, slow biodegradable and non-biodegradable components. Servias and Billen proposed a similar concept (Langlais et. al., 1991). It is assumed that there are no interactions or competition among these three components and they are assessed by dissolved organic carbon (DOC). The DOC changes of the biodegradable components in the biofilter are both governed by Equations 2, 3 and 5.

SENSITIVITY ANALYSES OF PARAMETERS

A sensitivity analysis was used to determine which parameters impacted the model results most. The parameters used in this model are listed in Table 1. For each parameter, the starting value for the sensitivity analysis was either selected from the literature or estimated from early experiments. Selected parameters were varied one at a time over a range of values to test the response of the model.

Table 1. Input Parameters for Sensitivity Analysis

		Fast component			Slow component		
X	K_L	V_{max1}	K_{s1}	f_1	V_{max2}	K_{s2}	f_2
mg cells/L	m/s	mg/mg cells/day	mg/L	-	mg/mg cells/day	mg/L	-
200	$1 \cdot 10^{-5}$	2.5	0.2	0.15	0.1	0.2	0.25

The maximum utilization rate determines how fast the substrate can be utilized by microorganisms in the filter. If the removal of organic matter by biofiltration process is controlled by the biodegradation rate, changes in the biodegradation rate will greatly affect the model prediction. The impacts of V_{max1} and V_{max2} on the model predictions of the substrate concentration change with the biofilter depth are shown in Figures 1 and 2, respectively. The model prediction is very sensitive to both of the maximum utilization rates. The DOC concentration at each filter depth decreases as the maximum utilization rates increase. However, the responses of the model prediction to the changes in V_{max1} and in V_{max2} are different. The change in V_{max1} greatly affects the model prediction in the top part of the biofilter. In the top part of the filter, the degradation of the fast degradable component dominates the DOC removal and the removal due to the degradation of the slow degradable component is negligible compared with the fast biodegradable component. The impact of biomass on the prediction is similar to those of the maximum utilization rates as the biomass and maximum utilization rate are coupled together in the model equations: XV_{max}. The increase in biomass in the biofilter provides more bioreaction sites on the surface of the support media and increases the biodegradation rate.

The response of the model prediction to the changes of K_{s1} is shown in Figures 3. The model is moderately sensitive to K_{s1}. As the K_{s1} value increases, the model

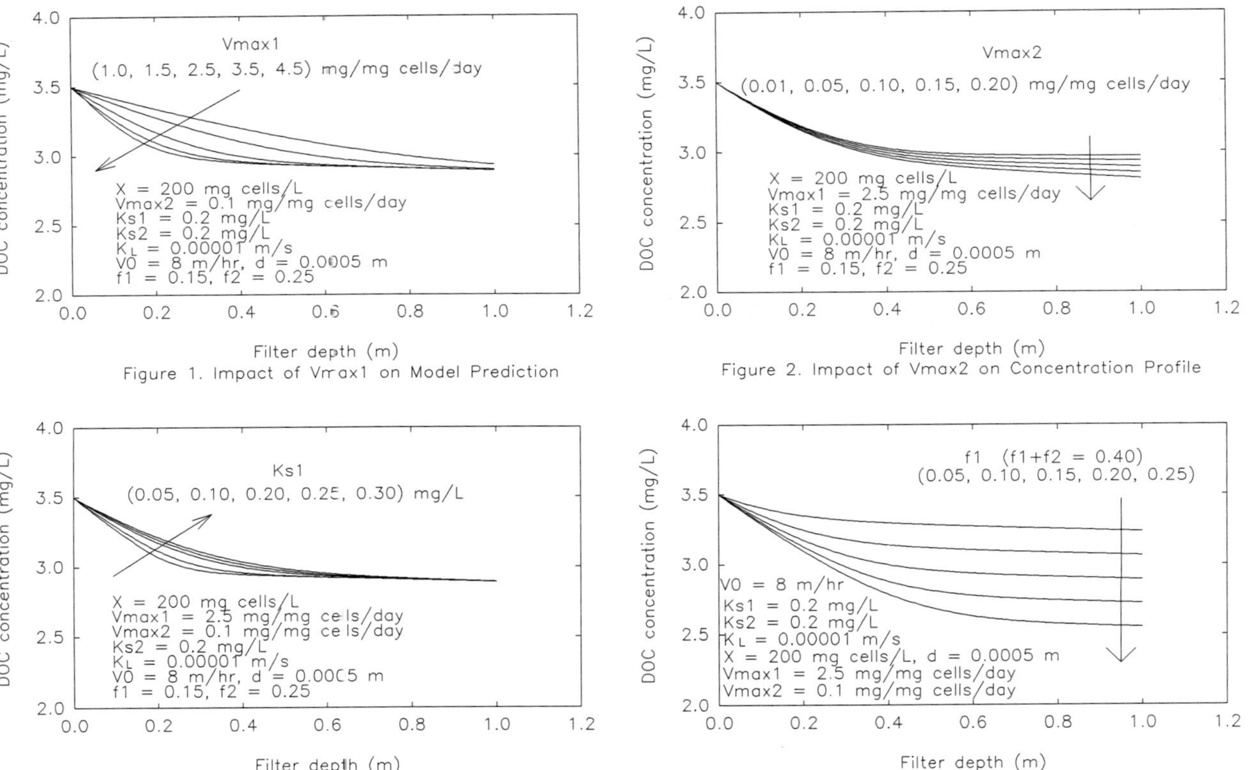

Figure 1. Impact of Vmax1 on Model Prediction

Figure 2. Impact of Vmax2 on Concentration Profile

Figure 3. Impact of Ks1 on Biofiltration

Figure 4. Impact of BDOC Fractions on Biofiltration

predicts a slower biodegradation rate in the top part of the filter. In the Monod type reaction, the reaction rate approaches zero-order when K_s is much smaller than the reactant concentration, and will be independent of the K_s value. When K_s is much larger than the reactant concentration, the reaction rate approaches first-order, and will be very sensitive to the changes of the K_s value. The sensitivity analysis results indicate that under drinking water treatment conditions, the biodegradation of the fast degradable component can not be simplified to zero-order or first-order reactions as the degradation rate is moderately sensitive to change in the K_{s1} value. However, the degradation of the slow degradable component can be simplified to zero-order reaction as it is insensitive to changes in the K_{s2} value. The concentration fractions of the fast and slow degradable components strongly impact the predicted concentration profile as shown in Figures 4. The DOC removal at the top of the filter increases as the fast biodegradable fraction increases.

The external mass transfer coefficient, K_L, has little impact on the model prediction as shown in Figure 5. When the external mass transfer coefficient is greater than $5 \cdot 10^{-6}$ m/s, the model is essentially insensitive to the changes. When the coefficient becomes smaller than $5 \cdot 10^{-6}$ m/s, the external mass transfer starts to affect the biodegradation rate and may become the rate-limiting step. However, at the loading rates commonly used in rapid filtration, the mass transfer coefficient is greater than $5 \cdot 10^{-6}$ m/s. Combining this result with the sensitivity analyses of the biodegradation rate, it can be concluded that the bioreaction on the surface of filter media, as compared to external mass transfer, is the rate-limiting step in the biofiltration of natural organic matter under drinking water treatment conditions.

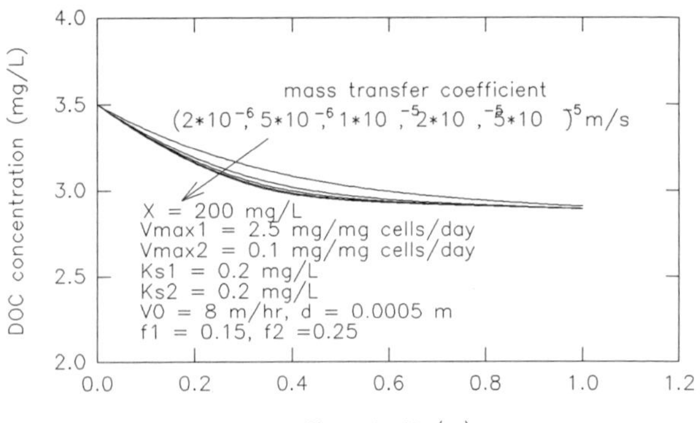

Figure 5. Impact of External Mass Transfer on Biofiltration

As the model is very sensitive to the parameters such as the maximum utilization rates, biomass concentration and the fraction of each component, these parameters need to be accurately evaluated for each individual source water by experiments before applying this model to prediction. For the insensitive parameters, the empirical values can be applied. For example, K_L can be estimated by Gnienlinski correlation (Sontheimer et. al., 1988).

MODEL VALIDATION

A continuous flow biofilter was designed to verify the biofiltration model. In the experiment, an ozonated humic substance solution was used as substrate and previously bioacclimated sand was used as filter media and biomass source. The humic substances were extracted from Germany groundwater by ion-exchange resins. The concentrated humic substance solution was diluted to about 35 mg DOC/L with chlorine- and organic-free tap water. The diluted solution was ozonated with an ozone dose of 0.35 mg O_3/mg DOC. The influent solution was prepared by diluting the ozonated humic substance solution to about 3 mg DOC/L with chlorine- and organic-free tap water. The pH value of the influent solution was adjusted to 7.2 to 7.5. The sand was bioacclimated with Ohio River water for more than a month. The biomass in the filter was measured as nmol of lipid phosphate per gram sand media (White et. al, 1979), then converted to mg cells per liter by a factor of 0.209 nmol phosphate/mg cells (Findlay et. al., 1989) and the filter bed density which was 1500 g/L. This yielded a biomass concentration of 200 mg cells/L. The biomass concentration with bed depth was kept uniform by mixing the media and repacking the bed before each run. V_{max} and K_s of both fast and slow degradable components were estimated as described elsewhere (Wang and Summers, 1994). These values are listed in Table 1. The total biodegradable fraction (f_1+f_2) of the ozonated humic substances was measured to be 45% by BDOC shaker method.

Figure 6 shows the experimental results and the model simulation at 5 m/hr. The error bars around the data represent the standard deviation of three to five samples. By best-fitting the experimental results, fractions of the fast and the slow degradable components were determined. The fast biodegradable component was estimated to be 15% and the slow biodegradable component 30%. The fast biodegradable component is completely utilized in the top 20 cm of the bed, while the slow biodegradable component is not fully utilized even at the end of the 1.0 meter bed. Comparisons of experimental results at 1.5, 8 and 15 m/hr with model predictions using the same parameter values are shown in Figures 7, 8 and 9, respectively. The results show that this model accurately predicts the biodegradation of the DOC of this ozonated humic solution at all filter depths for hydraulic loadings commonly used in practice. With increasing velocity, the easily biodegradable fraction penetrates deeper into the filter. However, at equal EBCTs, the same amount of DOC is removed. This suggests that the performance of the biofilter, in term of DOC removal, will not be affected by changing the filter velocity or the size of the filter media as long as the EBCT is the same. Therefore, the optimal operations of the biofilter for the removal of natural organic matter and particles may be achieved at the same time.

ACKNOWLEDGMENT

The authors would like to thank Ms. Debbie Moll at the Department of Civil and Environmental Engineering, University of Cincinnati for reviewing this manuscript. This study was supported by the USEPA, Drinking Water Research Division through

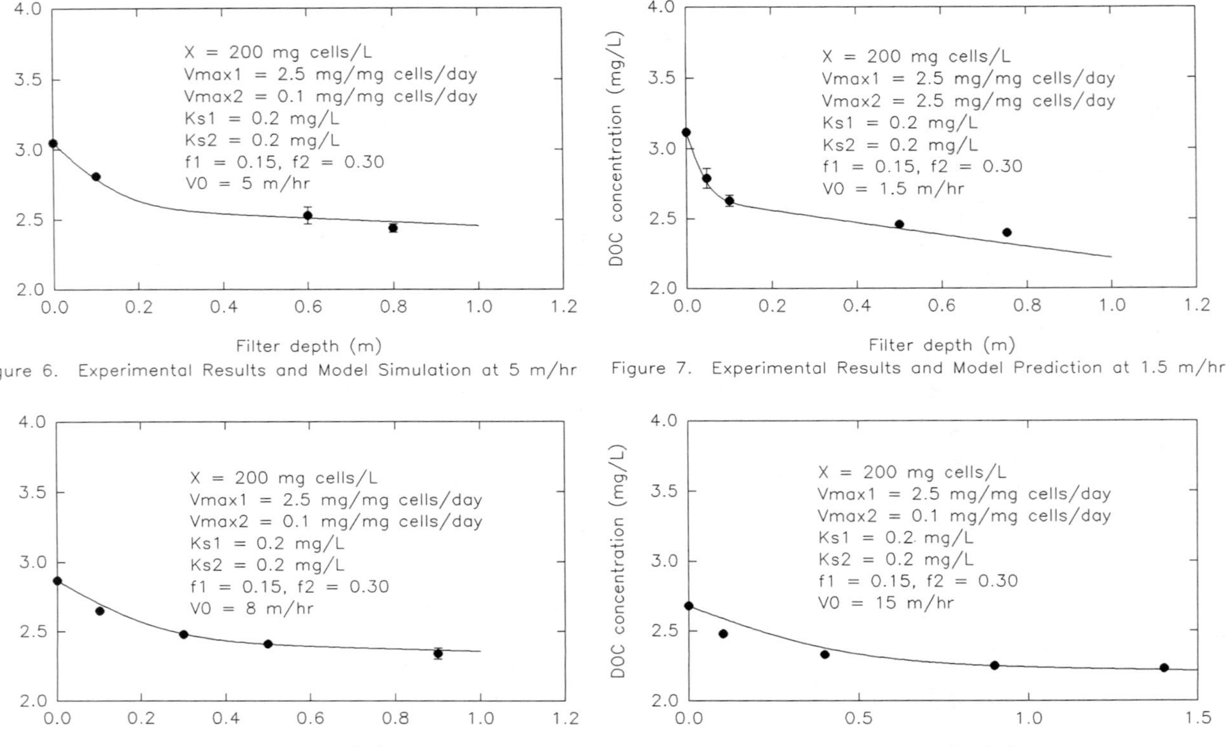

Figure 6. Experimental Results and Model Simulation at 5 m/hr

Figure 7. Experimental Results and Model Prediction at 1.5 m/hr

Figure 8. Experimental Results and Model Prediction at 8 m/hr

Figure 9. Experimental Results and Model Prediction at 15 m/hr

Cooperative Agreement CR-821891 with the University of Cincinnati. The views expressed in this paper are those of the authors and do not necessarily reflect the views of the USEPA.

REFERENCES

Carlson, M. and F. DiGiano, 1992, The use of a bioreactor to assess biostablization. *Proceedings*, 1992 AWWA-WQTC, Toronto, Canada.

Findlay, R. H. et. al., 1989, Efficacy of phospholipid analysis in determining microbial biomass in sediments. *Applied and Environmental Microbiology*, Vol. 55, No. 11.

Langlais, B. et. al. (eds), 1991, *Ozone in Water Treatment: Application and Engineering*. Lewis Publishers. pp. 273-285.

LeChevallier, M. W. et. al., 1992, Evaluating the performance of biologically active rapid filters. *J. AWWA*, Vol. 84, No. 4.

Lu, P. and P. M. Huck, 1993, Evaluation of methods for measuring biofilm thickness and density in biological drinking water treatment. Presented at 1993 AWWA-WQTC, Miami, FL.

Miltner, R. J. and R. S. Summers, 1992, A pilot-scale study of biological treatment. *Proceedings*, 1992 AWWA Annual Conference, Vancouver, B.C..

Mogren, E. M., 1990, Measurement of biodegradable dissolved organic carbon in drinking water. *Master Thesis*, University of Cincinnati, Cincinnati, Ohio.

Prévost, M. et. al., 1989, Full-scale evaluation of biological activated carbon for the treatment of drinking water. *Proceedings*, 1989 AWWA-WQTC, Philadelphia, PA.

Prévost, M. et. al., 1990, Comparison of biodegradable organic carbon (BOC) techniques for process control. *Proceedings*, 1990 AWWA-WQTC, San Diego, CA.

Rittmann, B. E. and P. M. Huck, 1989, Biological treatment of public water supplies. *CRC Critical Reviews in Environmental Control*, Vol. 19, No. 2.

Servais, P. et. al., 1991, Microbial activity in GAC filters at the Chiosy-le-Roi treatment plant. *J. AWWA*, Vol. 83, No. 2.

Sontheimer, H. et. al., 1988, *Activated Carbon for Water Treatment* (2nd edition). DVGW-Forschungsstelle, Universität Karlsruhe, pp. 273-274.

Wang, J. Z. and R. S. Summers, 1993, The evaluation of organic matter and disinfection by-product control in biofilters with biomass and bioactivity analyses. *Proceedings*, 1993 AWWA-WQTC, Miami, FL.

Wang, J. Z. and R. S. Summers, 1994, Biofiltration model application to NOM removal by drinking water biofilters. *International Seminar on Biodegradable Organic Matter*, Montréal, Québec, Canada. June 16-17, 1994.

White D. C. et. al., 1979, Determination of sedimentary microbial biomass by extractable lipid phosphate. *Oceologia,* Vol. 40, pp. 51-62.

Zhang, S. and P. M. Huck, 1993, Parameter estimation for kinetic modeling for biological water treatment. Presented at 1993 AWWA-WQTC, Miami, FL.

Biofilm Sorption of Natural Organic Material

Gary Carlson[1] and JoAnn Silverstein[2]

ABSTRACT

The biosorption of non-ozonated and ozonated natural organic matter (NOM) onto a dense biofilm was investigated at the University of Colorado Environmental Engineering laboratories. Bench-scale experiments were conducted using four different biomass concentrations produced by varying the amount of supplemental carbon substrate(glucose) to each individual reactor: reactor 1(R1) received 40mg/l as carbon/day, reactor 2 (R2) received 20mg/l as carbon/day, reactor 3 (R3) received 10mg/l as carbon/day, and reactor 4 (R4), the abiotic control reactor, received no carbon substrate. In the experiments influent natural DOC range varied from 0-24mg/l with an influent chloroform precursor concentration range of 0-800ug/L. The biosorption removal of NOM was measured as DOC and chloroform precursor concentrations. Statistically, ozonation did not change the amount of NOM removal by biosorption compared to non-ozonation.

INTRODUCTION

Natural organic material (NOM) can react with chlorine disinfectants to form toxic disinfection by-products (DBP's) such as chloroform ($CHCl_3$). Increasing regulation of disinfection by-products has lead to investigation of methods to remove NOM from water before disinfection. Addition of chemical coagulants has been found to reduce NOM precursors during the treatment process, but chemical coagulation produces sludges which are difficult to manage. Activated carbon has been found to be effective in DBP control, but this can be expensive and may require additional pretreatment. Recently there has been interest in using microorganisms to remove NOM from drinking water, by one or a combination of two mechanisms: (1) sorption of NOM onto biofilms, or biosorption, and (2) biodegradation of NOM compounds. In

---

[1] Graduate student, University of Colorado, Boulder, Colorado 80309-0428
[2] Associate Professor, University of Colorado, Boulder, Colorado 80309-0428

this project, the biosorption of NOM was examined using bench-scale packed bed reactors with dense biofilms.

BACKGROUND

Biosorption is the term given to the accumulation and uptake of materials by surface sorption and/or sorption into the cells of bacteria, fungi, and other biomass types. (Bell & Tsezos, 1988; Kuyucak & Volesky, 1988; Volesky, 1990; Pujol & Canler, 1992). The term sorption is used because it is difficult to separate chemical adsorption processes from physical adsorption processes in a biological system. In some cases it is not clear if the materials are sorbed onto or into the cell or cell membranes(Baugham & Paris, 1981). Furthermore, the surfaces of biomass are made up of biopolymers and lipids that usually become a swollen gel, differing greatly in chemical composition from place to place within the biomass.

There have been many attempts to quantify biosorption. Some researchers have examined biosorption by used killed biomass(Steen & Karickhoff, 1981; MacRae, 1985; Bell & Tsezos,1988) to eliminate biodegradation interference during experiments, and to provide more stable sorption by reducing growth in the biomass. Dead biomass was also used to eliminate active cell-mediated transport processes. Little research has been done on the use of a live biomass for use as an adsorbent for common raw water contaminants (Bell & Tsezos, 1988). Some research has shown that the removal of organic compounds using a dead microbial biomass is usually greater than the removal using a live biomass, while other researchers have reported no difference in removals. Contradictory findings may be the result of differences in the inactivation process which produce different new surface properties or even an increase in the surface area available for sorption. The absence of metabolic protection in dead cells may allow increased transport of hazardous organics changing the adsorptive uptake(Tsezos & Seto, 1986).

Biosorption of NOM is a specialized application which must consider the complex nature of both the biomass and NOM. One process thought to affect biosorption of NOM is ozonation. Ozonation of raw water decreases the apparent molecular weight of NOM, decreases the fraction of aromatic compounds, and increases both the acidity and the concentrations of aldehydes in solution(Edwards & Benjamin, 1992; Amy et al., 1991; Langlais et al., 1991). Overall, then, ozonation results in different sorbate characteristics that may influence biosorption.

APPROACH

This research is an effort to examine the effects of ozonation on changing NOM sorption by biofilms and to quantify biosorption by standard analytical techniques, e.g. isotherm modeling. Two isotherm models were selected to analyze the sorption data.

Freundlich Model $\log(x/M) = \log(K_f) + (1/n)\log(C_e)$ (1)

Linear Model $(x/M) = K_dC_e + \text{Intercept}$ (2)

where (x/M)is the specific mass sorbed(mass DOC or $CHCl_3$ precursor sorbed per mass of dry biofilm); C_e is the equilibrium concentration of DOC and $CHCl_3$ precursors; K_f, 1/n, and K_d are constants for a given sorbent and sorbate.

MATERIALS AND METHODS

Biomass Bench-Scale System. Four parallel packed bed biofilm reactors were used for sorption experiments. The reactors received daily supplements of oxygen, mineral nutrients and varying amounts of carbon substrate (glucose) to produce a range of 4 biofilm thicknesses: reactor 1 (R1) received 40mg/L as carbon/day; reactor 2 (R2) received 20mg/L as carbon/day; reactor 3 (R3) received 10mg/L as carbon/day; reactor 4 (R4), the abiotic control reactor, received no supplemental carbon. Glucose was used to grow biofilm which was not adapted to metabolizing the harder-to-degrade NOM so that biosorption could be studied separately from biodegradation. The time required for sorption experiments also was limited to the minimum needed to achieve equilibrium between the NOM and the biofilm, reducing the possibility of biodegradation of NOM. Reactor liquid was mixed by recirculating pumps to maximize contact between the water and biofilm. Table 1 summarizes the experimental system parameters.

Table 1. Experimental system parameters

Reactor volume (no biomass)	= 5.3 liters
Reactor volume R1 - R3 respectively (wet biomass)	= 4.0 - 4.9 liters
Media characteristics (Koch Flexirings)	
porosity	= 0.93
surface area	= 340 m^2/m^3
diameter	= 1.3 cm
Surface area per reactor	= 16000 cm^2
Recirculation rate (Mixing time)	= 48 min

Natural Organic Matter. Natural source water was collected from a mountain lake drinking water source for Boulder, Colorado. Data showed that this lake water had low DOC values of 1-5mg/L, depending on the seasonal variation. Preliminary experiments showed that at these low DOC concentrations, the sorption of DOC was difficult to assay. The background DOC from the biofilm reactors was approximately 3.0mg/L, which was often equivalent to the raw source water DOC concentration. The NOM was concentrated through a commercial nanofiltration system(NF-70, Filmtec Corp.) in order to develop isotherm models. The concentrate was then diluted to a specific DOC concentration for each experiment.

Biomass Measurements. Biomass was measured by removing a sample of plastic media from each reactor, drying the media at 105°C for 48 hours and then weighing dried media and biomass. The sample of media was taken randomly throughout the reactor bed to get a representative sample. The weight of this resulting sample was then compared to clean media weight, the difference being the amount of biomass(dry weight) for the given sample. The average weight per ring value was used with a count of the total number of rings in each reactor to calculate the total amount of biomass.

Experimental Procedure. Biosorption experiments were conducted by contacting non-ozonated and ozonated NOM solutions at varying concentrations with four different biomass concentrations. The influent TOC concentrations ranged from 3mg/L - 24 mg/l and the chloroform precursor concentration measured as chloroform ranged from

100ug/L - 800ug/L. Ozonation of NOM was done by adding an ozone stock solution to the raw water sample to achieve a O_3:DOC ratio of 1:1. The non-ozonated and ozonated NOM solutions were then placed in each reactor. The solutions were not buffered and preliminary experiments showed that the pH did not change during sorption experiments. Preliminary experiments also showed that the time required to reach equilibrium was reached within four hours. Furthermore, biodegradation may have occurred after 16-24 hours. Therefore for all sorption experiments, the equilibrium NOM concentration measurements were from the four hour samples. Samples were filtered through a 0.22um filter and duplicates were fractionated into two size categories: >10k daltons, and <10k daltons. Each resulting sample was then analyzed for DOC concentration and chloroform concentration. The concentration of each parameter determined by analysis of the initial solution before placement in biofilm reactors was taken as the initial concentration of each parameter. The amount of removal was calculated by the difference in the initial concentration and the equilibrium concentration for each reactor.

RESULTS AND DISCUSSION

Reactor Characteristics. From reactor 1(40mg/l of glucose per day) to reactor 3 (10mg/l of glucose per day) there was a definite range of biomass between the reactors given in Table 2 below:

Table 2. Reactor Characteristics

Reactor #	Biomass (dry g/reactor)	% Removal of $CHCl_3$ Precursors	
		Non-Ozonated	Ozonated
1	30 - 36	31± 7	31±10
2	16 - 21	29±10	31±14
3	9 - 11	24±10	27±12
4	0	0	0

Table 2 shows that the amount of biomass in each reactor increases with increasing substrate addition. There is also a trend for ozonated and non-ozonated NOM removal (measured as $CHCl_3$) of increasing percent removal as biofilm mass in the reactor increases. NOM percentage removal measured as TOC followed the same trend as chloroform precursors. The similarities in percentage removal for reactors 1 (R1) and 2 (R2) may be due to the high variance in removals between experiments for each reactor.

Sorption Isotherms. Isotherms were fitted to the data using the linearized Freundlich and linear isotherm equations with a least squares method. The data values for each individual reactor were plotted for each size fraction. Overall, it was found that significantly more NOM removal occurred in the three reactors with biofilm than in the abiotic control reactor which received only buffered oxygenated water. Statistical analysis showed that the Freundlich and linear isotherm parameters were not statistically different (95% confidence level) between the reactors for all size fractions. This suggests that the amount of NOM sorbed per gram of biomass for both thick and thin biofilms is not dependent on the amount of biomass, but on the liquid equilibrium concentration of NOM. Indicating that there may not be transport limitations in the thickest biofilm.

Because the specific sorption was the same for all biofilm containing reactors, the individual data values for all three reactors(R1, R2, and R3) were grouped together for fitting isotherm parameters. The 95 % confidence intervals for the 1/n and K_f

Freundlich parameters showed an extremely large variance for K_f values and acceptable variances for the 1/n values. The large variance in the K_f parameter values (3-6 orders of magnitude) was due to the sensitivity of the linearized log-log model. The large variance associated with K_f makes it impossible to draw any conclusions using the Freundlich Isotherm model in analyzing the biosorption data. Table 3 summarizes the parameters for the linear isotherm equation. The 95% confidence intervals for the parameter values from the least squares regression are shown to emphasize the variability associated with each parameter. The statistical analysis shows that the deviations of K_d are around 20-30% of the average value.

Table 3. Linear Isotherm Parameters

	TOC		$CHCl_3$ Precursors	
Size Fraction	K_d	95% CI*	K_d	95% CI*
No Ozone				
Total	0.084	0.042 - 0.126	0.102	0.040 - 0.164
>10k dalton	0.067	0.034 - 0.100	0.128	0.098 - 0.158
<10k dalton	0.103	0.048 - 0.159	0.131	0.053 - 0.210
Ozone				
Total	0.088	0.041 - 0.135	0.082	0.023 - 0.142
>10k dalton	0.062	0.026 - 0.099	0.057	0.003 - 0.111
<10k dalton	0.067	0.007 - 0.127	0.084	0.023 - 0.144

* Confidence interval limits for linear isotherm parameter K_d.

Figures 1 and 2 show the TOC linear isotherms for the total fraction for non-ozonated NOM and ozonated NOM respectively. Figures 3 and 4 show the $CHCl_3$ precursor linear isotherms for the total fraction for non-ozonated and ozonated NOM respectively. The error bars are the result of error associated with biomass measurements from each reactor. The error in biomass measurement increases as the biomass in each reactor decreases, thus increasing the error associated with the q values for the reactors with less biomass. The error bars are 95% confidence intervals for each value of (x/M). At higher equilibrium NOM concentrations, the specific sorption may be increasing more than that given by the slope of the regression line. This increase in sorption may be from increased diffusion into the biomass.

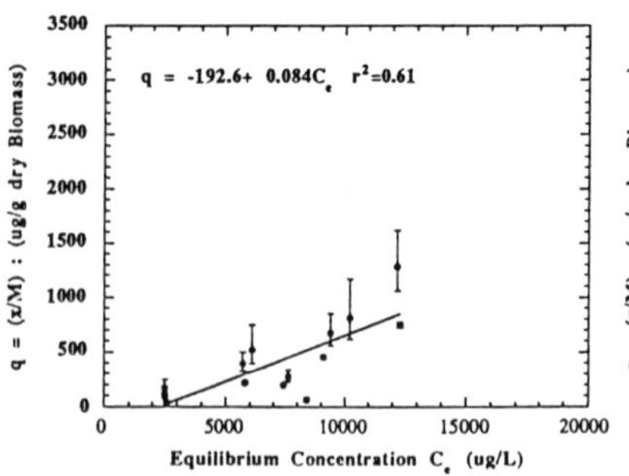

Figure 1. Linear Isotherm, total TOC fraction non-ozonated NOM.

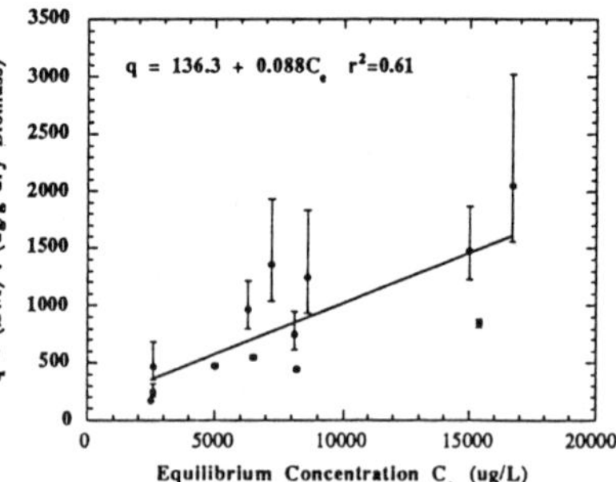

Figure 2. Linear Isotherm, total TOC fraction ozonated NOM.

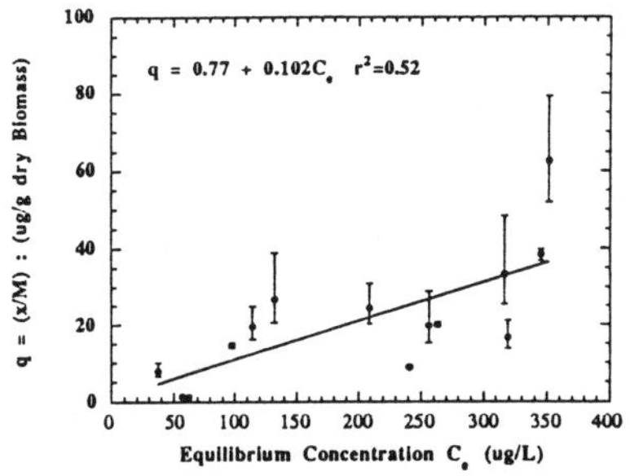

Figure 3. Linear Isotherm, total $CHCl_3$ precursor fraction non-ozonated NOM.

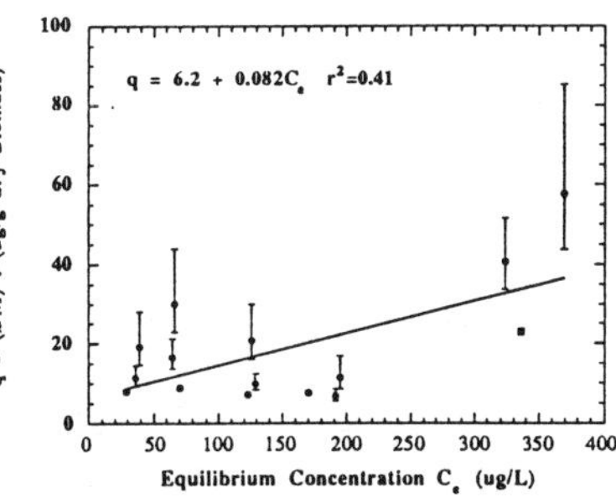

Figure 4. Linear Isotherm, total $CHCl_3$ precursor fraction ozonated NOM.

Statistical analysis (pairwise t-test and ANOVA) was used to show with a 95% level of confidence that the K_d values of the total, >10k fraction, and <10k fractions for TOC and chloroform precursors were not statistically different in each of the ozonated and non-ozonated sorption experiments. Therefore, the total, >10k fraction , and the <10k fraction showed similar biosorptive behavior.

The K_d values for non-ozonated NOM and ozonated NOM were then compared to see if there was a difference in isotherm parameter values. For both TOC and chloroform precursors, it was found with a 95% level of confidence that the K_d values were not statistically different between ozonated and non-ozonated NOM. Therefore, it appears that the ozonation of NOM did not change the partition coefficients of TOC and chloroform precursors, suggesting that the specific amount biosorbed at a specific equilibrium concentration of TOC and chloroform precursors is similar for both ozonated and non-ozonated NOM. A possible explanation is that ozonation had little effect on the resulting molecular size distribution. Prior to ozonation the >10k fraction was 63±3% of the total amount of NOM with the <10k fraction making up the remaining 34-6% measured as TOC. After ozonation the >10k fraction was 57±3% of the total amount of NOM with the <10k fraction making up 39±5% of the total amount NOM measured as TOC. Thus, ozonation increased the smaller molecular fraction (<10k daltons) while decreasing the larger molecular fraction (>10k daltons). However, this change in ozonated NOM is very small, 0 - 5% increase in the <10k fraction, compared to the non-ozonated NOM.

CONCLUSIONS

The dense biofilm studies have shown that:

- Significant amounts of NOM(TOC and $CHCl_3$ precursors) were removed in the biofilm reactors compared to the abiotic control reactor. The amount of removal ranged from 20%-30% of the influent concentrations depending on the biofilm mass.

- Specific sorption, the NOM sorbed per gram of biomass at equilibrium, measured as both TOC and $CHCl_3$ precursors, was dependent on the bulk liquid equilibrium concentrates following the assumptions of isotherm models.

- The linear sorption isotherm gave sorption capacity predictions that were more reliable than the Freundlich Isotherm, which had a very large variance associated with the K_f parameter. Therefore, comparison of molecular size and ozonation sorption isotherms were based on the linear isotherm parameter K_d.

- The sorption of the high molecular weight (>10k daltons) fraction and the low molecular weight (<10k daltons) fraction of TOC and $CHCl_3$ precursors did not differ statistically for both non-ozonated and ozonated NOM at 23^{0}C.

- Values K_d were not statistically different between ozonated and non-ozonated NOM isotherms. In general, the biosorption data of non-ozonated and ozonated NOM by a dense biofilm suggests that ozonation has little or no effect on biosorption of NOM at 23°C and at the ozone dose used.

REFERENCES

Amy, G.L., Tan, L., and Davis M.K.. *The Effects of Ozonation and Activated Carbon Adsorption on Trihalomethane Speciation.* Water Research. 25:2, 191-202, 1991.

Baugham, G.L. and Paris D.F.. *Microbial Bioconcentration of Organic Pollutants From Aquatic Systems - A Critical Review.* CRC Critical Reviews in Microbiology. 8, 205-228, 1981.

Bell, J.P. and Tsezos, M.. *The Selectivity of Biosorption of Hazardous Organics by A Microbial Biomass.* Water Research. 22:10, 1245-1251, 1988.

Edwards, M. and Benjamin, M.M.. *Effect of Preozonation on Coagulant-NOM Interactions.* J.A.W.W.A. 84:8, 63-72, 1992.

Kuyucak, N. and Volesky, B.. *The mechanism of Cobalt Biosorption.* Biotechnology and Bioengineering. 33:2, 823-831,1989.

Langlais, B., Reckhow, D.A., and Brink, D.R.. *Ozone In Water Treatment: Applications and Engineering.* Lewis Publishers, New York. 1991.

MacRae, I.C.. *Removal of Pesticides in Water by Microbial Cells Adsorbed to Magnetite.* Water Research. 19:7, 825-830, 1985.

Pujol, R. and Canler, J.P.. *Biosorption and Dynamics of Bacterial Population in Activated Sludge.* Water Research. 26:2, 209-212, 1992.

Steen, W.C. and Karickhoff, S.W.. *Biosorption of Hydrophobic Organic Pollutants by Mixed Microbial Populations.* Chemosphere. 10:1, 27-32, 1981.

Tsezos, M. and Seto, W.. *The Adsorption of Chloroethanes by A Microbial Biomass.* Water Research. 20:7, 851-858, 1986.

Volesky, B.(editor). *Biosorption of Heavy Metals.* CRC Press, Boston, 1990.

Investigations of Potential Radical Interaction in the Formation of Brominated Organic Compounds in Water

Rengao Song, Adam Eyring, and Roger Minear
Institute for Environmental Studies
University of Illinois, Urbana, IL 61801

A system has been designed that permits the examination of the formation of organic bromine compounds via a radical pathway. Hydroxyl radicals ($HO^{\bullet}$) generated by the Fenton reaction were reacted with aqueous bromide ion (Br^-) and natural organic matter (NOM) to form bromide radicals ($Br^{\bullet}$, $Br_2^{\bullet -}$, ...), which may in turn react with NOM and/or NOM radicals to form organic bromine compounds, measured as total organic bromine (TOBr). Preliminary experimental data show that in the presence of *tert*-butyl alcohol, an effective $HO^{\bullet}$ scavenger, the reaction was quenched. When the concentration of Br^- was increased, the amount of TOBr increased. TOBr also increased when the concentration of NOM was increased. These results suggest that a radical pathway, in addition to the well-established pathway involving HOBr, may also account for the formation of TOBr. However, further experiments is definitely necessary.

Water ozonation has proven to be a promising technology in drinking water treatment. It has the advantages of avoiding the use of chlorine and its associated organic disinfection by-product formation,[1] is more effective than chlorine against cysts and viruses, and it controls taste and odor. There are generally three problems that have come about with ozonation: the formation of assimilated organic carbon (AOC), which can cause biological regrowth in the distribution system; a short half-life for dissolved ozone, which then will not provide a disinfectant residual downstream; and the formation of bromate, which has been proven to be a toxic compound.[2] In addition, when bromide is present, water ozonation can produce organic disinfection by-products (DBPs) such as bromoform, bromoacetic acids (BAAs), and bromoacetonitriles (BANs). Some of these organic DBPs are regulated by the Environmental Protection Agency (EPA). Therefore, it is necessary to investigate the formation of organic bromine species. This study will also help us to understand bromate formation mechanisms and mass balance, which is a major issue of water ozonation.

In the formation of organic bromine compounds in water ozonation, hypobromous acid (HOBr) is generally believed to be the primary reactant with NOM, as shown in the following simplified mechanism:

$$O_3 + Br^- \text{----->} OBr^- + O_2$$
$$OBr^- + H^+ \text{----->} HOBr \qquad pKa=8.8 \text{ at } 20°C$$
$$HOBr + NOM \text{----->} TOBr\ (CHBr_3, BAAs, BANs, etc)$$

Recently, it has been difficult to explain some experimental data using only this mechanism. For example, during water ozonation, the TOBr was inhibited as alkalinity was increased.[3,4] It is well-known that bicarbonate ions are HO• scavengers. Thus, the $t_{1/2}$ of DO_3 was increased, which in turn should have increased TOBr formation.

Based on the alkalinity effects, it is postulated that a radical pathway is involved in the formation of TOBr. There is also evidence[5] in the literature that dichlorine radicals ($Cl_2^{\bullet -}$) can react with unsaturated organic compounds to produce chlorine adducts and it is thus presumed that other halogens may form such radicals that react in a similar fashion. A few pulse-radiolysis studies found that HO• could lead to Br•, $Br_2^{\bullet -}$, and $HOBr^{\bullet -}$. Therefore, some addition reactions between these radicals and unsaturated organic compounds may be occurring during water ozonation.

Since radical reactions are very fast and bromate formation also involves radical reactions, we have to eliminate bromate formation in order to investigate the TOBr formation with a radical pathway. Thus, reactions in this study were conducted in the absence of ozone. The Fenton reaction, in which HO• can be generated, was used to produce HO• to react with Br^- and NOM in an attempt to form organic bromine compounds. Known HO• scavenger, *tert*-butyl alcohol, was also added to the reaction to confirm the formation of HO•.

Experimental methodology

Reagents. All chemicals used in this experiment were of reagent grade or better and used without further purification. Potassium bromide was the source of Br^- and iron(II) sulfate provided Fe^{2+} for the Fenton reaction. Alkalinity was added in the form of sodium bicarbonate. Water purified with a Millipore Milli-Q system was used for solutions and blanks.

Analytical Equipment. The TOBr was measured with a Dohrmann DX-20A TOX Analyzer equipped with an AD-3 Adsorption Module. Both 2,4,6-trichlorophenol and bromoform were used to measure the percentage recovery (90% and 92%, respectively) and generate the standard curves for determining the concentrations of TOBr formed. Samples were run through columns containing granular activated carbon (GAC) to adsorb organics and washed with 5 ml of 5000 ppm KNO_3 to remove CL^-, Br^-, I^-, and other interfering anions. The GAC was combusted in a furnace at 800°C to form HBr, which was detected with a silver electrode.

A Shimadzu TOC-500 was used to measure the total organic carbon in the NOM fractions.

NOM Isolation. NOM was isolated and concentrated from raw water sampled at Lake Michigan. The sample was filtered with a 0.22 um tangential flow filter to remove particulate matter. The filtrate was pretreated with a cation exchange column in the sodium form to remove hardness, and then applied to a continuous flow system consisting of ultrafiltration and reverse osmosis (RO) membranes. The dissolved NOM recovery from 310 L of filtered water was 95%, based on dissolved organic carbon measurements. This allowed both concentration variation and study of different molecular size fractions. The molecular size fractions studied were 1000 (1 K) and 30,000 (30 K) Daltons. The levels of organic halides in the NOM were measured and they were taken into account when determining the extent of the reactions.

Reactor System. A true batch reactor was used to study the TOBr formation.[6] The reactor was a 500-mL graduated cylinder coated with aluminum foil and consisting of spigot one inch from the bottom to allow sample collection over time. A Teflon cover was kept on the solution surface to avoid loss of VOCs and keep light out of the reaction matrix. A magnetic stir bar was used to provide continuous mixing action. The reactor was recharged with 500 mL of solution for each new sample, since the concentrations of Br^-, NOM, etc. were changed. Each reactor batch consisted of specific concentrations of Br^-, NOM, and H_2O_2 in Milli-Q water. Sodium bicarbonate was added for the alkalinity analysis. The concentrations for Br^- and NOM were changed to examine any effects, while H_2O_2 was kept at the same concentration (1.4×10^{-2} M) for all runs.

The Fenton reaction was used to produce HO• by the one-to-one combination of iron(II) and hydrogen peroxide:

$$Fe^{2+} + H_2O_2 \text{ ------> } Fe^{3+} + HO\bullet + OH^-$$

The experiments were carried out in the dark in order to get better reproducibility by eliminating the HO• formed by photolysis of $Fe(OH)^{2+}$ and/or $Fe(OH)_2^+$.[7]

Reactions. Into 250 mL of water were dissolved known amounts of Br^- and H_2O_2, in that order. This solution was added to the reactor. The NOM fractions were diluted in another 250 mL of water and added to the reactor. Fe^{2+} was added to the reactor (for a concentration of 3.7×10^{-4} M) to start the reaction. It was verified that no reaction occurred without the presence of iron. Also, no reaction occurred without the presence of H_2O_2. Samples of 125 mL were collected after a 5-minute reaction period, quenched with 0.1 M ascorbic acid to quench the radical reactions, filtered with Whatman hardened filter paper (Cat. No. 1450042) to remove any precipitates, acidified to pH<2 with 3.8 M nitric acid, and 100 mL was run through the GAC columns. Precipitates were seen in some samples and they were most likely due to iron hydroxides. Acidification was necessary to keep any brominated acids undissociated. *tert*-Butyl alcohol and bicarbonate were added to the reactor with the NOM for testing the quenching effects.

Results and discussion

When examining the TOBr concentrations, there was no distinction made as to which bromine species was the reactant with the NOM to form TOBr. Br• and $Br_2^{\bullet -}$ are assumed to be the main reactants, but our main focus was to show that a radical pathway involving bromine is possible in the formation of TOBr. An orthogonal matrix of experiments allowed independent observation of the effects of the experimental variables. The baseline condition was set, in which bromide concentration is equal to 2mg/L and DOC = 3 mg/L

Effect of bromide concentration. Figure 1 shows the TOBr formed at bromide

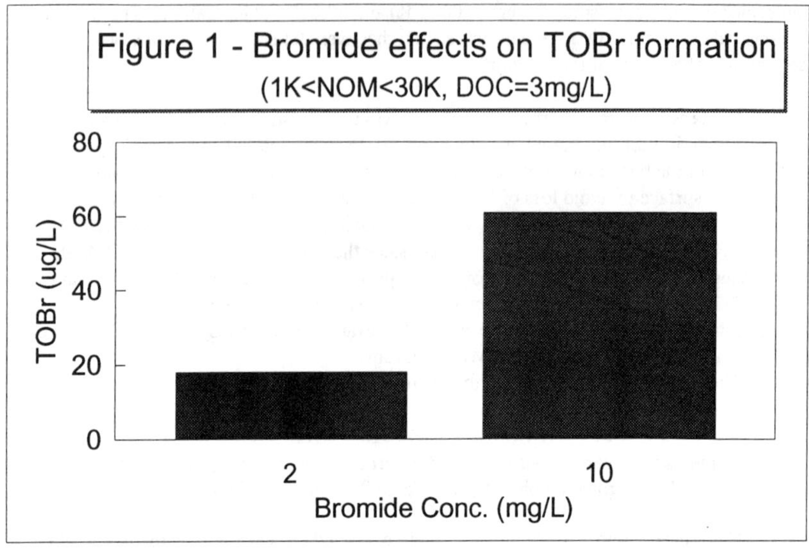
Figure 1 - Bromide effects on TOBr formation
(1K<NOM<30K, DOC=3mg/L)
TOBr (ug/L)
80
60
40
20
0
2
10
Bromide Conc. (mg/L)

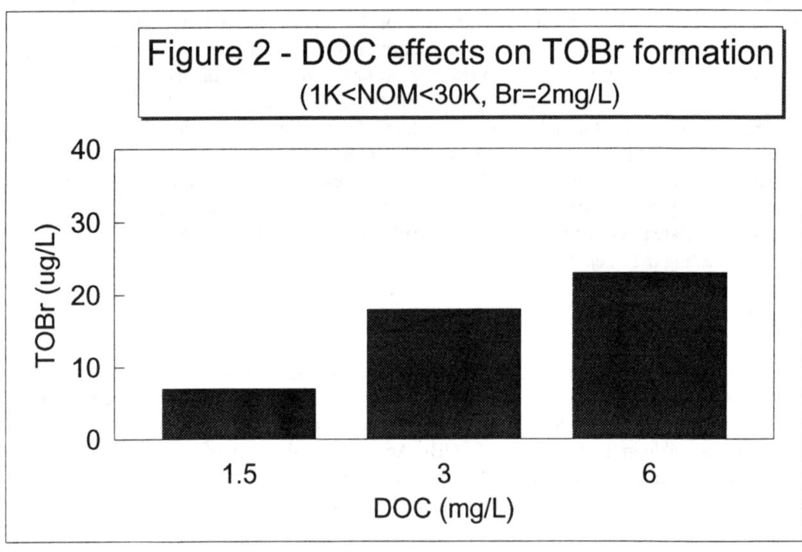
Figure 2 - DOC effects on TOBr formation
(1K<NOM<30K, Br=2mg/L)
TOBr (ug/L)
40
30
20
10
0
1.5
3
6
DOC (mg/L)

concentrations of 2 and 10 mg/L. There was an increase in the amount of TOBr formed as the concentration was increased. This may be due to higher concentrations of Br• and $Br_2^{\bullet -}$ from the higher amount of starting Br^-. As a result of the higher amounts of Br• and $Br_2^{\bullet -}$, there was a higher probability of reactions with NOM and/or NOM radicals, resulting in more TOBr.

Four types of reactions may have occurred between the bromine radicals and the natural organic matter: 1) An oxidation reaction in which NOM serves as an electron donor and bromine species are reduced to bromide ions; 2) An addition reaction in which Br• and $Br_2^{\bullet -}$ are added to double bonds to form adducts; 3) Electron abstraction; 4) radical-radical reactions in which Br• and $Br_2^{\bullet -}$ react with NOM• to form TOBr.

Effect of DOC. Figure 2 shows the TOBr formed at 1K NOM concentrations of 1.5, 3, and 6 mg/L. The total organic bromine concentration increased on increasing the NOM from 1.5 to 6 mg/L. Increasing NOM levels, which increased the probabilities of reactions between NOM and HO•, Br• and/or $Br_2^{\bullet -}$, lead to higher TOBr production if reaction types 2 and 4 do happen. Since we also found that HO• can destroy some background TOBr, it may suggest that the higher amounts of TOBr result from a higher frequency of reactions between NOM and HO• than the destruction of TOBr by HO•. Therefore, there could be a tradeoff between TOBr formation and destruction.

Effect of alkalinity. Figure 3 shows the TOBr formed at alkalinity concentrations of 1 and 4 mM, which correspond to total inorganic carbon (TIC) of 12 and 48 mg/L. The formation of TOBr decreased as the alkalinity was increased. True and Zafiriou proposed the following reactions during their investigation of reaction of $Br_2^{\bullet -}$ produced by flash photolysis of sea water with components of the dissolved carbonate system.[8]

$$Br_2^{\bullet -} + HCO_3^- \text{----}> 2Br^- + HCO_3^{\bullet}$$
$$Br_2^{\bullet -} + CO_3^{2-} \text{----}> 2Br^- + CO_3^{\bullet -}$$

However, the evidence and the reaction rates for theses two reactions are not available. In contrast, the $HCO_3^{\bullet}$ has a pKa of 7.5, and its conjunct base of $CO_3^{\bullet -}$. can react with bromide ions to form bromide radical ($k=5\times10^5\ M^{-1}s^{-1}$) The major evidence they based on is that there was a decrease in $Br_2^{\bullet -}$ and an increase in $HCO_3^{\bullet}$. this phenomena can be explained in the following way:

$$Br_2^{\bullet -} \text{----}> Br^{\bullet} + Br^-$$
$$Br^{\bullet} + OH^- \text{----}> HOBr^{\bullet -}$$
$$HOBr^{\bullet -} \text{----}> Br^- + HO^{\bullet}$$
$$HO^{\bullet} + HCO_3^- \text{----}> H_2O + HCO_3^{\bullet -}$$
$$HO^{\bullet} + CO_3^{2-} \text{----}> H_2O + CO_3^{\bullet -}$$

Thus, the role of alkalinity for TOBr formation through a radical pathway may be more complicated than expected before.

Effect of t-BuOH. Figure 4 shows the effects of the presence of 1 mM tert-butanol on TOBr formation. After the addition of 1 mM *tert*-butanol, there was a decrease in the yield of TOBr. This may be attributed to the following step:

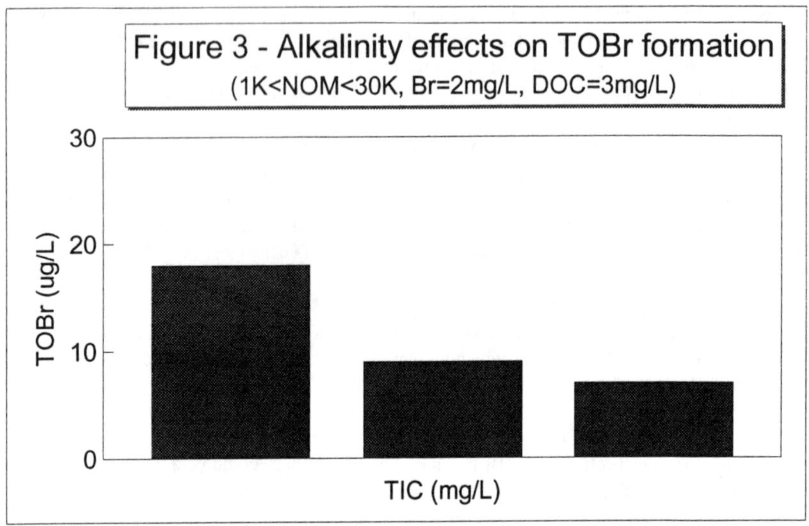

Figure 3 - Alkalinity effects on TOBr formation
(1K<NOM<30K, Br=2mg/L, DOC=3mg/L)

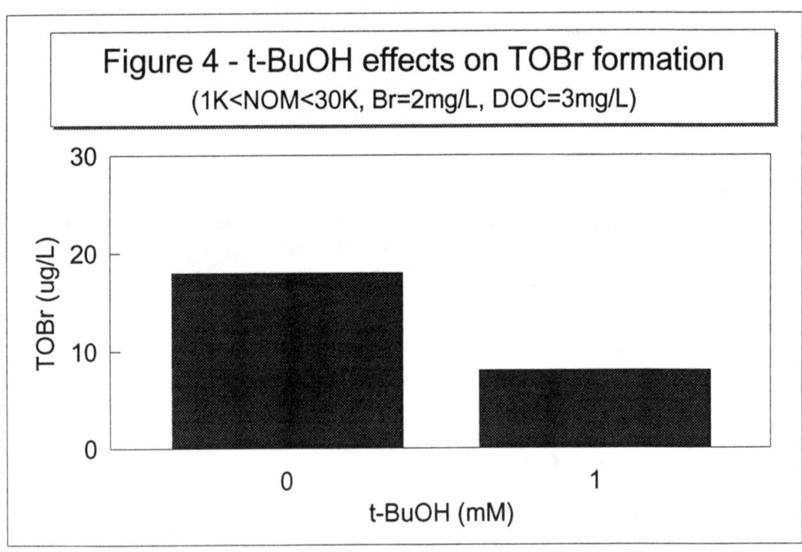

Figure 4 - t-BuOH effects on TOBr formation
(1K<NOM<30K, Br=2mg/L, DOC=3mg/L)

$$HO\bullet + t\text{-}BuOH \text{ ------> } H_2O + t\text{-}BuOH\bullet$$

The rate constant is $6.0\times10^8\ M^{-1}s^{-1}$. Thus, *tert*-butanol effectively competed with NOM and Br^- for HO•. Therefore, the formation of TOBr through the proposed radical pathway was decreased.

Summary and conclusions

Based on these preliminary results, it is likely that a radical pathway may contribute to TOBr formation. During water ozonation, HO• generated through a reaction involving ozone decomposition[9] can react with Br^- and NOM to form Br•, $Br_2^{\bullet -}$, and NOM radicals. Then the reactions of bromine radicals with NOM and/or NOM radicals would form bromine adducts. Thus, these radicals could work in conjunction with HOBr to form TOBr. The use of alkalinity and *tert*-butanol support the radical mechanism because other studies have concluded they have effects (i.e., radical quenching) that indicate the presence of radicals. Based on these results, a mechanism has been proposed with the following steps:

$$Br^- + HO\bullet \text{ ----->} HOBr^{\bullet -}$$
$$HOBr^{\bullet -} \text{ ----->} Br\bullet + HO^-$$
$$Br\bullet + Br^- \text{ ----->} Br_2^{\bullet -}$$
$$NOM(H) + HO\bullet \text{ ----->} NOM\bullet + H_2O$$
$$Br\bullet + NOM\bullet \text{ ----->} NOM\text{-}Br$$
$$Br_2^{\bullet -} + NOM\bullet \text{ ----->} NOM\text{-}Br + Br^-$$
$$Br\bullet\ (Br_2^{\bullet -}) + NOM \text{ ----->} \bullet NOM\text{-}Br + (Br^-)$$
$$\bullet NOM\text{-}Br + NOM \text{ ----->} NOM\text{-}Br + NOM\bullet$$

Since the radical pathway seems to be pretty much dependent on low radical flux, The formation of TOBr through this radical pathway could be very small to be compared in magnitude to the TOBr via HOBr pathway.

Even though these preliminary results support the proposed radical mechanism to some degree, there are some uncertainties. First, TOBr is not a good indicator analytically to verify this mechanism, especially at relatively high bromide and low TOBr concentrations.[10] Second, natural organic matter has some relatively high TOX background compared to the TOBr after reaction, which can cause some difficulties during data interpretation. Thus, a model compound and gas chromatograph technique is being used to investigate this mechanism right now. At the same time, bromide concentrations are monitored during the experiments. Also, ultraviolet photolysis of hydrogen peroxide is being investigated as another HO• producer.[11] Third, Fenton reaction has its own problems, too. For example, it is not clear what are the real reactive species;[12] the interference caused by Fe(III) precipitation which may adsorb some bromide reactants during experiment. The biggest problem with Fenton reagents is that no free HO radical is formed[13] during the reactions even though the reaction rates with some substrates such as NOM with Fenton reagents are similar to free HO radicals. Finally, the role of dissolved oxygen which is a diradical could be much more important[14].

In conclusion, although a radical mechanism is proposed for TOBr formation which can explain some experimental data, the uncertainties still remain. It is not clear whether this pathway is important practically if the mechanism does exist. Additional research is definitely necessary to clarify the issue.

References

1. Rook, J.J. Haloformes in Drinking Water, *Journal of American Water Works Association, 23(1):168-172 (1976)*

2. Amy, G., Siddiqui, M., Ozekin, K., and Westerhoff, P. Threshold Levels for Bromate Formation in Drinking Water, *Proceedings of the International Water Supply Association International Conference: Bromate and Water Treatment, Paris, November 22-24, pp. 169-180 (1993)*

3. Glaze, W.H.; Weinberg, H.S.; Cavanagh, J.E. Evaluating the Formation of Brominated DBPs During Ozonation. *Journal of American Water Works Association* 85(1): 96 (1993).

4. Song, R. Minear, R.A. Amy, G. Westerhoff, P. Comparison of Bromide-Ozone Reactions with NOM Separated by XAD-8 Resin and UF/RO Membrane Methods. *Preprints of the Environmental Chemistry Division Papers, (206th ACS meeting) 33:225-228 (1993).*

5. Neta, P. and Huie, E. R. Rate Constants for Reactions of Inorganic Radicals in Aqueous Solution. *Journal of Physical and Chemical Reference Data. Volume 17, No. 3 (1988).*

6. Westerhoff, P. Amy, G., Song, R. and Minear, R.A. Evaluation of Rate Constants for Dissolved Ozone Decay and Bromate Formation. *Preprints of the Environmental Chemistry Division Papers, (206th ACS meeting) 33:229-232 (1993)*

7. True, M.B. and Zafiriou, O.C. Reaction of Br_2^- Produced by Flash Photolysis of Sea Water With Components of the Dissolved Carbonate System. *Photochemistry of Environmental Aquatic Systems.* ACS, Washington, D.C. (1987).

8. Faust, B.C. and Hoigne,J. *J. Atmos. Environ. 24A, 79-89 (1990).*

9. Langais, B.; Reckhow, D.A.; Brink, D.R., eds. *"Ozone in Water Treatment - Application and Engineering." Lewis, 1991, p.17.*

10. Organic Halogen (Total). *Standard Methods, 16th Edition. pp 516-525 (1985).*

11. Peyton, G.R. and Glaze, W.H. In Photochemistry of Environmental Aquatic Systems; *ACS Symposium Series 327; American Chemical Society: Washington, DC. pp 76-88 (1986).*

12. Wailing, C. Fenton's Reagent Revisited. *Accounts of Chemical Research. Vol.8, pp 125-131 (1975).*

13. Sawyer, D.T., Kang C., Llobet, A., and Redman, C. Fenton Reagents (1:1 Fe^{II} Lx/HOOH) React via [$LxFe^{II}OOH(BH+)$] (1) as Hydroxylases(RH->ROH), not as Generators of Free Hydroxyl Radicals (HO). *J. Am. Chem. Soc., Vol. 115, No. 13 (1993).*

14. Larson, R. Personal Communications (1993).

Acknowledgments

The authors would like to thank Dr. Richard A. Larson of Institute for Environmental Studies, University of Illinois, for his valuable advice.

Treatment of Drinking Water Containing Bromate and Bromide Ions

Taha F. Marhaba[1], Ph.D., M. ASCE
Steven J. Medlar[2], P.E.

Abstract

Granular activated carbon and reverse osmosis nanofiltration and hyperfiltration were examined for both bromate and bromide ion removal using surface treated water in New Jersey. Bromate ion was formed when water containing bromide ion was ozonated. It was concluded that bromate ion was chemically reduced to bromide ion by the granular activated carbon. Also, reverse osmosis nanofiltration rejected bromate and bromide ions an average of 89% and 84%, respectively, whereas hyperfiltration rejected bromate and bromide ions an average of 97% and 89% respectively.

Introduction

The United States Environmental Protection Agency (USEPA) is developing regulations for disinfection and disinfection by-products in drinking water, referred to as the Disinfectant/Disinfection By-products (D/DBP) Rule (proposed March 1994). The D/DBP Rule will lower the Maximum Contaminant Level (MCL) of total trihalomethanes (TTHMs) from 100 to 80 ug/L in the first stage. Haloaceticacids (HAA5) are also a target as well as total organic carbon (TOC) (now referred to as "DBP precursors"). The limits are also expected to be lowered in a second stage in 1998. With all the disadvantages that lie ahead in using only chlorine as a disinfectant, the industry is likely to shift to other disinfectants or oxidants which could incorporate ozone for overall process optimization.

---

[1]Environmental Engineer, Stone & Webster Engineering Corp., 3 Executive Campus, Cherry Hill, NJ 08034. Lecturer, Dept. of Civil & Environmental Engineering, Rutgers University, Piscataway, NJ 08854.

[2]Senior Vice President, Camp Dresser & McKee, Raritan Plaza I, Edison, NJ 08818. Visiting Professor, Rutgers University, Piscataway, NJ 08854.

Ozone is not free from forming carcinogenic by-products. The ozonation by-product that is targeted for regulation is bromate ion (BrO_3^-), a suspected carcinogen, which can be formed when water containing bromide ion (Br^-) (usually >0.2 mg/L) is ozonated. The MCL during the first stage of regulation may be set at 10 ug/L which is at a higher cancer risk factor (10^{-4}) than the one normally used for setting limits (10^{-5}). The issue is that 10 ug/L is the current Practical Quantitation Limit (PQL) for field analysis. Another factor is that BrO_3^- control and removal have not been fully studied. The proposed MCL for BrO_3^- in the first stage should go into effect in 1997. A second ruling in 1998 is expected to lower the MCL provided that research in BrO_3^- control, removal, health effects, and analysis has provided that justification.

This study's main objective, as part of a research project (Marhaba, 1993) in conjunction with Rutgers University, Elizabethtown Water Company, and the New Jersey Department of Environmental Protection & Energy, was to investigate the removal of BrO_3^- and Br^-, the precursor, by the non-conventional processes of granular activated carbon (GAC) filtration/adsorption and reverse osmosis (RO) nanofiltration and hyperfiltration. The study used pilot plant equipment at Elizabethtown Water Company's Raritan-Millstone Water Treatment Plant (EWC) in Bound Brook, New Jersey. The equipment mainly included chemical feed, ozonation, contact storage, GAC, and RO systems. The influent to the pilot plant was from sources within the water treatment plant which treated surface water. Br^- addition was necessary since the Br^- levels in the water were insignificant to form BrO_3^- when ozonated.

BrO_3^- has not been reported to be present in natural water, and apparently is only formed by man-made chemical reactions. The most relevant of those reactions is the reaction of ozone with Br^-, the precursor.

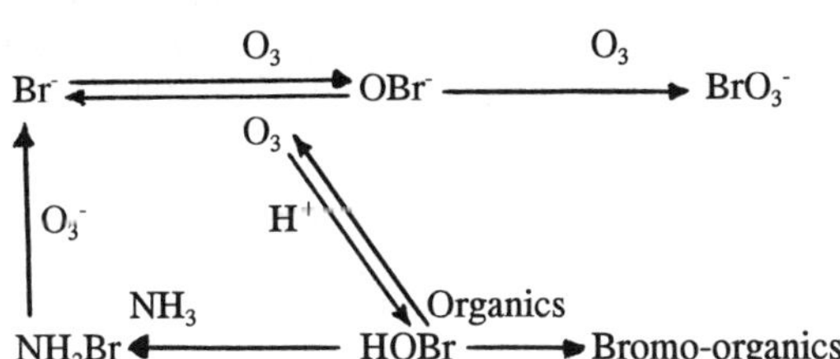

Figure 1. Conceptual Model of Dominant Reactions in BrO_3^- Formation

Figure 1 shows a conceptual model of the dominant reactions involved in the formation of BrO_3^- (Haag and Hoigne 1983 and 1984, and Von Gunten and Hoigne 1993). According to the figure, the concentrations of Br^- and ozone play major roles in the amount of BrO_3^- formed. These and other water quality conditions affect the overall formation potential. Therefore, the amount of BrO_3^- formed depends entirely on site-specific conditions.

Haag and Hoigne (1983), Krasner et al. (1991), and Siddiqui and Amy (1993) showed an increase in BrO_3^- formation at higher pH (from 7 to 9). Krasner (1992) showed that ozone dose plays a critical role in BrO_3^- formation. Siddiqui and Amy (1993) concluded that higher ozonation temperatures produced more BrO_3^-. This is probably due to the decrease in the pK_a of HOBr and OBr^- (Krasner et al. 1992), causing an increase in concentration of OBr^-. Krasner et al. (1992) also showed that even at higher pH levels, disinfection requirements can be attained, without increasing BrO_3^- formation, by keeping the ozone concentration low and extending contact time.

Methods and Equipment

Five different processes were evaluated; three ozonation for BrO_{3-} formation and two non-ozonation with the addition of BrO_3^-. The influent to all the pilot experiments was either from the EWC plant clearwell or sedimentation basin effluent. The different processes in each category (ozonation and non-ozonation) were developed to examine the effect of the presence of ammonia and temperature changes on BrO_3^- formation and removal. The non-ozonation processes were performed to provide data for high BrO_3^- concentrations to the GAC and RO systems which were run parallel following chemical feed, ozonation, and contact storage.

BrO_3^- and Br^- were analyzed by an outside laboratory using the method of ion chromatography (Novatek 1993) with the borate eluent (Revised USEPA Method 300.0). The samples were passed through a metal free column to remove dissolved metals and a guard column before separation on the analytical column. The limit of detection for BrO_3^- was 2-5 ug/L. Sample holding time was between 2 and 7 days. Ozone was dosed in the first contactor and the contactor's residual was measured immediately using the Indigo Colorimetric Method. Ozone dose was based on the minimum 3-log inactivation requirements of the USEPA Surface Water Treatment Rule.

The Calgon Filtrasorb-300 GAC was used as a medium in the study (two medium replacements). The DOW Filmtec NF-70 nanofiltration and FT-30 hyperfiltration membranes were used. Both membranes were thin film composites with spiral wound configurations.

Runs were performed with different initial conditions. That is, Br^-, BrO_3^-, pH, and ozone dose were varied for each run, except for repeats; Br^- concentrations up to 5 mg/L, BrO_3^- concentrations up to 300 ug/L, pH ranging from 6.8 to 9.3, and ozone dosages from 2 to 5 mg/L.

The general procedure for an experimental run was to run the pre-GAC and RO equipment until stabilization occurs in pH and ozone residual. Following stabilization the pre-GAC and RO systems tanks would be filled and the GAC and RO systems initiated.

Initiating a GAC run was performed by removing the water in the GAC filter and refilling it with the GAC influent (from the contact storage tanks) and backwashing with it an average of 10 minutes. This assures that the water going through the GAC is the homogenous influent. Mainly, two GAC medium replacements were performed. Runs using Medium 1 were for a duration of one hour each. Runs with Medium 2 were performed for a continuous period of 32 hours without backwashing between runs.

The duration of the reverse osmosis runs were approximately 10 minutes each. For a particular run, the influent to the RO system was from the same source as the GAC system.

Observations and Results

Removal By GAC

During the first 1,100 liters of Medium 1's throughput, Br^- was being adsorbed onto the GAC. It was observed that Br^- adsorption initially took place and then decreased until the 1,100 liter throughput mark. Following the 1,100 liter throughput point, desorption initiated. It was clearly observed that the amount of desorption was related to the influent Br^- concentration. Whenever a decrease in influent Br^- occurred, significant desorption of Br^- took place. The amount desorbed at a certain time was proportional to the accumulated Br^- mass and inversely proportional to the influent Br^- concentration.

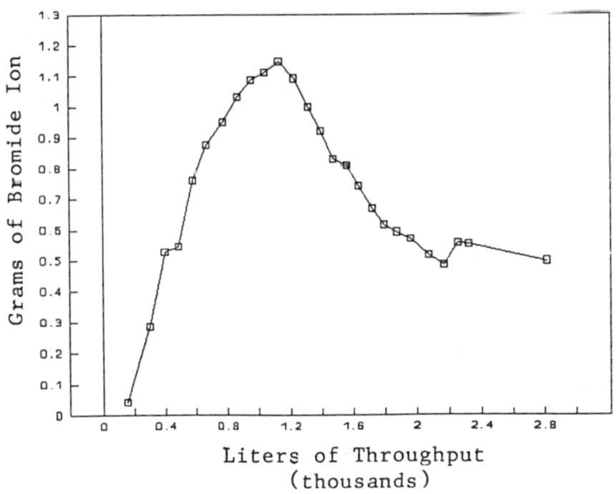

Figure 2. Accumulation of Br^- on GAC Medium 1

Figure 2 shows the amount of observed Br^- present on the GAC Medium 1 over throughput. The proportion of Br^- desorbed from the 1,100 liter throughput point and on, decreased due to the availability of less adsorbed Br^-. Hence the slope of the receding section of the curve is decreasing. The accumulation took into account BrO_3^- which was incorporated as Br^- ($[Br^-] = [BrO_3^-]*0.625$). A peak of about 1.2 grams of Br^- was adsorbed which correlated to 0.21 mg Br^- per gram of GAC, not including the mass retained from backwashing adsorption. The last run using Medium 1 was performed by passing water without the addition of BrO_3^- or Br^-. The results showed an effluent Br^- of 636 ug/L. This implied that Br^- was still being desorbed and true equilibrium was not reached.

The removal of BrO_3^- was observed to be dependent on the empty bed contact time (EBCT) and the influent BrO_3^- concentration. The EBCT in the system was varied between 5 and 12 minutes, but constant for each run. Table 1 shows the BrO_3^- removal runs for GAC Medium 1. The runs with the lower EBCT and high influent BrO_3^- concentrations showed a decrease from the 100% BrO_3^- removal level. Runs with higher influent BrO_3^- concentrations and similar EBCTs showed a decrease in BrO_3^- removal.

Table 1
GAC Medium 1 BrO_3^- Removals

EBCT (min)	Influent BrO_3^- (ug/l)	Effluent BrO_3^- (ug/l)	% BrO_{3-} Removal
5.0	24	4.0	83%
8.5	40	non-detect	100%
8.3	196	14.4	93%
8.1	69	non-detect	100%
9.3	258	8.0	97%
9.0	169	12.0	93%
9.0	3	non-detect	100%
6.6	290	14.6	95%
8.7	171	12.1	93%
10.2	285	22.5	92%

Runs through GAC Medium 2 were performed to examine the removal phenomenon of BrO_3^-. During Medium 1's runs, it was suspected that BrO_3^- was being reduced chemically by the GAC to Br^-, but this was not observed quantitatively because of the high concentrations of Br^- being desorbed. In the first two runs of Medium 2, only BrO_3^- was added at about 155 ug/L. The GAC filter was run continuously during the first 21 hours, accumulating 1,639 liters. The influent and effluent were sampled at hour 7 and hour 21. No BrO_3^- was observed in both runs mainly because the EBCT exceeds the estimated minimum EBCT at that influent concentration. Br^- was observed in the effluent, indicating that BrO_3^- was reduced to Br^-.

The effluent Br^- concentration increased during throughput on Medium 2. This was due to the adsorption and then desorption of Br^- on the GAC, as previously observed in Medium 1. It is suspected that all the Br^- produced by reduction of BrO_3^- will showup in the GAC effluent when Br^- adsorption/desorption reaches equilibrium. This would correlate to 96 ug/L effluent Br^-. Still after 21 hours of continuous running, a 36 ug/L effluent Br^- concentration was observed; 38% of what was expected.

The last run using Medium 2 incorporated the addition of BrO_3^- and Br^- to the influent. The EBCT was decreased from 6.3 minutes allowing BrO_3^- to showup in the effluent. The effluent Br^- concentration was 227% more than that of the influent, which was due to the desorption of Br^- adsorbed from the previous water treated.

According to the mass balance of BrO_3^- and Br^- calculated as Br^- in and out of the GAC filter for Medium 2, the accumulation of Br^- on the GAC decreased after about 1,600 liters of throughput. The maximum accumulation was about 0.1 gram or 0.02 mg per gram of GAC. This did not correlate with Medium 1's observations. Hence, the capacity of GAC to retain Br^- is highly dependent on the Br^- concentration being applied, either directly from the Br^- or from the chemically reduced BrO_3^-.

Removal By RO

The nanofiltration membrane was first operated at 40 psi (276 kPa) and then at 75 psi (517 kPa). As expected, the performance was better at the higher pressure. At the lower pressure, the variance was higher, especially for BrO_3^- removal. Hence, at the recommended pressure of 70 psi (517 kPa), performance was better and less variable.

The rejection of BrO_3^- was higher than Br^- since BrO_3^- has a larger molecular structure and a higher molecular weight than Br^-. The rejection ranges for Br^- and BrO_3^- were 43%-100% and 75%-100%, respectively (including data from both pressures) The lower rejections correlated with the higher influent concentrations and vice-versa. The average rejection for BrO_3^- and Br^- ions were 89%, and 84%, respectively, at pH from 6.8 to 9.3, influent Br^- concentrations from 35 to 5962 ug/L, and influent BrO_3^- concentrations to 263 ug/L.

Hyperfiltration membranes require higher pressures to perform properly. The operating pressure was at 115 psi (793 kPa). Average recoveries were at 7%, which was lower than the recommended. This was due to the limited capacity of the RO system pump. The average rejection for BrO_3^- and Br^- was 97% and 89%, respectively, at pH from 7.1 to 8.7, influent Br^- concentrations to 1372 ug/L, and influent BrO_3^- concentrations from 5 to 285 ug/L. The rejection of BrO_3^- and Br^- were higher than that by nanofiltration but not to a large extent. Hence, hyperfiltration may not be as cost-effective as nanofiltration for treating water with low total dissolved solids (<3000 mg/L).

Summary and Conclusions

The following conclusions were drawn on the basis of this study using the treated surface water at EWC in Bound Brook, New Jersey during the months of January to June, 1993:

1) The GAC used chemically reduced BrO_3^- to Br^-. The amount reduced was a function of the EBCT and influent BrO_3^- concentration. Hence, nearly complete removal is possible under the specific conditions cited. Breakthrough of BrO_3^- was not observed during the study, but the observed mass of BrO_3^- reduced was 0.03 mg/gr of GAC. Further research is needed to determine the reduction capacity of GAC in a dynamic GAC system. No change in the removals were observed when BrO_3^- was spiked into the influent. The minimum EBCT for the particular water conditions, was estimated at 10 minutes.

2) The GAC removed Br^-, but breakthrough began immediately, allowing Br^- to show up in the effluent above the limit of detection (8-44 ug/L). The retaining capacity was proportional to Br^- loading either directly from Br^- or indirectly from the reduced BrO_3^-. The release of Br^- was inversely proportional to the influent Br^- concentration and proportional to the accumulated Br^- mass at time, t.

3) Concentrations of ammonia up to 0.1 mg/L as NH_3-N were effective in preventing BrO_3^- formation at Br^- concentrations <5 mg/L, pH<8.1, and ozone dosages <3.7 mg/L.

4) RO Nanofiltration, using the specified thin film composite membrane, removed BrO_3^- and Br^- an average of 89%, and 84%, respectively, at pH from 6.8 to 9.3, influent Br^- concentrations from 35 to 5962 ug/L, and influent BrO_3^- concentrations to 263 ug/L, for this specific water quality.

5) RO Hyperfiltration, using the specified thin film composite membrane, removed BrO_3^- and Br^- an average of 97% and 89%, respectively, at pH from 7.1 to 8.7, influent Br^- concentrations to 1372 ug/L, and influent BrO_3^- concentrations from 5 to 285 ug/L, for this specific water quality.

Acknowledgement

The authors would like to thank Mr. Ed Mullen, Elizabethtown Water Company, for guidance the use of the pilot equipment and laboratory services. Thanks to Mr. Paul Schorr, P.E., New Jersey Department of Environmental Protection and Energy, Bureau of Safe Drinking Water, for his significant consultations, and funding the ion chromatography analyses. Special thanks to Dr. Rip G. Rice, RICE International Consulting Enterprises, for his invaluable expert advising, comments, and discussions. The authors also thank Professors Yong S. Chae and Q. Guo, Department of Civil & Environmental Engineering, Rutgers University, for their significant consultations.

Appendix

References

Haag, W., and J. Hoigne, "Ozonation of Bromide-Containing Waters: Kinetics of Formation of Hypobromous Acid and Bromate", *Environmental Science and Technology*, 17:5:261 (1983).

Haag, W., and J. Hoigne, "Kinetics and Products of the Reactions of Ozone with Various Forms of Chlorine and Bromine in Water", *Ozone Science & Engineering*, Vol. 6, pp. 103-114 (1984).

Krasner, S., J. Gramith, E. Means, N. Patania, I. Najm, and E. Aieta, "Formation and Control of Brominated Ozone By-products", Presented at the Annual Conference of the American Water Works Association, Philadelphia, PA (June 1991).

Krasner, S. , W. Glaze, H. Weinberg, P. Daniel, and I. Najm, "Formation and Control of Bromate During Ozonation of Waters Containing Bromide", Presented at the Pan American Committee Conf. of the Int. Ozone Assoc., Pasadena, CA (March 1992).

Marhaba, T., "Treatment/Removal of Bromate Ion and Its Precursor By Granular Activated Carbon and Reverse Osmosis From Drinking Water", Ph.D. Dissertation, Dept. of Civil & Environmental Engineering, Graduate School, Rutgers University, New Brunswick, NJ (Oct. 1993).

Novatek, "Novatek Application Note, Method No. C9-02001, The Measurement of Inorganic Species By Ion Chromatography", Oxford, Ohio (Jan 1993).

Siddiqui, M. and G. Amy, "Factors Affecting DBP Formation During Ozone-Bromide Reactions", *Journal of the American Water Works Association*, Vol. 85, pp. 63-72 (Jan. 1993).

Von Gunten, U., and J. Hoigne, "Factors Controlling the Formation of Bromate Ion During Ozonation of Bromide-Containing Waters", *Ozonews*, 21:3:21 (Mar. 1993).

Evaluation of Enhanced Coagulation for DBP Control

Richard J. Miltner[1], Sheila A. Nolan[1], S.M. ASCE,
Marcus J. Dryfuse[1], S.M. ASCE, and R. Scott Summers[2], M. ASCE

Many water utilities will have to meet the requirements of enhanced coagulation cited in the draft Disinfectant/Disinfection By-Product (D/DBP) Rule (Pontius, 1993). The mechanisms and pH dependency of coagulation of organic carbon have been discussed by Randtke (1988) and Quasim, et al. (1992). Utilities modifying coagulation in order to lower levels of total organic carbon (TOC) and DBP precursors must take care that the modifications do not compromise other water quality parameters or plant operation. A pilot scale study was conducted that examined the control of DBP precursors by conventional and enhanced coagulation. The objectives of the study were to assess: (1) the usefulness of TOC as a surrogate for DBP precursors; (2) the performance scale-up from jar testing to the pilot scale; (3) the differences in resulting DBP formation in conventional and enhanced plants; and (4) the differences in filter run times and particulate control in conventional and enhanced plants.

Experimental Design

Four surface waters having TOC concentrations exceeding 4 mg/L were jar tested. These preliminary jar tests were done following the protocols in the draft D/DBP Rule for enhanced coagulation. The jar test procedures have been described in detail (Tryby, et al., 1993). Based on jar testing, East Fork Lake water, was selected for piloting.

Two parallel pilot-scale treatment plants employing rapid mix, flocculation, and sedimentation were used to compare conventional and enhanced coagulation and are described in Figure 1. Following sedimentation, waters were chlorinated and sand filtered. On the conventional plant, pH was adjusted after the filter. On the enhanced plant, pHs were adjusted before one and after another parallel filter. All

---

[1] Environmental Engineers, USEPA, Drinking Water Research Division, Cincinnati, Ohio 45268

[2] Assistant Professor, Department of Civil and Environmental Engineering, University of Cincinnati, Cincinnati, Ohio 45221

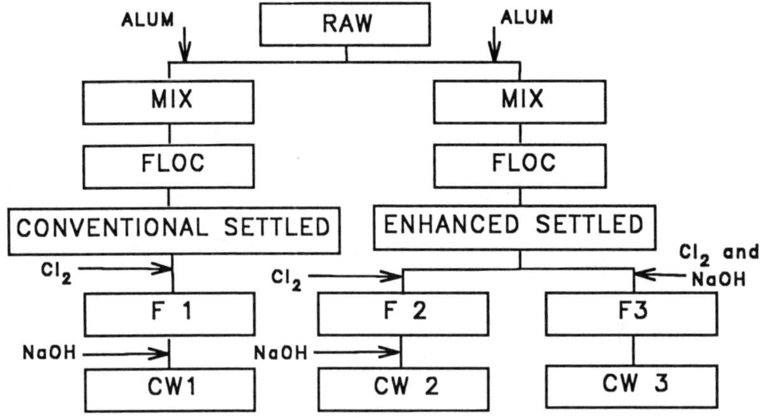

Figure 1. Pilot plant schematic diagram.

final pHs were adjusted to near 8 with NaOH. Water quality monitoring included: TOC, UV absorbance at 254 nm, trihalomethanes (THMs), haloacetic acids (HAAs), total organic halide (TOX), turbidity, particle counts, alkalinity, and THM, HAA, and TOX precursors. Operational data included: alum dose, pH, chlorine levels, headloss development, and filter run time. Preliminary jar testing was completed by 8/17/93; pilot operation began on 9/2/93 and was completed by 9/14/93.

DBP precursors were assessed under uniform formation conditions (UFCs), i.e., pH 8 ± 0.1, 20° ± 1°C, 24 ± 1 hours, and 1 ± 0.3 mg/L free chlorine residual. The term HAA6 refers to the sum of six HAAs: mono- di- and trichloroacetic acids, mono- and dibromoacetic acids, and bromochloroacetic acid. Both monohalogenated acids were not detected. Particle counting was done with a Hiac Royco model HRLD150 sensor and model 9064 counter.

Jar Testing

Figure 2 shows results of jar tests performed prior to and following the pilot plant operation. Conventional coagulation was defined as the alum dose that produced settled turbidities less than 1 NTU. It resulted in 26 percent removal of TOC. Enhanced coagulation resulted in 52 percent removal of TOC, and was defined as the alum dose where TOC concentrations reached a minimum plateau. For both 8/17 and 9/14 waters, this plateau was reached when sufficient alum was added to drop the pH to 6.6 or below. A utility, however, would not be required to operate on the plateau as, according to the draft D/DBP Rule, only 35 percent removal of TOC would be required given the TOC and alkalinity of this water. The jar tests also showed that control of TOC resulted in

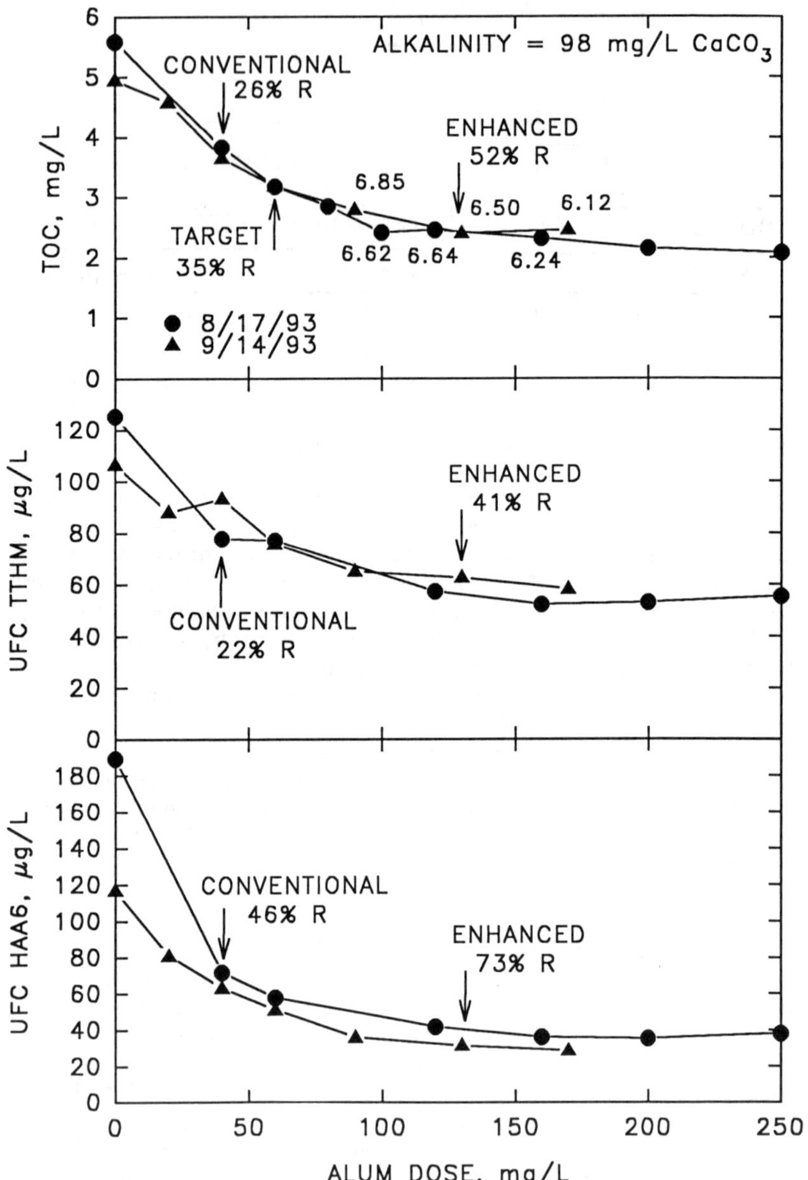

Figure 2. Jar test results for East Fork Lake water. All percent removals are based on 9/14 data. Select settled water pH values are given.

control of the precursors for THMs, HAAs, and TOX, and in control of UV absorbing compounds, indicating that TOC is a good surrogate (Table 1).

Table 1. Comparison of Jar Testing and Pilot Plant Operation.

	Percent Removal			
	Conventional		Enhanced	
	Jar Test[a]	Pilot Plant[b]	Jar Test[a]	Pilot Plant[b]
TOC, mg/L	26	29	52	54
UV Abs, 1/cm	42	39	65	64
UFC TTHM, μg/L	22	24	41	47
UFC HAA6, μg/L	46	44	73	73
UFC TOX, μg Cl/L	38	36	61	63
Alum Dose, mg/L	40	44	130	152
Turb, NTU	0.47	0.99	0.37	0.73
pH	7.23	7.34	6.50	6.57

a Percent removal based on 9/14/93 jar test data.
b Mean values of 6 sample days over two week operation.

Piloting and Scale-up

Figure 3 and Table 2 give mean results for the pilot plant operation. As with jar testing, conventional coagulation was defined as the alum dose that produced settled water turbidities less than 1 NTU. Enhanced coagulation was defined as the alum dose that gave pH 6.6 or below during rapid mixing. Mean removals of TOC during piloting were 29 and 54 percent, respectively, for conventional and enhanced coagulation. Based on piloting, the control of TOC predicted the control of THM, HAA, and TOX precursors and the control of UV absorbing compounds (Table 1). TOC's role as a surrogate has been demonstrated previously for Ohio River water (Tryby, et al., 1993).

Downstream of coagulation, there was no significant removal of TOC or DBP precursors with filtration. Figure 3 also shows the rapid in-plant formation of DBPs. Approximately half of the concentration of the DBPs formed in 24 hours was present after 20 minutes of chlorine contact during filtration, and all of what was formed in 24 hours was present after 12 hours of chlorine contact during clear well storage.

Table 1 summarizes the comparison between bench and pilot scale testing for both conventional and enhanced treatment. Except for turbidity, the jar tests were a good predictor of the pilot-scale performance, provided that the jar tests were conducted at the same time as the piloting. Percent removal data for the 8/17 jar test, conducted two weeks before piloting, slightly overpredicted DBP control (Figure 2). TOC is a conservative

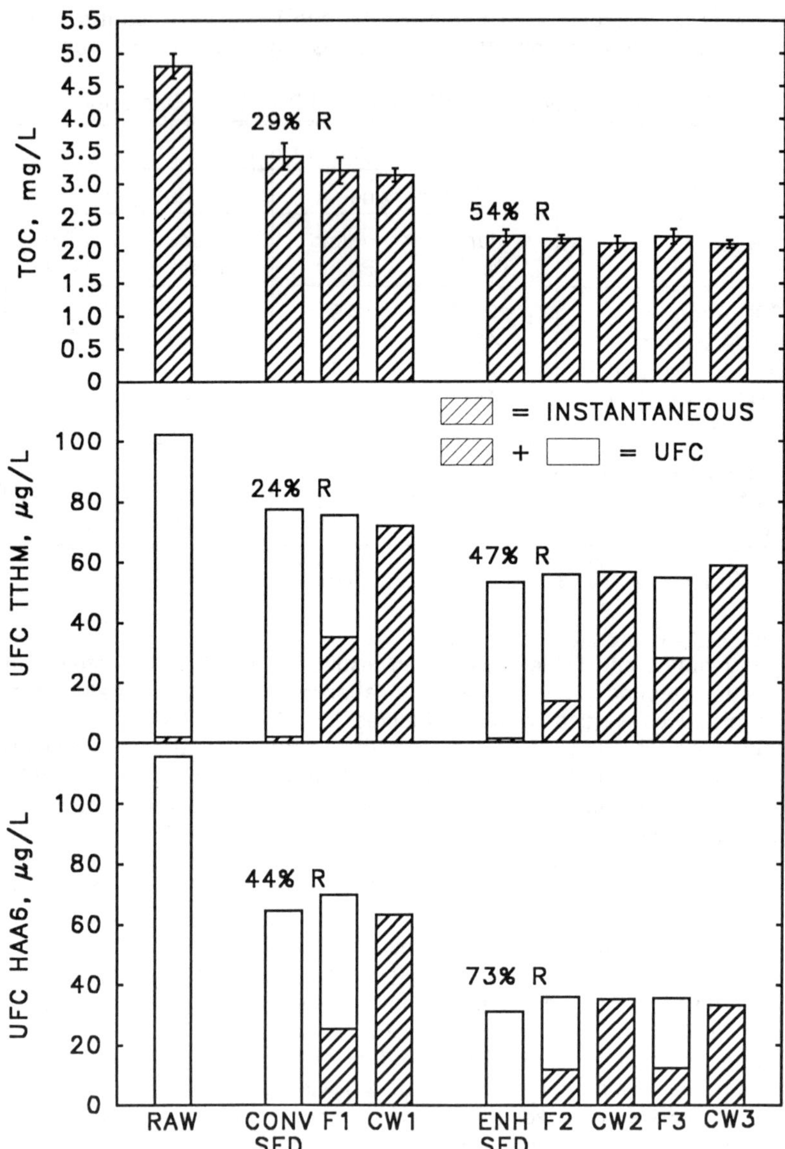

Figure 3. Pilot plant results for East Fork Lake water. Values represent mean of 6 sample days over two week operation. Error bars represent standard deviation.

Table 2. Mean* Values for Pilot Plant Operation.

	Turb NTU	Alum Dose mg/L	pH	Cl_2 Dose mg/L	Alk mg/L $CaCO_3$	UV Abs 1/cm
Raw	3.40		8.16		99	0.129
Conv Set	0.99	44	7.37		81	0.079
F1	0.24		6.90	4.81	92	0.067
CW1			7.96		90	
Enh Set	0.73	152	6.68		35	0.047
F2	0.17		6.90	3.99	40	0.040
CW2			7.78		74	
F3	0.18		7.95	4.07	67	0.043
CW3			8.11		69	

* Means over two weeks operation.

predictor of the removal of precursors of THMs, HAAs, and TOX, and of UV absorbing compounds.

Speciation

Figures 2 and 3 and Table 1 present the summed concentrations of individual THM and HAA compounds. Speciation, however, shows a shift to the more-brominated compounds. The shift is observable between raw and conventional treatment, and continues to shift as the alum dose is increased to reach enhanced coagulation (Table 3). As the organic precursor was removed, bromide remained, and thus the bromide-to-organic carbon ratio increased. Although the concentrations of formed HAAs and THMs decreased, the percentage of brominated species increased.

Particulate Control and Plant Operation

Settled and filtered turbidities (Table 2) on the enhanced plant were slightly lower than those on the conventional plant, but the differences were more distinguishable when particle count data were examined (Figure 4). Total particle counts (TPCs) were approximately one half log lower in settled waters with enhanced coagulation than with conventional coagulation, and near one log difference existed comparing conventional (F1) and enhanced (F2 and F3) filtered waters. Accordingly, with a higher particle load, conventional filter run times were shorter than those of enhanced filters. The particle count data and corresponding headloss data were collected in the middle of the pilot operation. Although the influent particle load to F2 and F3 were identical, F3 headloss build-up was slower because it was trapping fewer particles than F2. It is possible that this phenomenon can be explained by the fact that F2 was operated at pH 6.9,

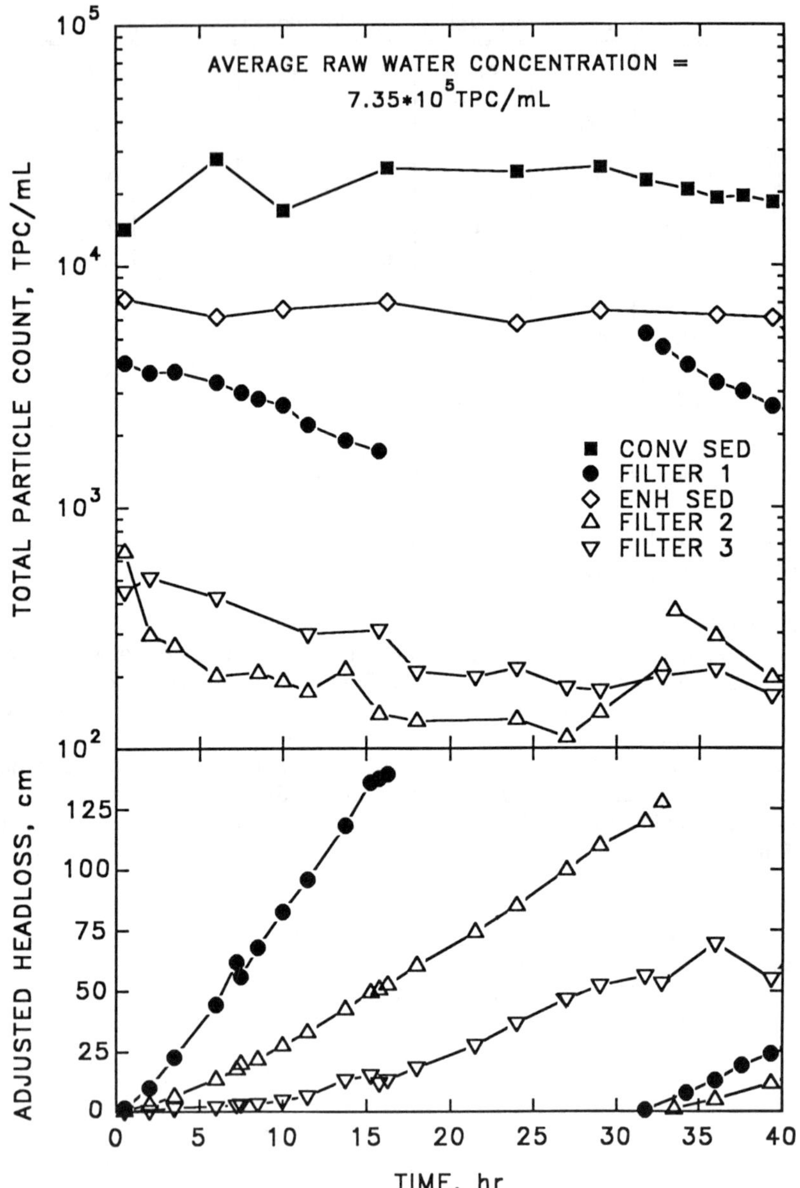

Figure 4. Total particle counts and headloss development during piloting for East Fork Lake water.

Table 3. UFC THM AND UFC HAA6 Speciation* During Piloting.

	TOC mg/L	UFC THM µmole/L	UFC THM-Br µmole/L	Ratio UFC THM-Br: UFC THM
Raw	4.81	0.814	0.079	0.097
Conv Set	3.42	0.620	0.072	0.117
Enh Set	2.21	0.424	0.064	0.150
	TOC mg/L	UFC HAA6 µmole/L	UFC HAA6-Br µmole/L	Ratio UFC HAA6-Br: UFC HAA6
Raw	4.81	0.794	0.0133	0.017
Conv Set	3.42	0.447	0.0126	0.028
Enh Set	2.21	0.216	0.0111	0.051

* Mean values for two week pilot operation.

while F3 was operated at pH 8, where aluminum is much more soluble. Thus, aluminum precipitating in F2 would result in faster headloss build-up. Aluminum data were collected and will be reported elsewhere. This highlights the importance of properly evaluating the trade offs in the selection of the point of final pH adjustment.

Acknowledgements

The authors thank the managers of the East Fork Lake State Park for their cooperation. The also acknowledge the significant contributions of University of Cincinnati undergraduate and graduate students and USEPA technicians in pilot operation and analytics. This work was funded through USEPA in-house funds and through cooperative agreement CR 816700 between the USEPA's Drinking Water Research Division and the University of Cincinnati. Although the research described was funded by the USEPA, it has not been subjected to agency review and does not necessarily reflect the views of the agency. Mention of commercial names does not constitute endorsement or recommendation by the agency.

References

Pontius, F.W. (1993). "D/DBP Rule to Set Tight Standards." *J. Am. Water Works Assoc.*, 85(11), 22.

Quasim, S.R., et al. (1992). "TOC Removal by Coagulation and Softening." *J. Envir. Engrg.*, ASCE, 118(3), 432.

Randtke, S.J. (1988). "Organic Carbon Removal by Coagulation and Related Process Combinations." *J. Am. Water Works Assoc.*, 80(5), 40.

Tryby, M.E., et al. (1993). "TOC Removal as a Predictor of DBP Control with Enhanced Coagulation." *Proceedings, Seminar on Enhanced Coagulation, Water Quality Technology Conference*, AWWA, Miami, Fla.

TRIHALOMETHANES IN CHLORINATED SACRAMENTO-SAN JOAQUIN DELTA DRINKING WATER SIMULATED WITH THE EPA WTP-THM MODEL

Russ T. Brown[1], Associate Member ASCE, Doug Brewer[1], Doug Owen[2], Zaid Chowdhury[3]

Abstract

Trihalomethane (THM) concentrations at a representative water treatment plant simulated with the EPA Water Treatment Plant THM (WTP-THM) model were used to assess potential drinking water quality impacts of a proposed water storage project in the Sacramento-San Joaquin Delta. Bromide and dissolved organic carbon (DOC) are the most important precursor water quality parameters. Bromide and DOC concentrations were estimated from a mass balance transport model of Delta inflows and agricultural drainage sources.

The EPA WTP-THM simulation model results were compared with data from a conventional water treatment plant using Delta water as its source. The raw water quality data were used in combination with unit process operating data to evaluate the accuracy of model predictions of THM concentrations. The EPA WTP-THM model was modified to accept 25 years of monthly raw water quality and to automatically estimate the chlorine doses necessary to satisfy requirements for disinfection and predict THM concentrations for each set of raw water quality conditions with the specified treatment process parameters (Jones & Stokes 1992).

The impact assessment strategy determined the incremental THM concentrations at a representative water treatment plant resulting from bromide and DOC concentrations changes attributable to the proposed water storage project. The risk of increasing the frequency, magnitude or duration of exceedances of the current EPA THM maximum contaminant

---

[1] Jones & Stokes Associates Inc. 2600 V Street Sacramento, CA 95818

[2] Malcolm Pirnie Inc. 703 Palomar Airport Suite 150 Carlsbad CA 92009

[3] Malcolm Pirnie Inc. 4636 University Dr. Suite 150 Phoenix AZ 85034

level (MCL) standard of 100 $\mu g/l$ (and the proposed 80 $\mu g/l$ MCL) were determined for each project alternative.

Description of EPA WTP-THM Model

The water treatment plant THM model was developed by Malcolm Pirnie Incorporated for EPA (EPA 1992) to support analyses of alternative Disinfection By-Products (DBP) Rules and the Surface Water Treatment Rule (SWTR). The primary purpose of the model is to simulate the formation of disinfection by-products (DBP) and disinfectant concentrations in water treatment plants and distribution systems based upon specified raw water quality, treatment process characteristics, and chemical dosages. The model predicts THM concentrations based on a series of regression equations developed to summarize THM data collected from various water treatment plants and experimental data on the kinetics of THM formation.

The model simulates the removal of organic THM precursor compounds (dissolved organic carbon [DOC] and ultra-violet absorbance at 254 nm [UVA]), and the adjustments in pH that occur as chemicals are added. When free chlorine is added, chlorine demand, chlorine decay, and THM concentrations are calculated. The calculations proceed until all treatment processes and disinfectant dosages are simulated.

The model predicts the total THM concentration, and then determines the concentration of individual THM molecules by estimating the relative concentrations from separate regression equations for each of the four THM molecules. The multiple-logarithmic regressions are each similar, but vary in the coefficient values for the independent variables (Amy, Chadik, Chowdhury 1987). For example, the total THM equation is:

$$\text{THM}\ (\mu g/l) = 0.3254 * \text{DOC}^{0.44} * \text{UVA}^{.351} * \text{Cl}_2^{0.409}$$

$$* \text{Hours}^{0.265} * \text{Temp}^{1.06} * (\text{pH}-2.6)^{0.715} * (\text{Br} + 1)^{0.516}$$

The magnitude of the coefficient for each independent variable indicates how THM concentrations will respond to change in that variable, other conditions remaining the same. For example, a temperature change from 15 to 25 degrees C will increase THM by about 75%. Doubling DOC from 3 to 6 mg/l will also double the UVA value because the two variables are linearly related (UVA = 0.003 * DOC), and this combination will increase THM by about 75%. Higher DOC may also require a higher chlorine dose. Increasing the chlorine dose from 2 to 3 mg/l will increase THM by about 20%. Increasing bromide from 0.2 to 0.4 mg/l will increase THM by about 10%, according to the THM regression equation.

Verification of EPA WTP-THM Model Results

The ability of the EPA WTP-THM model to accurately simulate water treatment plant operations and finished water quality and THM concentrations was tested by comparing model results with monthly measurements collected at the Penitencia WTP from January 1991 to August 1992. The Penitencia Water Treatment Plant, operated by the Santa Clara Valley Water District, was selected for verification because it uses chlorine for primary disinfection of Delta water. The Penitencia treatment plant uses conventional alum coagulation with flocculation, sedimentation, and dual-media filtration. Thirteen sets of samples were available for verification.

The raw water contained moderate to high levels of DOC and bromide, because of drought conditions in California. Figure 1 shows the raw water DOC and measured and predicted final DOC concentrations at the Penitencia WTP. Raw water DOC ranged from 3 to 7 mg/l. The alum coagulation effectively removed about 40% of the DOC, but after pre-chlorination. Because the WTP-THM model parameters were adjusted to match the final DOC value for each sample date, the modeled DOC removal was accurate. Unfortunately, pre-chlorination doses were not measured, so the chlorine residuals following sedimentation were used to estimate the pre-chlorination doses for each sample date.

Figure 2 shows the raw bromide concentrations and the measured and predicted bromide incorporation factor, defined as the moles of bromide incorporated per mole of THM, with a maximum value of 3 for pure bromoform (Hutton and Chung 1994). Raw water bromide ranged from 0.1 mg/l to 0.6 mg/l. The bromide incorporation factor ranged from 0.25 to 1.75, and the predicted bromide incorporation followed the measured pattern. The bromide incorporation is generally greater with increased bromide, but is reduced when DOC concentrations and the required pre-chlorination doses are increased, as occurred for the April 91 sample.

The bromide yield, defined as the ratio of bromide in the finished water THM compared to the raw water bromide concentration ranged from 5% to 15%, but was generally underpredicted by the WTP-THM model. Malcolm Pirnie Incorporated has recently completed a joint project with Metropolitan Water District and its member agencies to collect additional data and improve the brominated-THM regression equations used in the WTP-THM model (MWD 1994). These new equations produce substantially more brominated-THM concentrations for bromide concentrations in the range of 0.1 to 1.0 mg/l.

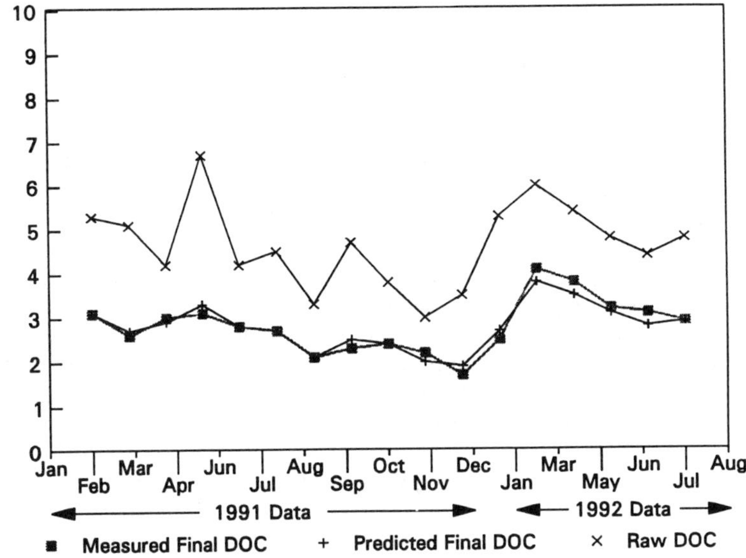

Figure 1. Comparison of Penitencia WTP Measured, Predicted, and Raw DOC Concentrations

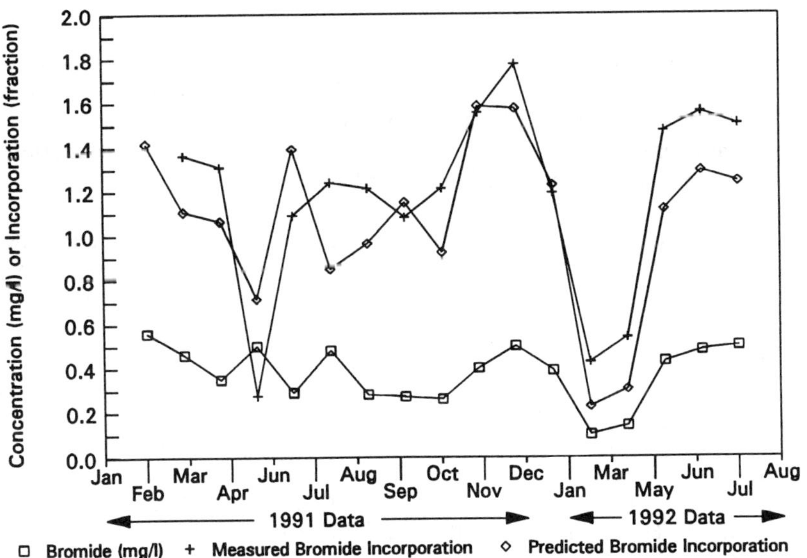

Figure 2. Comparison of Penitencia WTP Measured and Predicted Bromide Concentrations and Bromide Incorporation

Figures 3 and 4 compare measured and predicted concentrations for each of the four THM molecules. Measured total THM concentrations ranged from about 40 to 140 μg/l and averaged 72 μg/l, whereas the predicted total THM concentrations averaged ranged from 25 to 100 μg/l and averaged 50 μg/l. Some of the differences may be attributed to the relatively high bromide concentrations, which produce relatively large amounts of brominated-THM concentrations. The brominated-THM species are generally underpredicted by the WTP-THM model.

Other differences in THM concentrations occurred during winter when water temperatures were low. THM predictions were generally better during the summer months when water temperatures were warmer. Ambient water temperature is a strong determinant in the regression equation used for THM formation, and the model equations may be oversensitive to the effects of temperature.

Simulation of Delta Drinking Water THM Concentrations

THM concentrations are directly related to raw water DOC and prechlorination practices. Figure 5 shows simulated THM production as a function of estimated Delta DOC, using a treatment process characterized by a 0.65 mg/l prechlorination residual, 40 percent DOC removal, and 1.5 mg/l chloramine residual.

The modified EPA WTP-THM model results indicate that THM formation is directly related to raw water DOC, with higher raw water DOC resulting in higher THM formation. Variation in the predicted THM formation is caused by temperature and raw water ammonia concentrations, which both require increased pre-chlorination doses.

Figure 6 shows the simulated pre-chlorination dose as a function of estimated Delta DOC and ammonia concentrations. Ammonia oxidation is simulated to require a pre-chlorination dose of 7.6 times the ammonia concentration. The simulated pre-chlorination dose is about 40% of the DOC concentration.

Figure 7 shows the strong simulated effect of temperature on THM concentration. The simulated THM concentration (μg/l) was about 10 times the DOC concentration (mg/l) at 15 degrees C, and increased to about 20 times the DOC concentration (mg/l) at 28 degrees C. These results from the WTP-THM model suggest that the Penitencia WTP THM formation can be summarized by a very simple regression:

$$\text{THM } (\mu g/l) = 0.667 * \text{Temp(C)} * \text{DOC (mg/l)}$$

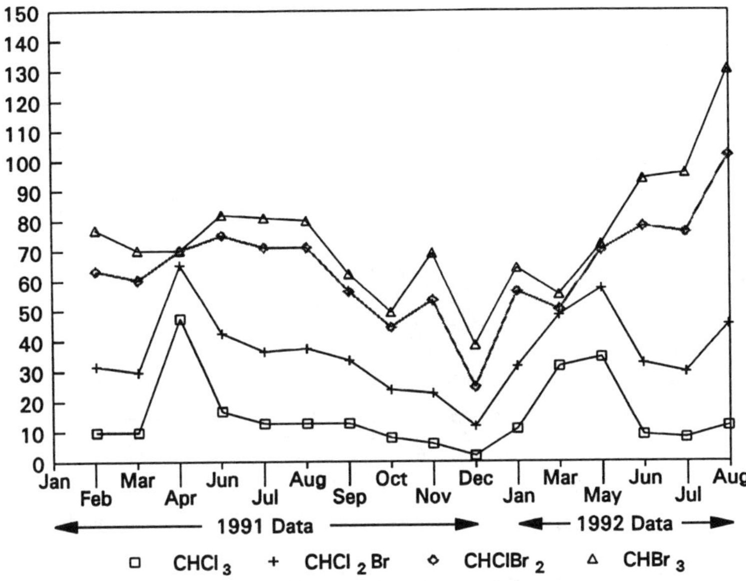

Figure 3. Measured Penitencia WTP THM Concentrations

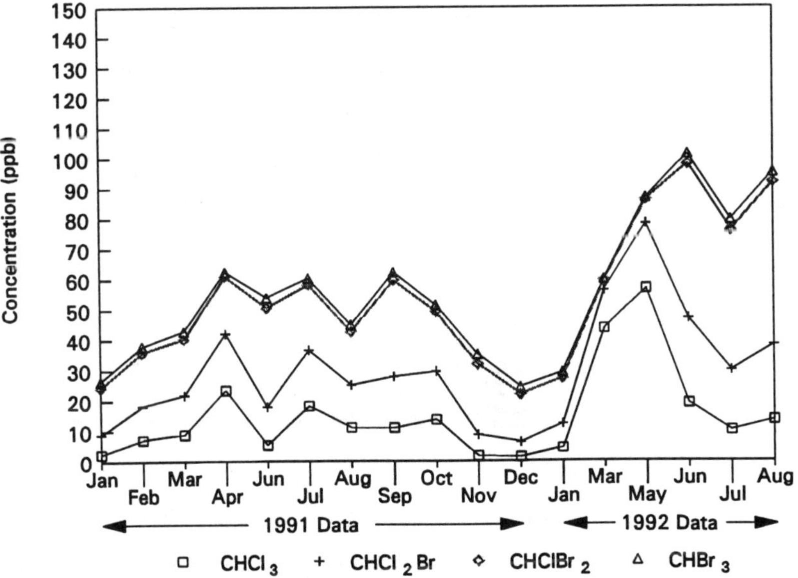

Figure 4. Predicted Penitencia WTP THM Concentrations

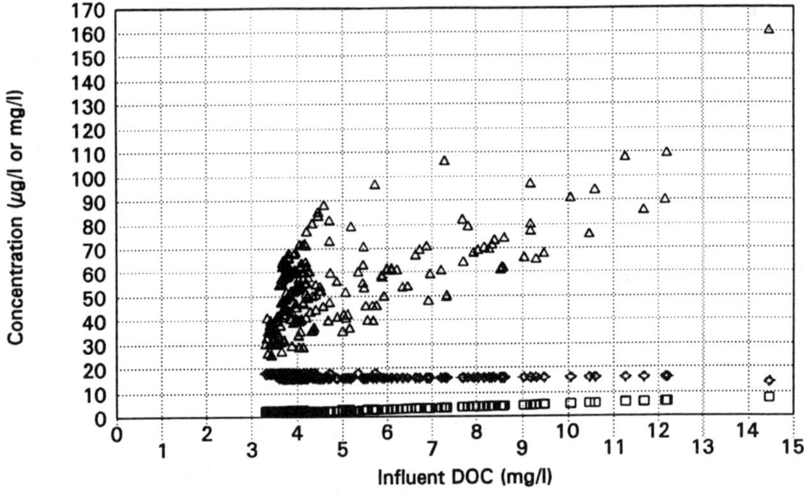

Figure 5. Simulated Total THM Concentration as a Function of Measured Influent DOC Concentration

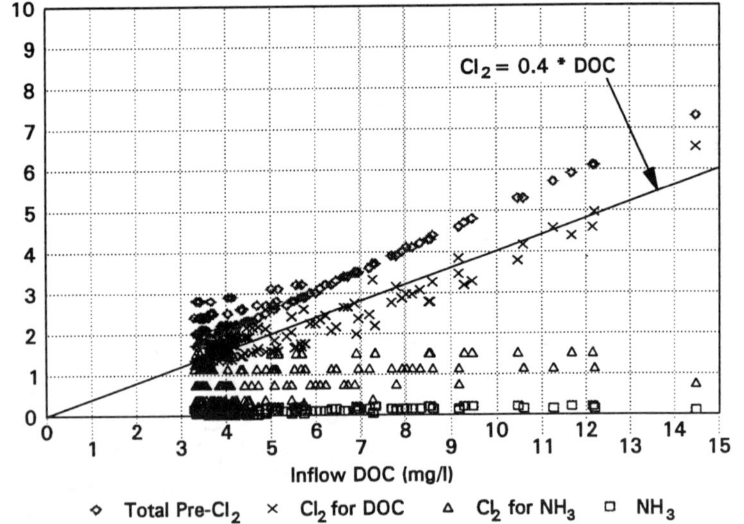

Figure 6. Pre-Chlorination Dose as a Function of NH_3 and DOC Concentrations

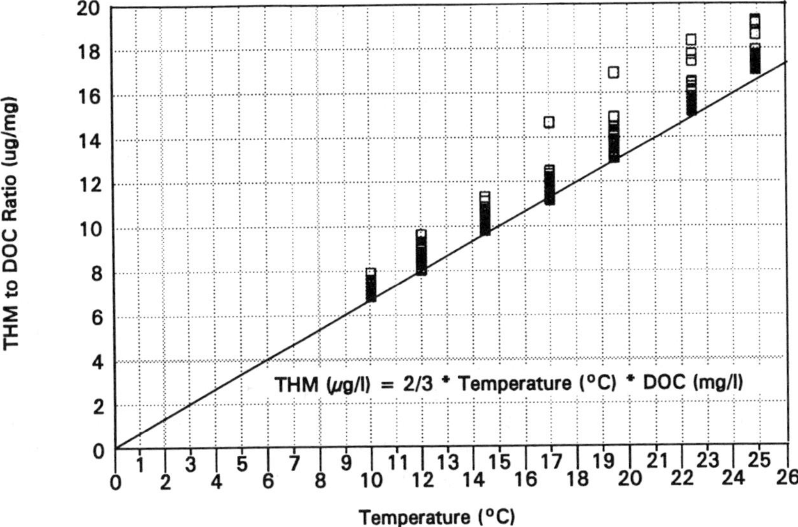

Figure 7. Simulated THM to DOC Ratio as a Function of Temperature

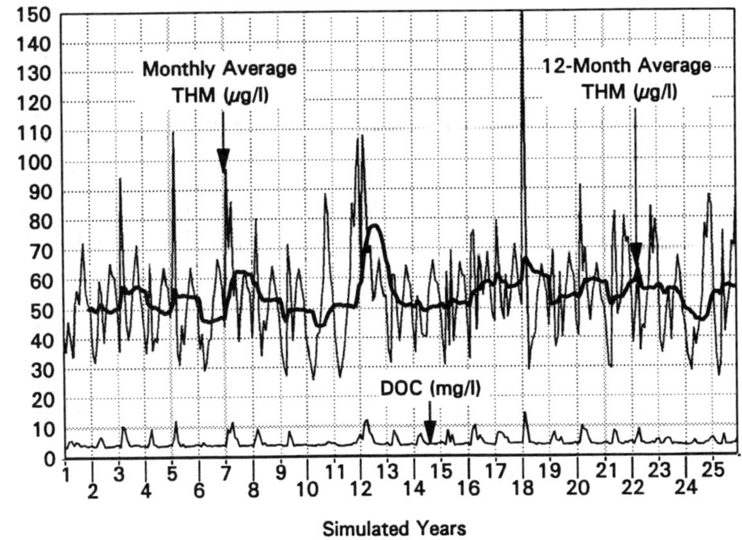

Figure 8. 25 Years of Simulated Monthly and 12-Month Average Total THM Concentrations

Figure 8 shows the monthly time series of simulated THM concentrations for the Penitencia WTP with conventional pre-chlorination treatment of Delta water, and the 12-month moving average of the THM concentrations that may correspond to the current EPA MCL criteria based on annual averages of quarterly sampling. The effects of raw water DOC concentrations and seasonal temperature effects are easily detected.

Acknowledgements

This work was performed as consultants to the California State Water Resources Control Board while preparing the Environmental Impact Report and Environmental Impact Statement for the proposed Delta Wetlands Project. The US Army Corps of Engineers is the lead Federal agency for the project. The entire effort was funded by Delta Wetlands Corporation. Technical review of the drinking water impact assessment strategy was provided by the California Department of Water Resources and Metropolitan Water District of Southern California staff.

References

G. L. Amy, P. A. Chadik, Z. K. Chowdhury (1987) Developing Models for Predicting Trihalomethane Formation Potential and Kinetics JAWWA 79(7):89-97

Environmental Protection Agency (1992) Water Treatment Plant Simulation Program Version 1.21 User's Manual. Office of Ground Water and Drinking Water Technology Transfer. Washington D.C. Prepared by Malcolm Pirnie Incorporated.

R. H. Hutton, F. I. Chung (1994) Bromide Distribution Factors in THM Formation ASCE Journal of Water Resources and Management 120(1):1-16

Jones & Stokes Incorporated (1992) Simulation of Water Quality Impacts on THM Formation for Delta Wetlands Impact Assessment. Sacramento California. Prepared by Malcolm Pirnie Incorporated.

Metropolitan Water District (1994) Bay-Delta Water Quality Modeling. La Verne California. Prepared by Malcolm Pirnie Incorporated.

A Gap in the Surface Water Treatment Rule: The Watersheds

Daniel A. Okun[1]

Abstract

Watershed control is required by the Surface Water Treatment Rule of the 1986 Amendments to the Safe Drinking Water Act for systems that do not provide filtration. Where filtration is available, no watershed protection is prescribed by the federal government. The oocysts of *Cryptosporidium*, which are difficult to remove by filtration and almost impossible to inactivate by chlorination, are highly infective. Measures are required to fill the gap in providing watershed control for the 90% of the surface water supplies in the U.S. that are now filtered.

Introduction

The philosophy that should guide the provision of safe water, where resources are being stretched and threatened by rapid development, is embodied in the traditional approach to the evaluation of raw water sources — the sanitary survey. Sanitary surveys, which were mandated by the 1962 U.S. Public Health Service Drinking Water Standards, emphasize the characterization of pollution sources, rather than merely the monitoring and analysis of the finished water. The operative statement in the standards with regard to drinking water source selection was that "...the water supply should be obtained from the most desirable source which is feasible, and effort should be made to prevent or control pollution of the source" (USPHS 1962). These strictures form the foundation upon which the U.S. Congress structured the Safe Drinking Water Act (PL 93-523, December 16, 1974).

---

1 Kenan Professor of Environmental Engineering, Emeritus, University of North Carolina at Chapel Hill, CB# 8060, Chapel Hill, NC 27599. Fellow, American Society of Civil Engineers.

The concern for quality of the source is found in the appendix to the EPA National Interim Primary Drinking Water Regulations of 1976 (US EPA, 1975): "Production of water that poses no threat to the consumer's health depends on continuous protection. Because of human frailties associated with protection, priority should be given to selection of the purest source. Polluted sources should not be used unless other sources are economically unavailable, and then only when personnel, equipment, and operating procedures can be depended on to purify and otherwise continuously protect the drinking water supply."

Recognition of the importance of sanitary surveys is also found in the preamble to the Regulations: "EPA encourages the states to conduct sanitary surveys on a systematic basis. These on-site inspections of water systems are more effective in assuring safe water to the public than individual tests taken in the absence of sanitary surveys." Surveys are not mandated.

The Current Dilemma

Unfortunately, the current drinking water regulations have never been compiled in a single document, presumably because they are evolving continuously. The National Interim Primary Drinking Water Regulations are no longer in print, so the admonitions concerning selection and protection of water sources do not challenge those who make decisions about water supply. All attention by the regulatory agencies is focused on establishing maximum contaminant levels (MCLs) for individual contaminants and by the water utilities on meeting these MCLs. Little attention has been given to source selection and protection.

This lack of concern is understandable, if not justifiable, because experience has shown that available technology, rapid sand filtration and chlorine disinfection, could render almost any surface water safe with respect to the transmission of infectious disease. The "chemical revolution" that accompanied World War II created new types of risks which were slow to be incorporated into regulations because their effects would not appear for decades if at all. The possibility of health effects, particularly cancer, and the plethora of synthetic organic chemicals found in drinking water sources led to passage of the SDWA. In addition, the emergence of the health significance of trihalomethanes (THMs) and other disinfection by-products (DBPs) led to what *The Milwaukee Journal* characterized as "Fatal Neglect" (Rowen & Behm, 1993). Water drawn from Lake Michigan and filtered and that met the applicable MCLs (if not the philosophy of the regulations) led to the worst water-borne epidemic of modern times with almost 400,000 people becoming seriously ill from cryptosporidiosis, with more than 50 deaths and countless numbers left terminally ill. Even with filtration, because of the human frailties mentioned, water source protection must be the first barrier.

Balancing the risks between the future health effects from exposure to trace chemical contaminants, and the immediate and readily observable health effects from microbial contaminants, has become the major sport engaging the drinking water cognoscente, with the refereeing as difficult as that for figure-skating at the Olympics. At a recent pre-Milwaukee international conference, "Safety of Water Disinfection: Balancing Chemical and Microbial Risks" (Craun, 1993), conferees informally leaned towards the latter being of greater concern today.

What can be agreed upon, however, is that selection of the best feasible source and its protection provides a barrier to the transmission of both chemical and microbial disease. Watershed protection limits the number and concentration of anthropogenic chemical contaminants in water and reduces the concentration of precursors of DBPs and as well reduces the concentration of parasites that are often difficult to remove by conventional filtration and may be impossible to inactivate by disinfection alone.

It was the concern for the parasites, *Giardia lamblia* in particular, that provided the impetus for the 1986 Amendments of the SDWA including a requirement for EPA to promulgate the Surface Water Treatment Rule (SWTR) which was finally issued in 1989 (USEPA, 1988). For the first time in federal regulations, site-specific watershed control requirements were established for water purveyors, but only for those without filtration who sought to rely on disinfection alone, which account for only about 10% of the systems and 10% of the population served by surface water. In addition to rigorous watershed control, unfiltered systems needed to meet source water coliform and turbidity limits as well as treated water disinfection residuals and comply with total THM MCLs. CT values also were to be met, based on the inactivation of *Giardia* cysts, although EPA in the SWTR indicated that it was "...conducting studies to determine whether additional regulations may be necessary to control for cryptosporidiosis" (USEPA, 1989).

Cryptosporidiosis is similar to giardiasis but much more severe. Therapy is not yet available. It is particularly threatening, and often fatal, to immuno-compromised individuals, such as those undergoing radiation or chemotherapy, AIDS patients and HIV-positives, the elderly and very young children. Moreover, the oocysts of *Cryptosporidium* are highly infective, only a few need to be ingested to cause disease. They are small and resistant, difficult to remove and/or inactivate. Monitoring their presence and concentration is very difficult and no surrogate for them is yet available. The absence of fecal coliform does not assure the absence of viable oocysts. They originate in animals, particularly cattle, and humans. The best approach for their control is to prevent their discharge into the watershed.

As important as watershed protection is, watershed protection with disinfection as the only treatment may rarely be sufficient to protect against the risks of parasites. An Expert Panel on New York City's Water Supply was established by EPA in 1992 to recommend whether New York City, with the largest water supply in the U.S., should be granted avoidance of filtration of its Catskill and Delaware waters under the SWTR. (New York had already consented to filter its Croton supply, which provides 10-15% of the water for the city.) The city's Department of Environmental Protection (DEP), which includes the Bureau of Water Supply, applied for avoidance of filtration based upon its plans for watershed control. The panel recognized that it might be feasible for some watersheds to be so well protected that disinfection alone might be adequate to reduce the risks of giardiasis. New York's Catskill and Delaware watersheds, however, were already so well developed, with 50 towns, 35 wastewater treatment plants, 200,000 summer population and some 400 dairy farms, that watershed protection alone was not deemed to be adequate by the panel (USEPA, 1993). Although the criteria for the control of *Giardia* might possibly be met, the control of the oocysts of *Cryptosporidium*, for which regulations had not yet been established would not be feasible with disinfection alone. CT values for *Cryptosporidium* would need to be several orders of magnitude greater than for *Giardia*. Nevertheless, the city continues to press for avoidance and EPA, on December 31, 1993, granted a three-year "time neutral" extension to the city, requiring the initiating of plans for filtration so that in the event the city could not meet the some 150 conditions placed upon it by the ruling, time for bringing filtration on line would not have been lost.

Not all cities are likely to fight filtration so strenuously. EPA's 1975 interim regulations called for a reduction in the turbidity standard from five to one NTU. In response, several Connecticut water companies knew that they would be obliged to provide filtration. So, ostensibly to help pay for the facilities, they sought to sell off, for development, watershed lands that they had purchased a half-century earlier and which had sky-rocketed in value, being within commuting distance of New York City. The State Health Department, at the initiative of their chief sanitary engineer, got the state legislature to declare a moratorium on further sale of watershed land for development and to establish a

Council on Water Company Lands. After considerable study, and public hearings, the Council recommended and the legislature adopted a policy of preventing watershed lands from being sold for development. The water companies sued the state, but the state was upheld in the federal courts on the basis that the state had the power to establish regulations for protection of the public health (U.S. District Court, 1977). Not incidentally, state and federal courts have always upheld local authorities that have instituted land use controls on watersheds. Unfortunately, too few local authorities have had the political will to challenge development interests.

However, as it now reads, the SWTR does not impose any restrictions on development on drinking water supply watersheds if filtration is provided.

In general, most upland watersheds, being located at some distance from the cities they serve, are rural and sparsely developed. Some cities have purchased all or a good part of their watersheds to maintain them as protected sources. However, in periods of financial exigency, these are attractive lands often seen as potential fruitful sources for the infusion of substantial funds into a community's exchequer and thereafter, because of their increased value when developed, a source of continuous income from property taxes. Decisions as to whether or not to continue to preserve watershed lands is largely local. Greenville, SC, with two wholly-owned mountain watersheds is unique in having taken steps, which required state legislation (South Carolina, 1991), that would prevent municipal officials in the future from selling off their watershed lands. In most cases, however, development interests in communities dominate land use decisions. Some watersheds are so fully developed that they are no longer suitable sources of water supply.

Agricultural, industrial, commercial and residential developments are allowed to proceed, largely unregulated, on most drinking water supply watersheds in the U.S. Controls may sometimes be exerted in those instances where the watersheds provide water to those who live on the watersheds. The far more common situation, where the watershed land lies in another jurisdiction, poses the most serious obstacle to watershed protection. In such cases, land use controls and the costs of so-called best management practices are perceived to be, and may very well be, a limit on economic development in that jurisdiction. The benefits of watershed protection would accrue to people in the urban areas being served by the watersheds who incur none of the costs.

Solutions

Among several possible approaches to addressing this problem are:

- Federal and state regulations mandating lands use controls for all drinking water watersheds, other than run-of-river supplies, in the interests of public health.

- Federal and/or state tax and other incentives for adopting land use controls on watersheds.

- State legislation that would require the benefited water purveyor or jurisdiction to reimburse the owners of the watershed lands, or the political jurisdictions involved, for losses in value. The amounts would be negotiated, mediated, or arbitrated by the state.

- Exchanges or sale of development rights.

- Tax payments by purveyors for watershed lands owned by others that would mitigate the loss of value resulting from land use controls.

- Integration of the administration of the Clean Water and the Safe Drinking Water Acts especially as they are directed at watersheds.

North Carolina has recently enacted a law requiring all local governments that have jurisdiction over drinking water supply watersheds, regardless of who uses the water, to prepare land-use plans for the watersheds to be approved by the state (North Carolina, 1992). However, in the final hectic hours of the 1993 legislative assembly, an assemblyman introduced an amendment to an unrelated bill that would exempt a county he represented from being required to make a land-use plan to protect a drinking water supply in a neighboring county. The battle over implementation of land use controls has begun.

The federal government and the states, together with professionals from a wide range of disciplines, need to develop strategies for watershed protection that would be economically and politically feasible. Filtration without watershed protection is not enough to assure protection of public health.

References

Craun, G.F. (ed.) 1993. Safety of Water Disinfection: Balancing Chemical and Microbial Risks. ILSI Press., Washington, DC, 690 pp.

North Carolina. 1992. Rules for Surface Water Supply Watersheds. Environmental Management Commission. Raleigh, NC.

Rowen, J. and D. Behm. 1993. Fatal Neglect. A Special Report, *The Milwaukee Journal*, September 19-26, 1993. 16 pp.

South Carolina. 1991. Chapter 8, Conservation Easement Act of 1991. Columbia, SC.

U.S. District Court. 1977. Bridgeport Hydraulic Co. et al. vs. The Council on Water Company Lands of the State of Connecticut et al, Civ. No. B-85-212. December 29, 1977.

U.S. Environmental Protection Agency (USEPA). 1975. National Interim Primary Drinking Water Regulations, 40 FR 59565, Dec. 24, 40 CFR 141, Washington, DC.

U.S. Environmental Protection Agency (USEPA). 1988. Evaluation of Specific Criteria of the Surface Water Treatment Rule. Washington, DC. 326 pp.

U.S. Environmental Protection Agency (USEPA). 1989. Drinking Water; National Primary Drinking Water Regulations; Filtration, Disinfection; Turbidity, *Giardia lamblia*, Viruses, *Legionella*, and Heterotrophic Bacteria; Final Rule. 40 CFR Parts 141 and 142. June 29, 1989, Washington, DC.

U.S. Environmental Protection Agency (USEPA). 1993. Report of the Expert Panel on New York City's Water Supply, April, 1993, Environmental Protection Agency, Region 2, New York, NY, 134 pp.

U.S. Public Health Service (USPHS). 1962. Public Health Service Drinking Water Standards - 1962. Publication No. 956, U.S. Department of Health, Education and Welfare, Washington, DC.

Field Test of Stormwater Best Management Practices

S. L. Yu[1], M. ACSE, S. L. Barnes[2], M. ASCE,
R. J. Kaighn, Jr.[3], M. ASCE, and S. L. Liao[3]

Abstract

A dry detention pond, hydraulically modified and draining a small and highly impervious area, and a highway median grassed swale were monitored for a fifteen-month period to test their ability in reducing stormwater pollution. Pollutants examined included total suspended solids, total phosphorus, chemical oxygen demand, and total zinc. The results indicate that for the modified dry pond (with a 7.6 cm or 3 in outflow orifice) an average removal rate of about 50% was obtained for total suspended solids; 40% for total phosphorus; and 30% for both chemical oxygen demand and total zinc. Slightly higher removal rates were obtained for the 31 m or 100 ft long swale with a check dam. Modeling analyses of the pond show that the pollutant removal rate is closely related to the total volume and the intensity of the rainfall events. Both the pond and the swale are fairly effective in removing pollutants when the rainfall volume is small and when storm intensity is low and duration is long.

Introduction

A field program was initiated in 1991 for testing the pollutant removal efficiency

---

[1]Professor, Department of Civil Engineering and Applied Mechanics, University of Virginia and Faculty Research Scientist, Virginia Transportation Research Council, Charlottesville, VA 22903

[2]Project Engineer, CH2M-Hill, 625 Herndon Parkway, Herndon, VA 22070

[3]Graduate Students, Department of Civil Engineering and Applied Mechanics, University of Virginia, and Graduate Research Assistants, Virginia Transportation Research Council, Charlottesville, VA 22903

of certain best management practices (BMPs). A dry detention pond and a highway median grassed swale in the Charlottesville, Virginia area were selected for the study. Manual as well as automatic sampling methods were used to monitor stormwater runoff into and out from the two facilities. Pollutants examined included total suspended solids (TSS), total phosphorus (TP), chemical oxygen demand (COD), and total zinc (Zn). The field monitoring work started in March, 1992 and continued until July, 1993. A total of eleven storm events were sampled at each BMP site.

Methods of Sampling and Analysis

The dry detention pond drains an area about 3.2 ha (7.9 acres), of which most is a parking facility. The outlet structure was modified in that a 7.6 cm (3 in) orifice was placed at its bottom. The low flow channels inside the pond were lined with riprap for erosion control. The pond bottom was heavily vegetated during the sampling period. The highway median swale is about 31 m (100 ft) long with a longitudinal slope of 5% and side slope of 1 vertical to 3 horizontal. A V-notch weir was installed at the outlet point of the swale for flow measurement. Wooden barriers were placed along both sides of the swale to prevent lateral runoff from flowing into the swale.

For each storm four to ten samples were taken at each sampling point. From the flow and the pollutant concentration data, the fluxes of materials coming into and going out of the facility were computed. The pollutant removal efficiency of the facility was then computed by a mass balance analysis.

Results and Discussion

Modified Dry Detention Pond. Table 1 presents a summary of pollutant loadings and removal efficiency, together with storm depth and intensity, antecedent dry days, and the inflow median particle size for each storm event for the modified dry pond.

The results show that on the average, the modified dry pond removed about 50% of TSS, 35% of COD, 40% of TP, and 30% of Zn, with an overall average of about 40% aggregate removal of all pollutants monitored. It can be seen from Table 1 that, for storms with small total volume of rainfall, e. g., less than 13 mm (0.5 in), the removal efficiency is generally higher, especially for TSS. The same trend (i.e., removal decreases with increasing volume of rainfall) is to a lesser extent also observed for TP and Zn. However, the removal of COD does not seem to relate to the volume of rainfall. An attempt was made to relate the removal efficiency of the pond to other pertinent parameters such as average rainfall intensity, antecedent dry days, detention time, particle size distribution, and

Table 1. Summary of Detention Pond Data*

Storm Date	Pollutant	Flux In (g)	Removal Efficiency (%)	Antecedent Dry days	Rain Intensity (mm/hr)	Total Depth (mm)	D50 (μm)	K (1/day)
04/30/92	TSS	411	64.4	3	1.80	3.6	2.6	128
	TP	9.6	39.6					73
05/05/92	TSS	512	89.9	4	1.30	2.0	3.5	210
	TP	14.5	87.4					210
03/31/93	TSS	2236	12.3	1	0.51	4.8		11
	COD	2023	50.6					52
	TP	27.4	76.5					108
	Zn	48.1	51.3					11
04/26/93	TSS	4698	81.3	4	2.12	6.4	32.0	89
	COD	4650	46.9					28
	TP	39.0	53.3					23
	Zn	31.5	19.4					4
05/29/92	TSS	18070	41.0	2	0.70	26.7		7
	TP	441	(33.1)					0
06/04/92	TSS	16460	38.4	4	1.90	50.8		17
	TP	343	(41.4)					0
	Zn	619	71.6					20
04/09/93	TSS	12522	41.0	2	2.73	36.8		18
	COD	14446	10.5					7
	TP	39.0	42.5					11
	Zn	117.4	21.9					3
07/03/93	TSS	18965	60.4	2	8.82	17.6	43.0	23
	COD	7715	46.1					14
	TP	61.7	59.8					9
	Zn	15.6	23.7					2

*Data for three other storm events were incomplete and therefore were not included drawdown time. However, the results are somewhat inconclusive and therefore no trend could be detected.

On the whole the dry pond tested in this study performed fairly well when compared with literature data. It is possible that the inflow channel riprap and also the heavy vegetation at the pond bottom provided certain removal through flow retardation, infiltration and adsorption / sedimentation.

Highway Median Swale. For the first six storms sampled in 1992, the side inflow barriers were not in place. Also leaks were discovered along the barriers during the first two events after the barriers were installed. Therefore, only the results obtained for the last four storms are listed in Table 2. Overall the swale removed a fairly high percentage of the pollutants examined. It should be noted that for the 7/10/93 storm, the rainfall volume was very small (5.1 mm or 0.2 in)and all the runoff infiltrated or was held by the check dam (weir) at the outlet. There was no pollutant mass flowing out and hence the 100% removal.

Literature information on field performance of grassed swales is scarce. It is, however, generally agreed that the ability of the swale to infiltrate the stormwater runoff is an important factor. Therefore, the swale length, longitudinal and side slope, and the density and height of the grass all will affect the swale pollutant removal efficiency. For the swale tested in this study, the removal rate is quite high for very small storms but decreases rapidly when rainfall volume exceeded 15 mm (0.6 in).

Table 2. Swale Removal Efficiencies and Rainfall Data

Storm Date	Efficiency (percent) TSS	COD	TP	ZN	Rainfall Intensity (mm/hr)	Volume of Rainfall (mm)
05/31/93	89	88	92	88	1.23	9.9
07/10/93	100	100	100	100	10.20	5.1
07/12/93	73	81	94	89	14.56	10.9
07/19/93	86	67	80	58	18.84	16.3

Modeling Analysis

Dry Detention Pond. The detention pond data was further analyzed using the VirginiA STorm (VAST) model. VAST (Tisdale and Yu, 1987; Yu and Das, 1987) is a watershed hydrology and runoff quality model. The VAST model uses the same quantity and quality algorithms found in such well known models as STORM, SWMM and HEC-1, but is much less data intensive. A subprogram of VAST called VAPOLL is a detention pond program which routes pollutants through a pond. The pollutant routing is done by solving the following advective diffusive transport equation numerically:

$$\frac{\partial c}{\partial t} = E\frac{\partial^2 c}{\partial x^2} - U\frac{\partial c}{\partial x} - Kc \tag{1}$$

where c is the pollutant concentration, E the dispersion coefficient, U the flow through velocity, K a first order decay coefficient, x the longitudinal distance, and t the time.

Data for eight storm events was analyzed using VAST. For each storm and for each pollutant, VAST runs were made to calibrate the values of E and K so that the measured concentrations and mass loadings at the pond outlet were matched. The computed K values are included in Table 1. The results show that E is not a sensitive parameter. K can also be called a removal coefficient because it represents mainly the removal of the pollutant through settling, adsorption, infiltration, etc., as the amount decayed is probably very small due to the relatively short time of simulation. Figure 1 depicts an example of the VAST calibration results.

The removal coefficient, K , was found to relate fairly well with the pond mass removal efficiency for TSS, TP, and Zn, but not for COD (Figure 2 shows both TSS and COD). This is expected because for particulate type of pollutants settling is the main removal mechanism (represented by K). COD is mostly in dissolved form and hence the lack of correlation with K. It was also found that K relates to the total volume of rainfall very well, as seen in Figure 3. Such a relationship can be useful in estimating the removal efficiency of a dry pond for different rainfall amounts.

Highway Swale Modeling. In order to develop a modeling approach for estimating swale pollutant removal efficiency, results from the present study and those available in the literature were examined. Important parameters affecting

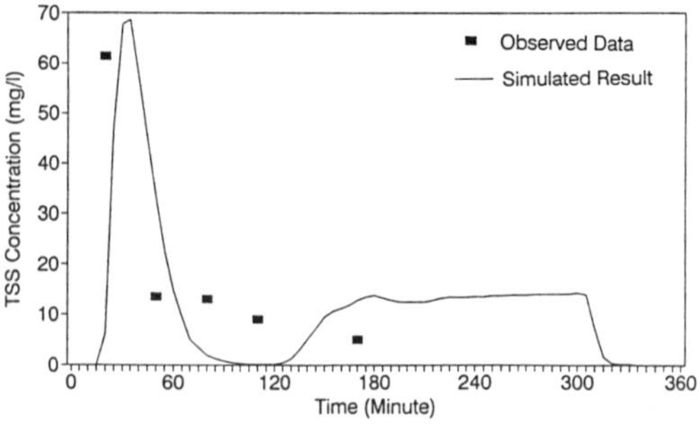

Figure 1. Example Pond Calibration Run

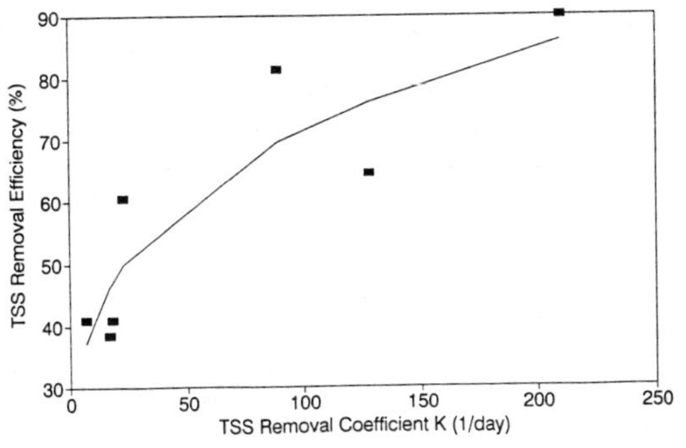

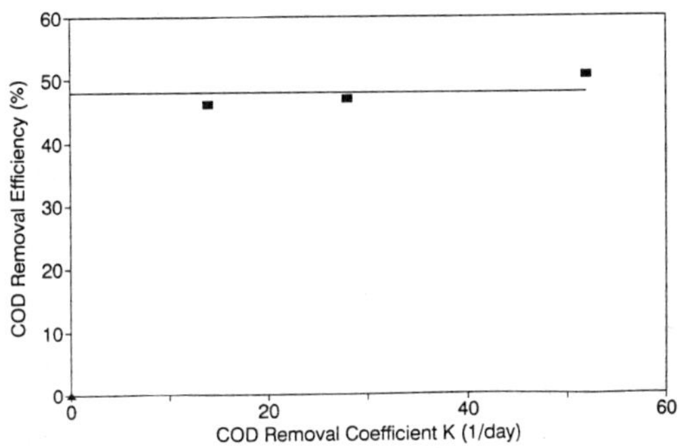

Figure 2. K versus Removal for TSS and COD.

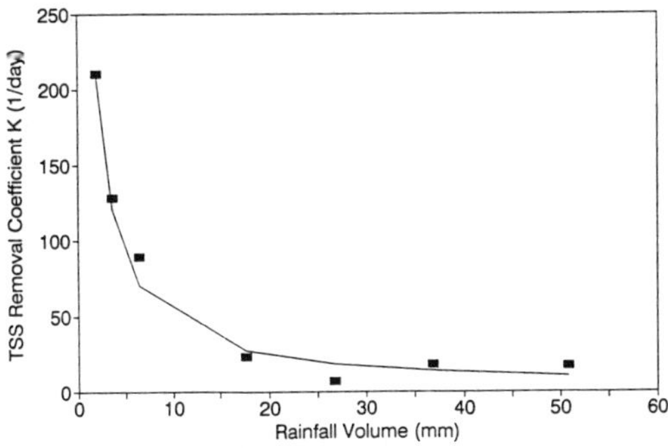

Figure 3. K versus Rainfall Volume for TSS.

swale efficiency include swale length, infiltration rate, grass characteristics, longitudinal and side slopes, etc. The swale length has been most commonly used as a design parameter. With swale data reported in New Hampshire (Oakland, 1986), Florida (Yousef et al., 1985), and buffer strip data from Virginia (Yu et al., 1987; and Dillaha, 1986) together with results from the present study, a regression analysis was made using Zn, which was monitored in each of the studies, yielding the following equation:

$$R_{ZN} = 8.806\,D^{0.510} \tag{2}$$

where R_{ZN} is the removal of zinc in percent, and D the swale length in meters. The results are shown graphically in Figure 4a. Figure 4b shows a similar analysis for total phosphorus. The scattering of the data prevented a regression of the points, and an estimated curve is included in the plot.

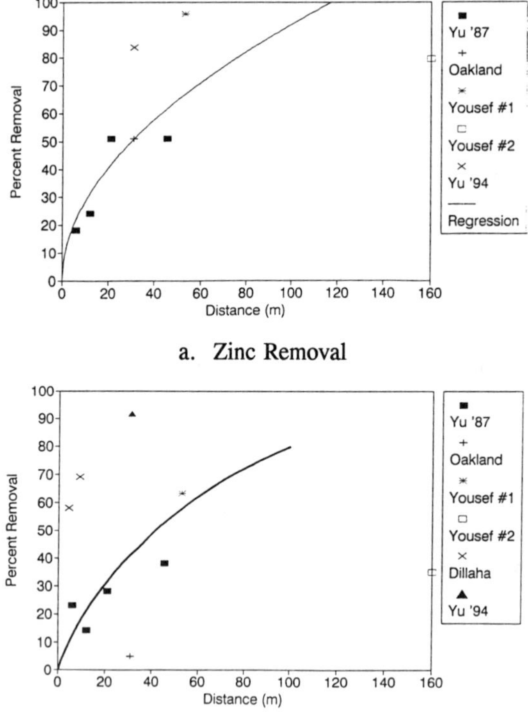

Figure 4. Removal Efficiency Versus Distance for Several Studies

At the present time (early 1994), the VAST model is being modified to include a subroutine which will estimate pollutant removal by a swale based on such parameters as the swale length, infiltration rate, etc.

Conclusions

Field test results for a modified dry detention pond and a highway median swale show that for both facilities, the total volume of rainfall is a key determinant in the pollutant removal efficiency. The small outflow orifice, the riprap inflow channel and the dense vegetation growth on the pond bottom all contributed to the moderate removal efficiency of the pond. Likewise the weir, which functions as a swale block, also helped raise the removal rate of the swale. Modeling analyses show that the pond removal rate relates closely to the rainfall volume, and that the swale length is an important parameter in determining swale efficiency.

References

Dillaha, T. A., et al. 1986. *Use of Vegetated Filter Strips to Minimize Sediment and Phosphorus Losses from Feedlots.* VPI-VWRRC-BULL 151. Blacksburg: Virginia Polytechnic Institute and State University.

Oakland, P. 1983. An evaluation of urban storm water pollutant removal through grassed swale treatment. In *1983 international symposium on urban hydrology, hydraulics and sediment control*, pp. 183-185. Lexington: University of Kentucky.

Tisdale, T. S., and Yu, S. L. 1987. *The VAST Model for Stormwater Runoff: Documentation and User's Guide*. Rpt. No. UVA/530376/ce87/101 Charlottesville: University of Virginia, Dept. of Civil Engineering.

Yousef, Y. A., Wanielista, M. P., Harper, H. H., Pearce, D. B., and Tolbert, R. D. 1985. *Removal of highway contaminants by roadside swales.* Report No. FL-ER-30-85. Tallahassee: Florida Department of Transportation.

Yu, S. L., and Das, K. C. 1987. "Micro-Computer Model for Regional Storwmater Management" *Procedings of the Fourth International Conference on Urban Storm Drainage*. Lausanne, Switzerland. pp. 327-328.

Yu, S. L., Norris, W. K., and Wyant, D. C. 1987. *Urban BMP Demonstration Project in the Albemarle/Charlottesville Area.* Rpt. No. UVA/ 530358/ce88/102 Charlottesville: University of Virginia, Dept. of Civil Engineering.

Evaluation and Management of Non-Point Source Pollutants in the Lake Tahoe Watershed

G. Fred Lee, PhD, PE, DEE (Member)[1] and Anne Jones-Lee PhD (Member)[2]

Abstract

Lake Tahoe, California-Nevada, one of the most oligotrophic lakes in the world, is experiencing decreased water clarity and increased periphyton growth, and water supplies drawing from the lake are experiencing increased algal-related tastes and odors. The growth of algae in Lake Tahoe is primarily limited by the nitrogen (nitrate and ammonia) loads to the lake, which have been increasing over the years. The nitrogen that is causing the increased fertilization of the lake is primarily derived from atmospheric sources through precipitation onto the lake's surface. A potentially highly significant source of atmospheric nitrogen in the Lake Tahoe Basin is automobile, bus, and truck engine exhaust discharge of NOx. The fertilization of lawns and other shrubbery, including golf courses, within the Lake Tahoe Basin is also leading to significant growths of attached algae in the nearshore waters of the lake. The fertilizers are transported via groundwater to the nearshore areas of the lake.

In order to prevent further deterioration of Lake Tahoe's eutrophication-related water quality, there is immediate need to control atmospheric input of nitrate and ammonia to the lake's surface, and to control use of fertilizers on lawns, shrubbery, and golf courses in the watershed. The states of California and Nevada, and the Tahoe Regional Planning Authority need to focus considerable attention on the determination of whether restricting NOx emissions from vehicular traffic within the basin would have a significant beneficial impact on Lake Tahoe's water clarity.

---

[1] President, G. Fred Lee & Associates, 27298 E. El Macero Dr., El Macero, CA 95618-1005

[2] Vice-President, G. Fred Lee & Associates

Introduction

The extremely high clarity of its waters and the natural beauty of its setting make Lake Tahoe, California-Nevada one of the unique lakes of the world. However, the water quality in this lake has been deteriorating at a significant rate over the past 20 years (Figure 1). The rate of deterioration matches the rate of increase in urbanization and use of the lake's watershed by permanent residents and visitors (Figure 2). Even with the diversion of all point-source wastewater discharges out of the Lake Tahoe Basin a number of years ago, the water clarity has been decreasing at an average rate of 1 to 2 ft/yr. While the Secchi depth was on the order of 30 m in 1970, today it is on the order of 20 m (see Figure 1). The decreasing water clarity has been caused by the increasing input of aquatic plant nutrients to the lake that stimulate planktonic algal growth.

Lee *et al.* (1978) and Rast and Lee (1978) developed a relationship between planktonic algal chlorophyll in lakes and reservoirs and Secchi depth (water clarity) where increased algae causes reduced light penetration. They also quantitatively related the Secchi depth to the normalized nutrient loadings to waterbodies; that relationship was expanded by Jones and Lee (1986). It is clear from the data of Goldman and others that while Lake Tahoe is ultra-oligotrophic and is one of the clearest lakes in the world, increased algal growth is occurring in this lake that is significantly reducing light penetration in the watercolumn.

It has been established that planktonic algal growth in Lake Tahoe is primarily limited by nitrogen in the forms of nitrate and ammonia. The increase in input of algal-available nitrogen, however, has resulted in a more balanced ratio of nitrogen to phosphorus with respect to the ratio needed by algae for growth. The result is that both nitrogen and phosphorus are often limiting algal growth in the lake today.

There has been considerable controversy over the years about how to best manage the non-point sources of nutrients that are causing the excessive growth of planktonic algae in Lake Tahoe. Several years ago the Tahoe Regional Planning Authority (TRPA) adopted the Individual Parcel Evaluation System (IPES) as a means of trying to control the deterioration of Lake Tahoe water quality based on an attempt to control nitrogen input to the lake from property development. IPES is a multi-parameter scoring system in which the property ground slope and other erosion-related parameters are assigned an arbitrarily developed score. The individual IPES parameter scores are summed to give a total score to rank properties for priority for development. A critical examination of the components of IPES and its application shows that it is an arbitrary approach for limiting development in the Lake Tahoe watershed, using a pseudo-technical approach for social engineering in the Lake Tahoe Basin. The underlying purpose of IPES was to control the rate of development of undeveloped properties in the Lake Tahoe watershed. It is clear that IPES is a technically invalid approach that will not

Figure 1. Decrease in Lake Tahoe's Annual Average Secchi Depth (After Goldman, 1988)

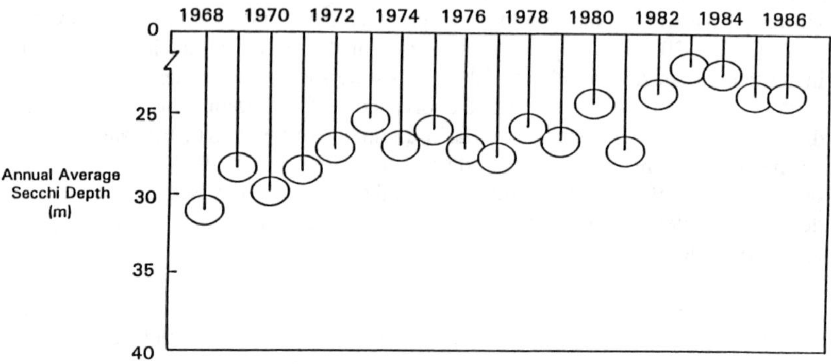

Figure 2. Increase in Lake Tahoe's Primary Productivity (After Goldman, 1988)

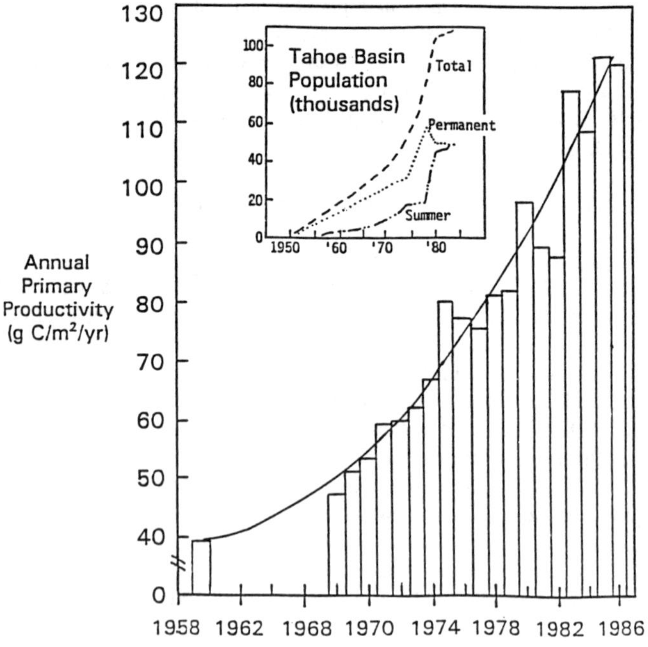

significantly control the input of available nitrogen and phosphorus to the lake to reduce the rate of decrease in water clarity due to planktonic algal growth.

While controlling property development will limit the input of algal nutrients to the lake to some extent, the IPES approach is basically flawed in addressing the primary causes of increased algal growth in the lake. In order to gain an understanding of the relative significance of various sources of nitrogen that are leading to increased growth of planktonic algae in Lake Tahoe, an estimate of the nitrogen sources for the lake was made and is presented in Table 1. It is apparent that the key to the increase in planktonic algal growth in the lake is the increase in atmospheric nitrogen that enters the lake through its surface. It is also clear that the development of property and the associated erosion that has occurred in the Lake Tahoe watershed has not been a significant factor in the increased nitrogen input to the lake relative to the inputs that have occurred from atmospheric sources.

In order to determine if the input of atmospheric nitrogen to the lake is potentially controllable, it is necessary to understand its origin. There are both in-basin and out-of-basin sources of atmospheric nitrogen. In 1987 the California Air Resources Control Board determined that about 2500 mt (metric tons) of nitrogen oxides (NOx) are emitted to the atmosphere in vehicular exhaust in the Lake Tahoe watershed (Table 2). That 2500 mt NOx/yr in-basin atmospheric input of nitrogen could contribute significantly to the 100 mt N/yr nitrogen input to the lake surface.

Managing Lake Tahoe's Water Quality

Presented below are a series of actions that should be considered for management of water quality in Lake Tahoe.

NOx-Nitrogen Control

There is an urgent need to immediately significantly curtail the nitrogen inputs to Lake Tahoe. Jones and Lee (1991) recommended that in order to begin to effectively slow the rate of deterioration of the lake's water quality that is related to algal growth in the open and nearshore waters of the lake, aggressive action should be immediately taken toward greatly reducing, if not essentially eliminating, the use of internal combustion engine-based automobiles, trucks, and buses within the Lake Tahoe watershed. This recommendation is based on the finding that the exhaust from these vehicles could be a significant source of atmospheric nitrogen for the lake. In order to evaluate the full significance of this source, there is need to conduct studies to properly define the relative roles of in-Tahoe-Basin atmospheric nitrogen sources and out-of-basin atmospheric nitrogen sources that lead to entrance into the lake of forms of nitrogen (nitrate and ammonia) that could stimulate algal growth within the waterbody.

Table 1. Estimated N Load to Lake Tahoe (tonnes N/yr)
(After Jones and Lee, 1991)

Source	*Pre-Development*	*Now*
Atmosphere - onto Lake Surface	2.5	~100
Surface Water Runoff	4	16
Groundwater	0.5	2
Total N Loads	7	118

Table 2. Estimated Contributions of NOX from Motor Vehicles
(After Jones and Lee, 1991)

	tonne NOX/yr
Automobiles	800
Light & Medium Trucks	630
Heavy Duty Trucks	1160
Total	2500

Source: Air Resources Control Board, 1987

Recently, a light rail system was proposed for the Yosemite Valley in order to reduce the traffic congestion and pollution of that area. A similar system needs to be adopted for the Lake Tahoe Basin in which automobiles and other vehicles are parked outside of the basin and people and goods are transported into the basin via a non-internal combustion engine-based transportation system.

Nearshore Lake Water Quality

Jones and Lee (1991) recommended that all lawns, including golf courses, and fertilized shrubbery be banned in the Lake Tahoe watershed. They feel that the basin should be allowed to return to native vegetation that does not require fertilization and/or irrigation. Adoption of this approach would significantly reduce nitrogen inputs to the nearshore waters of Lake Tahoe and thereby reduce the periphyton growth in some areas.

While at this time domestic wastewater disposal is not allowed within the Lake Tahoe watershed, i.e., the system is sewered with the wastewaters exported out of the watershed, it is highly likely that previous wastewater disposal practices could be significant sources of nutrients for some nearshore areas of Lake Tahoe, contributing to localized algal related problems in these areas. Nutrients derived from the previous use of septic tank wastewater disposal systems and wastewater spray irrigation disposal systems are, or could be, significant sources of nutrients which stimulate algal growth in some parts of the nearshore waters of Lake Tahoe.

Jones and Lee (1991) suggested that additional work needs to be done to determine the potential significance of past wastewater and solid waste (landfill) disposal practices within the Lake Tahoe Basin as a source of nutrients for nearshore water quality problems. If there is interest in controlling excessive periphyton growth in a particular part of the nearshore area of the lake where the nutrients contributing to the excessive growth in that region are significantly derived from past wastewater or solid waste disposal practices, it may become necessary to intercept the groundwater before it reaches the lake by pumping and treating the groundwater to remove the nutrients.

IPES Validity

Jones and Lee (1991) concluded that the Lake Tahoe Regional Planning Agency's Individual Parcel Evaluation System (IPES), which is being used in an attempt to control population growth in the basin, is technically invalid with respect to protecting the lake's water quality. The IPES score is a growth-limiting mechanism used by TRPA for the purpose of "protecting" lake water quality. However, the IPES score on a property is not related to the amount of nitrogen or, for that matter, other forms of algal-available nutrients that ultimately reach the lake from that property.

Erosion Control

It should be recognized that erosion control is very important in the Lake Tahoe Basin for reducing the scarring of the terrestrial resources of the area. The TRPA should abandon the use of the current IPES for regulating population growth in the basin. In its place, a more reliable approach for limiting erosion due to development should be formulated and used. It should further be recognized, however, that erosion control has little impact on the lake's water quality and that a significantly different approach, based on a proper evaluation of the nutrient sources for the lake that contribute algal-available forms of nutrients to the lake waters, is needed to address the lake's water quality problems.

Conclusions

The increased fertilization of Lake Tahoe is causing several highly significant water quality problems, including algal-related tastes and odors and decreased water clarity.

In order to reduce the frequency and severity of algal-related domestic water supply water quality problems in Lake Tahoe, it appears that it will be necessary to significantly curtail the use of automobiles and other vehicles powered by internal combustion engines in the Lake Tahoe watershed and to ban the use of lawn and shrubbery fertilizers and watering within the lake's watershed. It may be necessary to ban all watered lawns and shrubbery in the Lake Tahoe watershed and allow the basin to return to native vegetation. There is no doubt that unless action is taken to limit NOx emissions within the basin, the lake's open water clarity will continue to decrease, and ultimately, the highly unique character of Lake Tahoe (i.e., its water clarity) would be lost. It is imperative that highly aggressive action be taken now to reverse the changes in water quality that are occurring today.

The key issue that needs to be addressed to reverse and/or control the Lake Tahoe water clarity decrease is whether or not a restriction in use of internal combustion engines in the Lake Tahoe Basin could reduce the nitrogen input to the lake sufficiently to stop the deterioration of the lake's water clarity. It would be possible to require that most of the visitors to the lake park their vehicles out of the lake basin and use a light-rail or other system for transport to and within the lake watershed, and to control commercial traffic. The states of California and Nevada and the TRPA need to focus considerable attention on the determination of whether restricting vehicular traffic within the lake basin and the attendant reduction in the NOx emissions would have a significant beneficial impact on Lake Tahoe water clarity.

It is important to properly evaluate the impacts of contaminants in non-point sources on water quality in the waters receiving runoff from these sources. Failure to make such evaluations can readily lead to the initiation of inappropriate management programs for non-point sources that do not properly address the causes of real water quality problems, and can lead to substantial waste of public funds without an improvement in the designated beneficial uses of the receiving waters.

Acknowledgment

The authors wish to acknowledge the assistance of T. Durbin of Hydrologic Consultants, Davis, California, C. Mahannah of Water Research & Development, Reno, Nevada, and Dr. C. Goldman of the University of California, Davis in providing background information for this paper.

References

Air Resources Control Board, Annual Data, State of California, Sacramento, CA (1987).

Goldman, C. R., "Primary Productivity, Nutrients, and Transparency during the Early Onset of Eutrophication in Ultra-oligotrophic Lake Tahoe, California-Nevada," Limnology and Oceanography 33:1321-1333 (1988).

Jones, R. A., and Lee, G. F., "Eutrophication Modeling for Water Quality Management: An Update of the Vollenweider-OECD Model," World Health Organization's Water Quality Bulletin 11(2):67-74,118 (1986).

Jones, R. A., and Lee, G. F., "Evaluation of the Impact of Urbanization of a Forested Watershed on Eutrophication-Related Water Quality," Proc. "California Watersheds at the Urban Interface," Third Biennial Conference, Watershed Management Council and the University of California Water Resources Center, Report No. 75, Riverside, CA p. 173 (1991).

Lee, G. F., Rast, W., and Jones, R. A., "Eutrophication of Waterbodies: Insights for an Age-Old Problem," Environ. Sci. & Technol. 12:900-908 (1978).

Rast, W., and Lee, G. F., "Summary Analysis of the North American (US Portion) OECD Eutrophication Project: Nutrient Loading-Lake Response Relationships and Trophic State Indices," EPA 600/3-78-008, US EPA-Corvallis (1978).

Achieving Adequate BMP's for Stormwater Quality Management

Anne Jones-Lee, PhD (Member)[1] and G. Fred Lee, PhD, PE, DEE (Member)[2]

Abstract

There is considerable controversy about the technical appropriateness and the cost-effectiveness of requiring cities to control contaminants in urban stormwater discharges to meet state water quality standards equivalent to US EPA numeric chemical water quality criteria. At this time and likely for the next 10 years, urban stormwater discharges will be exempt from regulation to achieve state water quality standards in receiving waters, owing to the high cost to cities of the management of contaminants in the stormwater runoff-discharge so as to prevent exceedances of water quality standards in the receiving waters. Instead of requiring the same degree of contaminant control for stormwater discharges as is required for point-source discharges of municipal and industrial wastewaters, those responsible for urban stormwater discharges will have to implement Best Management Practices (BMP's) for contaminant control.

The recommended approach for implementation of BMP's involves the use of site-specific evaluations of what, if any, real problems (use impairment) are caused by stormwater-associated contaminants in the waters receiving that stormwater discharge. From this type of information BMP's can then be developed to control those contaminants in stormwater discharges that are, in fact, impairing the beneficial uses of receiving waters.

Introduction

The urban stormwater quality management program being developed by the US EPA evolved from the US EPA's 1992 report to Congress (US EPA, 1992) which stated,

---

[1] Vice-President, G. Fred Lee & Associates, 27298 E. El Macero Dr., El Macero, CA 95618-1005

[2] President, G. Fred Lee & Associates

"Based in part on national assessments conducted by the US Environmental Protection Agency (EPA) it is now recognized that nonpoint sources and certain diffuse point sources (e.g., stormwater discharges) are responsible for between one-third and two-thirds of existing and threatening impairments of the Nation's waters (US EPA, 1991)."

The US EPA (1992) report to Congress is based on an inappropriate assessment of the impact of urban stormwater-associated contaminants on receiving water quality (Lee and Jones-Lee, 1993a). In developing that assessment, the US EPA and states used the highly over-protective water quality standards equivalent to US EPA water quality criteria. There is considerable technical justification for not requiring that urban stormwater discharges be controlled so as to meet such state water quality standards at the point at which they enter a waterbody (lake, river, stream, or the ocean), because of the short-term, episodic nature of those discharges and because contaminants, such as heavy metals, in stormwater discharges are typically in chemical forms that are not available/toxic to aquatic life (Lee and Jones, 1991). Therefore for most contaminants, applying current water quality standards to stormwater discharges through the NPDES permit system used for wastewater discharges can lead to massive waste of public and private funds for contaminant control with limited improvement in the designated beneficial uses of the waters receiving the stormwater discharges.

The problems in achieving water quality standards in waters receiving stormwater discharges during the time of discharge have resulted in a relaxation of this requirement in favor of achieving stormwater contaminant control BMP's. The goal of the current US EPA stormwater quality management program for urban and industrial areas is to *"develop a comprehensive planning process which involves public participation and inter-governmental coordination to reduce the discharge of pollutants to the maximum extent practicable (MEP)."* The implementation of the regulations requires an estimate of the *"reductions in loadings of pollutants from discharges of municipal storm water constituents from municipal storm sewer systems expected as a result of the municipal storm water quality management program."* (WRCB, 1993).

The relaxation of the requirement of achieving water quality standards in waters receiving stormwater discharges and the focus on BMP's to control pollutants to the maximum extent practicable require the development of an approach by which state regulatory agencies can judge the adequacy of efforts to develop stormwater contaminant control BMP's. The California Water Resources Control Board and its Regional Boards (WRCB, 1993) have determined that the implementation of the federal regulations *"requires permittees to evaluate effectiveness of storm water management program by:*
Runoff: *Reduction of pollutants discharged in storm water to the MEP*
Receiving water: *Discharges do not impact beneficial uses*
Cause an exceedance in water quality objectives [standards]."

Stormwater quality BMP's range from non-structural contaminant control programs, such as the control of illegal connections and illicit discharges, to structural controls such as detention basins, grassy swales, and treatment works similar to those used for domestic and industrial wastewaters. The costs of the structural BMP's can be very high.

While for a period of time stormwater discharges will likely be legally exempted from regulation to meet water quality standards, the approach that is being used at the federal and state levels for definition of the control of stormwater-associated contaminants to the "maximum extent practicable" could readily result in a *de facto* implementation of current state water quality standards to stormwater discharges as the basis for judging when adequate BMP's have been implemented for a particular discharge. Such an approach could readily lead to substantial waste of public and private funds in the implementation of BMP's for stormwater-associated contaminants that have little or no impact on water quality in the receiving water.

There are two categories of exceedances of water quality standards. One category comprises exceedances of standards that reflect the presence of sufficient concentrations of toxic/available forms of contaminants to cause an impairment of the designated beneficial uses of waterbodies. The other category comprises "administrative" exceedances in which contaminants exist in the water in concentrations greater than water quality standards without the impairment of the designated beneficial uses of the waterbody. The existence of "administrative" exceedances reflects the highly overly protective nature of the water quality standards, including the incorporation of non-toxic, unavailable forms of chemicals in the assessment of the exceedance. It is important that the BMP's developed for stormwater quality management address real water quality problems and not merely the elimination of "administrative" exceedances of overly protective, numeric water quality standards of the type that have been adopted by many states.

The approach being adopted by the California Water Resources Control Board, of using exceedances of water quality standards (objectives) as a basis for developing the contaminant control goals of BMP's in determining contaminant control to the "maximum extent practicable" can readily lead to significant over-regulation of stormwater discharges in order to address "administrative" exceedances of water quality standards that do not reflect actual use-impairment of the surface waters. The process of "ratcheting down" the allowed concentrations of contaminants in urban stormwater discharges to achieve current state water quality standards is not technically valid and is not in the best interest of cost-effective management of real water quality problems associated with stormwater discharge.

Stormwater Quality Management

Stormwater quality management entities are in the process of developing contaminant control programs for Phase I stormwater discharges. Those programs typically initially focus on the implementation of "best management practice" (BMP). Numerous guidance manuals for BMP's for stormwater have been developed and are under development by various professional groups and regulatory agencies (MWCOG, 1992; APWA, 1993). Lee and Jones (1991) and Lee and Jones-Lee (1992a) discussed the importance of focusing BMP's on real water quality problems caused by the particular discharge for the particular site of focus. At the urging of environmental groups, Congress specified in the 1972 Clean Water Act that all municipalities provide a standard basic degree of treatment for domestic wastewaters ("secondary" treatment) irrespective of the indications of the need for such treatment to protect beneficial uses of a particular receiving water. This approach should not be followed to direct the construction of contaminant control systems for stormwater discharges.

One of the common BMP approaches to "controlling" contaminants in stormwater discharges is the construction of detention basins on the discharges to effect a decrease in concentration of chemical contaminants that are discharged to receiving water. While such facilities cause removal of some of the larger particulates, contaminant forms associated with those detained particulates are largely unavailable to cause toxicity to aquatic life; detention basins allow the passage of dissolved contaminants that could adversely affect aquatic life. Thus, as discussed by Lee and Jones (1991) and Lee and Jones-Lee (1992a), the construction of detention basins on stormwater discharges is largely ineffective in controlling real water quality problems that may be caused by stormwater-associated contaminants.

Need for Control of Particulate Forms of Contaminants

Some regulatory agencies are regulating the discharge/runoff of particulate forms of contaminants that could become associated with the sediment in receiving waters, under the mistaken belief that such control would protect "sediment quality." There is no technical foundation for these approaches. First, it is not possible to translate concentrations of total heavy metals in effluents/runoff or in ambient waters to sediment quality problems. Non-toxic/unavailable forms of heavy metals associated with particulates in watercolumns usually do not lead to the formation of toxic/available forms when those particulates settle to the sediments. Further, sediments typically detoxify toxic forms of chemicals. Understanding the aqueous environmental chemistry of toxicants is critical to the proper regulation of chemical contaminants. Site-specific biological effects studies on sediments should be used to determine exceptions to the expected detoxification reactions. Properly conducted ambient water and sediment toxicity measurements should be used to determine if heavy metals are causing real toxicity in receiving waters. "Short-term" chronic toxicity tests (using fish larvae and zooplankton) on ambient waters in conjunction with reliably determined concentrations of dissolved forms of contaminants provide a good indication of whether exceedances of discharge limitations for heavy metals and other contaminants, or of effluent toxicity are causing toxicity of significance to designated beneficial uses of the water.

Lee and Jones (1992b), Lee and Jones-Lee (1993b,d), and Jones-Lee and Lee (1993), discussed technical deficiencies in the current approaches for the development of sediment quality criteria and standards, and also alternative approaches for technically valid, cost-effective evaluation of the water quality significance of sediment-associated chemical contaminants. As they discussed, a properly conducted, non-numeric "weight of evidence" evaluation of ambient water conditions of aquatic life toxicity, bioaccumulation, and numbers and types of desirable organisms of concern to the public should override chemical concentration-based numeric criteria for the regulation of discharges/runoff and for determining the degree of clean-up of contaminated sediments needed to protect beneficial uses.

Detention Basin Sediment Quality

Not only are detention basins largely ineffective in controlling water quality problems in waters receiving stormwater drainage, but also there is increasing concern about the potential problems associated with management of particulate matter that is collected in them. Stormwaters from urban areas typically contain elevated concentrations of particulate forms of contaminants such as lead, some of which will settle out in a properly designed, operated, and maintained detention basin.

The assessment of lead-contaminated soil, sediment, or waste for its classification as "hazardous waste" was originally made based on the results of EP-Tox test and is now made based on the results of the US EPA's TCLP test. The prescribed basis for establishing the allowed level of lead that can be leached in the TCLP test procedure without the tested material's being classified as "hazardous waste" is the drinking water standard of 50 μg/L, multiplied by a factor of 100; on that basis, 5 mg/L lead is allowed to leach from the soil, sediment, or waste under the test conditions before the material is classified as a "hazardous waste" and in need of management as such.

The US EPA and some state regulatory agencies have recently reduced the accepted concentration of lead in drinking water to 15 μg/L. This change would be expected to cause a substantial reduction in the amount of leachable lead that would "pass" the TCLP, and cause more materials to be classified as "hazardous waste." It will likely be found that some soils that accumulate in stormwater detention basins will contain sufficient amounts of TCLP-leachable lead to be classified as "hazardous waste." Therefore those responsible for operation and maintenance of stormwater detention basins could find themselves in the position of having to manage the collected solids as "hazardous waste." This increases the cost of disposal from a few tens of dollars per ton to a few hundred dollars per ton for management in a "hazardous waste landfill."

Infiltration Basins

Another approach that is often considered to be BMP for stormwater-associated contaminants is the construction of infiltration basins to promote the passage of stormwaters into the groundwater aquifer system. In the past, the construction of infiltration basins has been done with little or no regard for the potential for groundwater pollution by contaminants in the stormwater. Lee and Jones-Lee (1993c) discussed the importance of the proper evaluation of the potential for groundwater pollution by contaminants in waters that are directly or incidently introduced into groundwaters.

Implementation of BMP's

The first step that should be taken in defining a BMP for a particular stormwater discharge is to define the specific water quality impairment being caused by stormwater-derived contaminants. The authors have yet to find a documented aquatic life toxicity problem caused by stormwater discharges to receiving waters that was not due to illegal or illicit disposal/connections. Thus, as discussed by Lee and Jones (1991) and Lee and Jones-Lee (1992a) if a real water quality problem is identified, the first "BMP" that should be undertaken is the careful examination of the system for illegal or illicit disposal or connections of industrial or commercial waste to the stormwater system. If rectifying those conditions does not resolve the water quality problem, a more detailed evaluation of the source of the contaminants causing the problem should be made to determine if they can be controlled at the source. Only after these actions have been taken without resolution of the water quality impairment should consideration be given to structural BMP treatment options for the stormwater runoff.

Stormwater treatment should focus on those components of the stormwater responsible for the specific water quality impairment of the particular receiving water. Because the stormwater-associated contaminant(s) that may cause an impairment of a beneficial use would be site-specific, and because of the rarity of aquatic life toxicity problems caused by urban stormwater runoff not due to illegal or illicit disposal or connections, a "standard" off-the-shelf BMP cannot and should not be prescribed for stormwater runoff treatment. For example, detention basins should not be constructed if there are no impairments of beneficial uses in the receiving waters caused by those materials that would be removed in such a facility at the site in question.

Far too often contaminant control entities construct treatment works to "solve" a what is defined as "problem" based on inappropriate regulatory requirements, rather than to challenge the validity of the regulatory requirements. This is especially important in the area of stormwater quality evaluation and management since it is well-known that the existing US EPA water quality criteria and standards that are applied to waters receiving such discharges are largely inappropriate for assessing the potential impact of stormwater-associated contaminants on beneficial uses of the receiving water. Thus, the entity making the water quality evaluation should take assertive action to ensure that any regulatory requirements imposed on the discharge are technically valid and will result in a significant, discernible improvement in the designated beneficial uses of the receiving water.

These recommendations are consistent with those made by the Water Environment Federation for changes in the Clean Water Act. WEF (1992) stated with regard to management of stormwater discharges,

> *"The permitting process, however, must take into account the inherent difference between conventional discharges and stormwater discharges."*
>
> *"Discharge-specific permits should be limited only to those discharges which are having a quantifiable and significant adverse impact on receiving water quality."*

Conclusion

The current approach for BMP implementation is often not technically valid and will result in substantial waste of public and private funds with little or no improvement in the designated beneficial uses in the receiving waters for stormwater discharges. The implementation of BMP's for control of stormwater-associated contaminants to the maximum extent practicable should be based on finding a real water quality problem - use-impairment that can be solved by the BMP.

References

APWA (American Public Works Association), "Best Management Practices for Storm Water Permit Compliance," American Public Works Association, Chicago, IL (1993).

Jones-Lee, A., and Lee, G. F., "Potential Significance of Ammonia as a Toxicant in Aquatic Sediments," Proc. First International Specialized Conference on Contaminated Aquatic Sediments: Historical Records, Environmental Impact, and Remediation, IAWQ, Milwaukee, WI, pp. 223-232, June (1993).

Lee, G. F., and Jones, R. A., "Suggested Approach for Assessing Water Quality Impacts of Urban Stormwater Drainage," IN: Symposium Proceedings on Urban Hydrology, American Water Resources Association Symposium, AWRA TPS-91-4, AWRA, Bethesda, MD pp. 139-151 (1991).

Lee, G. F., and Jones-Lee, A., "Development of Technically Valid, Cost-Effective Control Programs for Stormwater-Associated Contaminants," Presentation to American Public Works Association/State Water Resources Control Board Stormwater Quality Task Force, September (1992a).

Lee, G. F., and Jones, R. A., "Sediment Quality Criteria Development: Technical Difficulties with Current Approaches, and Suggested Alternatives," Report of G. Fred Lee & Associates, El Macero, CA, January 6 (1992b).

Lee, G. F., and Jones-Lee, A., "Water Quality Impacts of Stormwater-Associated Contaminants: Focus on Real Problems," IN: Proc. of First International IAWG Specialized Conference on Diffuse Pollution: Sources, Prevention, Impact and Abatement, Chicago, IL, pp. 231-240, September (1993a).

Lee, G. F., and Jones-Lee, A., "Sediment Quality Criteria: Numeric Chemical vs. Biological Effects-Based Approaches," Proc. Water Environment Federation National Conference, Surface Water Quality & Ecology, Anaheim, CA, pp. 389-400, October (1993b).

Lee, G. F., and Jones-Lee, A., "Water Quality Aspects of Incidental and Enhanced Groundwater Recharge of Domestic and Industrial Wastewaters - An Overview," Proceedings of Symposium on Effluent Use Management, TPS-93-3, pp. 111-120, American Water Resources Association, Bethesda, MD (1993c).

Lee, G. F., and Jones-Lee, A., "Sediment Quality Criteria: Numeric Chemical vs. Biological Effects-Based Approaches," Proc. Water Environment Federation National Conference, Surface Water Quality & Ecology, Anaheim, CA, pp. 389-400, October (1993d).

MWCOG (Metropolitan Washington Council of Governments), "A Current Assessment of Urban Best Management Practices; Techniques for Reducing Non-Point Source Pollution in the Coastal Zone," MWCOG, Washington, D.C. March (1992).

US EPA, "Environmental Impact of Stormwater Discharges: A National Profile," EPA 841-R-92-001, US EPA Office of Water, Washington, D.C., June (1992).

WEF (Water Environment Federation), "Recommendations - Reauthorization of the Clean Water Act," Water Environment Federation Clean Water Act Reauthorization Committee Report, WEF, Alexandria, VA, May (1992).

WRCB (Water Resources Control Board), Presentation to the State of California Storm Water Quality Task Force (1993).

MODELING POLLUTANT FATE AND TRANSPORT IN CONSTRUCTED WETLANDS

Russell T. Brown[1], Associate Member ASCE, Joanne J. Field[2], Michael J. Zanoli[2], and Ron W. Crites[3]

Abstract

The **C**onstructed **W**etlands **F**ate and **A**quatic **T**ransport **E**valuation Model (**CWFATE**) was developed to evaluate alternative design and operations of the constructed wetlands treatment system (CWTS) at the Sacramento Regional Wastewater Treatment Plant in Elk Grove, California. The model simulates the daily fate and transport of metals and other influent pollutants in an effort to predict CWTS effluent concentrations, removal efficiency, and long-term bioaccumulation of pollutants in sediment, vegetation, organic material, and aquatic organisms. Due to the difficulty of obtaining scientific information for model verification, final calibration is postponed until further field data become available.

Introduction

The CWTS is a pilot project designed to evaluate the effectiveness of further treatment of secondary effluent from the Sacramento Regional Wastewater Treatment Plant using natural wetland processes for removal of metals, organic materials, and other pollutants from the water column. The 17-acre wetland consists of 11 treatment cells and a habitat cell planted with common tule or alkali bulrush and operated as either plug flow, batch flow, recycled flow, or overland flow cells. CWFATE generally simulates pollutant fate and transport through plug flow cells, which are characterized by continuous inflow and outflow.

---

[1] Senior Environmental Scientist, Jones & Stokes Associates, 2600 V street, Sacramento, CA 95818

[2] Environmental Specialist, Jones & Stokes Associates, 2600 V Street, Sacramento, CA 95818

[3] Project Engineer, Nolte and Associates, 1750 Creekside Oaks, Sacramento, Ca 95833

CWFATE was envisioned primarily as a tool to interpret future CWTS field data and evaluate the performance of the constructed wetland. More specific model goals were to:

- Simulate the long-term bioaccumulation of metals and other pollutants in the sediment, vegetation, organic material, and aquatic organisms;

- Investigate the potential for more efficient design and operation of the CWTS; and

- Evaluate the ability of the effluent to satisfy proposed discharge water quality requirements.

CWFATE uses formulations from existing water quality models of lakes and streams, and was developed in Lotus 1-2-3 format to immediately integrate model input data, formulations, and results with graphical and statistical summaries (Jones & Stokes 1993).

Model Concept

The general organizing concept for the CWFATE model is a daily accounting of the water balance and mass budgets for inflowing nutrients (carbon and nitrogen) and metals within the water, sediment, and biomass components of the constructed wetlands. Figure 1 shows a conceptual diagram of a CWTS cell, identifying the water budget terms and the sediment and biomass components. Each CWTS cell is divided into a series of 25 connected segments, allowing the model to simulate longitudinal gradients within a CWTS cell.

The model can simulate pollutant fate and transport for an entire year, which allows the model to characterize and account for the seasonal factors influencing wetland hydrology, biology, and biochemistry. Long-term factors and bioaccumulation are simulated by modeling a sequence of annual cycles. CWFATE is formulated to simulate the fate and transport of a single metal. Various metals of potential interest can be simulated by using different inflow concentrations and partition coefficients for each biomass component.

Selection of Appropriate Model Variables

The model variables selected for use in CWFATE can be grouped according to four major mass budgets:

- Hydrological variables including inflow, rainfall, seepage, evaporation, transpiration, recycle, and water temperature (water budget);

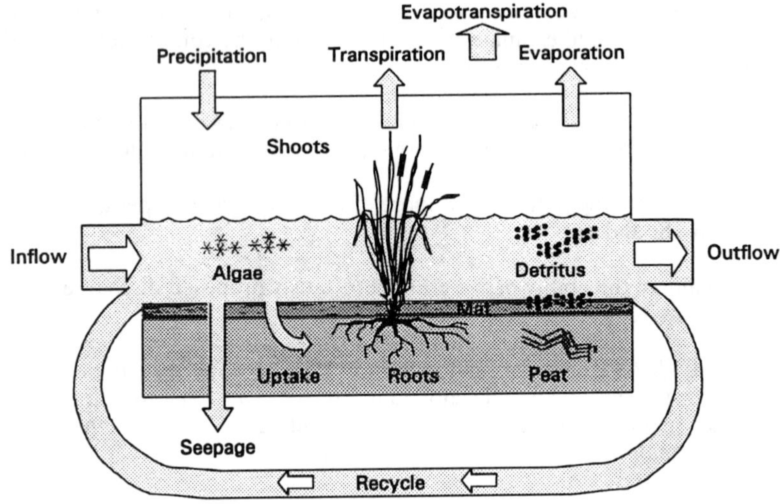

Figure 1. Diagram of a Model Segment Indicating Water Budget Terms and Biomass Components

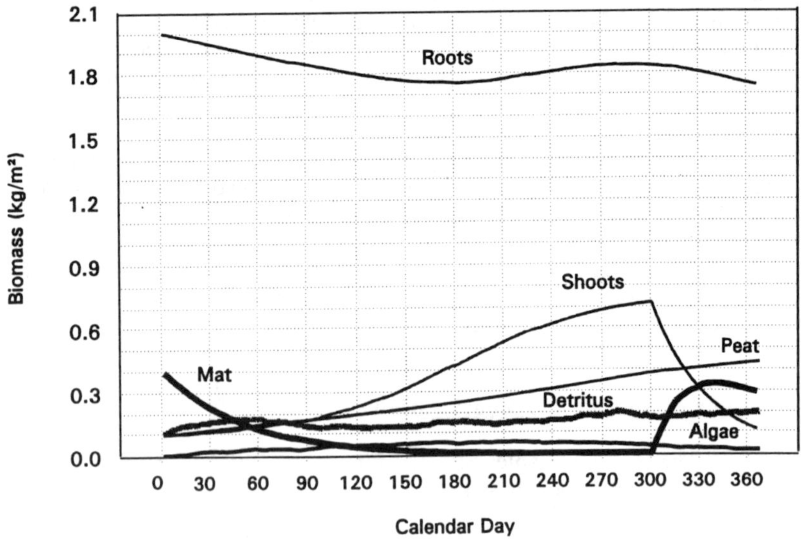

Figure 2. Simulated Seasonal Patterns of Shoots, Roots, Mat, Peat, Detritus, and Algae in Segment 1

- Aquatic plant growth and decay variables including roots, shoots, mat of dead vegetation, particulate detritus, and peat accumulation in the soil, as well as growth of attached algae (carbon budget);

- Nutrient cycling variables including both aerobic decay in the water column and anaerobic processes within the soil and peat (nitrogen budget); and

- Metals uptake, adsorption, immobilization, and general partitioning variables (metals budget).

Model Formulation

The water budget provides the basic framework for the other processes within the constructed wetland. Water temperature affects the rate of all modeled biological processes, including primary production of aquatic plants and algae, decay of organic material into detritus and peat, and denitrification. Water temperature is dependent on inflow temperatures, meteorological conditions, and shading and sheltering by the vegetation canopy.

The environmental features that make the constructed wetland a unique aquatic treatment facility are related to the growth of the aquatic plants (shoots and roots), the algae, and the accumulation of vegetative materials (mat, detritus, and peat) within the CWTS (Moshiri 1993). Estimating the seasonal magnitude of these plant biomass terms is an important requirement for the CWFATE model. Net primary production of the wetland plants are modeled as a function of solar radiation, water temperature, soil nitrogen content, and an assumed maximum canopy density. As the shoot biomass approaches the maximum canopy density, more of the primary production is diverted to the roots. The growth of the shoots is stopped on a specified date to simulate plant senescence caused by colder fall temperatures. Shoots and roots die at a temperature- dependent rate. Shoots first become part of the mat, then decay to detritus and peat. Roots decay directly to peat.

Algae primary production is modeled like the aquatic plant growth, but with shading from the shoot canopy and self-shading as the algae approaches a peak biomass level. A temperature-dependent loss rate is used to balance the primary production of algae by transferring the biomass to the particulate detritus term, which can then accumulate as peat or be transported out of the CWTS.

The detritus is assumed to be partially suspended in the water column, with the suspended fraction increasing with water velocity or wind mixing. The major long-term accumulation of organic materials is assumed to be the peat component, originating from decayed root and detritus deposited in the anaerobic layer at the soil-water interface.

The nitrogen budget is modeled to account for uptake, storage, and release from biomass components and to simulate the loss to denitrification, an important indicator of the magnitude of anaerobic processes in the peat-soil matrix. Denitrification may be limited by the available supply of organic carbon.

Metals are transported from segment to segment either dissolved in the water column or adsorbed onto detritus, and become distributed within the sediment and biomass components of the segment according to simple partitioning and passive uptake mechanisms. Atmospheric deposition may add some metals to the wetlands.

The CWFATE model assumes that there are essentially three connected pools of metal within the wetland. The first pool of metal is within the water column either as dissolved metal or adsorbed to the detritus, algae, or mat. The second pool is within the shoot and root plant tissue. The third pool of metal is within the peat-soil matrix, either dissolved in the pore water or adsorbed onto the peat. The partitioning of metal is calculated from adsorption coefficients and the biomass of each adsorbing component. For the CWFATE model, the three-way partitioning in the water column onto detritus, mat and algae has been estimated with the following equation:

DISSOLVED = TOTAL/(1+PARALG*ALG+PARDET*DET+PARMAT *MAT)

Where PARALG, PARDET, and PARMAT are the partitioning coefficients for the algae, detritus, and mat, and ALG, DET, and MAT are the biomass per unit area (g/m^2) for these components, respectively. DISSOLVED is the dissolved metal per unit area (g/m^2) and TOTAL is the total metal per unit area available for partitioning (g/m^2).

Metal enters the soil pool from root decay and from the water column as seepage and transpiration, and leaves the soil dissolved as seepage and plant uptake. The total metal is partitioned between the peat and the soil pore water based on a similar partitioning equation. The partitioning in the anaerobic soil-peat environment may be different from the water column. Metal enters the plants passively as dissolved metal from the soil water and is controlled by the transpiration rate. The metal is divided between shoots and roots in the same ratio as primary production is. Only when the dissolved metal in the soil water is high can large quantities of metal be taken into the plant. The possibility of metal exclusion coefficients has not been implemented in the model, but can be added once metal concentrations in shoots and roots are measured. Metal leaves the plant as it dies and decays. Metals are accumulated on a long-term basis only in the peat layer of the soil.

As additional information becomes available from the CWTS or similar facilities, these initial model formulations can be easily modified and the coefficient values changed to provide a better interpretation of the available field data.

Model Inputs and Outputs

Inputs required for the model are daily input variables, model coefficients, and initial values for all modeled variables. The required daily variables for CWFATE are: flow rate (gallons per minute), evapotranspiration (ET) estimate (mm/day), precipitation (mm/day), solar radiation (ly/day), mean wind speed (m/s), inflow and outflow temperatures (°C), biochemical oxygen demand (BOD) in inflow (mg/l), inflow of total nitrogen (mg/l), and metal concentration in inflow (μg/l).

Model coefficients include physical dimensions and characteristics of the wetland cell, biomass conversion and decay rates, and metal partitioning coefficients for each organic cell component. Suggested default coefficient values can be used for initial model studies, and adjusted as monitoring or experimental data becomes available for calibration. These coefficients can also be changed systematically to investigate the sensitivity of the simulated results.

The CWFATE model provides two types of outputs for selected model variables: time series and longitudinal profiles. Daily results from the first, middle, and last segments of the wetland cell are saved for a year to show seasonal variation and bioaccumulation. Longitudinal profiles are saved from each of the 25 segments at the end of each quarter so that patterns along the wetland cell can be evaluated. Standard graphs are provided in the spreadsheet, so simulation results can be easily reviewed or compared with field data.

Model Results

Figure 2 shows a simulated seasonal pattern of growth and decay for the six biomass components for the inflow segment of the constructed wetland cell. The largest biomass term is the roots, with a net decay of biomass in the first half of the year and a net gain in the second half. The shoots exhibit an increase in biomass in the spring, with fairly constant biomass throughout summer and a rapid decrease as the shoots die in the fall and become mat material.

The mat biomass is simulated as a storage of organic material that is more slowly transformed to detritus. Detritus is accumulating from

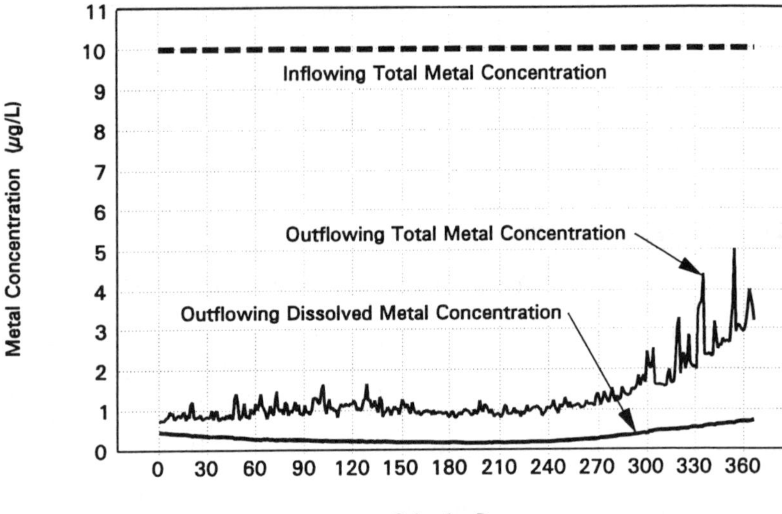

Figure 3. Simulated Inflow and Outflow of Metal Concentrations in Segment 1

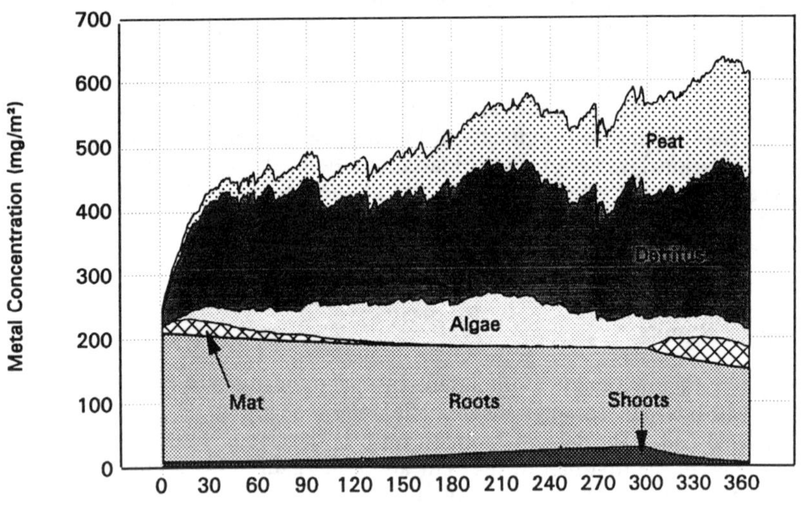

Figure 4. Simulated Metal Partitioning and Bioaccumulation in Segment 1

inflowing particulates and mat decay, and being decomposed to CO_2, transformed into peat, and transported to the next segment as suspended detritus. The algae biomass follows a seasonal pattern with peak values in the summer.

Figure 3 shows the simulated inflow and outflow of a selected metal. The inflow concentration for this unidentified metal was a constant 10 μg/l. For the default values of the model coefficients used for this simulation, the majority of the metal was retained in the constructed wetland. The partitioning of the metal shifts between the seasonal biomass components, and slowly accumulates in the peat and root components. Most of the outflow of metal was adsorbed onto detritus at the end of the year. Only low concentrations of dissolved metal were simulated in the outflow.

Figure 4 shows the simulated bioaccumulation of a selected metal in each of the six biomass components for the inflow segment. This segment has the greatest metal accumulation because the inflow of dissolved metal provides a constant source for adsorption and uptake. The partitioning of the metal shifts between the seasonal biomass components and slowly accumulates in the peat and root components.

Future Model Calibration

Adjustment of the coefficient values or modification of the model formulations to better match observations from the prototype constructed wetland must await the full operation of the facility and the collection of further field data. The CWTS was planted in 1993 with wetland plants, and initial operation using secondary effluent is scheduled to begin in 1994. Baseline monitoring of water quality and aquatic life was initiated in 1993, and performance monitoring of the CWTS will begin in 1995.

Acknowledgements

This work was conducted by Jones & Stokes, a subcontractor to Nolte and Associates, as part of Phase II of the overall design and construction of the wetland facility for Sacramento Regional County Sanitation District.

References

Jones & Stokes Associates, Inc. 1993. CWFATE: constructed wetland fate and aquatic transport evaluation model. (JSA 91-126.) July 1993. Prepared for Sacramento Regional County Sanitation District, Elk Grove, CA.

Moshiri, G. A. 1993. Constructed wetlands for water quality improvement. Lewis Publishers. Boca Raton, FL.

MIAMI INTERNATIONAL AIRPORT STORMWATER NPDES PLAN

Armando I. Perez [1], Jonathan Z. Goldman [2], Michael F. Schmidt [2] and Edward E. Clark[3]

Abstract

Miami International Airport (MIA) is endeavoring to essentially double its traffic volume by the turn of the century. This is a great challenge since the site is already highly developed. Space, safety and other constraints make it difficult to implement conventional detention/retention stormwater practices. Other practices were evaluated to control stormwater quantity/quality, since some of the downstream bodies of water are flood-prone or environmentally sensitive.

Introduction

Miami International Airport is one of the world's 10 busiest airports and a vital contributor to the local tourist and trade-oriented economy. A few years ago, conceptual plans to build a new regional airport near the Florida Everglades were set aside because of environmental concerns. Since then, the policy of the Dade County Aviation Department (DCAD) has been to maximize the throughput of the existing airport, which is already highly developed and surrounded by developed areas (see Figure 1).

(1) Member ASCE, Vice President, Camp Dresser & McKee Inc., Miami, FL

(2) Respectively, Environmental Engineer (Miami, FL) and Water Resources Engineer (Jacksonville, FL), Camp Dresser & McKee Inc.

(3) President, Clark Engineers-Scientists Inc., Miami, FL

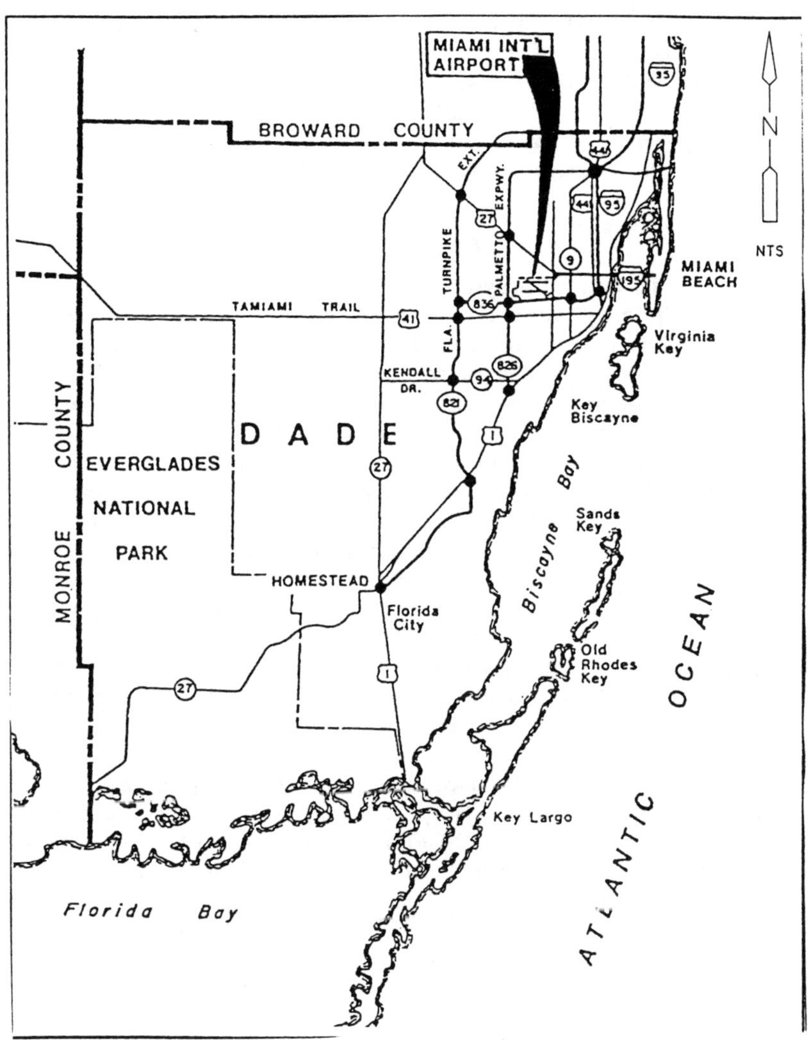

Figure 1. Location Map

In 1991, DCAD commissioned a stormwater master plan to address both water quantity and water quality control needs to allow redevelopment in conformance with DCAD's overall land use master plan. The stormwater master plan contained an element titled "Pollution Control Master Plan," which evaluated structural and nonstructural controls aimed at improving the quality of discharges, in anticipation of federal NPDES permit requirements.

Water Quantity Analysis

The water quantity analysis was directed at the principal conveyance systems (the "primary" system), which discharge at five outfalls permitted by the South Florida Water Management District (SFWMD). Over 10 years ago, SFWMD established permitted peak discharges at each outfall. SFWMD regulations require filing of a permit modification application when major changes are contemplated in the primary systems upstream of the outfalls. To accommodate the increased runoff from proposed new impervious area while meeting SFWMD limits, DCAD had to consider onsite storage. However, storage options were limited by: (1) the low elevations of some existing buildings, (2) Federal Aviation Administration (FAA) concerns about open storage ponds attracting wading birds, which could endanger aircraft engines and the production of fog by ponds over the runway approaches, and (3) concern on the part of Dade County's Department of Environmental Resources Management (DERM) about potential recharge to the shallow aquifer possibly causing migration of underground contaminant plumes, and (4) the high value of land for economically productive uses.

To date, the plan has resulted in successful permitting for the first outfall in the priority schedule. The proposed primary system was sized using the computer dynamic models RUNOFF and EXTRAN, which are components of the Storm Water Management Model (SWMM), and consists of both onsite and offsite components (the latter a former borrow pit). One of the plan outputs was a set of design standards, intended to ensure that firms engaged in primary system design preserve the desired flow and stage-storage relationships. Permitting efforts for other outfalls are proceeding using similar methodologies.

Water Quality Analysis

For the "industrial core areas", the pollution control element of the plan recognized the limited capacity to store and treat along the primary system, and therefore focused on source controls along the secondary system. That is, the approach is to use "housekeeping" practices to minimize exposure to rainwater, and capture of the "first flush" of wash off near its source, before flows become large and unmanageable. The combination of housekeeping and first flush treatment (including state-of-the-art coalescing plate-oil/water separators) is embodied in the "treatment train" concept (see Figure 3), which become part of the SFWMD permit modification conditions. The treatment train includes an oil skimmer device near the outfall as its final component. Treatment train guidelines were also included in the design standards for the benefit of firms conducting secondary system design. A typical layout of the special inlets, grit chambers and oil-water separators components of the treatment train is shown in Figure 4. The design standards specify a rate of flow corresponding to the runoff from 0.25 inches per hour of rainfall. Thus, the system will totally treat the typical storm in Miami (0.50 inches volume, 2-3 hour duration) as well as portions of larger storms.

Permitting Strategy

Regarding stormwater NPDES permitting, DCAD elected to file an individual (rather than group or general) permit for two reasons: (1) various local constraints suggested a tailored permit, and (2) a substantial amount of inventory data and conceptual plans had already been developed by the filing date. The water quality data collected for the permit filing was limited to the single storm event required. Water quality generally meets State of Florida criteria for the receiving bodies of water. This suggests that miscellaneous control practices implemented before the plan may have been beneficial. Nonetheless, DCAD has commissioned additional wet weather sampling to monitor water quality trends over time. Also, DCAD is conducting separate environmental audits of the various tenants at MIA.

Conclusions

As the airport continues to redevelop, there will be continued challenges to modify the stormwater system in a manner that maintains flood control onsite while keeping outfall discharges below permitted levels. The RUNOFF and EXTRAN models will be key tools in this regard. Also, the BMP

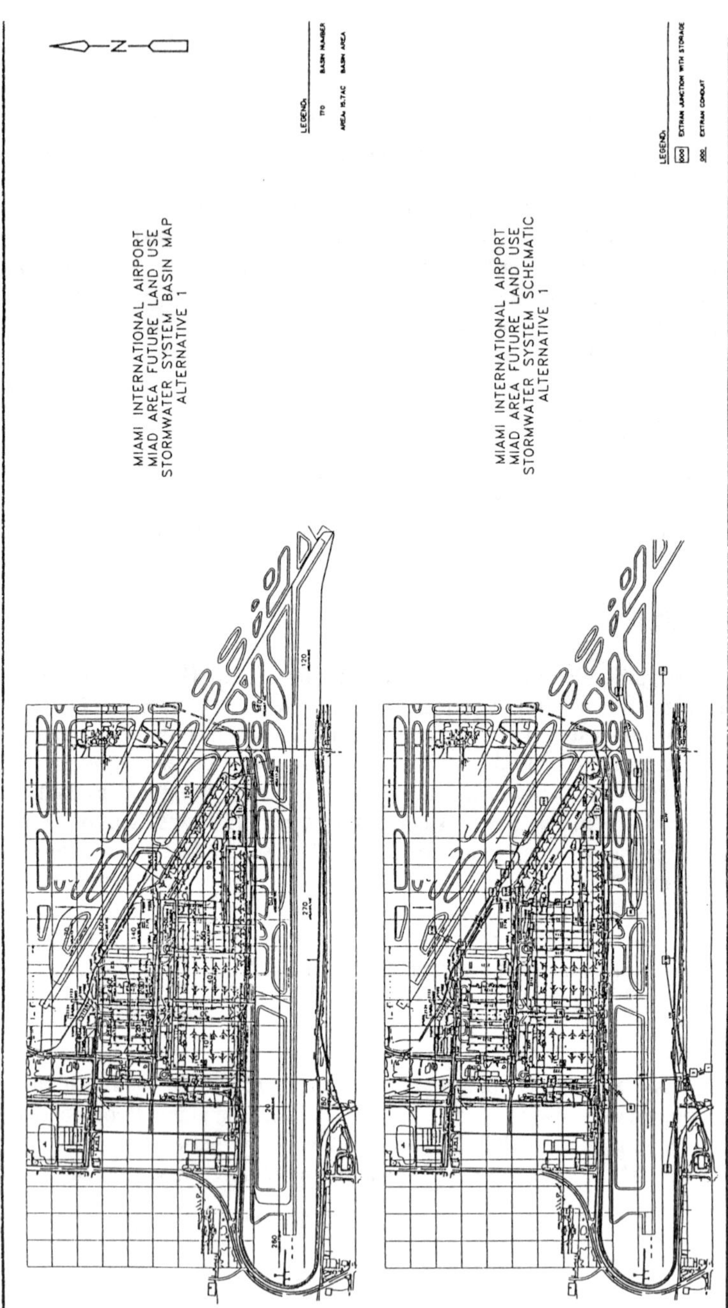

Figure 2. Runoff/Extran Model Layouts in MIAD Area

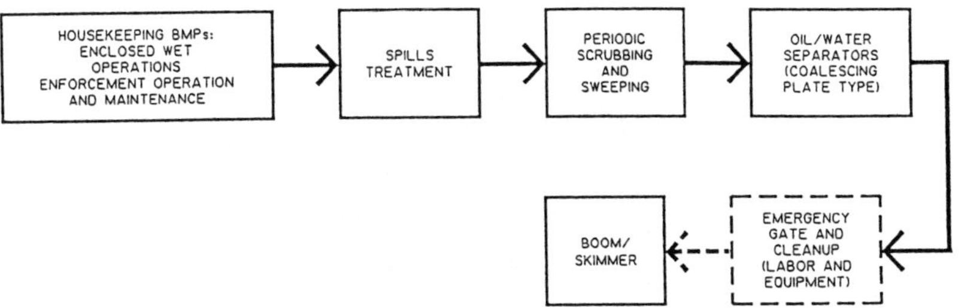

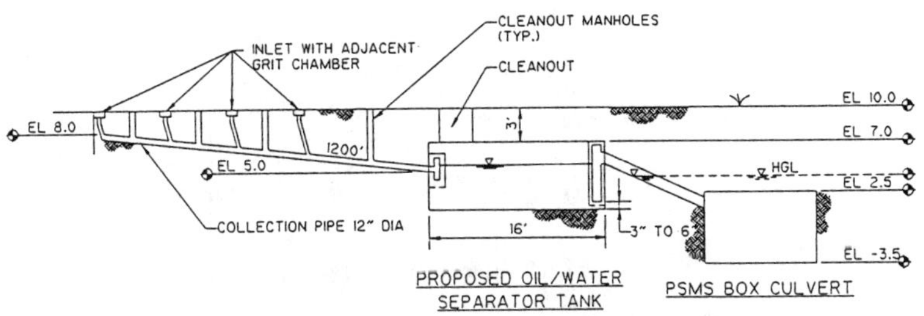

Figure 3. Treatment Train Concept

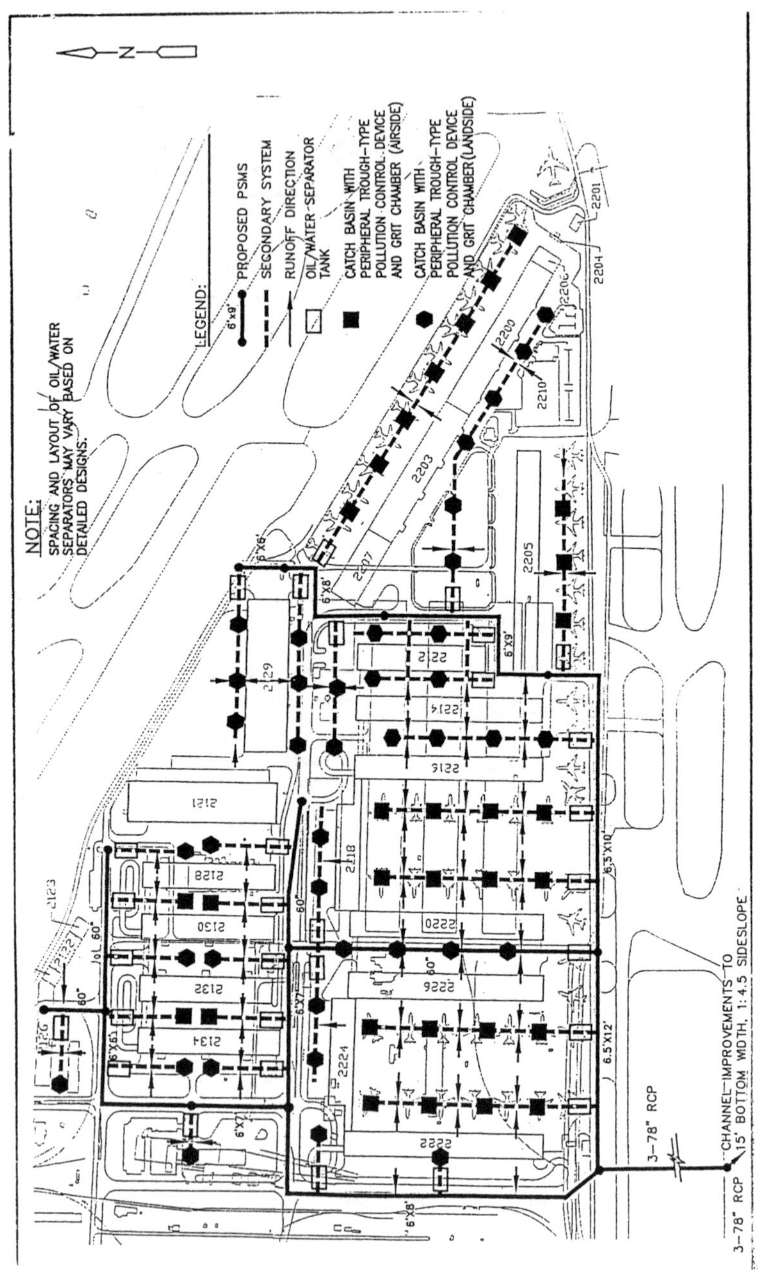

Figure 4. Layout of Inlets, Grit Chambers and Oil-Water Separators in MIAD Area

treatment train may require refinement as more water quality are collected or regulations evolve.

Control of Odor and VOC Emissions at Wastewater Treatment Plants -Boston Harbor Case Study-

Robert Getter[1], Cheryl Breen[2], and Mark Laquidara[3]

Abstract

Siting of the new wastewater treatment plant (WWTP) for the Massachusetts Water Resources Authority (MWRA) in Boston was based on an assumption of mitigation of total reduced sulfur (TRS) and volatile organic compound (VOC) emissions. Collection and treatment of exhaust streams from potential emission sources was recommended. Best Available Control Technology (BACT) for VOC control was conservatively suggested to consist of wet scrubbing followed by carbon adsorption based on initial sampling performed in 1988 during facilities planning, which estimated uncontrolled VOC emissions in excess of 1000 tons per year. This concept was carried forward to the design phase in 1990, concurrent with an extensive air emissions testing and pilot treatment program at the MWRA's existing primary treatment plant. Results of the pilot program, however, indicated source VOC concentrations well below what was expected as a result of the initial sampling study. Use of the 1990 pilot data in a top-down BACT analysis led to a recommendation to reconsider VOC control with carbon adsorption on the basis of prohibitive cost.

This paper summarizes the background and permitting approach for five new odor control facilities on Deer Island for the Boston Harbor Project, with emphasis on the new primary treatment facilities. The paper also presents results from the 1990 emissions characterization and pilot program, providing generally applicable ideas for solving the difficulties of characterizing and estimating emissions for WWTPs. Results from operation of the pilot facilities illustrate the effectiveness of wet scrubbing and carbon adsorption in removing TRS and VOCs from wastewater treatment exhaust air streams. In addition, pilot program results indicate the importance of flexibility in design of odor control systems to accommodate variations in concentrations of TRS and VOCs.

---

[1]Project Manager, and [2]Project Manager, Metcalf & Eddy, Inc., Wakefield, Massachusetts, [3]Manager, Process Control, Massachusetts Water Resources Authority, Boston, Massachusetts.

<u>Introduction</u>

The focus of the MWRA's Boston Harbor cleanup is a new 480 million gallon per day (mgd) pure oxygen secondary treatment plant on Deer Island, a 200 acre peninsula in Boston Harbor (see Figure 1). The new plant will replace existing primary treatment plants on Deer Island and Nut Island, and will provide wastewater treatment for a service area which includes 43 communities in the greater Boston area with a population of over 2.0 million (Willnow, 1992). Other major elements of the overall cleanup program include: a sludge pelletizing facility which began operation in December of 1991, thereby ending discharge of sludge to the harbor; and an ongoing program for control of combined sewer overflows.

Facilities planning, completed in 1988, projected design year (2020) flows and determined unit processes to be used in the new treatment plant. Following facilities planning, the MWRA selected Metcalf & Eddy to develop the concept design for the new plant, to perform detailed design for the preliminary, primary treatment and effluent disinfection facilities, and to coordinate detailed design by multiple firms for the remainder of the plant facilities. The plant was divided into approximately 20 major design packages, which were in turn subdivided into approximately 30 major construction packages for detailed design.

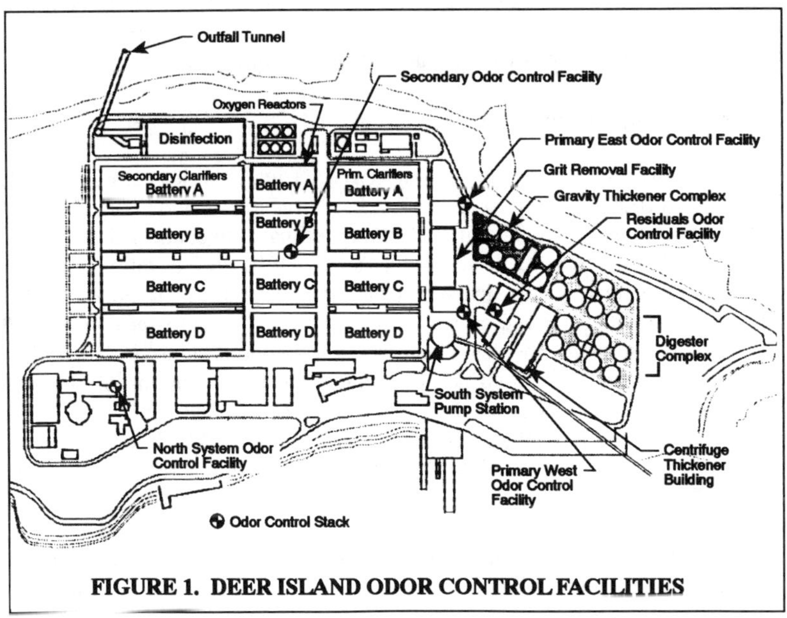

FIGURE 1. DEER ISLAND ODOR CONTROL FACILITIES

Metcalf & Eddy, together with architectural subconsultant Tso Associates, completed the design of the East and West Odor Control Facilities in 1991. These primary odor control facilities, which treat exhaust air from covered primary treatment unit processes, include packed tower wet scrubbers and carbon adsorbers for control of odors and VOCs, respectively. The residuals odor control facility includes carbon adsorbers and two-stage wet scrubbing for removal of ammonia and TRS. The secondary odor control facility includes carbon adsorption only for control of TRS and VOCs. In addition, construction of a 2 million gallon per day (mgd) permanent pilot plant, which includes carbon adsorption for odor control, has recently been completed on Deer Island. Construction of the primary and residuals odor control facilities is to be completed in mid-1994 for start-up of the first phase of the new treatment plant.

Covered and ventilated unit processes were selected for the new Deer Island treatment plant's wastewater pumping, grit removal, primary clarification, residuals handling, and pure oxygen secondary processes to minimize emissions of odors and VOCs. Ventilation rates for unoccupied spaces exposed to wastewater at the Deer Island plant were generally based on three air-changes per hour (ACH) to maintain negative pressure (Metcalf & Eddy, 1989). Mechanisms for generation of odors and VOCs within these spaces include aeration of liquid containing channels, gas mixing in the pure oxygen reactors, volatilization due to liquid drops over weirs, and volatilization from free liquid surfaces. Ventilation rates for occupied spaces exposed to wastewater at the Deer Island plant were based on 12 ACH. Collected gases are ducted to odor control facilities for treatment. Table 1 presents flowstreams resulting from the required ventilation for primary treatment unit processes. Figure 2 illustrates the typical equipment arrangement for the primary odor control facilities.

Characterization of Emissions

Emissions were initially characterized during facilities planning in 1988, and characterization was refined in 1991 based on a pilot study conducted at the existing treatment plant. Facilities planning estimated uncontrolled emissions of VOCs from gases to be collected at the new plant based on liquid stream VOC concentrations, theoretical calculations, and limited gas stream sampling at remote headworks facilities. Based on gas steam sampling, the increase in uncontrolled of VOCs emissions for the new plant was estimated to be in excess of 1000 tons per year. In addition, facilities planning projected violations of the state's Acceptable Ambient Levels for specific VOCs based on liquid stream concentrations and theoretical calculations. As a result, BACT for VOC control was conservatively suggested to consist of wet scrubbing followed by carbon adsorption.

In order to verify the emission characterization and assumed removal efficiencies, sampling and analysis of exhaust from a 300 ft by 15 ft section of

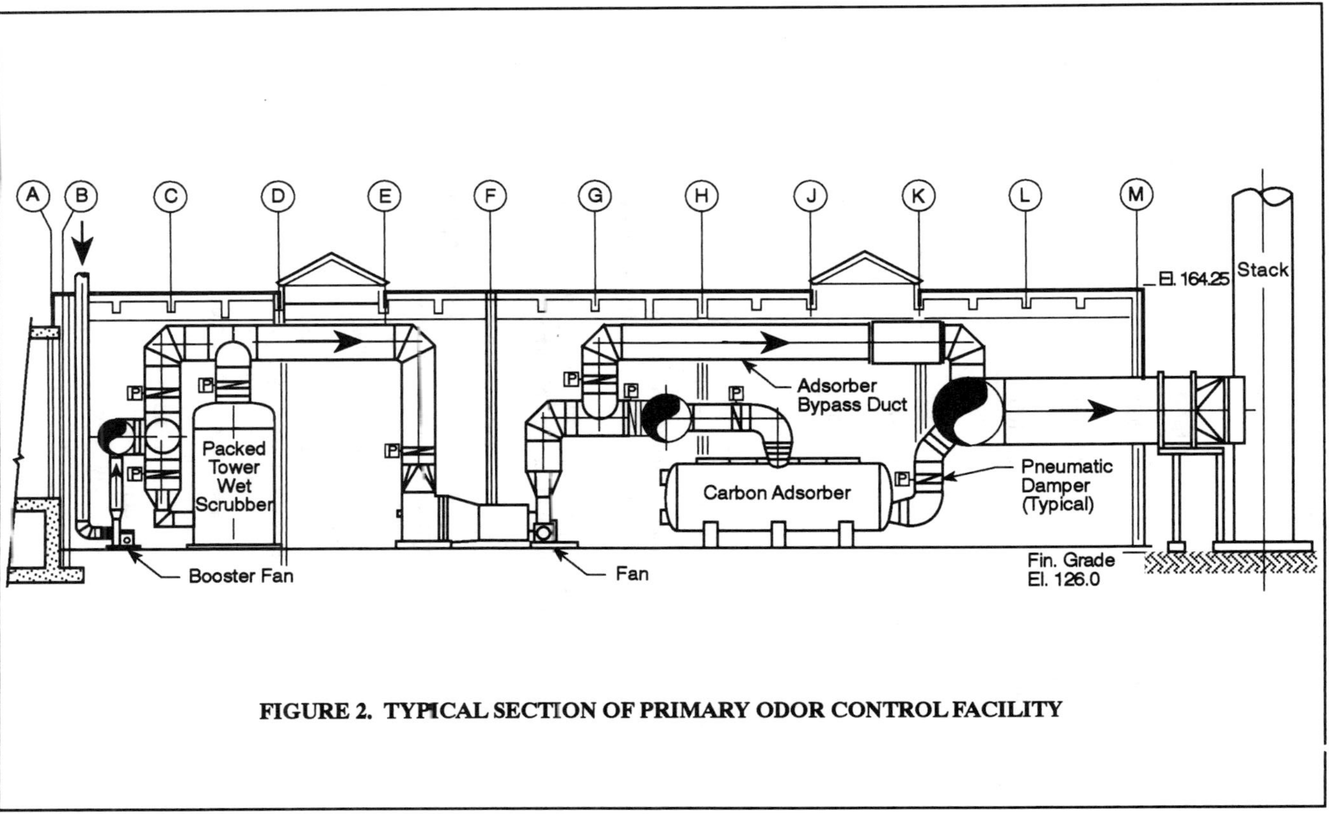

FIGURE 2. TYPICAL SECTION OF PRIMARY ODOR CONTROL FACILITY

TABLE 1. PRIMARY ODOR CONTROL FLOWSTREAMS

EAST ODOR CONTROL FACILITY:	cfm	#Fans	#Packed Tower Scrubbers	Scrubber Diameter, Ft.
Clarifier Maintenance[(1)]	44,000	1	0	-
Clarifier Normal Vent.	43,500	2	2	12
Grit Facility Vent.[(2)]	72,400	3	3	12
WEST ODOR CONTROL FACILITY:				
Clarifier Maintenance	44,000	1	0	-
Clarifier Normal Vent.	36,450	2	2	10
Grit Facility Ventilation	35,600	2	2	12
South Pump Station Wetwell	25,500	2	2	9

Notes:

1. Clarifier, maintenance ventilation provides 12 ACH for one clarifier during maintenance, with treatment by carbon adsorption.
2. One flowstream standby.
3. Each primary odor control facility includes eight, 10 ft. dia. x 20 ft. long horizontal carbon absorbers, each with a 3 ft. bed depth.

aerated influent channel at the existing primary treatment plant was conducted as part of the 1990 pilot program. A schematic of the pilot facilities is shown in Figure 3. The subject influent channel receives raw wastewater from a tunnel conveyance system, and therefore has the greatest potential for emitting TRS and VOC compounds in the highest concentrations. Diffused air is supplied at a rate of 3.0 to 3.5 scfm per foot of channel length at a depth of 8 feet to the 10-foot deep channel. The channel was enclosed by plywood covers and plastic sheeting, and the exhaust gases were drawn off at a rate of 1250 scfm, maintaining a slight negative pressure in the headspace above the water surface. Gases were ducted to a pilot treatment system consisting of a 3-foot diameter sodium hypochlorite wet scrubber followed by a carbon adsorption unit. Packing depth for the scrubber was 12 feet, and the activated carbon bed depth for the 4.4-foot diameter adsorber was varied between 1 and 3 feet.

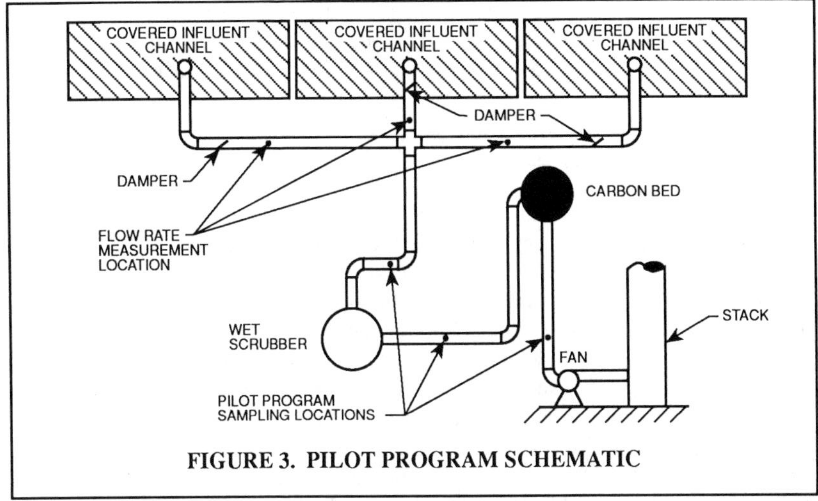

FIGURE 3. PILOT PROGRAM SCHEMATIC

A sampling and analysis program was conducted based on protocols developed by ENSR with input and approval from the Massachusetts Department of Environmental Protection. Grab samples were collected using both Tedlar bags and Tenax sorbent. All samples were prepared and analyzed by gas chromatography/mass spectrometry (GC/MS). Continuous emissions monitoring was provided for TRS and NMHC.

Analysis determined that non-methane hydrocarbon (NMHC) concentrations were dramatically lower than originally estimated, and that approximately 96 percent of the VOCs were methane, a non-regulated compound (ENSR, 1991). Average and maximum NMHC concentrations during piloting were 5.6 ppmv and 44 ppmv respectively, with an average molecular weight of 100. A total of 37 target compounds were also selected for analysis, including odorous compounds typically found in wastewater, such as mercaptans and sulfides, and compounds for which state air toxics regulations have been set. The target VOC compounds comprised only 15 percent of the total NMHC.

TRS concentrations during the 1990 pilot program were found to be highly variable, and significantly higher than anticipated, with concentrations ranging from 3 ppmv to over 200 ppmv, and averaging 78 ppmv. Similar results have been found during start-up of the new permanent pilot plant in the fall of 1993. However, data appear to indicate a seasonal trend, and levels at the full-scale plant are expected to be lower based on more dilute flow streams.

The 1990 emissions data were used to derive emission estimates for the new plant, using concentration factors of 0.25 to 1.0 for estimated concentrations from process areas where emissions are expected to be the least and greatest, respectively. For example, ventilation from aerated channels and weirs was assigned a concentration factor of 1.0, while ventilation from above quiescent tank surfaces was assigned a concentration factor of 0.25 (Metcalf & Eddy, 1991).

Pilot Results and BACT Analysis

Results of piloting indicated very effective TRS removal by wet scrubbing using sodium hypochlorite, with an average removal efficiency of 99.5 percent through the test period of 36 days, based on average inlet and outlet CEM measurements. Chemical feed rates were adjusted during the test period to maintain a pH of 10 and outlet TRS concentrations at less than 0.5 ppmv. Piloting also indicated an average total VOC removal by carbon adsorption of 80 percent, but suggested that full-scale operation would require frequent carbon replacement and high costs, with relatively minor VOC control benefits. Performance was significantly lower after only approximately 220 hours per foot of bed depth, with VOC removal efficiency deterioration from 95 percent to 75 percent in one test. In addition, air modeling indicated that the state's Acceptable Ambient Levels for VOCs would be met without carbon adsorption (ENSR, 1992). Therefore, priority was placed on odor control with wet scrubbing, instead of VOC control with carbon adsorption. Use of the pilot data and emissions estimates in top-down BACT analyses led to the recommendation that wet scrubbing without carbon adsorption be considered BACT.

Design of the primary treatment facilities was completed in 1991, and both sodium hypochlorite wet scrubbers and carbon absorbers are provided for treatment of emissions from both liquid and residuals treatment processes. The decision to implement the two systems for the primary and residuals facilities, in spite of the BACT analysis, was based on the expectation that the state's future VOC regulations will be more stringent in response to the Clean Air Act Amendments of 1990. Also, the selected processes provide the MWRA with flexibility and redundancy in odor control to ensure compliance with emission limits and mitigation commitments with its neighboring community. The MWRA intends to operate the carbon adsorption units only as needed based on monitoring of stack emissions. The BACT analysis is currently under regulatory review and the MWRA is preparing a testing protocol to provide a one-year emissions monitoring program in support of its air quality plans approval applications for intermittent operation of carbon adsorption.

Conclusions

The results of the emissions characterization and pilot program indicate the importance of flexibility in the design of odor and VOC control processes to accommodate variability and uncertainty in levels of odors and VOCs in exhaust streams. The arrangement of the packed tower scrubbers, fans and carbon absorbers for the primary odor control facilities, for example, allows bypassing the carbon absorbers in the event that VOC removal is not required. Two-speed fans are provided as required to allow bypassing the carbon absorbers or wet scrubbers as alternate modes of operation. System flexibility has proven to be particularly valuable given the potential for higher than expected levels of TRS. Each primary ventilation flowstream is provided with a redundant fan and scrubber. Parallel operation of both scrubbers in a flowstream is possible in the event that H_2S (which comprised >99 percent of TRS during the pilot program) concentrations exceed the 25 ppm design basis for H_2S removal of the scrubbers, as it did in piloting. The capability to either bypass the carbon adsorbers or operate both scrubbers in parallel will provide the MWRA with flexibility to meet treatment objectives more effectively and economically.

References

ENSR Consulting and Engineering (1992), "Air Quality Dispersion Model Analysis of the New Deer Island Wastewater Treatment Facilities," for the Massachusetts Water Resources Authority.

ENSR Consulting and Engineering (1991), "Deer Island Primary Treatment Facilities: Total Reduced Sulfur and VOC Best Available Control Technology Analysis," for the Massachusetts Water Resources Authority.

Lager, J.A., et.al. (1992), "Air Emissions Control for Wastewater Treatment Facilities - Boston Harbor Case Study," WEF Annual Conference.

Lager, J.A., et.al. (1991), "Measurement and Control of Volatile Organic Compounds and Reduced Sulfur Species in Wastewater Treatment Air Emissions," WPCF Annual Conference.

Metcalf & Eddy, Inc. (1991), "Methodology for the Derivation of Mass Air Emissions from the Deer Island Secondary WaStewater Treatment Plant," for the Massachusetts Water Resources Authority.

Metcalf & Eddy, Inc. (1989, "Boston Harbor Project-Deer Island Related Facilities Design package 10, North System Headworks, Concept Design Report," for the Massachusetts Water Resources Authority.

Willnow, L.D., et. al.(1992), "Hydraulic Modeling of the Deer Island Wastewater Treatment Plant for the Boston Harbor Project," WEF Annual Conference.

Quantification of the Mechanisms Controlling the Removal Rate of Volatile Contaminants by Air Sparging

Gretchen L. Hein[1], Neil J. Hutzler[2], M.ASCE, John S. Gierke[3], M. ASCE

Abstract

Air sparging is a technology for removing dissolved and volatile phase chemicals (VOCs) from the saturated zone. It is used in conjunction with soil vapor extraction (SVE) and sometimes pump and treat systems. The air sparging process involves the injection of air at some depth below the water table; as the air rises, VOCs transfer to the vapor phase and discharge to the unsaturated zone. In almost all applications, SVE is used to capture the sparging gas as well as to clean up the unsaturated zone. Several researchers have investigated air flow patterns, but relatively little ongoing research is being done to define the mass transfer rates that govern the performance of air sparging. This paper presents a model to quantify the removal mechanisms during air sparging. The model simulation is compared to an analytical solution and the dimensionless groups are analyzed to determine the dominant mass transfer processes. The model shows that mass transfer is dominated by gaseous dispersion and advection, and liquid diffusion.

Introduction

Existing models for air sparging are insufficient to adequately design sparging systems (Marley et al. (1992), Sellers and Schreiber (1992), Gvirtzman and Gorelik (1992), Ostendorf et al. (1993)). They are greatly simplified when compared to actual field conditions, both in regard to the porous media profile and the contaminant composition. Because these models have not been tested, they typically consider the mass removal very simply by assuming equilibrium between the flowing gas and the surrounding aqueous and sorbed concentrations. Nevertheless, Johnson (1993) and Ji (1993) observed that mass transfer can limit the removal rate. Ostendorf et al. (1993) model the behavior of sparging using

[1]PhD Cand., Dept. of Civ. & Env. Eng., Mich. Tech. Univ., Houghton, MI,49931
[2]Professor, Dept. of Civ. & Env. Eng., Mich. Tech. Univ.
[3]Assistant Professor, Dept. of Geo. Eng., Geo. & Geophysics, Mich. Tech. Univ.

a mass transport model that includes advective transport in the gas-phase and a "first order source term". They found that the process was mass transfer limited because at high air flowrates a decrease in removal efficiency occurred, but they did not quantify the mass transfer rates. Therefore, the purpose of this paper is to describe a relatively simple model that simulates the flow of air through a single air channel filled with porous media and adsorbed water. The relative importance of advection, liquid diffusion, gaseous dispersion and the mass transfer resistance at the air/water interface using this model are investigated. The numerical model is compared to an analytical solution presented by Van Genuchten, et al. (1984).

Model Development and Analytic Solution

Figure 1 shows a schematic of the sparging system within a cylindrical regime. Equations were derived to simulate chemical extraction by completing mass balances on the liquid and gaseous-phases, applying initial and boundary conditions and including mass transfer and equilibrium expressions. Then they were converted to dimensionless form to decrease the number of unknowns and permit the analysis of the dimensionless groups. Table 1 lists the mass transport dimensionless groups developed for a single sparging finger. For the initial development of the model equations, the following assumptions were made:

1. The porous media profile is homogeneous and isotropic.
2. Air flow is parallel to the z-coordinate and originates from the sparger (the z-coordinate follows the centerline of the sparging finger).
3. The sparging finger consists of porous media particles, adsorbed water, and air where the width of the sparging area ranges from 10 to 20 particle diameters (Ji et al., 1993).
4. Diffusion in the liquid phase occurs perpendicular to the z-coordinate and is the only transport mechanism in the liquid phase (Advective transport in the liquid phase is negligible because the flow of groundwater is much slower than the air flow.).
5. The time required for the groundwater to be displaced from the sparging regime will be small when compared to the period for remediation.
6. The mass flowrate remains constant throughout the sparging finger. Therefore, if the area of the sparging finger remains constant, the velocity of air will increase as the air rises in the profile.

The liquid mass balance was completed around a cylindrical shell at a distance from the center of the air finger which ranges from the radius of the air finger (R_P) to the radius of influence (R_{OI}). The reduction of accumulation of chemical in the liquid equals the net mass transport out due to liquid diffusion. This relationship assumes that the contaminant transport outside the air finger occurs in the radial direction and diffuses toward the finger while no transport occurs in the axial direction.

$$\frac{\partial c_W(z,r,t)}{\partial t} = \frac{Ed}{r}\frac{\partial}{\partial r}\left(r\frac{\partial c_W(z,r,t)}{\partial r}\right)$$

The air-phase mass balance is obtained by completing a mass balance around a cylindrical element of an incremental length with a radial dimension equal to R_P. Since the mass flowrate is assumed to be constant, the advection term is expressed as a function of the mass flowrate instead of air velocity. The accumulation in the air phase equals the amount of mass transferred to the air phase by dispersion, advection and mass transfer.

$$\frac{\partial c(z,t)}{\partial t} = \frac{1}{Pe}\frac{\partial^2 c(z,t)}{\partial z^2} - \frac{\partial c(z,t)}{\partial z} + 2St\left(c_W(z,r,t) - c(z,t)\right)$$

The initial conditions assume that the entering air is clean and that the contaminant in the surrounding porous media is at an initial concentration less than or equal to the contaminant's solubility. The first boundary condition for both the air and liquid phases assumes a zero concentration gradient. For the air phase, the Type II boundary condition is located at the upper boundary of the saturated zone. The liquid-phase zero gradient boundary occurs at the radius of influence. The second boundary condition for the air and liquid phases, respectively in the following two equations, requires a mass balance over the porous media column (Gierke et al, 1992). This boundary condition allows the numerical code to maintain conservation of mass.

These eqautions are solved using orthogonal collocation. This numerical method has been used in modelling SVE (Gierke, et al, (1992). Previous applications have verified the advection-dispersion portion of this code (Gierke, et al. (1992)) and have found it to be both stable and consistent with experimental results and analytical solutions. Therefore, the analytical solution developed by Van Genuchten, et al. (1984) is used to verify the liquid diffusion portion of the numerical model.

The numerical solution of the model was compared to a modification of an analytical solution developed by Van Genuchten, et al (1984). Their solution simulates the breakthrough of a chemical through a porous media macropore. For air sparging, the media is contaminated and the chemical removal is of interest. Therefore, the Van Genuchten solution is subtracted from 1.0 to show the elution of chemical. This solution is valid for the case where no gaseous dispersion occurs in the sparging finger and the region being remediated is infinite in the radial direction. The boundary and initial conditions used in the numerical method are valid for this solution. The analytical solution is described by the following equations:

$$c_m = \frac{c}{c_{WO}} = 1.0 - \exp\left(-\frac{\theta D_W L}{R_P^2 \theta v_m}\right) erfc\left[\frac{\theta L}{R_P \theta v_m}\left(\frac{D_P R_a}{t - R_m L/v_m}\right)^{0.5}\right]$$

Where:

$$v_m = \frac{G}{\rho_A A_P \theta}, \; R_m = 1.0, \; R_a = 1.0$$

And: $v_m > R_m L$. When $v_m \leq R_m L$, $c_m(L,t) = 1$.

The analytic solution approaches 1.0 when D_P is high because this equation assumes the system is infinite in the radial direction. Therefore, when diffusion through the porous media is fast, the concentration in the sparging finger never becomes depleted. The concentration is initially at 1.0 then it asymptoticly approaches 0.53. Figure 4 shows the shape of the analytic solution when $Ed_P = 0.000567$. The minimum value is dependent upon the liquid diffusion and the mass florate.

Results and Discussion

A sensitivity analysis was performed to determine the effect the different mass transfer groups had on the elution curve. The numerical model was found to be most sensitive to changes in Pe and Ed_P but not St. Therefore, the mass transfer was dominated by liquid and gaseous diffusion and advective transport. A numerical simulation was then compared to an analytical solution. These simulations were comparable during the initial sparging period but not for the overall elution pattern due to the infinite radial dimension in the analytic solution.

The model was not sensitive to changes in the Stanton number (St) (See Figure 2). The Stanton number characterizes the relationship between air/water mass transfer and advection. The three curves coincide, which also corresponds to previous work completed using SVE (Gierke, et al. (1992)). This finding supports the hypothesis that the mass transfer is controlled by the physical properties of microscale mixing and molecular diffusion. It is strengthened in the analysis of the Peclet number (Pe) and the pore diffusion modulus (Ed_P). When either of these two groups changed while holding the others constant, the time to remove the chemical changed and so did the shape of the elution curve (See Figures 3 and 4).

The Peclet number, which is the ratio of transport due to adevection to that of gaseous diffusion, shows the model sensitivity to changes in either the mass flowrate or the gaseous diffusion rate. As Pe increases or the effect of gaseous diffusion decreases, the elution curve becomes sharper and shifts to the left. As

Pe decreases, the curve became more spread out and the time required for complete removal of the chemical increases. The sensitivity of the model to Pe is indicative of field situations. At low gas flowrates, the time required for a volatile chemical to be removed via air sparging is long. The opposite is found at large flowrates until the air flowrate exceeds the rate at which the chemical can diffuse into the sparging finger. The removal time increases with decreasing Pe because the chemical has time to diffuse back through the finger.

The ratio of the rate of iquid diffusion compared to that due to to advection is described by Ed_P. The magnitude of this dimensionless group indicates the rate at which the chemical diffuses through the saturated sparging region to the air finger. As Ed_P increases, the time to remove the chemical decreases (Figure4). As the diffusion modulus decreased, two phenomena occurr. First, the amount of tailing and spreading increased. This behavior shows that the chemical requires more time to reach the air finger than with faster diffusion rates. Secondly, the three curves exhibited the typical elution curve form. The curvature results from the chemical inside the sparging finger at the initiation of the process. The model assumes that initially chemical is evenly distributed throughout the contaminated regime. Therefore, when sparging begins, a short period of time is required to displace the chemical laden air with cleaner air. Then the gaseous phase concentration quickly begins to decrease as the sparging process continues.

Since the region considered in the analytical solution is infinitely wide while the model simulates a finite area, the comparison between the model and the analytical solution is only valid for early times (Figure 4). The times prior to 1.3 hours indicate when the analytic solution and the numerical simulation (Ed_P = 0.00056) should correspond. After 1.3 hours, c/c_0 drops below 1.0 at R_{OI} for the numerical solution and the two curves begin to deviate from each other. The concentration decrease occurs both at the air inlet and at the end of the sparging regime. This behavior is due to the rapid air flowrate in comparison to the liquid diffusion. The air is travelling through the sparging finger such that the difference in removal at the R_{OI} for both locations is similar. If the length of the sparging regime was increased, the times for the these two values would not coincide. This point also corresponds to the location where the analytic solution crosses the numerical solution. The numerical solution and the analytical solution both reach a minimum concentration after approximately 14 hours. This result is expected because both are simulating a cylindrical configuration. The analytical solution approaches a concentration of about 0.56 due to the infinite radial dimension. The numerical solution approaches zero because a finite amount of chemical is in the porous media and the media has a finite area. Both curves also exhibit a short period where the concentration is approximately one at the start of sparging. The time required to displace the volume of air in the sparging finger is indicated by this time value.

Figure 1: Conceptual Picture of a Cylindrical Sparging Zone

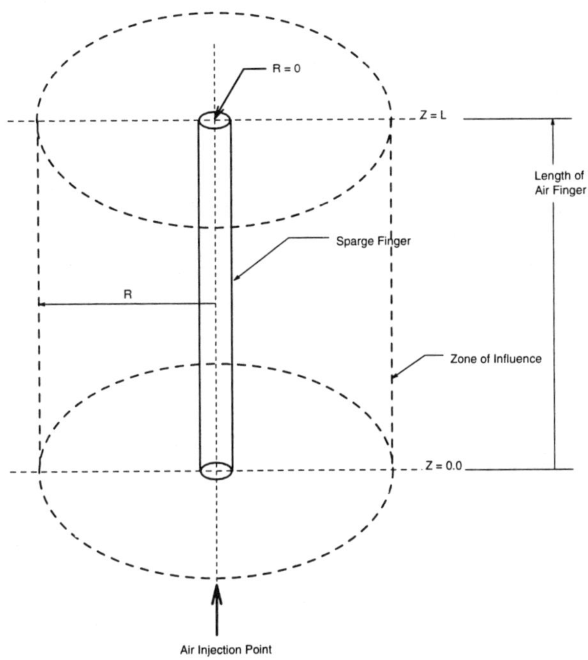

Table 1: Mass Transport Dimensionless Group Description

Dimensionless Group	Equation	Definition
Pe	$GL(\rho_A A_P D_A)^{-1}$	rate of transport by advection / rate of transport by axial dispersion
St	$K_L aL\rho_A A_P(2HG)^{-1}$	rate of transport by air/water mass transfer / rate of transport by advection
Ed	$\rho_A A_P D_W L(R_{OI}^2 G)^{-1}$	rate of transport by radial diffusion / rate of transport by advection

Figure 2: Effect of Varying The Stanton Number During Sparging

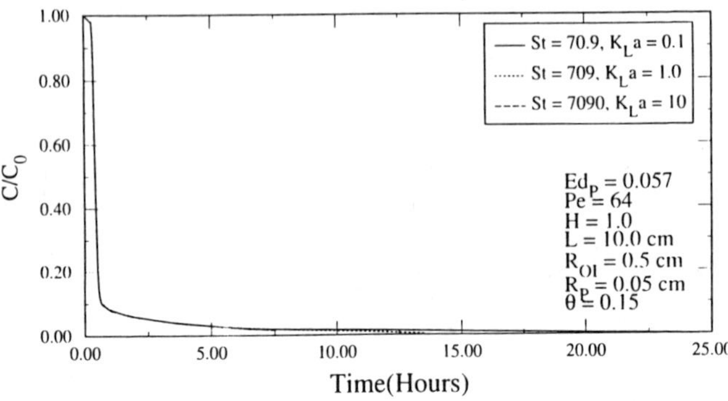

Figure 3: Effect of Varying The Pore Diffusion Modulus During Sparging

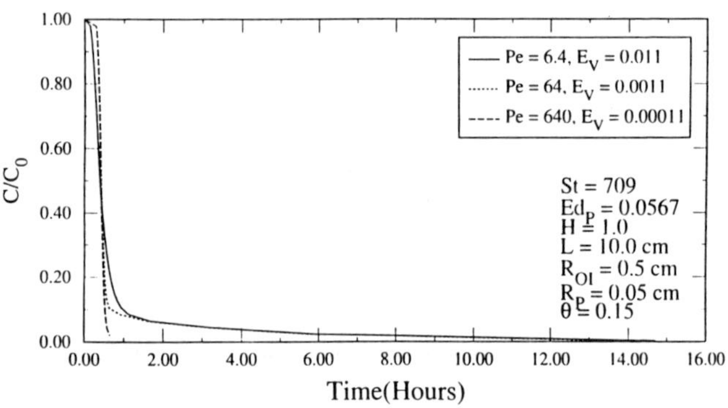

Figure 4: Effect of Varying The Pore Diffusion Modulus During Sparging

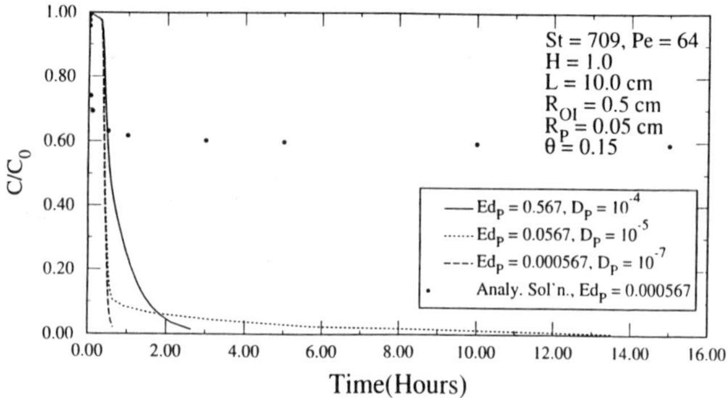

Conclusions

In air sparging, the gas flowrate and the liquid diffusion rate control the time required for remediation based on these preliminary simulations. The accuracy of the code to simulate experimental results will be tested to complete the model analysis.

References

Gierke, J.S., N.J. Hutzler, D.B. McKenzie, "Vapor Transport in Unsaturated Soil Columns: Implications for Vapor Extraction". Water Resources Research, **28**(2), 1992, pp. 323-335.

Gvirtzman, H., S. Gorelik, "The Concept of In-Situ Vapor Stripping for Removing VOCs from Groundwater". Transport in Porous Media, **8**, 1992, pp. 71-92.

Ji, W., A. Dahmani, D.P. Ahlfeld, J.D. Lin, E. Hill, III, "Laboratory Study of Air Sparging: Air Flow Visualization". GWMR, Fall, 1993, pp. 115-126.

Johnson, R.L., "In Situ Air Sparging for Removing Chlorinated Solvents from Groundwater". Nat'l. Groundwater Assoc., 45th Annual Convention, Chlorinated Vol. Org. Compounds in Ground Water, Oct. 17-20, 1993.

Marley, M., F. Li, S. Magee, "The Application of a 3-D Model in the Design of Air Sparging Systems". Petr. Hydro. & Org. Chem. in Groundwater: Prevention, Detection & Restoration, Houston, TX, Nov. 4-6, 1992b, pp. 377-392.

Ostendorf, D.W., E.E., Moyer, E.S. Hinlein, "Petroleum Hydrocarbon Sparging from Intact Core Sleeve Samples". Proc. of the 1993 Petr. Hydro. & Org. Chem. in Ground Water: Prevention, Detection, & Restoration, Houston, TX, Nov. 10-12, 1993, pp. 415-427.

Sellers, K., R. Schreiber, "Air Sparging Model for Predicting Groundwater Cleanup Rate". Petr. Hydro. & Org. Chem. in Groundwater: Prevention, Detection & Restoration, Houston, TX, Nov. 4-6, 1992, pp. 365-376.

Van Genuchten, M.T., D.H. Tang, R. Guennelon, "Some Exact Solutions for Solute Transport Through Soils Containing Large Cylindrical Macropores". Water Resources Research, **20**(3), 1984, pp. 335-346.

Variable Declaration: A = area of sparging region, cm^2; A_P = area of pore, cm^2; c = dimensionless gaseous concentration; c_m = dimensionless gaseous effluent concentration; c_W = dimensionless liquid phase concentration; C_{WO} = initial chemical concentration in the liquid phase, mg/L; D_A = gaseous diffusion coefficient, cm^2/s; D_W = liquid diffusion coefficient, cm^2/s; G = gaseous mass flowrate, g/s; H = dimensionless Henry's Law constant; K_La = overall mass transfer rate based upon liquid side, sec^{-1}; L = length of sparging regime, cm; L_{MTZ} = length of mass transfer zone, cm; r = dimensionless radial distance from the center of the sparge channel; R_a = retardation in sparging finger; R_m = retardation in sparging region; R_{OI} = radius of influence, cm; R_P = radius of sparging finger, cm; t = time; z = dimensionless distance from origin, depth of saturated soil; ρ_A = air density, g/cm^3; θ = porosity of porous media

A Probability Based Approach to the Design and Operation of Soil Vapor Extraction Systems

Kenneth K. Kebbell[1] and William F. McTernan, Member ASCE[2]

Introduction

Soil vapor extraction (SVE) or soil venting has increasingly been the operation of choice when the remediation of soils contaminated by volatile organic carbon is warranted. There are several computer codes available either in the public domain or through proprietary sources that can be employed to assist the engineer in the design of these systems. The key issues with regard to venting performance are air flow and mass removal rates. Equation 1 presents an analytical radial gas flow model that is a commonly accepted design basis employed in many of these codes.

$$Q = \pi H \frac{k_a}{\mu} P_w \frac{\left[1-(P_{atm}/P_w)^2\right]}{\ln(R_w/R_I)} \tag{1}$$

where:

Q	=	air flow rate
k_a	=	soil air permeability
μ	=	viscosity of air
P_w	=	absolute pressure at extraction well
P_{atm}	=	absolute ambient pressure
R_w	=	radius of vapor extraction well
R_I	=	radius of influence of vapor extraction well
H	=	thickness of well screen

[1]Graduate Student, School of Civil and Environmental Engineering, Oklahoma State University, Room 207 Engineering South, Stillwater, OK 74078
[2]Professor, School of Civil and Environmental Engineering, Oklahoma State University, Room 207 Engineering South, Stillwater, OK 74078

A molar mass balance for each chemical component is completed over a series of user defined venting time steps from the calculated air flow rate according to equation 2:

$$\frac{dM_i}{dt} = -\eta Q C_i^{eq} \quad (2)$$

where:

M_i	=	total number of moles of component i in the soil
Q	=	total gas flow rate through the contaminated zone
C_i^{eq}	=	equilibrium molar gas phase concentration of species i
η	=	efficiency factor to account for non-equilibrium effects

Air flow through the soil is mainly controlled by permeability which directly affects contaminant mass removal rates. Thus, from equations 1 and 2 the soil's air permeability becomes a critical design factor, varying within and among soils and also with soil moisture. Because of the extreme variability of these soil parameters, adequate estimates of air permeability are often difficult to make without specialized lab or field testing. Uncertainties in the permeability estimation lead to uncertainties in the overall venting design. A statistical linkage between the deterministic models used to describe soil vapor extraction efficiencies (Equations 1 and 2) and these highly varied soil properties was accomplished by Monte Carlo simulation.

Soil Air Permeability

Soil air permeability can be indirectly estimated from saturated hydraulic conductivity and relative permeability determinations according to:

$$k_i = \frac{\eta_w k_s}{\rho_w g(3600)} \quad (3)$$

where:

k_i	=	intrinsic permeability
η_w	=	viscosity of water
ρ_w	=	density of water
k_s	=	saturated hydraulic conductivity

Soil air permeability is related to intrinsic permeability by the following:

$$k_a = k_i k_r \quad (4)$$

where:

k_a	=	soil air permeability
k_i	=	intrinsic permeability
k_r	=	relative permeability of soil to air

Relative permeability was taken as air porosity which was estimated as total porosity minus water content.

Research Structure

The subject research focused upon the development of a probability based screening tool which utilized randomly generated soil properties to produce a normally distributed range of air permeability's. These were then used in a soil venting model in conjunction with incremental descriptions of contaminant spill volume and thickness, pressure drop, and radius of influence to produce probability curves representing Total Hydrocarbon (TH) and select constituent expected recoveries from a typical gasoline spill in the subsurface. These probability curves were generated by soil textural class for four (4) different soil types, varying from sand through sandy loam and loamy sand to loam. The design engineer need only know the soil texture at the spill location to access these curves, generating a prediction of the overall probability of success (defined by percent recovery of the TH and individual constituents). Additional disaggregation of the data sets was accomplished allowing a comparison of the effects of spill size, contaminated zone thickness and extraction vacuum for each of the performance criteria. Figure 1 presents the overall research structure used in this effort.

Monte Carlo Simulation

The basic Monte Carlo approach involves repeatedly solving a deterministic or single-value problem using inputs generated randomly from specific probability distributions. The techniques used in this study to generate the normally distributed soil properties were developed by Agricultural Research Service (ARS) and Environmental Protection Agency (EPA) personnel for other applications (Rawls and Brankensiek, 1985 and Carsel and Parrish, 1988). Based on soil textural class, they employed multivariant regression to develop normal distributions of correlated soil properties necessary to calculate soil air permeability. These properties included residual water saturation, van Genuchten water retention parameters and saturated hydraulic conductivity (van Genuchten et al., 1991). The saturated hydraulic water contents thus generated were employed as indicated by equation (3) while the van Genuchten parameters and the residual water saturation were used to calculate water contents. Water contents changed with each simulation of permeability and directly impacted soil air permeability according to equation (4).

Monte Carlo methods were utilized to repeatedly generate a statistically significant number of correlated sets of soil properties. Each set of randomly generated soil properties represented one realization of air permeability based on the described relationships. Soil air permeability realizations were input into an available deterministic or single value venting code and sequential simulations were

Figure 1. Flow chart showing research structure and overall range of spill site variables.

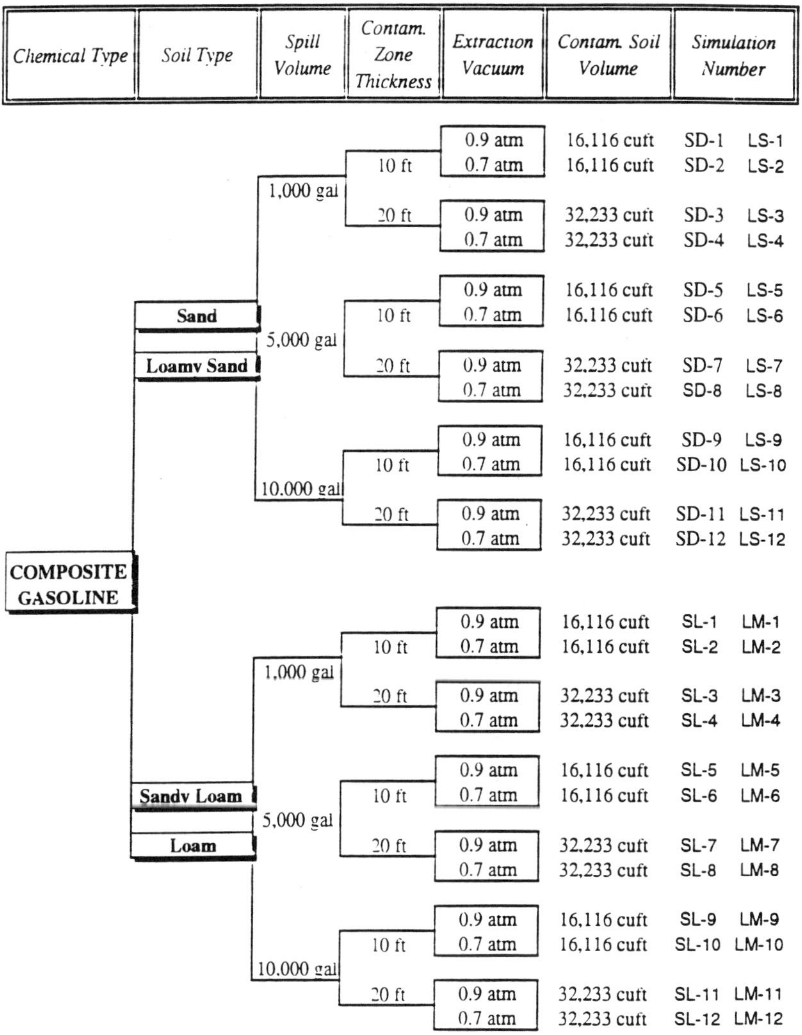

Chemical Type	*Soil Type*	*Spill Volume*	*Contam. Zone Thickness*	*Extraction Vacuum*	*Contam. Soil Volume*	*Simulation Number*	
COMPOSITE GASOLINE	**Sand** / **Loamy Sand**	1,000 gal	10 ft	0.9 atm	16,116 cuft	SD-1	LS-1
				0.7 atm	16,116 cuft	SD-2	LS-2
			20 ft	0.9 atm	32,233 cuft	SD-3	LS-3
				0.7 atm	32,233 cuft	SD-4	LS-4
		5,000 gal	10 ft	0.9 atm	16,116 cuft	SD-5	LS-5
				0.7 atm	16,116 cuft	SD-6	LS-6
			20 ft	0.9 atm	32,233 cuft	SD-7	LS-7
				0.7 atm	32,233 cuft	SD-8	LS-8
		10,000 gal	10 ft	0.9 atm	16,116 cuft	SD-9	LS-9
				0.7 atm	16,116 cuft	SD-10	LS-10
			20 ft	0.9 atm	32,233 cuft	SD-11	LS-11
				0.7 atm	32,233 cuft	SD-12	LS-12
	Sandy Loam / **Loam**	1,000 gal	10 ft	0.9 atm	16,116 cuft	SL-1	LM-1
				0.7 atm	16,116 cuft	SL-2	LM-2
			20 ft	0.9 atm	32,233 cuft	SL-3	LM-3
				0.7 atm	32,233 cuft	SL-4	LM-4
		5,000 gal	10 ft	0.9 atm	16,116 cuft	SL-5	LM-5
				0.7 atm	16,116 cuft	SL-6	LM-6
			20 ft	0.9 atm	32,233 cuft	SL-7	LM-7
				0.7 atm	32,233 cuft	SL-8	LM-8
		10,000 gal	10 ft	0.9 atm	16,116 cuft	SL-9	LM-9
				0.7 atm	16,116 cuft	SL-10	LM-10
			20 ft	0.9 atm	32,233 cuft	SL-11	LM-11
				0.7 atm	32,233 cuft	SL-12	LM-12

SD - sand
SL - sandy loam
LS - loamy sand
LM - loam

completed. The resultant outputs were pooled into probability density functions and plotted.

Composite Gasoline/Spill Constituents

Gasolines, fresh and weathered, vary widely in their major chemical components. For this effort a composite gasoline was employed consisting of benzene, toluene, ethylbenzene, MTBE, naphthalene, N-hexane and M-, O-, and P-xylenes as well as a "light and "heavy" end of unidentified hydrocarbons. The relative percentages of each were determined by comparisons with natural and other simulated gasolines as well as with simulated SVE performance data compiled preliminary to this effort. That is, the composite gasoline developed behaved similarly when modeled to larger, more complex mixtures reported in the literature (EPA 1988 and 1991,; Hartley and Ohanian 1990; Johnson et al., 1990; Weaver 1992, and Potter, 1990).

Results:

Figures 2-5 present probability plots of residual soil hydrocarbon concentrations remaining after two years of venting for sand, loamy sand, sandy loam and loam respectively. Three spill sizes of 1,000, 5,000 and 10,000 gallons respectively are shown on each plot as is a horizontal line drawn at 1,000 mg/kg of TH residual in the soil. This is a reference to a rather liberal treatment standard and is included only for discussion. It can be seen that this reference level is reached approximately 45% of the time in sand but only 12%, 5% and 0% of the time in the other soil types respectively. Extending venting times to 5 years increased the efficiencies to 45% and 15% for the loamy sand and sandy loam but exhibited no significant increase for the loam. These latter data are not included in this presentation for the sake of brevity but are available from the authors. Other data, also available include similar plots for the BETX plus series previously described and a series of analyses describing the probabilities of contaminant removal given the effects of the control or operational variables presented in Figure 1.

Summary

A technique linking Monte Carlo simulation of critical soil properties with deterministic modeling of soil vapor extraction systems for a series of process alternatives was accomplished. This produced a probability based nomograph available to design engineers capable of defining the chance of contaminant recovery from variously configured well systems.

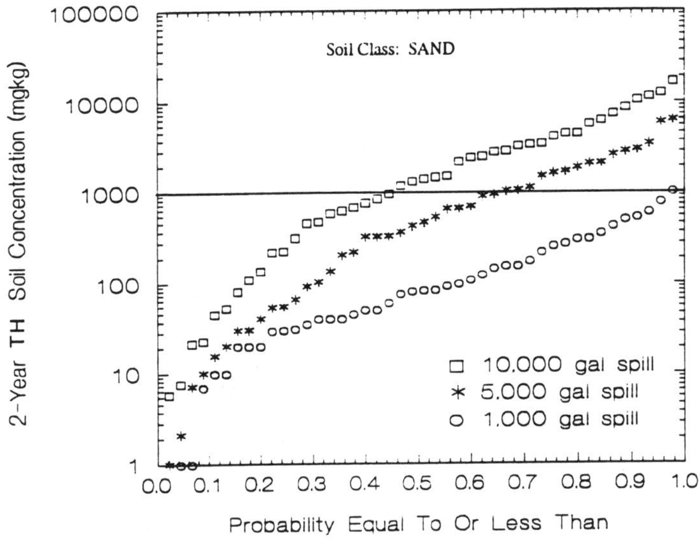

Figure 2. Probability of expected TH soil concentration after a two year venting period. Plot shows results of all sand soil simulations.

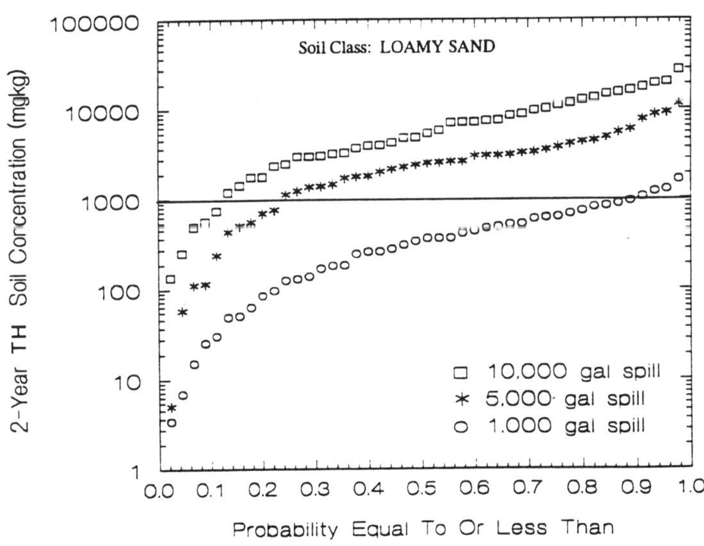

Figure 3. Probability of expected TH soil concentration after a two year venting period. Plot shows results of all loamy sand soil simulations.

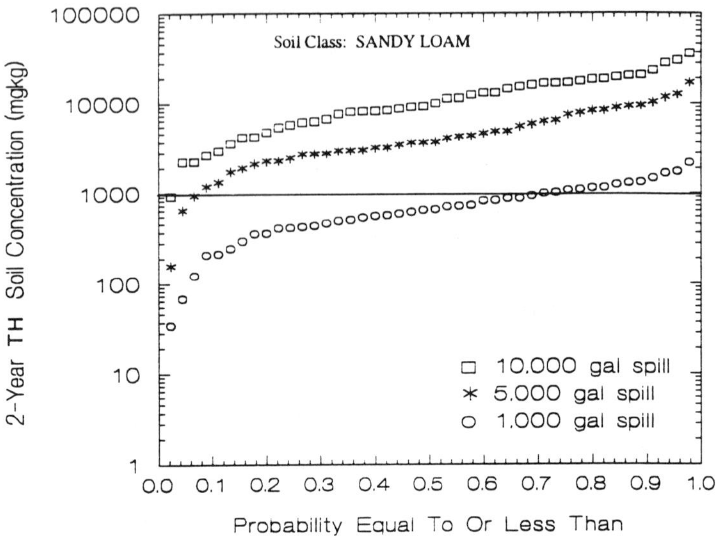

Figure 4. Probability of expected TH soil concentration after a two year venting period. Plot shows results of all sandy loam soil simulations.

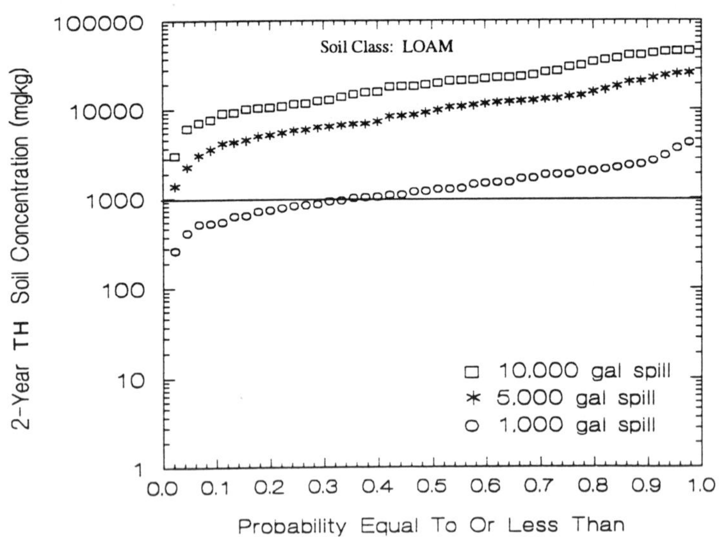

Figure 5. Probability of expected TH soil concentration after a two year venting period. Plot shows results of all loam soil simulations.

Selected References

Carsel, R.F. and Parrish, R.S. 1988. "Developing Distributions of Soil Water Retention Characteristics." Water Resources Research. 24:5:755-769.

Hartley, W.R. and Ohanian, E.V. 1990. "A Toxilogical Assessment of Unleaded Gasoline Contamination of Drinking Water." In: Petroleum Contaminated Soils. Volume 3. Lewis Publishers, Inc. pp. 327-340.

Johnson, P.C., Kemblowski, M.W., Colthart, J.D., 1990. "Quantitative Analysis for the Cleanup of Hydrocarbon-Contaminated Soils by In-Situ Soil Venting." Ground Water. 28:3:413-429.

Potter, T.L. 1990. "Fingerprinting Petroleum Products: Unleaded Gasolines." In: Petroleum Contaminated Soils. Volume 3. Lewis Publishers, Inc. pp. 83-92.

Rawls, W.J. and Brankensiek, D.L. 1985. "Prediction of Soil Water Properties for Hydrologic Modelling." In: Proceedings of Symposium on Watershed Management. American Society of Civil Engineers. New York, New York. pp. 293-299.

U.S. Environmental Protection Agency. 1988. Cleanup of Releases from Petroleum UST's: Selected Technologies. Office of Underground Storage Tanks. EPA/530/UST/-88/001. Washington, D.C.

U.S. Environmental Protection Agency. 1991. Guide for Conducting Treatability Studies Under CERCLA: Soil Vapor Extraction. Office of Emergency and Remedial Response. EPA/540/2-91/091A. Washington, D.C.

van Genuchten, M.Th., Leij, F.J. and Yates, S.R. 1991. The RETC Code for Quantifying the Hydraulic Functions of Unsaturated Soils. United States Environmental Protection Agency, RSKERL. EPA/600/2-91/065. Ada, OK.

Weaver, N.K., 1992. "Gasoline." Chapter 7 In: Hazardous Materials Toxicology: Clinical Principals of Environmental Health. Williams and Wilkins. Baltimore, MD.

In Situ Treatment of VOCs by Recirculation Technologies

O. F. Webb[1], R. L. Siegrist[2], M. R. Ally[3], W. E. Sanford[2], P. M. Kearl[4], and J. L. Zutman[4]

Abstract

Confronted with contaminated land from the world wars and the postwar industrialization period, German researchers and practicing professionals have worked to develop processes for effective environmental restoration. This presentation documents efforts by Oak Ridge National Laboratory (ORNL) researchers to (1) identify collaborators and German technologies exhibiting near-term potential for clean-up of volatile organic contaminated soil and groundwater at Department of Energy sites, (2) critically assess performance, and (3) inform interested agencies. The project was limited to identification and preliminary evaluation and included engineering computations, groundwater flow modeling, and treatment process modeling. Two processes were identified: (1) the vacuum vaporizer well/groundwater recirculation well (German: Unterdruck-Verdampfer-Brunnen/Grundwasser-Zirkulations-Brunnen, or UVB/GZB) and (2) the porous pipe/horizontal well (PP/HW). Both technologies induce a recirculation flow field in the aquifer and enable simultaneous down hole treatment of the aquifer and vadose zone. University of Karlsrue (Germany) researchers have demonstrated the UVB/GZB technology in shallow aquifers with moderately high saturated thickness and hydraulic conductivities. The PP/HW technology offers potential for VOC treatment in sites with thin aquifers or heterogeneities. This paper describes identified German technologies and includes critical evaluations of well performance, associated treatment processes, operating variables, and aquifer-well interactions.

---

[1] Oak Ridge National Laboratory, Chemical Technology Division, PO Box 2008 MS-6226, Oak Ridge, TN 37831-6038

[2]ORNL, Environmental Science Division, Oak Ridge, TN 37831-6038

[3]ORNL, Chemical Technology Division, Oak Ridge, TN 37831-6038

[4]Oak Ridge National Laboratory, Environmental Science Division, Grand Junction, CO 81502-2567

Introduction

Considerable research and development in environmental restoration and waste management have occurred in several foreign countries, most notably the former West Germany. Recognizing the potential gain from international collaboration, an International Technology Exchange Program (ITEP) was initiated within the U. S. Department of Energy (DOE) Office of Technology Development. Within ITEP it was envisioned that promising technologies would be transferred to an Integrated Program or Integrated Demonstration for further research and development. This paper describes German processes and technologies identified by Oak Ridge National Laboratory (ORNL) researchers that appear to have near-term potential for enhancing volatile organic compound (VOC) cleanup. In accordance with the intent of ITEP, and with DOE guidance, the project scope was limited to identification and preliminary evaluation of technologies that have near-term potential for improving DOE environmental restoration. Subsequent analyses consisted of traditional engineering analysis and groundwater flow modeling.

Materials & Methods

The project approach was to (1) identify potential collaborators and technologies; (2) investigate identified technologies to determine strengths, weaknesses, and benefits; and (3) inform interested agencies. Collaborator and technology identification occurred principally through written and telephone conversations between American and German researchers. Promising technologies were evaluated by a site visit, traditional engineering analysis, and a specially designed three dimensional groundwater flow subsurface model. Because the DOE complex represents a wide spectrum of conditions, contaminants, and regulatory environments, representatives from potential DOE sites were also involved. For the site visit, a team was selected based on their areas of expertise and their representation of sites where laboratory and field testing could readily occur. The agenda for the site visit included technical presentations, discussions, field test site and laboratory tours at University of Karlsruhe (UoK) (the German collaborating university), and visits to private industry vendors and their field sites.

Preliminary Findings

Three technologies were identified through the preliminary investigation; biological treatment that uses air injection and water infiltration (ODWR) for oxygen and nutrient delivery; specially designed vertical wells that recirculate subsurface fluids termed Unterdruck-Verdampfer-Brunnen (UVB) or Grundwasser Zirkulations Brunnen (GZB) systems; and shallow horizontal drilling systems with porous piping for recirculating subsurface fluids.

The ODWR technology employed a grout curtain extending to bedrock and an overhead enclosure to enable measurement of system flows, temperatures, nutrients, oxygen and carbon dioxide for estimating respiratory quotients and biodegradation rates. Nutrient delivery occurred by injection of air and aqueous solutions into the soil. The technique appeared appropriate for shallow contaminants strongly adsorbed to soils. The technology was not considered in detail because it appeared to demonstrate only limited applicability for DOE (more complex hydrology and deeper contamination are common at DOE sites).

The UVB/GZB technology depicted in Fig. 1 with an idealized capture zone demonstrated significant potential. The UVB/GZB consists of a vertical well with screened sections below and above the water table, a treatment section, and an above ground air blower. These vertical wells use an airlift effect (UVB) or electric pump (GZB) to remove groundwater through the bottom section, move it vertically, and reinject it. The upper and lower screened sections can be in different layers but the layers need to be hydraulically connected to form a circulation cell. The system has proven application for inducing recirculation in relatively homogenous aquifers and under ideal conditions can encompass the entire aquifer thickness. A blower, mounted on the outlet, moves air from the ground surface, through the well, and back to the well head. A separation plate prevents mixing of treated with entering

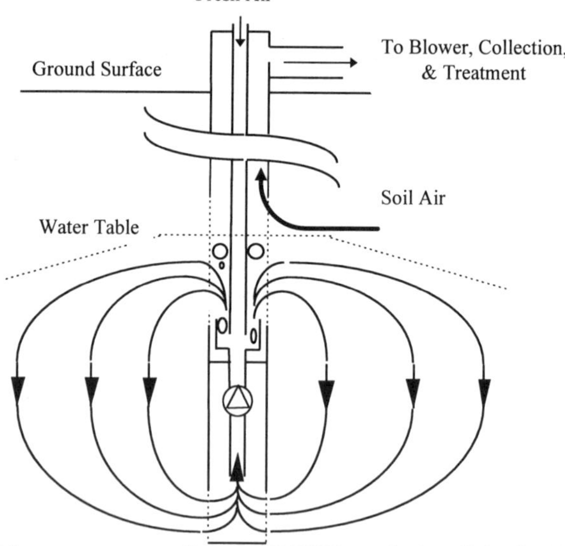

Fig. 1. Major components of the UVB/GZB vertical well for in situ recirculation and treatment of groundwater. (ORNL Drawing 93-10197)

water. Although less efficient than a countercurrent stripper (Treybal, 1980), this cocurrent design permits the advantages of a simple well design-only a blower for operation and well housing leaks have less potential of reducing effectiveness (Siegrist et al., 1993). Soil air receives treatment by entering the UVB/GZB at the second screened section due to vacuum developed by the blower. The gaseous waste stream is treated above ground prior to release (Herrling et al. 1991). No water is brought to the surface, thereby eliminating some problems associated with treatment, disposal, or reinjection. The technology also has potential for inducing strong vertical and horizontal gradients for aiding removal of nonaqueous phase liquids (NAPLs).

The porous piping horizontal well (PP/HW) system consists of horizontal well pairs (for withdrawal and injection of subsurface fluids) connected to a caisson that houses treatment and pumping equipment. Groundwater (and/or soil air) enters one well, passes through the caisson connecting the wells, and then is reinjected into the aquifer through a second well. KSK Microtunneling Technologies in Rastatt (manufacturer) and FlowTex Directional Drilling (installs the horizontal wells) have demonstrated soil vapor extraction of jet fuel at Frankfort airport using a horizontal, radial system. The KSK technology was uncomplicated and attractive because it was up-scaled from the utility industry, rather than down-scaled from the oil industry. The German drilling technique was demonstrated in fine soils, (rock and gravel may cause problems). KSK/FlowTex demonstrated horizontal well installation rates of 80 to 120 m/d at a cost of about $100 per meter (10 m depth). American companies manufacture similar equipment. Schumacher Environmental in Crailsheim has developed a porous plastic pipe (pore sizes of 10 to 100μm). UoK Institute of Applied Geology, Schumacher, and KSK/FlowTex are involved in a joint project for application of porous piping installed with horizontal drilling techniques for soil vapor extraction.

Comparison of German Recirculation Processes with Baseline Technologies

Conventional pump-and-treat (P&T) technologies are widely used for remediating groundwater (Palmer and Fish 1992). In this process groundwater is extracted, treated above ground, and disposed of and/or reinjected upgradient. Keely (1989) and Mercer et al. (1990) pointed out P&T limitations resulting from subsurface complexity and behavior of contaminants such as NAPLs. Travis and Doty (1990) cite 19 sites where P&T has continued for up to ten years and conclude that little success was achieved in reducing concentrations to target levels. Some authors participating in a recent National Research Council report suggest that restoration quality might not be achieved in less than 100 to 200 years (Abelson 1990). Currently, conventional P&T technology is widely recognized as plume control technology (Nyer 1990) but not as source reduction technology. Systems that reduce operating costs or provide some source treatment have an advantage over conventional P&T. Lepore et al. (1990), Palmer and Fish (1992), and Russell et al. (1992) and others discuss emerging technologies for P&T enhancement; however, these technologies require further research and field application prior to broad acceptance. Operating and capital costs are discussed by Thomsen et al. (1989) and Doherty (1992). For a P&T system installed for the Michigan Department of Natural Resources, capital costs of $3,000,000 and operating and maintenance costs of $145,000/year were incurred. Operation for 20 to 30 years will incur additional costs.

The UVB/GZB and the PP/HW technologies appear to offer the greatest near-term benefits due to potential technical and institutional drivers. Moving a gas from the deep subsurface poses fewer technical problems than moving groundwater. The recirculation technologies are capable of treating two soil phases simultaneously, whereas traditional technologies treat single phases. Although designed for stripping of VOCs, their use may support other treatments, such as down hole chemical oxidation or bioremediation. The UVB/GZB is well demonstrated at full-scale for air stripping of VOCs (trichloroethylene, trichloroethane, perchloroethylene, benzene, toluene, and xylene) in Germany (Herrling et al., 1991) and the process performance is reported to be good. Capital and operating costs are reported to be significantly less than for traditional methods. The PP/HW technology lacks significant testing. The well configurations of the PP/HW may more significantly affect circulation geometry than the UVB/GZB. Also, short-circuiting problems (between injection and extraction wells) are expected to be less significant with the PH/HW system. Extensive field testing in different hydrogeological settings, computer simulations, and detailed cost comparisons to other methods need to occur before these recirculation technologies can be implemented on a broad scale.

A unique application of horizontal recirculation wells at the Savannah River Site is described by Kaback 1989. The system consists of two horizontal wells completed in the contaminated aquifer and in the overlying vadose zone. Air and methane are injected into the aquifer well to strip VOCs and enhance microbial degradation. The vadose zone well collects the vapors. Initial reports indicate good VOC removal, but do not indicate system efficiency.

Although limited applications have occurred in groundwater programs, horizontal wells are used extensively in oil fields. Gilman and Jargon (1992) report that horizontal wells accelerated recovery in homogeneous, moderate permeability reservoirs (3 times greater) over vertical installations. For more permeable reservoirs, horizontal wells accelerate production with only slightly higher ultimate recoveries. Taber and Seright (1992) concluded that combinations of horizontal injectors and producers increased production rates by as much as 10 times and improve areal sweep efficiencies.

Herrling et al. (1991) published nomographs for determining influence radius and well spacing. The capture zone geometry of the recirculation systems depend upon many factors including pump rates, screen placement, aquifer thickness and permeability, confining zones, vertical/horizontal conductivity, and groundwater flow. Model assumptions include dominant convective transport; top and bottom well screens separated by one quarter of the aquifer thickness; radial symmetry in an homogeneous, anisotropic, confined aquifer; and horizontal to vertical hydraulic conductivity ratios of 1, 5, and 10. Groundwater modeling of the UVB/GZB supported the nomographs. The UVB/GZB has been applied in fairly homogeneous aquifers with horizontal hydraulic conductivity's of 10^{-1} to 10^{-3} cm/s. Little information exists on system effectiveness in varied hydrologic conditions with aquifer heterogeneity.

An analysis using published nonmographs (Herrling et al., 1991) leads to several conclusions. In general, well spacing at a constant pump rate is insensitive to aquifer thickness, except for small values of the dimensionless parameter. For highly permeable thick aquifers, low pumping rates require close well spacing (little circulation is developed because the screened section influence one another very little). Conversely, low permeability aquifers yield large well spacing but require very slow pump rates. The analysis suggests that the technique does not work well in thin aquifers. The technique appears best suited for aquifers about 10 m thick with fine to coarse sand materials. Groundwater modeling results indicates that capture zone width increases with aquifer thickness, mean hydraulic conductivities, circulation rates.

Chemical and biological changes caused by treatment may promote fouling of equipment, the well-aquifer interface, and sediments, adversely impacting operation. Precipitate formation is a function of available reactants, pH, and redox conditions. If ferrous ion is present, increases in pH and Eh may shift iron to an insoluble state. For example, addition of oxygen might drive the system to higher Eh values causing iron to precipitate. Carbon dioxide removal (from stripping) may also precipitate iron. Options are available for mitigating these effects gases (Siegrist et al., 1993). For example, recycling stripping offgas would reduce removal or addition of dissolved subsurface gases.

CONCLUSIONS

This presentation documents efforts by ORNL researchers to (1) identify collaborators and German technologies exhibiting near-term potential for VOC clean-up at DOE sites, (2) critically assess performance, and (3) inform interested agencies. Based on the analysis the following conclusions have been drawn: (1) Two German technologies, the UVB/GZB and PP/HW technologies; (2) The UVB/GZB technology has been researched and demonstrated in shallow aquifers of amenable hydraulic properties (i.e., thick, homogeneous, and conductive); (3) Application of the PP/HW technology exhibits significant potential but requires significant research and demonstration prior to broad acceptance; (4) Available nomographs predict UVB/GZB capture zone geometries; however, independent flow and transport models are need on a site by site basis; (5) the potential to couple stripping with other in situ technologies (e.g., advanced oxidation) appears promising; (6) precipitates from redox, pH, and biological changes may foul the well. A site by site analysis is needed to establish rate and impact of fouling. Gas recycling remains an alternative.

Further research and demonstration are required to fully develop these in situ recirculation technologies and document benefits. Pertinent issues for laboratory and field tests include: (1) verification of recirculation geometries and sensitivities to aquifer properties, (2) fouling of wells, (3) evaluation of VOC treatments, (4) development of alternative treatments, and (5) direct comparison with conventional systems.

ACKNOWLEDGMENTS

The International Technology Exchange Program within the US Department of Energy provided project sponsorship and oversight under contract DE-AC05-84OR21400 with Martin Marietta Energy Systems, Inc. This research also supported in part by an appointment to ORNL Postdoctoral Research Associate Program administered jointly by ORNL and the Oak Ridge Institute for Science and Education.

APPENDIX. REFERENCES

Abelson, P. H. 1990. Inefficient remediation of groundwater pollution. *Science*. ed., p. 733.

Doherty, R. E. 1992. *Pollution Engineering*, p. 61-64.

Gilman, J. R. and J. R. Jargon. 1992. *World Oil*. p. 55-60.

Herrling, B., W. Beurmann, J. Stamm, and M. Schoen "UVB/GZB Technique for In Situ Groundwater Remediation of Strippable Contaminants: Operation and Dimensioning of Wells", *Envirotech Vienna 1990*, W. Pillmann and K. Zirm, Eds., p.631, 1990.

Herrling, B., W. Beurmann, and J. Stamm, "Hydraulic Circulation System for In Situ Bioreclamation and/or In Situ Remediation of Strippable Contamination", *In situ Bioreclamation, Applications and Investigations for Hydrocarbon and Contaminated Site Remediation*, R. E. Hinchee and R. F. Olfenbuttel, Eds., p.173, 1991.

Herrling, B., W. Beurmann, and J. Stamm, "In Situ Groundwater Remediation of Strippable or Volatile Contamination Using the UVB-Method", *Proceedings of the European Conference Advances in Water Resources Technology*, G. T. Tsakiris, Ed., p.315, 1991.

Herrling, B., J. Stamm, E. J. Alesi, P. Brinnel, G. F. Hirshberger, and M. R. Sick. 1991. *In situ Groundwater Remediation of Strippable Contaminants by Vacuum Vaporizer Wells (UVB): Operation of the Well and Report About Cleaned Industrial Sites,* forum on Innovative Hazardous Waste Treatment Technologies, U. S. EPA, Dallas, Texas.

Kaback, D. S., B. B. Looney, J. C. Corey, L. M. Wright, III, and J. L. Steele, 1989. Horizontal Wells for In-Situ Remediation of Groundwater and Soils. *National Water Well Assoc.*, E. I. du Pont de Nemours & Co., Savannah River Lab., Aiken, S. C., and Sirrine Environmental Consultants, Greenville, S. C.

Keely, J. F. 1989. *Performance Evaluations of Pump-and-Treat Remediations.* EPA/540/4-89/005, U. S. Environmental Protection Agency. October.

Lepore, J. V., L. H. Turner, D. S. Kosson, and R. C. Ahlert. 1990. *Journal of Hazardous Materials* 25:289-307.

Mercer, J. W., D. C. Skipp, and D. Giffin. 1990. "Basics of Pump-and-Treat Ground-Water Remediation Technology", EPA-600/8-90/003, U. S. Environmental Protection Agency.

Nyer, E. 1990. *Groundwater Monitoring Review*, p. 70-72.

Palmer, C. D. and W. Fish. 1992. *Chemical Enhancements to Pump-and-Treat Remediation.* EPA/540/S-92/001, U. S. Environmental Protection Agency. January.

Russell, H. H., J. E. Mathews, and G. W. Sewell. 1992. *TCE Removal from Contaminated Soil and Groundwater.* EPA/540/S-92/002, U. S. Environmental Protection Agency. January.

Siegrist, R. L., O. F. Webb, M. R. Ally, W. E. Sanford, P. M. Kearl, J. L. Zutman, *In Situ Treatment of VOCs By Recirculation* Technologies, ORNL/TM-12317, June, 1993.

Taber and Seright. 1992. Horizontal Wells Inject New Life Into Mature Field. *Petroleum Engineer International*. p. 49-50.

Thomsen, K. O., M. A. Chaudhry, K. Dovantzis, and R. R. Riesing. 1989. *Groundwater Monitoring Review*, p. 92-99.

Travis, C. C. and C. B. Doty. 1990. *Science* 250(4982):733.

Treybal, R. E., Mass Transfer Operations, McGraw-Hill, 1980.

Emissions, Ambient Concentrations, and Potential Health Hazards of VOCs from Wastewater Treatment

Jin-Sheng Lin [1]
Lynn M. Hildemann [2] *(A.M., ASCE)*

Introduction

Traditionally, the design and operation of wastewater treatment plants have focused on the removal of conventional pollutants and toxic chemicals from the liquid stream. The emission of VOCs (volatile organic compounds) into the atmosphere has only recently been identified as a significant pollution source and thus has attracted increasing attention. The airborne VOCs not only pose potential heath hazards to the onsite workers and nearby residents, but also act as possible precursors for urban ozone formation. A few states (California and Massachusetts, for example) now require that wastewater treatment facilities (WWTFs) monitor and prepare VOC emission inventories.

The fate of contaminants in wastewater treatment facilities is influenced by biodegradation, adsorption, chemical reaction and volatilization processes. Emissions via volatilization occur in almost every stage of the treatment process. For example, volatilization is the dominant loss pathway for VOCs in primary treatment and also a major removal mechanism, in addition to biodegradation, for chlorinated VOCs in activated sludge systems (Mihelcic *et al.*, 1993). Although the most reliable information on how VOCs are transported and transformed during treatment processes is through field sampling and measurements, mathematical models are often much cheaper and also can be effective. A number of steady-state models have been developed for predicting the general fate of VOCs in liquid streams. What is lacking is an accurate dispersion model that can link emissions and dispersion, and is able to predict ambient VOC concentrations, thereby allowing an assessment of the human health impact.

---

[1] *Doctoral Student, and*
[2] *Assistant Professor, Civil Engineering Department, Stanford University, Stanford, CA 94305-4020*

To determine the extent to which compounds diffuse and are diluted in the atmosphere after being emitted from a pollution source, Gaussian-based plume models are usually utilized. The Gaussian model neglects the variations of wind speed and turbulent eddies with height, as well as the roughness of the downwind terrain and the dry deposition of pollutant components to the ground. The error due to the combined effects of all these oversimplifications tends to be greatest when the sources and receptor sites are close to the ground.

The purpose of this publication is to present a methodology for evaluating the potential health effects of the emissions from wastewater treatment facilities (aeration basins, in particular) on humans, a near-groundlevel "receptor". The emissions of a few selected chlorinated compounds from a typical basin are first estimated using an emission model (Chrysikopoulos *et al.* 1992a). The emission rate plus the prevailing meteorological conditions are then input into a newly-developed dispersion model to predict downwind concentrations. Finally, the degree of exposure level and health risks are assessed by comparing the model predictions with concentration levels for which health effects are known. The method presented herein is suitable for an environmental impact assessment of wastewater treatment facilities.

Emission Model

The predictive model presented in Chrysikopoulos *et al.* (1992a) is used in this paper for illustrative purposes. This model estimates emissions from aeration basins using liquid-phase concentrations and other easily-obtained operational data. It is based on two-film resistance theory, assuming that, for volatile compounds, the resistance in the liquid phase governs the rate of interphase mass transfer. Adsorption and biodegradation are not considered in the model. The mass transfer rate constant of a VOC is estimated from that of oxygen, which in turn is calculated from agitator operational data (brake power and oxygen transfer rate).

Dispersion Model

The dispersion model presented in Chrysikopoulos *et al.* (1992b) is linked with the above-mentioned emission model to predict ambient concentrations. The special feature of this model is the more accurate representation of dispersion phenomena within the atmospheric boundary layer, including realistic profiles of wind speed and eddy diffusivities, the surface roughness of the terrain, and dry deposition as a sink mechanism. The model can predict concentrations downwind of an area-shaped source, such as an aeration basin.

Potential Health Hazard

The predicted ambient concentrations of a few selected chlorinated compounds are compared with two concentration levels to evaluate the potential health hazards due to the emissions. The Carcinogenicity Risk Concentration (CRC) is the concentration that would likely result in a 10^{-6} probability of developing cancer in a person's lifetime. The Chronic Inhalation Reference Concentration (RfC), on the other hand, is the concentration that is likely to have no appreciable risk of deleterious health effect during one's lifetime. Readers interested in where/how to obtain and estimate these two concentrations are referred to Eklund *et al.* (1991).

Application and Discussion

For illustration, we consider a typical aeration basin located upwind of a terrain containing root crops (surface roughness $q_{z_0} = 0.1$ m). The volume of the basin is $V = W \times L \times H = 20 \times 40 \times 4 = 3,200$ m^3. Oxygen is supplied into the tank by a 30 kW surface aerator with oxygen transfer rate of 2.13 kg$_{O_2}$/kW hr. The saturation concentration of oxygen at 20°C is 9.17×10^{-3} kg/ m^3. Under these conditions, the mass transfer rate for oxygen $(k_L a_T)_{O_2}$ is approximately 6×10^{-4} s^{-1}. The atmosphere is assumed to be stable (stability class F, a worst case scenario) with wind speed $\overline{U_X}(q_z^*) = 1.5$ m/s and vertical diffusivity $K_{ZZ}(q_z^*) = 0.025$ m^2/s at a reference height $q_z^* = 10$ m.

The ambient concentration C(x) downwind of this hypothetical aeration basin under the stated conditions can thus be obtained using the emission and dispersion models, provided that the Henry's law constant for the compound of interest is known. Figure 1 shows how the **dispersion factors**, defined as the concentration ratio of air to liquid ($C(x)/C\ell$), vary with downwind distance for four chlorinated compounds, dichlorodifluoromethane (CCl_2F_2), carbon tetrachloride (CCl_4), trichloroethylene ($CHCl{=}CCl_2$), and chloroform ($CHCl_3$). Dichlorodifluoromethane has the highest dispersion factors, due to its high Henry's law constant. The most highly impacted area occurs at about 100 m downwind.

For example, if a compound's concentration in the aeration basin, $C\ell$, is 0.01 mg/l, its ambient concentration at a specific distance downwind is simply the product of the dispersion factor and 0.01. Table 1 tabulates the maximum ambient concentrations of the compounds considered and compares them with CRC and RfC values. Since the maximum exposure levels exceed RfC values, this indicates that, under the worst case scenario, a certain health risk is posed by this aeration basin.

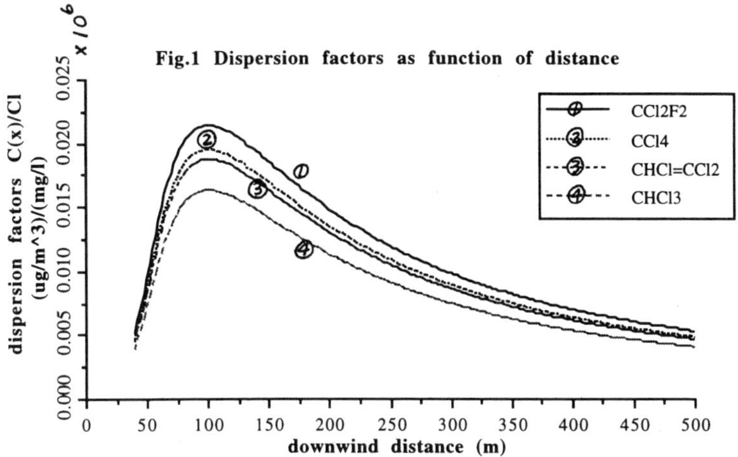

Table 1. Estimated Maximum Ambient Concentrations and Potential Health Hazards ($C\ell$ = 0.01 mg/l)

Compound	$H_C^{\ddagger}$ (-)	$C_{max}(x)$ ($\mu g/m^3$)	CRC ($\mu g/m^3$)	RfC ($\mu g/m^3$)
CCl_2F_2	9.30	213	–	200
CCl_4	0.94	188	0.067	$2^{\dagger}$
$CHCl{=}CCl_2$	0.42	187	0.590	–
$CHCl_3$	0.24	163	0.043	$40^{\dagger}$

Notations: H_C : Henry's law constant @293° K
$C_{max}(x)$: Maximum ambient concentration (x ≈ 100 m)
CRC : Carcinogenicity Risk Concentration
RfC : Chronic Inhalation Reference Concentration
‡ : from Roberts *et al.* (1984)
† : no RfC is available, estimated from Chronic Oral Reference Dose (RfD) (Eklund *et al.*, 1991)

Conclusions

A method is presented to evaluate the potential health effects due to emissions from wastewater treatment facilities. The ambient concentrations, and hence, the health risks involved, can be estimated from liquid-phase concentrations and other easily-obtained data. The method is suitable for environmental impact assessment of wastewater treatment facilities.

Acknowledgements

Funding for this study was provided by the U.S. Department of Energy through the U.S. Environmental Protection Agency-supported Western Region Hazardous Substance Research Center, under agreement R-815738-01. The contents of this publication do not necessarily represent the views of these organizations.

References

Chrysikopoulos, C. V., Hildemann, L. M., and Roberts, P. V. (1992a). "Modeling the Emission and Dispersion of Volatile Organics from Surface Aeration Wastewater Treatment Facilities," *Water Research*, Vol. 26, No. 8, pp. 1045 – 1052.

Chrysikopoulos, C. V., Hildemann, L. M., and Roberts, P. V. (1992b). "A Three Dimensional Atmospheric Dispersion-Deposition Model for Emissions from a Ground-Level Area Source," *Atmospheric Environment*, Vol. 6A, pp. 747 – 757.

Eklund, B., Smith, S., Durham, J. F., and Touma, J. S. (1991). "Estimation of Emissions, Ambient Air Concentrations, and Health Effects from Air Stripping of Contaminanted Water," presented at 84th Annual Meeting and Exhibition in Vancouver, British Columbia, published by the *Air and Waste Management Association.*

Mihelcic, J. R, Baillod, C. R, Crittenden, J. C., and Rogers, T. N. (1993). "Estimation of VOC Emissions from Wastewater Facilities by Volatilization and Stripping," *Air and Waste*, Vol. 43 , pp. 97 – 104.

Roberts, P. V, Munz, C., and Dändliker, P. (1984). "Modeling Volatile Organic Solute Removal by Surface and Bubble Aeration," *Journal Water Pollution Control Federation*, Vol. 56, No. 2, pp. 157 – 163.

Destruction of Gas Phase TCE Using a Plasma Reactor

Ho-Sik Yoo[1], John N. Veenstra[2],
Ven-Yen Tsai[3], Arland H. Johannes[4]

Abstract:

A plasma reactor was used to treat off-gas contaminated with trichloroethylene. The plasma reactor was powered by an alternating current with frequencies up to 1000 Hz. The study showed that over 95% conversion of gas-phase trichloroethylene can be achieved. An optimum frequency for the alternating current existed for maximum power input and conversion. The optimum frequency was dependent on the reactor geometry and the primary voltage applied. Conversion rate did not decrease for increased inlet concentration.

Introduction

Concern about ground water quality has increased as a result of the numerous times that various organic pollutants have been discovered in ground water in the United States. Since ground water constitutes a large portion of water used in United States, a potential health threat to the public exists.

There are several technologies commonly used to treat ground water contaminated with volatile organic chemicals (VOCs). The study of Hand *et al.* (1989) showed that two methods, granular activated carbon (GAC) adsorption

---

[1]Assistant Professor, Department of Environmental Engineering, College of Engineering, Kyonggi University, Kyonggi-Do, Korea 440-760.
[2]Professor of Civil and Environmental Engineering, Oklahoma State University, Stillwater, OK. 74078.
[3]Research Associate, School of Chemical Engineering, Oklahoma State University, Stillwater, OK. 74078.
[4]Professor of Chemical Engineering, Oklahoma State University, Stillwater, OK. 74078

and packed tower air stripping (AST), are the most viable methods for VOC removal. In the GAC system, however, the adsorbent must be regenerated and the regenerant must be processed to isolate the original adsorbates. These processes increase the total cost of the system. Adsorption typically costs more than an AST, depending on the contaminant (Clark and Eilers, 1982; ESE, Inc., 1985). Therefore, ASTs are often the least expensive and most popular method to remove organics provided the pollutants are volatile. Trichloroethylene (TCE) is a very good candidate for air stripping in this sense. TCE is a common contaminant found in ground water and was used as the target compound in this study. Additional reasons to choose ASTs are: a) easy to operate with minimum skill, b) less operation and management cost than carbon adsorption, and c) aerating to remove one specific contaminant will also reduce concentrations of other VOCs.

With all the advantages listed above, however, air stripping does not permanently remove the VOCs from the environment. It removes the VOCs from the liquid-phase and transfers it to the vapor-phase. So, water pollution is just transferred to air pollution. Regulations concerning air pollution have been becoming more stringent (CAAA, 1990). It is necessary to develop off-gas control devices for ASTs. Gas-phase adsorption onto activated carbon is currently considered the standard technology. However, gas-phase GAC has the same disadvantage as liquid-phase GAC. The development of a method to incinerate and ultimately dispose of the toxic wastes in an air stream is the goal of this research. This research investigates the possibility of using a plasma reactor as the off-gas control technology of an AST.

In general, plasmas can be thought of as an ionized gas mixture. Gases are normally good electrical insulators. Under the influence of an applied electric field of sufficient strength, however, gas molecules can be ionized. Electric conduction then takes place and an electrical discharge occurs. The most familiar type of gas discharge which can occur is a hot localized arc. Under the right current conditions, an incomplete breakdown of the gas can occur and a silent discharge results. This discharge can be maintained and its properties are very different from an arc. It is cooler and more diffuse and actually fills the gap between the electrodes with a soft glow.

Non-isothermal plasma reactors utilize electrical energy to create a relatively low temperature plasma (electric discharge) in a reactor cavity. The ionized species and electrons are accelerated to high speed by electromagnetic energy. When organic material flows into the plasma, their chemical bonds are broken by collision with electrons, causing dissociation to occur. The resulting fragments recombine along the pathways to form simple reaction products.

The use of a plasma reactor to destroy volatile organic compounds has been studied by several investigators. Some studies (Sheinson *et al.*, 1987; Fraser *et al.*, 1985; Piatt, 1988) were done to investigate the destruction of methane in a plasma reactor. Sheinson *et al.*, (1987) showed increasing methane destruction with increasing power input in a plasma reactor. They also found the existence of an optimum frequency for the alternating current that yields maximum power input. Barat and Bozzelli (1989) studied the reactions of chlorinated hydrocarbons including TCE with water vapor or molecular hydrogen in a low-pressure microwave plasma. All these studies have not been intended to treat the off-gases from ASTs.

Method and Materials

The experiment consisted of two phases. In the first phase, TCE destructive tests were done on a bench scale. These tests were designed to determine the destruction efficiency of the plasma reactor. TCE-contaminated gas was generated in a saturation device to simulate the AST off-gas. In addition, electrical characteristics and power requirements of the plasma reactor were estimated. In the second phase, feed gas to the plasma reactor was supplied from a pilot scale AST. In this stage, TCE was not added to the feed water of the AST. Therefore, a TCE-free off-gas was fed to the plasma reactor. The only reason for using the pilot-scale AST was to generate large quantities of off-gas with the appropriate temperature and humidity.

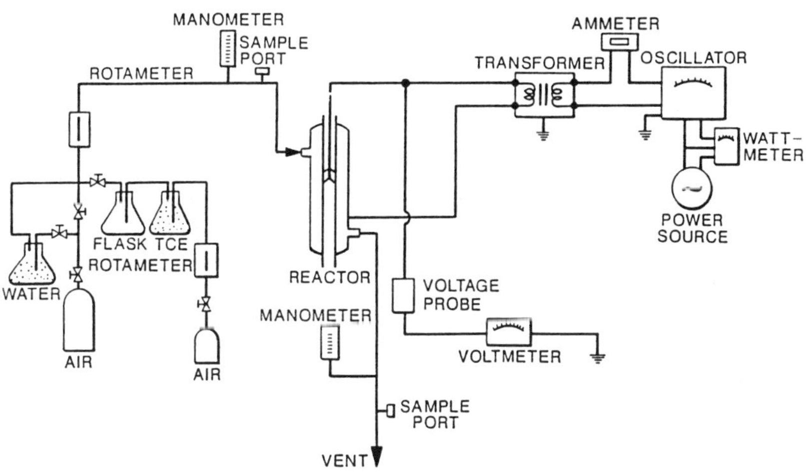

Figure 1. Schematic of Plasma Reactor System

The experimental apparatus used in the first stage is shown in Figure 1. The apparatus includes a power source (California Instruments Model 161T oscillator with a range of 0 to 120 volts and frequency generation of 40 to 5,000 Hz), transformer (Jefferson Electric with a maximum secondary voltage of 15,000 volts), high voltage test probe, ammeter, gas carrier system, reactor, and an outlet gas analysis system. The reactor was constructed using Pyrex glass and consisted of coaxial glass tubes. The inside of the inner tube and outside of the outer tube were coated with inorganic silver paint that served as the electrodes. The length of the reactor was 37.5 cm with a 20.2 cm long effective discharge zone. The outer diameter of the inner tube was 2.0 cm and the inner diameter of the outer tube was 2.64 cm. The gap between inner and outer tube was 0.32 cm. When an electric potential was applied across these electrodes, the glass walls served as a dielectric causing the current to diffuse into a plasma or "glow" in the annulus. Using this reactor design, plasmas can be generated at atmospheric pressure using frequencies below 1000 Hz. For the measurement of the secondary voltage, a Simpson AC high voltage probe was used in conjunction with a Simpson 620 multimeter. The primary power input (total power input to plasma reactor) was measured using a General Electric wattmeter. Gas flow rate to the reactor was measured using a calibrated rotameter. The TCE-contaminated air was prepared by passing air over the surface of pure liquid TCE. This stream was mixed with another air stream in a tee-connection to obtain an approximately constant concentration of TCE in the feed gas. The mixing air was humidified by passing air through a flask which contained water (Figure 1).

In the first phase destructive study (bench scale), destruction efficiency was monitored using gas chromatography (GC) equipped with a flame ionization detector. The GC column was packed with 10% SP-1000 on 80/100 Supelcoport. The gas-phase concentration, ppb or ppm, was expressed on a volume/volume basis and not by a weight/weight basis in this study. The following operating conditions of the GC were used: oven temperature, 75°C; run time, 3 min; injector temperature, 100°C; detector temperature, 250°C; carrier gas (nitrogen), 30 cc/min.

In the second phase, feed gas to the plasma reactor was supplied from a pilot scale AST instead of the TCE-saturation device. The air stripping tower was composed of a glass column with an inside diameter of 7.52 cm and a length of 1.83 m. The tower consists of three sections, a 1.4 m long center piece, open at both ends, and two 0.35 m long end pieces each sealed at one end. The packing height was 1 m with a bed porosity of 0.75. The ceramic packing used in this study was 0.64 cm Intalox Saddles with a packing factor of 725 and a specific surface area of 984 m^2/m^3 (Treybal, 1980). The AST operating condition were: water (TCE-free) flow rate, 1 L/min; water temperature, 22°C; air flow rate, 30 L/min; air temperature, 24°C; pressure drop across the tower, 80 $N/m^2/m$.

In the second phase non-destructive study, a small portion of the off-gas flow from the AST was separated from the main flow using a tee-connection. The amount of the slip stream was adjusted by using a control valve to obtain the desired flow rate. The branch stream was directed to the top port of the reactor and exited from the bottom.

Results and Discussion

Figure 2 shows the TCE conversions as a function of frequency and flow rate of air/TCE mixture. The word, total, is used in this paper concerning power because the net power input to sustain the plasma will be a portion of the total power. The other remaining portion of the total power will be dissipated in other electrical devices such as the frequency generator and transformer. The net power input to the plasma reactor was not measured due to the lack of available measuring devices. The total power input to the plasma reactor is actually what the reactor electrical cost is based on. The frequency of the current was varied from 60 Hz to 1000 Hz in 100 Hz increments. High conversion was maintained and then fell off at 700 Hz. At the frequency range of high conversion, high power input was maintained (105 W - 110 W). Power input decreased to 92.5 W at the frequency of 700 Hz and to 86.5 W at 800 Hz. That is, high conversion occurs in the range of high power input as expected. It is seen that the system can be tuned by varying the frequency to get high power input and high conversion. Volumetric flow rate of the feed gas determines the residence time of the gas in the discharge zone. Conversion decreased as gas flow rate increased from 2910 to 4410 ml/min. About a 3% difference in conversion efficiency was seen between the two gas flow rates at the optimum frequency, 0.5 kHz. At frequencies other than the optimum, a larger difference in conversion efficiency was seen between the two gas flow rates.

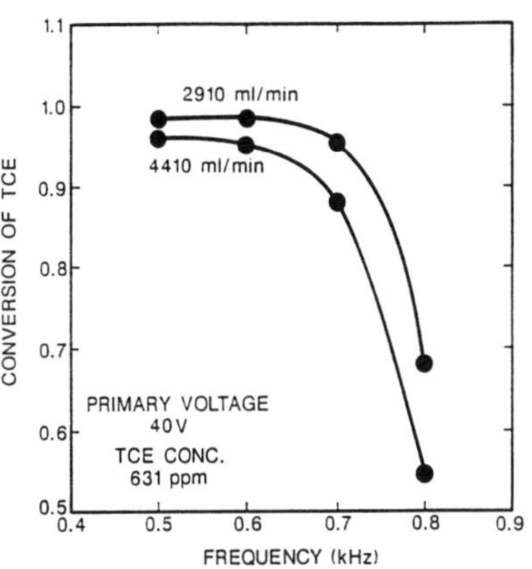

Figure 2. Effect of Flow Rate and Frequency on Conversion of TCE in Dry Air Discharges

Figure 3 shows the variation of conversion with TCE concentration. The percentage removal did not decrease until an extremely high concentration of TCE was introduced into the reactor (15 kppm). Therefore, the plasma reactor will have a definite advantage for a heavily contaminated air stream, over the gas-phase GAC process in which carbon

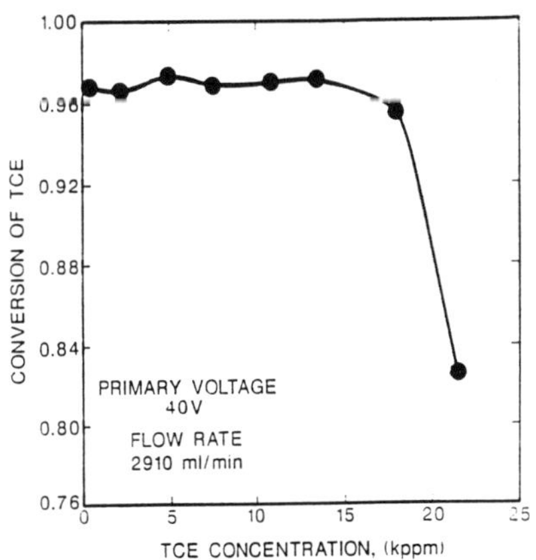

Figure 3. Effect of Concentration on Conversion of TCE in Dry Air Discharges

usage rate will be a function of the contaminant concentration.

The humidification of the source air did not decrease the conversion. In fact, a slightly higher conversion (1%) was measured with humidified gas. The humidity was obtained by bubbling the air through a flask.

The plasma reactor was connected to the AST to investigate the electrical characteristics of the total system with respect to power, frequency, and voltage. A series of breakdown tests with TCE-free air (non-destructive tests) were performed to determine the breakdown voltages and frequencies of the plasma reactor. Here, breakdown means that the flowing gas is ionized so that a plasma is established in the reactor cavity. Breakdown could be easily identified because of an audible noise from the reactor and the sudden increase in power input to the system. In darkness, a blue-colored glow could be seen during breakdown.

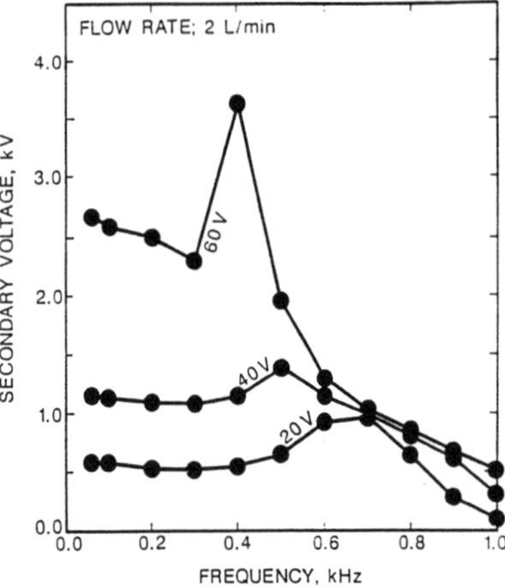

Figure 4. Effect of Frequency and Primary Voltage on Secondary Voltage

Figure 4 shows the variations of secondary voltage with frequency at different primary voltages. The voltage remains fairly constant until a noticeable increase of voltage is observed at breakdown or corona starting frequency followed by a gradual decrease. As the primary voltage increases from 20 V to 60 V, the frequency that draws maximum secondary voltage decreases.

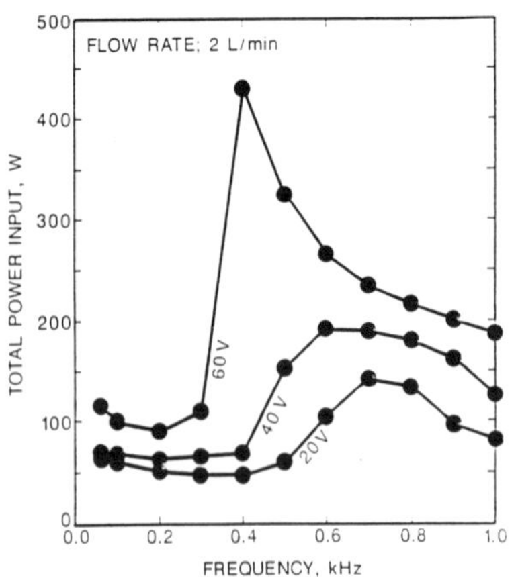

Figure 5. Effect of Frequency and Primary Voltage on Total Power Input

Total power input for the reactor is shown in Figure 5. The power input reaches a maximum immediately after the plasma is established. Total power input to the system is dependent on frequency. As the frequency increases, the power input remains fairly constant until a steep increase up to a maximum is observed followed by a gradual decrease. As showed earlier in the destructive tests, the maximum power input yielded the maximum conversion of the target contaminant. Therefore, the frequency that draws the maximum power input may be called the optimum frequency. The optimum frequency was also dependent on the reactor geometry and the primary voltage applied. These results are discussed in more detail by Tsai (1990). As the primary voltage increases from 20 V to 60 V, the optimum frequency decreases. The optimum frequency roughly coincides with the frequency that draws maximum secondary voltage in Figure 4. The system can be tuned by varying the frequency to obtain higher power input. A higher primary voltage increases secondary voltage and total power input.

Flow rates were varied from 1 L/min to 12 L/min. Power input did not change as the flow rates changed at each frequency (60-1000 Hz) and at each primary voltage (20 V, 40 V, and 60 V). That is, the shape of Figure 5 did not change at varying flow rates. There were some fluctuations in power input at the optimum frequency as the flow rate increased. This power fluctuation over time was a typical phenomenon of the plasma even at constant flow rates. Since it was shown that increased flow rate did not draw more power to the system, increased flow rate reduced the power demand by the reactor per unit volume of carrier gas. This fact helps explain the decreased conversion efficiency at increased flow rate in Figure 2.

The results of this preliminary study show the plasma reactor to have potential as an off-gas control device for ASTs. It should be noted that in this work the experiments were not conducted on a fully optimized design. Potentially, scaling up may increase the power efficiency that is actually transferred to the plasma reactor.

Conclusions

A preliminary evaluation of a plasma reactor was conducted to investigate its potential as a feasible off-gas control technology for an AST. The specific findings of the study are summarized below:

1. Destruction tests conducted on an air stream containing TCE showed a plasma reactor was capable of greater than 95% destruction efficiency.

2. Studies conducted to investigate the electrical characteristics of the plasma reactor showed that an optimum frequency for the alternating current exists for maximum power input. The optimum frequency is dependent on the primary voltage applied.

3. The total power input to the plasma reactor did not increase as the air flow rate to the unit increased from 1 to 12 L/min. Therefore, the amount of power per unit volume of air decreased as flow rate increased. This fact partially explains the drop in TCE destruction efficiency as the flow rate to the plasma increases, but the effect of decreased residence time in the reactor also potentially contributed to lower conversion.

Selected References

Barat, R.B., and Bozzelli, J.W. 1989. "Reaction of Chlorocarbons to HCl and Hydrocarbon in a Hydrogen-Rich Microwave-Induced Plasma Reactor." Environ. Sci. Technol.. 23 (6), 666-671.

Clark, R.M. and Eilers R.G., 1982. "Treatment of Drinking Water for Organic Chemical Contamination: Cost and Performance." Proceedings, Atlantic Workshop on Organic Chemical Contamination of Ground Water. AWWA/IWSA, Nashville, TN.

Clean Air Act Amendments, Public Law 101549, Title I - XI, November 15, 1990.

Environmental Science and Engineering Inc., 1985. Techniques and Costs for the Removal of VOCs from Potable Water Supplies. ESE Draft Report No. 84-912-0300, Gainesville, FL.

Fraser, M.E., Fee, D.A., and Sheinson, R.S. 1985. "Decomposition of Methane in an AC Discharge." Plasma Chemistry and Plasma Processing. 5(2), 163-173.

Hand, D.W., Crittenden, J.C., Miller, J. M., and Gehin, J. L. 1989. Performance of Air Stripping and GAC for SOC and VOC Removal from Ground Water. Project Summary, EPA/600/S2-88/053.

Piatt, M.A., 1988. Methane Destruction in an Alternating Current Plasma Reactor. Master Thesis, Chemical Engineering, Oklahoma State University, OK.

Sheinson, T.S., Smyth, N.S., Piatt, M.A., and Wills R.A., 1987. "Plasma Induced Destruction of Hydrocarbons." In: Proceedings of the International Congress on Hazardous Materials Management. Chattanooga, TN.

Treybal, R.E., 1980. Mass-Transfer Operations. 3rd Edition, p198, McGraw-Hill Chemical Engineering Series. p. 198.

Tsai, V., 1990. Conceptual Design and Performance Analysis of Frequency-Tuned Capacitive Discharge Reactors. Ph.D. Dissertation, Chemical Engineering, Oklahoma State University, OK.

Effect of Roughness and Thickness of Biofilms on External Mass Transfer Resistance

Tian C. Zhang, Paul L. Bishop, M. ASCE, and James T. Gibbs[1]

Abstract

The effect of roughness and thickness of biofilms on the dissolved oxygen (DO) concentration boundary layer thickness (DO CBLT) and the external mass transfer process in a biofilm system were investigated using a microelectrode technique. The experimental results indicated that (a) an increase of biofilm roughness would reduce the external mass transfer resistance; and (b) the absolute roughness increased with an increase in the biofilm maximum thickness, but this increase diminished when the biofilm thickness exceeded 1200 μm. The microelectrode technique is a useful tool to study the external mass transfer process.

Introduction

In the overall process of substrate utilization by biofilms, the first step is substrate transport from the bulk liquid to the liquid-biofilm interface. This step is very important because, under the steady state condition, it is equal to the sum of the overall reaction resulting from internal diffusion and substrate utilization. The aim of the present study is to use the microelectrode technique to study biofilm systems, with an emphasis on the external mass transfer process. Topics examined include determination of the concentration boundary layer, the magnitude of the mass transfer coefficient, and factors that affect the external mass transfer process. Previous studies by our group showed that three main factors affect the external mass transfer resistance: substrate loading rate, fluid streamwise velocity, and roughness and thickness of the biofilm (Zhang and Bishop, 1994a). While the

[1]Respectively, Post-Doctoral Assistant, Professor of Environmental Engineering, Graduate Research Assistant, Department of Civil & Environmental Engineering, University of Cincinnati, Cincinnati, OH 45221-0071.

first two factors are easy to investigate, the third factor deserves much more attention. Therefore, the particular objectives of this paper are: (a) to introduce the microelectrode technique for measurement of the DO CBLT near the fluid-biofilm interface; and (b) to investigate the effect of roughness and thickness of biofilms on external mass transfer resistance.

Materials and Methods

Test chamber and biofilms Fig. 1 shows the flow system used in this study. The flow chamber is an open channel which was designed to be suitable for measurement of the DO CBLT. The test biofilm (usually with a size of 0.9 cm x 1.5 cm) was sampled from a laboratory-scale rotating drum biofilm reactor (RDBR), and then mounted flush with the flat plate (by adjusting the thickness with double sided tape).

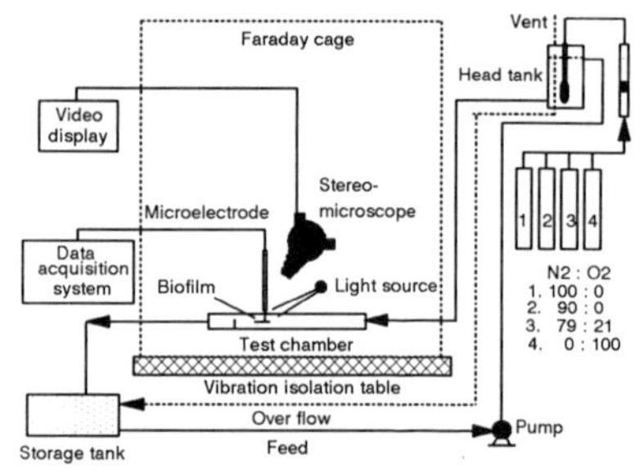

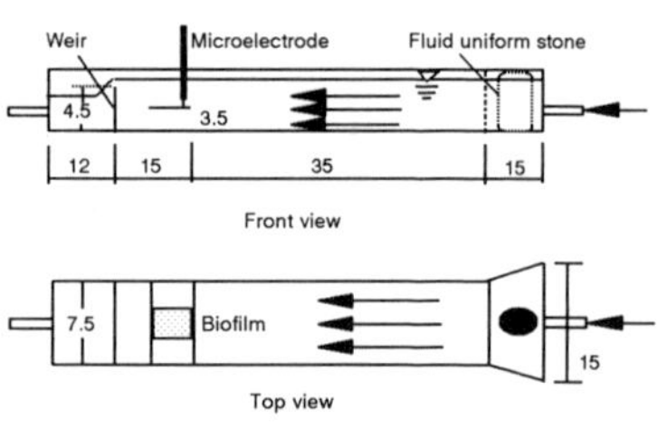

Fig.1 Experimental apparatus. (Top) Flow system; and (Bottom) Test chamber.

Artificial biofilms were used to control the "global" roughness of the surface as well as the demand for substrate/nutrient. The artificial biofilms were obtained using the following procedure: (a) take the effluent of the RDBR and centrifuge it to collect 200-300 mg biomass (based on total suspended solids (TSS)); (b) mix the biomass collected in step (a) with 10-15 ml of 3% agarose (gel type 7, Sigma) at a temperature of about 55°C; (c) pour the agarose to make the artificial biofilms with thicknesses around 1000 μm; (d) cut the artificial biofilm into two pieces of suitable size; (e) press a piece of sandpaper (e.g., Flint # 2½, 804, Armour's, U.S.A.) onto the surface of one piece of the artificial biofilm to create pre-defined roughness, and keep the other artificial biofilm as a smooth biofilm.

Measurement method Polarographic recessed cathode gold O_2 microelectrodes (O_2 microelectrodes) were used to measure the DO concentration profiles, the biofilm thickness, and surface contour (so that the roughness of the biofilm could be calculated). All of the microelectrodes had tip diameters of less than 5 μm. The test solution was DI water with 4 mM KCl and various glucose concentrations.

The bulk oxygen concentrations were controlled using air, 10% O_2 and N_2 gases to aerate the substrate tank. The whole system was then situated on a Micro-g Series 63-500 high-performance vibration isolation table (63-527-01, TMC, MA) with Faraday cage (TMC, MA). A three-dimension micro-manipulator, controlled from outside of the cage, was used to position the tip of the microelectrode. A light source over the top of the biofilm allowed us to very clearly focus on the microelectrode and biofilm surface through the microscope. A more complete description of the microelectrode technique can be found in Zhang (1994).

Results and Discussion

Biofilm roughness and its effect on external mass transfer resistance Up to now, studies on biofilm roughness have been conducted by borrowing the concept of equivalent sand roughness (e_s) in fluid mechanics. By measuring the flow rate and pressure drop, Picologlou et al. (1980) calculated the e_s for the biofilm, and found that e_s generally increases with biofilm thickness. Using the e_s concept, it is difficult to directly and microscopically investigate the characteristics of the biofilm surface because the information obtained totally depends on phenomenon-observations outside of the biofilm. In order to study the biofilm surface on a microscopic scale, two new definitions of the biofilm roughness are introduced in this study, both based on the concept of the mean line (Thomas, 1982). First, the absolute roughness of the biofilm, H_a, is the height that gives the same area as that represented by the cross-hatched area in Fig. 2, where H_m is the maximum thickness of the biofilm and H_b is the smallest thickness of the biofilm. Second, the biofilm relative roughness, H_r, is equal to $H_a/(H_a + H_b)$. These definitions of biofilm roughness are consistent with the concepts of the base biofilm and the surface biofilm (Characklis and Marshall, 1990). Using the microelectrode, it is easy to identify the surface and the bottom of the biofilm if we choose the surface of the flat plate as the reference plane (zero). Then, the H_a of the biofilm can be easily measured using a cutting and weighing method.

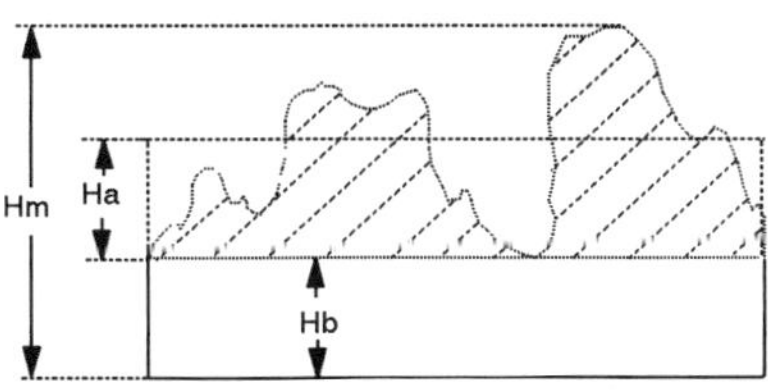

Fig. 2 Schematic of biofilm roughness

Substrate degradation effectiveness of a biofilm system greatly depends upon mass transfer resistance. One factor affecting the degree of mass transfer resistance is the thickness of the hydrodynamic boundary layer (HBL), through which mass is transferred only by diffusion mechanisms. Diffusion and consumption by microbes are balanced in the system to maintain a concentration gradient. In this study, the region where the concentration of nutrient is lower than 99% of that in the bulk solution is defined as the concentration boundary layer

(CBL).

Fig. 3 shows the effect of biofilm roughness on the DO CBLT. From Fig. 3a, b, and c, it can be seen that the microelectrode technique can be used to measure the biofilm contour on the microscopic scale. This is very important because this type of examination makes it possible for us to obtain information on: a) characteristics of the biofilm roughness, even though the contour of the biofilm surface can exhibit self-similarity, that is, the profile is the same whatever the magnification; and b) the substrate/nutrient CBL (e.g., DO CBLT). Since the early 1960s, electrochemical methods have been used in chemical engineering to study transport phenomena (Reiss and Hanratty, 1962, 1963; Jolls and Hanratty, 1969; Mizushina, 1971). This technique has been developed such that microelectrodes can be used to determine flow regimes in packed bed reactors (Latifi, et al., 1989, 1992; Pauli et al., 1991), and to measure the mass transfer boundary layer around a spherical biocatalyst particle (Hooijmans, et al., 1990). To our knowledge,

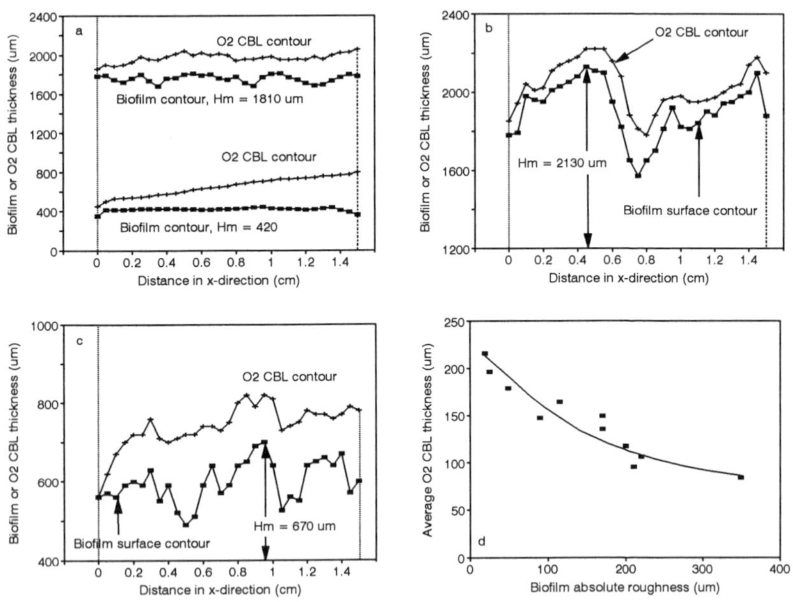

Fig. 3 Effect of the biofilm roughness on the dissolved oxygen concentration boundary layer thickness. a. For artificial biofilms. The biofilm with Hm = 1810 um is a rough biofilm, while the biofilm with Hm = 420 is a smooth one; x-direction is the direction of flow. At x = 0, the flow meets the biofilm. During the measurements, the location at which the microelectrode pierced the biofilm was change along the x-direction for a certain streamwise velocity; b. and c. For real biofilms; and d. Effect of biofilm absolute roughness on average DO concentration boundary layer thickness.

however, the microelectrode technique has not been used to simultaneously obtain information on the biofilm surface characteristics and the substrate/nutrient CBL. The direct experimental measurements shown in Fig. 3 demonstrate that the microelectrode technique is a useful tool to study the biofilm roughness and the external mass transfer resistance. The results in Fig. 3 also indicate that the size of the biofilm sample used in this study (0.9 cm x 1.5 cm) usually includes the typical characteristics of the biofilm surface being studied, and therefore is suitable for this study.

The concepts of the absolute roughness and the relative roughness of the biofilm introduced in this study effectively describe the characteristics of the biofilm surface. Although the idea of roughness is well-formed conceptually, there is no rigorous mathematical or geometric definition of the biofilm roughness. In order to quantify a variation from the mean elevation of a surface, the definition and derivation of that mean must first be defined. There are two dominant definitions in the standards used by industry: the first is defined as the line through and parallel to the geometrical profile such that the area of solid above is equal to the area of void below that line. The second is a more statistical definition employing the concept of standard deviation and is referred to as the Root Mean Square (RMS) mean line. This line minimizes the square of the areas of solid and void above and below it (least squares method). It may be noted that these two definitions are equivalent. Many methods of measuring surface roughness are based on these mean line definitions (Thomas, 1982). Another method of measuring surface roughness is the E-system. This method involves the profiles that are created by graphically rolling circles of varying radii along the surface profile. This method has been incorporated into the German, Italian, and Swiss national standards and has been fairly rigorously defined (Thomas, 1982). However, neither the concept of the mean line nor the E-system are currently ready to be used in investigations of biofilm roughness. The definitions of roughness introduced in this study enable us to characterize the biofilm surface in a simple but quantitative way. Based on this biofilm surface information, the effect of the biofilm roughness on fluid flow and the external mass transfer resistance can be investigated.

For any thickness of the biofilm, there is a DO CBL (see Fig. 3a, b, and c). Fig. 3d shows the effect of the biofilm roughness on the average DO CBLT. This average DO CBLT was calculated by adding all of the measured DO CBLT, and then dividing by the number of measurements. The two-dimensional modification of the measured DO CBLT was conducted using the same method in Zhang and Bishop (1994a), that is, the thickness perpendicular to the local biofilm surface was used as the DO CBLT in each point. From Fig. 3d, it can be understood that the DO CBLT decreases with an increase of biofilm roughness. Therefore, an increase in biofilm absolute roughness would result in an increase of external mass transfer, since the thickness of substrate/nutrient CBL is often considered as a representative of the external mass transfer resistance.

Relationship among biofilm thickness, roughness and external mass transfer resistance Fig. 4 shows the relationship among thickness, absolute roughness and the relative roughness of the biofilm. From Fig. 4, it can be seen that the absolute roughness increased with an increase in the biofilm maximum thickness, H_m, but this rate of increase became small when the biofilm was thicker than 1200 μm. The relative roughness of the biofilm does not exhibit a very clear trend as a function of the biofilm thickness. Over the whole range of biofilm thicknesses studied, this relative roughness always remained as 10-20% of the biofilm maximum thickness. Therefore, the biofilm thickness does not greatly affect the biofilm relative roughness.

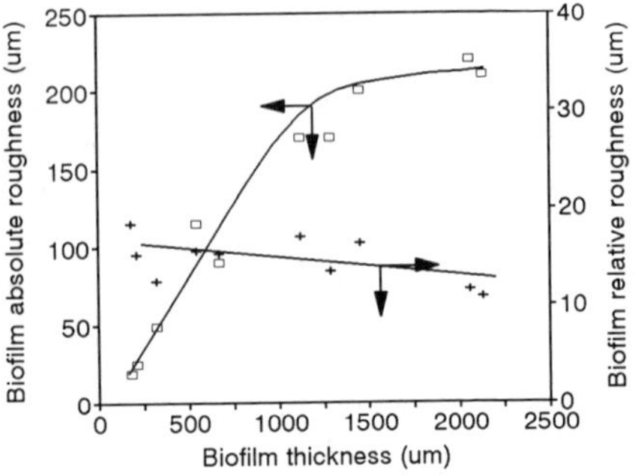

Fig. 4 Effect of biofilm thickness on biofilm roughness

Recently, it has been reported that biofilm thickness greatly affects spatial distributions of biofilm properties, such as biofilm densities and porosity. It was found that the average density decreases with an increase of biofilm thickness, and the biofilm porosity increases with an increase in biofilm thickness. In addition, gradients in biofilm properties in the direction away from the substratum were also reported (Zhang and Bishop, 1994b). These results are most important in the investigation of the effect of the biofilm thickness on the biofilm roughness. There are two reasons to explain why the thicker the biofilm, the rougher the biofilm surface. First, since the upper layers of a biofilm are less dense and more porous, daughter cell formation forces material up toward the surface of the biofilm. At the surface there is a greater variety of possible directions in which this material can expand. This multidirectional expansion is the method by which the biofilm becomes rougher. In fact, increasing roughness can be a strategy for microorganisms to survive because roughness decreases the diffusional length, permitting greater respiratory activity in the lower depths of the biofilm. Second, the cohesive (mechanical) strength of the biofilm decreases with an increase in biofilm thickness because the average density decreases and porosity increases with an increase in biofilm thickness. As a result of this, the detachment (erosion and/or sloughing) occurs more easily for the thicker biofilm than for the thin biofilm. This statement is supported by Wanner and Gujer (1986), who reported that the erosion rate is proportional to the biofilm mass concentration and also the square of the biofilm thickness.

Fig. 5 shows the effect of the biofilm thickness on the average DO CBLT. it can be seen that, based on a two-dimensional modification of the calculation of

DO CBLT, the DO CBLT decreases with an increase in the biofilm thickness. Therefore, the rougher and thicker the biofilm, the less the external mass transfer resistance.

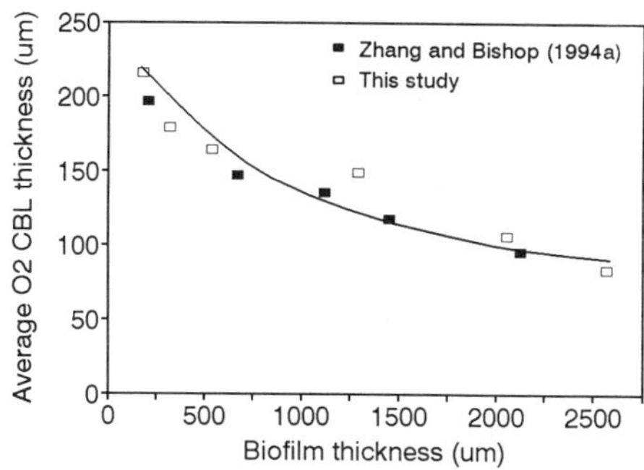

Fig. 5 Effect of biofilm thickness on average oxygen concentration boundary layer thickness

It is well known that mass is transported by two mechanisms: (a) *diffusion*, which is the transport of mass within a phase from a high concentration to a low concentration in a solid or liquid, and (b) *advection* (sometimes termed convection), which results from bulk fluid motion and influences both internal and external mass transport. In turbulent flow, most of the resistance to mass transport occurs near the interface. As a consequence, the biofilm roughness strongly influences the external mass transfer resistance. It was reported that mass transfer at a rough surface can be as much as three times higher than that at a smooth surface (Davies, 1972). Using the artificial biofilm as the test materials, we found that biofilm roughness created fluid velocity fluctuations which could be detected by current fluctuations at the O_2 microelectrode. This velocity fluctuation in turn affected the thickness of the DO concentration boundary layer (Zhang and Bishop, 1994a).

Conclusions

The following conclusions are derived based on the experimental measurements obtained in this study. (a) The microelectrode technique employed in this research is a useful tool to study the external mass transfer resistance. (b) It is useful to introduce the concepts of absolute and relative roughness of the biofilm. (c) The absolute roughness increases with an increase in the biofilm maximum thickness, but the rate of this increase decreased when the biofilm thickness exceeded 1200 μm. However, the biofilm thickness shows little influence on the biofilm relative roughness. (d) The roughness of the biofilm surface has an positive effect on external mass transfer. The rougher the biofilm surface, the less the external mass transfer resistance.

Acknowledgements

The authors wish to think M. Kupferle and S. Fitzgerald of the University of Cincinnati for their assistance with laboratory management and reactor operation. This work was funded by the National Institute of Environmental Health Sciences (NIEHS), U.S.A..

References

Characklis, W. G., and Marshall, K. C., "Biofilms: a basis for an interdisciplinary approach," in Biofilms, ed. by Characklis, W. G., and Marshall, K. C., pp. 3-15. John Wiley & Sons, Inc., Now York. (1990).

Davies, J. T., Turbulence Phenomena, Academic Press, New York, (1972).

Hanratty, T., and Campbell, J. A., "Measurement of wall shear stress," in Fluid Mechanics Measurements (Edited by R. J. Goldstein), pp. 572-574. Hem. Publ. Corp., New York. (1983).

Hooijmans, C. M., Geraats, S. G. M., Potters, J. J. M., and Luyben, K. Ch. A. M., "Experimental determination of the mass transfer boundary layer around a spherical biocatalyst particle," Chem. Enging. J. **44**, pp. B41-B46. (1990).

Jolls, K. R., and Hanratty, T. J., "Use of electrochemical techniques to study mass transfer rates and local skin friction to a sphere in a dumped bed," A.I.Ch.E. J., **15**, pp. 199-205. (1969).

Latifi, M. A., Midoux, N., and Storck, A., "The use of micro-electrodes in the study of the flow regimes in a packed bed reactor with single phase liquid flow," Chem. Enging. Sci., **44**, pp. 2501-2508. (1989).

Latifi, M. A., Rode, S., Midoux, N., and Storck, A., "The use of microelectrodes for the determination of flow regimes in a trickle-bed reactor. Chem. Enging. Sci., **47**, pp. 1955-1992. (1992).

Mizushina, T., "The electrochemical method in transport phenomena," in Advances in Heat Transfer, **7**, pp. 87-161. (1971).

Pauli, J., Sobolik, V., and Onken, U., "Oxygen as depolarizer in electrodiffusion diagnostics of flow," Chem. Enging. Sci., **46**, pp. 3302-3304. (1991).

Picologlou, B. F., Zelver, N., and Characklis, W. G., "Biofilm growth and hydraulic performance," J. Hyd. Div. ASCE, **106**, pp. 733-746. (1980).

Reiss, L. P., and Hanratty, T. J., "Measurement of instantaneous rates of mass transfer to a small sink on a wall," A.I.Ch.E. J., **8**, pp. 245-247. (1962).

Reiss, L. P., and Hanratty, T. J., "An experimental study of the unsteady nature of the viscous sublayer," A.I.Ch.E. J., **9**, pp. 154-160. (1963).

Thomas, T. R., Rough Surfaces, ed. by Thomas, T. R., Longman, New York. (1982).

Wanner, O., and Gujer, W., "A multispecies biofilm model," Biotech. & Bioengng., **28**, 314-328, (1986).

Zhang, T. C., and Bishop, P. L., "Experimental determination of the dissolved oxygen boundary layer and mass transfer resistance near the fluid-biofilm interface," accepted for publication in Wat. Sci. Tech.. (1994a).

Zhang, T. C., and Bishop, P. L., "Density, porosity and pore structure of biofilms," accepted for publication in Wat. Res., (1994b).

Zhang, T. C., "Influence of biofilm structure on transport and transformation processes in biofilms," Ph.D. Dissertation, University of Cincinnati, Cincinnati, OH 45221. (1994).

PROCESS SELECTION FOR TREATMENT OF SOC CONTAMINATED WATERS

Bruce I. Dvorak[1], Desmond F. Lawler[2], and Gerald E. Speitel Jr.[3] M.ASCE

Introduction

The selection of the least-cost treatment option for treating synthetic organic chemical (SOC) contaminated wastewaters is often a complex endeavor. There are too many potential treatment processes and series-of-processes to perform a detailed evaluation of all alternatives. This research was undertaken to simplify the selection of the least expensive treatment process(es) for a given set of conditions. Mathematical process performance and cost models were developed for eight treatment processes. Four aqueous treatment processes were considered: air stripping, liquid-phase adsorption, fixed-film biological oxidation, and biodegradation within a carbon adsorption column. Because off-gases from air stripping towers are frequently regulated, four off-gas treatment processes also were considered: gas-phase adsorption (both on- and off- site regeneration), thermal incineration, and catalytic oxidation. The least-cost design for each process was identified for a set of wastewaters typical of contaminated groundwaters, drinking waters, and industrial wastes. The results were synthesized to create generalizations concerning process selection. The specific objective of this research was to develop analytical tools to aid engineers faced with complex decisions concerning process selection for the treatment of SOC contaminated waters.

Approach

Sophisticated mathematical models for process performance were combined with cost models incorporating capital and operations and maintenance costs to describe each process in this study. These mathematical models have been coded into FORTRAN computer programs. For each treatment process, the models were used to find the optimal design for each wastewater in the matrix of test conditions. A variety of test conditions (e.g., different chemicals, various concentrations, different regulatory constraints) were included in the waste matrix. Treatability

[1]Graduate Research Assistant, [2]Professor, and [3]Associate Professor, Environmental and Water Resources Engineering, Dept. of Civil Engineering, University of Texas at Austin, Austin, Texas 78712-1076.

parameters for each compound were obtained from the technical literature. Both single and multiple component cases were considered, including chemicals such as aromatic hydrocarbons, chlorinated alphatics, chlorinated hydrocarbons, and phenolics typically found in contaminated groundwaters, drinking waters, and industrial wastewaters. Single component cases have been used to develop generalizations concerning process selection.

Eight processes were studied in the research. The multiple component plug flow homogeneous surface diffusion model (HSDM developed by Crittenden and Weber, 1978) was used to describe the process performance of liquid-phase carbon adsorption. This model describes both adsorption equilibrium and kinetics. The cost model for liquid-phase adsorption was based upon Adams and Clark (1991). Combined biodegradation and adsorption was modeled by an equilibrium model developed by Erlanson (1993). The cost model for biodegradation in an adsorption column was a modification of the liquid-phase adsorption cost model.

A packed bed biofilm reactor was selected to be representative of biological treatment at low concentrations. Aerobic fixed-film biodegradation was modeled by the pseudoanalytical solution for steady-state biofilms by Saez and Rittmann (1988 and 1992). The fixed-film cost model was developed using a combination of the liquid-phase adsorption cost model, cost models developed for other biological treatment processes, and recommendations from vendors.

Henry's Law and the Onda et al. (1968) correlations were used to describe equilibrium and kinetics of air stripping in order to predict the required tower size. Air stripping costs were modeled following Adams and Clark (1991). Since off-gases from aqueous-phase treatment processes are regulated in some cases, four off-gas options have been considered. Thermal incineration and catalytic oxidation process performance and costs were modeled following QAQPS (1990). Gas-phase adsorption (both on and off-site regeneration) was modeled using equilibrium models of Crittenden et al. (1989) and cost equations from Adams and Clark (1991).

Synthesis of Results

Figure 1 shows a comparison of cost contours for air stripping alone (dashed lines) and air stripping followed by gas-phase adsorption (solid lines). For air stripping alone, the optimal (least-cost) design is in the region of the closed contour; for air stripping followed by gas-phase adsorption, the optimal region is shifted up and to the left, i.e., to higher liquid loading rates and lower stripping factors ($S = \frac{Q_G H}{Q_L}$), where Q_G is the volumetric gas flow rate, Q_L is the liquid flow rate, and H is the Henry's constant. This example illustrates the necessity of designing the two processes jointly. Similar results held for all chemicals and types of off-gas treatments, because all off-gas treatment options are significantly less expensive when the gas-phase concentration is maximized and the gas flow rate is minimized. Typically an air stripper followed by off-gas treatment uses a stripping factor between 2.0 and 4.0 depending upon the target compound's volatility.

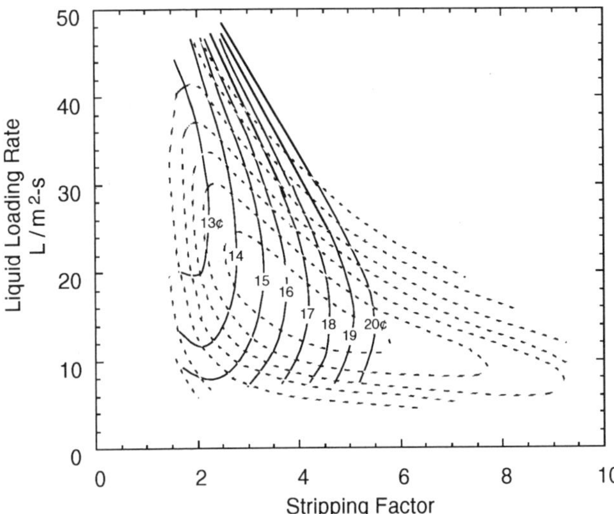

Figure 1. Cost Contours for Air Stripping followed by Gas-Phase Adsorption of 1,2-DCA. (Dashed contours are the cost of air stripping alone, costs are in cents/m^3 treated, costs are for July 1991, removal efficiency = 99.75%, liquid flow rate = 5,540 m^3/day, temp. = 20°C, and packing = 51mm plastic tellerettes.)

Four processes are often considered for treating air stripping off-gases. Figure 2 shows the flow and concentration conditions for each process when it is least-cost for benzene. The second (outside) axis on Figure 2 shows the flow rate and liquid-phase concentration for a stripping tower with a stripping factor of 2.0. The least-cost regions for other relatively adsorbable chemicals are at similar gas flow and gas concentration values (inside axes) as shown in Figure 2. For other chemicals, the second (outside) axes vary in magnitude.

Henry's constants vary greatly for SOCs; the Henry's constant determines the quantity of air required to strip a given compound for a given stripping factor. Factors such as gas concentration and waste heat of combustion are unimportant under typical air stripping concentrations in defining thermal incineration and catalytic oxidation costs. Other than the presence of chlorinated compounds that may poison catalysts, thermal incineration and catalytic oxidation costs are primarily a function of the volume of gas to be treated. Since the Henrys' constant of a compound determines the required volume of air required, the Henry's constant is the most important parameter in defining thermal incineration and catalytic oxidation costs when the liquid flow rate is the same.

Liquid-phase adsorption is nearly always more expensive than air stripping alone. Only when the Henry's constant is low (less than approximately 0.002) is liquid-phase adsorption the better alternative, as listed in Table 1. When off-gas treatment is required for air stripping, liquid-phase adsorption is more competitive,

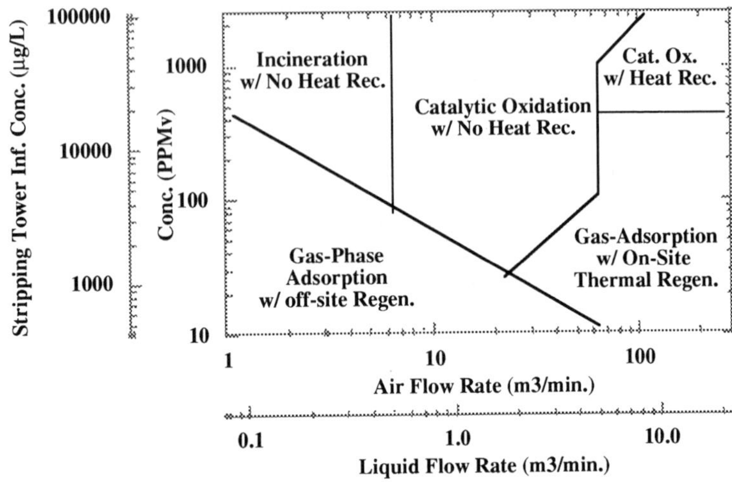

Figure 2. Least-Cost Process for Off-Gas Treatment of Benzene. (Design Life of 5 years.)

Table 1. Estimated Costs of Treating Single-Component Wastewaters. Costs are in cents/m³ treated, for December 1993. Assumed flow rate of 3780 L/min. (1.44 MGD), temperature is 20°C, influent concentration of 2 mg/L, effluent concentration of 50 µg/L, and design life of 20 years.

Compound	Air Stripping Alone	Air Stripping w/off-gas GAC	Air Stripping w/off-gas Thermal Incin.	Liquid-Phase GAC Adsorption	Packed Bed Biofilm
Phenol	NA	NA	NA	18.7	3.65
4-Nitrophenol	NA	NA	NA	5.98	4.02
Napthalene	9.17	24.7	137.	7.97	4.27
DBCP	5.17	16.4	103.	8.52	NA
o-Dichlorobenzene	3.79	9.52	33.3	6.32	6.67
1,2-DCA	3.34	11.1	25.1	57.3	NA
Methylene Chloride	2.53	83.9	18.8	109.	NA
Chloroform	1.81	6.12	14.6	38.5	NA
Benzene	1.83	6.14	11.9	12.4	3.91
Toluene	1.72	5.3	12.2	10.5	3.88
TCE	1.47	4.53	9.22	14.3	NA
Vinyl Chloride	1.24	604.	5.87	NA	NA

NA - Process is not technically feasible.

and the selection of the least-cost treatment process appears more difficult. However, from the wide variety of designs made in this research, a relatively simple selection diagram has been developed as explained subsequently.

Partitioning parameters are useful for analyzing separation processes like stripping and adsorption. The Henry's constant of a compound is a good measure of the ease of using air stripping for treating the waste. Adsorption is a more complex process than stripping. A single dimensionless partitioning parameter does not exist for adsorption, although partitioning can be approximated by the solute distribution parameter:

$$D_g = \frac{\text{Mass of adsorbate in the solid phase at equilibrium}}{\text{Mass of adsorbate in the fluid phase at equilibrium}} = \frac{\rho_a q_e(1-\varepsilon)}{\varepsilon C_o}$$

where: ρ_a = adsorbent density,
q_e = mass of adsorbed pollutant per mass of adsorbent (GAC) at equilibrium with C_o,
ε = porosity, and
C_o = influent concentration.

The use of a "volatility" and "adsorbability" parameter as an axis on a graph creates a tool to compare stripping and adsorption costs. This type of graph can help identify cases when either air stripping with off-gas treatment or liquid-phase adsorption is most likely to be the least cost treatment. Figure 3 is an example of this graph. Five regions exist in Figure 3. In the upper left is a region where liquid-phase adsorption is clearly the best alternative; just to the right is a region where liquid-phase adsorption and air stripping with gas-phase adsorption costs are similar (within 30%), and further detailed study is required to identify the less expensive option. Farther to the right is a region where air stripping with gas-phase adsorption is clearly the best. Below is a region where air stripping followed by either gas-phase adsorption or thermal incineration is cost competitive (within 30%). Finally, at the bottom is a region where adsorption costs in either phase are large, and incineration off-gas process with air stripping should be considered.

When multiple component wastes are considered, selection of a treatment option is more complex than for single component wastes. Figure 3 can be used in many cases for multiple component wastes, although caution should be used in interpreting the recommendations. Since chemicals act independently of each other in stripping, the Henry's constant of the most restrictive contaminant (requiring the tallest tower) should be used to estimate the "volatility" parameter for the waste. The "adsorbability" of the waste can be estimated by calculating the D_{gL} for each component in the waste as if it were in a single component system, and using the minimum D_{gL}. Caution in interpreting the results is particularly warranted when the minimum Henry's constant and the smallest D_{gL} are for different chemicals.

Biological treatment of SOCs is an attractive treatment alternative because it can be inexpensive, it can result in no hazardous residuals, and it emits no off-gases when in-line oxygen addition is used. As a treatment alternative, a packed bed

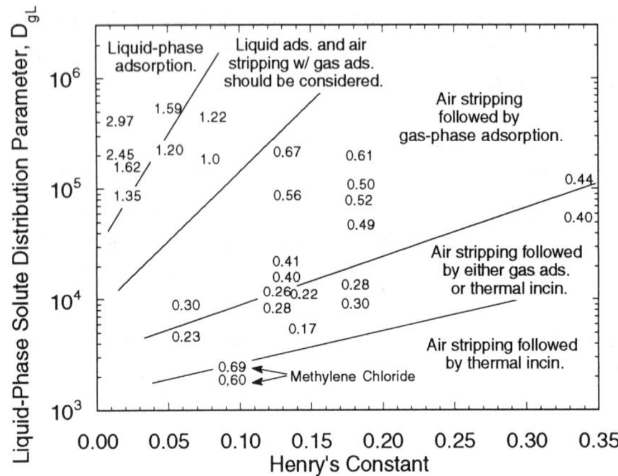

Figure 3. Process Selection Diagram: Liquid-Phase Adsorption vs. Stripping with Off-Gas Treatment. (Numbers on the diagram represent the ratio of the costs of air stripping followed by gas-phase adsorption; numbers are plotted on the diagram at the Henry's constant and solute distribution parameter value of the waste tested; the constraints are the same as those for Figure 1.)

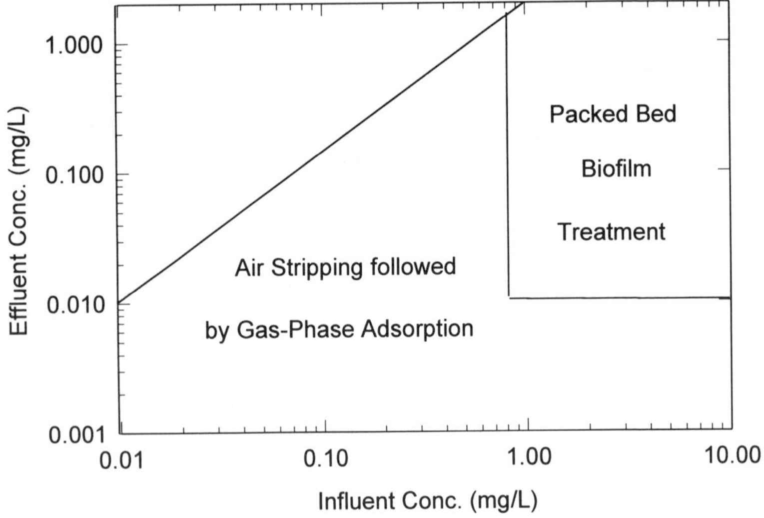

Figure 4. Least-Cost Process for Treatment of a Toluene Contaminated Water. (Costs are in cents/m^3, liquid flow rate=5,540 m^3/day, and temp.=20°C.)

biofilm reactor will be cost competitive with air stripping with off-gas treatment and liquid-phase adsorption only for the most biodegradable compounds such as phenol, toluene, and benzene, as listed in Table 1, and when both the influent and effluent concentrations are both relatively high. Low effluent concentrations are difficult to achieve and costly with biological treatment. The concentration ranges where the biological option and air stripping with off-gas adsorption are the least-cost treatment for toluene is shown in Figure 4; liquid-phase adsorption and air stripping with off-gas thermal incineration were not economically competitive in this scenario. Performance of a biological process can be more difficult to predict than that of a physical/chemical process; thus the biofilm region in Figure 4 represents situations in which a biological option deserves further investigation. The location of these least-cost regions will vary greatly depending upon the treatability characteristics of each compound.

Biodegradation within an adsorption column can greatly increase the bed life. Combined biodegradation and adsorption is of most interest when a wastewater has both a biodegradable and non-biodegradable component. As shown in Figure 5, the increase in the adsorption column bed life caused by biodegradation is affected by three factors: which component (biodegradable or non-biodegradable) is more

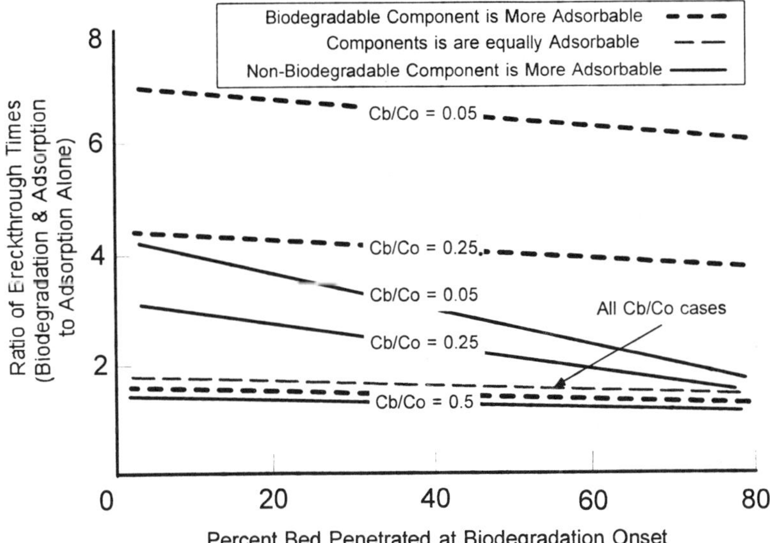

Figure 5. Increase in Adsorption Column Bed Life from Microbial Degradation of One Component of a Bicomponent System. (C_b/C_o represents the fraction of the biodegradable component that is not microbially degraded in the column.)

adsorbable, the fraction of the initial concentration not biodegraded (C_b/C_o), and the percentage of the bed penetrated at the onset of biodegradation. The percentage of the bed penetrated at the onset of biodegradation is a function of three factors: bed size, time required for microbial acclimation to the SOC, and flow rate. Figure 5 shows that biodegradation can increase the bed life in any case, but the greatest benefit occurs when the biodegradable component is the most adsorbable within the mixture.

Conclusions

Generalizations for rapid identification of the least-expensive treatment option(s) have been developed in this research. Individual process optimization generalizations have been developed, such as the effect shown in Figure 1 concerning the need for joint optimization of air stripping and off-gas treatment. Process selection aids also have been synthesized from the results, as shown in Figures 2, 3, 4, and 5. Tools such as these can aid engineers in making intelligent decisions concerning preliminary treatment process selection, with the combined result of better, cost-effective designs and more rapid selection.

References:

Adams, J.Q., and Clark, R.M. (1991). "Evaluating the Costs of Packed-Tower Aeration and GAC for Controlling Selected Organics," *J. Am. Water Works Ass.*, 83 (1), 49-57.

Crittenden, J.C., and Weber, W.J., Jr. (1978),"Predictive Model for Design of Fixed-Bed Adsorbers: Parameter Estimation and Model Development," *J. Environ. Eng. ASCE*, 104(2), 185-197.

Crittenden, J.C., Rigg, T.J., Perram, D., Tang, S., and Hand, D.W. (1989), "Predicting Gas-Phase Adsorption Equilibria of Volatile Organics and Humidity," *J. Environ. Eng. ASCE*, 115(3), 560-573.

Erlanson, B.C. (1993), "Multicomponent Equilibrium Model for Simultaneous Biodegradation and Adsorption in Granular Activated Carbon Columns," Thesis presented to the University of Texas at Austin in partial fulfillment of the requirements for the degree of Master of Science.

Onda, K., Takeuchi, H. and Okumuto, Y. (1968), "Mass Transfer Coefficients Between Gas and Liquid Phases in Packed Columns," *Journal of Chemical Engineering of Japan*, 1(1), 56-62.

Saez, P.B. and Rittmann, B.E. (1988), "Improved Pseudoanalytical Solution for Steady-State Biofilm Kinetics," *Biotechnology and Bioengineering*, 32(7), 379-385.

Saez, P.B. and Rittmann, B.E. (1992), "Accurate Pseudoanalytical Solution for Steady-State Biofilms," Communications to the Editor, *Biotechnology and Bioengineering*, 39, 790-793.

QAQPS (1990), "QAQPS - Control Cost Manual," 4th ed. Economic Analysis Branch, Office of Air Quality Planning and Standards, U.S. EPA Report *EPA-450/3-90-006*, Research Triangle Park, N.C.

Anaerobic Biodegradation of Chloroform under Methanogenic Conditions

Devesh[1], Munish Gupta[1], Makram T. Suidan[2], Gregory D. Sayles[3]

Abstract

The degradation of chloroform is studied for two different methanogenic cultures grown on acetate and methanol exclusively, as the primary substrates. In chemostats, chloroform was fed at different concentrations as high as 2000 μg/l along with the primary substrate and chloroform degradation greater than 98 % was observed. The kinetics of degradation of chloroform and the primary substrate were investigated using BMP tests and it was seen that the methanol-fed methanogenic culture exhibited higher rates of chloroform degradation than the acetate-fed methanogenic culture. Besides, chloroform inhibited acetate degradation at any concentration while methanol inhibition was observed only for chloroform concentrations higher than 800 μg/l.

Introduction

Chloroform is a priority pollutant that poses a great threat to public health as a suspected carcinogen. Cost effective technology is needed to degrade this compound because of its presence in industrial effluents, in landfill leachates and at Superfund sites.

Bouwer *et al.*(1981) showed that chloroform is non-degradable under aerobic conditions. Recent research has shown that chloroform at low concentrations,

---

[1]Grad. Res. Asst., Dept. of Civil and Environmental Engineering, University of Cincinnati, Cincinnati, OH 45221-0071.

[2]Professor, Dept. of Civil and Environmental Engineering, University of Cincinnati, Cincinnati, OH 45221-0071.

[3]Biochemical Engineer, Risk Reduction Engineering Laboratory, U.S. EPA, Cincinnati, OH 45268.

can be anaerobically degraded by methanogenic bacteria in the presence of a primary substrate (Bouwer and McCarty, 1983). Bouwer *et al.* (1981) conducted batch tests on a mixed methanogenic culture which was grown on acetate at 35°C. Initial chloroform concentrations of up to 40 μg/l was completely degraded by the culture, while degradation at higher initial concentration (157 μg/l) was not very conclusive. Mikesell and Boyd (1990) studied the degradation of chloroform using pure cultures of two different species of Methanosarcina. It was seen that chloroform did not degrade in the absence of a primary substrate. However, chloroform was rapidly biotransformed in the presence of methanol and 65-78 % of methylene chloride was formed.

Chloroform at high concentrations has been found to be inhibitory to methanogenesis. Toxicity of chloroform to methanogenesis is important since it is the rate-limiting step in anaerobic digestion. Narayan *et al.* (1993) have shown that chloroform concentrations of more than 200 μg/l inhibited the degradation of acetate and acetone in a methanogenic reactor, while methanol was completely utilized. Yang and Speece (1986) have shown that unacclimated cultures are more prone to inhibition by chloroform as compared to acclimated ones.

The present study investigates the anaerobic biodegrability of chloroform in methanogenic environments utilizing acetate and methanol as the primary substrates. Acetic acid and methanol were chosen for this study since they are important precursors to methane production in anaerobic systems. In addition, the effect of initial chloroform concentration on the degradation rate of chloroform and the primary substrates (acetate and methanol) is also studied using biological methane potential (BMP) tests.

Materials and Methods

Experimental Set Up

The experiments were conducted in two ten-liter stainless steel jacketed chemostats (see schematic in Figure 1.). The chemostats were maintained at 35°C by recirculating heated water in their outer jackets. The contents of the chemostat were kept completely mixed with magnetically coupled variable speed mixers. The chemostats had a detention time of 10 days. The seed culture was obtained from a mixed culture methanogenic reactor operated at the USEPA Test and Evaluation Facility at Cincinnati, Ohio. Each chemostat was equipped with two constant speed (2rpm) Masterflex pumps to feed solutions of nutrients and a buffer. The nutrient solution consisted of essential inorganic salts and vitamins necessary for the growth of microorganisms. A buffer solution containing sodium hydroxide and sodium carbonate was fed to maintain a constant pH of 7.2 in the system. Chloroform was dissolved in the primary substrate and was injected into the nutrient line using a high precision single syringe infusion pump. Continuous flow of nutrient was kept for an uninterrupted feed of chloroform and primary substrate to the chemostat. To achieve steady-state, the chemostats were operated for a period of 160 days with out the addition of chloroform. After achieving steady-state, chloroform at different concentrations was fed to the chemostats keeping the feed concentration of primary substrate constant. BMP tests were then performed on both the chemostats when the chemostats were operating at steady-state.

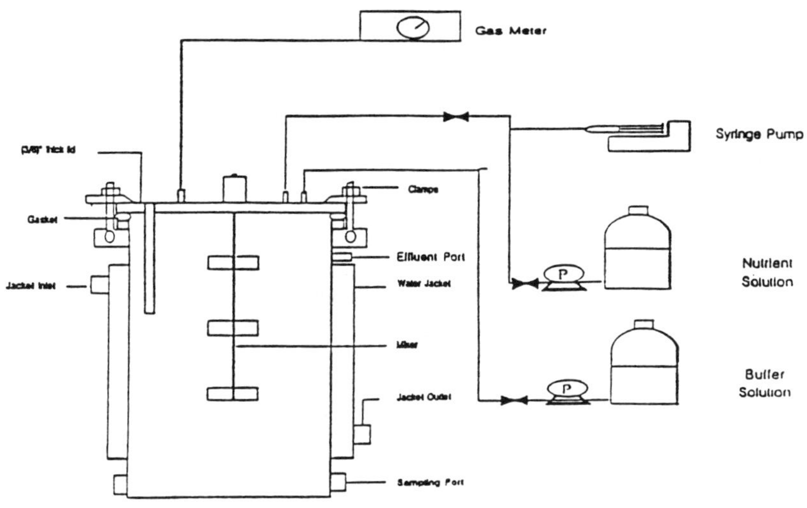

Figure 1. A Schematic of the Chemostat.

Daily measurements of nutrient and buffer feed rates, total gas production, effluent pH and operating temperature were performed on the chemostats. A wet gas tip meter was used to measure the total gas production. The gas samples were analyzed weekly for the composition of the gas. Liquid samples were collected weekly from the sampling port and filtered through 0.45 μm Magna Nylon supported plain filters to analyze for chemical oxygen demand (COD) and volatile fatty acids (VFA). Effluent liquid and gas samples were analyzed twice a week for chloroform and methylene chloride. Chloromethane was not measured in this study.

Biological Methane Potential (BMP) Tests

The chemostats operating at steady-state were used as a consistent source of culture for BMP tests. A 160 ml serum bottle was used for performing BMP tests. Nutrient and vitamin solutions were added to the serum bottle to ensure the availability of essential nutrients. In addition, reducing conditions were maintained by adding 125 mg/l of each sodium sulfide and cysteine hydrochloride. Resazurin was used as an indicator of anaerobic conditions. The serum bottle was purged with a gas mixture of 70% nitrogen and 30% carbon dioxide to remove any oxygen dissolved in the solution and present in the headspace of the serum bottle. Effluent from the chemostat was then collected in the serum bottle. A known chloroform and primary substrate concentration (50 mg/l) was then spiked in the serum bottle and it was immediately sealed with a teflon cap and aluminium crimp. Concentrations of chloroform, methylene chloride and primary substrate (acetate and methanol) were then monitored with time.

Results & Discussion

Steady-State

The chemostats were run without chloroform for a period of 160 days. This state was taken as a reference to compare the changes in chemostat performance for different chloroform loadings. Steady-state was confirmed by stable effluent concentrations of VFA, VSS and COD. Feed concentration of acetate to the chemostat was 2.51 g/l while that of methanol was 1.93 g/l. After obtaining steady-state, chloroform at 100 μg/l was introduced in the chemostats to ensure a gradual acclimation of the microorganisms to chloroform. The chemostats were operated for at least three additional turnovers (30 days) after reaching stable conditions to confirm steady-state. Successive steady-states were obtained for increased chloroform loadings of 500, 750, 1000 and 2000 μg/l. The steady-state values of effluent chloroform and methylene chloride concentrations for various influent chloroform concentrations are shown in Table 1. The standard deviation of the data is shown in the parenthesis.

Biological Methane Potential (BMP) Tests

BMP Tests were conducted when the chemostats were operating at steady-state. The concentration of the primary substrates (acetate and methanol) was kept constant at 50 mg/l. BMP tests were performed with different initial chloroform concentrations and at these different concentrations, initial rate of degradation of chloroform and the primary substrate was calculated. The BMP tests were continued at increasing initial chloroform concentrations until inhibition with respect to chloroform and primary substrate degradation was observed. The tests performed with inactivated cells (by adding mercuric sulfate and autoclaving the culture) showed no degradation or loss of chloroform or the primary substrate over a period of time.

Table 1. Effluent steady-state methylene chloride (MeCl) and chloroform concentrations for various loadings of chloroform.

	Acetate-fed Chemostat		Methanol-fed Chemostat	
Influent Chloroform Concentration	MeCl	Chloroform	MeCl	Chloroform
100	2.2	0.6 (0.003)	< 0.01	0.6 (0.08)
500	< 0.01	< 0.01	< 0.01	< 0.01
750	0.24 (0.17)	0.11 (0.01)	0.16 (0.02)	3.63 (0.49)
1000	0.28 (0.01)	0.15 (0.04)	0.17 (0.07)	6.1 (0.51)
2000	0.25 (0.3)	0.52 (0.47)	0.19 (0.24)	32.1 (9.25)

Acetate fed Methanogenic Culture - BMP tests were performed at initial chloroform concentrations of 20, 40, 100, 150, 180, 325, 400, 500 and 650 μg/l and 50 mg/l of acetate as the primary substrate. Figure 2 shows the results for an initial chloroform concentration of 180 μg/l. This culture showed a maximum chloroform degradation rate of 30-32 μg/l/hr at an initial chloroform concentration of 150 μg/l. The chloroform degradation rate decreased on further increasing the initial chloroform concentration, indicating chloroform inhibition. The rate of acetate degradation decreased with the addition of chloroform showing inhibition by chloroform. No acetate degradation was seen for chloroform concentrations higher than 320 μg/l and a constant rate of chloroform degradation was observed.

Methanol fed Methanogenic Culture - BMP tests for this culture were conducted at initial chloroform concentrations of 15, 70, 135, 335, 460, 500, 660, 800, 1100, 1400 and 1900 μg/l. An example of the test results for this culture at an initial chloroform concentration of 335 μg/l is shown in figure 3. The initial methanol concentration was kept constant at 50 mg/l. This culture showed higher chloroform degradation rates compared to the methanogenic fed acetate as primary substrate. The maximum rate of chloroform degradation was calculated to be 54 μg/l/hr at an initial chloroform concentration of 460 μg/l. No inhibition with respect to the methanol degradation was observed for chloroform concentrations less than 800 μg/l. The higher initial chloroform concentrations inhibited the degradation of methanol. The rate of chloroform degradation at different initial chloroform concentrations is shown in figure 4.

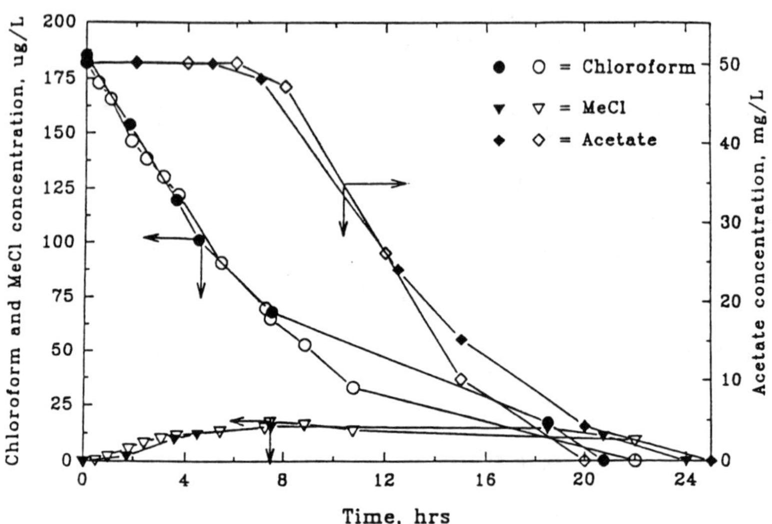

Figure 2. A BMP test for the acetate-fed Methanogenic culture

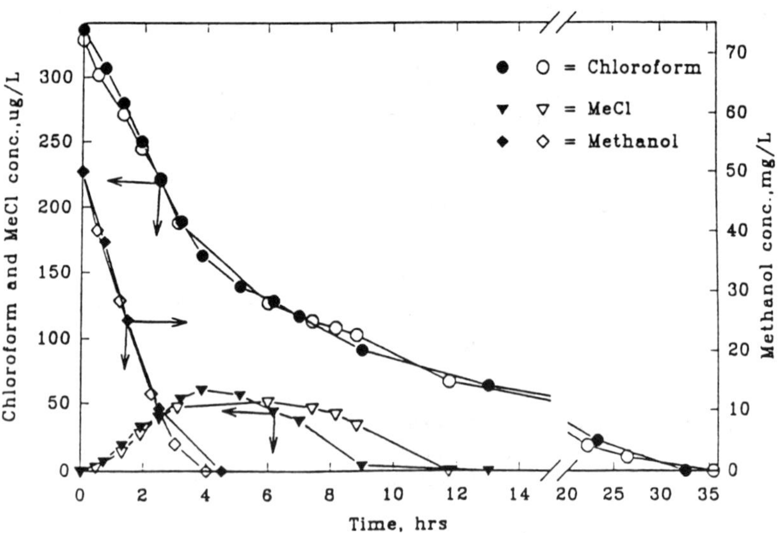

Figure 3. A BMP test for the methanol-fed Methanogenic culture

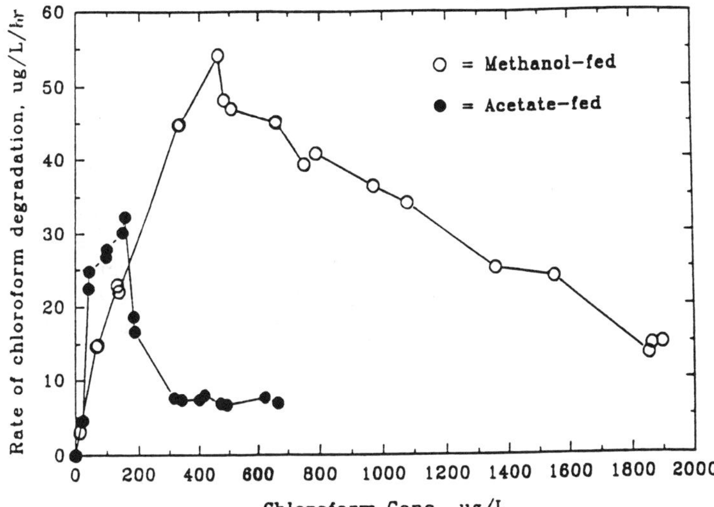

Figure 4. Rate of Chloroform degradation at various initial Chloroform concentrations for Methanogenic cultures.

Methylene chloride was observed as one of the products of degradation of chloroform in both the cultures. It accumulated transiently but was degraded in due course of time. Egli *et al.* (1990) showed that the anaerobic degradation of chloroform can occur either by reductive dehalogenation leading to the formation of methylene chloride or by an oxidative pathway leading to carbon dioxide by a series of unknown reactions.

Comparison of the Acetate fed and Methanol fed Methanogenic Cultures - The comparison of the two cultures indicates that the inhibition of chloroform and its degradation in methanogenic environment depends on the type of primary substrate. Chloroform at any concentration inhibited the degradation of acetate while the methanol degradation was not inhibited for initial chloroform concentrations up to 800 μg/l. At chloroform concentrations less than 200 μg/l, the acetate-fed methanogenic culture exhibited higher chloroform degradation rates while at chloroform concentrations more than 200 μg/l, the methanol-fed methanogenic culture had higher rates of chloroform degradation. The difference became more pronounced at increasing initial chloroform concentrations. These results can be used for interpreting inhibition patterns and in designing reactors treating chloroform.

Conclusions

The degradation of chloroform was studied in methanogenic cultures with acetate and methanol as the primary substrates. The chloroform degradation efficiency of more than 98 % was seen for influent chloroform concentrations up to 2000 μg/l in the chemostats. In the BMP tests, the acetate fed methanogenic culture showed a maximum rate of chloroform degradation of 30-32 μg/l/hr at an initial chloroform concentration of 150 μg/l. Chloroform inhibited acetate degradation and for concentrations higher than 320 μg/l, no acetate was degraded by microorganisms. The methanol fed methanogenic culture exhibited higher chloroform degradation rates for chloroform concentrations more than 200 μg/l and the maximum chloroform degradation rate for this culture was 48-52 μg/l/hr at an initial chloroform concentration of 500 μg/l. Methanol degradation was not inhibited for initial chloroform concentrations up to 800 μg/l.

Acknowledgements

This research was supported by the U.S. Environmental Protection Agency under COE-UC/RREL Cooperative agreement CR-81670 project activity no. WG-2.

References

Bouwer, E.J. and McCarty, P.L.,"Transformations of 1- and 2-carbon halogenated aliphatic organic compounds under methanogenic conditions," Appl.Env.Micro.,45(4):1286 (1983).

Bouwer, E.J., Rittmann, B.E. and McCarty, P.L.,"Anaerobic degradation of halogenated 1- and 2-carbon organic compounds,"Envir.Sc.Tech.,15(5):596(1981).

Egli, C., Stromeyer, S., Cook, A.M. and Leisinger, T.,"Transformation of tetra- and trichloromethane to CO by anaerobic bacteria is a non-enzymatic process,"FEMS Micro. Lett. 68:207 (1990).

Mikesell, M.D. and Boyd, S.A.,"Dechlorination of chloroform by *Methanosarcina* strains" Appl. Env. Micro., 56(4):1198 (1990).

Narayan, B., Suidan, M.T., Gelderloos, A.B. and Brenner, R.C.,"Treatment of VOC's in high strength wastes using an anaerobic expanded-bed GAC reactor,"Wat. Res. 27(1):181 (1993).

Yang, J. and Speece, R.E.,"The effect of chloroform toxicity on methane fermentation,"Wat. Res. 20(10):1273 (1986).

Effects of Molecular Oxygen and pH on the Adsorption of Aniline to Activated Carbon

Peter Fox[1]
Kamalesh Pinisetti[2]

Abstract

This paper examines the influence of molecular oxygen and pH on the adsorption of aniline to F-300 Calgon Carbon. Molecular oxygen increased the adsorptive capacity of GAC for anilines by 250-400% at pH 3, 30-83% at pH 5, 17-42% at pH 9, and 8-45% at pH 11 (higher than those obtained in the absence of molecular oxygen). At pH 7, some of the products formed are poorly adsorbed as evidenced by an increase in UV absorbance in the oxic isotherms as compared to the other isotherms. Oxygen uptake measurements revealed significant consumption of molecular oxygen during the adsorption of aniline compounds. It is speculated that the increase in the GAC adsorptive capacity under oxic conditions was due to the polymerization of these adsorbates on the carbon surface.

Introduction

The effect of molecular oxygen on the adsorption of aromatic compounds to activated carbon can be very significant. Recent studies by Vidic et al. (1990) and Vidic and Suidan (1991) revealed that the adsorptive capacity of granular activated carbon for several phenolic compounds is highly influenced by the presence of molecular oxygen in the test environment. The GAC adsorptive capacity for o-cresol that was attainable under oxic conditions was as much as 200% above that obtained in the absence of molecular oxygen (Vidic et al., 1990; Vidic

[1]Assistant Professor, Department of Civil Engineering, Arizona State University, Tempe, AZ-85287.

[2]Environmental Specialist, Development Engineering Inc., 5110 N. 40th St., Suite 201, Phoenix, AZ-85018.

and Suidan, 1991; and Vidic et al., 1992). A similar phenomenon was determined for the adsorption of phenol, 2-chlorophenol, and 3-ethylphenol (Vidic and Suidan, 1991). Molecular oxygen appears to catalyze the formation of phenolic dimers, trimers, and polymers on the activated carbon surface. These products are strongly adsorbed or occupy less surface area than phenol (Vidic et al; 1992). The influence of molecular oxygen on adsorption has been overlooked in both isotherm studies and column studies. Many compounds contain phenolic groups or have chemical properties similar to phenol and therefore, the adsorption of these compounds could be strongly influenced by molecular oxygen.

Aniline is an aromatic compound that has similar chemical properties as phenol and would be expected to react in a manner similar to phenol. Aniline is considered to be more susceptible to electrophilic substitution than phenol and aniline has a critical oxidation potential (COP) of 1.135 V while the COP of phenol is 1.089 V (Fieser, 1930). Both the COP and amenibility to electrophilic substitution appear to correlate with the ability of molecular oxygen to affect the adsorption of aromatic compounds.

Methodology

This research has studied the influence of molecular oxygen and pH on the adsorption of aniline to F-300 Calgon Carbon (12x40 US Mesh). The bottle-point technique was used for all adsorption isotherm tests. Adsorption isotherm tests were done under anoxic conditions, with aerobic conditions with air as the headspace and under oxic conditions with pure oxygen as headspace. The oxic procedure is similar to the traditionally used techniques and involves placing accurately weighed portions of GAC (+0.1 mg) into a series of 100 ml bottles. The bottles are then purged with pure oxygen for a short period of time (5-10 min.). Air dissolved in the aniline solution is eliminated by stripping with pure oxygen gas for 1 hour. The bottles are then filled with 90 ml of adsorbate solution and sealed with rubber stoppers and aluminium caps. The air from the headspace is replaced by pure oxygen (1-2 min. purging). The described procedure facilitates interference of molecular oxygen from three sources: oxygen entrapped within the pores of GAC particles, the DO present in the aniline solution and the headspace in the bottles.

The anoxic procedure, on the other hand, requires the absence of molecular oxygen from isotherm bottles. This was achieved by displacing the air associated with carbon particles with nitrogen gas, stripping the air from the aniline solution with nitrogen, and purging the headspace in the bottle with nitrogen gas prior to sealing.

Isotherm bottles were placed on a shaker and allowed to equilibrate for at least ten days. Each set of bottles is accompanied by a blank without carbon. All the experiments were performed at room temperature (25°C).

The effects of pH was studied by doing both oxic and anoxic isotherms at pHs of 3, 5, 7, 9, and 11. For pHs of 5 and 9, molecular oxygen was measured using a DO probe immediately before isotherm preparation and after 14 days on the shaker.

Aniline concentrations were measured by UV absorbance using wavelengths of maximum absorbance with a HP 8452A diode-array spectrophotometer. A complete spectrum (190nm-820nm) was recorded for each sample analyzed.

Results and Discussion

The carbon adsorption data was plotted according to the Freundlich isotherm equation, $q_e=k.c^{1/n}$. Freundlich coefficients k and 1/n for both oxic and anoxic adsorption isotherms are listed in Table 1.

pH	Isotherm Type	k	1/n	R
3	Oxic	20.04	0.41	0.990
	Anoxic	8.26	0.33	0.890
5	Oxic	57.24	0.24	0.994
	Anoxic	31.34	0.29	0.990
7	Oxic	2.33	0.84	0.985
	Anoxic	46.98	0.26	0.992
9	Oxic	54.69	0.24	0.995
	Anoxic	38.48	0.26	0.996
11	Oxic	59.44	0.25	0.996
	Anoxic	40.91	0.29	0.999

Table 1. Freundlich Isotherm Equation Parameters

The presence of molecular oxygen in the test environment had a significant impact on the adsorptive capacity of GAC for aniline. Molecular oxygen increased the adsorptive capacity of GAC for anilines by 250-400% at pH 3, 30-83% at pH 5, 17-42% at pH 9, and 8-45% at pH 11 (higher than those obtained in the absence of molecular oxygen). The adsorption isotherms for aniline for oxic, and anoxic cases (pH 5) are presented in Figure 1.

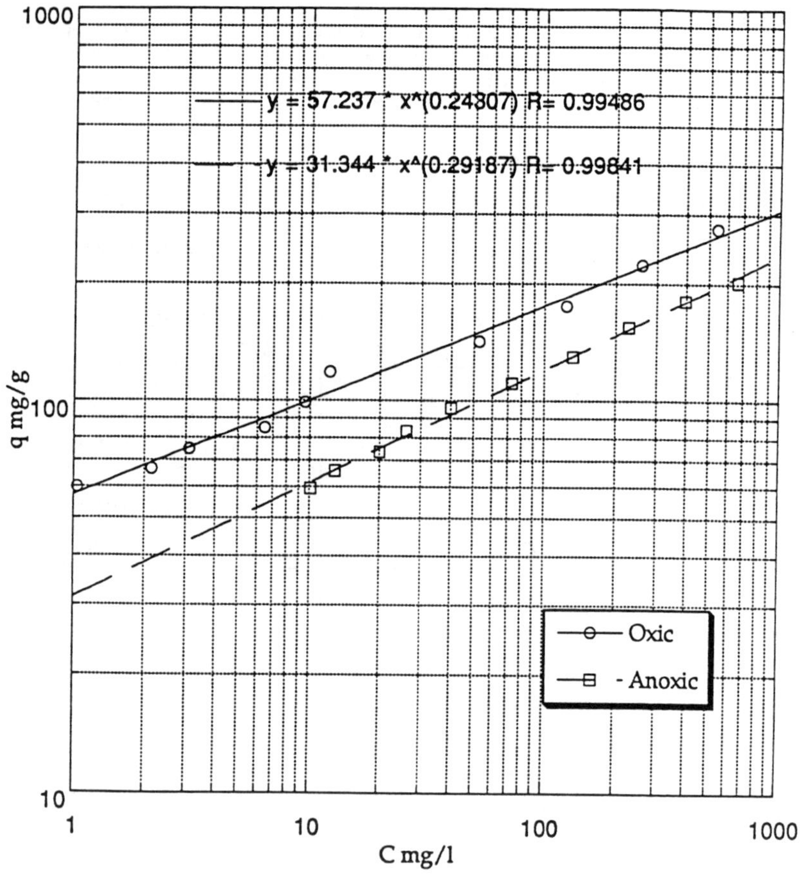

Figure 1. Adsorption Isotherms for Aniline (pH 5)

For pHs of 5 and 9, molecular oxygen was measured using a DO probe immediately before isotherm preparation and after 14 days on the shaker. Figure 2, shows the relationship between the dissolved oxygen consumed and the mass of carbon for both aerobic and oxic conditions. As is apparent from Figure 2, the amount of oxygen consumed during the adsorption of aniline is significantly higher for the oxic condition.

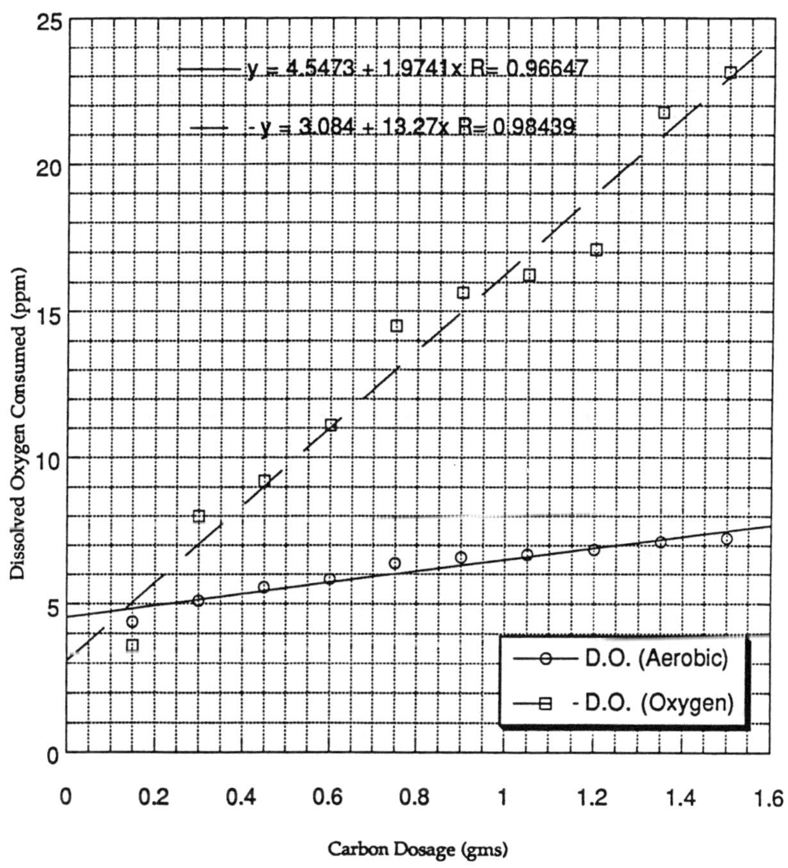

Figure 2. Dissolved Oxygen Consumed v/s Mass of Carbon (pH 5-Aerobic/Oxic)

At pH 7, some of the products formed are poorly adsorbed as evidenced by an increase in UV absorbance in the oxic isotherms as compared to the other isotherms. At pH 7, the observed UV absorbance spectrum of oxic samples was very different than the spectrum of aniline (Figure 3). The absorbing compounds were likely products of the reactions of aniline. This phenomena was not observed at any other pH and it appears that there was a unique reaction at pH 7. The unique reaction at pH 7 might indicate that there is both an acid catalyzed intermediate and a base catalyzed intermediate and that these two intermediates only reacted with one another at pH 7 to form weakly adsorbed products.

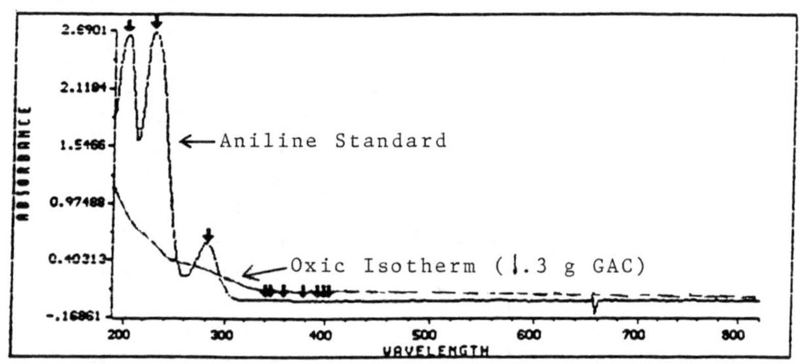

Figure 3. Comparison of Spectroscopic Scans (pH 7)

The pK_b of aniline is 9.3 and at pHs of 7, 9, and 11 the aniline is completely ionized. At pH 5, the majority of the aniline is ionized (66%) and at pH 3, the aniline is present in the unionized state. The presence of molecular oxygen in the test environment exerted a stronger influence on the adsorptive capacity of GAC for aniline when present in the ionized form.

According to Pillai et al. (1982), azocompounds (N=N dimers) have been reported as one class of the major transformation products of substituted anilines and aniline based herbicides under a variety of environmental conditions. According to Iwan et al. (1976); and Voudrias (1985), free radicals on the carbon surface, produced by its oxidation with chlorine, presumably interact with anilines to produce free anilino radicals ($C_6H_5NH.$). Symmetrical pairing of the anilino radicals leads to the formation of azobenzenes (N=N dimers).

Since oxygen is also considered to be strongly electrophilic, it can be speculated that the increase in the GAC adsorptive capacity under oxic conditions is a result of adsorbate polymerization on the carbon surface.

Conclusions

The general effects of molecular oxygen on the adsorption of aniline were similar to previous observations made with phenols. Weakly adsorbed products produced at pH 7 indicate that molecular oxygen might not always increase adsorptive capacity. Oxygen uptake measurements revealed significant consumption of molecular oxygen during the adsorption of aniline compounds. For aniline, pH had a strong effect on the degree to which molecular oxygen influenced adsorption.

References

Fieser, L. F., (1930) An Indirect Method of Studying the Oxidation Reduction Properties of Unstable Systems, Including those from the Phenols and Amines. Journal of American Chemical Society, **52**.

Iwan, J., Hoyer, G.A., and Goller, D., (1976) Transformations of 4-chloro-o-toluidine in Soil: Generation of Coupling Products by One-Electron Oxidation. Pesticide Science, **7**, 621.

Pillai, P., Helling, C.S., and Dragun, J., (1982) Soil Catalyzed Oxidation of Aniline. Chemosphere, **11**, 299.

Vidic, R.D., Suidan, M.T., Traegner, U.K., and Nakhla, G.F., (1990) Adsorption Isotherms: Illusive Capacity and Role of Oxygen. Water Research, **24**, 1187.

Vidic, R.D., and Suidan, M.T., (1991) Role of Dissolved Oxygen on the Adsorptive Capacity of Activated Carbon for Synthetic and Natural Organic Matter. Environmental Science and Technology, **25**, 1612.

Vidic, R.D., Suidan, M.T., and Brenna, R.C., (1992) Oxidative Coupling of Phenolics on the GAC Surface. Water Environment Federation, 65th Annual Conference and Exposition, New Orleans, LA, Sept. 20-24.

Voudrias, E.A., Larson, R.A., and Snoeyink, V.L., (1985) Effects of Activated Carbon on the Reactions of Free Chlorine with Phenols. Environmental Science and Technology, **19**, 441.

The Evaluation of Empty Bed Contact Time on the Biodegradation of Pentachlorophenol using an Anaerobic GAC Fluidized Bed

Gregory J. Wilson[1], Julie A. Wagner[2], Amid P. Khodadoust[3], Makram T. Suidan[4], and Richard C. Brenner[5]

Abstract

The objective of this research is to evaluate the effects of decreasing the empty-bed contact time (EBCT) on pentachlorophenol and its intermediates utilizing an anaerobic granular activated carbon (GAC) fluidized bed. The bioreactor A was fed an 100 mg/l influent concentration of PCP. EBCTs of 1.167 to 0.292 days for reactor A resulted in a PCP reduction greater than 99%. In addition, an equimolar conversion of PCP to monochlorophenol was observed.

Introduction

[1]Graduate Research Assistant, University of Cincinnati, Department of Civil and Environmental Engineering, Cincinnati, OH, 45221-0071

[2]Industrial Process and Air Specialist, CH2MHILL, Milwaukee, WI

[3]Graduate Research Assistant, University of Cincinnati, Department of Civil and Environmental Engineering, Cincinnati, OH, 45221-0071

[4]Professor, Environmental Engineering, University of Cincinnati, Cincinnati, OH, 45221-0071

[5]Chief, Biosystems Engineering Section, Risk Reduction Engineering Laboratory, U.S. EPA, Cincinnati, OH, 45268

Pentachlorophenol has been used worldwide by the wood industry as a preservative and by the agricultural industry as a pesticide and herbicide (Crosley, 1981). Relatively water insoluble and slowly biodegradable, PCP is a priority pollutant found at many wood preserving sites through out the United States. Due to its wide usage, toxic concentration levels of PCP have been observed in lake sediments, industrial and municple sewage, and soils (Bryant, 1991). Both state and federal legislation mandate removal and of PCP from contaminated sites. Previous studies have focused on the aerobic biodegradation including the use of white rot fungi. Anaerobically, much attention has been given to the degradation pathways using various cultures acclimated to chlorinated phenols (Mikesell *et al.*, 1986; Nicholson *et al.*, 1992). However, existing biological treatment technologies have not yet proven to be cost effective, anaerobic GAC fluidized beds provide a promising alternative. The ability to absorb chlorinated compounds as well as providing an excellent attachment surface for biofilms makes GAC an attractive media. Wagner *et al.* (1992) have demonstrated PCP can be degraded using fluidized beds; however, to make this technology cost effective, an investigation into reducing the EBCT was required.

Materials and Methods

An anaerobic fluidized-bed granular activated carbon (GAC) bioreactor (Reactors A) was used to study degradation of PCP in this study. The bioreactor has a total volume of 10 L including the recycle loop and consists of a water jacketed main column, an influent header, and an effluent header. The inner jacket consists of a 96.5-cm long Plexiglas tube with a 10.2-cm internal diameter, surrounded by an outer Plexiglas tube. The influent header, comprised of Plexiglas, rests at the bottom of the main column. Marbles are packed within the influent header to aid in the uniform distribution of the fluidizing water across the column, while preventing media entering the column's recycle line. The Plexiglas effluent header provides a sampling port for the effluent as well as a gas collection port. The reactor has a side arm extending below the recycle inlet allowing for carbon replacement, if necessary.

The feed solutions were introduced into the suction side of the recycle loop. The feed solutions consisted of an inorganic buffer solution and a nutrient solution, fed separately to minimize any external biological activity. Each feed reservoir was equipped with a constant-speed pump controlled through a programmable timer. The PCP solution was dissolved in ethanol and fed into the suction side of the recycle loop using a Model 11 high precision syringe infusion pump (Harvard Apparatus, Inc., South Natick, MA) with a 10-mL fixed needle syringe (Hamilton Company, Reno, Nevada) via 1/16-in. A316 stainless steel tubing. Initially, 1 kg of 16 x 20 U.S. Mesh F400 GAC (Calgon Corporation, Pittsburgh, PA) was placed in each of the bioreactors. The recycle rate was set to achieve a 30% expansion from the initial carbon height. To maintain mesophilic conditions, a constant temperature bath provided 35°C water circulating through the outer jacket.

PCP (99%, Aldrich Chemical Co., Milwaukee,WI) and ethanol (95%, USP Grade, Midwest Grain Products, Weston, MO) comprised the synthetic solution fed by the syringe pumps. The nutrient solutions for the bioreactor provided the essential vitamins and salts needed for microbial growth. The inorganic buffer solution for reactor A consisted of sodium carbonate, sodium sulfide, and some ethanol. The role of the buffer solution was to ensure a relatively constant pH while providing another vector for introducing the remaining ethanol. SuperQ quality grade water was used for all solutions.

Daily monitoring of the bioreactor included gas volume production, buffer and nutrient flow rates, syringe flow rates, and pH. Gas production was measured using wet tip gas meters (Environmental & Water Research Engineering, Nashville, TN). The pH was measured using an Orion Model 720A pH meter (Orion Research Co., Boston, MA). Weekly samples were taken for effluent (including PCP and its degradation products), gas composition, chemical oxygen demand (COD), chlorides, volatile fatty acids (VFAs), and alcohols. The effluent concentrations of PCP, tetra-, tri-, and dichlorophenols were analyzed on a Hewlett-Packard 5890 Series II gas chromatograph (Hewlett Packard, Palo Alto, CA) using an electron capture detector (ECD). The ortho-, para-, and metachlorophenols and phenol were analyzed on the same gas chromatograph using a flame ionization detector (FID). A Hewlett-Packard 5890 Series II gas chromatograph (Hewlett-Packard, Palo Alto, CA) equipped with a thermal conductivity detector was used to determine the percentage of carbon dioxide, oxygen, nitrogen, and methane in the product gas. CODs were determined using the Hach low range digestion vials (0-150 mg/L) and a Hach COD reactor Model 45600 (Hach Co., Loveland, CO). The digested vials were measured for transmittance on a Bausch & Lomb Spectronic 70 Spectrophotometer (Bausch and Lomb, U.S.A.). Chloride concentrations in effluent samples were determined using an Orion Model 9617BN chloride selective electrode. VFA and alcohol concentrations were analyzed on a Hewlett-Packard 5890 Series II gas chromatograph (Hewlett-Packard, Palo Alto, CA) equipped with a flame ionization detector. Carbon extractions were conducted periodically using the procedure outlined by Fox *et al.* (1989).

Results and Discussion

Bioreactor A has been operating in excess of 950 days, however this study began on the 480th day. Wagner *et al.* have previously reported research in the first 480 days of operation including acclimation and isotherms. The goal of this study is to evaluate the effects of decreasing the EBCT in the anaerobic GAC fluidized beds while maintaining a constant influent PCP concentration of 100 mg/L. Decreased operating cost is the advantage of optimizing the EBCT. In order to reduce the EBCT by 50% requires the PCP mass loading to be doubled as seen in Table 1.

Phase	Days of Operation	PCP Feed (g/d)	Ethanol (g/d)	Flow Rate (L/d)	EBCT (hours)
I	480-606	0.60	4.28	6.0	28.01
II	607-824	1.20	8.33	12.0	13.99
III	825-present	2.40	16.66	24.0	7.01

Table 1. Bioreactor Operation Summary

Flow rates for Reactor A was arbitrary selected to be 6.0 corresponding to a PCP loading of 0.60 g/day during phase I. More than 99% of the PCP was degraded during phase I as indicated by Figure 1. Additionally, the intermediates are represented in Figure 1 including mono-, di-, and trichlorophenols. Concentration are expressed in mmol/L to show nearly an equimolar conversion of PCP to monochlorophenol throughout all phases of operation. Influent PCP and effluent (no biological activity) concentrations were calculated using a model developed in our lab.

Figure 2 show a COD balance on the reactor. Influent COD loadings account for the PCP, ethanol, and trace vitamins present in the nutrient solution. The effluent and gas COD represent 75 to 80% of the influent COD throughout the three phases with the remaining 20 to 25% going to biomass production. In addition, the effluent gas accounted for the majority of total effluent COD loadings. However, at higher PCP loadings, a lower percentage of influent COD going to effluent gas and biomass to both reactors resulted in suggesting PCP toxicity is inhibiting the conversion. Figures 2 and 3 show stable performance during each phase of operation as measured by the COD, volatile fatty acids (VFA), and alcohol concentrations. Fluctuations in VFA and alcohols did occur during periods of increased loadings returning, however, to stable levels.

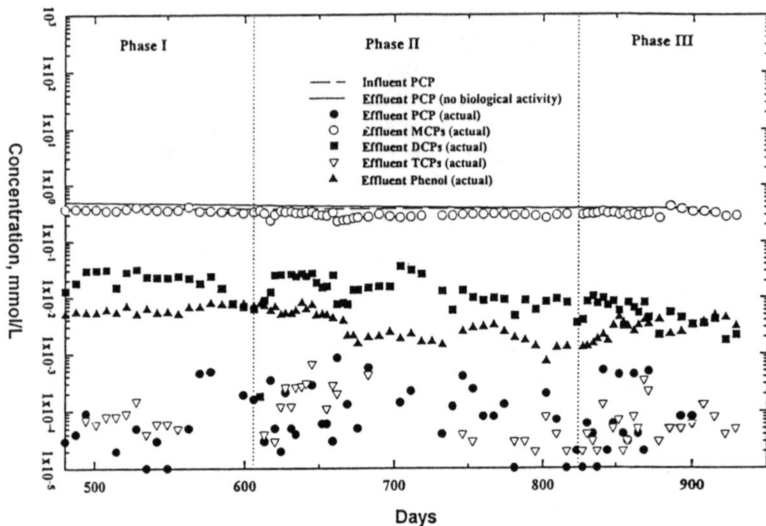

Figure 1. PCP and Intermediates Effluent Concentrations.

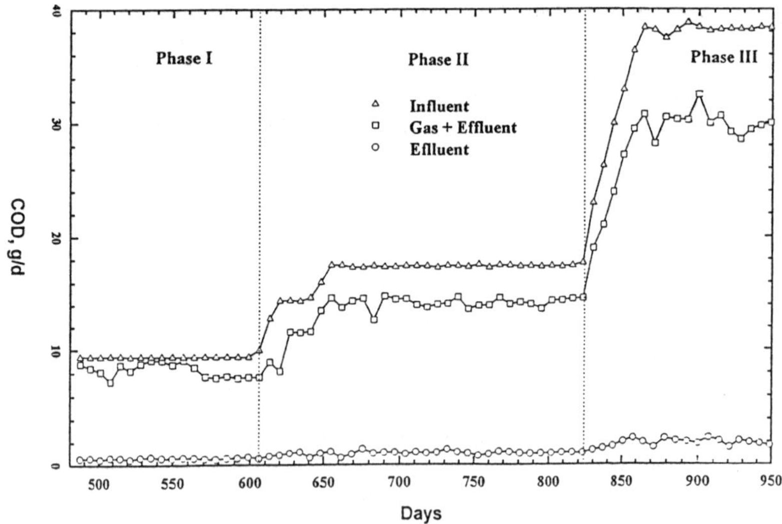

Figure 2. COD Balance.

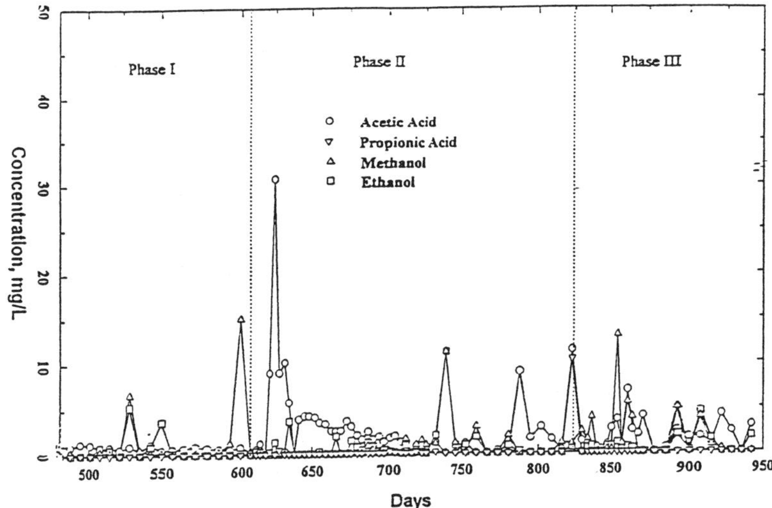

Figure 3. Volatile Fatty Acids and Alcohols.

Figure 4 represents chloride balances for all phases of operation. The chloride potential is defined as the equimolar amount of chloride from all potential sources (i.e., all chlorinated phenols in the feed). The delta chloride represents the difference between the measured effluent chloride concentration and concentration of chloride in the influent. Effluent concentrations lower than the chloride potential indicate PCP or its derivatives are either adsorbing on the GAC or incomplete dechlorination is occurring. Throughout phase 1, the delta chloride remained above the chloride potential in the bioreactor suggesting possible desorption of chlorinated compounds from the GAC medium. The chloride concentration was used as an indicator of performance throughout the study.

Carbon extractions were performed periodically to determine the mass of adsorbed phenols. The data in Figure 5 suggest that the total adsorbed monochlorophenols remained relatively constant in Reactor A; however, the distribution between 3- and 4-chlorophenol was changing. Metachlorophenol (3CP) became the dominant monochlorophenol absorbed on the carbon as the EBCT decreased. Dichlorophenol decreased with a reduction in the EBCT while tri-, tetra-, and pentachlorophenol remained relatively constant. The total monochlorophenols decreased with 3-chlorophenol being the dominant species as the EBCT decreased.

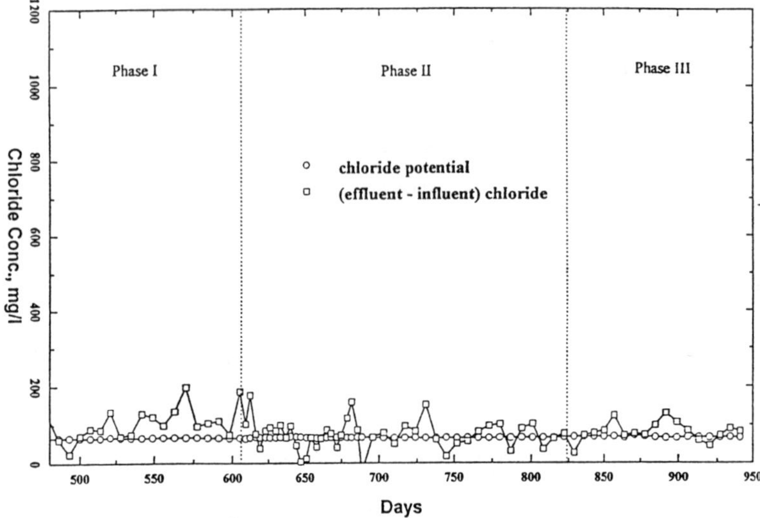

Figure 4. Chloride Balance.

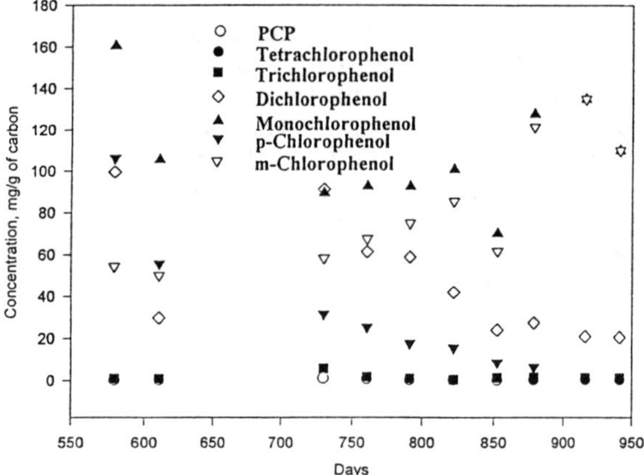

Figure 5. Carbon Extractions.

Conclusion

The anaerobic GAC fluidized-bed bioreactor has been demonstrated to effect greater than 99% removal of the influent PCP concentration. Additionally, equivalent molar concentrations of monochlorophenols were observed in the effluent through all phases of operation with 3-chlorophenol and dichlorophenols adsorbing on the GAC the most. Decreasing the EBCT threefold did not inhibit degradation of PCP and its intermediates, thus allowing removal of PCP at much lower detention time providing a significant cost advantage.

Acknowledgments

This research was made possible through Cooperative Agreement CR821029 from the U.S. Environmental Protection Agency. The findings and conclusion expressed in this paper are solely those of the authors and do not necessarily reflect the views of the Agency.

References

Bryant, F.O., D.H. Hale, and J.E. Rogers, "Regiospecific dechlorination of pentachlorophenol by dichlorophenol-adapted microorganisms in freshwater, anaerobic sediment slurries," Appl. Environ. Microbiol. 57: 2293-2301 (1991).

Crosby, D.G., "Environmental chemistry of pentachlorophenol," Pure Appl. Chem. 53: 1051-1080 (1981).

Edgehill, R.U. and R.K. Finn, "Microbial treatment of soil to remove pentachlorophenol," Appl. Environ. Microbiol. 45: 1122-1125 (1982).

Fox, Peter, "Innovative reactor design for the treatment of biologically inhibitory wastewater," Ph.D Thesis - Environmental Engineering, Department of Civil Engineering, University of Illinios at Urbana-Champaign (1989).

Guthrie, M.A., E.J. Kirsch, R.F. Wukasch, and C.P.L. Grady, Jr., "Pentachlorophenol degradation-II," Water Res. 18: 451-461 (1984).

Mikesell, M.D. and S.A. Boyd, "Complete reductive dechlorination and mineralization of pentachlorophenol by anaerobic microorganisms," Appl. Environ. Microbiol. 52(4): 861-865 (1986).

Nicholson, D.K., S. Woods, J.D. Istok, and D.C. Peek, "Reductive dechlorination of chlorophenols by a pentachlorophenol-acclimated methanogenic consortium," Appl. Environ. Microbiol. 58(7): 2280-2286 (1992).

Wagner, J.A. Khoudadoust, A.P., Suidan, M.T., and Safferman, S.I., "Treatment of PCP containing wastewater using anaerobic fluidized-bed GAC bioreacters," Proceedings of the 1993 Water Environment Federation Conference, Water Environment Federation, Los Angles (1993).

Removal of Arsenic by Enhanced Coagulation and Membrane Technology

S. David Chang[1], Hector Ruiz[1], William D. Bellamy[2], Carl W. Spangenberg[3], and Debra L. Clark[3]

Background

The U.S. EPA is promulgating a new MCL for arsenic based on epidemiology studies. The concern from the water industry is that the new MCL will be significantly lower than the current MCL of 50 μg/L. The new MCL, some speculate may be as low as 2 μg/L or less, is expected to have a significant impact on water utilities that depend on groundwater supply or use aquifer storage and recovery technology for potable water resources.

A questionnaire sent to the participants at the Arsenic Workshop (May, 1993) sponsored by Association of California Water Agencies and the Metropolitan Water District of Southern California indicated that 66% of the 50 water agencies responded will not be in compliance if the new MCL is set at 5 μg/L. The issues concern most of the water utilities are compliance cost and treatment technologies. There are very few, if any, water utilities that are treating water with the objective of removing arsenic to levels less than 2 μg/L. Some technologies such as reverse osmosis are capable of removing arsenic to a very low level, however, there is no data available due to the analytical limitation. All these information are critical to the development of treatment technologies and compliance cost.

Experimental Approach

A pilot study was conducted at the Irvine Ranch Water District (IRWD) in Santa Ana, California to develop local groundwater resources. Arsenic removal is evaluated simultaneously with natural organic material removal. Treatment technologies evaluated were conventional treatment processes and membrane technology (microfiltration, nanofiltration, and reverse osmosis).

---

[1]CH2M HILL, 2510 Red Hill Avenue, Santa Ana, CA 92705
[2]CH2M HILL, 6060 South Willow Drive, Englewood, CO 80222
[3]Irvine Ranch Water District, 15600 Sand Canyon Avenue, Irvine, California 92718

The IRWD pilot facility takes water from two confined aquifers with different organic contents. The water quality of these two aquifers are shown in Table 1. The pilot facility has the capability of blending the water from these two aquifers to evaluate the impact of organic content on arsenic removal. Arsenic was spiked into the groundwater at 20 to 50 μg/L in the form of arsenate (+5).

Conventional Treatment

Bench-scale Testings. Jar tests, with both alum (at pH 6) and ferric chloride as coagulant, has conducted to determine chemical dosages required to achieve different levels of effluent quality. Following coagulant addition, the water was rapid-mixed at 100 rpm for 60 seconds, flocculated at 35 rpm for 15 minutes, and allowed to settle for 30 minutes. The supernatant was then filtered through a 0.45-μm membrane filter. To confirm that test results of water spiked with arsenic can be applied to waters with naturally existing arsenic, groundwater contaminated with arsenic, Albuquerque, New Mexico and Reno, Nevada, were tested to compare the results to the bench-scale testings.

Pilot-scale Testings. The conventional treatment train includes a 150-liter-per-minute (40 gpm) package treatment plant (rapid mixing, flocculation, and sedimentation) followed by filter columns. Enhanced coagulation with ferric chloride was conducted to confirm jar test results.

Membrane Technology

The membrane testing facilities at the IRWD pilot facility include a pilot-scale microfiltration unit (Memcor Microfiltration System, Memtec America Corporation), a membrane screening unit, and a full-scale membrane train.

Memcor Microfiltration System. The Memcor is a self-cleaning, continuous microfiltration system. The Memcor unit utilizes nominal 0.2 micron hollow fiber membranes to remove suspended solids. An advanced gas backwash technique, developed by Memtec, is used to remove accumulated solids from the membrane surface and maintain membrane performance (Vickers, 1992). In drinking water applications, the Memcor system can be operated under direct filtration mode to meet the SWTR filtration requirements and replace conventional flocculation, sedimentation, and filtration processes. This is especially attractive in wellhead groundwater treatment applications. Memcor system requires chemical (coagulant) addition to be effective for arsenic removal. Ferric chloride was used as coagulant. The Memcor system was tested in parallel with the conventional treatment train to

Table 1. Raw Water Quality

Aquifer	Depth (m)	pH	Turbidity (NTU)	Color (CU)	TOC (mg/L)	Arsenic[A] (μg/L)
Shallow	60	7.5	0.5	<5	0.5	20-50
Deep	600	8.5	1.5	180	11	20-50

A: Spiked arsenate concentration

compare the arsenic removal efficiency.

Membrane Screening Unit. The screening unit is a single 2-1/2-inch element test unit. It allows multiple element types to be evaluated independently in a relative short period of time. All the membranes were screened at the flux rate of 0.61 m^3/m^2/day (15 gallons/ft^2/day).

Full-Scale Membrane Train. The full-scale unit is an 18-element membrane test unit composed of standard 4-inch production type membrane elements. It includes six pressure vessels arranged in a 2-2-1-1 array that allows operation at hydraulic recovery rates of a full-scale facility.

Results and Discussion

Jar Tests

Jar tests, with both alum (at pH 6) and ferric chloride, were conducted to determine chemical dosages required to achieve different levels of arsenic removal.

Jar Tests with Spiked Arsenate. Figure 1 shows the jar test result of shallow aquifer groundwater (TOC=0.4 mg/L) spiked with 50 μg/L of arsenate. As indicated in Figure 1, ferric chloride is more effective than alum (pH = 6) in removing arsenic. One major observation during the jar test was that no settlable floc formed through the 15-minute flocculation period for tests with alum dosage under 80 mg/L and ferric chloride dosage under 40 mg/L. This is because of the low particulate (turbidity) content of groundwater. Figure 2 shows the results of a jar test which was performed with a blended groundwater (TOC of 3.5 mg/L). A comparison of both jar tests indicate that, TOC impaired the arsenic removal for coagulant dosages (both alum and ferric chloride) under 20 mg/L, however, TOC enhanced the arsenic removal for coagulant dosages above 40 mg/L. The existence of organic materials (TOC of 3.5 mg/l) seemed enhanced floc formation during the jar tests and improved the removal of arsenic.

Jar Tests with Groundwater with Naturally Occurring Arsenic. Groundwater from Albuquerque, New Mexico and Reno, Nevada was tested to confirm the jar results from spiked groundwater. Figure 3 shows the results for the groundwater from Albuquerque. The water had an arsenic content of 19 μg/L and a TOC content of 0.3 mg/L. The arsenic content in the groundwater is 100% arsenate. The levels of filtered arsenic in Figure 3 are similar to that in Figure 1. This indicates the results obtained from spiked groundwater can apply to groundwater that originally contains arsenate. No visible floc formation was observed during the jar tests.

Jar tests were also performed on groundwater from Reno. The removal of arsenic was approximately 25%. The speciation of arsenic in the Reno groundwater is not known at this time. It is suspected that the majority of the arsenic in the Reno groundwater may be in the form of arsenite.

Lack of settlable floc formation with low turbidity and low TOC groundwater indicate that conventional sedimentation may not be effective in removing the very

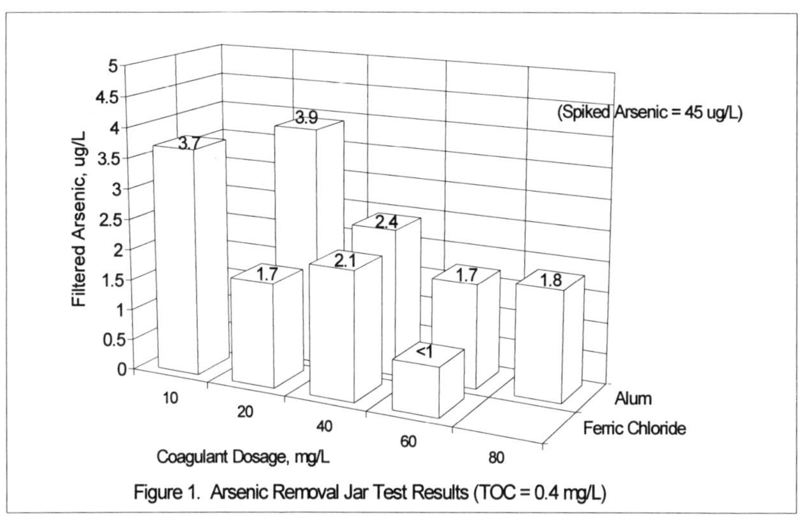

Figure 1. Arsenic Removal Jar Test Results (TOC = 0.4 mg/L)

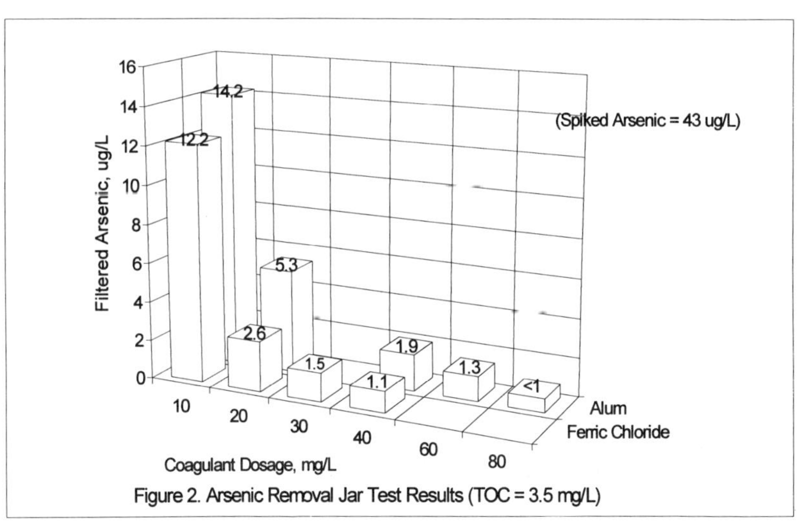

Figure 2. Arsenic Removal Jar Test Results (TOC = 3.5 mg/L)

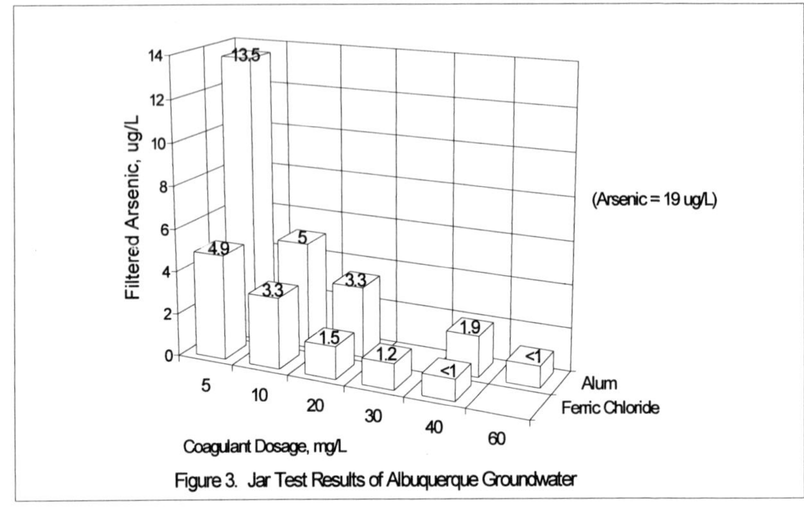

Figure 3. Jar Test Results of Albuquerque Groundwater

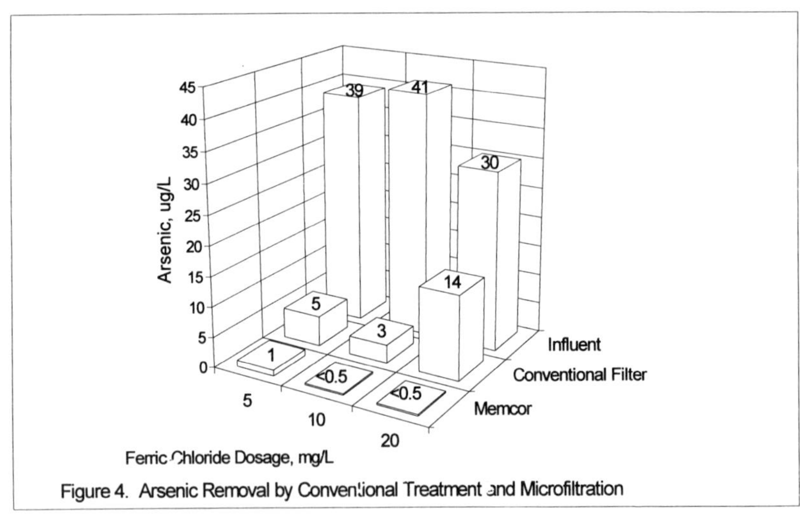

Figure 4. Arsenic Removal by Conventional Treatment and Microfiltration

fine floc formed by enhanced coagulation and filter will have to remove all the solids. This will lead to short filter run time unless some nuclei are added to the water to enhance floc formation and settling.

Conventional Treatment Train and Memcor System Testings

The conventional package unit was used to conduct two phases of testings. The first phase tested the organic and arsenic removal using water from the deep aquifer. The second phase was a parallel testing of conventional treatment and Memcor System. Ferric chloride was used as the coagulant because the jar test results suggested that it is more effective than alum.

Arsenic Removal in High-TOC Groundwater. Arsenate was spiked to high-TOC (11 mg/L) water from the deep aquifer at a level of 40 μg/L. The objective was to remove the TOC content of the groundwater to a level of less than 2 mg/L and observe the arsenic removal that can be achieved. The conventional treatment train was operated under the following conditions: flocculation time of 20 minutes, sedimentation basin (with tube settler) hydraulic loading of 0.06 $m^3/m^2/min$ (1.5 gpm/ft^2), and a filtration rate of 0.33 $m^3/m^2/min$ (8 gpm/ft^2). The ferric chloride dosage required to achieve the TOC removal objective was 120 mg/L with 10 mg/L of cationic polymer. The arsenic content in all filtered water samples were less than the detection limit (1 μg/L). The removal efficiency was more than 97%. As the results indicate, for groundwater with high TOC content, arsenic will be removed simultaneously with organic matter by enhanced coagulation.

Conventional Treatment and Memcor Microfiltration System. The jar tests results indicate that a direct filtration system, like the Memcor System, can achieve high particle removal efficiency and handle high solids loading may be effective to remove the very fine floc formed during the enhanced coagulation process of low turbidity groundwater. A series of parallel testings of conventional treatment and Memcor System was performed to compare the arsenic removal efficiency. The shallow aquifer groundwater (TOC=0.4 mg/L) spiked with approximated 40 μg/L of arsenate was used for the tests. After ferric chloride addition, a flow of 7.6 liters per minute (2 gpm) was diverted from the flocculation basin to the Memcor System. The Memcor System was operated at the flux rate of 2.7 $m^3/m^2/day$ (67 $gallons/ft^2/day$) and the recovery was 95% (5% reject or waste stream). The feed pressure required was 207 kilopascal (30 psi). Figure 4 summarizes the test results. Memcor system consistently out-performed the conventional filter. The arsenic concentration in the permeate was 1 μg/L or less for all the conditions tested.

Membrane Screening and Full-Scale Testings

Five different membranes were tested with the membrane screening unit and the full-scale membrane unit. The results are summarized in Table 2. As shown in Table 2, TS-60 membrane can not remove arsenic.

Nanofiltration membrane such as NF-70 shows high arsenic rejection (>97%) at 15% recovery (15% permeate flow and 85% reject flow). This is consistent with the bench-scale test results of Thompson and Chowdhury (1993) with the NF-70

Table 2
Membrane Testing Results

Membrane	Type	Feed Pressure (psi)	Flux (gfd)	Recovery (%)	Arsenic Concentration (μg/L)		Rejection (%)
					Feed	Permeate	
Membrane Screening Unit							
TriSep TS-60	UF/NF	17	15	15	29	29	0
Filmtec NF-70	NF	37	15	15	44	1	98
		39	15	31	47	8.3	82
TriSep TS-80	NF	48	15	15	46	<1	>97
		49	15	30	47	<1	>97
TriSep ACM1	LPRO	88	15	15	42	<1	>97
		85	15	31	55	2	96
Fluid System TFCL-HR	LPRO	142	16	15	47	1	98
		150	16	30	50	3	94
Full-scale Membrane Unit							
Hydranautics Nitto-Denko NTR 7450	UF/NF	90	20	90	18	15	17
		90	20	90	45	38	16
TriSep TS-80	NF	113	15	65	20	7	65
		105	20	90	19	16	16

UF: Ultrafiltration Membrane NF: Nanofiltration Membrane LPRO: Low Pressure Reverse Osmosis Membrane

membrane. But, the arsenic rejection by NF-70 deteriorated significantly at 31% recovery as shown in Table 2.

Nitto-Denko 7450 membrane was tested for arsenic removal at full-scale in parallel with organic removal. It is effective in removal the natural organic materials in the deep aquifer groundwater, but is not effective in rejecting arsenic under the test conditions.

TS-80 membrane which shows high rejection (>97%) of arsenic at both 15% and 30% recovery with screening unit fail to achieve high level of rejection when tested in full-scale. The arsenic rejection is 65% when operated at 75% recovery and further dropped to 16% when operated at 90% recovery. This is indicates nanofiltration membranes and are not tight enough to remove arsenic although high percentage of arsenic rejections were observed under low recovery test conditions. Low pressure reverse osmosis membrane may be required for effective arsenic removal at high recovery levels.

The ACM1 and TFCL-HR membranes both show high arsenic rejection with screening unit. The full-scale testing of low pressure reverse osmosis membranes can not be performed due to schedule conflict with the organic removal testings at the IRWD Pilot Facility.

Conclusions

- Ferric chloride is more effective than alum in arsenic removal for the groundwater tested. To achieve same level of arsenic removal, ferric chloride dosage required is about half of the alum dosage required.

- Microfiltration is more effective than conventional filter in removing fine particles formed during enhanced coagulation and ,thus, achieved higher arsenic removal than conventional filter.

- Nanofiltration membranes can reject arsenic at low recovery levels (15 to 30%), but is ineffective when operated at high recoveries (75 to 90%). Low pressure reverse osmosis membrane may be required for arsenic removal at high recovery level.

References

Association of California Water Agencies and Metropolitan Water District of Southern California, Arsenic Workshop Questionnaire Report, 1993 Arsenic Workshop for Utility Managers and Water Quality Specialists, May 1993.

Thompson, M. A. and Z. K. Chowdhury, Evaluating Arsenic Removal Technologies, 1993 American Water Work Association Annual Conference, San Antonio, Texas, June 1993.

Vickers, J. C., Memcor Microfiltration System Design - An Engineering Perspective, Memtec America Corporation, March 1992.

ARSENIC REMOVAL VIA SOFTENING

Laurie McNeill and Marc Edwards[1]

Abstract

The removal of arsenic during softening was investigated using synthetic solutions. Arsenic removal is facilitated by formation of a variety of solids including $Mg(OH)_2$, calcite and $Mn(OH)_2$. At pH > 11.0 arsenic removal was independent of magnesium concentration for initial concentrations of 20 or 50 mg/L Mg^{+2}, but arsenic removal was improved at lower pHs at the higher initial magnesium concentration. For systems initially containing only Mn^{+2}, arsenic removal appears to be through sorption onto $Mn(OH)_2$ solids rather than due to formation of a $Mn_3(AsO_4)_2$ precipitate. The presence of trace amounts of orthophosphate hindered overall removal of arsenic from synthetic groundwater solutions.

Introduction

Arsenic (As) is a common trace inorganic contaminant in drinking water supplies that has recently been identified as a significant health risk. The current assessment, which is subject to considerable debate, ranks arsenic in drinking water as a risk comparable to that posed by exposure to tobacco smoke and indoor radon gas [Waterweek, 1993]. Accordingly, the US Environmental Protection Agency (EPA) is expected to lower the current limit on arsenic in drinking water from 50 μg/L to a range of 0.5 - 5 μg/L.

Although a variety of highly effective arsenic removal techniques exist, their implementation often involves considerable cost and/or modifications to existing water treatment processes. Fortunately, previous studies indicate that arsenic may be effectively removed during water softening and coagulation processes, two treatment steps that are currently in place at many existing facilities. Because higher levels of arsenic often seem to occur in regions of the US with high hardness waters, improving arsenic removal by softening may be an efficient and cost-effective means of meeting the new standard for many utilities.

---

[1]Department of Civil Engineering, 4-28; University of Colorado; Boulder, CO 80309

Not much is currently known regarding arsenic removal during softening processes. Logsdon and Sorg (1978), who have conducted the bulk of the experimental work in this area, illustrated that As(V) may be removed in significant amounts (>90%) above pH 10.5. However, their studies began with initial arsenic concentrations above 300 μg/L and fundamental removal mechanisms were not identified. Additional research is obviously needed to improve fundamental understanding and to allow the rational optimization of low-level arsenic removal in drinking water sources.

Our experiments were aimed at elucidating arsenic removal mechanisms during water softening. Planning these experiments required some understanding of arsenic chemistry. Arsenic is a naturally-occurring toxic metal with four stable oxidation states (depending on the redox conditions) of -3, 0, +3, and +5. In oxygenated waters, arsenic is principally present in the +5 state as arsenate (arsenic acid, H_3AsO_4). Under reducing conditions arsenic typically occurs in the +3 state as arsenite (arsenious acid, H_3AsO_3). If sulfide is present, one of two solid forms may be stable including realgar (AsS) and orpiment (As_2S_3). Two other arsenic solids of interest include $Ca_3(AsO_4)_2$ and $Mn(AsO_4)_2$. Although the formation of the latter solids have not been considered previously, it is possible that these solids might be important in removing arsenic via precipitation reactions during water softening. In addition, arsenic might either sorb or co-precipitate with other solids formed during softening including $Mg(OH)_2$, $Mn(OH)_2$, and $CaCO_3$.

Procedure

The basic experimental procedure was to conduct simple softening experiments in batch to study arsenic removal from synthetic groundwater solutions. Solids formed by softening were removed from the water by centrifugation (@300 rpm for 10 minutes) followed by filtration through a 0.45 μm pore size membrane filter. Arsenic remaining in solution was determined using the colorimetric method of Johnson et.al (1977). We estimate that our accuracy with this technique is +/- 5 μ g/L. Initial arsenic concentrations were varied from 75 μg/L to 150 μg/L in a "synthetic groundwater" that also contained the following ions: 300 mg/L alkalinity as $CaCO_3$, 2 mM Ca^{+2}, 32 μg/L PO_4^{-3}, 1 mg/L Mn^{+2}, 20 mg/L Mg^{+2}, and 2 mg/L F^-. This synthetic groundwater was then softened with incremental dosing of lime (CaO) to predetermined final pH values. Results were compared to similar experiments in which components of the groundwater were altered. The initial experiments were aimed at isolating mechanism(s) of arsenic removal during softening and examining potential competition from trace orthophosphate.

Results

Softening of the synthetic groundwater solution indicated that arsenic removal is nearly 95% above pH 10.5 (Figure 1). This experiment was repeated for an identical water without ortho-phosphate. Total arsenic removals in these two systems (Figure 1) were approximately the same, but there is somewhat greater removal in the system without orthophosphate at high pH. On the basis of these results it appears that trace orthophosphate can compete with arsenic for removal during softening. This observation is not surprising given the similar chemistry of orthophosphate and orthoarsenate.

It has been hypothesized that arsenic removal during softening is due to sorption/co-precipitation onto $Mg(OH)_2$. To test this hypothesis arsenic was added to two solutions, one containing only 20 mg/L Mg^{+2} and the other containing 50 mg/L Mg^{+2}. After the waters were softened (Figure 2) using NaOH (to avoid confounding effects of calcium in the results) both waters exhibited about 90% removal of arsenic at pH 11.5. Calculations based on solubility of $Mg(OH)_2$ predict that > 97% of the Mg^{+2} is precipitated at pH 11.0 in each system. Since little additional solid can form above this pH, it is not surprising that additional improvements to arsenic removal did not occur at pHs greater than 11.0. As a final point, the water with the higher magnesium concentration had somewhat higher removals at the lower pH values tested. Since more $Mg(OH)_2$ is predicted to form at lower pH values in the system with higher initial Mg^{+2}, this result is consistent with arsenic removal occurring via $Mg(OH)_2$ precipitation/co-precipitation.

Waters with high levels of arsenic often have high concentrations of manganese [Hounslow, 1980; Korte, 1991; Mattess, 1981; and Masscheleyn et.al., 1991]. Our preliminary calculations indicate that solids of manganese ($Mn_3(AsO_4)_2$ and $Mn(OH)_2$) may also be influential in arsenic removal. To test this prediction, we examined two waters containing only 1 mg/L Mn^{+2}, one with 75 μg/L initial As(V) and the other with 150 μg/L As(V). After softening with NaOH to pH > 10.5 (Figure 3) there was about 50% removal in the system with 75 μg/L As(V) versus about 30% removal in the 150 μg/L solution. However, the total amount of arsenic removed (in μg/L) is approximately equal for both conditions (Figure 3). This behavior is more consistent with an adsorptive rather than precipitative arsenic removal mechanism. If removal occurred via precipitation as $Mn_3(AsO_4)_2$, equilibrium predictions suggest that the amount of arsenic remaining in solution would be approximately constant regardless of initial arsenic concentration (over the initial arsenic concentration range tested).

Other experiments have indicated that at low concentrations (75 μg/L) arsenic is not significantly removed by formation of the calcium arsenate, $Ca_3(AsO_4)_2$, at least at calcium concentrations as high as 2 mM Ca^{+2}. Future studies will determine if arsenic is removed during softening by co-precipitation with $CaCO_3$ or apatite ($Ca_5(PO_4,AsO_4)_3OH$), and to examine the removal of As(III) by softening processes.

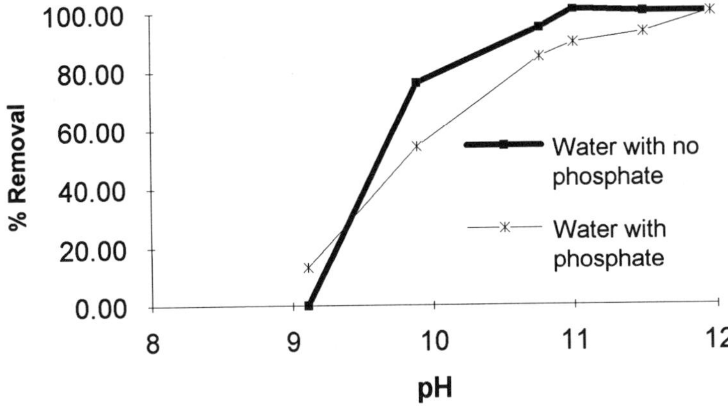

Figure 1: Removal of arsenic (V) from a synthetic groundwater (300 mg/L alkalinity as $CaCO_3$, 2 mM Ca^{+2}, 32 μg/L PO_4^{-3}, 1 mg/L Mn^{+2}, 20 mg/L Mg^{+2}, and 2 mg/L F^-). Initial [As] = 75 μg/L. Note that the presence of orthophosphate hinders arsenic removal at higher pHs even though the same final removal (@ pH 12) is obtained.

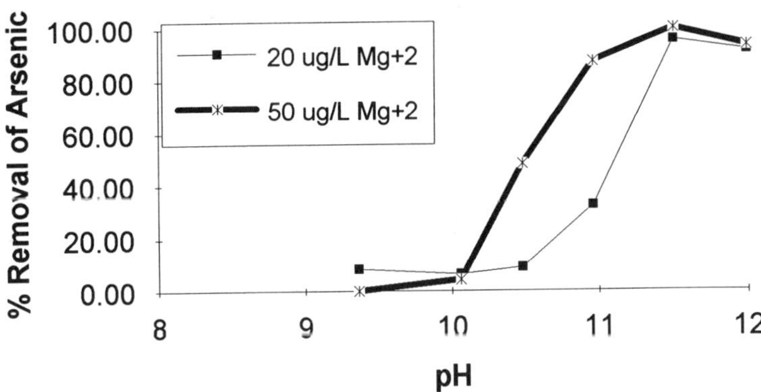

Figure 2: Removal of As(V) from a system initially containing 20 μg/L Mg^{+2} or 50 μg/L Mg^{+2}. Initial [As] = 75 μg/L. Arsenic removal was approximately 90% at higher pH in both waters, but the system with the higher initial concentration of Mg^{+2} removes more arsenic at the lower pH values.

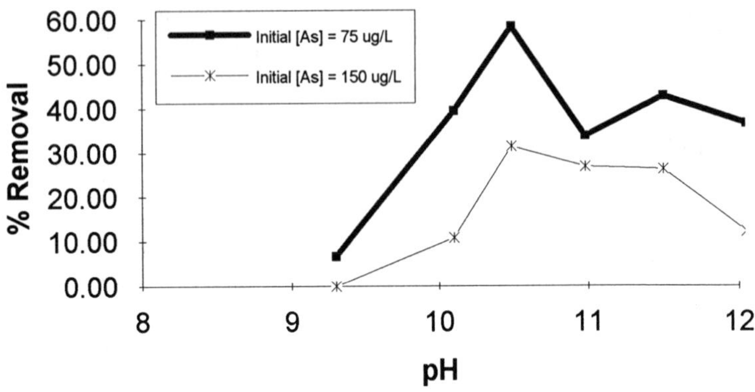

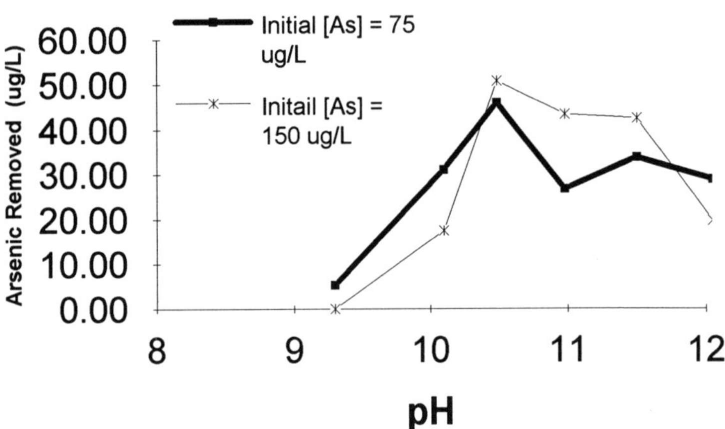

Figure 3: Removal of As (V) from a synthetic water containing only 1 mg/L Mn^{+2} and the indicated initial arsenic concentration. Percentage arsenic removal was greater for the lower initial arsenic concentration (above) but the total mass of arsenic removed in each system is approximately the same (below), suggesting removal occurred via an adsorption mechanism .

References

Hounslow, A.W., "Ground-water Geochemistry: Arsenic in Landfills," *Goundwater*, V.18, No. 4, 331-333 (1980).

Johnson, D.L., "Simultaneous Determination of Arsenate and Phosphate in Natural Waters," *EST,* V. 5, No. 5, 411-414 (1973).

Korte, Nic, "Naturally Occurring Arsenic in Groundwaters of the Midwestern United States," *Env. Geol. Water Sci.,* V. 18, No. 2, 137-141 (1991).

Masscheleyn, P.H. R.D. Delaune, and W.H. Patrick, Jr, "Effect of Redox Potential and pH on As Speciation and Solubility in a Contaminated Soil," *EST*, V. 25, No. 8, 1414-1418 (1991).

Mattess, G. "In-situ Treatment of As Contaminated Groundwater," *The Science of the Total Env.*, 21, 99-104 (1981).

Sorg, T.J., and G.S. Logsdon, "Treatment Technology to Meet the Interim Primary Drinking Water Regulations for Inorganics: Part 2." JAWWA, V. 70, No. 7, 379 - 393 (1978).

Waterweek, p. 6, March 15, 1993.

ARSENIC REMOVAL BY MANGANESE GREENSAND FILTERS

Thon Phommavong[1] and T. Viraraghavan[2]

Abstract

Some of the small communities in Saskatchewan are expected to have difficulty complying with the new maximum acceptable concentration (MAC) of 25 μg/L for arsenic. A test column was set up in the laboratory to study the removal of arsenic from the potable water using oxidation with $KMnO_4$, followed by manganese greensand filtration. Tests were run using water from the tap having a background arsenic concentration of <0.5 μg/L and iron concentration in the range of 0.02 to 0.77 mg/L. The test water was spiked with arsenic and iron. Results showed that 61% to 98% of arsenic can be removed from the potable water by oxidation with $KMnO_4$ followed by manganese greensand filtration.

Introduction

The two most common valence states for arsenic are As^{3+} and As^{5+}. In aerobic or well oxygenated water, arsenate is the most stable form while in the deep ground water or in the deep lake sediments, arsenite is the most common form. Four aqueous forms are: H_3AsO_4, $H_2AsO_4^-$, $HAsO_4^{2-}$, and AsO_4^{3-}. Gulledge and O'Connor (1973) stated that the predominant ionic form of arsenic(V) is pH dependent. The acute effects of the inorganic arsenic compounds (mostly As^{3+}) are well documented (World Health Organization, 1981). Chronic effects on the peripheral nervous system in humans due to long-term exposure to arsenic-contaminated water, has been reported. Hyperpigmentation, warts and hyperkeratosis of the palms and soles were reported to be the most common occurrences. Smith et al. (1992) reported that the population cancer risks due to arsenic in U.S. water supplies may be comparable to those from evnironmental tobacco smoke and radon in homes. The epidemiological investigations indicate that inorganic arsenic could be carcinogenic to man. However, the carcinogenic property of the arsenic has not been confirmed. The conclusive animal bioassay and data are not available at this time.

[1]Environmental Engineer, Saskatchewan Environment, Regina, Saskatchewan, Canada

[2]Professor, Faculty of Engineering, University of Regina, Regina, Saskatchewan, Canada

Arsenic Removal

Laboratory experiments and pilot plant studies indicate that arsenic can be removed by the conventional alum or iron coagulation (Gulledge and O'Connor, 1973), activated alumina (Bellack, 1971; Elson et al., 1980), ion exchange (Fox, 1989), reverse osmosis (USEPA, 1977), and lime softening (USEPA, 1985). Major operational factors such as flow rate, down time, and media clogging also affected the removal of arsenic (Shen, 1973; Hathaway and Rubel, 1987). A pilot study conducted at Kokomo, Indiana by Cooper and Thomson (1990) showed that arsenic as well as iron and manganese can be removed by using a high-rate contact adsorption clarification-mixed media filtration process. The study demonstrated that arsenic, iron and manganese can be reduced to <0.0005 mg/L, <0.05 mg/L and <0.02 mg/L respectively. The above reductions were achieved at a flow rate of 10 gpm/ft^2 on the contact adsorption clarifier and 5 gpm/ft^2 on the filter; pH was 7.3; $KMnO_4$ was used as an oxidant to achieve a "greensand effect." Other chemicals used were ferric sulphate and non-ionic polymer.

Objectives and Scope of Study

Based on the historical data, some communities in Saskatchewan may need to modify their water treatment plants in order to comply with the new MAC for arsenic (25 μg/L). The objectives of this study are as follows:

1) To determine if oxidation with $KMnO_4$ followed by manganese greensand filtration can effectively remove arsenic; and
2) To study the effect of iron concentration on the arsenic removal efficiency.

Methodology

Tests were conducted using 660 mm (26") of manganese greensand filter media and 305 mm (12") of support gravel in the polyethylene column of 153 mm (6") in diameter. Prior to the study, the greensand filter was packed and conditioned (the media was allowed to sit in the $KMnO_4$ solution overnight) in accordance with the recommendations of the Department of Saskatchewan Environment. For all runs, the flow was maintained in the range of 1893 to 2270 mL/min (2.5 to 3 gpm/ft^2) by using the flow indicator, and control valves.

Free chlorine, total chlorine, total iron, and manganese concentrations were determined by using a spectrophotometer (Hach Co. DR 2000). The pH of the water was measured by using a portable Hewlett Packard pH meter. Arsenic samples were preserved with nitric acid and analyzed by the Saskatchewan Research Council (SRC). Arsenic samples were digested with nitric/sulphuric acid followed by persulphate oxidation. Arsenic was reduced to arsine using sodium borohydride. Arsine was aspirated into an argon/hydrogen flame and/or heated quartz tube and quantitated by atomic absorption techniques.

Batches of raw water were spiked with an inorganic arsenic standard solution which was obtained from Fisher Scientific. In addition to the arsenic, powdered ferric chloride (about 20% of iron by weight) was also added to some batches of raw water to raise the iron concentration. Trivalent iron (ferric) was chosen so that oxidation of iron during stirring and exposure to air will be minimized. Six experimental filter runs were conducted. In all the runs, the tap water was spiked with arsenic. In the first run, no ferric chloride was added. In the second, third, fourth and sixth runs, different amounts of ferric chloride were added to maintain different arsenic to iron ratios. In the fifth run, the background iron concentration was 0.77 mg/L, without the addition of ferric chloride. In all the runs, the required amount of $KMnO_4$ was added.

Results and Discussion

Table 1 shows the results of the six experimental runs. Table 2 shows the arsenic and iron removal efficiencies after two hours of filtration.

Table 1. Test Results for Arsenic, Iron, and Manganese.

Sample No. and Description	Arsenic µg/L	Iron mg/L	Manganese mg/L	$KMnO_4$ Dosage mg/L
# 1-Tap Water	<0.5	0.02	0.090	0.5
# 2-Raw As Water-Run1	110.0	0.02	0.090	
# 3-Treated @ 0:30	4.4	0.01	0.055	
# 4-Treated @ 1:00	6.2	0.02	0.060	
# 5-Treated @ 1:45	9.8	0.01	0.093	
# 6-Raw As Water-Run2	110.0	0.58	0.090	1.0
# 7-Treated @ 0:30	9.0	0.05	0.066	
# 8-Treated @ 1:00	9.8	0.02	0.058	
# 9-Treated @ 1:30	9.2	0.02	0.058	
#10-Treated @ 2:00	9.2	0.04	0.066	
#11-Raw As Water-Run3	110.0	1.05	0.077	1.5
#12-Treated @ 0:30	14.0	0.10	0.075	
#13-Treated @ 1:00	11.0	0.06	0.092	
#14-Treated @ 1:30	9.0	0.04	0.084	
#15-Treated @ 2:00	9.0	0.02	0.092	
#16-Raw As Water-Run4	135.0	1.05	0.077	1.5
#17-Treated @ 0:30	7.5	0.02	0.109	
#18-Treated @ 1:00	8.6	0.01	0.096	
#19-Treated @ 2:00	8.6	0.01	0.100	
#20-Raw As Water-Run5	228.0	0.77	0.058	1.3
#21-Treated @ 1:00	22.0	0.04	0.035	

Table 1 (continued)

Sample No. and Description	Arsenic µg/L	Iron mg/L	Manganese mg/L	$KMnO_4$ Dosage mg/L
#22-Treated @ 1.45	28.0	0.04	0.070	1.3
#23-Treated @ 2:15	24.0	0.03	0.104	
#24-Treated @ 3:00	45.0	0.02	0.091	
#25-Treated @ 3:45	62.0	0.01	0.120	
#26-Treated @ 5:00	56.0	0.02	0.139	
#27-Treated @ 6:00	81.0	0.01	0.116	
#28-Raw As Water-Run6	105.0	5.77	0.058	5.9
#29-Treated @ 0:30	41.0	0.05	0.056	
#30-Treated @ 1:00	36.0	0.04	0.042	
#31-Treated @ 2:00	36.0	0.01	0.090	
#32-Treated @ 3:00	28.0	0.03	0.058	New Batch
#33-Treated @ 4:00	25.0	0.01	0.064	New Batch
#34-Treated @ 6:00	23.0	0.00	0.091	

Table 2. Comparison of Arsenic and Iron Removal Efficiency After 2 Hours of Filtration.

Run No.	Raw Water		Treated Water		% Removal	
	Fe,mg/L	As,µg/L	Fe,mg/L	As,µg/L	Fe	As
1	0.02	110.0	0.02	6.8	0	94
2	0.58	110.0	0.03	9.3	94	92
3	1.05	110.0	0.05	10.5	95	90
4	1.05	135.0	0.01	8.2	99	94
5	0.77	228.0	0.04	24.7	95	89
6	5.77	105.0	0.01	37.7	99	64

Run #1

The $KMnO_4$ feed rate was set at approximately 0.9 mg/L. The filter was operated for 1 hour and 50 minutes. The results showed that an average of 94% of arsenic was removed; no iron was removed; manganese removal was poor. This test demonstrated that the arsenic was retained in the filter. It appears that most of the arsenic was oxidized by $KMnO_4$ and adsorbed by the greensand media. A study on lake sediments conducted by Oscarson et al (1980, 1981) indicated that Arsenic (III) can be oxidized to arsenic (V) and simultaneously adsorbed by the MnO_2. However, the adsorption capacity was low. The low adsorption could be due to the incomplete arsenic oxidation.

Run #2

The arsenic removal averaged at about 92%. Additionally, the iron removal (94%) was also good. This test run showed a slight reduction in the manganese concentration. The arsenic removal efficiency is practically the same as the first run. An increase of 0.56 mg/L in iron concentration showed no effect on arsenic removal.

Run #3

The arsenic removal through the filter averaged 90%. Iron removal averaged at about 95%. A slight manganese removal was observed at 30 minutes into the test run. No manganese removal was recorded at 1, 1:30, and 2:00 hours into the test run. The higher manganese concentration observed could be due to a slight over feeding of $KMnO_4$, low pH, or insufficient detention time.

The arsenic removal obtained in this test run is not significantly different from the test run #1. The iron concentration ratio of test run #1 to test run #3 is about 1:52. Although the iron concentration was increased significantly, the iron to arsenic ratio for this run was only 10:1. A study done by Cooper and Thomson (1990) on arsenic removal using high-rate contact adsorption clarification-mixed media filtration process, showed that the arsenic removal efficiency increased with the iron to arsenic ratio. They observed that most of the arsenic removal was achieved by adsorption on the ferric hydroxide; the best removal was achieved at the highest iron to arsenic ratio of 28.3:1.

Run #4

The analysis of the three treated samples taken at 30 min, 1:00 and 2:00 hours respectively indicated fairly constant arsenic and iron removals. The arsenic and iron removals were recorded at 94% and 99% respectively. Once again, no manganese removal was observed. The manganese concentration in the treated water (0.109, 0.096, and 0.100 mg/L) was consistently higher than that of the raw water (0.077 mg/L).

During this run, the water (before entering the filter) was once again having a faint pink colour. No pink colour was observed in the filtrate. At this time, the manganese may be breaking through. Additionally, it is possible that $KMnO_4$ was overfed. As the result, higher manganese concentration was found in the treated water samples.

Run #5

Prior to this run, the filter was backwashed. The iron to arsenic ratio was reduced from (run #4) 7.7:1 to 3.4:1. The arsenic removal was in the range of 90% to 64% from the beginning to the sixth hour respectively. Although arsenic was being removed at the sixth hour, arsenic concentration in the treated water sample was 81 μg/L. This concentration is much higher than the MAC of 25 μg/L. Based on this result, it appears that the treatment process of oxidation with $KMnO_4$ followed by filtration through manganese greensand media may not be a suitable process for treating water that contains >100 μg/L of arsenic. Further study is required.

Once again, iron removal was as expected (averaged at about 95%). The manganese removal was poor. It appears that the arsenic is replacing the manganese. That is the arsenic is adsorbing on the surface of the manganese oxide rather than the Mn^{2+}. The removal of manganese was evident at the 1:00 hour mark. After that however, the manganese concentration steadily increased.

High manganese concentration in the treated water may be caused by the reduction of manganese oxide. Arsenic can be oxidized by manganese oxide. One of the products of this reaction is Mn(II). A study conducted by Huang et al (1982), showed that Mn(II) was released into the water when arsenic was oxidized by MnO_2. The feasibility of the oxidation of arsenic was indicated by the positive log K value, where K is an equilibrium constant at 25°C. The reaction takes the following form:

$$HAsO_2 + MnO_2 + 2H^+ = H_3AsO_4 + Mn^{2+} \quad \text{Log K} = 23.1$$

The equation above indicates that Mn^{2+} is being released while the arsenite is oxidized to arsenate. As a result, the manganese concentration in the treated water was elevated.

Run #6

The filter was backwashed; the arsenic concentration was reduced to about 100 μg/L. This test run suggested that increasing iron concentration may not necessarily improve the arsenic removal as many past studies had pointed out. This test indicated that the arsenic removal efficiency was in the range of 61% to 78%. As stated earlier, the $KMnO_4$ dosage was 5.9 mg/L. Theoretically, this dosage is just enough to oxidize the iron and manganese. Lack of $KMnO_4$ for oxidizing arsenic many cause this low arsenic removal.

The arsenic removal improved with time. A removal in the range of 61% to 66% was achieved during first batch of water. This low removal may have something to do with low or lack of chlorine residual in the water since the water tank was filled and left overnight. Knocke et al (1990) stated that chlorine residual will help in the oxidation reaction and enhance the performance of the manganese greensand filter.

Although the rest of this run was performed as soon as the tank was filled and thoroughly mixed, the arsenic removal did not improve significantly. The arsenic removal was in the range of 73% to 78% (last two batches) compared to 61% to 66%

(first batch). This removal efficiency is less than that obtained from the previous five tests.

Manganese removal efficiency was poor once again. Manganese removal was observed for the first two hours of the test. It is interesting to note that the manganese concentration in the treated water increased with time. Again, it appears that arsenic is competing with manganese and dominated the adsorption process.

Conclusions

The following conclusions are drawn based on this study:

- Arsenic can be removed by oxidation with $KMnO_4$ followed by manganese greensand filtration;
- As low as 61% and as high as 98% of arsenic removal efficiency was recorded during this study;
- No definite conclusions regarding the effect of iron concentration on the arsenic removal efficiency can be made at this time as the results of this study did not generate enough concrete evidence. It appears that increasing iron concentration will not significantly improve the arsenic removal efficiency;
- Adsorption of arsenic by manganese oxide seems to be a dominant mechanism since increasing iron concentration did not significantly improve the arsenic removal;
- Elevated manganese concentration in the treated water samples could be caused by 1) replacement of manganese adsorption by sorption of arsenic onto the greensand 2) $KMnO_4$ may be overfed at times and/or insufficient detention time and 3) the release of Mn^{2+} (as a result of the reduction of manganese oxide) when the arsenic was oxidized by the manganese oxide.

Acknowledgements

The authors wish to thank the Water Quality Branch of Saskatchewan Environment and Public Safety for providing most of the test equipment and funding for arsenic analysis, Gord Will of the Water Quality Branch for his assistance and valuable discussions and Darling Duro, Division of Watergroup Canada Ltd., for the greensand and anthracite.

References

Bellack, E., (1971). "Arsenic Removal from Potable Water." J. Am. Water Works Assoc. , 63 (7), 454.

Cooper, K.H., and Thomson, R.E. (1990). "Iron, Manganese and Arsenic Removal Using High-Rate Contact Adsorption Clarification-Mixed Media Filtration Process." AWWA Annual Conference Proceedings, Cincinnati, Ohio.

Elson, M.C., Davies, H.D., and Hayes, R.E. (1980). "Removal of Arsenic from Contaminated Drinking Water by a Chitosan/Chitin Mixture". Water Research, 14 (1), 1307.

Fox R.K. (1989). "Field Experience With Point-of-Use Treatment Systems for Arsenic Removal." J. Am. Water Works Assoc., 81(2), 94.
Gulledge H. J. and O"Connor T. J. (1973). "Removal of Arsenic (V) From Water by Adsorption on Aluminum and Ferric Hydroxides." J. Am. Water Works Assoc., 65 (8), 511.
Hathaway, W.S., and Rubel, F. Jr. (1987). "Removing Arsenic From Drinking Water". J. Am. Water Works Assoc., 79 (8), 61.
Huang, P.M., Oscarson, D.W., Liaw, W.K., Hammer, U.T. (1982). "Dynamics and Mechanisms of Arsenite Oxidation by Freashwater Lake Sediments". Hydrobiologia, 91, 315-322.
Knocke, W., Occiano, S., Hungate, R. (1990). "Removal of Soluble Manganese From Water by Oxide-Coated Filter Media". AWWA Research Foundation, Denver, CO.
Oscarson, D.W., Huang, P.M., and Liaw, W.K. (1980). "The Oxidation of Arsenite by Aquatic Sediments". J. Environ. Qual., 9 (4), 700-703.
Oscarson, D.W., Huang, P.M., Liaw, W.K. (1981) "Role of Manganese in the Oxidation of Arsenite by Freshwater Lake Sediments". Clays and Clay Minerals., 29 (3), 219-225.
Shen, U.S. (1973). "Study of Arsenic Removal From Drinking Water". J. Am. Water Works Assoc., 65 (8), 543.
Smith, A.H., Hopenhayn-Rich, C., Bates, M.N., Goeden, H.M., Hertz-Picciotto, I., Duggan, H.M., Wood, R., Kosnett, M.J., and Smith, M.T. (1992). "Cancer Risks from Arsenic in Drinking Water." Environmental Health Perspectives, 97, 259-267.
USEPA. (1977). EPA-600/8-77-005, Manual of Treatment Techniques for Meeting the Interim Primary Drinking Water Regulations. Water Supply Research Division, Cincinnati. .
USEPA. (1985). EPA/600/2-85/094, Pilot Study for Removal of Arsenic from Drinking Water at the Fallon, Nevada Naval Air station.
World Health Organization. (1981). "Environment Health Criteria 18: Arsenic." Geneva.

Removal of Bromate After Ozonation By Electron Beam Irradiation

M. Siddiqui[1], G. Amy[2], W. Cooper[3], C. Kuruz[4] , and T. Waite[5]

Abstract

Drinking water regulations related to disinfection, disinfectant by-product (DBP) control, DBP precursor removal and contaminated source water pre-treatment will become increasingly stringent as the regulation/negotiation process proceeds. Of critical concern will be the development of new process that can address one or more of these critical areas. This paper discusses the use of an innovative treatment process, high energy electron beam irradiation (HEEB), in drinking water treatment to remove bromate, an ozonation disinfection by-product and total organic carbon (TOC), precursors for organic DBPs. Preliminary bench-scale and pilot-scale studies indicate that the process is capable of removing bromate and TOC. Reduction of bromate is dependent on HEEB dose, initial bromate concentration, pH of the source water and alkalinity of the influent water.

Introduction

Draft drinking water regulations in the U.S. will specify a maximum contaminant level (MCL) of 10 µg/L for bromate (BrO_3^-) and a best available treatment (BAT) of pH adjustment. To date, most BrO_3^- control strategies have involved inhibiting or minimizing BrO_3^- formation through acid or ammonia addition. Adjustment to pH 6.0 prior to ozonation will significantly reduce BrO_3^- formation; however, acid addition may not be viable or cost effective for high alkalinity waters. Ammonia addition can theoretically tie up bromine as monobramamine; however, the complexity of bromamine chemistry has yielded mixed BrO_3^- formation results in lab and pilot studies. The work reported herein involves removing BrO_3^- after its formation, when

---

[1] Assistant Professor Adjunct, University of Colorado at Boulder

[2] Professor, University of Colorado at Boulder

[3] Professor and Director, DWRC, Florida International University

[4] Associate Professor, University of Miami

[5] Professor, University of Miami

other control strategies are not cost effective and/or reliable. If the proposed BrO_3^- MCL in the U.S. is lowered further, a combination of minimized production and subsequent removal may be required as a BrO_3^- control strategy. While several researchers have done preliminary studies on BrO_3^- formation (Siddiqui and Amy, 1993; Krasner et. al., 1993), there is a paucity of data related to BrO_3^- removal in drinking waters.

High energy electron beam (HEEB) process can potentially be used at various points in the process train, as either a pre-oxidant or a post-disinfectant. In one scenario, HEEB could be used in place of pre-ozonation; here, it would provide CT credit, and reduce TOC (including synthetic organic compounds) while not contributing to the formation of DBPs. In another scenario, HEEB would be used later in the process train following a pre-ozonation step; here it can be used to provide additional CT requirements while destroying ozonation by-products such as BrO_3^-. HEEB irradiation can destroy halogenated disinfection by-products including THM and HAA species and also may provide effective inactivation of microbes of present or future regulatory interest, including enteric viruses and Giardia.

The primary objective of this research was to evaluate HEEB as an oxidant/reductant which can decompose DBPs such as BrO_3^- formed by other disinfectants.

Experimental Methods and Analytical Procedures

The high energy electron beam (HEEB) unit selected to process the samples (bench-scale) was a 8 MeV Linatron machine rated at 1kW (Nutek Corp., Palo Alto, CA). The samples were placed in a steel wire basket on three sides and irradiated. The conveyor speed and machine pulse rate was set to give the required dose to the samples. The samples were given incremental doses to produce the specified doses. Radiochromic dosimeters were placed on the samples to determine the doses. The dosimeters were read with a spectrophotometer. BrO_3^- varying in concentration from 100-320 µg/L was spiked into MQW and different source waters and transferred to 35 ml screw-cap vials. The bottles were kept at 4 ^{o}C before and immediately after irradiation to minimize thermal decomposition of BrO_3^-. Doses ranging from 0.15 to 0.75 Mrads were applied. This range reflects the lower and upper limits of dose as measured by dosimeters placed external to the product and does not represent dose delivered within the sample containers.

Full-scale experiments were performed at two different pH levels (4.5 and 9.0) and five different doses (0, 50, 100, 150, 200, 400 krads). The influent water had a TOC of 4.6 mg/L. The facility consists of an horizontal 1.5 meV insulated core transformer electron accelerator whose beam current can be varied from 0 to 50 mA to achieve absorbed doses of 0-800 krads at water flows of 120 gpm. The influent stream is presented to a scanned electron beam in a falling stream about 114 cm wide and 0.4 cm thick. This full-scale electron beam facility (EBRF) has been described in detail elsewhere (Kuruz et. al., 1991).

Br^- and BrO_3^- measurements were accomplished by ion chromatography (IC) using a Dionex 4500i (Dionex Corp., Sunnyvale, CA) series system with an IonPac AS9-SC column and a suppresser column. A 2mM Na_2CO_3/0.75 $NaHCO_3$ eluent was used for

Br^- determination and a 40mM H_3BO_3/20mM NaOH eluent was employed for BrO_3^- determination. Minimum detection limit for BrO_3^- using borate eluent at the time of this research was 2 μg/L. For samples with high chloride ion (Cl^-) content, a silver cartridge was used to remove Cl^- prior to IC analysis to minimize its interference with BrO_3^-measurement. Both Colorado River Water (CRW; DOC=3.2 mg/L) and State Project Water (SPW; DOC=3.8 mg/L) contain up to 150 mg/L as Cl^-. Conductivity detector response was almost perfectly linear ($r^2 \geq 0.99$) for standards ranging from 25 to 500 μg Br^-/L and from 5 to 50 μg BrO_3^-/L.

Results and Discussion

High-energy electron-beam irradiation has been shown to be efficient for removing trihalomethanes (THMs) (Cooper et. al., 1993). Irradiation of water by electron beam results in the formation of reducing species such as aqueous electrons (e^-_{aq}), hydrogen atoms (H•), and oxidizing species such as hydroxyl radicals (OH•). BrO_3^- reacts with reactive intermediates formed when aqueous solutions are irradiated. The e^-_{aq} is a powerful reducing agent with an E_o of -2.77. The hydrogen atom accounts for approximately 10% of the total free radical concentration in irradiated water (Buxton et. al., 1988). The H• undergoes two general types of reactions, hydrogen addition and hydrogen abstraction. The OH• radical can undergo several types of reactions with chemicals in aqueous solution.

The types of reactions that are likely to occur are addition and hydrogen abstraction. The reaction of these intermediates with BrO_3^- has been reported and the relevant reactions are listed below (Buxton et. al., 1988):

$e^-(aq) + BrO_3^- + 2H^+ \Longrightarrow BrO_2\cdot + H_2O \quad k \approx 10^9\ M^{-1}s^{-1}$ [1]

$e^-(aq) + BrO_2^- \Longrightarrow BrO\cdot + O^{2-} \quad k \approx 10^{10}\ M^{-1}s^{-1}$ [2]

$e^-(aq) + BrO^- \Longrightarrow O\cdot^- + Br^- \quad k \approx 10^{10}\ M^{-1}s^{-1}$ [3]

$H\cdot + BrO_3^- \Longrightarrow BrO_3\cdot + H^+ \quad k \approx 10^7\ M^{-1}s^{-1}$ [4]

$OH\cdot + BrO_3^- \Longrightarrow BrO_3\cdot + OH^- \quad k \leq 10^6\ M^{-1}s^{-1}$ [5]

The $BrO_3\cdot$ radical is highly reactive in aqueous solution and additional reactions of $BrO_3\cdot$ with intermediates will lead to the formation of Br^-.

Bench-scale Experiments: CRW, SPW, and MQW (NOM-free) source waters were irradiated with doses of 250, 500, and 750 krads respectively with an initial BrO_3^- of 320 μg/L (Figure 1), and 150, 300, 450 krads with an initial BrO_3^- of 100 μg/L (Figure 2). Br^- and free bromine (HOBr/OBr^-) concentration were measured after irradiation and Br^- mass balances showed almost complete recovery of initial Br (note that BrO_3^- is 63% Br by wt.) in MQW water. Complete recovery of initial Br^- in BrO_3^- was not observed for CRW and SPW, ostensibly due to the formation of organic bromine (TOBr). Reduction of BrO_3^- in CRW was higher than MQW and SPW presumably due to the presence of carbonate alkalinity. The $CO_3\cdot^-$/$HCO_3\cdot$ radicals are produced by the reaction of e^-_{aq}, OH· and H· with bicarbonate ions and dissolved carbon dioxide in water. These radicals are strongly reducing species, with a redox

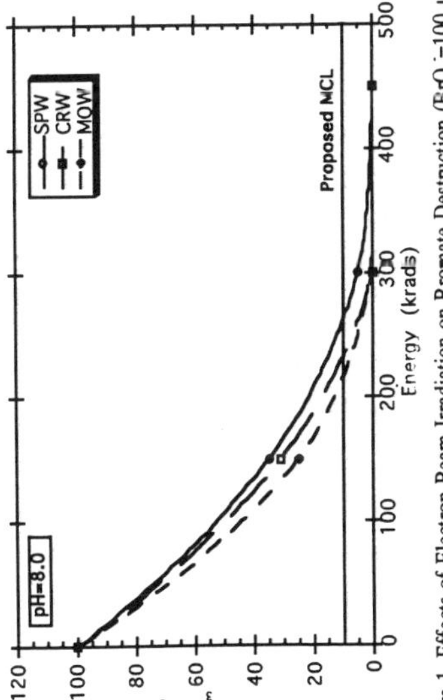

Figure 1. Effects of Electron Beam Irradiation on Bromate Destruction (BrO_3^-=100 µg/L)

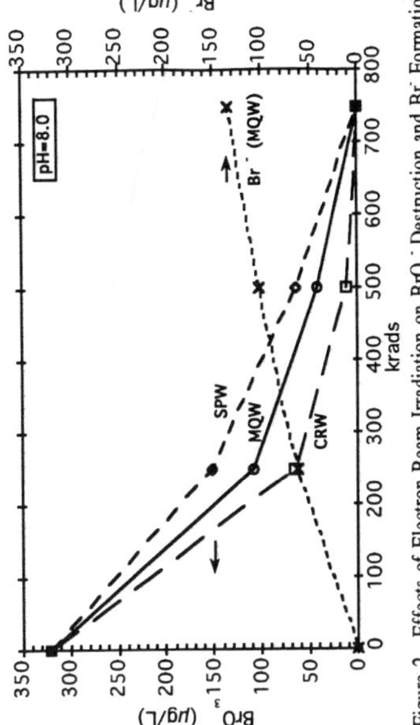

Figure 2. Effects of Electron Beam Irradiation on BrO_3^- Destruction and Br^- Formation

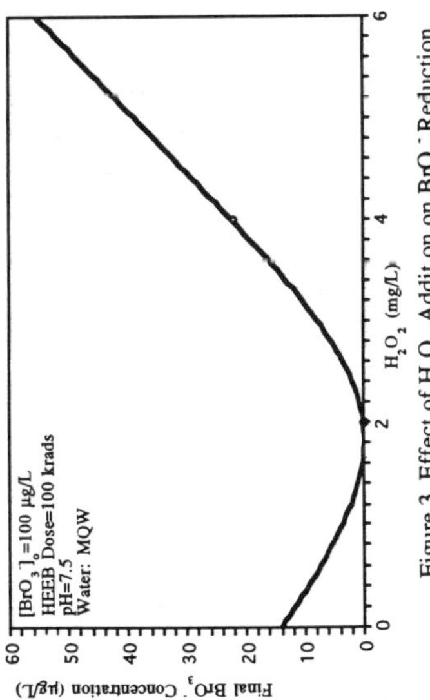

Figure 3. Effect of H_2O_2 Addition on BrO_3^- Reduction

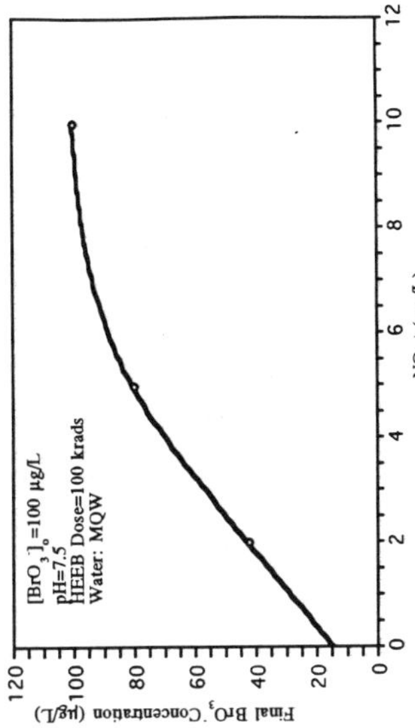

Figure 4. Effect of Adding NO_3^- on BrO_3^- Reduction

potential of -2 V. The carbonate radicals acts predominantly as electron acceptors; hydrogen abstraction by the carbonate radicals is generally very slow. The $CO_2\cdot^-$ radical is present in this form throughout most of the pH range and only protonates in strongly acidic solutions ($pK_{CO_2\cdot^-}$=1.4)(Buxton et. al., 1988). Increasing alkalinity further (≥ 200 mg/L as $CaCO_3$) from 50 mg/L in MQW, BrO_3^- reduction by HEEB irradiation slightly decreased ostensibly due to competition between e_{aq}^- and carbonate radicals for bromine ions.

Experiments performed by spiking H_2O_2 showed a relatively strong inhibition effect on BrO_3^- reduction indicating that H_2O_2 is acting as a scavenger of both OH· radicals and e^-_{aqs} (Figure 3) at concentrations greater than 2 mg/L. BrO_3^- reduction enhanced when H_2O_2 concentration was less than 2 mg/L indicating that there is a threshold level of H_2O_2 above which it simply acts as a sink for aqueous electrons and OH. radicals. This threshold level of H_2O_2 is depended on HEEB dose, pH, and background characteristics of source water. The relevant reactions are shown below:

$e^-_{aq} + H_2O_2 \Longrightarrow OH^- + OH\cdot$ $\qquad k = 1.1 \times 10^{10}$ [6]

$OH\cdot + H_2O_2 \Longrightarrow H_2O + HO_2\cdot$ $\qquad k = 2.7 \times 10^{9}$ [7]

Also reactions involving hydrated electrons produced during HEEB irradiation with protons can lead to the formation of hydrogen peroxide (H_2O_2) in the presence of dissolved organic matter (DOC). Cooper and Zika (1983) have reported sunlight-induced hydrogen peroxide accumulation rates, normalized to DOC, of 0.2-0.3 μmol of H_2O_2/(mg of DOC)/hr in natural water samples. Thus H_2O_2 acts as OH· radical scavenger or promoter of excited state species. It should be noted that these levels of H_2O_2 fall within the threshold levels described above for waters containing TOC of ≤ 5 mg/L.

In the presence of tert-butanol which is a strong OH· radical scavenger, reduction of BrO_3^- was not impaired suggesting that e^-_{aq} is mainly responsible for BrO_3^- reduction and OH· radicals are not strong reducing species and are consumed by hydrated electrons ($k=3\times10^{10}\ M^{-1}\ s^{-1}$) faster than they can react with bromide species. To further confirm this behavior, solutions containing BrO_3^- were irradiated in the presence of high concentrations of electron scavengers (nitrate, and chloroacetate); the destruction of BrO_3^- was found to decrease significantly indicating that e^-_{aqs} are in fact responsible for BrO_3^- reduction during HEEB irradiation (Figure 4). 10 mg/L as NO_3^- was sufficient to completely inhibit the reduction of BrO_3^- with an initial BrO_3^- concentration of 100 μg/l (HEEB dose=200 krads). However, nitrate has some rather complex radiation chemistry associated with it. That is, the reaction with e^-_{aqs} will form nitrite (NO_2^-), which can further react with hydroxyl radicals and in the presence of an aromatic ring can actually form compounds like nitrobenzene (Cooper, 1994). The reaction between aqueous electron and NO_3^- is given below.

$e^-_{aq} + NO_3^- \Longrightarrow O\cdot^- + NO_2^-$ $\qquad k = 0.7 \times 10^{9}\ M^{-1}\ s^{-1}$ [8]

Full-scale Experiments: Full-scale HEEB irradiation experiments were conducted in a source water containing TOC = 4.6 mg/L, alkalinity=45 mg/L and UV_{254} = 0.09

cm^{-1} (continuous flow of water at 150 gpm) by spiking BrO_3^- at concentrations ranging from 200 to 800 µg/L and at two different pH levels (pH=4.5 and 9.0) and the results are shown in Figures 5 and 6. Better removal of BrO_3^- was observed at pH=9.0 than at pH levels of 4.5 ostensibly due to higher production of OH· radicals and the possible interaction of alkalinity in the production of $CO_3^{\cdot -}$ radicals. TOC concentration and UV_{254} absorbance levels decreased by 26% and 42% at pH=4.5 and 25% and 32% at pH=9.0 respectively on increasing HEEB doses from 50 to 200 krads (Figures 7 and 8). pH decreased on increasing HEEB doses as expected because more OH^- ions are consumed to produce OH· radicals which subsequently react with TOC and BrO_3^-.

Economics: The cost of treatment using HEEB irradiation depends on many factors, such as the dosage required, the volume of water to be treated, and the size of the treatment facility (Cooper et. al, 1993)). Cooper et. al. (1993) estimated a cost of $1.52/1,000 gallons to reduce a THM concentration from 80 µg/L to less than 20 µg/L requiring a dose of 100 krads. Since the formation of BrO_3^- during drinking water treatment is less than 50 µg/L, destruction of BrO_3^- requires less energy than THMs and the cost of reducing BrO_3^- concentration from 50 µg/L to <10 µg/L would be less than $1.52/1,000 gallons. Moreover if electron beam is employed to destroy THMs, BrO_3^- and other DBPs, the cost would be more economical.

Conclusions

(1) Solutions irradiated with HEEB reduced BrO_3^- to Br^- with OBr^- as an intermediate specie. A HEEB dose of 60 krads is sufficient to reduce 70% of BrO_3^- from an initial concentration of 100 µg/L, the highest concentration normally encountered in drinking water treatment.

(2) Presence of hydrogen peroxide or NO_3^- (strong electron scavenger) significantly reduced reduction of BrO_3^- whereas addition of radical scavenger t-butanol did not affect the reduction of BrO_3^- indicating that e_{aq}^- are stronger reducing species than other excited species. Hence in source waters with significant levels of NO_3^-, HEEB use may not be feasible and cost effective.

(3) The presence of natural organic matter (NOM) did not significantly change BrO_3^- reduction efficiency. Also HEEB irradiation reduced TOC concentration by 26% for a HEEB dose of 200 krads. This reduction efficiency is dependent on the type of TOC in source water.

Acknowledgments: The authors wish to thank Wenyi Zhai for performing the bromide ion analyses.

References

Buxton, G, C. Greenstock, "Critical Review of Rate Constants for Reactions of e^-_{aq}, H· and HO· in Aqueous Solutions" *J. Phy. Chem. Ref. Data*, 17(2), 1988.

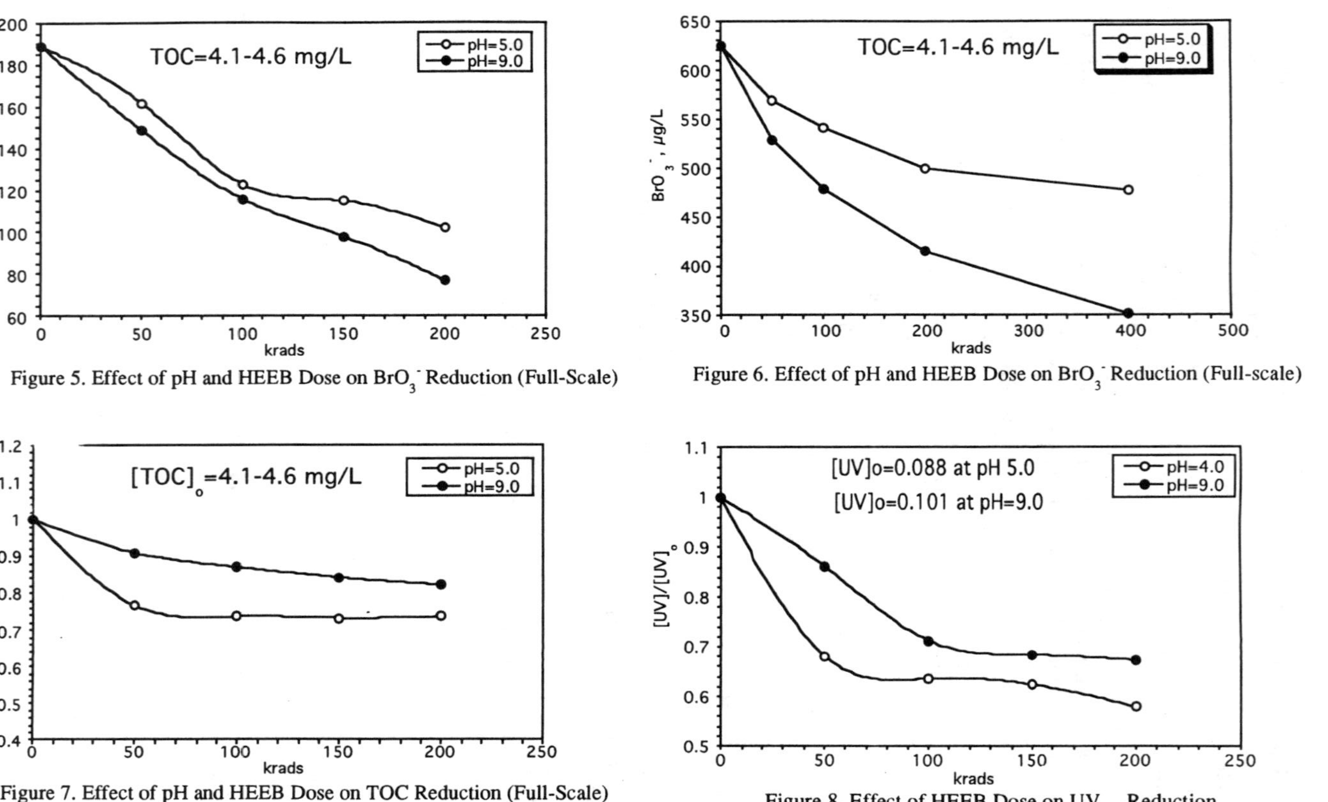

Figure 5. Effect of pH and HEEB Dose on BrO_3^- Reduction (Full-Scale)

Figure 6. Effect of pH and HEEB Dose on BrO_3^- Reduction (Full-scale)

Figure 7. Effect of pH and HEEB Dose on TOC Reduction (Full-Scale)

Figure 8. Effect of HEEB Dose on UV_{254} Reduction

Cooper, W., E. M. Cadavid, M. Nickelson, K. Lin, C. Kuruz, and T. Waite, "Removing THMs From Drinking Water Using High-Energy Electron-Beam Irradiation", *JAWWA*, 85(8), 1993

Cooper, W., Zika, R. G. *Science* (Washington, D.C.), 220:711-712, 1983a.

Gunten, V.U., and J. Hoigne, "Factors Affecting the Formation of BrO_3^- During Ozonation of Bromide-Containing Waters", *J. Water SRT-Aqua* 41:5, 1992.

Krasner, S. W., William H. Glaze, Howard S. Weinberg, Phillippe A. Daniel, and Issam Najm, "Formation and Control of BrO_3^- During Ozonation of Waters Containing Bromide", *JAWWA*, 85:96-103, 1993.

Kuruz C. N. et. al. "High-Energy Electron-beam Irradiation of water, Wastewater and Sludge", Advances in Nuclear Science and Technology, Plenum Press, New York, 1991.

Siddiqui, M. and Gary Amy "DBPs Formed During Ozone-Bromide Reactions In Drinking Waters" *J. AWWA*, 1993, 85, 63-72, 1993

CONTRIBUTION OF WATERBORNE RADON TO HOME AIR QUALITY

Arun K. Deb, F. ASCE[1]

INTRODUCTION

Radon-222 is a member of the uranium decay chain and is formed from the decay of radium-226. Radon and its decay products emit alpha particles during the decay process. If radon is inhaled, alpha particles emitted from inhaled radon and its daughters increase the risk of lung cancer. Radon is soluble in water; thus when radon comes in contact with groundwater it dissolves. The radon concentration in groundwater may range from 100 pCi/L to 1,000,000 pCi/L. When water with a high radon level is used in the home, radon is released from the water to the air and thus can increase indoor air radon concentration. Considering the estimated health risk from radon in public water supply systems, EPA has proposed a maximum contaminant level (MCL) of 300 pCi/L for radon in public drinking water supplies.

To address the health risks of radon in water and the proposed regulations, the American Water Works Association Research Foundation (AWWARF) initiated a study to determine the contribution of waterborne radon to radon levels in indoor household air.

OBJECTIVE

The objective was to perform a controlled experiment to explore the contribution of waterborne radon to home air quality and the effect of radon removal from a centralized water system on indoor radon levels in three communities with widely variant waterborne radon contamination levels.

---

[1] Vice President, Roy F. Weston, Inc., 1 Weston Way, West Chester, PA 19380

STUDY SETUP

Community Description

Three small communities in New Hampshire served by groundwater systems have been selected for this study.

Homeowner participation was solicited from three communities: 19 homes from Community A, 85 homes from Community B, and 15 homes from Community C participated in this study. The indoor air radon testing program was carefully developed so that airborne radon contributions from waterborne radon could be estimated. In participating homes, indoor air radon concentrations in at least three locations and water radon concentrations were measured. Packed tower aeration (PTA) systems were installed in all three community water treatment systems to remove radon from water. Measurements of airborne radon and waterborne radon in each of the participating homes were repeated to determine post-treatment air and water radon concentrations.

Airborne Radon Measurements

Airborne radon measurements were made using electret-passive environmental radon monitors (E-PERMs). An E-PERM is a device that uses an electrostatically charged Teflon disc, called an electret, to measure the radon concentration in air. Once activated, the electret surface, which has a positive charge, attracts negative ions produced by the alpha particles generated from decaying radon. The radon enters a filtered ionization chamber housing the electret, and negatively charged ions are drawn to the surface of the electret, thereby reducing its surface voltage. The electret surface voltage is read using a portable digital meter (surface potential electret reader, or SPER) before and after exposure to radon. The resulting differential voltage is then used to calculate the average radon concentration present during the exposure period. For a 7-day exposure period, the lower level of detection (LLD) of airborne radon measurements using E-PERM is approximately 0.3 pCi/L (EPA 1989).

Radon Monitor Placement

Release of waterborne radon to indoor air depends on waterborne radon concentrations, water temperature, and the degree of agitation of water. Thus, the largest releases of waterborne radon in the home are due to activities such as showers and washing dishes or clothes.

Radon monitor locations were selected to best differentiate between airborne radon contributions from the soil and those from household water. Radon monitors were placed within each house in areas with varying potentials for radon/water separation. These areas included the following:

Area 1: High potential for release of radon from water, and high potential for airborne contribution from soil gas (e.g., basement laundries).

Area 2: Low potential for release of radon from water but with high potential for airborne contribution from soil gas (e.g., basement family rooms).

Area 3: High potential for release of radon from water and low potential for airborne contribution from soil gas (e.g., first-floor kitchens or bathrooms and second-floor laundries).

For each selected house, E-PERMs were placed in each of the three areas for a continuous 7-day period before and after the installation and operation of the community-wide radon removal water treatment systems. A 7-day period for indoor air radon measurements was selected to cover the weekly cycle of household water-related activities. The voltage of the electret was measured at the beginning and end of the 7-day period to calculate the 7-day average radon concentration.

Water Sample Collection and Analysis

Samples of household water were collected for radon analysis concurrently with the placement of the airborne radon monitors for both the pretreatment and post-treatment testing phases. Water samples were collected only at the time monitors were placed. The analysis of radon in water was completed using the scintillation method.

Waterborne Radon Removal System

A Packed Tower Aeration (PTA) system was used to remove radon from water in the three communities. PTA systems were installed at the well-head supply for each community. A general schematic of the PTA radon removal system used in Communities A, B, and C is shown in Figure 1.

DATA ANALYSIS

Airborne radon and waterborne radon data collected were entered into a relational computer database system.

Average waterborne radon concentrations in household water in Communities A, B, and C during pre- and post-treatment conditions were 1,292 and 126 pCi/L; 2,411 and 241 pCi/L; and 21,295 and 520 pCi/L, respectively.

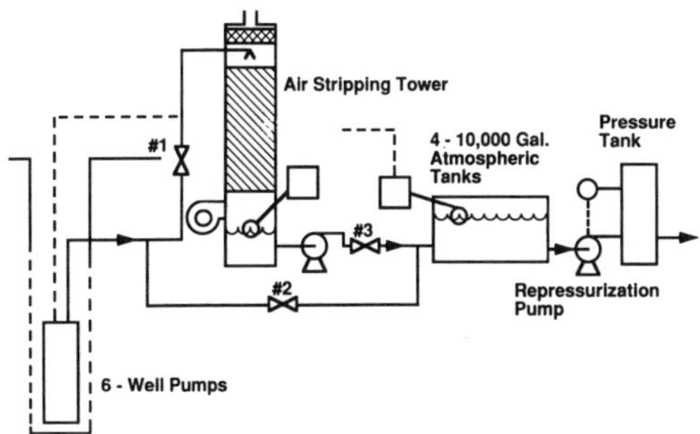

Figure 1. Schematic of PTA Radon Removal Systems

Matched pair statistical regression analysis of pre- and post-treatment indoor air data was conducted for each community (Figures 2 through 4) and also separately for locations in each community.

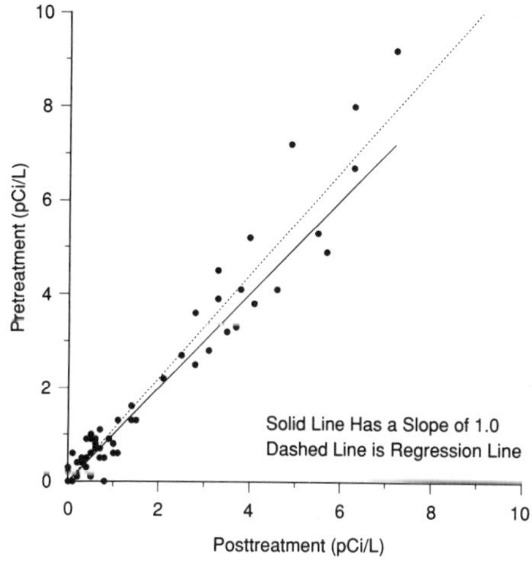

Figure 2. Matched Pair Analysis, Community A

The slopes of the regression lines of the matched pair data for Communities A, B, and C were 1.11, 1.03, and 1.87, respectively, indicating small indoor air radon reductions in Communities A and B compared with Community C. Background indoor air radon concentrations for the three communities were estimated from post-treatment indoor air radon concentrations in basement family room (BF) and basement laundry room (BL) locations. Background radon concentrations for Communities A, B, and C were estimated at 3.7 pCi/L, 1.0 pCi/L, and 1.6 pCi/L, respectively. Community A had a high

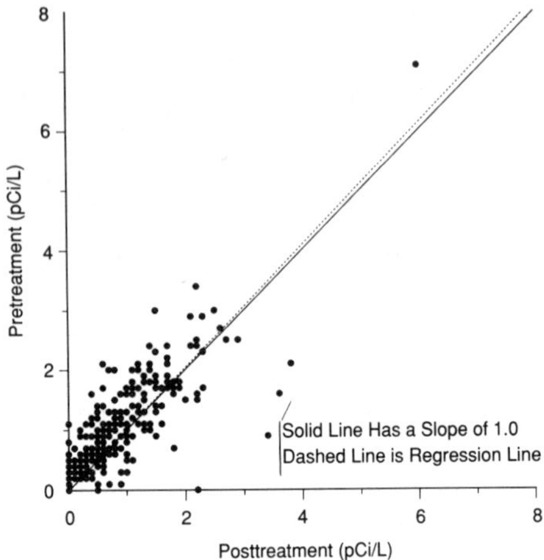

Figure 3. Matched Pair Analysis, Community B

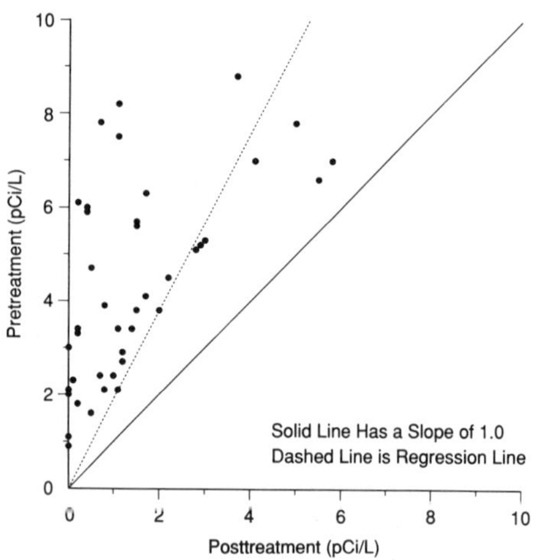

Figure 4. Matched Pair Analysis, Community C

background radon concentration (3.7 pCi/L) and a relatively low waterborne radon concentration (1,292 pCi/L). Community B had a low background indoor radon concentration (1.0 pCi/L) and a moderate waterborne radon concentration (2,411 pCi/L). Community C had a low background indoor radon concentration (1.6 pCi/L) and a very high waterborne radon concentration (21,295 pCi/L). Analysis of the reduction of indoor air radon concentration between pre- and post-treatment periods indicated that very little average reduction of indoor air radon occurred in Community A (0.1 pCi/L) and Community B (0.2 pCi/L); a large average reduction of indoor air radon occurred in Community C (2.9 pCi/L).

Analysis of Radon Reduction Data

A regression analysis was conducted of the reduction of all household waterborne radon concentrations (ΔRn_{water}) and the corresponding reduction of household airborne radon concentrations (ΔRn_{air}) for all three communities combined. A plot of all data, the regression line, and the 95% confidence lines appear in Figure 5. The fitted regression equation with a correlation coefficient (R^2) of 0.71 is:

$$\Delta Rn_{air} = (0.00013)\Delta Rn_{water}$$

This equation gives an average ratio of radon concentration in air to radon concentration in water of 1.3×10^{-4}.

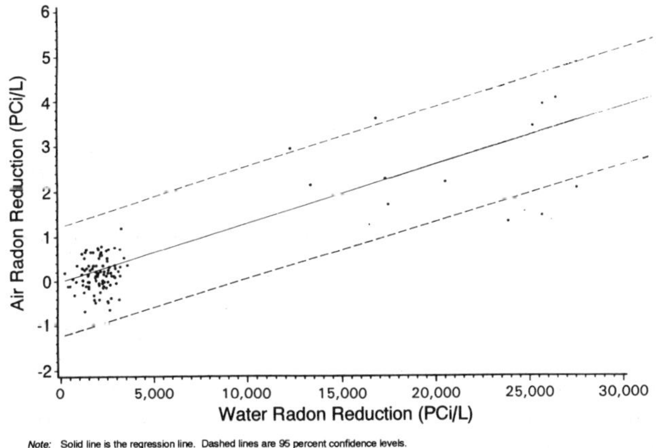

Note: Solid line is the regression line. Dashed lines are 95 percent confidence levels.

Figure 5. Air Radon Reduction Versus Water Radon Reduction

Analysis of Distribution of Data

To obtain an overall, community-wide distribution of airborne radon concentration reduction due to reduction of waterborne radon for each community, a cumulative distribution analysis of pre- and post-treatment airborne radon concentrations was conducted. A cumulative distribution of measured airborne radon concentration values (both before and after treatment) provides a picture of the overall reduction of radon in each community. Figures 6, 7, and 8 show the cumulative distribution of pre- and post-treatment measured radon concentrations in Communities A, B, and C, respectively.

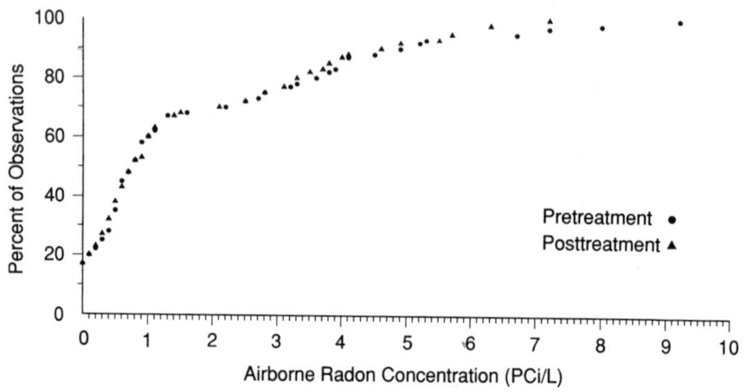

Figure 6. Airborne Radon Cumulative Distribution, Community A

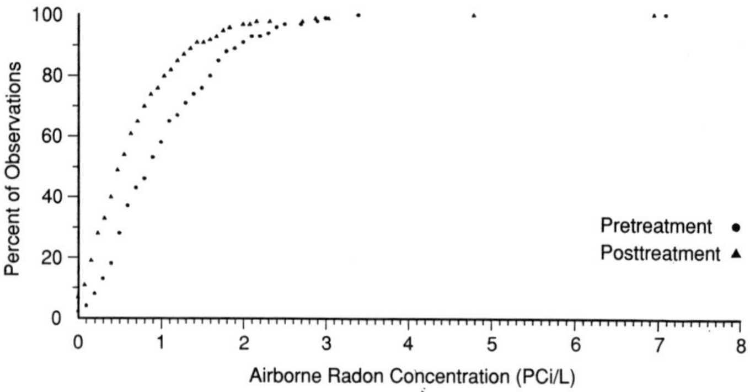

Figure 7. Airborne Radon Cumulative Distribution, Community B

Community A essentially shows very little difference in the cumulative distribution of measured airborne radon values before and after treatment of water for removal of radon. The difference of mean of pre- and post-treatment airborne radon data in Community A is 0.1 pCi/L.

In Community B, there is a small but discernible difference in the cumulative distributions. Approximately 95% of the airborne radon observations in Community B were less than 2 pCi/L, and the concentration difference in the mean of pre- and post-treatment radon concentration in air is approximately 0.2 pCi/L. Community C shows a greater difference in the cumulative distributions. The difference in the mean concentration of pre- and post-treatment airborne radon for Community C is 2.9 pCi/L.

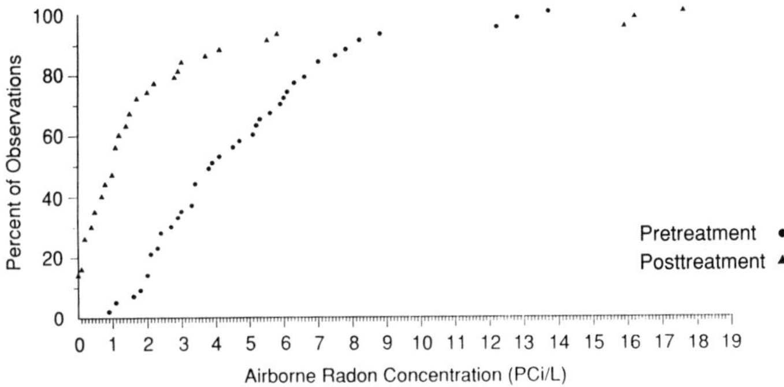

Figure 8. Airborne Radon Cumulative Distribution, Community C

CONCLUSIONS

1. In Community A, the contribution of waterborne radon to indoor air radon is small relative to the contribution from soil gas and other sources. Thus, the water treatment system for radon removal had little effect on indoor air radon in Community A. This situation could create a false sense of safety for Community A. Residents could be led to believe that the radon problem is solved by the installation of a water treatment system.

2. In Community B, average pretreatment indoor air radon concentrations in all locations are low. The small difference in pre- and post-treatment indoor air radon concentrations indicates a small contribution of waterborne radon to indoor air radon.

3. In Community C, a relatively large reduction of indoor air radon between pre- and post-treatment measurements at all locations indicates a large contribution of waterborne radon to indoor air radon and the effectiveness of a water treatment system in indoor air radon reduction.

4. The effect of reducing waterborne radon on indoor air radon concentrations has been found to be a reduction of 1.3×10^{-4} pCi/L of indoor air radon for every reduction of 1 pCi/L of waterborne radon.

REFERENCE

1. USEPA. 1989. Indoor Radon and Radon Decay Product Measurement Protocols, Las Vegas; Office of Radiation Programs, Eastern Environmental Radiation Facility.

ACKNOWLEDGMENT

The sponsorship of the American Water Works Association Research Foundation (AWWARF) of this project is acknowledged.

Nation-Wide Bromide Occurrence and Bromate Formation Potential in Drinking Water Supplies

P. Westerhoff, M.Siddiqui, J. Debroux, W. Zhai, K.Ozekin
and G. Amy
University of Colorado, Boulder

Abstract
The average bromide concentration obtained in a survey of 101 nation-wide drinking water utilities was almost 100 μg/l. Whereas previous studies have targeted ozonation studies on waters containing high levels of bromide, a broad ozonation study herein illustrates that, under typical treatment conditions, a significant portion of utilities may exceed the proposed bromate MCL of 10 μg/l. A bromate prediction diagram has been developed to assist utilities in assessing the in potential for bromate formation.

Introduction
Bromide occurs ubiquitously in surface and ground water supplies throughout the nation, and can pose health concerns by forming brominated chlorination disinfection by-products (DBPs) and bromate. Levels of organic and inorganic DBPs are being lowered by the USEPA to protect human health as more epidemiological data and analytical techniques become available. Upon chlorination of waters containing bromide, bromide can substitute for chlorine and react with organic material to form such organic DBPs as brominated THMs or HAAs depending upon pH and other water quality characteristics (Glaze et al, 1993; Pourmoghaddas et al, 1993; Symons et al, 1993). Of particular importance to utilities is the fact that brominated organics equate to a higher mass of halogenated organics than chlorinated DBPs since the molecular weight of bromide (79.9 g/mol) is twice that of chloride (35.5 g/mol). Upon ozonation, bromide can quickly oxidize to bromate (via hypobromite), an inorganic DBP with a proposed USEPA MCL of 10 μg/l. Hypobromite/hypobromous acid can also lead to the formation of regulated organic DBPs.

Drinking water utilities have primarily used chlorine to disinfect water (potable supplies or cooling tower water) for many years. Chlorination acts as a disinfectant and oxidant in these supplies, but recently the benefits attributed to microbial inactivation and metal oxidation have been contrasted to the formation of organic DBPs. Sine regulation of DBP concentrations in drinking water by the USEPA, alternative disinfectants/oxidants have been considered for chlorine with the intent of decreasing the DBPs of health concern. As new regulations require higher finished water quality, ozone stands at the forefront of alternative disinfectants/oxidants, yet some unanswered questions persist related to DBPs formed during ozonation (e.g., bromate and others).

While the literature indicates that bromide is of concern to several major utilities, it has not been clear as to whether bromide, on a national basis, is an isolated or extensive problem. The first intent of this study is to describe the national occurrence of bromide ion in drinking water sources. Second, as utilities consider ozonation as an alternative disinfection process to chlorination, the potential for bromate formation may become a factor in the selection. This study examines the results of controlled ozonation studies of over thirty nation-wide drinking water supplies to provide a broad view for the potential of bromate formation based on varying water quality conditions.

Methods and Procedures

Upon receipt, samples were characterized by analyzing for bromide, chloride, pH, alkalinity, hardness, and ammonia nitrogen. The impetus for measuring chloride is to discern differences between coastal or salt impacted regions and other geographic regions. Hardness, alkalinity, DOC, and pH were measured to provide geochemical inferences for bromide occurrence. pH is also an important parameter because of its effects on chlorination and ozonation DBPs. Ammonia can play a role in bromate formation during ozonation, or react with free chlorine. Dissolved organic carbon (DOC) was measured as an indicator of organic DBP precursors and oxidant demanding constituents.

Bromide and chloride were measured by ion chromatography (IC) using a Dionex (DX300) IC with an IonPac AS9 column with a minimum detection limit of 5 μg/l. Bromate was analyzed using a borate IC eluant and an Ionpac AS9SC column with a minimum detection limit of 2 μg/l; Dionex silver filters were used to remove chloride during bromate analyses. Alkalinity, hardness, ammonia and pH were all measured according to standard methods. DOC measurements were performed using a Shimadzu TOC-5000 low-level analyzer.

Ozonation was performed in a true batch mode. A concentrated ozone stock solution (~40 mg/l at 4°C) of Milli-Q water was obtained by ozonating 500 ml for several hours with an Orec Ozone Generator and pure oxygen. Aliquots of the stock ozone solution were quickly transferred to either 72 ml serum vials in order to completely fill the vials, or to larger modified graduated cylinders if larger volumes were involved. The reactors were headspace free to prevent any loss of ozone to the atmosphere, and bromate was analyzed after 24 hours (after residual ozone had completely decayed).

Bromide Sources and Occurrence

Over approximately an 18-month period, a total of 164 samples were evaluated from over 100 participating drinking water utilities (Amy et al, 1993). These samples included seasonal variations for some utilities. The survey broadly selected random utilities, except for 10 known utilities with bromide-related problems. The database has been segregated geographically into targeted, large (serving >50,000 people) and small (serving < 50,000 people) utilities (Figure 1).

Bromide may enter water supplies via salt water intrusion, weathering of geological formations, anthropogenic sources (e.g. fuel additives), or aerosol uptake from ocean waters, and is influenced by geographical location, climatological events, and population and industrial density. The study found that large utilities are influenced more than small utilities (Figure 2). Randomly selected large utilities had an average bromide concentration of 78 μg/l (range: 0 to 896 μg/l); targeted utilities had an average of 397 μg/l (range: 33 to 887 μg/l); small utilities had an average of 6 μg/l (range: 0 to 30 μg/l). Table 1 summarizes the results of bromide occurrence by source.

Table 1 - Bromide Occurrence by Source

	Lakes	Rivers	Ground Water
Average	38 μg/l	101 μg/l	168 μg/l
Median	23	63	62
Range	2-322	4-426	2-2690
Std Dev	54	115	428
N	34	29	36

As can be observed, bromide occurrence has a large range, but generally tends to be lower in lakes and rivers than in ground water. The entire survey of 101 drinking water supplies showed an average national bromide concentration of almost 100 μg/l. Seasonal variations in bromide concentrations were minimal, compared to the seasonal fluctuations in the ratio of bromide to DOC. This important ratio influences oxidant doses and DBP formation. The fluctuations in DOC can be attributed to seasonal DOC cycles due to lake turnover and runoff (range: 0.2 to 11 mg/l). Ammonia concentrations ranged between 0.040 to 0.690 mg/l. Poor correlations were observed between bromide and either hardness (range:), alkalinity (range:), or pH (range: 6.89-8.83), suggesting that there is no simple geochemical explanation for bromide occurrence. However, the strongest correlation was found between chloride and bromide. Seawater exhibits a chloride to bromide ration of 292:1 (mg/mg). Some of the water supplies exhibited values approaching this ratio, suggesting that either connate seawater or salt deposits may be a source of bromide.

Surveys by European countries have found similar trends in bromide occurrence as in America. Dutch drinking water utilities sources have bromide concentrations ranging from 100 to 400 μg/l (Kruithof and Meijers, 1993). Finnish waters have bromide concentrations into the hundreds of micrograms per liter, and 30 to 100 μg/l are rather common (Hisvirta, 1993). Legube et al (1993) surveyed 25 European water utilities and observed bromide concentrations in the range of 0 to 1,040 μg/l (average = 130 μg/l).

Bromate Formation - A Synopsis

Haag and Hoigne (1983) proposed that bromate formation is described by the following oxidation pathway, portraying the roles of bromide, ozone, and an intermediate species, hypobromite ion:

$$O_3 + Br^- \rightarrow O_2 + OBr^-$$

$$2O_3 + OBr^- \rightarrow 2O_2 + BrO_3^-$$

The pH dependency of bromate formation, lower bromate formation under lowered pH conditions, is reflected by:

$$H^+ + OBr^- \rightleftarrows HOBr \qquad (pK_a = 8.7 \text{ @ } 20°C)$$

NOM affects on bromate formation, with lower bromate formed in the presence of DOC, are embodied by the competition of NOM for molecular ozone and for bromine species to form organo-bromine:

$$HOBr + NOM \rightarrow Org\text{-}Br$$

Ammonia may reduce bromate formation by combining with the intermediate species according to the reaction (Krasner et al, 1993):

$$HOBr + NH_3 \rightarrow NH_2Br + H_2O$$

Other researchers suggest that $OH^{\cdot}$ radical pathways, initiated by ozone decomposition, can play a major role in bromate formation (Von Gunten and Hoigne, 1993; Yates and Stenstrom, 1993; Westerhoff et al, 1993). Through radical reactions carbonate alkalinity scavenges $OH^{\cdot}$ radicals and decreases the rate of ozone decomposition (Reckhow, Legube and Singer, 1986). It has also been postulated that the radical carbonate ion formed during this process may itself play a further role in oxidizing bromide to bromate.

Bromate Formation Potential

Thirty three waters from the above survey, and some local waters, have been examined for their potential to form bromate upon ozonation. The average bromide concentration in the surface waters examined is 87 μg/l (n = 22) and 272 μg/l for the ground waters (n = 11). These waters were ozonated under various conditions to examine the role of water source on bromate formation (DOC, pH, ammonia, alkalinity, bromide).

One set of experiments exposed all the waters to the same transferred) ozone dose of 3 mg/l to simulate a common ozone dose, despite the variation of DOC (ozone demanding substances) of 0.2 to 12 mg/l and pH (influences the distribution of HOBr/OBr^- and ozone decomposition rates) of 6.9 to 8.9, and other variable water quality characteristics. The results indicate that detectable (> 2 μg/l) bromate formation occurred in virtually all waters (Figure 3 and 4), and many utilities using these source waters may potentially violate the proposed MCL of 10 μg/l.

In a separate set of experiments, all the waters were adjusted to pH 7.0 with HCl or NaOH and then dosed with 3 mg/l of ozone. The average ambient pH was approximately 8.0 before adjustment, and since a suggested bromate control option includes reducing pH, a constant pH was examined (Figure 5). pH depression generally led to a reduction in pH, while an increase led to increased bromate formation. Exceptions to these observations may be the result of the complex nature of bromate formation. Decreasing the pH decreases the bromate driving force, OBr^-, but increases the driving force due to molecular ozone (molecular ozone decomposes slower and is therefore more available for bromide oxidation at lower pHs due to the free radical initiation dependency on OH^- ions).

In combination with a series of experiments conducted at varying pH and ozone doses, generalized relationships to predict the potential for bromate formation can be obtained with the aid of statistical analysis. Bromate prediction diagrams can be generated to estimate bromate formation based on water quality parameters and ozone doses (Figure 6). This type of an approach is intended to allow a utility to quickly assess the level of concern for bromate formation upon ozonation. As will be discussed below, this diagram is an approximate prediction and bromate formation is highly dependent upon all the water constituents and source of DOC.

Three of the waters examined herein turned pinkish during ozonation, and further analyses indicated the presence of manganese which was being oxidized during ozonation. Controlled studies on one water resulted in decreasing bromate formation in the presence of increasing manganese concentrations. It can be speculated that the manganese exerted an ozone demand, and a manganese concentrations of 0, 0.2, 0.4 and 1.0 mg/l resulted in bromate concentrations of 42, 32, 30, 22 μg/l after ozonation, respectively.

Summary

With the advent of new USEPA regulations limiting the levels of halogenated organic DBPs which result primarily from chlorination, many utilities are examining the use of ozone as an alternative disinfectant. Bromide ion concentration can play a critical role in the disinfection/oxidation process selection due to the possible formation of brominated organic DBPs and bromate. This study provides a useful overview of the extent of bromide occurrence and its potential to form bromate upon ozonation via bromate prediction diagrams.

References

Amy, G., Siddiqui, M. Zhai W., DeBroux, J. and Odem, W. (1993). *AWWA Annual Conference Proceedings - San Antonio*, June.

Glaze W.,Weinberg, H.,and Cavanaugh, J.(1993).*JAWWA*, 85:1:96-103.

Haag, W. and Hoigne, J. (1983). *ES&T*, 17:5:261-267.

Hisvirta, K. (1993). *Proceedings of the International Water Supply Association International Conference: Bromate and Water Treatment*, Paris, November, pp.125-134.

Krasner, S., Glaze, W., Weinberg, H., Daniel, P, and Najm, I. (1993). *JAWWA*, 85:1:73-81.

Kruithoff, J. and Meijers, R. (1993). *Proceedings of the International Water Supply Association International Conference: Bromate and Water Treatment*, Paris, November, pp.125-133.

Legube, B., Bourbigot, M., Bruchet, A., Deguin, A., Montiel, A., and Matia, L. (1993). *Proceedings of the International Water Supply Association International Conference: Bromate and Water Treatment*, Paris, November, pp.135-150.

Pourmoghaddas, H., Stevens, A., Kinman, R., Dressman, R., Moore, L, and Ireland, J. (1993). *JAWWA*, 85:1:63-72.

Reckhow, D., Legube, B., and Singer, P. (1986). *Wat Res*, 20:8:987-998.

Siddiqui, M. and Amy G. (1993). *JAWWA*, 85:1:63-72.

Symons, Krasner, Simms, and Sclimenti, (1993). *JAWWA*, 85:1:51-62.

Von Gunten, U., and Hoigne, J. (1993). *Proceedings of the International Water Supply Association International Conference: Bromate and Water Treatment*, Paris, pp.51-56.

Westerhoff, P., Amy, G., Minear, R., Siddiqui, M., Song, R., and Ozekin, K. (1993). *AWWA Annual Conference Proceedings-San Antonio*, June.

Yates, R. and Stenstrom, M. (1993). *AWWA Annual Conference Proceedings-San Antonio*, June.

Figure 1. Geographical Distribution of Utilities According to EPA

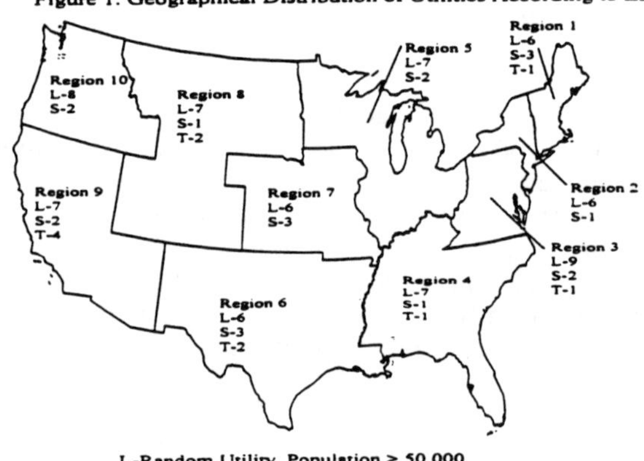

L-Random Utility, Population > 50,000
S-Random Utility, Population < 50,000
T-Targeted Utility

Figure 2. Seasonal Averages of Bromide Concentration (ug/L) According to EPA Region

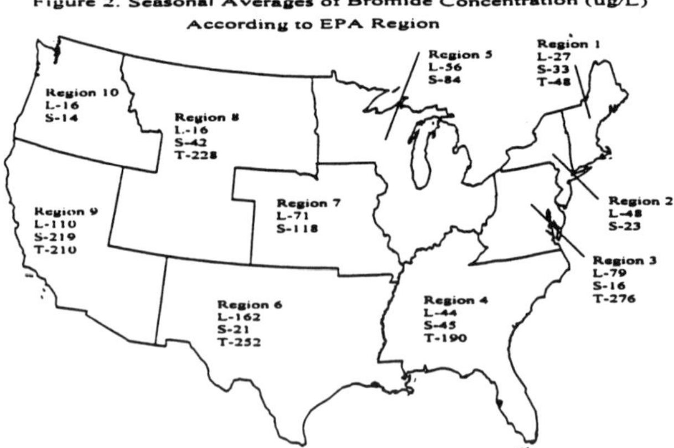

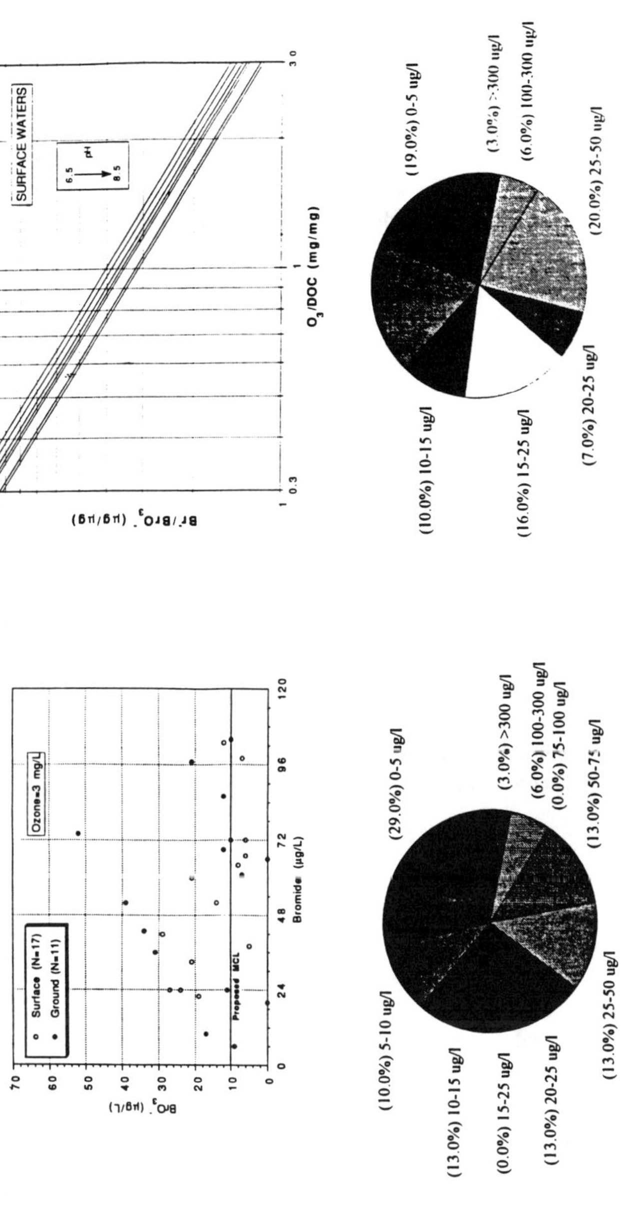

Figure 3 - Bromate Formation in U.S. Drinking Water Supplies

Figure 4 - Bromate Formation Distribution (3mg/l Ozone; pH = Amb)

Figure 5 - Bromate Formation Distribution (3mg/l Ozone; pH = 7.0)

Figure 6 - Bromate Potential Prediction Diagram for Surface Waters

Computer Aided Design of Trickling Filters

Bruce E. Logan[1]

Abstract

Computer programs based on mechanistic mass transfer models have been developed by the author to predict sBOD removal, ammonia removal, and oxygen transfer in plastic media trickling filters. Although the use of these models requires the numerical solution of a set of differential equations, the availability and speed of modern personal computers makes these programs accessible to design engineers. Application of the program that predicts sBOD removal (TRIFIL2), using a default molecular weight distribution, predicted lower removal (68%) than observed (95%) for a trickling filter in Tucson AZ. Size distributions of sBOD were therefore measured using two ultrafiltration membranes with molecular weight cutoffs of 1,000 and 10,000 daltons. Using the observed sBOD size distributions as input into the computer model we calculated 93% removal which compared very well to the observed removals. We conclude that high sBOD removals at the Tucson plant are due to the high concentration of low molecular weight organic compounds in the wastewater.

Introduction

The recent WEF Manual on the design of wastewater treatment plants (WEF Manual of Practice 1992) describes many approaches to designing trickling filters for BOD removal from wastewater. Several of these design equations are based on the assumption that substrate removal is first order with detention time in the reactor, or that

---

[1]Associate Professor, Department of Chemical and Environmental Engineering, University of Arizona, Tucson, AZ, 85721.

$$\frac{dc}{dz} = -k_1 c \qquad (1)$$

where c is the substrate (sBOD) concentration, t is the detention time in the reactor, and k_1 is a first-order overall rate constant. This approach is flawed, however, since in order to derive an equation of this form (using a shell balance) is must be true that the fluid above the biofilm is completely mixed. The requirement of this solution for complete mixing is justified below.

If we have a control volume of fluid of height δ and width w above a biofilm, the rate wastewater organics enters and leave the control volume are $v\delta wc_z$ and $v\delta wc_{z+\Delta z}$. Since the reaction term is not in the fluid, but in the biofilm, the only way to obtain a differential equation of the necessary form is to use the chemical flux out of the control volume as if it was a reaction in the control volume. For first-order biofilm kinetics, the flux into the biofilm is $c_s(kD)^{1/2}$. Since the flux occurs across the cross-sectional area $w\Delta z$, the overall rate is $c_s(kD)^{1/2}w\Delta z$. Combining all terms, and expressing this equation in terms of a limit as $\Delta z \to 0$, we have

$$\lim_{\Delta z \to 0} \frac{c|_{z+\Delta z} - c|_z}{\Delta z} = -\frac{c\,(k\,D)^{1/2}}{v\,\delta} \qquad (2)$$

Taking this limit produces the differential equation

$$\frac{dc}{dz} = -\left[\frac{(k\,D)^{1/2}}{v\,\delta}\right] c \qquad (3)$$

Thus, Eq. 3 is the same as Eq. 1 if the term in brackets in Eq. 3 is equal to the overall first order rate constant k_1.

In order for Eq. 3 to be derived we had to assume that the flux from the fluid was not a function of concentration gradients in the fluid, or that the flux was not mass transfer limited. The only justification for the flux to be assumed to occur without a gradient is to assume either complete mixing or very slow kinetics. Complete mixing is not possible since fluid flow is laminar. Slow kinetics are not justified since computer solutions indicate biofilm kinetics must be rapid to account for observed removal rates.

Investigators have tried to adjust integrated solutions to the first-order equation by adding additional empirical "correction" terms in Eq. 1 to make the solution agree with observed BOD removals in trickling filters. However, such approaches have no rational theoretical basis and cannot be accepted since the basic formulation of a completely mixed fluid layer is incorrect.

Computer models can more accurately model BOD removal in trickling filters since the governing equations describing mass transfer and biofilm kinetics can be solved using numerical techniques. Different assumptions are made by different investigators while constructing these computer models. Logan et al. (1987a, b) assumed that sBOD removal was independent of oxygen transport and that fluid flow was dictated by the geometry of the plastic media modules. Hinton and Stensel (1989, 1991) have argued that oxygen limits sBOD removal and that biofilm "stalactites" dictate fluid flow patterns within the media. Both issues are still being debated. Whatever the mechanism responsible for fluid patterns in plastic media, side-by-side tests have shown that media of similar specific surface area can differ in BOD removal (Logan et al. 1987a).

Methods

Experiments conducted at the University of Arizona over the last 7 years have been designed to understand the fate of macromolecules in fixed film wastewater treatment systems. The purpose of this research was to investigate whether a measured size distribution of dissolved organic matter in wastewater could be used in computer models to predict sBOD removal in trickling filters. The computer model used to predict sBOD removal was TRIFIL2 (Logan 1993) the second version of the model TRIFIL developed by Logan et al. (1987). This and other trickling filter models (TFO2, NTF2 and LTF) are available from the author upon request.

All wastewater samples analyzed in this part of our project were obtained from the Roger Road Wastewater Treatment Plant located 7 miles from the University of Arizona. The plant has two 28 ft high trickling filter towers 1900 m^2 (21,380 ft^2) filled with BF Goodrich vertical-flow 30 ft^2/ft^3 (VFb-30) media designed to treat 30 MGD. The plant usually operates at 1.14 gpm ft^{-2} with a variable recycle.

Molecular Weight Distributions. The molecular weight distributions of amino acids, dissolved organic carbon (DOC), and sBOD were examined using molecular size separations based on ultrafiltration techniques (Logan and Jiang 1990). Wastewater samples obtained from the Roger Rd. plant were sequentially prepared by centrifugation (<4000 g), and filtration through pre-rinsed (500 ml) 5 μm and 0.45 μm pore diameter cellulose acetate filters. Samples for analysis of amino acids were filtered twice through 0.2 μm polycarbonate filters. Molecular weight separations were performed using two ultrafiltration membranes with 1,000 and 10,000 nominal molecular weight (NMW) cutoffs in 200 ml stirred ultrafiltration cells (Amicon, Model 8200) under nitrogen. Permeate concentrations were corrected for membrane rejection using the permeation model of Logan and Jiang (1990).

The TRIFIL2 trickling filter model (Logan 1993) uses an assumed 5-component size distribution based on data originally developed by Levine et al. (1985). In developing this model it was assumed sBOD was equally distributed into these 5 components with diffusion coefficients of 112, 80, 65, 50 and 30 $\mu m^2\ s^{-1}$. The experimental protocol used here produced molecular weight distributions of <1,000 daltons, <1,000 to <10,000 daltons, and >10,000 daltons to <0.45 μm. Using either chemical molecular weight (Polson 1950, Frigon et al. 1983) or molecule size and the Stokes-Einstein equation (Bird et al. 1960) produced diffusion coefficients of 500, 100 and 10 $\mu m^2\ s^{-1}$ used in the model for these three size fractions.

Amino acid concentrations in the wastewater were investigated to see if treatment efficiency was correlated to amino acid removal. The total concentration of dissolved amino acids (DTAA, e.g. dissolved polypeptides, proteins and amino acids) was measured in unconcentrated domestic wastewater at a detection limit of 10 to 100 nM (coefficient of variation $\leq$0.10 at 100 nM) using precolumn *ortho*-phthaldialdehyde (OPA) derivatization, reverse phase HPLC separation, and fluorometric detection (Confer et al. 1994).

Non-purgable dissolved organic carbon (DOC) was measured using a Shimadzu 5000 DOC analyzer, autosampler, and a two point calibration curve with blank correction. Standards and samples were acidified to pH 2.5 and purged of inorganic carbon (CO_2) prior to analysis. sBODs were calculated using either 60-ml BOD bottles or 22-ml tubes filled to capacity and crimp-sealed with teflon stoppers. Bottles were prepared in triplicate using a synthetic seed (Polybac) at two dilutions (Wagenseller 1994).

Results

Molecular Weight Distributions. Molecular weight separations indicated that the majority of DTAA were in the combined form and therefore had large molecular weights (Figure 1). Primary clarifier effluent contained the largest concentrations (31 μM) of DTAA with 31% as free amino acids. Removal percentages during biological treatment were 89% for amino acids and 47% for combined (large molecular weight) amino acids. Overall, DTAA were a small percentage (8 to 13%) of DOC, and DTAA removals were not well correlated with DOC removals (Confer et al. 1994).

sBOD removal in Trickling Filters. Most of the sBOD applied to the Roger Road trickling filter was contained in the small molecular weight fraction (76% <1,000 amu), with only a small percent (5%) in the largest (>10,000 amu) size fraction. Typical sBOD removals average around 95% removal (Fig. 2) DOC concentrations in the wastewater were also mostly in the small size fraction (72%).

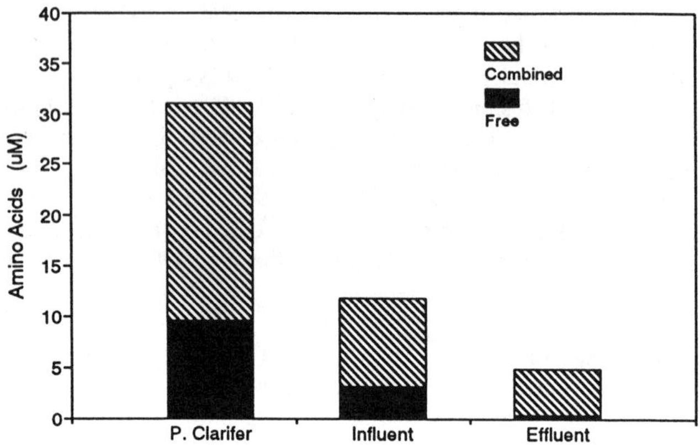

Figure 1. Combined and free amino acids in trickling filter wastewater.

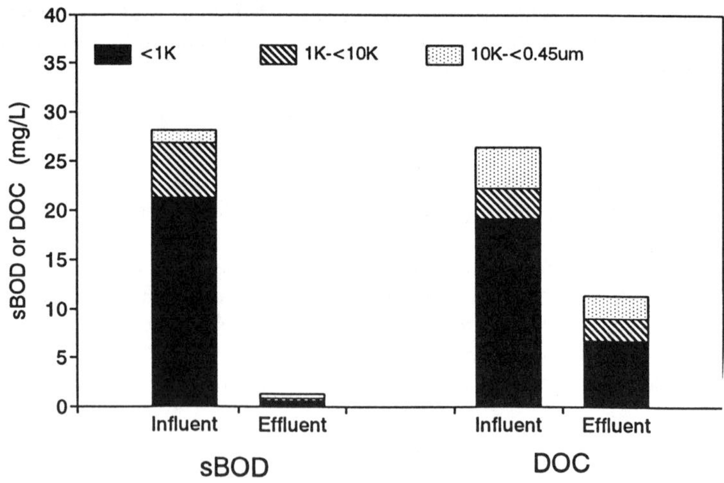

Figure 2. sBOD and DOC concentrations in trickling filter wastewater entering and leaving the biotower.

Only 57% of the DOC was removed, indicating that not all of the DOC was biodegradable (Wagenseller 1994).

At an influent sBOD of 28.2 mg l^{-1}, we measured an effluent sBOD of 1.4 mg l^{-1} (95% removal). When the 5-component size distribution was used in the TRIFIL2 model the predicted sBOD removal was 68%. However, when we used the observed input distribution of 21.3 mg l^{-1} (500 μm^2 s^{-1}), 5.6 mg l^{-1} (100 μm^2 s^{-1}) and 1.3 mg l^{-1} (10 μm^2 s^{-1}), the model prediction was 93% removal. This compares favorably with the observed removal indicating that the model does not require recalibration, and further, that sBOD removal can be predicted from measured sBOD molecular weight distributions.

Discussion

The design of trickling filters has historically proceeded based on the assumption that BOD removal was kinetically limited in the treatment process. Consequently, pilot scale plants have been required to address efficient design for wastewaters of varying composition. However, mass transfer models such as the one used here have been gaining wider support based on their fundamental approach to media geometry and wastewater properties. Based on the above calculations, it may now be possible to anticipate the performance of a given design on different wastewaters by measuring the molecular size distributions of the sBOD components and using these in computer models. This level of design is made possible by the availability and low cost of computers and ultrafiltration separation devices.

In the development of any model, there are always questions concerning assumption made in the model derivation. One recurring question is whether oxygen can limit trickling filter performance since it is assumed in the TRIFIL2 model that lack of oxygen does not restrict sBOD removal. The model incorporates this assumption since there is a lack of evidence to support oxygen-limited sBOD removal in trickling filters. Logan et al. (1987a, 1989) have shown that sBOD removals observed at full scale trickling filters can exceed the calculated maximum rate of oxygen transfer to the biofilm, and this data has never been contested. Laboratory investigations by Hinton (1989) have shown that substrate removal by biofilms can be accompanied by decreased oxygen uptake; however, it has not been shown that oxygen was the cause of substrate-limited uptake. An alternative explanation for the Hinton data is that substrate transport exceeded the uptake rate of the microbial biofilm. A limitation of the uptake rate at the biofilm surface would limit the substrate gradient and therefore the sBOD flux. Given an oxygen flux still below its own maximum rate, and substrate uptake coupled to oxygen uptake, the oxygen uptake would also flatten out. Thus a plateau in substrate update does not *require* that oxygen limited substrate uptake. The ability of the TRIFIL2 model to reproduce sBOD removal data at a variety of pilot and full scale sites still remains as the best evidence to support the model assumption that oxygen does not limit sBOD removal in trickling filters.

A point of agreement among trickling filter computer models is that the size of molecules is an important factor in sBOD removal by biofilms. The chemical flux into a biofilm is proportional to chemical diffusivity, chemical gradient, and fluid mechanics. Further development of the models and the techniques to collect size distribution data can help design engineers design the most cost effective treatment systems.

References

Bird, R.B., Stewart, W.E., Lightfoot, E.N. (1960). *Transport phenomena*. John Wiley and Sons, New York, N.Y.

Confer, D.R. Logan, B.E. Aiken, B.S. and Kirchman, D.L. (1994). "Measurement of dissolved free and combined amino acids in unconcentrated wastewaters using HPLC". *Wat. Environ. Res.* In press.

Frigon, R.P., Leypoldt, J.K., Uyeji, S. and Henderson, L.W. (1983). "Disparity between Stokes radii of dextrans and proteins as determined by retention volume in gel permeation chromatography." *Anal. Chem.*, 55, 1349-1354.

Hinton, S.W. (1989). "A mechanistic model for the trickling filter process which considers the effects of both substrate and oxygen availability on substrate uptake kinetics," Doctor of Philosophy thesis presented to the University of Washington, Seattle.

Hinton, S.W. and Stensel, H.D. (1989). Discussion of "A fundamental model for trickling filter process design," by B.E. Logan, S.W. Hermanowicz and D.S. Parker. *J. Water Pollut. Control Fed.*, 61(3), 363-364.

Hinton, S.W. and Stensel, H.D. (1991). "Experimental observations of trickling filter hydraulics." *Wat. Res.*, 25(11), 1389-1398.

Levine, A.D., G. Tchobanoglous, and T. Asano. (1985). "Characterization of the size distribution of contaminants in wastewater: treatment and reuse implications." *J. Wat. Pollut. Cont. Fed.*, 57(7):805-816.

Logan, B.E. (1993). "Oxygen transfer in trickling filters." *J. Environ. Engin.* 119(6):1059-1076.

Logan, B.E., Hermanowicz, S.W. and Parker, D.S. (1987a). "A fundamental model for trickling filter process design." *J. Water Pollut. Control Fed.*, 59(12), 1029-1042.

Logan, B.E., Hermanowicz, W.W., Parker, D.S. (1987b). "Engineering implications of a new trickling filter model." *J. Water Pollut. Control Fed.*, 59(12), 1017-1028.

Logan, B.E., Hermanowicz, S.W., Parker, D.S. (1989). Reply to Discussion of S.W. Hinton and H.D. Stensel. *J. Water Pollut. Control Fed.*, 61(3), 364-366.

Logan, B.E. and Q. Jiang. 1990. "A Model for determining molecular size distributions of DOM." *J. Envir. Engin. Div., ASCE*, 116(6):1046-1062.

Polson, A. (1950). "Some aspects of diffusion in solution and a definition of a colloidal particle." *J. Phys. Chem.*, 54, 649-652.

Wagenseller, G.A. (1994). "Molecular size distributions of BOD and DOC in trickling filter wastewaters", M.S. thesis presented to the University of Arizona, Tucson.

WEF Manual of Practice, No. 8 (1992). "Suspended-growth Biological Treatment", Chap. 11. In: *Design of Municipal Wastewater Treatment Plants*, Water Environment Federation, Alexandria, Virginia.

DEVELOPMENT OF A MODEL FOR THE ABF/AS PROCESS

Dr. Julian Sandino[1]
Dr. Ross E. McKinney[2], Honorary Member

Abstract

A modelling approach has been established for the activated biofiltration/activated sludge (ABF/AS) process. This approach indicates that ABF/AS systems can be considered as a variation of the activated sludge process, in which the biocell functions as a high-rate reactor, providing oxygen transfer and contact between soluble organics in the wastewater and the biomass recycled from the downstream aeration basin. This modelling approach provides for the first time a tool for the rational (rather than empirical) design of new ABF/AS facilities, as well as for the operation of existing systems.

Introduction

The activated biofiltration/activated sludge (ABF/AS) process (Figure 1) is one of the most popular fixed-film/suspended growth biological processes in wastewater treatment. The most important feature of the ABF/AS process is the recycling of settled biomass from the final clarifier through the biocell. Similarly to other "combined" systems, the ABF/AS process is thought to provide the stability of the fixed-film process while achieving the higher effluent quality of the suspended growth system.

---

[1]Process Engineer, Black & Veatch, 8400 Ward Parkway, Kansas City, MO 64114

[2]Process Consultant, 2716 Oxford Road, Lawrence, KS 66044

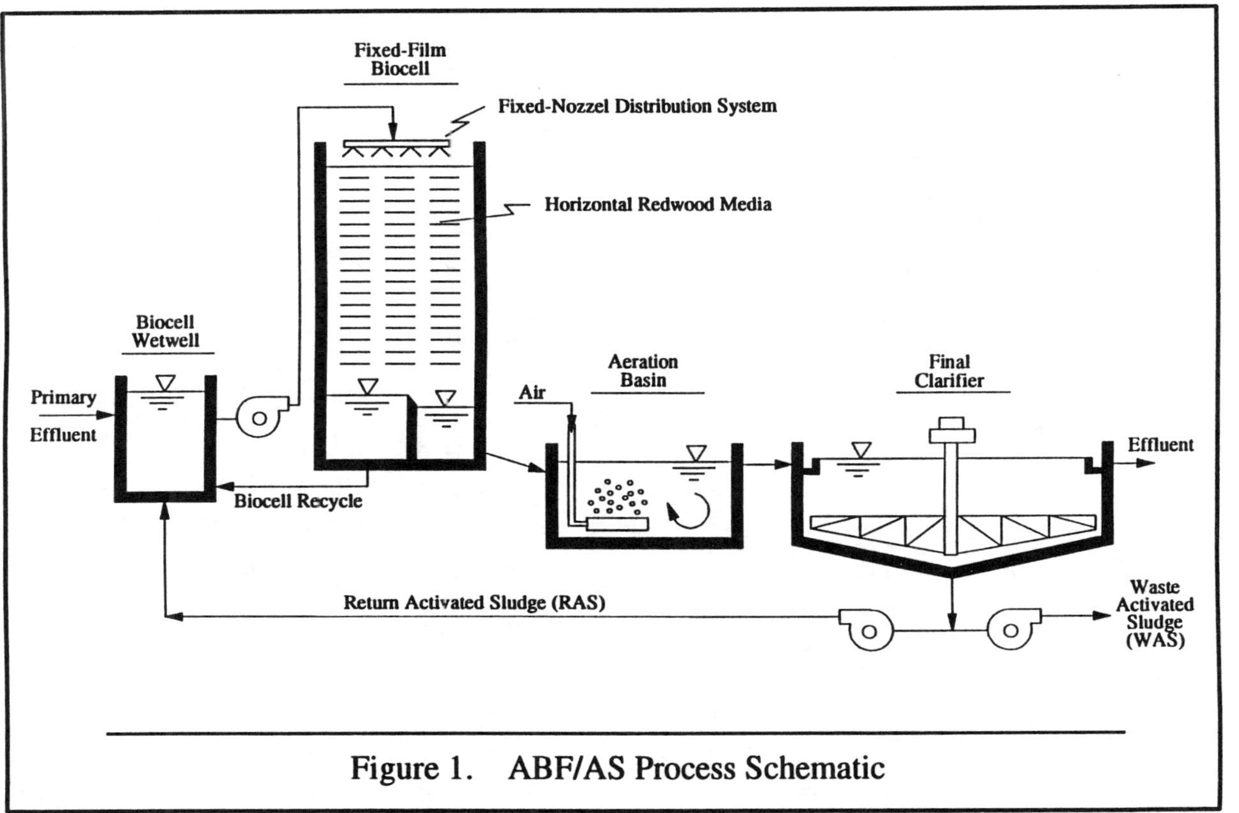

Figure 1. ABF/AS Process Schematic

Traditionally, ABF/AS facilities have been designed based on a series of guidelines empirically derived from the results of several pilot studies. This approach, however, has proven inadequate when considering applications that do not match closely the conditions under which the empirical relationships were derived. In addition, failure to recognize the actual biological principles governing the process has limited the use of these guidelines for purposes of operating the system. This lack of understanding of the fundamental process mechanisms involved can help explained in part the incidence of failures of plants using the ABF/AS process, as well as the overall decline in popularity of this system in recent years. The objectives of this research were to develop an understanding of the fundamental biological principles governing this process and to establish a rational modelling approach that could be used to size and operate ABF/AS facilities.

Modelling Approach and Methodology

In developing a rational model for the design and performance evaluation of the ABF/AS process, the biocell was regarded as an extension of the aeration basin in an activated sludge process. The basic role assumed for the biocell was that of a high-rate reactor, providing oxygen transfer and contact between the soluble organics and the biomass (both attached to the media and recycled from the clarifier).

The approach taken in developing a model for the combined system was to initially evaluate separately the biocell from the aeration basin. Once relationships were established that accurately described the biocell performance, they were coupled with previously established completely mixed activated sludge (CMAS) kinetic equations.

Data collected from the Helena Wastewater Treatment Plant (WWTP) in Montana, and the T.E. Maxson WWTP in Memphis, Tennessee were used to derive a relationship for soluble organics removal in the biotower. The differences between these treatment plants in terms of wastewater composition, ambient conditions, and equipment configuration allowed the establishment of generalized modelling relationships that can be considered applicable to other ABF/AS facilities.

The establishment of the biocell performance model was carried out in four steps. In the first step, variables thought to determine performance were identified on a preliminary basis. This mostly intuitive approach was based on the consideration of previous studies and observations on ABF/AS facilities, attached growth systems (mainly trickling filters), and biological treatment processes in general. By considering the basic components in biological treatment processes, namely substrate, microorganisms, contact time, and temperature, a group of variables thought to affect biocell performance were identified. The next step corresponded to a factor analysis of these preliminary identified variables utilizing the previously described data base. The purpose of this statistical analysis was to identify the existence of possible correlations between variables. In the third step, variables shown by the factor analysis to have a high correlation with biocell performance were used as the basis for a dimensional analysis. This analysis provided a procedure to simplify the determination of a relationship between these variables by establishing a number of dimensionless products of these parameters. From the dimensional analysis, the following relationship for biocell performance (in terms of soluble organics removal) was established:

$$SBOD_5r = 0.243\,(D^{3.380})\,(Q_t^{0.586})\,(Q_p^{-0.575})\,(\theta^{(T-20)})\,(RASvs^{0.01})$$

Where:

$SBOD_5r$	=	Soluble BOD_5 removal across the biocell as expressed in terms of plant flow, mg/l
D	=	Biocell media depth, m
Q_t	=	Biocell total hydraulic loading, $L/m^2 \cdot s$
Q_p	=	Plant raw wastewater flow expressed as a hydraulic loading, $L/m^2 \cdot s$
θ	=	Modified Arrenhius constant, estimated to be 1.013
T	=	Wastewater temperature, °C
RASvs	=	Return activated sludge volatile solids load to biocell, estimated from return activated sludge flow and volatile solids concentration in the return activated sludge stream, kg/d

Previously established fundamental biochemical energy-synthesis principles and completely mixed activated sludge (CMAS) relationships (Mckinney, 1962, 1984) were used to predict biomass generation in the biocell as well as to simulate the aeration basin stage of the ABF/AS process. By combining the biocell performance relationship with McKinney's biomass generation relationships and CMAS equations, the complete simulation of the process was accomplished. A complete listing of all the proposed ABF/AS modelling relationships is included at the end of this paper.

In summary, the steps involved in simulating the integrated ABF/AS process are as follows:

1. Establishment of biocell influent characteristics

In summary, the steps involved in simulating the integrated ABF/AS process are as follows:

1. Establishment of biocell influent characteristics
2. Simulation of biocell soluble BOD_5 removal
3. Estimation of biocell active mass generation
4. Establishment of aeration basin influent characteristics
5. Simulation of aeration basin operation

The adequacy of the modelling approach above described was verified by conducting a series of statistical analysis that included the two-sample student's t-test, the one-way analysis of variance (ANOVA), and the paired comparison t-test. The sample population analyzed corresponded to the measured and the calculated values for each of the monthly average data points in the Helena and Memphis data base. In addition, model predictions were compared against actual performance values recorded at three other full-scale ABF/AS facilities: Concord, NH; Southbridge, MA; and Twin Falls, ID. Results from this analysis indicated that the proposed ABF/AS model estimates can be considered statistically equivalent (at a 95 percent confidence level) to the observed values at the full-scale facilities considered.

Discussion of Results

The proposed model for the ABF/AS process considers the biocell as an extension of the aeration basin in the activated sludge process. The biocell provides oxygen transfer to a mixture of wastewater and returned activated sludge, converting readily biodegradable soluble organics into additional biomass. Biocell performance, in terms of soluble organics removal, was found to be a function of media configuration (surface area and depth), amount of biomass recycled, hydraulic loading conditions, and wastewater temperature.

Given the short detention times experienced in the media, the amount of biomass endogenation and hydrolysis of the influent suspended organic fraction can be considered negligible within the biocell. This results in a considerable amount of the system's total oxygen demand being exerted in the aeration basin of the ABF/AS process. Previous ABF/AS modelling approaches failed to fully establish the extent of the aeration basin oxygen demand by neglecting to recognize that soluble organics were converted in the biocell into biomass rather than destroyed. Problems experienced in many ABF/AS facilities can be attributed to this lack of understanding of the process.

Another important result of this research relates to the establishment of previously developed biochemical energy-synthesis principles as well as more general activated sludge simulation relationships as applicable also to the ABF/AS process. This fact establishes once more that aerobic wastewater treatment processes follow for the most part the same basic principles regardless of the configuration of the system components.

Conclusions

The model here presented for the ABF/AS process can be considered as a combination of empirical and rational approaches; empirical in that the simulation of soluble organics across the biocell is accomplished through an equation derived from a multiple regression analysis of data collected from two full-scale facilities; rational in that well-established energy-synthesis relationships and activated sludge kinetic equations were incorporated as well. However, it is important to note that even the empirically derived component of the model is based on well understood and widely accepted fundamental aerobic wastewater treatment principles. The proposed modelling approach should serve not only as a tool in the design of new treatment facilities, but also in the operation of the many ABF/AS installations currently in service.

References

McKinney, R.E., "Mathematics of Complete Mixing Activated Sludge", Journal of the Sanitary Engineering Division, ASCE, 88, 87 (1962).

McKinney, R.E., "Activated Sludge Treatment Systems", University of Kansas, Lawrence, Kansas (1984).

SUMMARY OF ABF/AS MODELLING RELATIONSHIPS

a. BIOCELL INFLUENT PARAMETERS.

Flows (expressed as hydraulic loading rates)

Q_P, L/m^2· s: Plant raw wastewater flow

Q_T, L/m^2· s: Total biocell flow (including return activated sludge and biocell effluent recycle flows)

Primary Effluent (PE) Characteristics

PE $TBOD_5$, mg/l:	Total five-day biochemical oxygen demand
PE TSS, mg/l:	Total suspended solids
C_i:	Inert organic fraction of the volatile suspended solids (typically 0.4 for domestic sewage)
C_{ii}:	Inert inorganic fraction of the suspended solids in influent (VSS/TSS ratio)
T, °C:	Wastewater temperature

b. BIOCELL PERFORMANCE.

$SBOD_5$ Removal

$$SBOD_5r\text{, mg/l} = 0.243\,(D^{3.580})(Q_T^{\,0.586})(Q_P^{\,-0.575})(1.013^{(T-20)})(RASvs)^{0.01}$$

Where:

$SBOD_5r$	=	$SBOD_5$ removed across biocell, mg/l
D	=	Biocell media depth, m
RASvs	=	Return activated sludge volatile loading to the biocell, kg/d

Biomass Generation

$$\text{ABF } M_a\text{, mg/l} = 0.84\ SBOD_5r$$

Where:

ABF M_a	=	Biomass generated in biocell, mg/l VSS

c. AERATION BASIN OPERATION.

Substrate Removal

$$F\text{, mg/l} = F_i/(K_M K_T\ t + 1)$$

Where:

F	=	$SBOD_5$ in effluent, mg/l
F_i	=	Influent $TBOD_5$ = PE $TBOD_5$ - $SBOD_5r$
K_m	=	Metabolism factor, 15/hr @ 20 °C
K_T	=	Temperature correction factor at temperature T, °C, equal to $1.072^{(T-20)}$
t	=	Hydraulic retention time in aeration basins based on plant raw wastewater flow only, hrs

Suspended Solids

$$M_a\text{, mg/l} = K_S K_T F/(K_e K_T + 1/t_S) + \text{ABF } M_a/(t/t_S + t\, K_e K_T)$$

Where:

M_a = Active microbial mass, mg/l VSS
K_s = Synthesis factor, 12.6/hr @ 20 °C
K_e = Endogenous factor, 0.02/hr @ 20 °C

$$M_e\text{, mg/l} = 0.2\, K_e K_T M_a t_s$$

Where:

M_e = Endogenous mass, mg/l VSS

$$M_i\text{, mg/l} = SS_{inf}\, C_i\, (1\text{-}C_{ii})\, t_S/t$$

Where:

M_i = Inert organic mass, mg/l VSS
SS_{inf} = Influent suspended solids = PE TSS, mg/l

$$M_{ii}\text{, mg/l} = SS_{inf} C_{ii} t_S/t + 0.1\, (M_a + M_e)$$

Where:

M_{ii} = Inert inorganic mass, mg/l TSS

$$MLVSS = M_a + M_e + M_i$$

Where:

MLVSS = Mixed liquor volatile suspended solids, mg/l VSS

$$MLSS = MLVSS + M_{ii}$$

Where:

MLSS = Mixed liquor suspended solids, mg/l TSS

Oxygen Demand

$$dO/dt = 0.57\, F_i/t + 1.1\, K_e K_T M_a$$

Where:

dO/dt = oxygen uptake rate (OUR), mg/l-hr

NITRIFICATION IN SINGLE-STAGE TRICKLING FILTERS

By Rao Y. Surampalli[1], O. Karl Scheible[2], and Shankha K. Banerji[3]

ABSTRACT: An assessment of nitrification of municipal wastewater in trickling filters was performed using operating data from wastewater treatment facilities currently in operation. A survey of trickling filter use in the United States, found twenty-seven plants accomplishing some degree of nitrification. Of the twenty-seven plants ten were identified as having operating data sufficient for further analysis. Although most of the plants evaluated were not designed as single-stage systems they all provided sufficient data to evaluate trickling filter performance as a single-stage system. The evaluated plants were generally meeting their permit requirements, including ammonia-nitrogen limits when applied. Analysis of the data compared favorably with the BOD loading rates suggested by the EPA Process Design Manual for Nitrogen Control.

KEYWORDS: trickling filters, nitrifying trickling filters, nitrification, BOD removal, single stage-nitrification

INTRODUCTION

Many municipalities may have ammonia limits added to their permits in the near future because of stringent water quality based ammonia limits. For the large number of facilities that include trickling filters in their treatment train, modifications to the filters would frequently be the most cost-effective solution to this additional treatment need. This manuscript evaluates the use of trickling filters for nitrification of municipal wastewater.

A survey was conducted to identify the extent to which trickling filters are used at municipal facilities in the United States to accomplish nitrification. This was not meant to be an exhaustive search, but of sufficient coverage to assess the state-of the-art, and to determine the availability of performance data.

1Senior Env. Engr, U.S. Env. Protection Agency, P.O. Box 172141, Kansas City, KS 66117.

2 Engr, HydroQual, Inc., Mahwah, NJ 07430

3Prof., Dept. of Civil Engineering, Univ. of Missouri, Columbia, MO 65211

The data was collected from full-scale treatment facilities and used to evaluate process performance and aid in understanding the effect of various operating parameters. Twenty-seven trickling filter plants that are accomplishing some degree of nitrification were identified. The design and performance data for the plants were obtained directly from the facility operators.

This study originally focused on single-stage trickling filters, a biological process application wherein carbon oxidation and nitrification are accomplished within the same unit without separation of the biomass used to accomplish these operations. Multiple-stage systems were added to the study due to the limited number of single-stage facilities. The multiple-stage systems evaluated in this report all had performance data that were measured after the first stage so they could be compared to single-stage systems.

A total of ten plants are practicing single-stage nitrification. Of these, six have the solids contact modification to enhance particulate BOD removal. Seventeen plants have separate stage nitrification, six of which use an activated sludge or stabilization pond in conjunction with the trickling filter.

Design flows range as high as 42.0 mgd with most plants between 50 and 100% of their design capacity. The majority of plants use plastic media. Twelve are exclusively plastic, while there is one slag media plant and two rock media plants. The rest have combinations of rock and plastic media filters. Depths of the rock filters range between 4 and 10 feet. The plastic media filters are generally deep, typically between 20 and 40 feet. Shallower filters with plastic media are typically retrofits of old rock filters.

None of the plants are experiencing problems with meeting permit requirements for BOD_5, suspended solids, and ammonia removal (if required), particularly during warmer temperature seasons. Problems have been noted at some plants with ammonia removal during cold temperature periods.

MATERIALS AND METHODS

Of the twenty-seven facilities that were identified as trickling filter plants accomplishing nitrification, ten were selected for further evaluation. These had sufficient data for analysis, which were made available by the individual plant operators. The following discussions present a description and assessment of the plants performance.

The Palm Springs, California wastewater treatment plant utilizes a single stage trickling filter system with four filters in parallel. The plant was operating at approximately 70 percent of its design flow of 10.9 mgd. Influent temperatures were moderate year-round, ranging between 23 to 28°C. The hydraulic loading rate to the trickling filters is relatively low at 125 gpd/ft^2 of filter cross-sectional area. The plant maintains a recirculation ratio of 1:1. Overall the Palm Springs secondary filter generates a consistent effluent quality, accomplishing high levels of ammonia removal. Although not required to nitrify, the loadings imposed on the system are consistent with those generally imposed for ammonia removal.

The Amherst, Ohio treatment plant utilizes two trickling filters placed in series without intermediate clarification. The trickling filters are each 40 feet wide, 90 feet long and 17 feet deep, with plastic cross-flow media. Currently, the plant is operating at an average flow equivalent to its design flow of 2.0 mgd. Influent temperatures for October through May ranged between 8 and 15°C, while the summer month temperatures ranged between 17 and 20°C. BOD and ammonia loadings to the trickling filters were relatively low. Recirculation is not practiced. The Amherst plant has consistently met ammonia removal requirements at loadings generally associated with nitrification design practices. Lower temperatures did cause lower removal rates.

The Chemung County wastewater treatment plant is in its first year of operation. The plant utilizes two trickling filters in series without intermediate clarification. The trickling filters are each 135 feet in diameter and 6 feet deep, with rock media. Influent temperatures ranged from 11 to 16°C. The average flow to the plant was 5.8 mgd. BOD and ammonia loadings to the plant's trickling filters are somewhat higher than the preceding plants. A recirculation ratio of approximately 3:1 is utilized. The plant is meeting its BOD_5 permit limit. No ammonia limit has been established.

The Wauconda wastewater treatment plant utilizes a single-stage trickling filter system with two filters in parallel. The trickling filters are each 50 feet in diameter and 28 feet deep, with plastic media. The media surface area is 30 ft^2/ft^3. The average flow to the plant for 1987 and 1988 was approximately 0.7 mgd, or 50 percent of its design capacity. Cold month influent temperatures ranged between 11 and 16°C, with a range of 17 to 21°C during the warmer months. BOD and loadings are consistent with those generally imposed for ammonia removal. Overall the plant is generating a high quality effluent and meeting its effluent requirements for both BOD_5 and ammonia-nitrogen.

The Ashland, Ohio facility is a single-stage trickling filter plant with two biotowers in parallel and a solids contact tank. The biotowers are 80 feet in diameter and 30 feet deep, with plastic cross-flow media. The media surface area is 30 ft^2/ft^3. Average design flow for the plant is 5.0 mgd. Mean monthly flow ranged from 54 to 66 percent of the 5.0 mgd design flow, with higher flows occurring in the colder months. Influent temperatures ranged from 17 to 22°C in the warm months and 13 to 16°C in the cold months. The mean flow, trickling filter BOD and ammonia influent levels varied considerably, with high levels in the warm months and low levels in the cold months. Because higher BOD and ammonia influent levels were offset by lower flow levels, loadings were very consistent for the one year period. Recirculation is practiced, but the rates are not measured. The plant has consistently met its effluent limits. The effluent ammonia-nitrogen levels are somewhat anomalous with high levels in the October through December period (average 8.6 mg/L), and 4.6 mg/L in January. Levels were consistently lower in the April through September period preceding this and February and March afterward.

The Bremen, Indiana wastewater treatment plant is a two-stage trickling filter process. The first-stage biotowers are each 35.3 feet in diameter, and 32 feet deep, with plastic media (34 ft^2/ft^3). The second-stage rock filter is 60 feet in diameter and 6 feet deep. Current flow is 85 percent of the 1.3 mgd design flow. Temperatures ranged from 10 to 14°C for the months November through May, and from 17 to 19°C for the remaining months. BOD loadings to the first-stage biofilters are relatively low, because of low effluent concentrations from the primary clarifiers. Loadings to the second-stage rock filter were not greatly different than the first stage loading due to differences in volume and media surface area. The plant was consistently in compliance with its discharge permit requirements for BOD_5 and ammonia.

The Allentown plant utilizes a two-stage trickling filter design. The first stage trickling filters are each 100 feet in diameter and 32 feet deep, with plastic media. The second stage is a large rock trickling filter, 8 feet deep and covering an area of approximately 8 acres. Normally 2 of the first-stage filters and 75 percent of the second stage filter are in service. The average flow was 82 percent of the 40.0 mgd design flow. The influent temperature ranged from 17 to 19°C for the warmer months and 11 to 16°C for the colder months. First-stage hydraulic, BOD_5 and ammonia-nitrogen loadings are very high, more typical of roughing filters. The second stage loadings at Allentown were more in line with those shown for the preceding plants and are consistent with design loadings for nitrifying plants. Recirculation is practiced only on the second stage, with a target ratio of 0.2:1. The plant consistently met its BOD_5 effluent permit requirements, but had trouble meeting the ammonia-nitrogen requirements during the summer months.

Cibolo Creek operates three parallel treatment plants. All utilize a two-stage trickling filter system. All of the trickling filters utilize plastic media. The first-stage filters for Plants A and B are 8 feet and 7 feet deep ,respectively, and 55 feet in diameter. The media surface area for both is 32 ft^2/ft^3. The second-stage filters for Plants A and B are each 7 feet deep, 55 feet in diameter, and use media with a surface area of 64 ft^2/ft^3. The first-stage of Plant C is 82 feet in diameter and 16 feet deep, while the second-stage has a diameter of 82 feet and a depth of 12 feet. The packing in both units is comprised of alternative layers of vertical (27 ft^2/ft^3) and cross-flow (48 ft^2/ft^3) plastic media. The design flow for the treatment facility as a whole is 6.2 mgd, which is split to the three plants (A - 29%, B - 16%, C - 55%). The current average flow for the facility is 35 percent of the design flow. Influent temperatures are moderate year-round, ranging between 20 and 27°C. Loadings to each plant differ to a degree. Loadings for BOD_5 and NH_3-N, for both stages, range from low in Plant C to highest in Plant A. Recirculation is practiced in both stages of all three plants. The recirculation ratio ranges from 2.4:1 to 6.0:1. The plants are all producing high quality effluent. The facility is consistently in the low flow range and meeting permit requirement for both BOD_5 and NH_3-N.

RESULTS AND DISCUSSION

The data that were received from various plants were reviewed as a whole, assessing the general operational characteristics for accomplishing nitrification. These analyses must necessarily be of a general nature, given the limits of the data and the narrow range of operating conditions experienced by the individual plants.

Figure 1 presents the ratio of the ammonia-nitrogen removal to the BOD_5 removal as a function of the BOD removal rate. As would be expected, the ratio decreases with increasing BOD removal rates, shifting to a process dominated by carbonaceous BOD removal with ammonia-nitrogen removal limited to that required for cell growth. If a nitrogen requirement for active ammonia-nitrogen removal systems is assumed to be 0.8 to 0.12 (lbs NH_3-N / lbs BOD_5), then the Allentown first stage, Bremen first stage and Ashland plants are considered carbonaceous removal processes with marginal ammonia removal activity outside that needed for cell growth. The remaining units show higher ratios (in particular the second-stage units for Bremen, Cibolo Creek and Amherst), indicating nitrification activity. The transitional BOD removal rate appears to be in the range of 0.2 to 0.4 lbs BOD_5/d-1,000 ft^2.

The effluent ammonia-nitrogen concentrations are compared to the equivalent period effluent BOD_5 levels accomplished by the system in Figure 2. This suggests that ammonia levels less than 2 to 4 mg/L NH_3-N will be reached when the effluent BOD_5 concentration is at levels less than 15 mg/L and preferably less than 10 mg/L.

Figure 3 presents effluent ammonia-nitrogen concentration as a function of the media surface area BOD loadings. The variability is somewhat high, but the data indicate a surface area loading less than 0.25 to 0.30 lbs BOD/1,000 ft^2-d is needed in order to yield effluent ammonia levels less than 2 to 4 mg/L. When BOD loading is expressed on a volumetric basis (Figure 4), the variability is reduced. This also shows that the BOD loadings need to be less than 10 lbs/1,000 ft^3-d for effective ammonia removal.

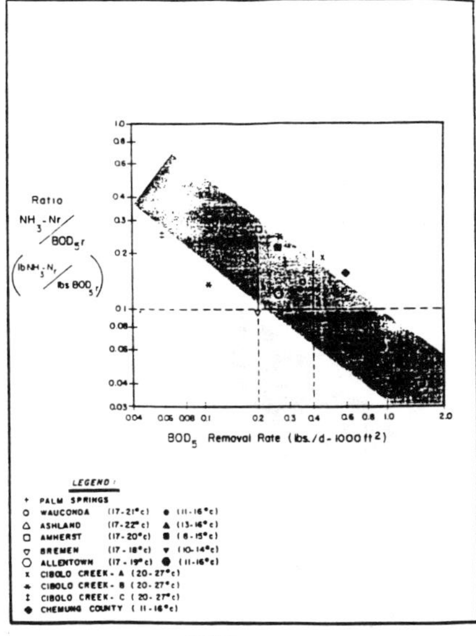

FIGURE 1

RATIO OF NH_3-N REMOVED TO BOD_5 REMOVED AS A FUNCTION OF THE BOD REMOVAL RATE

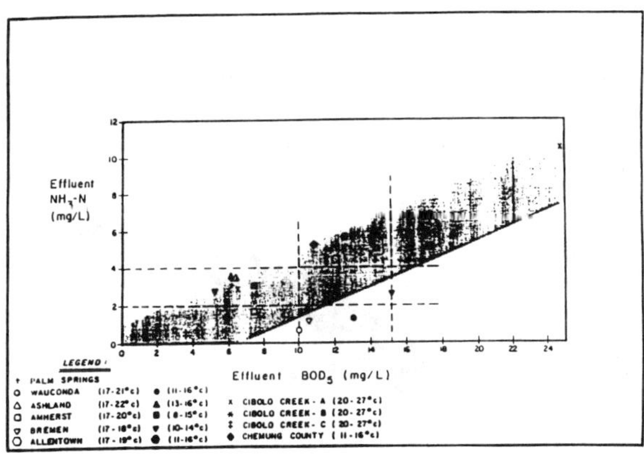

FIGURE 2

EFFLUENT NH_3-N LEVELS COMPARED TO EQUIVALENT BOD_5 EFFLUENT LEVELS

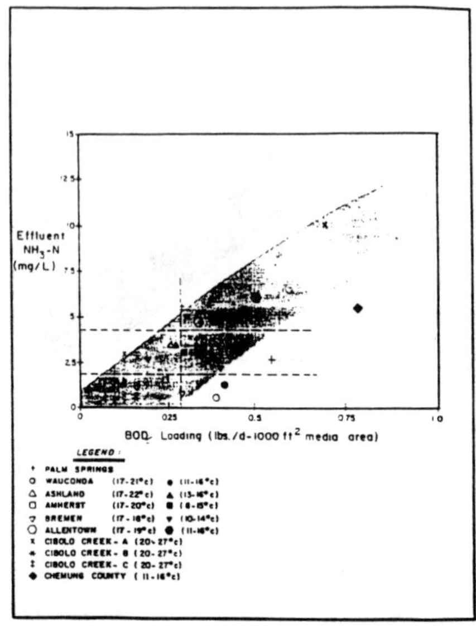

FIGURE . 3

EFFLUENT AMMONIA-NITROGEN CONCENTRATION AS A FUNCTION OF BOD MEDIA SURFACE LOADING

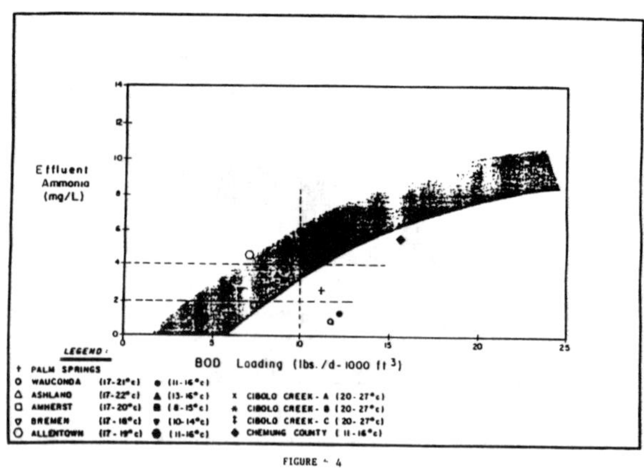

FIGURE - 4

EFFLUENT AMMONIA-NITROGEN CONCENTRATION AS A FUNCTION OF BOD VOLUMETRIC LOADING

These loadings conform to those suggested by the USEPA Process Manual for Nitrogen Control (1975), which recommends an organic loading of 10 to 12 lbs BOD_5/1,000 ft^3-d for nitrification in a single-stage trickling filter. On an areal basis, the USEPA suggested design loadings are 0.1 to 0.3 lbs BOD_5/1,000 ft^2-d, depending on temperature and effluent targets. These compare favorably to the loadings suggested in Figure 3.

SUMMARY AND CONCLUSIONS

The extent to which single-stage trickling filter nitrification is practiced is very limited. Ten single-stage plants were identified, with six of these utilizing the solids-contact process in conjunction with the trickling filter. Several other plants use separate two-stage processes for carbonaceous/nitrogenous BOD removal. They have either two-stage trickling filters with intermediate clarifiers or a trickling filter in series with an activated sludge process.

The evaluated plants were generally meeting their permit requirements, including ammonia-nitrogen limits when applied. Several plants exhibit some increase in effluent ammonia and BOD levels during cold weather months, although the differences are relatively small and there is no direct correlation apparent with temperature. Both plastic media and rock filters are represented by the plants. There are no apparent differences in performance related to the media; the reactor sizings are different because of the various specific area characteristics of the media.

Nitrification requires relatively low organic loadings. These can be expressed on a volumetric or media area loadings basis, and are generally set to yield BOD_5 levels less than 10 to 15 mg/L. Operating at these levels will assure an environment in which the autotrophic nitrifying bacteria can compete with the faster growing heterotrophic bacteria responsible for carbonaceous BOD removal. Loadings that are less than 10 lbs BOD/1,000 ft^3-d or 0.3 lbs BOD/1,000 ft^2-d will allow for nitrification and yield effluent ammonia-nitrogen levels less than 4 mg/L. These loadings, based on a review of combined data from 10 selected plants, compare favorably with the organic loading guidelines suggested by the USEPA for proper design and operation of single-stage trickling filters (1975 Process Design Manual for Nitrogen Control).

ACKNOWLEDGMENTS

The views/opinions expressed in this paper are those of the authors and should not be construed as opinions of the United States Environmental Protection Agency.

APPENDIX. REFERENCES

Process Design Manual for Nitrogen Control. (1975). U.S. Environmental Protection Agency.

MODELING FULL-SCALE DIFFUSED AERATION SYSTEMS

Terry L. Johnson[1], Member
Ross E. McKinney[2], Honorary Member

Abstract

The purpose of this research was to show the direct relationship of the oxygen mass transfer coefficient (K_La_T) to mixing, and to investigate the impacts of mixing on dissolved oxygen saturation concentration (Cs) in diffused aeration systems. Furthermore, the findings reveal a high correlation of K_La_T to mixing and to the type and orientation of diffused aeration systems, whereas Cs is dependent upon only the theoretical surface saturation and diffuser depth.

Introduction

The design of diffused aeration has largely been based on manufacturers' information, designers' experience, and, to a limited extent, models specific to only one aerator type and configuration. Transfer rates are often expressed as a simple function of air flow rate per unit volume or per diffuser. This design approach is not routinely accurate. At the same time, many researchers, including Hutchinson (1937), Eckenfelder (1952), Haney (1954), Elmore (1961), Kalinske (1965), Bewtra (1970), Smith (1970), Suschka (1971), Imhoff (1972), Camp (1974), Hunter (1978), and Metzger (1978) have indicated that diffused aeration performance is related to mixing due to bubble surface renewal and dispersion of the dissolved oxygen to the bulk liquid. However, no one has quantified or directly correlated oxygen transfer performance parameters for diffused aeration to mixing. This research was performed to relate diffused aeration performance to mixing, basin geometry, and diffuser layout. The results of this research were a series of models relating K_La_T to velocity gradient and critical geometry of the installation by diffuser type.

---

[1]Director Wastewater Treatment Technology, Black & Veatch, 8400 Ward Parkway, Kansas City, MO 64114.

[2]Process Consultant, 2716 Oxford Road, Lawrence, KS 66044.

Description of Data Base

The data base developed for this research includes 342 full-scale clean water tests on diffused aeration equipment. These tests were conducted on both porous and nonporous diffusers at various treatment plants around the country. The types of diffusers tested included Kenics and Polcon static mixer-aerators; Sanitaire, Parkson, PCI, and EDI wide-band coarse bubble diffusers; drilled pipe coarse bubble diffusers; Sanitaire box coarse bubble diffusers; M-S Spargers; Enviroquip coarse bubble diffusers; EDI plate diffusers; Norton ceramic domes; Sanitaire and Parkson ceramic disks; Eimco, Sanitaire, Roediger, and Wilfley Webber membrane disks; and Parkson and EDI membrane tube diffusers.

Of the 342 test results, 161 were available for coarse bubble units while the remaining 181 were for porous and membrane units, the latter being grouped together and referred to herein as fine bubble. These installations represented a broad range of critical physical and performance features as summarized in Table 1.

Table 1. Range of Physical and Performance Features

Parameter	Coarse Bubble		Fine Bubble	
	Minimum	Maximum	Minimum	Maximum
Test Volume, m^3	20.8	8,510	20.4	12,190
Sidewater Depth, m	3.0	7.6	2.3	7.6
Diffuser Depth, m	2.7	7.3	2.0	7.4
Number of Diffusers	2	1,152	6	2,736
Air Lateral Spacing, m	0.8	9.5	0.4	3.7
Total Air Flow Rate, L/s	9.7	11,760	1.5	5,340
K_La_T, hr^{-1}	1.12	37.5	0.80	50.6
Water Temperature, C	4.0	27.5	6.9	27.9
Measured Cs, mg/L	8.3	13.6	7.7	15.6

Correlation of Mass Transfer Coefficient to Mixing

Quantification of mixing by diffused aeration can be accomplished by determining the root-mean-square velocity gradient (G) for each testing condition. This relationship was reported by Das (1991) and is shown below as Equation 1.

Figure 1 shows the rough correlation of K_La_T to G for fine bubble and coarse bubble aeration devices. This observed correlation was verified by dimensional analysis based on the Buckingham Pi theorem, which confirmed the relationship of

K_La_T to G and other parameters. A similar investigation of oxygen saturation phenomena to G did not show a direct correlation and indicates that saturation for any diffused aeration system is independent of the mixing intensities normally encountered in wastewater treatment.

$$G = \sqrt{\frac{Q_a \cdot \gamma \cdot h}{V \cdot \mu}} \tag{1}$$

where,

Q_a = air flow rate, m^3/s
γ = liquid specific weight, $kg/m^2 \cdot s^2$
h = diffuser depth, m
V = aeration basin volume, m^3
μ = absolute viscosity, Pa·s

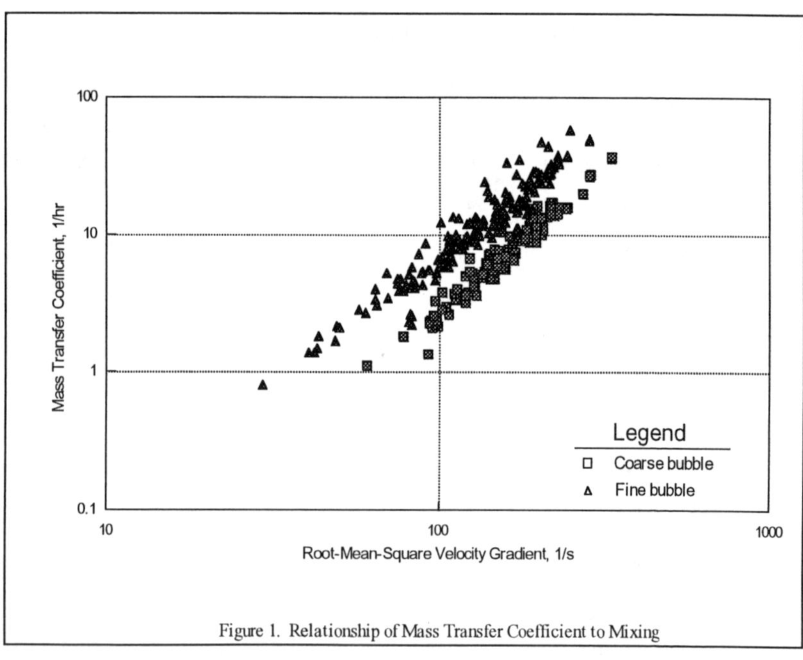

Figure 1. Relationship of Mass Transfer Coefficient to Mixing

Model Development

Based on the dimensional analysis, model forms were developed for predicting K_La_T for both coarse bubble diffused aeration systems and fine bubble aeration systems. The general form of the model as determined from the dimensional analyses is as follows:

$$K_L a_T = K \cdot G^a \cdot FC^b \cdot \frac{Q}{V}^c \cdot LS^d \tag{2}$$

where,

K = constant
G = root-mean-square velocity gradient, s^{-1}
FC = diffuser floor cover, %
Q/V = air flow rate per unit volume, $m^3 \cdot s/m^3$
LS = air lateral spacing, m
a-d = exponents derived from multiple regression

The procedure used to develop these models once the general form was determined from dimensional analysis was to perform a multiple regression on the logarithmic form of the equation, thereby directly determining the appropriate constants. The "K" and "a" through "d" constants determined by this method are shown in Table 2 for Equation 2. In each instance the correlation coefficient from the multiple regression is also listed.

Table 2. Models for Mass Transfer Coefficient ($K_L a_T$)

Type of Diffuser	Corr. Coeff.	K x 10^4	a	b	c	d
Coarse	0.974	2.04	2.13	0.08	0.06	0.07
Static Mixer	0.991	2.56	2.00	-	-	-
Fine	0.983	1,190	1.24	0.17	0.25	-0.05
Membrane Disk	0.985	2,080	1.11	0.13	0.24	-0.02

It was determined that the static aerator-mixer coarse bubble systems could be modeled reasonably accurately as a function of G only. In this case, the multiple regression correlation coefficient was 0.991 whereas the correlation coefficient for the more detailed model was only 0.993. Other types of coarse bubble systems are similar in that their performance can be reasonably modeled with the simplified equation similar to that for static aerator-mixers. It proved more reliable, however, to use the detailed model with the constants as stated in Table 2 to accurately predict all other types of coarse bubble system performance. Conversely, fine bubble systems can be accurately modeled only if terms for diffuser floor cover, air flow rate per unit aeration volume and air lateral spacing are included along with G. Attempts to simplify the models for fine bubble systems resulted in significant loss of accuracy indicating a higher dependence upon diffuser configuration and basin geometry.

Figures 2 and 3 demonstrate the comparison of model predictions against tests conducted at Los Angeles for the EPA. These tests were reported by Yunt (1980, 1988). Figure 2 is a plot of predicted $K_L a_T$ for coarse bubble aeration against actual

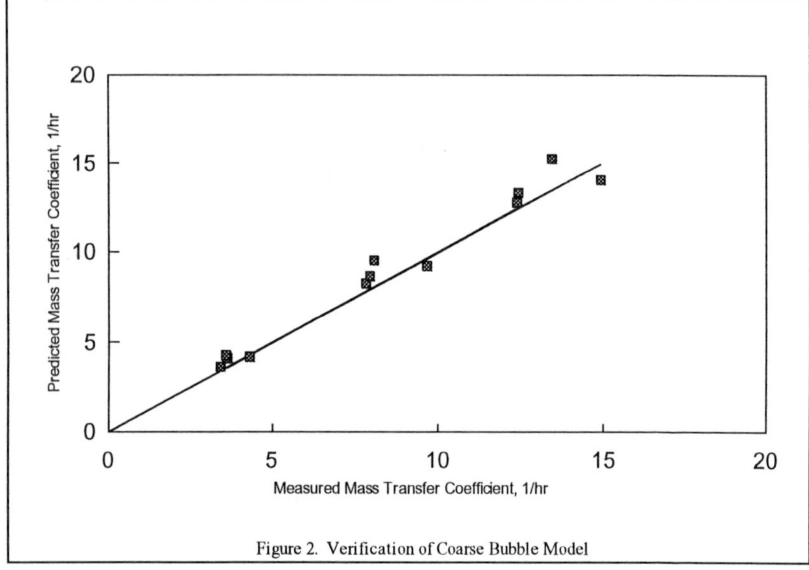

Figure 2. Verification of Coarse Bubble Model

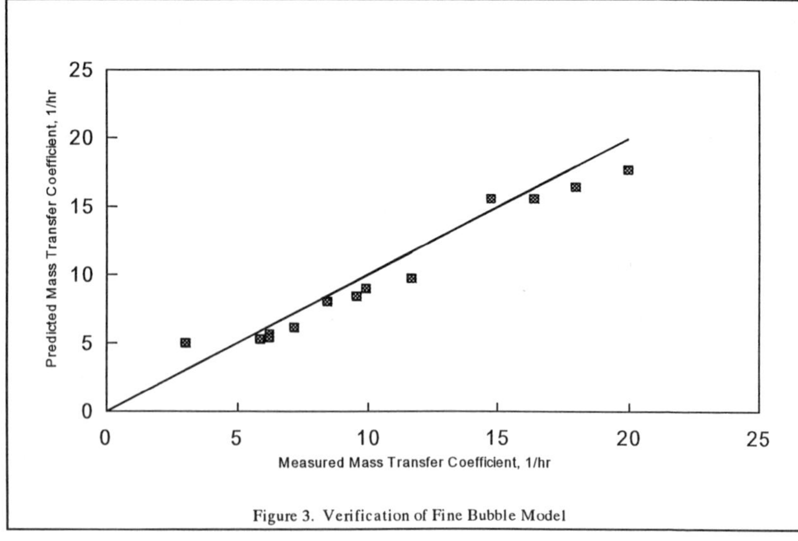

Figure 3. Verification of Fine Bubble Model

measured results for Sanitaire wide-band diffusers. Average error for the 12 tests was 9.2 percent. Figure 3 is a plot of predicted K_La_T for fine bubble aeration against actual measured results for Norton ceramic domes. In this case, the average error for 13 tests was 14.3 percent. When the results for both of these comparisons were further developed into oxygen transfer efficiency, the average errors were 9.4 and 11.0 percent for the coarse and fine bubble systems, respectively.

Concerning model development for Cs, rigorous data analysis showed that Cs was related to theoretical surface saturation and diffuser depth. No other parameters had significant impact on Cs. The models derived from multiple regression for all coarse bubble diffusers and for all fine bubble diffusers are shown below as Equations 3 and 4. The correlation coefficients from the regressions were 0.936 and 0.976, respectively.

$$Cs = 1.22 \cdot ThCs^{0.913} \cdot DWD^{0.074} \quad (3)$$

$$Cs = 0.715 \cdot ThCs^{1.146} \cdot DWD^{0.105} \quad (4)$$

where,

Cs = oxygen saturation concentration, mg/L
ThCs = theoretical surface oxygen saturation concentration at water temperature and site barometric pressure, mg/L
DWD = diffuser depth below water surface, m

Discussion of Results

The literature is replete with articles regarding diffuser geometry, air flow rates, depth, and temperature effects on diffused aeration performance. The models developed in this research and their proven accuracy promote convenient investigation of these changes. Although many comparisons are made possible by these models, Table 3 contains a summary of the impacts on K_La_T resulting from changes in key operating and design parameters.

Doubling either air flow rate alone or air flow rate per unit volume increases K_La_T more rapidly than the rate of increase for fine bubble systems. The effects of doubling diffuser water depth are also more dramatic on coarse bubble K_La_T than on fine bubble. The effect of temperature changes is directly related to the water viscosity at those respective temperatures, and the values in Table 3, when compared to commonly accepted temperature correction techniques, show that undercorrection is normal for coarse bubble and overcorrection is normal for fine bubble when correction is made to 20 C from temperatures less than 20 C. The converse is true if corrections are being made for temperature reduction. Finally, neither floor covering nor air lateral spacing have significant impacts on K_La_T, but floor coverage is more significant in the case of fine bubble units than it is for coarse.

Table 3. Impacts of Parameter Variation on $K_L a_T$

	Response of $K_L a_T$	
Parameter Doubled	Coarse Bubble	Fine Bubble
Air Flow, m^3/s	2.18	1.83
Air Flow per Unit Volume, $m^3/s{\cdot}m^3$	2.18	1.83
Diffuser Depth, m	2.09	1.54
Temperature (10→20), C	1.33	1.18
Diffuser Floor Cover, %	1.06	1.13
Air Lateral Spacing, m	1.05	0.96

Conclusions

Comparison of the results from these models and the results from diffused aeration manufacturers' curves reveals that these models predict $K_L a_T$ with reliable accuracy. The importance of mixing and its quantified impact on $K_L a_T$ has finally been achieved. Consequently, by using these models diffused aeration systems can be designed more accurately than in the past. Further conclusions are supported by these findings as follows:

- $K_L a_T$ is higher for fine than that for coarse bubble systems at the same mixing energy.

- Fine bubble diffused aeration performance can be accurately modeled by G in combination with diffuser floor cover, air lateral spacing and air flow rate per unit volume. The first of these three address the surface renewal rate in these systems while the last parameter provides an estimate of the quantity of air-liquid interface area per unit volume of basin.

- Temperature correction of $K_L a_T$ can most accurately be accomplished by using the relative liquid viscosities at the respective temperatures.

- Increasing air flow rate per diffuser has a detrimental affect on fine bubble diffused aeration system performance.

- Basin depth considerations have more importance to the performance characteristics of coarse bubble diffused aeration systems than it does to fine bubble systems.

References

Bewtra, J. K., W. R. Nicholas, and L. B. Polkowski, "Effect of Temperature on Oxygen Transfer in Water", Water Research, **4** (1970).

Camp, Thomas R. and R. L. Meserve, Water and Its Impurities, Dowden, Hutchinson and Ross, Inc., Stroudsburg, Pennsylvania (1974).

Das, Debankur, Thomas M. Keinath, Denny S. Parker, and Eric Wahlberg, "Floc Breakup in Activated Sludge Plants", Presented at the 1991 WPCF Conference, Toronto, Ontario, October (1991).

Eckenfelder, W. Wesley, Jr., "Aeration Efficiency and Design - I. Measurement of Oxygen Transfer Efficiency", Sewage and Industrial Wastes, **24**, 10 (1952).

Elmore, Harold L. and William F. West, "Effect of Water Temperature on Stream Reaeration" reported by Committee on Sanitary Engineering Research, Sanitary Engineering Division, ASCE (1961).

Haney, Paul D., "Theoretical Principles of Aeration", Journal American Water Works Association, **46**, 4 (1954).

Hunter, John S., III, "Accounting for the Effects of Water Temperature in Aerator Test Procedures", Proceedings - Workshop Toward an Oxygen Transfer Standard, Asilomar Conference Grounds, Pacific Grove, California (1978).

Hutchinson, M. H. and T. K. Sherwood, "Liquid Film in Gas Absorption", Industrial Engineering Chemistry, **29**, 836 (1937).

Imhoff, K. R. and D. Albrecht, "Influence of Temperature and Turbulence on Oxygen Transfer in Water", Proceedings of 6th Conference of Water Pollution Research Association, Jerusalem (1972).

Kalinske, A. A., "Evaluation of Oxygenation Capacity of Localized Aerators", Journal WPCF, **37**, 11 (1965).

Metzger, Ivan, "Effects of Temperature on Stream Aeration", Journal of the Sanitary Engineering Division, Proceedings of the ASCE, **94**, SA6 (1978).

Smith, Daniel W., "Modelling Oxygen Transfer in Diffused Aeration Tanks", PhD Dissertation, The University of Kansas, Lawrence, Kansas (1970).

Suschka, J. "Oxygenation in Aeration Tanks", Journal WPCF, **43**, 1 (1971).

Yunt, Fred W., Tim O. Hancuff, Dick Brenner and Gerry Shell, "An Evaluation of Submerged Aeration Equipment - Clean Water Test Results", Presented at WWEMA Industrial Pollution Conference, Houston, Texas (1980).

Yunt, Fred W. and Tim O. Hancuff, "Aeration Equipment Evaluation: Phase I - Clean Water Test Results", Prepared for the U.S. Environmental Protection Agency, EPA/600/2-88/022, March (1988).

Biodegradation of 2,4-Dinitrotoluene in a Two Stage System

S.L. VanderLoop[1], M.T.Suidan[1], M.A. Moteleb[1], S.W. Maloney[2]

ABSTRACT

An anaerobic/anoxic fluidized-bed GAC bioreactor in series with an activated sludge reactor was used to treat 2,4-Dinitrotoluene (2,4-DNT). A simulated high strength wastewater solution of 2,4-DNT, ethanol, and ethyl ether as well as carbonate buffer and nutrient solutions were fed to the anaerobic/anoxic reactor. The environment in the fluidized-bed reactor was varied to determine its effect on 2,4-DNT biodegradation. The effluent from this reactor was treated further in an activated sludge system. Methanogenic operation of the fluidized-bed resulted in stoichiometric transformation of 2,4-DNT to 2,4-diaminotoluene (2,4-DAT). The 2,4-DAT was completely mineralized by the activated sludge. The system failed to transform the 2,4-DNT under anaerobic conditions without addition of a primary substrate. The effects of operating the first stage under nitrate reducing conditions with a primary substrate is currently being investigated.

INTRODUCTION

Dinitrotoluene (2,4-DNT) is used in the manufacture of polyurethane foam and is a byproduct of propellant production. As a known carcinogen, NPDES permit levels have been proposed at 113 μg/l for the Radford Army Ammunition Plant (RAAP) in Radford, VA, U.S.A. The wastewater evaluated in this study was modeled after RAAP effluent from the water-dry process.

The most effective treatment technology currently available is adsorption of the propel-

---

[1]Department of Civil and Environmental Engineering, University of Cincinnati, Cincinnati, Ohio 45221-0071, U.S.A.
[2]U.S. Army Construction Engineering Research Laboratories, Champaign, Illinois 61826-9005

lants onto activated carbon followed by incineration. Bioremediation is potentially a more effective and less costly alternative to incineration. The biodegradation of 2,4-DNT can proceed through several different transformation pathways depending on the microbial environment. Liu *et al.* (1984) reported that anaerobic reduction of TNT and DNT proceeds through nitroso and proposed hydroxylamino intermediates to amino compounds, consuming one mole of hydrogen in each step. The instability of the hydroxylamino intermediate precludes direct measurement, but it often exists long enough in aerobic environments to undergo oxidative coupling. McCormick *et al.* (1977) found that nitrotoluene dimers can result from the aerobic reduction pathway indicating that oxidative coupling does occur. A number of pure bacterial cultures have transformed 2,4,6-TNT in sulfate reducing or denitrifying environments (Boopathy *et al.*, 1992; Preuss *et al.*, 1993). It is likely that 2,4-DNT is also capable of undergoing such transformations under similar conditions.

Previous work has shown that 2,4-DNT can be stoichiometrically reduced to 2,4-DAT under methanogenic conditions if supplemented with a primary substrate. This reduction may increase the toxicity of the wastewater, but it also opens up new pathways for mineralization. Although no further transformations are possible anaerobically, aminotoluenes are much more susceptible to enzymatic oxidation, and therefore ring cleavage, than their nitrotoluene precursors.

This research evaluated the effectiveness of an anaerobic/anoxic fluidized-bed GAC (AFBGAC) bioreactor in series with an activated sludge unit in treating a simulated high strength 2,4-DNT wastewater. The AFBGAC bioreactor has been proven effective in the treatment of toxic and inhibitory wastes such as coal gasification wastewaters and wastewaters containing chlorinated hydrocarbons and phenols (Suidan *et al.*, 1983, 1991; Flora *et al.* 1993). The GAC serves as a superior microbial attachment surface as well as a buffer against shock loads or buildup of toxic intermediates. The effluent from the first stage is fed to the activated sludge for mineralization of anaerobic/anoxic intermediates and for effluent polishing with respect to COD.

MATERIALS AND METHODS

Anaerobic/Anoxic Bioreactor: A schematic of the system is shown in Figure 1. The AFBGAC reactor was constructed of Plexiglas and consisted of a main body which was jacketed to maintain a constant temperature of 35°C; an influent header packed with marbles to more evenly distribute the flow over the 81.7 cm^2 cross sectional area; and an effluent header which routed the liquid effluent to the activated sludge unit and the gas to the collection system. The effluent header also had a sampling port which allowed withdrawal of GAC samples from any depth in the reactor using a small vial. The column was charged with 1.0 kg of 16 x 20 U.S. mesh F400 GAC (Calgon Corporation, Pittsburgh, PA).

The GAC was introduced through a side arm which extended into the reactor below the recycle withdrawal port to prevent uptake into the recycle stream and subsequent

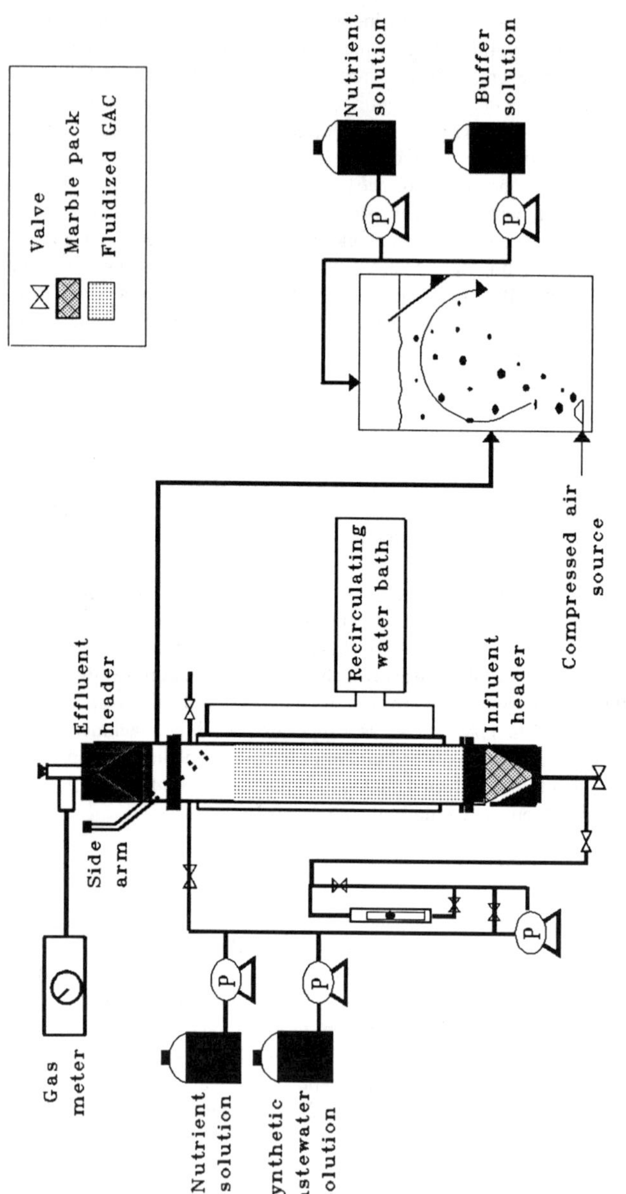

Figure 1. Two-Stage System Schematic

grinding through the pump. The recycle lines, made of polyvinyl chloride piping, increased the total reactor volume to ten liters. The recycle flowrate was controlled by two valves to maintain a bed fluidization of 30%. Synthetic wastewater and nutrient solutions were supplied to the recycle lines via neoprene and tygon tubing. Feed rates were controlled with fixed rpm peristaltic pumps wired to on/off timers to adjust the flows.

Activated Sludge Reactor: The activated sludge unit consisted of a 17.5 liter Plexiglas tank (7.75 x 8.0 x 17.5 inches) with an upflow clarifier and a 7 inch long, 0.25 inch diameter stainless steel diffuser. Clarification was achieved by means of a Plexiglas plate oriented approximately 25° from vertical which was raised or lowered to adjust the width of the opening between the plate and the reactor wall. The reactor was seeded with mixed liquor from a municipal wastewater treatment plant in Cincinnati Ohio.

Growth Media: The synthetic wastewater feed to the first stage was based on the composition of DNT production wastewater following the water dry process at RAAP. Combined feed concentrations of 110 mg/l 2,4-DNT and 100 mg/l ethyl ether were employed. Ethanol was provided as the primary substrate at 600 mg/l. A solution of essential vitamins and nutrients was supplied to each reactor. Sodium carbonate and sodium hydroxide buffers were used to maintain the pH at 7.3 and 8.0 in the anaerobic and aerobic reactors respectively.

System Operation: Synthetic wastewater and nutrient solutions were supplied to the AFBGAC reactor at a rate of 5.1 l/d and 1.05 l/d respectively, resulting in a hydraulic retention time of 1.6 days. The activated sludge received 1.05 liters of nutrients and 0.27 liters of buffer per day for a hydraulic retention time of 2.3 days. The sludge age was maintained at 8 days.

Analytical Methods: The anaerobic reactor was monitored daily for total gas production, feed flow rates, temperature, and effluent pH. Feed flow rates and effluent pH were measured daily in the continuous sludge system. Effluent liquid and gas samples were withdrawn weekly from the anaerobic bioreactors to analyze for total and soluble COD, alcohols, volatile fatty acids, and off gas composition. Mixed liquor volatile suspended solids in the aerobic reactor and in the clarifier as well as final effluent COD were also determined weekly. Effluent liquid samples were withdrawn twice weekly from each reactor to analyze for 2,4-DNT, 2-A-4NT, 4-A-2NT, 2,4-DAT, ammonia, and nitrate. A nitrogen balance around each reactor confirmed complete biotransformation or mineralization in each stage. The analytical methods for the various analyses follow:

i. Gas composition: Effluent gas samples from the anaerobic reactors were analyzed for nitrogen, oxygen, carbon dioxide and methane with a model 900 Perkin Elmer Gas Partitioner.

ii. Chemical oxygen demand: Filtered samples were acidified with 85% o-phosphoric acid to a pH of 2 and purged with prepurified nitrogen for 10 minutes to strip out the sulfide. Prepared Hach COD glass vials (range 0 - 150 mg/L) and Hach COD reactor Model 45600 (Hach Co., Loveland, CO) were used for the analysis. The percent transmittance of the digested samples was read using a Bausch and Lomb Spectronic 70 spectrophotometer.

iii. Volatile fatty acids and alcohols: Volatile fatty acids (acetic and propionic acids) and alcohols (methanol and ethanol) were analyzed with a Hewlett Packard 5890 gas chromatograph equipped with flame ionization detectors. VFA's were injected onto a 2 mm ID, 1.83 m glass column packed with 4% Carbowax on a 80/120 Carbopack B-DA. Alcohols were analyzed using a 2 mm ID, 1.83 m glass column packed with 5% Carbowax 20M on a 60/80 Carbopack B (Supelco, Inc., Bellefonte, PA).

iv. 2,4-Dinitrotoluene and its biodegradation products: The pH of the aqueous sample was first raised to 12 with 10 M NaOH and extracted using a 5:1 ratio of sample to ether. The ether extract was then injected onto a DB-1 capillary column (J & W Scientific, Folsom, CA). Detection limits for 2,4-DAT, 2-A-4-NT, 4-A-2-NT, and 2,4-DNT were 0.3, 0.1, 0.1, and 0.03 mg/L, respectively.

Final effluent levels of 2,4-DAT were below the detection limit of 0.3 mg/l. In order to reach a lower detection limit, the ether extract was condensed from 3 to 0.1 ml before it was injected. The detection limit for 2,4-DAT by this method is 50 μg/l.

RESULTS AND DISCUSSION

The AFBGAC bioreactor was operated in 3 stages, each of which examined the effect of a particular microbial environment on 2,4-DNT transformation. During stage I, which continued for a period of 8 weeks, the AFBGAC reactor was operated anaerobically with no primary substrate. During this time, DNT in the effluent rose to detectable levels and DNT adsorbed on the GAC increased to 15 mg/g. This was part of a previous study which determined that a primary substrate was required to effect the reduction of 2,4-DNT to 2,4-DAT anaerobically (Berchtold, 1994).

It was apparent that the system was failing so a moderate ethanol feed concentration of 600 mg/l was implemented in stage II of anaerobic operation. During this period stoichiometric reduction of 2,4-DNT to 2,4-DAT resumed and the resulting 2,4-DAT was completely mineralized in the activated sludge system. The steady-state GAC capacity was 38 mg/g of 2,4-DAT.

The influent COD was transformed mainly to gaseous end products (Figure 2). Gas composition was typically 87-93% methane with the balance as carbon dioxide. The difference between the influent COD and the sum of the soluble effluent and gas COD's can be attributed to biological growth. The anaerobic reactor was able to achieve an almost 90% reduction in soluble COD. 2,4-DAT accounts for approximately 75% of the 200 mg/l COD in the anaerobic reactor effluent. The balance is due to effluent ethyl ether, volatile fatty acids, and alcohols. In the activated sludge the DAT is mineralized and the final effluent COD is reduced to 15-32 mg/l.

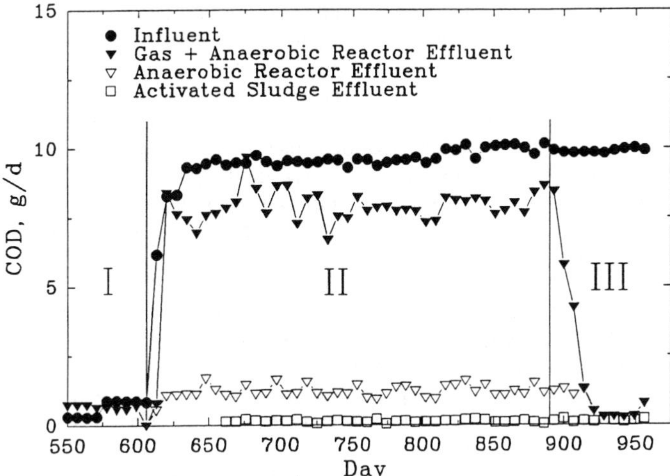

Figure 2. Two Stage System COD Balance.

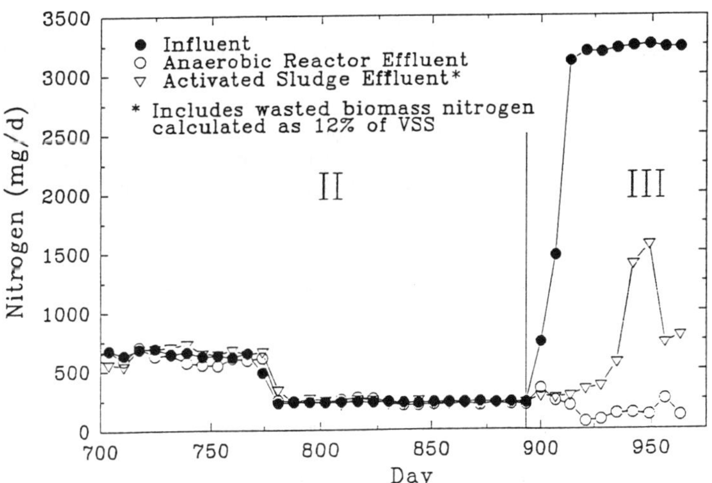

Figure 3. Two Stage System Nitrogen Balance

A nitrogen balance around the system confirms that the 2,4-DNT is completely transformed anaerobically and the resulting 2,4-DAT is mineralized in the activated sludge system. Figure 3 shows the total nitrogen content of the feed, the AFBGAC reactor effluent, and the final effluent. The primary influent nitrogen sources are ammonia and 2,4-DNT. The ammonia in the anaerobic reactor nutrient feed was decreased on day 773 to reduce the background nitrogen and allow small variations in performance to be discerned. Ammonia and DAT are the main sources of nitrogen in the AFBGAC reactor effluent. These are transformed to nitrate and biomass in the activated sludge system. A 12% biomass nitrogen content was used to calculate the nitrogen equivalent of the wasted sludge.

The possibility of anoxic 2,4-DNT degradation prompted the investigation of nitrate reducing conditions. Excess nitrate, based on influent COD, was supplied to the AFBGAC reactor to promote denitrifying conditions in stage III. Methane production ceased and off gas composition stabilized at approximately 90% N_2 with the balance as CO_2. The anoxic effluent 2,4-DAT concentration dropped below detectable limits while effluent 2,4-DNT levels rose to 2.5 mg/l but did not increase further. It is possible that a greater excess of nitrate is required to facilitate transformation of the remaining 2,4-DNT. Effluent nitrate levels were lower than those calculated based on influent COD indicating that nitrate was being reduced by organic matter stored on the GAC. Total effluent nitrogen from the activated sludge follows a typical regeneration curve with a lag time of five weeks (Figure 3). This surge of nitrogen was not detected in the AFBGAC reactor effluent. The form of organic nitrogen present in the anoxic reactor has not yet been identified.

Low anoxic reactor effluent COD values may seem inconsistent with the proposed high organic nitrogen content, but there is much ambiguity over which form of nitrogen results from COD tests. Possibilities include NO_3, NH_3, N_2, and N_2O. It is likely that it depends on the initial form of nitrogen. Preliminary results from COD tests on DNT suggest that N_2 or N_2O is the preferred form. This means that highly oxidized forms of organic nitrogen will result in depressed COD measurements, giving artificially low indications of organic content.

At this point, the extent of anoxic 2,4-DNT transformation is uncertain. Attempts to identify anoxic effluent constituents continue. In the interim additional nitrate is being supplied to force more complete oxidation of the 2,4-DNT in the anoxic reactor.

CONCLUSIONS

The microbial environment has a strong impact on the extent and pathway of 2,4-DNT biotransformations. This two stage treatment scheme was highly effective in treating 2,4-DNT and its derivatives under methanogenic/aerobic conditions. It was also effective in reducing the total COD of the simulated wastewater. However, the system failed to effect any transformations of 2,4-DNT anaerobically when a primary substrate was absent.

Denitrifying conditions promote 2,4-DAT transformations to highly oxidized organic nitrogen compounds that are readily mineralized in the activated sludge system. Further evidence for similar 2,4-DNT transformations is required.

ACKNOWLEDGEMENTS

This research was supported by the U.S. Army Construction Engineering Research Laboratory, Champaign, IL 61821.

REFERENCES

Berchtold, S.R., Suidan, M.T., Maloney, S.W., and VanderLoop, S.L. (1994). "Treatment of 2,4-dinitrotoluene using a two stage system: fluidized-bed anaerobic GAC and aerobic activated sludge reactors." Submitted for publication in J. WEF.

Boopathy, R. and Kulpa, C.F. (1992). "Trinitrotoluene (TNT) as a sole nitrogen source for a sulfate-reducing bacterium *Desulfovibrio* sp. (B strain) isolated from an anaerobic digester." *Current Microbiol.*, 25, 235-241.

Flora, J.R.V., Suidan, M.T., Wuellner, A.M., and Boyer, T.K. (1993). "Anaerobic treatment of a simulated high-strength industrial wastewater containing chlorophenols." Accepted for publication in *Water Env. Res.*

Liu, D., Thomson, K. and Anderson, A.C. (1984). "Identification of nitroso compounds from biotransformation of 2,4-Dinitrotoluene." *Appl. Environ. Microbiol.*, 47, 1295-1298.

McCormick, N.G., Feeherry, F.E., and Levinson, H.S. (1976). "Microbial transformation of 2,4,6-trinitrotoluene and other nitroaromatic compounds." *Appl. Environ. Microbiol.*, 31, 949-958.

McCormick, N.G., Cornell, J.H., and Kaplan, A.M. (1977). "Identification of biotransformation products from 2,4-dinitrotoluene." Appl. Environ. Mic.., 35, 945-948.

Preuss, A., Fimpel, J., and Diekert, G. (1993). "Anaerobic transformation of 2,4,6-trinitrotoluene (TNT)." *Arch. Microbiol.*, 159, 345-353.

Suidan, M.T., Siekerka, G.L., Kao, S.W., and Pfeffer, J.T. (1983). "Anaerobic filters for the treatment of coal gasification wastewater." *Biotechnol. Bioeng.*, 55, 1581.

Suidan, M.T., Wuellner, A.M., and Boyer, T.K. (1991). "Anaerobic treatment of a high strength industrial waste bearing inhibitory concentrations of 1,1,1-trichloroethane." *Water Sci. Technol.*, 23, 1385.

Measurement and Automatic Control of Chlorination

Robert Hill[1], Member, ASCE, and James Martin[2]

Abstract

Chlorination is perhaps the most commonly utilized process for wastewater effluent disinfection. In Houston's plants sodium hypochlorite solution (bleach) is mixed with secondary effluent for disinfection. The chlorinated effluent is then held in tanks for a minimum of 20 minutes (at maximum flow) to assure nearly complete disinfection. Most of Houston's plants also require dechlorination before discharging their effluent.

At Houston's 69th Street wastewater treatment plant, a chlorination control system using a polarigraphic type total chlorine analyzer was implemented. Several deficiencies of the instrument were corrected before good control was achieved. Dual analyzers were used to improve reliability and alarm instrument failures. The chlorine measuring point was located approximately 1 minute from the point of chlorine addition. A relatively simple cascaded control strategy using one residual chlorine loop and one flow control loop was implemented. A process and instrument diagram for the process is presented.

Savings from using this control strategy at the 69th Street WWTP are estimated to be greater than $125,000 per year. Savings in the first three months of operation have exceeded the estimate.

----

[1]Chief Engineer, Public Works & Engineering, City of Houston, 2525 S/Sgt Macario Garcia, Houston, Texas 77020.

[2]Section Chief, Public Works & Engineering, City of Houston, 2525 S/Sgt Macario Garcia, Houston, Texas 77020.

Past Performance of Chlorination Control Systems

Measurement of residual chlorine concentration and control of chlorination have traditionally been very difficult problems at wastewater treatment facilities. Virtually all of the automatic chlorination systems installed in Houston plants in the 1970s and 1980s failed to perform. The leading cause of failure was inaccurate and unreliable residual chlorine measurement by on-line instruments. Other significant problems included inadequate mixing at the point of chlorine addition, ammonia interactions (breakpoint chlorination), a variable chlorine demand, and control problems due to a long and variable dead time. Solutions to each of these problems will be discussed in later sections.

Selection of On-Line Residual Chlorine Analyzers

On-line analyzers for measuring residual chlorine are often perceived as the weak link in chlorination control. The perception is that chlorine analyzers are inaccurate, unreliability, and require "excessive" maintenance (even if these terms aren't defined with specific numeric values).

In order to evaluate the accuracy, reliability, and maintenance requirements of commercially available residual chlorine analyzers, the City of Houston utilized the services of the Instrumentation Testing Association of North America (ITA). ITA is a not-for-profit corporation which pools its members money to perform detailed instrument testing (Hill and Schuk 1992). On-line residual chlorine analyzers were thoroughly tested by ITA in 1985 and again in 1990. They tested both free and total chlorine analyzers from three leading manufacturers in 1985 and six in 1990. The testing was conducted by an independent laboratory according to a rigorous, peer-reviewed test protocol. The protocol included a bench test under carefully controlled laboratory conditions, 45-day field tests at water and wastewater plants, and a general analysis of design.

Based upon the 1985 ITA test results, the City of Houston selected the Hach model CL17 total residual chlorine analyzer for implementation. Results using the Hach analyzers have been very successful and were presented by Garrett et al. (1993). After using the Hach analyzers for some years, however, the City decided to experiment with the EIT model 5151T total chlorine analyzer at its larger facilities where they could be closely observed. Sample preparation, environmental controls, and other conditions necessary for reliable and accurate measurements will be presented in the next section.

Use of the EIT Polarigraphic Chlorine Analyzer

The EIT total chlorine analyzer utilizes a polarigraphic probe which measures the diffusion of chlorine across a thin, gas-permeable membrane. The analyzer uses no chemicals (with the exception of the electrolyte in the probe itself). The analyzer was selected for field evaluation based upon ITA's reported acceptable accuracy and the potential for less maintenance than the Hach analyzers. The three areas in which the EIT analyzer appeared to require less maintenance were:

(1) The probe appeared to be suitable for direct immersion without the need for a pumped sample. Sample pumping systems often require as much maintenance as the analyzer. This potential benefit was not realized.

(2) The EIT analyzer required essentially no chemicals.

(3) The EIT analyzer did not use small diameter (1/16" ID) tubing which requires filtering of particulates to prevent plugging. Sample filters typically require daily cleaning.

The EIT chlorine probes were initially installed at the end of a 3 meter conduit for direct immersion in the effluent channel. It quickly became apparent that this configuration did not provide accurate measurements. The apparent residual value could be easily changed by changing the depth of submergence. Also, the water velocity in the effluent channel affected the instrument reading. Both the water depth and velocity in the effluent channel changed constantly with changes in the flow rate. A sample pump and flow through cells were installed to overcome this problem.

Several of the analyzer electronic units were installed near the effluent overflow weir for almost a year before the contractor commissioned them. During this time, two of the analyzer's circuit boards exhibited mild corrosion while one analyzer was completely ruined. Although the instrument enclosures were rated NEMA-4, they failed to seal correctly due to apparent shrinkage of the plastic enclosure. One unit sustained substantial water damage beyond repair. This problem was resolved by first cleaning the remaining circuit boards and coating them with a corrosion resistant sealer. The entire electronics unit was then installed inside another larger NEMA-4 enclosure with an instrument air purge. This arrangement has protected the analyzers from further corrosion and water damage. The probes and flow through cell were installed in a separate adjacent enclosure to isolate the electrical and wet components.

After these modifications were made, the analyzers were placed in monitoring service for several weeks. It was observed that they tended to read high during the night and low during the day. This effect was eventually attributed to ambient temperature sensitivity. These problems were overcome by adding a vortex type cooler and a small strip heater to control the temperature inside the electronics enclosure to within approximately +/- 10 degrees Fahrenheit. Additionally, the entire enclosure was shielded to avoid direct exposure to sunlight.

Maintenance on the EIT analyzers has been very minimal. It has averaged 2 to 4 hours per month per analyzer for calibration and corrective maintenance.

Control Problems Due To Dead Time

A major control problem associated with chlorination control is the long and variable dead time experienced in the chlorine detention tanks. These tanks are typically sized for maximum flow rates, and the actual detention times may vary from 20 minutes to several hours depending on the influent flow rate. Control based solely upon an effluent feedback signal will typically be far out of phase and will generally lead to poor control results.

Fortunately chlorine demand seems to have two major components; an immediate demand which reacts very quickly, and a long term demand which reacts at a fairly slow and more constant rate. If the immediate chlorine demand can be satisfied, the effluent chlorine residual will be only slightly affected (+/-0.2 mg/L) by changes in detention time. Since the immediate chlorine demand is satisfied shortly after the point of chlorine injection, the residual measurement used for control should be taken within a few minutes detention time of this point. The first edition of MOP 21 (1978) recommended a maximum delay of 1 to 3 minutes. If longer dead times are necessary due to physical restrictions, they can be corrected by using a Smith-predictor (1957) control algorithms as described by Garrett et al. (1993).

Initial Mixing Requirement

Good dispersion and overall mixing at the point of injection is critically important to ensure that all of the immediate chlorine demand is satisfied. Good dispersion also prevents breakpoint chlorination and other undesirable reactions from occurring due to localized high chlorine concentrations. Two induction-type mixers were installed to provide the initial dispersion. These units were later modified to include a water purge to prevent forming a vacuum on the sodium hypochlorite feed line. The vacuum interfered with flow control.

Although the induction mixers provided excellent dispersion, their small power input (2-1/2 HP) did not provide adequate mixing for the entire flow through the entire chlorine channel (4 m x 2.5 m). An existing 15 HP mixer located downstream of the induction mixer was retained for more complete mixing throughout the channel.

Chlorination Control Strategy

A process and instrument diagram (P&ID) for a chlorination controller is shown as Figure 1. The sampling point for the chlorine analyzers is within 1 to 2 minutes detention time of the point of chlorine injection. This particular controller is based upon using two residual chlorine analyzers to increase the reliability of the chlorine system. Such design decisions are often justified in chlorination and dechlorination systems since even a momentary failure can result in a permit violation. System reliability is further increased by alarming if the two analyzers disagree more than a predetermined value (0.25 mg/L).

Under normal conditions, the average of the two chlorine analyzers is used as the process variable for a feedback control loop. Alternately, the plant operators can use a software switch to choose either of the single analyzers as the process variable if they feel one of the analyzers is questionable. This feature is often used when an analyzer is being maintained.

The actual control strategy is very simple. A magnetic flow meter and a control valve maintain the

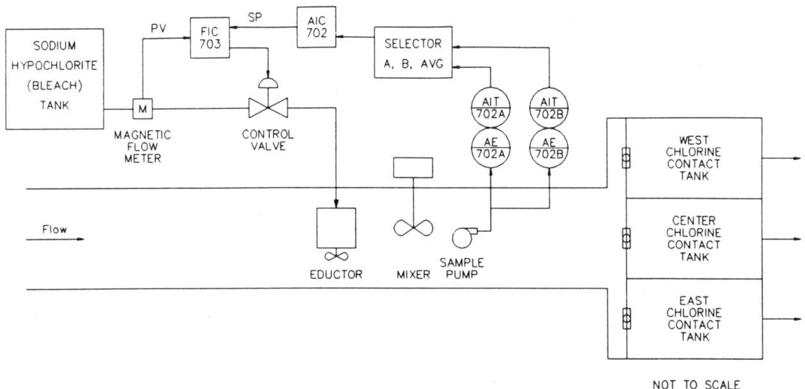

FIGURE 1 PROCESS AND INSTRUMENT DIAGRAM FOR CHLORINATION CONTROL

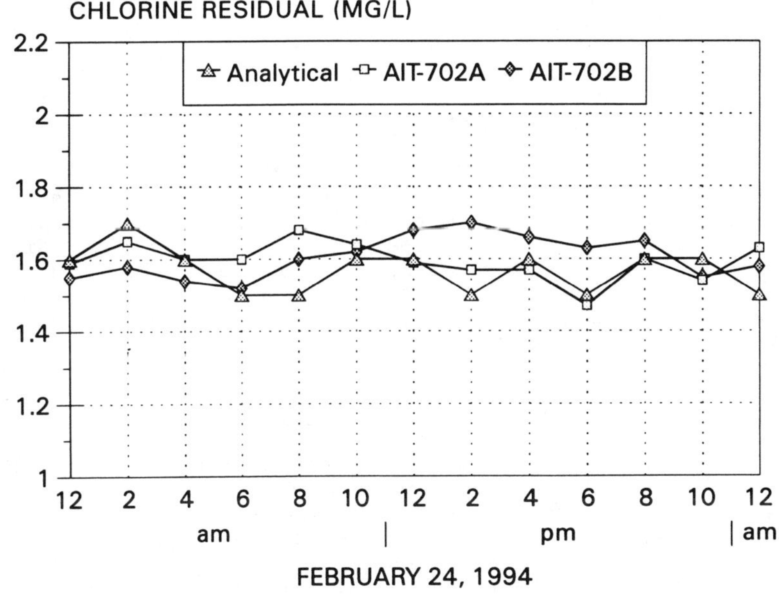

FIGURE 2 QUALITY CONTROL CHART FOR CHLORINE ANALYZERS

desired chlorine solution flow rate. The setpoint for the flow loop is cascaded from the residual control loop using the chlorine residual as the process variable. The plant operators enter the setpoint for the residual loop.

The chlorination control system has been designed for "graceful" degradation when equipment failures occur. In case of a failure of both analyzers, the plant operators can put the flow loop in AUTO mode and enter a chlorine flow setpoint. In case of a failure of the flow meter, the operators can set the valve position and read the chlorine flow manually in the field. In case of a failure of the automatic valve, the operators can use a manual bypass valve.

Benefits of Chlorination Control

The new chlorination control system has performed very well and allows effluent residual chlorine concentration to be controlled to within +/- 0.2 mg/L of the desired setpoint most of the time. This tight regulatory control has permitted lowering the chlorine setpoint from a nominal value of 2.0 mg/L during manual control to 1.5 mg/L with automatic analyzer control. The plant's discharge permit requires a residual of 1.0 mg/L after 20 minutes. Therefore, the 1.5 mg/L operating point still leaves a substantial safety margin.

The reduced residual chlorine setpoint as well as the improved efficiency of the induction-type mixer have resulted in a 20 to 30 percent reduction in chlorine usage. For the 69th Street WWTP, the savings are estimated to be over $125,000 per year for sodium hypochlorite alone. Additional savings in the bisulfite (used for dechlorination) are also anticipated but have not yet been fully documented.

Instrumentation Quality Control Testing

An important part of any instrument-based control strategy is a strong quality assurance/quality control (QA/QC) program to ensure the integrity of the on-line instruments. The QA/QC program consisted of periodically comparing the analyzer output with that of a portable spectrophotometer (Hach DR 100 colorimeter) utilizing the DPD methodology for total residual chlorine concentration. During the first few months of operation, QA/QC samples were taken every two hours. Based upon the first few months of operation and use of dual analyzers, one conformance test per shift would be very adequate. A typical set of conformance tests for a single day is shown in Figure 2.

Acknowledgements

We gratefully acknowledge the help of Instrument Control Service for installing, calibrating, and maintaining the instruments and control elements discussed in this paper. We also acknowledge the continued assistance of the 69th Street wastewater treatment plant operators and supervisors who carefully watched over the control system during startup and performed all QA/QC procedures.

Appendix I. References

1. Garrett, Jr., M. Truett, Ahmad, Zaki, and Young, Shelley (1993). "Experience With The Relay Procedure For Tuning Controllers In Automatic Control of Chlorination." 6th IAWQ Workshop on INSTRUMENTATION, CONTROL, AND AUTOMATION OF WATER AND WASTEWATER TREATMENT AND TRANSPORT SYSTEMS, Banff and Hamilton, Canada, June 17-25, 1993, pp. 377-386.

2. Hill, Robert and Schuk, Walter (1992). "The Instrumentation Testing Association." Instrument Society of America, Transactions, Applying Instrumentation and Automation in Environmental Engineering: Water and Wastewater, Vol. 31, No. 1.

3. Instrumentation In Wastewater Treatment Plants (1978). Manual of Practice No. 21, Water Pollution Control Federation, p. 80.

4. Smith, Otto J. M (1957). "Close control of loops with deadtime." Chemical Engineering Progress, Vol. 53, pp. 217-219.

5. The Water and Wastewater Instrumentation Testing Association of North America (1985). Report Nos. 85-1, 85-2, and 85-3, 1225 Eye Street, N.W., Suite 300, Washington, D.C.

6. The Water and Wastewater Instrumentation Testing Association of North America (1990). "Performance Evaluation of Residual Chlorine Analyzers For Water and Wastewater Treatment Applications." Report No. CH-1, 1225 Eye Street, N.W., Suite 300, Washington, D.C.

THE SENSITIZED PHOTOCATALYSIS OF A MIXED REACTANT SYSTEM OF 4-CHLOROPHENOL AND 4-NITROPHENOL

Melissa S. Dieckmann[1] , Kimberly A. Gray[1] and Prashant V. Kamat[2]

Abstract

Sensitized photocatalysis has been demonstrated with colored pollutants such as nitrophenols. With visible light excitation of an adsorbed nitrophenolic compound, charge injection into the conduction band of the semiconductor occurs and subsequent transformation of nitrophenols to a variety of products is possible. These phenomena have been observed in powder systems of TiO_2, but have not been clearly detailed in aqueous systems. This work reports the results of aqueous TiO_2 slurry experiments in which the degradation of 4-nitrophenol and 4-chlorophenol is studied in single reactant and mixed reactant systems. The primary objectives of this work are to establish the reaction pathway occurring in sensitized photocatalysis and to determine if sensitization can induce/drive the degradation of non-colored pollutants. The influence of organic concentration, oxygen and light energy are discussed and byproduct data are presented.

Introduction

Heterogeneous photocatalysis is capable of degrading many classes of compounds (Ollis, et al., 1991), but requires ultraviolet light, and thus may be rather energy intensive. Our previous research (Dieckmann, et al., 1992; Dieckmann, et al., 1993) has shown that colored compounds can be used in a single reactant system to sensitize the photocatalytic process with charge injection from the excited state of the colored pollutant, thereby undergoing degradation using visible light as the energy source (Kamat, 1991). The sensitized photocatalytic process has three possible advantages over direct photocatalysis: 1) it extends the range of excitation energies into the visible range, making fuller use of solar energy; 2) it may promote selective removal of colored pollutants, and in systems with low concentrations of colored pollutants may increase the sensitivity of the photocatalytic process for target removal of colored pollutants; and 3) it may

---

[1] Department of Civil Engineering and Geological Sciences, University of Notre Dame, Notre Dame, IN, 46556.

[2] Radiation Research Laboratory, University of Notre Dame, Notre Dame, IN, 46556.

have the ability to drive other reactions, which may make sensitized photocatalysis a more attractive large-scale process.

The basic mechanism of *direct* photocatalysis is well established (Kamat, 1991). An input of ultra-bandgap energy to the semiconductor particle causes a valence band electron to be promoted to the conduction band, causing charge separation. The conduction band electrons and valence band holes can then migrate to the surface and participate in oxidation-reduction reactions. The degradation of an organic pollutant is attributed to indirect reaction at the positive hole where adsorbed water or a hydroxyl group is oxidized to hydroxyl free radical (•OH), which then reacts with the pollutant.

In *sensitized* photocatalysis, colored compounds are adsorbed onto semiconducting surfaces; the adsorbed compound is promoted to an excited state by the input of visible radiation. This excited state can then inject an electron into the conduction band of the semiconductor and become oxidized to a cation radical. Sensitized photocatalysis, then, does not involve charge separation in the semiconductor or production of •OH; rather, it entails charge injection from an excited state of the colored compound. This cation radical may be very reactive and therefore is susceptible to further reaction, but is also extremely susceptible to recombination if the injected electron is not effectively swept from the conduction band. The mechanisms of direct photocatalysis and sensitized photocatalysis are demonstrated in Figure 1.

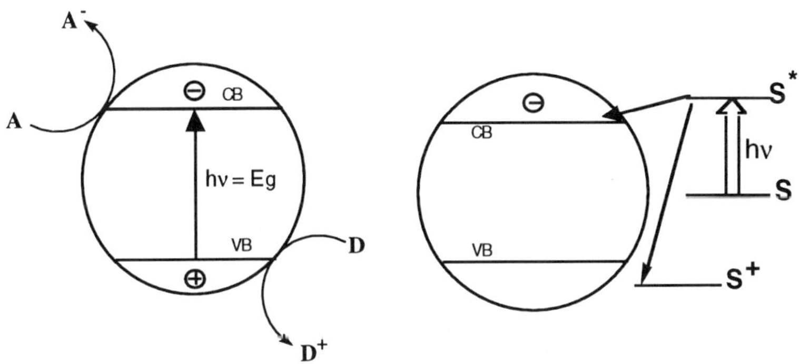

Direct photocatalysis **Sensitized Photocatalysis**

Figure 1. Comparison of direct and sensitized photocatalysis.

This paper explores the use of a model colored pollutant, 4-nitrophenol (4-NP), to sensitize its own degradation in a single reactant system, and to drive the degradation of an uncolored compound, 4-chlorophenol (4-CP). Previous studies have shown that 4-NP is capable of autocatalyzed degradation via sensitized photocatalysis (Dieckmann, et al., 1992) and that 4-CP can be degraded photocatalytically (see Al-Ekabi and Serpone, 1988; Mills, et al., 1993; Gray, et. al., 1994 and references therein). Specifically, we consider the kinetics of parent compound loss and extent of mineralization to determine the effect of the pollutant concentration, TiO_2 surface area, oxygen concentration and light energy on the degradation rate and mechanism. The pathway of 4-NP degradation is elucidated by considering organic and inorganic byproducts. Similar studies are conducted on the mixed reactant system to determine the extent to which sensitization of 4-NP can drive the degradation of 4-CP. These data are also considered in order to investigate mutual rate inhibition in this mixed reactant system, since preliminary studies indicate that mutual rate inhibition is a concern in direct photocatalytic reactions (Turchi and Ollis, 1989).

Results and Discussion

Experiments were conducted on single component and two-component TiO_2 slurries; the ratio of organic concentration to TiO_2 surface area, light energy, and O_2 content were varied. Organic byproduct analysis was conducted using reverse phase HPLC; inorganic byproduct analysis was conducted using colorimetric methods and ion selective electrochemistry. Slurry studies were conducted using a 450 W medium pressure mercury lamp and an 850 mL annular reactor with a borate buffer solution to maintain a pH of 8.5. A filter solution was used for light energy control and as a cooling solution. UV light conditions were achieved using a filter solution of $CuSO_4$ and 2,7-dimethyl-3,6-diazacyclohepta-1,6-diene perchlorate (DDDP) with a range of light $\geq$ 340 nm; visible light conditions were achieved using a filter solution of $CuSO_4$, $NaNO_2$, NH_3, and DDDP that filters out all wavelengths below 375 nm. The semiconducting material was Degussa P-25 TiO_2.

Single Reactant Studies

The most important factor affecting the ability of 4-nitrophenol to sensitize photocatalysis is the concentration of 4-NP relative to the surface area of the semiconducting particles. This is a result of the necessity for direct interaction between a short-lived excited state of 4-NP and the TiO_2 particle to achieve charge injection into the conduction band. As the concentration of 4-NP increases (for a constant TiO_2 amount), the rate of 4-NP degradation decreases, contradicting the expected first-order kinetics of solution chemistry, and indicating that sensitization (a surface reaction) occurs. This necessity for direct interaction between semiconductor and pollutant concentration may be a limitation in process design for a full-scale operation.

Another important factor in the sensitized photocatalysis of 4-NP is the presence of oxygen in the system. In direct photocatalysis, oxygen has been shown to be the major electron acceptor of the conduction band electron, and is hypothesized to participate in the degradation process via production of $^{\bullet}OH$. Oxygen is also thought to be necessary for complete mineralization to occur. In sensitized photocatalysis, oxygen is also a major electron acceptor, but additional roles of O_2 have not been previously explored. Therefore, we wished to explore the role of oxygen in the degradation and mineralization of 4-NP. In a deoxygenated system, 4-NP degradation is achieved but the reaction rate is slower; this is most likely the result of an increased recombination rate of the injected electron with the cation radical. Degradation of 4-NP in a deoxygenated system implies that 4-NP becomes the primary electron acceptor as well as the electron injector; therefore the reaction between the 4-NP cation radical created by charge injection and the 4-NP anion radical created by electron sweeping will produce ground-state 4-NP, which decreases the overall rate of 4-NP degradation. The effect of O_2 on mineralization is discussed later in the paper.

A preliminary reaction mechanism has been determined for the direct photocatalytic degradation of 4-NP. The detectable organic degradation products are 4-nitrocatechol (4-NC), hydroquinone (HQ) and 1,2,4-benzenetriol (BT). In an oxygenated system, the production of 4-NC is proposed to proceed via reaction with •OH. The attack of •OH on 4-NP leads to a hydroxyl adduct of 4-NP; 4-NC is the preferred substitution product. 4-NC production occurs primarily when a large pollutant:surface area ratio exists; this implies that 4-NC is produced in solution rather than while adsorbed at the surface. Hydroquinone (HQ) can be produced via direct substitution of the ntiro group by •OH. Hydroquinone and 4-NC are produced simultaneously, but HQ production dominates when the pollutant:surface area ratio is low. Indirect evidence of 1,2,4-benzenetriol appearance suggests that it is a byproduct of both 4-NC and HQ, and is proposed to be a result of •OH attack on 4-NC and HQ. In systems where O_2 is available, conversion of 4-NC and/or HQ to BT occurs early in the overall degradation process. This is confirmed by an analysis of NO_2^- and NO_3^- in the system during the reaction. In a deoxygenated system, the degradation rate decreases to 20% of the oxygenated system, but the intermediate analysis is not yet complete.

The mechanism of degradation via sensitized photocatalysis has also been studied mechanistically, and a preliminary reaction pathway has been proposed. The degradation products are identical to those identified for direct photocatalysis (4-NC, HQ, and BT)., but are produced under different conditions and in different proportions. In an oxygenated system, 4-NC may be produced by nucleophilic addition of hydroxide ion to the 4-NP cation radical. The borate buffer may encourage this hydroxylation since OH^- is abundant in a pH 8.5 solution. However, the overall proportion of 4-NC produced is much smaller than in direct photocatalysis, which is an expected consequence of producing a highly reactive 4-NP cation radical at the surface where it can quickly react, thereby inhibiting reactions occurring in solution Hydroquinone is also formed by reaction with 4-NP cation radical, and can be produced via attack of OH^- on the cationic carbon directly attached to the nitro group. The production of HQ/mol 4-NP degraded is much larger for sensitized photocatalysis than for direct photocatalysis; this supports the hypothesis that HQ production occurs on the TiO_2 surface. Benzenetriol production is proposed to occur by reaction of HQ and 4-NC with the activated oxygen species created when O_2 sweeps the conduction band electron. The actual activated oxygen species involved in this transformation is not yet known; however HQ transformation to BT appears to be more dominant than 4-NC transformation to BT. In a deoxygenated system, HQ production is much slower, 4-NC production dominates, and BT production is strongly inhibited. Finally, in the oxygenated system, there is an overall decrease in TOC which is not evident in deoxygenated systems; ring cleavage leading to mineralization is thought to occur immediately after denitration.

The reaction pathway for direct photocatalysis of 4-CP has been determined in a single reactant system (Stafford, et al., 1994). 4-Chlorophenol is degraded to 4-CC, HQ and non-aromatic intermediates via reaction with •OH and direct electron transfer. The spectral characteristics of 4-CP in solution indicate that sensitized photocatalysis of 4-CP in a single reactant system is impossible. Although 4-CP in solution does not absorb light below 375 nm, in the slurry system it has the ability to sensitize photocatalytic reactions illuminated by light where $\lambda \geq 375$ nm since its spectral characteristics are changed by its adsorption onto TiO_2 particles (Stafford, et al., 1993). The ability for sensitized photocatalysis to occur with pollutants that are normally uncolored in solution can expand the range of polluted waters which can be treated using this treatment process. The degradation of 4-CP via sensitized photocatalysis is mechanistically similar to 4-NP photocatalytic degradation: 4-chlorocatechol (4-CC), HQ and/or benzenetriol are the major reaction intermediates.

Mixed Reactant Studies

In Figure 2, results are presented from a series of experiments investigating 4-NP sensitization of TiO_2 photocatalysis in the presence of various electron acceptors. Since 4-NP degradation requires that the injected electron be swept by an acceptor to prevent recombination with the 4-NP cation radical, the effect of acceptor type on the rate of 4-NP disappearance has been evaluated. In the absence of O_2, 4-NP degradation occurs via charge injection from an excited 4-NP and the electron is swept by another 4-NP molecule; since 4-NP is not a very effective electron acceptor, the degradation rate is low. With addition of O_2, the rate of 4-NP degradation is increased, since O_2 is a more effective electron acceptor than 4-NP. In a deoxygenated mixed reactant system, 4-CP appears to be as effective an electron acceptor as O_2; after 1.5 hr., the degradation rate of 4-NP begins to slow as a result of lower availability of 4-CP due to degradation, further indicating the electron sweeping ability of 4-CP. When both 4-CP and O_2 are present in the system, the degradation of 4-NP is 2.5 times more rapid than with 4-CP or O_2 alone as the electron acceptor. This suggests that mutual rate enhancement in the 4-CP/4-NP system is a result of a synergistic interaction between 4-CP and O_2.

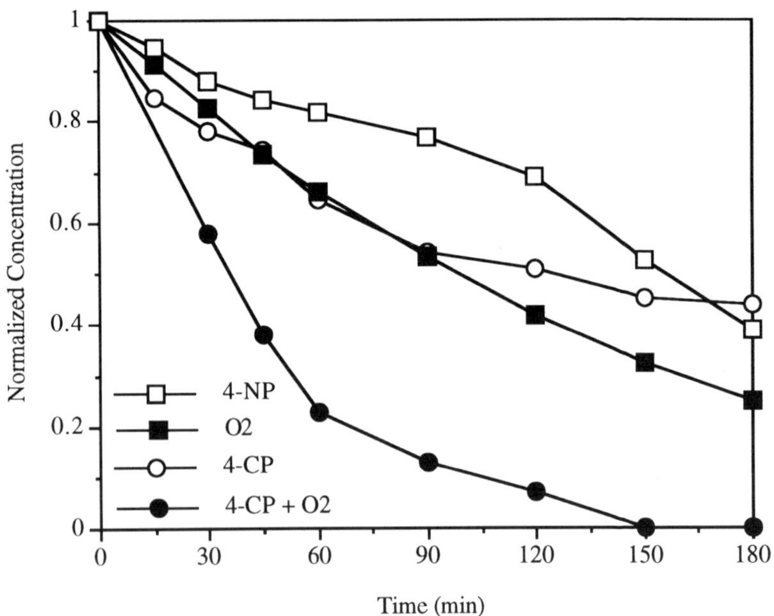

Figure 2. Degradation of 4-NP using various electron acceptors.

Since 4-CP is also capable of sensitizing photocatalytic reactions, there was some uncertainty in the roles of 4-CP and 4-NP in the degradation pathway. A kinetic analysis of the mixed reactant system and a mechanistic study are being conducted to determine the role of 4-CP and 4-NP in the mixed reactant system. Preliminary results indicate that 4-NP is the primary sensitizer which drives the degradation process while 4-CP acts primarily as an electron acceptor in conjunction with O_2. The total pollutant concentration:surface area ratio is particularly important in the mixed reactant system, since this seems to strongly affect the degradation rate of both compounds, and also seems to affect to a lesser extent the role of the 4-NP and 4-CP in the system. The suppression of the degradation rate as a function of available surface area in the mixed system has a much more drastic effect on 4-NP degradation than on 4-CP degradation. This seems to indicate that interaction of 4-NP with the surface is more crucial to its degradation pathway, and that 4-CP degradation is probably an indirect oxidation reaction occurring in solution.

Conclusions

In the single reactant aqueous system, it has been confirmed that 4-NP does sensitize the photocatalytic process. The ability for sensitization to occur is limited by the need for direct interaction between the 4-NP and TiO_2, and a maximum pollutant concentration:surface area ratio exists at which sensitization is effective. Analysis of organic intermediates has determined that the byproducts of direct and sensitized photocatalysis are identical, although the proportions and rates of appearance are different, indicating that different mechanisms dominate in sensitized and direct photocatalysis. The identified degradation byproducts are 4-nitrocatechol, hydroquinone, and 1,2,4-benzenetriol. In direct photocatalysis, 4-NP degradation predominantly occurs via $^{\bullet}OH$ attack; in sensitized photocatalysis, the attack of the cation radical 4-NP molecule with hydroxyl ion appears to be the predominant pathway of 4-NP degradation. In both direct and sensitized photocatalysis, mineralization requires O_2. Similar results are obtained for 4-CP degradation, and reaction byproducts were identified as 4-chlorocatechol, hydroquinone, and possibly 1,2,4-benzenetriol.

The introduction of a second component to the system results in very complex reaction kinetics. The kinetics of the mixed reactant system are strongly affected by the ratio of organic concentration:surface area of TiO_2. The mechanism also becomes more complicated since 4-CP can actively participate in the sensitization of the photocatalytic process. Both rate enhancement and inhibition are observed; the propensity to enhance or inhibit reaction kinetics is dependent on the pollutant concentration:TiO_2 surface area ratio. Most importantly, the spectral characteristics of the pollutant change with adsorption onto the semiconducting surface, allowing compounds such as 4-CP, which are normally uncolored in solution, to sensitize the photocatalytic process. This phenomenon suggests that sensitized photocatalysis may extend the range of photocatalysis to longer wavelengths since surface active pollutants which are uncolored can now participate in the degradation process at these wavelengths.

References

Al-Ekabi, H., Serpone, N. (1988) *J. Phys. Chem.*, *92*, 5726.

Dieckmann, M.S., Gray, K.A., Kamat, P.V. (1992)*Wat. Sci. Technol.*, *25*, 277.

Dieckmann, M.S., Gray, K.A., Zepp, R.G. (1994) *Chemosphere*, in press.

Gray, K.A., Stafford, U. (1993) *Res. Chem. Intermed.*, submitted.

Kamat, P.V. (1991) *Kinetics & Catalysis in Microheterogeneous Systems*, Mercel Dekker, Inc., New York, 275.

Mills, A., Morris, Davies, R. (1993) *J. Photochem. Photobiol. A: Chem*, *70*, 183.

Ollis, D.F., Pelizzetti, E., Serpone, N. (1991) *Environ. Sci. Technol.*, *25*, 1522.

Stafford, U., Gray, K.A., Kamat, P.V., Varma, A. (1993)*Chem. Phys. Let.*, *205*, 55.

Stafford, U., Gray, K.A., Kamat, P.V. (1994) *J. Phys. Chem* ., accepted.

Turchi, C.S., Ollis, D.F. (1989) *J. Catal.*, *119*, 483.

Innovative Treatment of Soil Contamination: Radiolytic Destruction of Dioxin and Co-Contaminants by Cobalt-60

Roger J. Hilarides and Kimberly A. Gray[1]

Abstract

Recent work in our laboratory has demonstrated that gamma radiolysis is a feasible method by which 2,3,7,8-tetrachlorodibenzo-p-dioxin (TCDD) can be converted to products of negligible toxicity. A standard soil has been artificially contaminated to a level of 100 ppb TCDD and destruction to a level less than 1 ppb has been achieved at a radiation dose of 800 KGy and with the addition of certain soil amendments (water and surfactant). By-product analysis has illustrated that the destruction occurs via step-wise reductive dechlorination and mass balance on carbon has been demonstrated. The presence of co-contaminants at much higher levels does not interfer with TCDD destruction. These results in combination with scavenger studies and target theory calculations indicate that *direct radiation effects* account for the major route of destruction. Process efficiency has been verified using real contaminated soils and sediments. A reactor design is proposed and an economic analysis is presented to show that radiolysis is technically feasible and economically competitive.

Introduction

There are approximately 500,000 tons of soil in the U.S. which are contaminated with dioxin, specifically 2,3,7,8-tetrachlorodibenzo-p-dioxin (TCDD) and require treatment (OTA, 1991). Current dioxin destruction techniques include thermal and non-thermal methods. Incineration is the most common and effective thermal treatment, and is used for dioxin destruction at many dioxin contamination sites under remediation. The capital and operating costs of incinerators, coupled with the administrative costs for permitting, have made them a less attractive treatment alternative. Moreover, the extreme public opposition to incineration has motivated many industries to search for feasible alternative treatment technologies. Some non-thermal dioxin destruction techniques include photolysis, radiolysis

[1] Department of Civil Engineering and Geological Sciences, University of Notre Dame, Notre Dame, IN 46556

(gamma radiation and electron beam) and chemical treatments. For the treatment of contaminated soil, gamma radiation has a major advantage over electron beams and photolysis because it has greater penetration into soil. The purpose of this paper is to explain those factors which influence the rate and extent of radiolytic destruction of TCDD and to discuss how this understanding can be integrated into a full scale reactor design.

Materials and Methods

These experiments have been conducted using an artificially contaminated 'standard soil' prepared by U.S. EPA (Synthetic Soil Matrix Blending Systems; SSM-91). This soil is a mixture of clay, silt, sand, top soil and gravel and is described fully elsewhere (Tabak et al. 1991; Hilarides et al. 1994a). 2, 3, 7, 8-TCDD was obtained from Cambridge Isotope Laboratories. Artificial soil contamination to approximately 100 ppb TCDD was accomplished in an homogeneous manner by a mixing procedure using hexane (Hilarides et al. 1994a). Analysis for chlorinated dioxins was developed and conducted by the Dioxin Laboratory of Occidental Chemical Corporation, Technology Center, Grand Island, NY. All irradiations were performed using a 10,000 Curie Shepherd 109 Cobalt-60 irradiator, which is a concentric, well type source that provides uniform dose distribution.

Results and Discussion

Destruction of 2,3,7,8-tetrachlorodibenzo-p-dioxin (TCDD) destruction on artificially contaminated soil using Cobalt-60 (^{60}Co) gamma radiation is feasible. In the presence of 25% water, 2% surfactant (RA-40^{TM}-nonionic) and high irradiation dose (800 KGy) greater than 90% TCDD destruction has been achieved (Hilarides et al. 1994 a,b). High irradiation doses are not a significant problem for ^{60}Co sources because no external power is required. Additionally, high doses can be achieved at costs that are competitive with current technologies (Hilarides et al. 1994b). The factors that influence the effectiveness of ionizing radiation in degrading TCDD on soil include radiation dose, soil moisture content, surfactant type and concentration, soil-TCDD age, and soil-surfactant equilibration time.

The role played by a surfactant in the destruction of TCDD adsorbed to soil is essential and complex. Large differences in destruction have been observed as a function of surfactant type and concentration. It is suspected that surfactants may play a specific role in the chemistry of destruction (i.e., hydrogen donor in radiolytic dechlorination), as well as to mobilize the dioxin to an interface where it is more susceptible to radiolytic destruction. Furthermore, depending on the pathway of destruction, the surfactant is expected to play different roles. Scavenger as well as kinetic studies have been performed to determine if the primary route of destruction is via direct or indirect effects.

In Figure 1 data (open and solid circles) are presented showing the rate of TCDD destruction in soils containing 25 % water and no surfactant (solid circles), and 2% nonionic surfactant (RA-40^{TM}, open circles). The presence of surfactant enhances both the rate and extent of destruction. The square symbols in this figure are data acquired from long irradiation experiments and provide a comparison with projected destruction based on initial kinetic behavior (solid and dashed lines). Theoretical calculations based on target theory have also been made and predict that a dose between 800 KGy and 1 MGy is required for complete destruction of TCDD by direct radiation effects (Lea, 1955; Hilarides et al. 1994a). There is a general agreement between the observed destruction and that predicted by target theory. Furthermore, the log relationship between TCDD destruction and radiation dose (time) shown in Figure 1 and results from scavenger studies indicate strongly that the initial transformation of TCDD to products occurs via direct radiation effects.

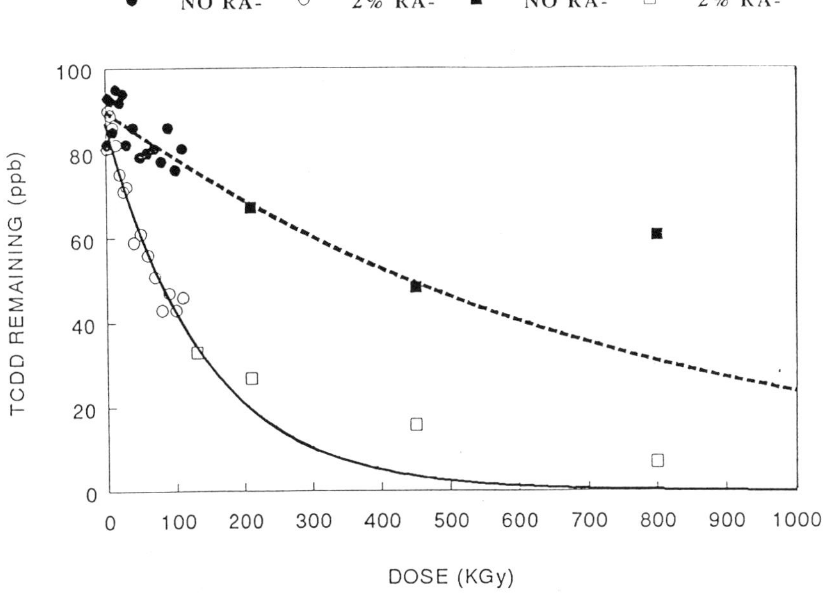

FIGURE 1. TCDD destruction with irradiation dose for soil/water systems in the presence and absence of surfactant (RA-40).

Beyond radiation doses of 200 KGy, the rate of TCDD destruction appeared to diminish as seen in Figure 1. Deterioration of the surfactant caused by the radiation flux is thought to account for this decrease in TCDD destruction. In order to maintain initial kinetics intermittent surfactant addition has been explored. These experiments have illustrated that a 2% threshold level of surfactant is required initially and that with incremental additions of 0.5% surfactant four times over the course of irradiation destruction to a 1 ppb TCDD residual was achieved. Preliminary investigations have indicated that the head group of the surfactant, which is an ethoxylated fatty alcohol, is cleaved radiolytically. Experiments are being conducted to test the effectiveness of other more resistant surfactants.

The effect on TCDD destruction of addition of high levels of co-contaminants to this artificial soils has been studied. Hexachlorobenzene and 4-chlorophenol have been added to the dioxin contaminated artificial soil each at a level of 100 ppm. Irradiation of the soil (25 % water and 2 % surfactant) at a dose of 800 KGy has produced the same destruction as shown in Figure 1. These results are consistent with the direct radiation pathway of destruction.

By-product analysis has been performed for all chlorinated dioxins (tri, di and mono-CDD) on soils irradiated at doses of 75, 150 and 450 KGy. At 75 KGy,

84 % of the initial TCDD had been converted to lesser chlorinated dioxins and at 150 KGy an 87 % conversion was measured. These results illustrate that TCDD loss is accounted for by successive reductive dechlorination and it is presumed that carbon mass balance at higher irradiations will be achieved with the analysis of non-chlorinated compounds and possibly oxidation products such as chlorophenols and biphenyls. By-products have also been analyzed under selected scavenger conditions. Under either oxidizing or reducing conditions or in the absence of scavengers, similar TCDD destruction and reation by-products were observed. This fact, combined with closing mass balance on carbon with reduction products even under oxidizing conditions, further confirms the fact that gamma radiolysis of TCDD on soils occurs via direct radiation effects.

Conclusion

These findings have been tested in real contaminated soils and sediments to determine the general applicability of model soil results. Based on the outcome of these studies a simple batch reactor system has been designed and cost estimates per ton of soils have been determined. One of the major components of operating expenses for radiolysis is surfactant cost . Preliminary estimates have shown that the operating and maintenance of gamma radiolysis for TCDD destruction on soil would be in the range of $ 200/ton of soil which is a fraction of the cost of incineration. The results of this research provide a promising basis for adapting an established technology to a new application and indicate that Cobalt-60 irradiation may be a feasible alternative to incineration for dioxin destruction in soils.

Acknowledgements

This research was sponsored by Occidental Chemical Corporation. The authors with to thank Drs. Jim Duffy and Larry Patterson for their comments and suggestions. The authors would also like to acknowledge the University of Notre Dame's Radiation Laboratory and the Center for Bioengineering and Pollution Control for use of their facilities.

References

Hilarides, R., Gray, K., Guzzetta, J., Cortellucci, N., Sommer, C., *Environ. Sci. Tech.,* submitted.

Hilarides, R., Gray, K., Guzzetta, J., Cortellucci, N., Sommer, C., (1994) *Env. Prog.,* in press.

Lea, D. E. (955) *Actions of Radiation on Living Cells,* Cambridge University Press, New York.

U. S. Congress Office of Technology Assessment, (1991) *Dioxin Treatment Technologies---Background Paper,* OTA-BP-O-93 (Washington, D.C.: U.S. Government Printing Office).

Tabak, M. E., Glynn, W., Traver, R. P., (1991) "Evaluation of EPA Soil Washing Technology for Remediation at UST Sites," U.S. EPA Report.

NCASI Experiments Related to Validation of Equilibrium Partitioning (EP) Theory

Steven W. Hinton[1]

Abstract

A review of NCASI's technical studies related to evaluating equilibrium partitioning (EP) theory as it is commonly applied in regulatory setting is presented. Although interim results are insufficient to conclude that EP theory is without flaws, they are supportive of the hypothesis that non-polar hydrophobic compounds bind to colloidal materials in proportion to the concentration of DOC present and that EP theory appears to be an adequate predictor of chemical distribution within stationary sediment beds. Ongoing experiments and future research directions are also described.

Introduction

One of the most common assumptions used in environmental models is EP theory which assumes that the dissolved to sorbed chemical concentration ratio for non-polar hydrophobic compounds is constant within aqueous mixtures containing dissolved and/or suspended organic carbon matter. Although currently used by US EPA for sediment criteria development and justification, field scale validation of EP theory with traditional measurements is lacking and estimates of binding coefficients for compounds typically found in receiving waters and paper industry treated effluents are subject to large uncertainties for solids matrices and are virtually non-existent for colloidal materials.

The purpose of this paper is to describe the interim results for two technical studies being conducted to evaluate EP theory as it is applied in regulatory settings.

---

[1] Research Engineer, National Council of the Paper Industry for Air and Stream Improvement (NCASI), Tufts University, Civil Engineering Dept., Anderson Hall, Medford, MA 02155.

Background and Theory

When organic-bearing solids and colloidal materials are exposed indefinitely to non-polar hydrophobic compounds in the presence of water, a thermodynamic equilibrium is believed to occur (Chiou *et al.*, 1979). The equations describing this equilibrium are:

$$C_s = S\ C_s' = S\ K_p\ C_w \tag{1}$$

$$C_b = DOC\ C_b' = DOC\ K_b\ C_w \tag{2}$$

C_s = solids sorbed chemical concentration (g_chem/liter soln)
C_s' = solids sorbed chemical concentration (g_chem/kg_solids)
C_b = DOC bound chemical concentration (g_chem/liter soln)
C_b' = DOC bound chemical concentration (g_chem/kg_carbon)
C_w = aqueous chemical concentration (g_chem/liter soln)
K_p = solids-water partition coefficient (g_chem/kg_solids per g_chem/liter soln)
K_b = DOC-water partition coefficient (g_chem/kg_carbon per g_chem/liter soln)
DOC = dissolved organic carbon concentration (kg_carbon/liter soln)
S = suspended solids concentration (kg_solids/liter soln)

In the above, dissolved organic carbon (DOC) concentration is used as a surrogate measure of the amount of truly dissolved (aqueous) and colloidal organic materials present in a mixture. Generally, DOC is operationally defined as the organic carbon present in a liquid after either filtration through filters retaining material greater than 0.45 μm or centrifugation removal of similar sized materials. Consequently, chemical concentration measurements of filtrate samples (Cm) represent the sum of the aqueous (C_w) and DOC bound chemical (C_b) concentrations.

$$Cm = C_w + C_b = C_w\ (1 + DOC\ K_b) \tag{3}$$

Equations 1 and 2 dictate that the relative distribution of hydrophobic compounds into aqueous, colloidal and solids sorbed phases at equilibrium is unaffected by the total amount of chemical present in the mixture and solely a function of the relative magnitudes of the partition coefficient values and sorbent concentrations. The partition coefficient values K_p and K_b are unique to the specific sorbents and sorbates involved, and are generally obtained from measurement or estimation procedures.

For non-polar hydrophobic compounds, equilibrium partitioning to solids has been shown to be highly correlated to the amount of carbon present (Chiou *et al.*, 1979; Karickhoff *et al.*, 1979), allowing measurements of partition coefficient values for a particular solid to be used in estimating another solid's partition coefficient value by correction for differences in fraction organic carbon (FOC) content. Equation 4 shows the relationship typically used for normalizing existing data and subsequently, estimating solids-specific partition coefficient values in the absence of measured values.

$$K_p = K_{oc}\ FOC \tag{4}$$

K_{oc} = carbon solids-water partition coefficient (g_chem/kg_carbon per g_chem/liter soln)
FOC = fraction organic carbon (kg_carbon/kg_solids)

Many empirical relationships are available for estimating K_{oc} based on the octanol/water partition coefficient value (K_{ow}) or solubility (Fetter, 1993; Lyman *et al.*, 1990). Relationships employing K_{ow} generally take the form of $LOG(K_{oc}) = A + B\ LOG\ (K_{ow})$ with typical coefficient values of $-0.8 < A < 0.62$ and $0.5 < B < 1.0$. For PCDD/Fs having $LOG(K_{ow})$ values in the 6 to 7 range, isomer specific K_{oc} values can be assumed approximately equal to the corresponding K_{ow} value (Eqn. 5) due to the substantial uncertainty associated with K_{ow} values in that range (EPA, 1989; Lyman *et al.*, 1990).

$$K_{oc} \approx K_{ow} \tag{5}$$

A consensus view of the magnitude of non-polar hydrophobic compound partitioning to dissolved and colloidal carbonaceous materials, however, does not presently exist. For estimating the value of K_b, Ambrose *et al.* (1988) recommend the general assumption of $K_b \approx K_{oc}$ for all compounds while Endicott *et al.* (1990) state that K_b is only 1 percent of the corresponding K_{oc} value for 2,3,7,8-TCDD. The available measurements of TCDD partitioning to DOC suggest K_b values in the range of 10^5 to $5 \cdot 10^5$ L/kg (Servos & Muir, 1989; Kukkonen, 1992) but these experimental determinations were limited in number and made under conditions atypical of aquatic environments, casting doubt on the precise value of K_b for TCDDs.

Current Projects and Interim Results

Two experiments designed to test the validity of EP theory using realistic sample volumes and conventional laboratory analyses methods are described briefly here. Contaminated solids materials were sieved into known size classes and analyzed for the non-polar hydrophobic compounds known as poly-chlorinated dibenzo dioxin/furans (PCDD/Fs) and for fraction of organic matter (FOM). Subsequently, two size classes of particles were exposed in separate reactors to a high grade distilled water, reagent humic acid, and bleached kraft effluent samples of three different dissolved organic carbon (DOC) concentrations. Aqueous exposures occurred in completely-mixed, brown-glass 4-liter reactors for 90 days. At the conclusion of the exposure period, the entire reactor contents were filtered through 1.2 μm and 0.7 μm glass fiber filters and subsequently analyzed for aqueous PCDD/F content by high resolution GC-MS methods. The following paragraphs describe typical examples of the laboratory results from these experiments.

1. Partitioning to Solids

To test the validity of EP theory for solids, sediment materials from industry impacted receiving waters were wet sieved into different size classes and analyzed for selected chlorinated organic chemicals and organic matter content. Given their intimate and infinite exposure to the intersticial pore water, EP theory would predict that the concentration of non-polar hydrophobic chemicals present in each size class should be approximately equal. Figure 1 shows the analysis results for 2,3,7,8-TCDD at one typical site. When normalized to organic matter content according to EP theory, chemical concentrations are within ±45 percent of the average. Given the wide variation in organic matter contents between size class (1.4-13.3% FOM), EP theory appears to be an adequate predictor of chemical distribution within stationary sediment beds. Preliminary results for a second site indicate a similar tendency and should be finalized during Autumn 1994.

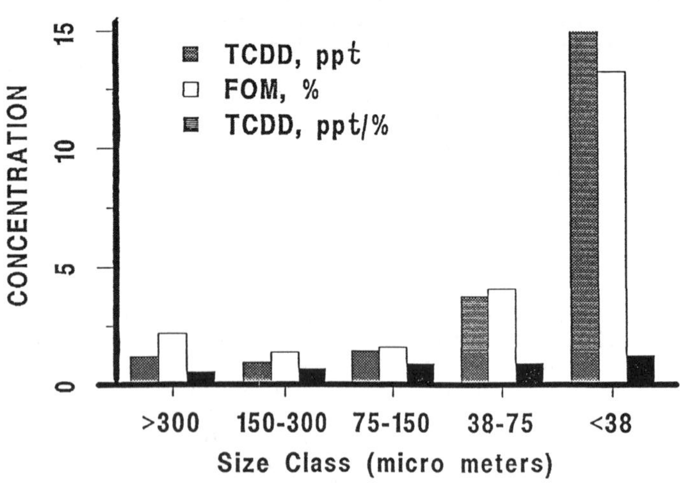

Figure 1. 2,3,7,8-TCDD concentrations and Organic Matter Percents vs. Sediment Particle Size Class.

2. Partitioning to Colloids

The results of exposing 1,2,7,8-TCDD containing solids in the 4-liter batch reactors to reagent water and two different colloidal materials at three dilution levels are shown in Figure 2. For both the treated kraft mill effluent and the reagent humic acid tests, the concentration of TCDD increased with increasing concentration of the colloidal sorbents. The dashed line approximates the expected theoretical relationship. Although the data are insufficient to conclude that EP theory is without flaws, it is supportive of the hypothesis that non-polar hydrophobic compounds bind to colloidal materials in proportion to the concentration of DOC present. Additional colloidal sorbents are currently being tested.

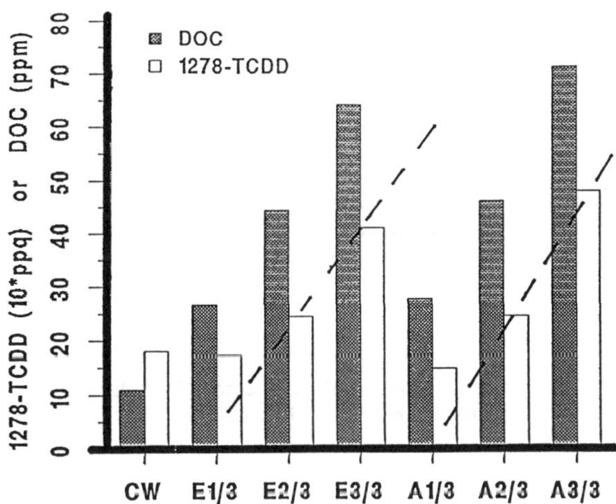

Figure 2. 1,2,7,8-TCDD Concentrations in Clean Reagent Water (CW), and 1/3, 2/3, 3/3 Dilutions of BKTP Effluent (E) and Humic Acid (A).

Summary And Recommendations

Based on the interim results from two experiments using realistic sample volumes and conventional laboratory analyses methods, there is insufficient information to conclude that EP theory is completely flawless. However, the results support the hypothesis that non-polar hydrophobic compounds bind to colloidal materials in proportion to the concentration of DOC present and that EP theory appears to be an adequate predictor of chemical distribution within stationary sediment beds.

References

Ambrose, R.B., Wool, T.A., Connolly, J.P. and Schanz, R.W. (1988). *WASP4, A Hydrodynamic and Water Quality Model--Model Theory, User's Manual, and Programmer's Guide*. US Environmental Protection Agency, Athens, Georgia.

Chiou, T.C., Peters, L.J. and Freed V.H. (1979). A Physical Concept of Soil-Water Equilibria for Nonionic Organic Compounds. *Science*, 206, 831-832.

Endicott, D.D., Richardson, W.L. and DiToro, D.M. (1990). Lake Ontario TCDD Modeling Report. In: *Lake Ontario TCDD Bioaccumulation Study Final Report*. Cooperative study including U.S. Environmental Protection Agency, New York State Department of Environmental Conservation, New York State Department of Health, and Occidental Chemical Corporation.

EPA. (1989). *Briefing Report to the Science Advisory Board on the Equilibrium Partitioning Approach to Generating Sediment Quality Criteria*, NTIS No. PB92-231521. U.S. Environmental Protection Agency, Washington DC.

Fetter, C.W. (1993). *Contaminant Hydrogeology*. Macmillan Publishing, New York, NY, pp 134.

Karickhoff, S.W., Brown, D.S. and Scott, T.A. (1979). Sorption of Hydrophobic Pollutants on Natural Sediments. *Wat. Res.*, 13, 241-248.

Kukkonen, J. (1992). Effects of Lignin and Chlorolignin in Pulp Mill Effluents on the Binding and Bioavailability of Hydrophobic Organic Pollutants. *Wat. Res.*, 26, 1523-1532.

Lyman W.J., Reehl, W.F. and Rosenblatt, D.H. (1990). *Handbook of Chemical Property Estimation Methods*. American Chemical Society, Washington DC.

Servos, M.R. and Muir, D.C.G. (1989). Effect of Dissolved Organic Matter From Canadian Shield Lakes on the Bioavailability of 1,3,6,8-Tetrachlorodibenzo-p-Dioxin to the Amphipod *Crangonyx Laurentianus*. *Envir. Tox. and Chem.* 8, 141-150.

THE USE OF IRON SALTS TO CONTROL DISSOLVED SULFIDE IN DISTRICTS' TRUNK SEWERS

Navnit A. Padival[1,2], William A. Kimbell[3], and John A. Redner[4]

ABSTRACT

Headspace H_2S removal via the precipitation of dissolved sulfide was investigated using iron chlorides. Full-scale experiments were conducted in a 25-mile sewer with an average flow of 200 MGD. Results were sensitive to Fe dosages and Fe^{3+}/Fe^{2+} blend ratios injected. A concentration of 16 mg/L Fe and a blend ratio of 1.9:1 (Fe^{3+}:Fe^{2+}) reduced dissolved sulfide levels by 97%. Total sulfide and headspace H_2S were also reduced by 63 and 79%, respectively. Liquid and gas phase sulfide reductions were largely owed to the effective precipitation of sulfide and the limited volatilization of H_2S gas, respectively. The catalytic oxidation of sulfide in the presence of Fe^{2+} and minute amounts of O_2 may have occurred. When compared to previous work, a combination of Fe^{3+} and Fe^{2+} proved more effective than either salt alone. No specific relation between the concentration of Fe or Fe^{3+}/Fe^{2+} blend ratio and crown pH was inferred. Iron salts may retard crown corrosion rates by precipitating free sulfide, thus reducing its release to the sewer headspace as H_2S and/or by inhibiting responsible bacteria.
KEYWORDS: wastewater, collection system, concrete, corrosion, industrial waste, hydrogen sulfide, ferric/ferrous chloride, sulfide control

INTRODUCTION

Concrete corrosion. Biogenic corrosion of sewage collection systems—concrete sewers and structures and treatment plant concrete structures—is a concern for wastewater agencies [ASCE, 1989; Parker, 1947; and Thistlethwayte (Ed.), 1972]. Severe problems are encountered in the US, Germany, Japan, Soviet Union, and many other parts of the world (Holmstrom and Wilander, 1977; Rozhanskaya *et al.*, 1977; Sand *et al.*, 1983; and Witsgall *et al.*, 1989). Corrosion is especially severe within the Los Angeles County Sanitation Districts sewer system. At the Districts, rehabilitation activities attributable to corrosion has already cost about $50 million (Redner, 1992).

The corrosion of sewers is a two-step, biological process. The first step occurs below the sewer hydraulic grade-line, where anaerobic conditions can exist. There, sulfate-reducing bacteria reduce sulfate in the wastewater to sulfide in the anaerobic slime layer on continuously wetted pipe walls. Sulfide diffuses into the bulk fluid. At near neutral pHs, an appreciable fraction of sulfide so generated volatilizes into the sewer headspace as H_2S gas. The second step occurs above the sewer hydraulic grade-line, typically at the pipe crown, where the available H_2S gas is biologically oxidized to sulfuric acid by sulfur-oxidizing bacteria. This acid attacks the cement binder in the concrete leading to corrosion, deterioration, and eventual collapse (ASCE, 1989).

---

[1]Project Engineer, [2]Corresponding author, [3]Senior Civil Engineer, and [4]Sewerage System Superintendent—of County Sanitation District of Los Angeles County, 1955 Workman Mill Road, P. O. Box 4998. Whittier, CA 90607-4998.

The Los Angeles County sewerage system. The Districts own and operate 1,200 miles of trunk sewers ranging from 6 to 144 inches in diameter. These sewers are primarily constructed of reinforced concrete and vitrified clay, but include some nonreinforced concrete, and clay-tile lined reinforced concrete. The Districts have attempted to control sulfide corrosion, from the initial use of clay-tile liners, through the design of adequate velocities along with sulfide control treatment in its tributaries. The philosophy was to control sulfide into the main collection sewers and to provide sufficient reaeration in the system to maintain aerobic conditions (Jin and Won, 1989).

Wastewater quality in the Districts' sewers significantly changed after 1975, when (i) "the ocean plan" limited toxic metals concentrations and (ii) USEPA implemented categorical pretreatment regulation for industrial discharge to municipal sewers in the early 1980s. The Districts implemented these regulations with it's own industrial waste pretreatment program (Martyn and Kremer, 1987). These regulations produced large reductions in concentrations of toxic metals contributed by metal-finishing and electroplating industries. Sulfide control became a major concern after pretreatment programs reduced the overall metals concentration by at least 65% (Redner *et al.*, 1989).

Current, full-scale attempts to impede corrosion include (i) NaOH shock dosing, (ii) treatment with $FeCl_3/FeCl_2$, (iii) air injection in force mains, and (iv) NaOH and Na_2CO_3 spraying of sewer crowns. Experimental evaluations of liquid phase sulfide oxidation using H_2O_2 and O_2 have been also performed. Restriction of S^{2-} transport to the sewer headspace by continuous pH adjustment (using NaOH) was successfully tested, but can be prohibitively expensive (Khan, 1992). The costs associated with the Districts' ongoing treatment programs are very high and have not been fully justified based on the effectiveness of treatment demonstrated. High-pH spraying of the crowns is an exception.

Iron and sulfur chemistry. Domestic and industrial wastewaters contain one or several species of sulfur compounds. Of the thirty-plus inorganic and ionic sulfur species that exist, only six are thermodynamically stable in aqueous solution at 25°C and 1 ATM: bisulfate, elemental sulfur, sulfate, hydrogen sulfide, hydrosulfide, and sulfide. Other forms of inorganic sulfur such as polysulfides, polythionates, and thiosulfate also occur in the natural environment, but are considered to be thermodynamically unstable and generally not found in significant concentrations. Sulfite and thiosulfate are found in some industrial wastewaters (Garrles and Naeser, 1958).

The salts of many metals, including those of iron, zinc, and copper, react with dissolved sulfide to form metal-sulfide precipitates, thus reducing H_2S release to the sewer headspace. For this to be fully effective, the metal-sulfide formed must be highly insoluble. Addition of iron salts to sewage to control dissolved sulfide levels has been practiced in a number of places since the 1920s (Pomeroy and Bowlus, 1946). Zinc was once used for sulfide control by both the City of Los Angeles (U.S. EPA, 1974) and the Districts (Won, 1988). The use of zinc is no longer recommended because of it's potential negative effects on secondary treatment processes and on final effluent quality. Copper is relatively costly and environmentally controlled.

Iron salts of chloride, nitrate, and sulfate have been added to wastewater either in the ferric or ferrous forms. Iron has generally been thought to become tightly bound to sulfide and other radicals in wastewater. Both Fe^{3+} and Fe^{2+} iron are effective in sulfide removal. Several distinct iron-sulfide complexes may form after iron addition, including pyrrhotite (varies from FeS to Fe_4S_5), ferric sulfide (Fe_2S_3), smythite (Fe_3S_4), pyrite (FeS_2), and marcasite (also FeS_2) (U.S. EPA, 1974). Fe^{3+} can also remove sulfide by directly oxidizing it to sulfur (Pomeroy and Bowlus, 1946) by reducing to Fe^{2+} (Dohnalek and FitzPatrick, 1983). The Fe^{2+} formed, is then precipitated as ferrous sulfide (FeS). It is supposed that these reactions proceed as follows:

$$2Fe^{3+} + S^{2-} \rightarrow 2Fe^{2+} + S^{o} \tag{1}$$

$$Fe^{2+} + HS^{-} \rightarrow FeS\downarrow + H^{+} \tag{2}$$

In accordance with Equation 2, 7 parts of iron should precipitate 4 parts of sulfide. In this reaction, the H^+ ion concentration has no significant effect within the pH range normally encountered in sewage. However, the reaction is greatly influenced by the presence of traces of dissolved O_2 (Pomeroy and Bowlus, 1946). Further, the presence of O_2 or even any weak oxidizing agents assist the formation of FeS_2. This is suggested by Equation 3:

$$Fe^{2+} + 2HS^{-} + O \rightarrow FeS_2\downarrow + H_2O + 2H^{+} \tag{3}$$

Not only does O_2 accelerate the precipitation of sulfide, but it also catalyses the oxidation of sulfide even beyond the formation of FeS_2. In many instances, the decrease in total sulfide can be more than one-half the decrease in dissolved sulfide, thus more than would be accounted for merely by the formation of FeS_2 (Pomeroy and Bowlus, 1946).

If a mixture of Fe^{3+} and Fe^{2+} is added to sulfide-bearing water, better results may be obtained than with either salt alone (Equation 4):

$$Fe^{2+} + 2Fe^{3+} + 4HS^{-} \rightarrow Fe_3S_4\downarrow + 4H^{+} \tag{4}$$

Pomeroy and Bowlus (1946) report that a mixture of Fe^{3+} and Fe^{2+} is more effective than either alone. By using only a few mg/L of excess Fe^{3+}, sulfide can be reduced to levels of 0.2 - 0.3 mg/L in the absence of aeration or to 0.1 mg/L, with slight aeration. The most effective results were obtained with solutions containing about two-thirds of the iron as Fe^{3+}. This suggests the precipitation of Fe_3S_4, but lacks evidence. It may exist as an intermediate step in the transition pathway to FeS_2.

From the above equations and hypotheses, Fe^{3+} is expected to be more effective than Fe^{2+}. However, Jameel (1989) reports that in practice, there is no such difference. In the complete absence of O_2, Fe^{3+} is better than Fe^{2+}. But a little O_2 greatly improves the effectiveness of Fe^{2+} and is without effect on the Fe^{3+}, so that under these conditions the relationship is reversed. Thus, Fe^{2+} is better when added to a free-flowing sewer, where some O_2 is always present. The principal effect of Fe^{3+}, when added to sewage of low sulfide content, seems to be oxidation of the S^{2-} at the time of mixing; very little precipitation occurs unless the initial S^{2-} concentration exceeds 10 mg/L (Jameel, 1989).

The San Jose Department of Water Pollution Control reports that when $FeCl_2$ is added to raw wastewater, some of Fe^{2+} is rapidly oxidized to Fe^{3+} in oxygenated-environments, at a favorable pH (7.1-7.6). Thus, adding $FeCl_2$ to the system is equivalent to adding a mixture of Fe^{2+}/Fe^{3+}. Likewise, in a reducing environment, Fe^{3+} is reduced to Fe^{2+}, and addition of $FeCl_3$ amounts to adding a similar mixture of Fe^{2+} and Fe^{3+} (Dezham *et al.*, 1988). In the same study, the authors report that when iron encounters highly-reducing environments, where sulfide concentrations are relatively higher than the other anions (especially hydroxides), iron can more effectively precipitate sulfide. However, both these factors—the reducing environment and the relative predominance of sulfide—tend to make iron available for precipitation as $FeS_{(s)}$. The reduction of Fe^{3+} to Fe^{2+} can cause dissolution of iron, because both $Fe(OH)_{2(s)}$ and $Fe_3(PO_4)_{2(s)}$ are more soluble with respect to iron concentration than their ferric counter parts.

Jameel (1989) reports that since January 1986, 95% control of dissolved sulfide has been provided in the City of Mesa, AZ interceptors. For the Baseline Road Interceptor, a dose of 12 mg/L of $FeCl_2$ has controlled the dissolved sulfide concentration at the monitoring point to 0.2 mg/L. For the Southern Avenue Interceptor, a $FeCl_2$ dose of only 2.4 mg/L, has controlled these concentrations by 83%. Following these studies, two additional facilities were constructed in March, 1987 at a cost of $200,000: the Alma School Road and the Noche De Paz Stations. Station operation has been trouble-free and is controlling dissolved sulfide in the above mentioned sewers to 0.5 mg/L.

The Districts completed a full-scale evaluation of $FeCl_2$ to control sulfide in a large diameter sewer. The study indicated that: (i) complete control of dissolved sulfide was difficult, although effective control was achieved for at least 17 miles downstream of the addition point; (ii) the dosage required for 90% sulfide control is 7:1 ($FeCl_2$ to dissolved sulfide) if the dissolved sulfide is 4 mg/L or greater, 15:1 if the dissolved sulfide is between 1 to 4 mg/L, and 100:1 if the dissolved sulfide is less than 1 mg/L; (iii) odor and headspace H_2S levels were reduced by 40 to 70% ; (iv) the pipe crown pH increased from 1 to 2. The study concluded that the cost to control dissolved sulfide concentrations to a level of 0.5 mg/L in this 25-mile sewer carrying 200 MGD was $1,500/day (Won, 1988).

The Greater Vancouver Regional District (GVRD) studied the effectiveness of $FeCl_2$ to control odor and corrosion in their Annacis Island wastewater collection system. Their objective was to limit dissolved sulfide to a low enough level to eliminate H_2S gas in the headspace. The dosing strategy is based on the premise that, although iron treatment has some impact on the environment, dosage will be maintained to provide acceptable effluent iron concentrations. They report the following conclusions: (i) sulfide is produced year-round in the sewers in spite of -10°F winters, without iron treatment, (ii) uncontrolled dissolved sulfide levels up to 2.5 mg/L are encountered during warm periods, (iii) immobilization of dissolved sulfide reduced H_2S in the sewer atmosphere significantly and controlled sewer odors with approximated theoretical iron concentrations exceeding 10 mg/L Fe in the wastewater, and (iv) addition of iron to the forcemains effectively controlled H_2S gas release at force main/gravity sewer intersections (So and Merry, 1993).

MATERIALS AND METHODS

Calculating chemical dosages. Using uncontrolled sulfide levels and relying on iron-sulfide chemistry, the amount of chemical to be injected was calculated. Experiments were varied so as to bracket the optimum iron concentration and blend ratio required for maximum sulfide control. Five Fe concentrations—10, 12, 14, 16, and 18 mg/L; and for each Fe concentration of iron, four blend ratios (Fe^{3+}:Fe^{2+})—1:1, 1.3:1, 1.6:1, and 1.9:1 were tested. Each dose and blend ratio was tested for one week.

Chemical feed system. The feed system was designed to function with minimum maintenance. $FeCl_3$ and $FeCl_2$ was delivered separately in 4000-gallon trucks at concentrations of 39 to 45% and 27 to 33%, respectively. It was unloaded via on-line pumps from the trucks into the 12,000-gallon fiberglass-reinforced plastic tanks, through a quick-connect coupling. The chemical was injected into the sewer at a predetermined blend ratio via a plant sewer. The control valves (Collins Instrument Company Inc., Model 1060), Programmable Logic Controller (PLC) (Allen Bradely Inc.,), and Magnetic Inductive Flowmeters (Krohne America Inc., Model K-480AS) adjusted the feed rates automatically according to the diurnal flow variation at the most downstream end of the study reach.

Study reach. From the San Jose Creek Water Reclamation Plant (SJCWRP), the 25-mile reach consisted of two separate trunk sewers—Joint Outfalls (J.O.) "B" and "H". At the upstream end, the sewer carries a part of SJCWRP flow and primary/secondary biosolids. Approximately 1/2 mile downstream, the District 21 Interceptor discharges it's flow from the Pomona Water Reclamation Plant and the Chino Basin. About 17 miles downstream of SJCWRP, at "the cluster", flow inputs from J.O. "E", "F", and "H" provides a 4x dilution. Here, the 200 MGD flow is tributary to J.O. "B" which terminates 8 miles downstream of the cluster in the Joint Water Pollution Control Plant (JWPCP).

Analytical and chemical procedures. Sewage was monitored for total and dissolved sulfides, crown and sewage pHs, total and soluble iron, and headspace H_2S gas. Total and dissolved sulfides were measured in the field by methylene blue test visual estimation "Standard Methods," (1990). Indicator pH strips (American Scientific Inc.) dictated crown and sewage pHs. Stain, length-type MSA (Mine Safety Appliances Co.,) tubes furnished H_2S data. Iron analysis was performed using AA spectrophotometry.

RESULTS AND DISCUSSION

Dissolved Sulfide. Dissolved sulfide increased steadily with distance downstream in the presence and absence of iron. At the addition point, levels were below detectable limits 100% of the time for all Fe levels and Fe^{3+}/Fe^{2+} blends tested. In the upper 17 miles of the reach excellent control was achieved because: (i) sewage is fairly aerobic and (ii) excess iron is present at all times. At points 17 and 25-miles downstream, respectively, reductions averaged 94% and 84% at most concentrations of Fe and Fe^{3+}/Fe^{2+} ratios. Reductions with 16 mg/L Fe were higher than the other four Fe concentrations tested. At the most remote end, a Fe^{3+}/Fe^{2+} blend of 1.9:1 (16 mg/L Fe) delivered a maximum reduction of 97%. In contrast to this, a blend ratio of 1:1 delivered a 70% control at the same Fe concentration. The data, in general, indicated that iron reacted with sulfide very well and fitted the stoichiometry: 7 parts of iron precipitated 4 parts of sulfide. In fact, 16 mg/L Fe precipitated more sulfide than predicted by Equation 2. This may be attributable to the presence of Fe^{3+}, and the reduction of Fe^{3+} to Fe^{2+} iron via the formation of S° and FeS (Dohnalek and FitzPatrick, 1983 and Pomeroy and Bowlus, 1946). Jameel (1989) reports similar findings in the City of Mesa wastewaters.

Won (1988) concluded that complete and consistent control of dissolved sulfide was difficult to maintain using $FeCl_2$. In comparing Won's study and the study reported here, the latter indicates better control using a $FeCl_3/FeCl_2$ blend in lieu of the $FeCl_2$ alone. Pomeroy and Bowlus (1946) also report that by using only a few mg/L of excess Fe^{3+}, sulfide levels can be reduced more than with Fe^{3+} or Fe^{2+} iron alone. They report that effective results were obtained with about 66% of the iron in the Fe^{3+} form. In this study, effective dissolved sulfide control was observed when Fe^{3+} used was almost twice Fe^{2+}.

Total sulfide. From the treatment point, total sulfide steadily increased with distance downstream, whether or not iron was present. Above the cluster, iron seems to retard the rate at which S^{2-} accumulated in the wastewater. Below the cluster, accumulation, in the presence and absence of treatment appeared to be similar. At the injection point, reductions for various Fe doses and blend ratios averaged 73%. At points 17- and 25-miles downstream, these reductions averaged 68% and 45%, respectively. At a concentration of 16 mg/L Fe and blend ratio of 1.9:1, reductions maximized at 63%.

Total sulfide reductions can be attributable to two factors—catalytic oxidation of S^{2-} in the presence of Fe^{2+} and inhibitory/catalytic effects on SRB. If catalytic oxidation occurred, some of the S^{2-} can be oxidized to FeS_2. The kinetics associated with this oxidation reaction is unclear (Pomeroy and Bowlus, 1946). Such reactions could occur in the upstream sections of the study reach where the sewage is relatively aerobic. The oxidation rates may be directly proportional to the amount of dissolved O_2, thereby delivering better reductions of total sulfide in the upper reaches. They also can be dependent on: pH, S^{2-} level, transition metals, and organic content (Chen, 1974). At micromolar levels, transition metals and organics catalyze the oxidation of S^{2-} (Chen *et al.*, 1972). The potential of domestic and industrial wastewater being supplemented with transition metals and trace organics is good (Martyn and Kremer, 1987).

SRB may be experiencing inhibitory/catalytic effects. The data suggests, however, that some toxic effect may in fact be exerted by iron in the upstream reaches. This toxicity may be relieved in the lower 8 miles due to the dilution provided by other flow inputs. Saleh *et al.* (1964) and Jack and Thompson (1983) have provided lists of metals that could inhibit SRB activity. Iron is one of them. If iron induced a catalytic effect, it would be expected that SRB from systems treated with iron, would produce more sulfide at higher rates, than SRB in untreated systems. This needs further investigation and can be determined by generation rates of the same SRB, but reacting to two different conditions— treated vs. untreated. SRB can also be in competition with other bacteria in the sewage. Iron could exert a selective pressure against the competitors of SRB allowing SRB to expand their niche. This could result in higher sulfide output. Iron addition could have permitted the establishment of a unique type of SRB in the pipe slime layers. This uniqueness might have shutoff its sulfate-reduction pathway (in Fe presence) and somehow triggered it's energy derivation from the metabolism of Fe (Weiss, 1992).

Hydrogen Sulfide. Background sewer headspace H_2S ranged from 0 ppmv at the injection point to 40 ppmv at the remote end. In certain instances, these values were as high as 100 ppmv irrespective of the location and pipe configuration. In the absence of iron, high values were more frequently reported in the lower 8 miles of the study reach when compared to upstream reaches. Results indicate that Fe concentrations were directly proportional to observed reductions, but was less than the control provided over dissolved sulfide. However, no apparent relationship between the various combinations of blend ratio and the H_2S reductions at a stable Fe concentration was seen. Average headspace H_2S reductions was the greatest at 16 mg/L Fe with a blend ratio of 1.3:1. At JWPCP, a 79% reduction was observed at 16 mg/L Fe and a Fe^{3+}/Fe^{2+} blend of 1.9:1. In contrast to this, at the same concentration of Fe, a blend ratio of 1:1 offered a reduction that was half as good. Reductions in H_2S gas and blend ratios at 16 mg/L Fe did not correlate. Won (1988) and Weiss (1991) document similar reductions in H_2S.

Total and soluble iron. Without treatment, total iron concentrations averaged 7.2 mg/L Fe. Most of this iron is contained in the SJCWRP discharge. During treatment, total iron at the remote end was in the expected range. That is, it was approximately the sum of the iron added at the Sulfide Control Facility and that discharged from SJCWRP. Background soluble iron at JWPCP ranged from 0.25 to 1.53 mg/L Fe. As iron dosage increased, so did the average values of soluble iron measured at JWPCP. Depending on the dosage, the average soluble iron at the remote end ranged from 1.86 to 4.52 mg/L. The availability of soluble iron at JWPCP might lead one to believe that excess iron is injected. Although some amount of soluble iron is present, it may not be available to precipitate sulfide because certain charged, complexed species of $Fe(OH)_3$ and $Fe(OH)_2$ iron are believed to exist in a soluble form (Stumm and Morgan, 1970).

Sulfide control costs. The sulfide control station has enabled the Districts to control sulfide at reasonable capital, operating, and chemical costs. The San Jose Creek Sulfide Control Facility was built at a capital cost of about $175,000 in 1986. The station's power consumption and utility costs are minimum. At a current dose of 16 mg/L Fe and a blend ratio of 1.9:1, the chemical cost is $6,500/day. Station operation has been virtually trouble-free, and sulfides have remained well under 95% control. The only continuing labor cost is a few hours/week for maintenance of the station.

CONCLUSIONS

Iron precipitates free sulfide, thus significantly restricting H_2S in the sewer headspace. Iron treatment is effective to at least 17 miles downstream of the application point, albeit complete and consistent sulfide control is difficult. Comparing with previous work, a Fe^{3+}/Fe^{2+} was found to be more effective than either salt alone. By using excess Fe^{3+} (when compared to Fe^{2+}), sulfide can be reduced to undetectable levels. A dosage of 16 mg/L Fe and a blend ratio of 1.9:1 (Fe^{3+}:Fe^{2+}) provided the maximum sulfide control during the study reach. An average dissolved sulfide reduction of 95% provided only a 70% sewer headspace H_2S control. With treatment, H_2S and dissolved sulfide reductions were not proportional.

The catalytic effect of iron on sulfide in the upper reaches may be an important factor for sulfide removal in the system studied. The sensitivities of sulfide removal in the lower end of the system are not clearly established. Oxygen can accelerate precipitation of sulfide and can also catalyze sulfide oxidation. In many instances, decreases in total sulfide were more than one-half the decrease of dissolved sulfide. This could be accounted for by factors other than iron-sulfide precipitation.

In the upper 17-mile reach, iron toxicity on sulfate-reducing bacteria may have occurred. Below the cluster, this inhibition seemed to be relieved by dilution provided by other flow inputs. This needs to be looked into from a microbiological standpoint. Iron chloride may retard crown corrosion rates by limiting the availability of H_2S gas to sulfur-oxidizing bacteria. However, no convincing results were observed. This does not eliminate the possibility that iron addition provides some corrosion control. Additional studies, which more precisely define the relationship between headspace H_2S and crown deterioration rate, are needed to determine whether iron treatment retards corrosion.

REFERENCES

1. APHA, AWWA, and WPCF (1990). *Standard Methods for Water and Wastewater.*
2. Sulfide in Wastewater Collection and Treatment Systems (1989). Manuals and Reports on Engineering Practice No. 69, ASCE, New York, NY.

3. Chen, K. Y. *et al.* (1972). Oxidation of Sulfide by O_2: Catalysis and Inhibition. *Jour. San. Engrg. Div.*, ASCE, 98: SA1:215.
4. Chen, K. Y. (1974). Chemistry of Sulfur Species and Their Removal From Water Supply. *Chemistry of Water Supply, Treatment, and Distribution* (A. J. Rubin, editor). Ann Arbor Science, Ann Arbor, MI.
5. Dezham, P. *et al.* (1988). Digester gas H_2S Control Using Iron Salts. *Journal of Water Pollution Control Federation*, Washington, D.C., 60:514-517.
6. Dohnalek, D. A. and FitzPatrick, J. A. (1983). The Chemistry of Reduced Sulfur Species and Their Removal from Ground Water Supplies. *Journal of the American Water Works Association*, 75:298-308.
7. Garrels, R. M. and Naeser, C. R. (1958). Equilibrium Distribution of Dissolved Sulphur Species in Water at 25°C and 1 Atm. Total Pressure. *Geochem. Cosmochim. Acta.*, 15:1-2:113.
8. Holmstrom, H., and Wilander, A. (1977). Control of Hydrogen Sulfide Production in Sewers. *Vatten* (Sweden), 33, 4, 394.
9. Jack, T. R., and Thompson, B. G. (1983). Patents Employing Microorganisms in Oil Production in Microbial Enhanced Oil Recovery. J. E. Zajic *et al.* (Eds.), Penn Well Books, Tulsa, Oklahoma.
10. Jameel, P. (1989). The Use of Ferrous Chloride to Control Dissolved Sulfides in the Interceptor Sewers at Mesa, Arizona, Brown and Caldwell Report.
11. Jin, C., and Won, D. (1989). Master Plan, CSD.
12. Khan, G. A. (1992). CSD.
13. Martyn, P., and Kremer, J. (1987). Implementation of an Industrial Wastewater Treatment Program. *Proc. 41st Ind. Waste Conf.*, Purdue University, IN.
14. Parker, C. D. (1947). Species of Bact. Assoc. with the Corrosion of Concrete. *Nature*, 159, 439.
15. Pomeroy, R. D. and Bowlus, F. D. (1946). Progress Report on Sulfide Control Research. *Sew. Works Journal*, 18, 597.
16. Process Design Manual for Sulfide Control in Sanitary Sewerage Systems (1974). EPA 625/1-74-005, U. S. EPA, Cincinnati, OH.
17. Redner, J. A. (1992). Sewerage System Superintendent, CSD.
18. Redner, J. A., *et al* (1989). Abstract for the 62nd Annual WPCF Conf., San Francisco, California.
19. Rozhanskaya, A. M., *et al.* (1977). Bacterial Analysis of Corrosion of Reinforced Concrete Constructions. *Pro. 10th Int. Cong. on Metal Corrosion* (India), 1, 713.
20. Saleh, A., *et al.* (1964). The Effect of Inhibitors on Sulphate Reducing Bacteria. *Jour. Appl. Bacteriol.*, 27, 281.
21. Sand, W., *et al.* (1983). Simulation of Concrete Corrosion in a Strictly Controlled H_2S Breeding Chambers. Recent Progress Biohydrometallurgy (Italy), 1, 667.
22. So, S. S. and Merry, C. (1993). Senior Engineer and Assistant Project Engineer. Greater Vancouver Regional District, Burnaby, Canada. Personal Communication.
23. Stumm, W. and Morgan, J. J. (1970). *Aquatic Chemistry: An Introduction Emphasizing Chemical Equilibria in Natural Waters*, Wiley Interscience.
24. The Control of Sulfides in Sewerage Systems (1972). D.K.B. Thistlethwayte (Ed.), Ann Arbor Science Publishers, MI.
25. Weiss, J. W. (1991 and 1992). CSD. Memorandums Related to In-house Work.
26. Witzgall, R. A., *et al.* (1989). Sulfide Corrosion Case Histories. Prepared for Water Pollution Control Federation Conference, San Francisco, CA.
27. Won, D. (1988). CSD.

Photocatalytic Transformations and Degradation of 2,4,6-tirnitrotoluene (TNT) in TiO2 Slurries

Daniel C. Schmelling and Kimberly A. Gray[1]

Abstract

The photodegradation of TNT in a TiO_2 slurry reactor was studied as both a potential treatment technique for the remediation of water with munitions contamination and to gain further insight into the behavior of a nitroaromatic compound in a photocatalytic system. Photocatalytic and direct photolytic reactions were compared by evaluating rates and extent of TNT transformation and mineralization in the presence and absence of oxygen. Nitrate, nitrite, and ammonium ion concentrations were determined and mass balances on carbon and nitrogen were performed for the catalytic system.

Introduction

TNT (2,4,6-trinitrotoluene) is a widely used secondary explosive and constitutes one component of an extensive explosives contamination problem [1-3]. Current technologies used for the remediation of TNT contamination such as activated carbon and incineration are considered expensive and have poor public approval and this has motivated extensive research on alternative treatment techniques [3,4]. The photocatalytic mineralization of many organic compounds in aqueous systems using titanium dioxide as a photocatalyst has been reported. Moreover, such destruction has been achieved using near UV and solar radiation. It therefore appeared likely that photocatalysis might prove an effective and economical process for the remediation of TNT contamination.

The work reported herein presents an analysis of the degradation of TNT in a TiO_2 (Degussa P25) slurry reactor. Photocatalyis is compared to direct photolysis using both low and high frequency UV radiation and in the presence and absence of oxygen. This comparison is made by contrasting rates and extent of TNT trans-

---

1. 156 Fitzpatrick Hall, Dept. of Civil Eng., University of Notre Dame, Notre Dame, IN 46556

formation and mineralization. The organic and inorganic products of these photoreactions are considered and the inorganic nitrogen species are quantified and evaluated. Additionally, the possible role played by the colored photodegradation products of TNT in sensitization of TiO_2 is assessed. These studies were designed to both indicated the potential of TiO_2 photocatalysis as a process for the remediation of TNT contaminated water and to give further insight into the photocatalytic behavior of nitroaromatics.

A series of eperiments was performed in which TNT was dissolved in a slurry of TiO_2, sparged with oxygen, and exposed to a medium pressure mercury lamp in a closed reactor for a period of two hours. The light was filtered so that only wavelengths >340 nm were transmitted. Experiments were also run with no TiO_2 and under an atmosphere of nitrogen. Additionally, experiments were performed using a λ >400 nm filter to examine semiconductor sensitization and with a λ >190 nm filter to assess the effect of high energy UV radiation. TNT concentration and some by-product information were determined by HPLC analysis. Dissolved organic carbon (DOC), dissolved inorganic carbon, and insoluble carbon were quantified using a TOC analyzer. Production of CO_2 gas was measured by gas chromatography. Nitrite, nitrate, and ammonium ions were monitored by ion chromatography.

Results and Discussion

In the presence of the TiO_2 photocatalyst, oxygen, and radiation with λ >340 nm, the destruction of TNT was rapid and complete (see Figure 1). After a two hour reaction there was a 95% reduction in DOC and a subsequent carbon balance accounted for 90% of the carbon as having been mineralized to CO_2. Analyses for inorganic nitrogen showed that after two hours approximately 55% of the total nitrogen existed as nitrate ion while 35% was present as ammonium ion thus demonstrating that both oxidation and reduction of TNT were occurring (see Figure 2). When this same experiment was performed without the TiO_2, TNT was also completely transformed although not as rapidly as with the photocatalyst. However, the reduction in DOC was less than 15% and only minimal amounts of inorganic nitrogen ions were observed.

The TiO_2 photocatalyzed reaction with TNT utilizing λ >340 nm radiation was repeated under an atmosphere of nitrogen as was the non-catalyzed control reaction. Under these deaerated conditions, the photocatalyzed reaction still exhibited an enhanced rate of TNT transformation and an increased amount of DOC reduction. Additionally, the predominant inorganic nitrogen species observed in the photocatalyzed reaction was ammonium ion whereas in the non-catalyzed reaction it was nitrite ion. This evidence demonstrates that the nitro groups on TNT and/or its reaction metabolites can serve to sweep the electron from the catalyst surface.

Although TNT does not absorb visible light it does form highly colored photolytic decomposition products. Due to the presence of colored products there was a possibility that sensitization of the semiconductor played a role in the

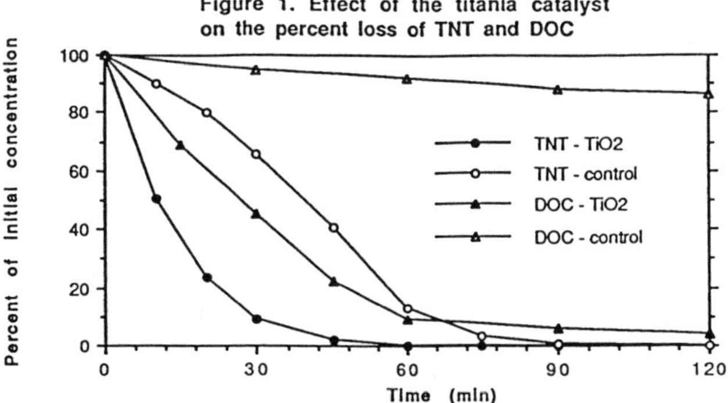

Figure 1. Effect of the titania catalyst on the percent loss of TNT and DOC

(Initial TNT conc.=50 mg/L; Initial oxygen conc. =40 mg/L; radiation>340 nm)

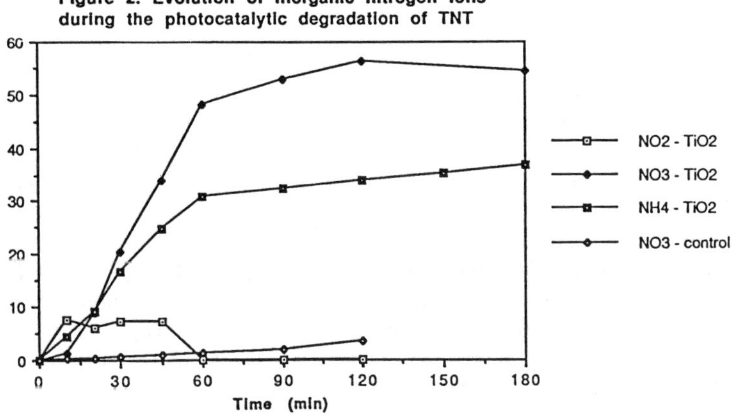

Figure 2. Evolution of inorganic nitrogen ions during the photocatalytic degradation of TNT

(Initial TNT conc.=50 mg/L; Initial oxygen conc.=40 mg/L; Radiation>340 nm; TiO2=250 mg/L) enhanced degradation observed with the TiO_2 photocatalyst. However, experimental analysis in which these colored products were photocatalytically reacted with TiO_2 using radiation with λ >400 nm revealed that sensitization was not significant.

The effects of employing high energy UV radiation were evaluated by running a series of catalyzed and non-catalyzed experiments using ultra pure water which has a UV cutoff of 190 nm as a filtering solution. The TiO_2 photocatalyzed reaction utilizing high energy UV occurred at a higher rate in both transforming and mineralizing TNT than the photocatalyzed reaction utilizing only radiation >340 nm. However, the extent of the destruction was equal in both caes. For the reaction conducted using high energy UV with no photocatalyst the percent loss of DOC was much greater than in the non-catalyzed reaction in which radiation <340 nm was blocked; approximately 79% loss versus 15% loss after 120 minutes. Nevertheless, the reaction run with the photocatalyst and 340 nm filter was still substantially faster and proceeded to a greater extent than the non-catalyzed reaction utilizing high energy UV.

Conclusions

These results have shown that TiO_2 photocatalysis using near UV radiation may be highly effective in the remediation of TNT contaminated waters. This process can achieve almost complete mineralization of TNT and it has the potential to be adapted to solar radiation which could offer an economic advantage over other advanced oxidation processes. The photocatalytic degradation of TNT appears to involve both oxidative and reductive steps. Ammonium ion accounts for approximately one third of the total nitrogen species produced under aerated conditions and appears to be formed predominantly via interfacial reactions at the catalyst surface and to a lesser extent by direct photoreduction. Data indicate that TNT is capable of sweeping electrons from the catalyst surface but not to the extent of fully reducing all aromatic nitro groups. Future work is expected to concentrate on illustrating degradation pathways through the identification of intermediate species and analyzing the specific role of the catalyst.

Acknowledgements

Notre Dame Center for Bioengineering and Pollution Control
U.S. Dept. of Education GAANNP
U.S. Dept. of Energy Radiation Lab at Notre Dame
National Science Foundation

References

1. Won, W.D., R.J. Hedkly, D.J. Glover, and J.C. Hoffsommer, (1974),

Metabolic disposition of 2,4,6-trinitrotoluene, *Appl. Microbiol.*, vol. 27, pp. 513-516.

2. Keither, L.H., and W.A. Telliard, (1979), Priority pollutants. I. A perspective view, *Environ Sci Technol*, vol 13, pp. 416-423.

3. Tri-service environmental quality strategic plan program, developed by: The tri-service reliance joint engineers-environmental quality tech area panel, Draft program user review, F. Belvoir, VA, 7-11 December 1992.

4. Wujcik, W.J., W.L. Lowe, and P.J. Marks, (1992) Granular activated carbon pilot treatment studies for explosives removal from contaminated groundwater, *Environ Prog*, Vol. 11, No. 3, pp. 178-189.

Activated Carbon as a Catalyst for Polymerization of Phenolic Compounds

Jacob A. Bourdeau[1] and Radisav D. Vidic[2]

Abstract

Previous studies by the principal investigator showed that the increase in the adsorptive capacity of activated carbon for phenolic compounds can be attributed to polymerization (oxidative coupling) of these compounds in the adsorbed phase which is catalyzed by the carbon surface in the presence of molecular oxygen (oxic conditions). This study was conducted to determine the origin of the catalytic properties of activated carbon surface towards these reactions. Particular emphasis was directed towards various metals and metal complexes imbedded in the activated carbon graphite crystalline structure. Adsorption isotherm studies performed in the presence and absence of oxygen using the same adsorbate (o-cresol) and different types of activated carbons were utilized as a preliminary screening tool in determining surface characteristics that influence the catalytic properties of activated carbon. The presence of Fe on the carbon surface is related to the ability of that activated carbon to promote polymerization of 2-methylphenol. Furthermore, the adsorptive capacity of activated carbon for molecular oxygen correlates well with the irreversible adsorption promoted by the GAC surface.

Background

Adsorptive capacity of activated carbon for organic compounds of interest is a key parameter governing the service life of a carbon bed and consequently the cost of this efficient technology for water purification. Studies by Vidic *et al.* (1991, 1993) revealed that the presence of oxygen promotes a significant increase in granular activated carbon (GAC) adsorptive capacity for many aromatic compounds as well as naturally occurring organic matter. The increase in GAC adsorptive

---

[1]Graduate Research Assistant, Department of Civil and Environmental Engineering, 949 Benedum Hall, University of Pittsburgh, Pittsburgh, PA 15261.

[2]Assistant Professor, Department of Civil and Environmental Engineering, 949 Benedum Hall, University of Pittsburgh, Pittsburgh, PA 15261

capacity was attributed to polymerization of these organic compounds through oxidative coupling reactions catalyzed by the activated carbon surface. Figure 1 shows that the oxic adsorptive capacity of a bituminous coal based activated carbon for 2-methylphenol can be almost three-fold the capacity that can be obtained in the absence of oxygen (anoxic conditions). Figure 2 shows that more than 90% of the adsorbate loaded onto the GAC surface under anoxic conditions can be recovered by solvent extraction. On the other hand, only 5-20% of 2-methylphenol was recovered from the GAC surface using the same procedure when the adsorption was performed in the presence of oxygen.

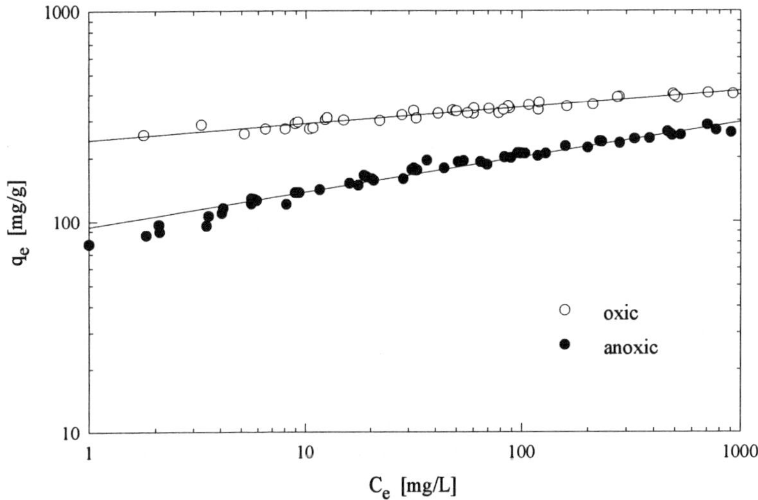

Figure 1. Adsorption Isotherms for 2-Methylphenol

Gas Chromatographic/Mass Spectroscopic (GC/MS) analyses revealed the presence of dimers, trimers, and even tetramers of 2-methylphenol in the extracts from the GAC used in the oxic adsorption isotherm experiments. This finding lead to the conclusion that polymerization of adsorbate under oxic conditions induces an increase in GAC surface loading and is responsible for incomplete recovery of the adsorbed compound. The amount of adsorbate still remaining on the carbon upon solvent extraction is therefore, denoted as irreversible adsorption.

In the past, irreversible adsorption of organics has been associated with the presence of oxygen containing functional groups on the carbon surface (Mattson *et al.*, 1969; Yonge *et al.*, 1985). Ishizaki *et al.* (1974) suggested that specific metals like copper and iron may lead to catalytic properties of activated carbon surface towards oxidation of mercaptans to disulfides. Lim et al. (1983) used cuprous chloride as a catalyst in a novel dephenolization scheme for coal gasification wastewater. Grant and King (1990) and Sorial *et al.* (1993) implicated the presence

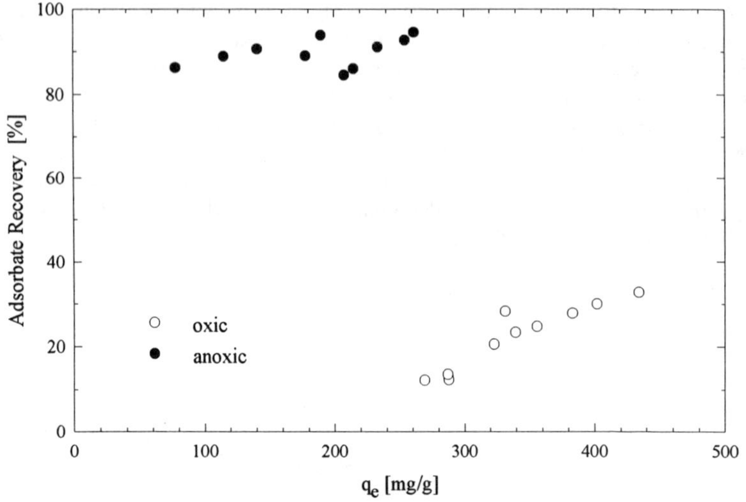

Figure 2. Recovery of 2-Methylphenol

of Mn as an important component of the catalytic properties of activated carbon towards polymerization of aromatics in the presence of oxygen.

The presence of oxygen during the adsorption of organic compounds leads to the formation of phenoxy radicals through the following reactions (Shibaeva, 1968):

$$PhOH + O_2 \rightarrow PhO^{\bullet} + HO_2^{\bullet}$$

In addition, the phenolate ion can also react with oxygen:

$$PhOH \leftrightarrow PhO^- + H^+$$

$$PhO^- + O_2 \rightarrow PhO^{\bullet} + O_2^-$$

$$O_2^- + H^+ \rightarrow HO_2^{\bullet}$$

The radicals formed from the above reactions can then react with other phenol molecules:

$$PhOH + HO_2^{\bullet} \rightarrow PhO^{\bullet} + H_2O_2$$

Hydrogen peroxide reacts with phenol molecules according to the following equations:

$$PhOH + H_2O_2 \rightarrow PhO^{\bullet} + H_2O + HO^{\bullet}$$

$$PhOH + HO^{\bullet} \rightarrow PhO^{\bullet} + H_2O$$

These phenoxy radicals can, depending on their resonance structure, react with other phenolic molecules or other phenoxy radicals to form different types of dimers, trimers, and tetramers of the parent molecule. The above reactions were demonstrated to take place at elevated temperatures (180-210 °C) and pressures (35 atm), thereby indicating high activation energies (Shibaeva, 1969). Joschek and

Miller (1966) were able to promote such reactions at lower temperatures and pressures, but in the presence of catalysts and steady irradiation, while Hay *et al.* (1959) reported oxidative coupling to take place at room temperatures and in the presence of copper (I) salt as a catalyst. Studies by Vidic *et al.* (1991, 1993) discovered that such reactions are also feasible at room temperatures with activated carbon surface acting as a catalyst.

The main goal of this study was to determine whether an overall parameter like ash content or a more specific measurements of certain metals on the GAC surface could explain the difference in the extent of irreversible adsorption of organic compounds observed for different types of activated carbons. In addition to ash and metal content of the carbon, dissolved oxygen (DO) uptake was utilized in explaining the catalytic properties of activated carbon surface.

Experimental Methods

Characteristics of activated carbons used in this study are outlined in Table 1. Letters B, W, and C denote bituminous coal based carbon, wood based carbon, and coconut based carbon, respectively. W2 is an extruded wood based carbon that incorporates some proprietary binder to facilitate production of 3.7 mm diameter pellets. C+Cu and C+Fe denote acid washed coconut based activated carbon loaded with CuCl and $FeCl_2$, respectively using the procedure similar to that used in the adsorption isotherm experiments. GAC particle size used in all experiments was 16x20 US Mesh size. Activated carbons were washed with deionized water to remove all the fines, dried at 105 °C for a period of 48 hours, and then stored in a desiccator until use. The model compound chosen for the study of the effect of GAC characteristics on oxidative coupling reactions was 2-methylphenol. All aqueous solutions were prepared using a deionized water buffered with 0.1 M phosphate buffer and the pH was adjusted to 7 by the addition of predetermined amount of 10 M NaOH solution. All adsorbate concentration measurements in the liquid phase was performed using a Perkin Elmer Lambda 2B UV/VIS spectrophotometer.

Two different experimental procedures for determining the adsorptive capacity of activated carbon, denoted henceforth as "oxic" and "anoxic", were utilized in this study. Details of both isotherm procedures are provided by Vidic and Suidan (1991) and will not be repeated here. The same reference also provides the description of the soxhlet extraction procedure utilized to recover adsorbates loaded onto the GAC surface in the presence and absence of molecular oxygen.

Acid washing was performed on the coconut based GAC to remove the metals and metal oxides imbedded in the graphite crystalline structure of activated carbon. Acid washing was conducted in the batch mode by successively contacting activated carbon with a 100 mL of fresh 2 N HCl solution. The iron content in the acid solution was used as an indicator of metal removal in each washing step. It was established that after 4 washes which took 17 days to complete, no additional Fe removal could be detected by flame atomic absorption spectrophotometer. The carbon was then thoroughly washed in distilled deionized water to remove any acid residual, dried at 105 °C for two days and stored in a desiccator until use.

Table 1. Activated Carbons Used in the Study

Carbon Type	Raw Material/ Treatment	Ash Content (%)	Metal Content (mg/kg)					
			Al	Ca	Cu	Fe	Mg	Mn
B1	Bituminous Coal	7.26	86	325	21	2984	101	5
B2	Bituminous Coal	7.80	166	665	29	6203	175	57
B2+	Acid Washed B2	5.37	15	7	10	699	26	72
W1	Wood	2.73	48	9822	53	140	838	603
W2	Wood + Binder	4.63	111	14063	23	614	1410	905
C	Coconut Based	0.80	6	105	13	182	365	9
C+	Acid Washed C1	0.24	2	10	DL	18	12	DL
C+Cu	C1+ with CuCl	--			2488			
C+Fe	C1+ with $FeCl_2$	--				1162		

DL = Detection Limit

Metal content of an activated carbon was determined by ashing 2 g of GAC in a muffler furnace at 550 °C for 24 hours. The residual ash was then dissolved in 25 mL of concentrated HCl and filtered through an acid-resistant, ash-free cellulose acetate filter (0.45-µm pore size) to retain the insoluble portion of the ash. The filtrate was then diluted with distilled deionized water and analyzed for metals using a Perkin Elmer 1100B atomic absorption spectrophotometer equipped for flame and graphite furnace analysis.

Measurements of oxygen uptake by activated carbons were performed according to the procedure described by Vidic and Suidan (1991).

Results and Discussion

Freundlich isotherm equation parameters describing the oxic and anoxic adsorptive capacity of nine different types of activated carbons together with the extraction efficiencies (% adsorbate recovered from the GAC surface) are shown in Table 2. The bituminous coal based carbons exhibited lower anoxic but higher oxic adsorptive capacity for 2-methylphenol than coconut and wood based carbons. Such behavior indicates that using bituminous coals as raw material for the production of activated carbon enhances the ability of that carbon to promote polymerization of organic compounds when compared to wood or coconut as raw materials. Adsorbate recovery by solvent extraction ranged from 80 - 100 % for all nine activated carbons when the loading process was performed in the absence of oxygen suggesting predominantly reversible adsorption under these conditions. On the other hand, extraction efficiencies achieved for the carbons loaded in the presence of oxygen ranged from 6-48%, indicating irreversible adsorption under oxic conditions. It is important to note that adsorbate recovery for bituminous coal based activated carbons was somewhat lower than those observed for the other adsorbents confirming the previously stated observation that bituminous coal based activated

Table 2. Freundlich Isotherms and Adsorbate Recovery

Carbon Type	Isotherm Type	K $(mg/l)(l/mg)^{1/n}$	1/n (--)	Adsorbate Recovery (%)
B1	oxic	282.2	0.059	6-26
	anoxic	79.0	0.189	84-93
B2	oxic	234.8	0.086	13-27
	anoxic	96.2	0.172	85-93
B2+	oxic	251.6	0.072	15-27
	anoxic	100.6	0.157	80-90
W1	oxic	77.1	0.097	41-48
	anoxic	76.4	0.109	81-91
W2	oxic	291.2	0.052	10-28
	anoxic	110.8	0.171	88-94
C	oxic	215.6	0.090	21-35
	anoxic	128.7	0.137	84-100
C+	oxic	225.8	0.085	15-40
	anoxic	128.4	0.148	80-92
C+Cu	oxic	209.0	0.137	18-29
	anoxic	115.0	0.176	80-90
C+Fe	oxic	221.0	0.103	20-31
	anoxic	118.5	0.174	83-88

carbons have more pronounced catalytic properties towards oxidative coupling of organic compounds.

Wood based carbon, W1, exhibited no measurable difference between the oxic and anoxic adsorptive capacity and yet only 41-48% of adsorbate was recovered from the carbon used in the oxic isotherm experiments while almost complete recovery was accomplished for the carbon used in the anoxic isotherm tests. Furthermore, a yellow tinge was observed in the oxic extracts indicating the presence of phenolic polymers. Such behavior indicates that even the wood based activated carbon exhibits some catalytic properties towards oxidative coupling of phenolics but to a such a small extent that no measurable effect on the exhibited adsorptive capacity can be observed.

This study further showed that 2-methylphenol can also undergo polymerization in the aqueous phase but in the presence of CuCl acting as a catalyst. A 20% reduction in the initial aqueous phase 2-methylphenol concentration of 111 mg/L was observed at room temperature over a period of 35 days in the presence of 54 mg/L CuCl, 16 mg/L dissolved oxygen, and at pH 5.3. The first indication of polymer formation was a color change from maroon to light brown and the formation of a brown precipitate. GC/MS analyses confirmed the existence of four dimers of o-cresol in the aqueous phase. However, the results of the isotherm experiments with acid washed coconut based activated carbon (carbon C+) indicated that the presence of Cu on the carbon surface is not a key factor in promoting increased loading under oxic conditions since this carbon contained almost no Cu (see Table 1) and yet exhibited identical oxic capacity to that of a virgin coconut based GAC (carbon C).

Irreversible adsorption, defined as the amount of adsorbate which can not be recovered from the GAC surface by solvent extraction, can be obtained from the data on the oxic GAC adsorptive capacity and adsorbate recovery. It was established that the irreversible adsorption remained constant over the entire range of adsorbate loading achieved in adsorption isotherm experiments. Relative irreversible adsorption is defined as the irreversible adsorption divided by the anoxic Freundlich isotherm coefficient K to facilitate comparison between activated carbons that exhibit different adsorptive capacities. Both absolute and relative irreversible adsorption are shown in Table 3 together with ash content, iron concentration, and oxygen uptake determined for the nine activated carbons used in this study.

The results shown in Table 3 indicate that activated carbons with a lower ash content (decreased presence of metals and metal oxides) exhibit lower irreversible adsorption. Such behavior is an indication that some of the metals that are imbedded in the graphite crystalline structure of activated carbon could be responsible for catalyzing oxidative coupling of organic compounds on the GAC surface. Furthermore, the iron content correlates quite well with the amount of adsorbate that is irreversibly adsorbed onto the GAC surface in the presence of oxygen suggesting that some of the iron containing compounds associated with activated carbon surface could be partially responsible for oxidative coupling of organic compounds.

Apart from iron there could be other metals on the GAC surface that can catalyze polymerization reaction as indicated by the fact that acid washing generally reduces the amount of irreversible adsorption. Furthermore, acid washed coconut based activated carbon loaded with Cu (C+Cu) promoted higher irreversible adsorption when compared to the acid washed carbon (C+) even though the ultimate loading under oxic conditions was identical for both carbons.

Oxygen uptake experiments were utilized to establish a comparison in the activated carbon surface activity as well as determine the extent to which oxygen

Table 3. Irreversible Adsorption

Carbon Type	Irreversible Adsorption [mg/g]	Relative Irreversible Adsorption	Ash Content [%]	Fe Content [mg/kg]	Oxygen Uptake [mg/g]
B1	287.6	3.64	7.3	2984	3.7
B2	276.7	2.88	7.8	6203	3.7
B2+	255.7	2.54	5.4	699	3.5
W1	72.65	0.95	2.7	140	0.6
W2	267.2	2.41	4.6	614	2.9
C	227.6	1.77	0.8	182	2.7
C+	190.6	1.48	0.2	18	NA
C+Cu	233.3	2.02	NA	18	NA
C+Fe	209.6	1.77	NA	1162	NA

may be available for oxidative coupling reactions. A good correlation exists between the activity of the carbon surface measured by O_2 uptake and the relative irreversible adsorption.

Acknowledgments

This research was funded by the University of Pittsburgh Central Research Development Fund. The conclusions expressed in this publication are solely those of the authors and do not necessarily reflect the views of the funding agency.

References

Grant, T.M. and King, C.J. (1990) "Mechanism of Irreversible Adsorption of Phenolic Compounds by Activated Carbon." *Ind. Eng. Chem. Res.*, 29, 264.

Hay, A.S., Blanchard, H.S., Endres, G.F., and Eustance, J.W. (1959) "Polymerization by Oxidative Coupling." *J. Am. Chem. Soc.*, 81:23, 6335.

Ishizaki, C and Cookson, J.T., Jr. (1974) "Influence of Surface Oxides on Adsorption and Catalysis with Activated Carbon," in Chemistry of Water Supply, Treatment, and Distribution, A.J. Rubin, Ed., Ann Arbor Science Publisher Inc.

Joschek, H.I. and Miller, S.I. (1966) "Photooxidation of Phenols, Cresols, and Dihydroxybenzenes." *J. Am. Chem. Soc.*, 88:14, 3273.

Lim, P.K., Cha, J.A., and Patel, C.P. (1983) "Aerobic Coupling of Aqueous Phenol Catalyzed by Cuprous Chloride: Basis of a Novel Dephenolization Scheme for Phenolic Wastewaters." *Ind. Eng. Chem. Process Design Division*, 22, 477.

Mattson, J.S., Mark, H.B., Malbin, M.D., Weber, W.J., Jr., and Crittenden, J.C. (1969) "Surface Chemistry of Active Carbon: Specific Adsorption of Phenols." *J. Colloid Interface Sci.*, 31, 116.

Puri, B.R. "Carbon Adsorption of Pure Compounds and Mixtures from Solution Phase." in Activated Carbon Adsorption of Organics from the Aqueous Phase, Suffet, I.H. and McGuire, M.J., Eds., Ann arbor Science, Ann Arbor, MI, 1980.

Shibaeva, L.V., Metelitsa, D.I. and Denisov, E.T. (1969) "Oxidation of Phenol with Molecular Oxygen in Aqueous Solutions 1. The Kinetic of the Oxidation of Phenol with Oxygen." *Kinetics and Catalysis*, 10:5, 832.

Sorial, G.A., Suidan, M.T., Vidic, R.D., Brenner, R.C. "Effect of GAC Characteristics on Adsorption of Organic Pollutants." *Water Environment Research*, 65:1, 53-75, 1993.

Vidic, R.D. and Suidan, M.T. (1991) "Role of Dissolved Oxygen on the Adsorptive Capacity of Activated Carbon for Synthetic and Natural Organic Matter." *Environ. Sci. Technol.*, 25:9, 1612.

Vidic, R.D., Suidan, M.T., and Brenner, R.C. "Oxidative Coupling of Phenols on Activated Carbon - Impact on Adsorption Equilibrium." *Environ. Sci. & Technol.*, 27:10, 2079-2085, 1993.

Yonge, D. R., Keinath, T. M., Poznanska, K., and Jiang, Z. P. (1985) "Single-Solute Irreversible Adsorption on Granular Activated Carbon." *Environ. Sci. Technol.*, 19, 690.

An Evaluation of the Impact of Recent Flooding on the Operation of a Groundwater Extraction and Treatment System at a Superfund Site

Kerry L. Gavett, Michael J. Fiore, and Eric J. Meyer[1]

Abstract

A groundwater extraction and treatment system was installed in 1987 at the Des Moines TCE Superfund Site. The purpose of the system is to prevent groundwater contaminated with chlorinated volatile organic compounds (VOCs) from migrating toward an infiltration gallery system which supplies drinking water to the City of Des Moines, Iowa. The extraction system was not operating for a three week period in July and August of 1993, when the system was flooded by the nearby Raccoon River. Data collected as part of a monitoring program have been evaluated to assess the affect of flooding on the operation of the system.

Records indicate that the flood did not have a long-term impact on the performance of the system. An examination of groundwater levels show that groundwater elevations receded quickly after the flood, similar to patterns observed after other periods of heavy precipitation. In fact, data collected nine weeks after the extraction system was returned to service indicate that the system continues to meet its containment objective. Water quality records indicate that the affect of the 1993 flood was similar to trends observed after earlier periods of heavy precipitation. Trichloroethene concentrations in the treatment system influent and in wells located in the vicinity of suspected source areas increased as a result of rising groundwater levels, and infiltration through residual contamination in the unsaturated zone. Groundwater quality in areas beyond suspected source areas does not appear to have been affected by the 1993 flood.

---

[1] ECKENFELDER INC., 1200 MacArthur Boulevard, Mahwah, New Jersey 07430.

Introduction

The purpose of this paper is to evaluate the impact of the 1993 flooding in the midwest on the operation of a groundwater extraction and treatment system at the Des Moines TCE Site, a Superfund site on the USEPA's National Priorities List. Specifically, this paper addresses the affect of flooding on the size and shape of the hydraulic capture zone, groundwater quality and treatment system influent quality.

Site Background

The City of Des Moines, Iowa and several nearby communities receive their drinking water from an infiltration gallery system that consists of a horizontal permeable conduit, over three miles long, buried in the alluvium of the Raccoon River. Groundwater from this system is treated at the Des Moines Water Works (DMWW) treatment facility. In 1975, the chlorinated VOCs trichloroethene (TCE), 1,2-dichloroethene (1,2-DCE), and vinyl chloride were detected in the groundwater. Initial studies indicated that contaminated groundwater was entering the northern portion of the gallery, located along a large meander of the Raccoon River. Groundwater contamination has been associated with several potential source areas located east of the Raccoon River. When the gallery was in operation, an elongated area of influence was created which locally reversed the natural gradient, causing contaminated groundwater to migrate from east to west beneath the river and enter the gallery system.

Seven extraction wells were installed along the eastern bank of the Raccoon River, opposite the northern gallery to prevent contaminated groundwater from migrating toward the infiltration gallery. The extraction system began to operate on December 17, 1987. Groundwater is pumped to an air stripping tower and treated effluent from the tower is routed underground to an outfall on the Raccoon River. One well was subsequently removed from operation in 1988, because of problems associated with iron encrustation. The hydraulic capture zone created by the extraction system is successfully preventing VOCs from migrating toward the infiltration system.

The extraction and treatment system was shut down for a three week period during July and August of 1993, when the Raccoon River breached the levee and inundated the site. Data collected as part of a monitoring program for the extraction and treatment system have been evaluated to assess the affect of the flooding on the operation of the system.

Effects of Flooding on Groundwater Elevations

Hydrographs have been prepared (Figure 1) to depict the change in the groundwater elevations using measurements made in three representative wells between December 1988 and 1993. Well NW-20 and NW-10 are located east and west, respectively, of the Raccoon River; well ERW-7 is an extraction well located east of the Raccoon River. Water level measurements between July and September 1993 are missing because of the disruption caused by the flooding of the Raccoon River during this period. However, the height of the groundwater table in August is estimated to have reached approximately 246.1 meters (m) near the Raccoon River based on the peak elevation of the river recorded at the DMWW. Total monthly precipitation recorded for Des Moines, as well as the change in stage of the Raccoon River at Van Meter, Iowa are plotted in Figure 1, for the period of record. As shown in Figure 1, similar fluctuations of groundwater elevations have been observed during the period of record in extraction wells and nearby monitoring wells located on both sides of the river. The fluctuations are related to the change in stage of the adjacent Raccoon River, which is in turn affected by the amount of precipitation.

As shown in Figure 1, the total annual precipitation in 1988 and 1989 was 56.1 and 73.9 centimeters (cm), respectively, which was lower than the 30-year average of 81.3 cm. For the years 1990 through 1993, the total precipitation for each year was higher than the annual average, especially in 1993, when total rainfall recorded in Des Moines was 142.0 cm.

As shown in the hydrographs, groundwater elevations were generally lowest in 1988 and 1989, corresponding to low precipitation during these years. Between January 1990 and December 1993, the groundwater elevations rose sharply three times. The first two peaks correspond to periods in which total monthly rainfall was greater than 19 cm for two consecutive months. The third peak occurred during the flooding in 1993, when the total monthly rainfall ranged between 19 and 31 cm for four consecutive months. During each of these periods of heavy precipitation, the groundwater table rose by as much as 3 to 10 m, but then dropped again within one to two months. Therefore, although the groundwater table was probably higher during the peak of the 1993 flood than any other time during the operation of the extraction system, the affect of the flood on groundwater elevations was similar to other periods of heavy precipitation in that groundwater elevations dropped quickly after precipitation subsided.

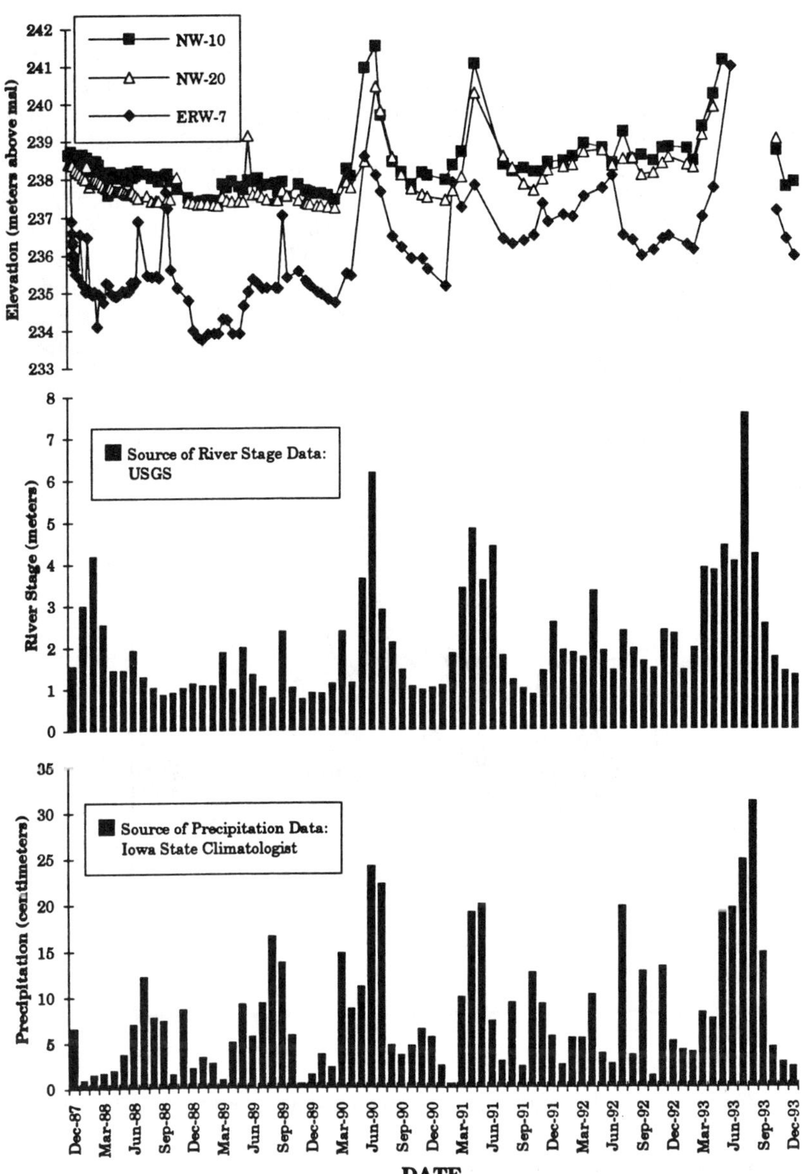

Figure 1. Hydrographs of selected wells, stage of Raccoon River at Van Meter, Iowa, and precipitation data for Des Moines, Iowa

The size and shape of the hydraulic capture zone were not affected over the long-term by the 1993 flood. A contour map was prepared (Figure 2) to show the configuration of the groundwater table on October 25, 1993, approximately nine weeks after the system resumed pumping. As shown in the figure, the extraction system is continuing to prevent contaminated groundwater from migrating beneath the river toward the infiltration gallery. An earlier study (Soukup and Huber, 1990) examined the ability of the system to achieve the required degree of containment during periods when the groundwater table is higher than normal. The results indicated that the overall shape and size of the capture zone did not vary greatly between wet and dry periods, although the absolute elevations varied by as much as 3 meters. Data collected after the 1993 flood confirm this.

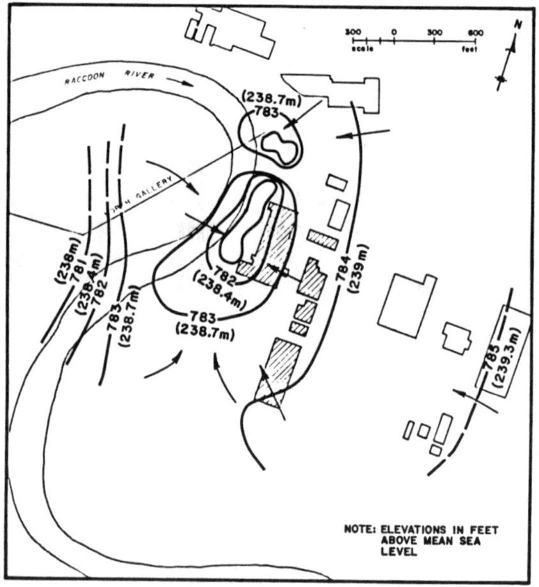

Figure 2. Configuration of groundwater table on October 25, 1993

<u>Effects of Flooding on Groundwater Quality</u>

Groundwater at the Des Moines TCE Site has been contaminated by chlorinated VOCs, principally TCE, which have densities greater than water, and therefore are categorized in their pure form as Dense Non

Aqueous Phase Liquids (DNAPLs). When a DNAPL is released, it migrates vertically downward through the unsaturated and saturated zones leaving residual liquid in the soil. If a sufficient volume is released, the DNAPL will continue to migrate downward until a relatively impermeable barrier, such as clay, is encountered. If the DNAPL is relatively immiscible in water, it may accumulate in pools at the top of the barrier. As groundwater migrates through the area of residual DNAPL and above the DNAPL pools, VOCs are dissolved into the groundwater and carried away.

The concentration of TCE in the treatment system influent is plotted in Figure 3 to illustrate the affect of flooding on groundwater quality. The plot shows that the concentration of TCE in the influent generally declined during the relatively dry period of 1988 and 1989. (The sharp increase in the concentration of TCE in April 1989 is not the result of an increase in the mass of TCE, but rather reflects a temporary decrease in the quantity of water from two extraction wells which add proportionately small TCE mass relative to the other extraction wells.) Since early 1990 the concentration of TCE has fluctuated, with increases in TCE corresponding to rises in groundwater elevations. The concentration of TCE most noticeably increased after the two sharp rises in groundwater elevations which were observed in 1990 and 1991, and after the flooding in 1993. Minor increases in TCE concentration were also observed in 1992 which correspond to smaller rises in groundwater elevations.

The increase in TCE concentration is probably the result of both the rising groundwater table coming into contact with residual DNAPL and also from infiltration through residual DNAPL in the unsaturated zone. This can be illustrated by examining water quality plots from individual wells. The concentrations of TCE are plotted in Figure 3 for wells ERW-7 and NW-7 located in the vicinity of a suspected source area. As shown in the plot, the concentration of TCE was generally declining asymptotically in both wells during the relatively dry period of 1988 and 1989. This pattern of decline is generally indicative of the presence of residual DNAPL in the saturated zone or pooled DNAPL continuously dissolving chemicals into the groundwater. Through time, the concentration of DNAPL dissolved in the groundwater declines, but does not reach zero, as long as residual DNAPL is present. Since early 1990, the concentration of TCE has fluctuated, and increases in TCE also appear to correspond to increases in precipitation and corresponding rises in groundwater elevations.

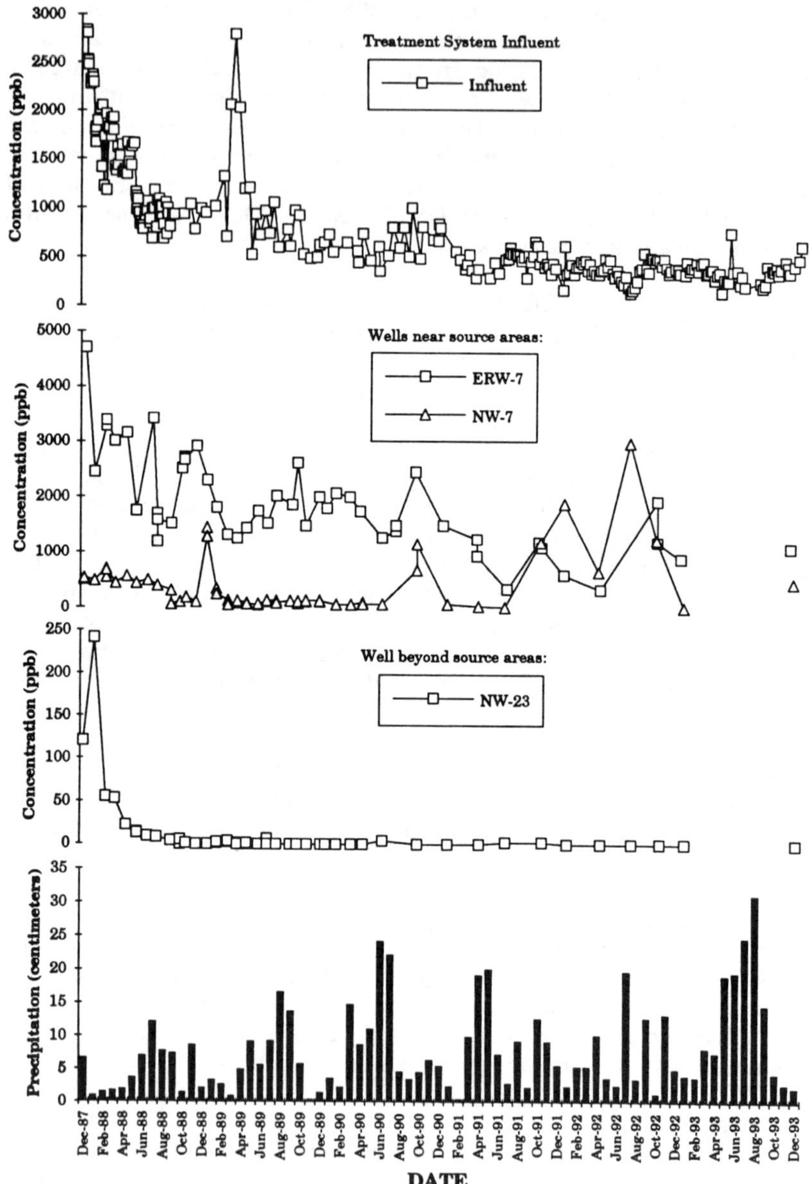

Figure 3. Concentrations of trichloroethene in treatment system influent and selected wells

The concentration of TCE in well NW-23, which is located beyond the source area, is also plotted in Figure 3. As shown in the plot, TCE in well NW-23 declined rapidly from the start of pumping until the end of 1989. Since that time, TCE has generally not been detected in the well, or has been present at extremely low levels (e.g. less than 10 parts per billion (ppb)). Increases in precipitation and corresponding rises in groundwater elevations have not affected water quality in well NW-23. This pattern is indicative of the decline in contaminant concentration during clean-up where only dissolved-phase material is present.

In summary, the concentration of TCE increases in wells in the vicinity of suspected source areas and therefore in the treatment system influent, when precipitation increases and groundwater elevations rise. A similar pattern was observed after each of the three periods when the groundwater table rose sharply between 1990 and 1993. Therefore, the affect of the 1993 flood on groundwater quality was similar to trends observed after earlier periods of heavy precipitation.

Conclusions

Based on an examination of data collected over a six year period, the 1993 flood did not have a long-term impact on the size and shape of the hydraulic capture zone of the extraction system. Data collected nine weeks after the extraction system was turned back on indicate that the system continues to prevent groundwater containing VOCs from migrating toward the infiltration gallery. An examination of groundwater levels indicated that groundwater elevations dropped quickly after the flood, similar to patterns observed after other periods of heavy precipitation. An examination of groundwater quality indicated that the affect of the 1993 flood was similar to trends observed after earlier periods of heavy precipitation. The concentration of TCE increased in the treatment system influent and in wells located in the vicinity of suspected source areas as a result of the rising groundwater table coming in contact with residual DNAPL and from infiltration through residual DNAPL in the unsaturated zone. Groundwater quality in areas beyond suspected source areas was not affected by the 1993 flood.

References

Soukup, W.G., and M.J., Huber, 1990. "Case Study: A Post-Audit of Actual Versus Predicted Groundwater Recovery System Performance", Proceedings of the Petroleum Hydrocarbon and Organic Chemicals in Groundwater Conference, NWWA, Houston, Texas.

PREDICTING THE IMPACT FROM SIGNIFICANT STORM EVENTS ON A HAZARDOUS WASTE SITE

By Udai P. Singh[1], M.ASCE, Neal P. Dixon[2], M.ASCE, and J. Stephen Mitchell[3]

ABSTRACT: The Stringfellow Hazardous Waste Site is a former Class I industrial waste disposal facility located near the community of Glen Avon in southern California. In response to community concerns regarding flooding and possible exposure to contaminants via the surface water pathway, a study was performed to evaluate the potential effect significant/episodic storm events may have on the site and its engineered structures as they exist during present day conditions. Specific storm events such as significant recorded historic storms as well as synthetic design storms were considered and the impact on the onsite area and surface channels in Pyrite Canyon downstream of the site was evaluated. Conclusions were reached, and recommendations were made to minimize the potential flood impacts and exposure to contaminants via the surface water pathway in the areas downstream of the site.

INTRODUCTION

The Stringfellow Hazardous Waste Site is a former Class I industrial waste disposal site located approximately 50 miles (80 km) east of Los Angeles in southern California. The disposal facility was sited at the northern end of Pyrite Canyon on the southern slopes of the Jurupa Mountains in Riverside County. Pyrite Canyon is 5 miles (8 km) northwest of the City of Riverside near the community of Glen Avon. For purposes of organizing information, the site, including its contaminated plume of groundwater, has been divided into four geographic zones. Zone 1 includes the original 17-acre (7 ha) disposal area in the northern part of Pyrite Canyon, southward to approximately 600 feet (183 m) downgradient of the subsurface barrier. Zone 2 encompasses the portion of Pyrite Canyon that extends from the southern edge of Zone 1 to the existing mid-canyon extraction wells. Zone 3 extends from the mid-canyon extraction wells down to the lower canyon extraction system north of U.S Highway 60. Zone 4 includes the area south of Highway 60 to the leading edge of the plume of site-related contaminated groundwater, approximately 12,000 feet (3,660 m) from Zone 1.

---

[1]Senior Environmental Engineer, CH2M HILL, 1111 Broadway, Suite 1200, Oakland, CA 94607
[2]Senior Water Resources Engineer, CH2M HILL, 2525 Airpark Drive, Redding, CA 96001
[3]Hydrologist, CH2M HILL, 7 West 6 Avenue, Suite 614, Helena, MT 59601

During the operation of the site from 1956 to 1972, approximately 34 million gallons (129 million L) of industrial wastes primarily from metal finishing, electroplating, and DDT production) were placed in unlined evaporation ponds located throughout the 17-acre (7 ha) disposal area. Release of onsite wastes to surface water downstream occurred from heavy rains in 1969 and 1978. From 1980 through 1982, the following interim abatement actions were initiated: (1) the removal of onsite surface liquids and some contaminated soil (Zone 1); (2) construction of a subsurface clay-core barrier on the downgradient end of the 17-acre site (Zone 1); (3) grading and installation of a surface cap/cover onsite Zone 1; (4) construction of a surface water diversion system around the 17-acre site (Zone 1); and (5) development of several onsite and downgradient extraction wells (Zones 1 and 2).

The site was added to EPA's Superfund National Priorities List (NPL) in 1983 as California's highest priority toxic waste site. Recent (1983 to 1990) remedial actions include: (1) construction of a french drain system immediately downgradient of the subsurface barrier; (2) expansion of the groundwater extraction system in the mid-canyon (Zone 2) and lower canyon (Zone 3); (3) construction and operation of a pretreatment plant near the site in Pyrite Canyon for treating extracted groundwater from onsite, mid-canyon, and lower canyon areas (Ullensvang and Singh, 1990); and (4) completion of the surface water diversion channels around the onsite area and reconstruction (rechannelization) of Pyrite Creek downstream of the onsite area to the box culvert under Highway 60. During the 1990s, the following remedial actions are in the process of being implemented: (1) installation of a groundwater extraction and treatment system in the contaminant plume area of the Glen Avon community (Zone 4); and (2) installation of a dewatering system in the onsite areas (Zone 1).

In response to community concerns regarding flooding and possible exposure to contaminants via the surface water pathway, a study was initiated in 1990 to evaluate the potential effect significant/episodic storm events may have on the site and its engineered structures as they exist during present day conditions (1990). Specific storm events such as significant historic storms and synthetic design storms, and the impact on the onsite area and surface channels in Pyrite Canyon down stream of the site were evaluated. The potential damage that was considered based on specific storm events analyses includes overtopping of channels, resulting in flow across and erosion of capped areas and engineered structures, and erosion of the onsite capped area because of ponding and overtopping of the subsurface barrier. Impacts only in the Pyrite Canyon north of Highway 60 were evaluated, as Zone 4 was not a part of this study (CH2M HILL, 1993).

DEFINITION OF EVENTS

The Stringfellow Hazardous Waste Site is located in Pyrite Canyon (Figure 1), a small box canyon with a drainage area at the Riverside Freeway of approximately 1.2 square miles (3.1 sq. km). Elevations on the watershed vary from approximately 865 feet (264 m) above mean sea level (msl) at the freeway to approximately 2,225 feet (678 m) msl at the highest point on the divide. The main water course extends about 1.1 miles (1.8 km) from the divide to the freeway with an average slope of 4.5 percent. The upper drainage is steep with overland slopes of up to 50 percent. As a result, the upper channels respond quickly to intense rainfall. Measurement of rainfall and runoff in Pyrite Canyon was started in November 1986 and is continuing to this time (CH2M HILL, June 1989).

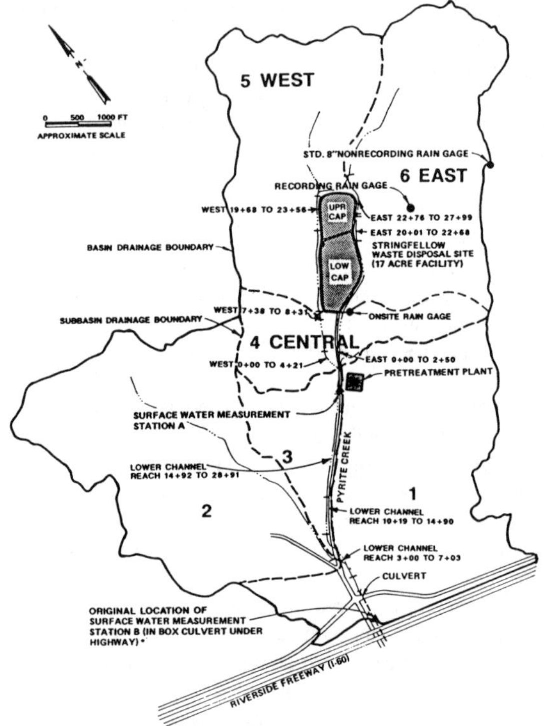

* Moved approx. 200 yds. upstream in November 1989.

Figure 1. Pyrite Canyon Showing Stringfellow Site, Subbasin Drainage Boundaries, and Surface Channel Reaches

Onsite precipitation gages were established in 1985-86 and 1986-87, respectively, by CDM and CH2M HILL. The precipitation data from these gages were compared with longer term records from nearby stations (RCFCD office gage, Mira Loma gage, and Glen Avon gage). Table 1 shows the long-term to onsite precipitation ratios by water year for each of these gages. The RCFCD office gage (Weather Station 2S/5W-14P01), approximately 3 miles (5 km) east of the site, was selected as being most representative of onsite precipitation. The RCFCD gage record covers the period from 1882 to the present. Records for 1882 through 1896 are total annual precipitation only. Daily rainfall records are available for the remainder of the record period. In addition, starting with a storm in February 1963, the RCFCD has 5-minute incremental precipitation records for over 100 major precipitation events through 1989.

Table 1
Ratio of Data from Nearby Precipitation Gages to Onsite Gage Records
(1985-86 Through 1987-88)

Water Year	CDM Onsite Rain Gage			CH2M HILL Recording Rain Gage		
	RCFCD	Mira Loma	Glen Avon	RCFCD	Mira Loma	Glen Avon
1985-86	0.94	1.25	--	--	--	--
1986-87	1.22	1.08	1.23	1.20	1.19	1.29
1987-88	--	1.21	--	--	1.19	--

For these reasons, the historic records from the RCFCD rain gage were used in this evaluation. During the operational and post-operational period of the Stringfellow facility, three major storm-related discharges to Pyrite Creek occurred in 1969, 1978, and 1983. Historic storms that likely caused surface water to leave the onsite area were identified as follows:

- Storm event beginning on February 23, 1969: total precipitation 5.52 inches (140.2 mm); duration 59 hours.
- Storm event beginning on February 28, 1978: total precipitation 2.98 inches (75.69 mm); duration 52 hours.
- Storm event beginning on March 2, 1983: total precipitation 2.24 inches (56.90 mm); duration 42.5 hours.

Actual 5-minute incremental precipitation records for each of these storms were used in the evaluation to determine runoff for present conditions.

In addition, design storms from the Riverside County Hydrology Manual (Riverside County, April 1978) were used in this evaluation to assess the capability of existing surface drainage facilities to handle the design storms. The design storms are identified as follows:

- 100-year, 1-hour precipitation = 1.30 inches (33.02 mm)
- 100-year, 3-hour precipitation = 2.20 inches (55.88 mm)
- 100-year, 6-hour precipitation = 3.20 inches (81.28 mm)
- 100-year, 24-hour precipitation = 6.0 inches (152.4 mm)

EVALUATION APPROACH

The U.S. Army Corps of Engineers' computer program HEC-1 was used to calculate runoff from Pyrite Canyon for the present upgraded drainage system for historic and design storms. The HEC-1 program is designed to simulate the surface runoff response of a drainage basin to precipitation (Corps of Engineers, 1987). Peak flows at selected locations in the basin were determined for the seven storms defined above, using an updated HEC-1 computer model changed from previous work to reflect existing site conditions. The highest peak flows from the seven storms were then used to determine if the present improved channels are capable of passing these flows. Manning's formula was applied and the current channels' geometry, slope, shape, and roughness coefficients were used in this hydraulic evaluation to determine potential overflows.

CONSTRUCTION OF THE HEC-1 MODEL

Rainfall and associated runoff events in the Stringfellow watershed have been recorded since November 1986. Precipitation and streamflow gaging stations were established at locations shown on the subbasin drainage boundary map (Figure 1).

A modeling study was conducted in 1989 to develop a HEC-1 model to represent the rainfall-runoff relationship for the Stringfellow watershed. The model was constructed to represent existing (1986-1989) conditions. Recorded precipitation and streamflow data from the onsite gages were used to calibrate the model. The calibration process involved inputting data describing a specific storm event, followed by selecting model parameters for the watershed representation so that the calculated hydrographs closely approximate the recorded hydrographs from the streamflow gage record. Using this calibrated model, design storms of various return frequencies were input to the model and the resultant runoff was calculated. Design storms used were for return periods of 2, 5, 10, 25, 50, and 100 years, and 15-minute, 60-minute, and 24-hour durations. The results of this study were published in a July 1989 report (CH2M HILL, July 1989).

For this analysis, the HEC-1 model from the 1989 study was modified to reflect existing conditions.

HYDROLOGIC (HEC-1) MODELING RESULTS

Using the updated HEC-1 model, historic recorded precipitation for the storms beginning February 23, 1969, February 28, 1978, and March 2, 1983, were input and runoff from each of these storms was calculated. The resultant runoff values for selected locations on the watershed are shown in Table 2.

Table 2
Calculated Runoff from Historic Storms Using New Site Conditions*

Storm Date	Peak Flow (cfs) at Node Location						
	6-East	5-West	Uprcap	Lowcap	4-Cntrl	Gage A	Gage B
2/23/69	67	56	3	4	27	157	284
2/28/78	28	24	2	3	11	67	84
3/2/83	80	63	6	10	35	194	224

*Note: 1 cfs = 0.028 m^3/sec

The updated HEC-1 model was also used to calculate potential runoff from the RCFCD design storms identified above. The time distribution of rainfall was taken directly from the previous rainfall/runoff simulation study. The results of this analysis are summarized for selected points in Table 3.

These calculations were based on clear water flows. In reviewing the Riverside County Flood Control District Hydrology Manual (1978) and from visual observations at the site, it is evident that the effects of sediment or debris flow should also be considered. The above HEC-1 results do not include debris flow and therefore reflect only the amount of clear water runoff from all storms.

Table 3
Runoff from Design Storms Using New Site Conditions*

Storm Duration	Frequency	Peak Flow (cfs) at Node Location 6-East	5-West	Uprcap	Lowcap	4-Cntrl	Gage A	Gage B
100-Year	1 Hour	186	147	21	34	97	438	512
100-Year	3-Hour	215	171	23	36	111	512	629
100-Year	6-Hour	264	209	24	37	134	622	811
100-Year	24-Hour	356	279	25	39	176	824	1,154

*Note: 1 cfs = 0.028 m^3/sec

HYDRAULIC EVALUATION

Surface Channel Capacity: The capability of the improved drainage channels to accommodate flows from the runoff event producing highest flows (100-year design flow) was assessed by applying Manning's formula to capacity limiting reaches, as represented in the construction documents for channel improvements. Results of the analysis for clear water hydraulics are shown in Table 4. The channel reaches listed in this table are shown on Figure 1.

Table 4
Surface Channel Hydraulic Capacity*
(Clear Water Analysis)

Channel	Reach	Minimum Channel Depth (ft)	Flow (cfs)	Normal Depth (ft)	Freeboard (ft)
Lower	3+00 to 7+03	6	1,154	2.7	3.3
Lower	10+19 to 14+90	5	824	3.5	1.5
Lower	14+92 to 28+91	5	824	3.6	1.4
West	0+00 to 4+21	4	400	2.5	1.5
West	7+38 to 8+31	4	294	2.7	1.3
West	19+68 to 23+56	3	250	2.2	0.8
East	0+00 to 2+50	4	400	2.3	1.7
East	20+01 to 22+68	4	356	2.2	1.8
East	22+76 to 27+99	3	267	2.1	0.9

*Note: 1 ft = 0.3 m; 1 cfs = 0.028 m^3/sec

A cursory analysis of the debris flow case was also conducted. Flows used in this analysis were derived by assuming the 100-year runoff event for the Stringfellow watershed to carry 120,000 cubic yards per square mile (35,400 m^3 per square km) of debris from the watershed. This debris volume used is somewhat arbitrary. It appears, however, that the Stringfellow watershed with shallow permeable soils, minimal vegetative cover, and steep slopes (35 percent average, ranging up to 50 percent or more) has a relatively high potential for debris flows. Methodologies for predicting volume and timing of debris flows are not well established at this time.

Dividing the 120,000-cubic-yard debris volume by the total volume of the clear water hydrograph shows that the debris flow could amount to as much as 80 percent of the clear water runoff. In the debris flow analysis, therefore, peak clear water flow rates at each location were multiplied by a factor of 1.8. No effort was made to take into account possible effects of the sediment load on the viscosity of the flowing water. The results of this analysis also are shown in Table 5.

Table 5
Surface Channel Hydraulic Capacity*
(Debris Flow Analysis)

Channel	Reach	Minimum Channel Depth (ft)	Flow (cfs)	Normal Depth (ft)	Freeboard (ft)
Lower	3+00 to 7+03	6	2,080	3.8	2.2
Lower	10+19 to 14+90	5	1,480	4.7	0.3
Lower	14+92 to 28+91	5	1,480	4.9	0.1
West	0+00 to 4+21	4	720	3.3	0.7
West	7+38 to 8+31	4	530	3.6	0.4
West	19+68 to 23+56	3	450	2.8	0.2
East	0+00 to 2+50	4	720	3.1	0.9
East	20+01 to 22+68	4	640	2.9	1.1
East	22+76 to 27+99	3	540	2.8	0.2

*Note: 1 ft = 0.3 m; 1 cfs = 0.028 m^3/sec

In several cases the channel is virtually bank full. If viscosity effects were considered, and if any sediment deposition were to occur or if significant wave action were to be generated, overflow would probably occur. The potential for damage from debris flows is mitigated in some jurisdictions by the construction of debris basins. Debris basins, if constructed at inlets to the east and west channels near the onsite area and to Pyrite Creek in the lower canyon area, should be able to mitigate the damage potential under these rare flow conditions.

There is small potential for overtopping in the upper reaches of the system. If overtopping were to occur, water would flow to the culverts and back to the west channel. Overtopping, if it were to occur, could also cause water to flow across the Stream "A" storage tank system location and could conceivably result in some local erosion.

Flow Measuring Structures: Through March 1991 there has been one temporary and one permanent flow-measuring weir in the main channel downstream from the confluence of the east and west channels. Because of new construction in the channels, the old streamflow measurement Station B was taken out from the box culvert under the Riverside Freeway and moved approximately 200 yards (183 m) upstream from the old station for temporary use only. The permanent flow-measuring weir in Pyrite Creek was installed in 1990 at approximately the same location as Station A. These structures represent significant restrictions to flow in the channel. In these steep, lined channels, all unobstructed flow is supercritical, characterized by relatively shallow depth of flow and high velocities. Hydraulic analysis of data provided for review indicates that the permanent structure will cause an upstream hydraulic jump at flows up to approximately 400 cfs (11 m^3/sec). At higher flows, the depth of flow over the weir is not great enough to support an upstream hydraulic jump. Theoretical water depths are below the top of bank. However, because flows in the Froude range experienced here (Froude No. between 2 and 3) are inherently unstable, considerable wave action and turbulence should be expected. Under higher flows, some overtopping by waves and/or spray may occur.

In the case of the upstream (permanent) weir, it appears that the overflow will be directed mostly over the right bank, flow through the quarry yard parallel to the channel, then re-enter the channel in the vicinity of station 7+00 (assuming flow is not backed up by the downstream temporary weir). The overflow would not cross the capped area, and damage from it would probably not be significant.

At the location of the downstream (temporary) weir, however, it appears that overflow would occur mostly on Pyrite Street. A significant portion of this overflow would be captured by the storm inlets and re-enter the channel in the box culvert under the freeway, while the remainder would find its way along Pyrite Street under the freeway.

An analysis of ponding on the capped areas from obstruction of the pipe drains was conducted using HEC-1. Hydrographs for each capped area were calculated and routed through the pipe drains. Two separate runs were made for the capped areas: one assuming a completely open pipe, and one assuming a completely blocked pipe. Based on these calculations, it appears that if the pipes are not blocked, overflow will not occur. If the pipes are completely blocked, the inlet basin to the pipe drain serving the upper capped area will not overflow. The lower capped area drain inlet, however, would pond, and then overflow into the roadway area, eventually reentering the channel in the vicinity of the confluence. The peak overflow rate would be about the same as the peak rate of runoff from the lower capped area (routing effects of the lower intake basin would be minimal). An overflow depth of about 0.3 foot (0.1 m) is estimated.

If this flow were to continue as sheet flow, it would have the capability of moving sediment particles about as large as 0.25 inch (6 mm). Any concentration of flow would increase its carrying power. The cap material is a moderately fine-grained, non-cohesive material and would probably be subject to erosion under sheet flow conditions. If allowed to continue long enough, the sheet erosion would probably lead to rill, and eventually to gully erosion.

Reasonable attention to maintenance of the pipe-drain inlet, and repair of damaged areas if overtopping were to occur, should prevent the cap from eroding and exposing the underlying material.

CONCLUSIONS AND RECOMMENDATIONS

- Out of the seven storms (three historic and four design) used in this study, the one that will produce the most runoff and the largest peak flows is the 100-year, 24-hour storm.

- The surface drainage system, as designed and constructed, will pass the clear water flow, as calculated at the various locations shown on Figure 1 for the 100-year, 24-hour flood. Two possible exceptions are the channel reaches just upstream from the weirs, which could overflow under these rare flow conditions.
- The downstream (temporary) weir should be removed to eliminate the overtopping hazard at this location. (Based on this recommendation and due to silting problems, the weir was removed in March 1991.)
- The pipe drains carrying flow from the onsite capped areas have sufficient capacity to carry runoff from these areas if kept clear. The ongoing program of observation and maintenance should be continued and emphasized to prevent runoff as a result of clogged drains.
- Diking around stockpiles of sand and gravel in the downstream areas and/or covering the piles would help reduce the sediment load in the lower channel area. Again, the current program of removing debris and silt from channels after major storms should be continued and emphasized.
- Consideration should be given to construction of debris basins at inlets to east and west channels near the onsite area and to Pyrite Creek in the lower canyon area to mitigate the damage potential from debris flows under these rare flow conditions.

ACKNOWLEDGEMENT

This study was funded by the U.S. Environmental Protection Agency, Region IX. Karen Ueno and Dante Rodriguez were the EPA project managers.

APPENDIX 1.—REFERENCES

CH2M HILL (June 1989). "Stringfellow Hazardous Waste Site, Riverside County, California. 1986-89 Rainfall/Runoff Monitoring Program, Summary Report." Prepared for U.S. EPA.

CH2M HILL (July 1989). "Stringfellow Hazardous Waste Site, Riverside County, California. Rainfall/Runoff Simulation Final Report." Prepared for U.S. EPA.

CH2M HILL (April 1993). "Stringfellow Hazardous Waste Site, Riverside County, California. Significant Storm/Seismic Events Impact Evaluations." Prepared for U.S. EPA.

Corps of Engineers (1987). "HEC-1 Flood Hydrograph Package-Users Manual." U.S. Army Corp of Engineers, Water Resources Center, The Hydrologic Engineering Center. Davis, California. Revised March 1987.

Hydrology Manual (1978). Riverside County Flood Control District, Riverside County, California.

Ullensvang, B.J. and U.P. Singh (1990). "Treatment and Discharge to a POTW: The Stringfellow Experience." *Water Environment and Technology*, Vol. 2, No. 1.

Environmental Effects of Flooding at a USCG Base

Virginia L. Bretzke, P.E., Member, ASCE[1]
Michael J. Donnelly[2]

Abstract
In July of 1993, the U.S. Coast Guard (USCG) Base St. Louis facility was flooded. As the flood waters receded and cleanup efforts were initiated, a number of environmental issues had to be addressed. This paper presents some of the environmental and health concerns that were identified and the measures taken to evaluate and clean up potential hazards.

Introduction

The USCG maintains Base St. Louis in an industrial area south of downtown St. Louis, Missouri, directly adjacent to the Mississippi River. The facility, which was primarily constructed in the 1940's, services the mid- and upper-Mississippi valley area by assisting with navigation of freight and other related services. The buildings at the facility include a boat storage facility, a maintenance shop (industrial building), administrative offices, a barracks, an officers club, and a boat dock. The facility covers approximately one acre.

As the Mississippi River levels began to rise in July of 1993, the USCG personnel attempted to keep the facility operational by sandbagging along the river and pumping water that accumulated on the facility side of the sandbags. As the river continued to rise, the USCG moved critical equipment offsite and turned off utilities. The site was evacuated, and USCG operations were moved to temporary facilities elsewhere. Eventually, the river rose to a level where it

1 Civil Engineer, BLACK & VEATCH Waste Science, Inc., 4717 Grand Avenue, Suite 500, Kansas City, Missouri 64112.

2 Geologist, BLACK & VEATCH Waste Science, Inc., 1415 Elbridge Payne Road, Suite 200, Chesterfield, Missouri 63017.

could no longer be controlled, and the site was flooded. Water levels at the site ranged from 0.3 to 2.3 meters (1 to 7 feet) deep.

The USCG realized that an expeditious cleanup would be critical to getting the facility operational after the flood waters receded. The site restoration activities were coordinated by the USCG Civil Engineering Unit in Cleveland, Ohio. The USCG Emergency Response Team from New Orleans performed much of the general cleanup, demolition, and emergency utility restoration activities onsite. The USCG also retained the engineering services of Black and Veatch and its subsidiary, BLACK & VEATCH Waste Science, Inc., to assist with the restoration efforts. This paper describes the various aspects of environmental concern that were evaluated and implemented by the USCG and the Black & Veatch team during the flood cleanup.

Regulatory Guidance

While the site was still flooded, the USCG began obtaining direction from the Missouri Department of Natural Resources (MDNR) regarding the disposal of material left onsite from the flood. The verbal direction received from the MDNR included the following:

- Mud, silt, sand, or other material from the river could be returned to the river if there was no obvious contamination.
- Potentially contaminated materials were to be collected, tested, and properly disposed of.
- Organic materials, such as logs, limbs, vegetation, and paper, could be disposed of in solid waste landfills.
- Initial flushing of the site could be conducted using river water.

The MDNR later issued some information sheets related to handling the wastes generated from the flood. Special procedures were established for the disposal of hazardous wastes by businesses that do not normally generate hazardous wastes. Because testing indicated that none of the wastes at Base St. Louis were hazardous, these procedures did not need to be implemented. However, this demonstrates the importance of keeping in touch with regulatory agencies when responding to flooding or other emergencies to be aware of any special provisions or guidance that they may be issuing.

Inherent Flood Hazards

Some of the initial areas of environmental, health, and safety concern addressed were those that are inherently associated with flooding. The following is a list of some of these hazards and measures that can be taken to minimize them:

- Biological hazards exist from food products, dead animals, sewage, and other organic wastes from the flood. Cleanup personnel should be properly immunized, especially for tetanus. Good personal hygiene, such as washing promptly with soap and clean water and laundering clothes in hot water and detergent after each wearing can help minimize the biological hazards. Food that was in contact with flood water should be properly disposed of. All materials that were in contact with flood waters should be thoroughly cleaned and disinfected.

- Because of the potential danger of electrocution and explosions, electric and gas utilities should be turned off until they can be properly restored.
- Potable water lines may be contaminated by flood water and should not be used until the water source is determined to be clean and the lines are flushed.
- The structural integrity of a building should be verified before entering the building. Water damaged walls and ceilings can suddenly fall. Basements should be drained slowly and carefully over a period of days to prevent structural damage caused by the pressure of residual high groundwater.
- Domestic and wild animals are displaced during flooding and should be approached with caution.
- Mosquito and other insect populations will rapidly increase in the wet conditions following flooding. Insecticides and insect repellents can be used, and standing water should be drained to eliminate insect breeding areas.

Personnel involved in the cleanup activities at USCG Base St. Louis were advised of these types of hazards. The site was covered with 5 to 25 centimeters (2 to 10 inches) of silt, but there was minimal organic wastes such as food or dead animals. The utilities had been turned off before the site was evacuated, and the utilities were properly restored as part of the site cleanup efforts. Water receptacles, such as sinks and drinking fountains, were marked to indicate that the water was unsafe to drink until services were restored. Buildings were inspected for safety before the cleanup workers were allowed to enter them. The cleanup included the removal of walls, flooring, and other building materials damaged by the flood.

Asbestos and Lead Paint Survey

Because of the age of the buildings onsite, the Black & Veatch team conducted a survey to determine whether there was asbestos or lead paint present in the buildings that would require special handling during the cleanup. Flooding can also make non-friable or encapsulated asbestos material friable, which could require removal in cases where it was not required previously. Thirty-two samples of building materials were collected for asbestos analysis, and 10 paint samples were collected for lead analysis.

The asbestos survey indicated that asbestos/cement wallboard was used in the industrial building, vinyl asbestos floor tiles were used in the industrial building and barracks, and some of the pipe and hot water tank insulation in the industrial building and barracks contained asbestos. Therefore, these materials were not disturbed during the cleanup effort.

Lead paint was determined to be present in only two locations--in the primer on the structural steel in the boat storage building and an exterior paint on a concrete masonry wall at the industrial building. The initial cleanup efforts did not involve disturbing these painted areas, but future restoration efforts may require that proper procedures be used for removing and disposing of the lead paint.

PCB Testing of Electrical Equipment Oils

During the restoration of the electrical utilities onsite, a transformer and a switch were identified as pieces of equipment that could contain polychlorinated biphenyl (PCB) oils. The Black & Veatch team collected one oil sample from the transformer and one wipe sample from the switch and had these samples analyzed for the presence of PCBs. The oil sample from the transformer was also analyzed for general oil quality.

The transformer oil contained 48 parts per million PCBs (Aroclor 1260) and, therefore, was classified as a non-PCB transformer (as defined in 40 CFR 761). This meant that the transformer could continue to be used and serviced without restrictions. Disposal of the drained transformer carcass is not restricted, but the use of the PCB containing dielectric fluid is restricted.

The liquid inside the switch was determined to be water, not oil (probably flood water). The wipe sample from the switch indicated that the PCB concentration was less than 2 micrograms on the 100-square centimeter wipe sample, so no special disposal procedures were required.

Environmental Cleanup Activities

As a part of the initial cleanup of the facility, cleanup activities were conducted for materials that were potentially contaminated due to the flooding at the site. These activities included collecting the potentially contaminated material, sampling and analyzing the material to determine how it should be disposed of, and properly disposing of the material. The Black & Veatch team procured and oversaw the work of a subcontractor, who performed the waste collection and disposal activities. The team collected samples of waste for analysis and evaluated the sampling results to identify the proper disposal method. Most of the samples were analyzed for toxicity characteristic leaching procedures (TCLP); total petroleum hydrocarbons (TPH); and benzene, ethylbenzene, toluene, and xylene (BETX) concentrations. A brief description of each type of material handled during the cleanup efforts follows.

Waste Motor Oil Spill. During the flooding, a 132-liter (500-gallon) storage tank containing approximately 50 liters (200 gallons) of waste motor oil spilled near the northeast corner of the site, adjacent to the dock. The directions provided by the MDNR regarding flood cleanup indicated that silt and mud with visible contamination should be collected and properly disposed of, as compared with uncontaminated silt and mud, which could be returned to the river. The waste motor oil was visibly evident as small droplets of brown colored fluid with diameters of approximately 1 to 5 millimeters. The oil was spread throughout the silt and could not be skimmed or contained in any way other than the gross containment and removal of the affected silt.

Much of the waste motor oil at the site was in the silt in a corridor approximately 1.3 meters (4 feet) wide between a building and a fence line. Other debris, such as sandbags and lumber, had also accumulated in this area. Efforts were made to segregate potentially contaminated material from uncontaminated material to minimize the amount of material disposed of as special wastes. The covers of the sandbags were cut off and disposed of as either construction debris or special wastes, depending on whether there was oil on the bag. The clean sand within the sandbags was then spread on the ground after the silt had been removed. The lumber was washed and disposed of as construction debris. The decontamination water from washing the lumber was collected.

One of the factors affecting the cleanup effort was difficulties with materials handling of the saturated silt material. If the silt was exposed to the sunlight, it dried out quickly. However, inside the boat storage shed, the silt was never exposed to sunlight and did not appear to be drying. Because of the high moisture content, this material could not pass the paint filter test required for landfill disposal. After evaluating various options for drying the silt in the boat shed, it was decided to mix agricultural lime with the silt to solidify it.

A sample of the soil and silt with the waste oil was collected and analyzed, and it was determined that this material could be disposed of as special wastes.

Paint Spill. During the flood, a drum of paint had spilled near the southwest portion of the site. Because there was a possibility of another flood peak occurring, USCG personnel collected the silt with visible paint and put it in drums. The Black & Veatch team collected and analyzed a sample of this material and determined that it could be disposed of as special wastes.

Hazardous Material Storage Building. A hazardous material storage building is located near the southwest portion of the facility. During the waste motor oil cleanup, some soil with oily material that appeared to have different characteristics than the waste motor oil was identified. This material was collected and containerized separately from materials with waste motor oil. It appeared that the oily material may have originated from the hazardous waste storage building. When this building was opened, approximately 28 drums were inside, of which approximately 8 drums were overturned. An HNu photoionizing detector was used to monitor the air near the doorway, and readings as high as 250 to 300 parts per million were recorded. The building was allowed to ventilate. After three weeks, the HNu readings had decreased to background levels. The drums were then moved from the building and staged nearby, with the exception of two drums. These two drums were bulging, but not leaking, and were left inside the building. The liquids in the drums included gasoline, waste oil, waste paints, and trichloroethylene (1 drum). The silt from the floor of the building was collected. A sample of the oily silt material collected from outside the storage building was analyzed. This material, along with the material from the floor of the building, was disposed of as a special waste.

Decontamination Water. During the cleanup of the waste motor oil, water was used to wash down some of the structures that had oil on them and to decontaminate equipment used for the cleanup activities. This decontamination water was collected, and a sample of the water was analyzed. Based on the laboratory analysis, the decontamination water was disposed of in the Metropolitan St. Louis Sewer District (MSD) sanitary sewer.

Liquids from Lift Station. As part of the examination of the facilities onsite, an onsite lift station had to be accessed. The lift station contained liquids that were potentially contaminated based on a visible thick, clear oily sheen, a strong chemical odor, and the presence of several floating unlabelled containers. These liquids needed to be removed in order for personnel to enter the lift station. These liquids were pumped from the lift station and containerized onsite. Based on analytical results of a sample of the liquid, these liquids were disposed of in the MSD sanitary sewer.

Liquids from Pit and Air Tank in Industrial Building. During the cleanup activities in the industrial building, the USCG identified a condensate pump pit and an air tank that had oily liquids in them. These liquids were collected and containerized in drums. The Black & Veatch team collected a sample of each of these liquids and analyzed the samples to determine how to properly dispose of the liquids. These liquids were originally analyzed for only TPH and BETX because the source of the liquids appeared to be known. Additional analyses were performed, as requested by the disposal facility, for chlorine in water, flash point, and PCBs in water. The liquids were disposed of at a solvent recycling facility.

Monitoring Well Restoration

There are nine existing monitoring wells onsite that were installed to monitor groundwater contamination from adjacent properties. The Black & Veatch team inspected these wells and measured the groundwater levels and the total depth of the wells. By comparing the well depths measured after the flood with the depths recorded when the wells were installed, it was determined that silt had accumulated in the bottom of some of the wells as a result of the flooding. There was also physical damage to the covers of most of the wells. A technical specification was prepared by the Black & Veatch team for the redevelopment and repair of these wells.

Schedule

Following flooding, it is imperative that cleanup activities be conducted as quickly as possible. Mud and silt that remain after flooding are much easier to clean up while they are still wet. Wet building materials need to be removed or dried out before they begin to mold and mildew, and food or other organic matter must be disposed of before it begins to rot.

Accordingly, the USCG conducted the cleanup of Base St. Louis on an expedited schedule. Before the flood waters had even subsided, the USCG had obtained guidance from regulatory officials on the requirements for cleanup and had contracted the cleanup crews and technical support needed to perform the cleanup activities. An initial inspection was performed immediately after the flood waters had receded from most of the site. The initial cleanup of the site, including the collection of potentially contaminated soil and silt, was completed within approximately two weeks of when the flood waters subsided. Utilities were reestablished within a month of the recession of the water.

Conclusions

Based on the flooding and subsequent cleanup activities at the USCG Base St. Louis, the following procedures are recommended for dealing with environmental concerns associated with flooding:

- Implement preventative measures before flooding occurs, such as removing hazardous materials stored at the site and turning off utilities.
- Contact regulatory agencies for guidance on handling the waste materials generated due to flooding and to find out if any special procedures are being instituted as a result of the flooding.
- Expedite cleanup efforts to minimize further damage and health risks.
- Be aware of inherent health and environmental risks associated with flooding.
- Evaluate potentially hazardous building materials that could be disturbed or require special handling during flood cleanup.
- Collect, test, and properly dispose of any potentially contaminated materials.

By implementing these steps, the USCG Base St. Louis was successfully cleaned up with minimal risks to health and the environment.

The ICR; $130 Million of Water Quality Monitoring

Albert Ilges[1], Fredrick W. Pontius

Abstract

To develop a Disinfectant/Disinfection By-products (D/DBP) Rule, EPA used a formal regulation negotiation (reg-neg) process which included representatives from water utilities, state and local agencies, environmental groups, consumer groups, and EPA. This reg-neg process resulted in agreement to propose three rules in 1994, 1) an information collection rule (ICR), 2) an "interim" enhanced surface water treatment rule (ESWTR), and 3) Stage 1 of a two stage D/DBP regulation. A "final" ESWTR and a Stage 2 D/DBP Rule will be negotiated based on the $130M of data that over 1,500 utilities will collect to comply with the ICR. A national database will be created to store, analyze, and allow public access to all the data collected under the ICR.

Introduction

The ICR was developed to obtain both microbial and DBP occurrence, exposure, and treatment data and requires a segment of public water utilities to invest an estimated $130M over the next 3.5 years to collect this data . The rule has three major components which require, 1) microbial monitoring, 2) DBP monitoring, and 3) bench- and pilot-scale research on DBP precursor removal.

The ICR data would form the basis by which utilities could establish levels of treatment that properly control microbial risk while complying with new D/DBP regulations. To this end, additional data collection requirements will include information on the occurrence of disinfectants, DBPs, potential surrogates for DBPs, source-water and within-treatment conditions that affect DBP formation, and

---

[1]Project Manager, AWWA Research Foundation, 6666 W. Quincy Ave., Denver, CO 80235

the treatability of DBP precursor removal (to be used in developing Stage 2 of the D/DBP rule).

ICR Component Details

Microbial Monitoring - The purpose of microbial monitoring is to provide data that can be used to:

- Assess pathogen occurrence through monitoring of *Giardia, Cryptosporidium, E. coli* or fecal coliforms, and "total culturable viruses";
- Improve understanding of microbial health risks and treatment effectiveness of pathogen removal;
- evaluate the adequacy of the SWTR's requirements for microbial removal;
- determine whether relationships exist among the occurrence, removal, or both of protozoan pathogens, e.g., *Giardia* and *Crypto*, and viruses, and
- evaluate whether different levels of treatment should be required based on source-water quality.

Surface water (SW) systems serving >100,000 (~233 systems) must monitor influent water to each plant for *Giardia* cysts, *Crypto* oocysts, viruses, fecal coliforms or *E. coli*, and total coliforms monthly for 18 months. If during the first 12 months any pathogen exceeds one per liter, the finished water must also be tested for the entire set of pathogens and indicator organisms during subsequent samplings. (Note - virus monitoring may be avoided if specified water quality conditions are met).

Monitoring requirements differ for the 700-plus SW systems serving 10,000 to 100,000 and are NOT required to monitor for viruses or to take finished water samples if raw-water pathogens exceed 1/L. Monitoring is required every two months for 12 months. Systems serving <10K are not required to monitor. Microbial monitoring is estimated to impact 1,725 SW plants serving >10K and cost $11.76M.

EPA has solicited public comment on, and is considering monitoring requirements for *Clostridium perfingens* and coliphages (viruses that infect *E. coli*).

DBP Monitoring - A two staged D/DBP rule was proposed by the reg-neg team due to the lack of data to address most of the DBPs on EPA's Priority List. Stage 1 sets MCLs for the following DBPs: trihalomethanes (THMs), haloacetic acids (HAAs), chlorite and bromate. EPA believes that DBPs not addressed in Stage 1 would be "controlled" if 1) systems met the MCLs for THMs and HAAs, and 2) conventional treatment plants implemented optimized coagulation to remove as much organic material as possible before disinfection, thereby minimizing the formation of all DBPs. Total organic carbon (TOC) has been designated as the surrogate for organic precursor material.

Stage 2 of the D/DBP rule will be formulated based on ICR field data designed to, 1) characterize source-water parameters that influence DBP formation, 2) determine the concentrations of DBPs in drinking water, 3) refine models for predicting DBP formation based on treatment and water quality parameters, and 4) establish cost effective monitoring requirements that are protective of public health. Ground Water (GW) and SW systems >100K are required to conduct monitoring as specified in the ICR in addition to reporting treatment plant operational data. (Note - systems using alternative disinfectants have disinfectant specific DBP monitoring requirements). GW systems serving 50-100K must monitor TOC levels at the distribution entry point but are not required to monitor for DBPs. DBP monitoring will affect some 292 SW and GW systems >100K. Total DBP monitoring cost is estimated at $56.53M, ranging from $26.5K to $50K per treatment site.

Another aspect of the DBP monitoring requirements includes treatment process information in order to characterize the various forms of treatment in current use by systems >100K. The data will be used to evaluate options available to large utilities to monitor and reduce DBP formation. This data, once in the national database, would be used to upgrade the Water Treatment Plant (WTP) Model (Harrington, et. al. 1992) (which was used in development of D/DBP rule Stage I) by including additional processes, prediction of other DBPs, and to better calibrate the model. Utilities must report detailed treatment process information as outlined in the rule. Reporting of process parameter for ~1,725 SW treatment plants >10K related to microbial treatment and the related DBP formation data from ~440 plants >100K is estimated at $3.88M.

EPA will develop software for utilities to report all required monitoring data and treatment process information. The national database would contain all data reported to EPA and be available to the public. Processed database output would target the requirements being considered for Stage 2 of the D/DBP rule and the ESWTR. Examples of what could be extracted from the database are, 1) the national distribution of bromide, TOC, factors that affect DBP formation, etc., 2) distribution of HAAs, chloral hydrate, etc. in distribution system waters, 3) treatment processes and operating conditions associated with minimum DBP levels, 4) bromate concentrations produced under different ozonation conditions.

Bench-/Pilot- Scale Testing - SW systems >100K and GW systems >50K must conduct bench or pilot studies on DBP precursor removal using GAC or membrane filtration unless certain water quality conditions are met or if their full-scale use is already in place. The purpose of these studies are, 1) obtain more information on the cost effectiveness of GAC and membranes for removing DBP precursors and reducing DBP levels, and b) to decrease the time that systems would need to install such full-scale technology, if required under Stage 2 of the D/DBP rule.

SW systems could be exempted from bench/pilot studies by one of two options, 1) systems using chlorine had an annual average of $<40\mu g/L$ for total THMs and $<30\mu g/L$ for total HAAs, or 2) source-water TOC before disinfection is <4.0 mg/L based on a one year monthly average. GW systems would be exempted if TOC in their finished-water is <2.0mg/L based on an average of monthly measurements for one year.

GAC bench-scale tests - Rapid small-scale column tests using at least two empty-bed contact times will be used for the GAC bench-scale tests. Tests are to be conducted quarterly for one year. Should results from the first quarter indicate GAC to be a poor treatment candidate according to rule guidelines, the last three quarters would be used to bench-test one membrane.

Membrane bench-scale tests - A minimum of two membranes with nominal molecular weight cutoffs of <1,000 are to be tested quarterly for one year.

GAC pilot-scale tests - These studies are to be continuous flow using columns with a minimum ID of 2 inches and incorporate at least two empty bed contact times. The hydraulic loading rate and GAC particle size are to be representative of those in full-scale practice.

Membrane pilot-scale tests - Continuous-flow tests will use membrane modules with a minimum 4.0 inch diameter and must be designed to assess flux loss. Testing will be conducted throughout the year and be of sufficient length to evaluate seasonal variations.

All pilot- and bench-scale testing have extensive water quality analyses and design reporting requirements. Simulated Distribution System (SDS) testing with chlorine is also required of all bench- and pilot-scale studies which must be analyzed for THMs, HAAs, total organic halide (TOX), and chlorine demand. The total costs for pilot and bench studies is put at between $45M and $76M. Bench-scale cost/facility is estimated at $150K and at $750K/facility for pilot studies.

The information reported here is based on the Proposed Rule published in the Federal Register on Thursday February 10, 1994. Copies of Proposed and Final Rules may be obtained through the Safe Drinking Water Hot Line (SDWHL) @ 1-800-426-4791. (The Water Treatment Plant model is also available through the SDWHL.)

Removal Of Chromium From Groundwater
Using Permeable Barriers: An Aquifer Simulation Study

Mark D. Schmidt, Stephen P. Shelton[1]

Abstract

Previous efforts to remediate groundwater contaminated by chromium-bearing industrial wastes have involved post-extraction methods, whereby groundwater is pumped to the surface treated and returned to the aquifer. This practice has proven effective for removing soluble pollutants. However, it is often costly and labor intensive and requires treating large volumes of water. Also, institutional obstacles such as ground and surface water discharge permits and groundwater rights must be considered.

An alternative to conventional remediation methods is the in situ permeable barrier process, which intercepts soluble contaminants from solution but allows groundwater to flow through. Trench-based barriers, backfilled with reactive media, result in the direct adsorption of chemical species or the oxidation or reduction of chemical species followed by precipitation.

Laboratory studies were conducted to determine the technical feasibility of using trench-based media to remove chromium from groundwater. In batch tests various doses of candidate media and 10g of silica sand were added to glass vials containing a chromium solution. Candidate media included powder-activated carbon, ferric oxide and agricultural limestone. Adsorption isotherms were plotted from batch test results. In aquifer simulation tests a chromium-containing solution was passed through an aquifer simulation model containing silica sand and a vertical barrier of candidate media.

Removal of soluble hexavalent chromium (Cr(VI)) to concentrations less than the maximum contaminant level (MCL) for total chromium in drinking water (0.05 mg/l) was demonstrated with all candidate media in the aquifer simulation model. Adsorption appeared to be the principle mechanism of removal for all candidate media considered.

Information gained from experience with the physical model was used in developing a computer generated, 1-D solute transport model to predict the movement of hexavalent chromium in an aquifer system.

[1]Mr. Schmidt is a graduate student in the Department of Civil Engineering, University of New Mexico, Albuquerque, NM 87131. Dr. Shelton is Interim Director of the Army Environmental Policy Institute, P.O. Box 6569, Champaign, IL 61826

Introduction

Discharge of chromium-bearing industrial wastes has contaminated groundwater at numerous United States sites (Griffen et al., 1977). Current technologies for removing chromium from groundwater use post extraction techniques, whereby groundwater is pumped to the surface, treated and returned to the aquifer. These methods are often costly and labor intensive, and require treating large volumes of groundwater. Furthermore, they may not achieve long term remediation goals. Also, institutional obstacles such as ground and surface water disposal permits and groundwater rights must be considered. These challenges have intensified the need to examine alternative strategies, such as in situ methods, for remediating contaminated groundwater.

The purpose of this study is to consider an in situ method, a permeable barrier, as a feasible alternative to conventional remediation strategies for removing chromium from groundwater. A permeable barrier consists of a chemically reactive media placed in a trench downgradient from a contaminant plume. It allows water to pass through but can prevent migration of pollutants. This approach is largely passive, in that a barrier functions with little or no maintenance for long periods.

Chromium is an essential nutrient to animals and plants. However, it can harm humans and the environment at high concentrations. The U.S. EPA specifies the maximum contaminant level (MCL) for chromium in drinking water as 0.05 mg/l (Federal Register, 1992). Of the various forms in which chromium can exist, hexavalent chromium (Cr(VI)) is considered the most toxic. Cr(VI), which also is soluble and very mobile in groundwater, is generally indicative of industrial waste contamination (Pattersen, 1975). Its mobility allows Cr(VI) to migrate long distances downgradient from a contaminant source.

In natural aquatic systems, Cr^{3+} and Cr^{6+} are the most common chromium species (Artiole and Fuller, 1979, Richard and Bourg, 1991). Under oxidizing conditions, aqueous chromium exists in an anionic form as $HCrO_4^-$ or CrO_4^- depending upon the pH. Under reducing conditions, aqueous chromium exists predominantly in a cationic form as Cr^{3+}. Below a pH of 3.6, the Cr^{3+} cation prevails. At a pH above this, various insoluble oxides, such as $Cr(OH)^{2+}$, $Cr(OH)_3$ and $Cr2O_3(s)$, are found (Rai et al., 1986, 1987).

Adsorption has long been used as a mechanism for removing Cr(VI) from solution. Huang and Wu (1977) used powder-activated carbon in batch studies while Griffen et al. (1977), James and Bartlett (1983), and Rai et al. (1986) examined chromium adsorption by various clay soils, soils containing $Fe(OH)_3$, and soils containing hematite, respectively. Artiole and Fuller (1979) examined agricultural limestone barriers for chromium attenuation in various soils. Hsia et al. (1992) investigated adsorption of hexavalent chromium by amorphous iron oxide and Stollenwerk and Grove (1985) considered the characteristics of adsorption of hexavalent chromium on aquifer alluvium.

Cr(VI) normally assumes the anionic form (i.e. CrO_4^- or $HCrO_4$-) in water. It is believed that two primary mechanisms affect adsorption of anions. These are specific and non-specific adsorption which control the process at solid solution interface (Hingston et al., 1967, Stollenwerk and Grove, 1985). Specific adsorption involves ligand exchange of mineral surface ions with negatively charged anions while nonspecific adsorption, refers to the balancing of the anions on the diffuse double layer opposite that of the positively charged adsorbent surface (Jury et al., 1991).

In addition to adsorption, Cr(VI) also may be affected by other chemical reactions. Because Cr(VI) is a strong oxidizing agent in acidic solutions, it can be reduced to the Cr(III) state if electrons are available in the solution. Both organic and inorganic compounds can donate electrons for this reaction. In groundwater systems, the ferrous ions, Fe(II), can provide these electrons. Fe(II) is found in magnetite and hematite, minerals common in soils and sediments. The reduction of Cr(VI) to Cr(III) by hematite is as follows (Rai et al., 1988).

$$3FeO + 6H^{+} + Cr^{6+}_{(aq)} = Cr^{3+}_{(aq)} + 3Fe^{3+}_{(aq)} + H_2O$$

In neutral to alkaline solutions, the end products are often hydroxide solids such as $Cr(OH)_3(s)$, $(FeCr)(OH)_3(s)$, and $Fe(OH)_3(s)$, (Rai et al., 1988). This indicates that the mobility of Cr(VI) may be limited due to reduction by Fe(II) minerals and resulting precipitation of Cr(III) hydroxide solids.

Laboratory experiments presented here examine the effectiveness of selected barriers for removing Cr(VI) in a simulated groundwater system.

Materials

Powder-Activated Carbon

A commercial activated carbon, Darco G-60 (-100 mesh), was used. This carbon has a surface area of 410 m^2/g and for the zero point of charge (ZPC) a pHZPC = 6.2 (Corapcioglu and Huang, 1987). The carbon was used as received with no additional purification.

Ferric Oxide

Ferric Oxide (Fe_2O_3) is common rust taken from weathered steel. The ferric oxide used was 200 mesh, Baker analyzed, reagent-grade. It was used as received with no additional purification.

Agricultural Limestone

Unsieved, commercial-grade limestone (53.94% $CaCO_3$ and 45.39% $MgCO_3$) from Bonham, Texas, developed for agricultural soil application, was used. When used as a leachate barrier, it was compacted to an average bulk density of 1.79 g/cm^3 at a moisture content of approximately 7%.

Silica Sand

Unsieved, washed silica sand from Wedron, Ill., was used as an aquifer medium and as a mixture with the barrier material. Table 1 shows the grain size distribution. The sand was compacted in the model to an average bulk density of 1.69 g/cm^3. Estimation of the pH for the silica sand was pHZPC = 2.0 (Stumm and Morgan, 1978).

Sieve Size	Percent Passing
#25	100
#50	45
#100	12
#200	2.3

Table 1. Grain Size Distribution of Silica Sand

Influent Solution

Simulated groundwater was prepared by filling a 200 liter nalgene tank with deionized water (>16.0 µmohm) and enough stock potassium chromate to result in a 1 mg/l solution. The solution was acidified to a pH of 2.5 (±.1) with 0.2N H_2SO_4 to simulate a highly acidic waste stream. The containers were stirred to ensure that the solutions were completely mixed.

Methods

Batch Studies

Laboratory batch studies were conducted to determine the maximum adsorption capacity of various media for removing chromium from solution and to determine the effects of pH on adsorption and the ionic form of chromium in solution (e.g. Cr(VI) vs. Cr(III)).

A continuous mixed batch system was employed for all batch adsorption experiments. Twenty ml aliquots of dionized water and chromium stock were piped into a series of 40 ml, borosilicate glass vials. After adjusting the initial pH with 0.2N H_2SO_4 or $Ca(OH)_2$, various doses of candidate media and 10 grams of silica sand were added to each solution. Blank vials, without candidate media, were also prepared and used to calibrate initial concentration. The vials were sealed with plastic caps, hand shaken, and placed on a spinner for 24 hours (±4 hrs). At the end of the reaction period, the samples were hand shaken and filtered through 45 µm filter paper to separate the media from the supernatant. Immediately after the filtration, the equilibrium pH, total chromium in solution and state of the chromium present were determined in the supernant. The amount of chromium adsorbed was determined from the difference between the concentrations before and after the reaction.

The first set of batch experiments was conducted to determine the maximum adsorption capacity of various media for removing chromium from solution. The chromium concentration and pH were kept constant while the amount of media was varied. The second set involved adding constant doses of media to a chromium solution while observing the effects of pH on adsorption and the ionic form of the element in solution (e.g. Cr(VI) vs. Cr(III)).

Laboratory Aquifer Simulation Model

An experimental aquifer model was designed and built to simulate migration of a contaminant in an unconfined, saturated porous media. This model is an intermediate step between more common column studies and actual field investigations. Figure 2 illustrates laboratory aquifer model profiles.

Figure 2. Laboratory Aquifer Simulation Model

The aquifer model was constructed of 0.5-inch Lexan polycarbonate. The Lexan material was clear and allowed for easy observation inside the model. The polycarbonate was assumed to have no effect on oxidation/reduction or retention of the chromium. Studies confirmed these assumptions. The overall dimensions of the aquifer model are 1.83 meters in length, 1.22 meters in height and .10 meters wide. These dimensions were selected to minimize the effect of any reactions between the chromium and polycarbonate and to avoid wall effects.

One constant head tank, 0.2-meters wide, is located at each end of the model. The sand and media mixture, which was compacted in approximately 8-cm lifts (with care being taken to avoid layering), was supported inside the box by 10 cm wide drilled Lexan panels. Pyrex fiberglass (Silver 8 Micron) was used as a filter between the media and the drilled Lexan panels at each end of the media bed.

A gradient was established across the bed by lowering the downgradient constant head tank while maintaining a constant flow rate in the upgradient constant head tank. A flow rate of 40 ml/min was maintained in the laboratory model and gave good approximations for actual groundwater velocities.

In the first experiment, 10% PAC was thoroughly mixed with silica sand at a moisture content of approximately 7%, then a barrier of limestone and the 10% PAC/ silica sand was placed vertically across the flow path. The limestone was not mixed with the silica sand but was wetted to a moisture content of approximately 7% and compacted separately. The media was compacted to an average bulk density of approximately 1.8 g/cm^3. The influent solution concentration was 1.0 mg/l Cr at a pH of 2.5. It was fed to the laboratory model at a flow rate of 40 ml/min. The initial calculations predicted a breakthrough on day 13.

A second experiment, using 10% ferric oxide instead of 10% PAC, was performed using the same methodology as Experiment 1. The initial calculations predicted a breakthrough on day 13.

Analytical Methods

All samples collected from the model and in batch studies were placed in 40-ml, borosilicate glass vials and capped prior to analysis. The pH measurements were conducted using a pH reference electrode and a Corning pH and Temp Meter 4. Total chromium and other inorganic metals in the sample solutions were analyzed by the Atomic Adsorption Spectroscopy procedure (Standard Methods for the Examination of Water and Waste water, 1991). The amount of hexavalent chromium was determined by the diphenyl-carbazide procedure (Standard Methods for the Examination of Water and Wastewater,1991).

Results and Discussion

Batch Studies

PAC and silica sand mixture

Figure 3 gives the Freundlich adsorption isotherm for 10 mg/l Cr(VI) on PAC. The resulting equilibrium equation and isotherm parameters are derived from the "best fit" line determined by regression analysis of the data and are presented in Table 2. A linear distribution was observed with little scatter of data ($r^2 = 0.981$). The data demonstrate that Darco G-60 (-100 mesh) Powder Activated Carbon is an effective adsorbent for Cr(VI) removal.

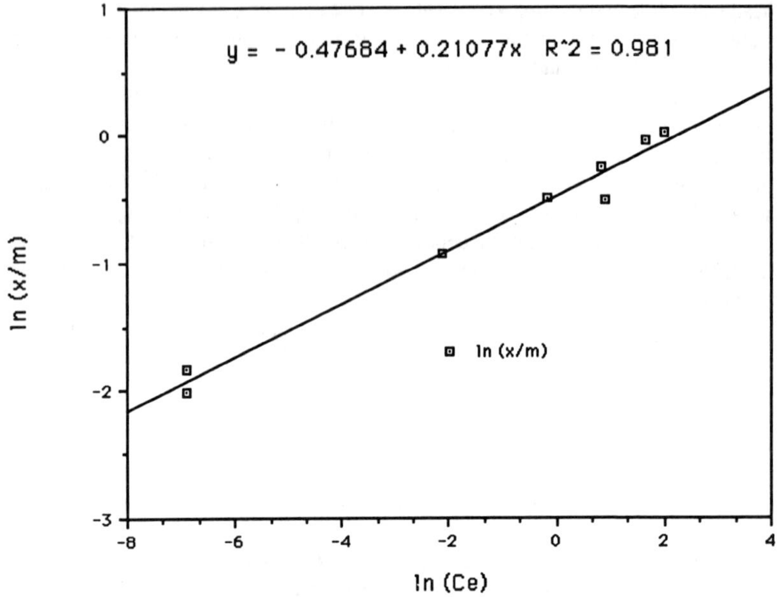

Figure 3. Freundlich Adsorption Isotherm for 10 mg/l Cr(VI) on PAC

K = 0.624
1/n = 0.21
q = x/m = 0.624 C 1/0.21

Table 2. Freundlich Adsorption Isotherm Parameters for Cr(VI) on PAC

Results indicate that adsorption density of Cr(VI) increases slightly with increasing pH to a maximum value near a pH of 5 and then declines with increasing pH. This finding is in agreement with other work (Huang and Wu, 1977). Results also demonstrate that Cr(VI) is reduced to Cr(III) in the presence of activated carbon but is dependent on pH. At pH above approximately 4, only Cr(VI) was found in the effluent. In the absence of activated carbon, all chromium remained as Cr(VI).

Ferric oxide and silica sand mixture

Figure 4 gives the adsorption isotherm for 10 mg/l Cr(VI) on ferric oxide (Fe_2O_3) and limestone.

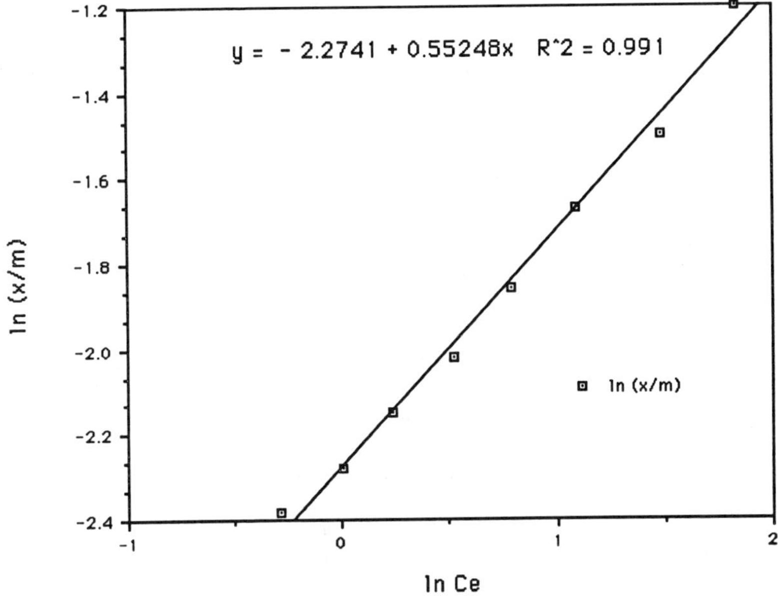

Figure 4. Adsorption Isotherm for 10 mg/l Cr(VI) on Ferric Oxide (Fe_2O_3) and Limestone

The resulting equilibrium equation and isotherm parameters are derived from the "best fit" line determined by regression analysis of the data and are presented in Table 3. A linear distribution was observed with little scatter of data ($r^2 = 0.991$). The data demonstrate that ferric oxide (Fe_2O_3) is also an effective adsorbent for Cr(VI) removal. However, by comparing Freundlich constants for values of K, it is seen that Darco G-60 (-100 mesh) Powder Activated Carbon is a more effective adsorbent for Cr(VI) than Ferric oxide (Fe2O3).

$K = 0.102$
$1/n = 0.55$
$q = x/m = 0.624\ C\ 1/0.55$

Table 3. Freundlich Adsorption Isotherm Parameters for Cr(VI) on Fe_2O_3

The adsorption density of Cr(VI) remains fairly constant with increasing pH to a maximum value of a pH of 8 and then declines with increasing pH. Cr(III) was not found in the effluent. Cr(VI) may have reduced to Cr(III) in the presence of Fe_2O_3, with the resulting Cr(III) ions forming precipitate $Fe(OH)_3$. In the absence of Fe_2O_3, all chromium remained as Cr(VI).

Limestone and silica sand mixtures

Agricultural limestone has little to no adsorption potential for Cr(VI) and caused little to no reduction of Cr(VI) to Cr(III). The limestone, however, reacts with the influent solution, producing HCO^{3-} and H_2CO_3 with an increase in the OH^- concentrations, increasing pH.

In summary, the batch experiments on removing hexavalent chromium from solution indicate that:

1) Darco 60 Powdered Activated Carbon and ferric oxide are effective adsorbents for Cr(VI) removal, while agricultural limestone has no adsorption potential for Cr(VI) removal.
2) Adsorption density of Cr(VI) increases with increasing pH for PAC to a maximum value and then decreases with further increase in pH. Adsorption density of Cr(VI) remains fairly constant with increasing pH for ferric oxide up to pH = 8 and then decreases as pH increases above 8.
3) Adsorption data for PAC followed the Freundlich isotherm. The ferric oxide Freundlich isoterem was not linear; however, the ferric oxide/limestone mixture was well simulated by the Freundlich isotherm.
4) Darco G-60 (-100 mesh) Powder Activated Carbon is a more effective adsorbent for Cr(VI) than Ferric oxide (Fe_2O_3) when compared by the Freundlich isotherms.
5) Cr(VI) is reduced to Cr(III) in the presence of PAC, Fe_2O_3, and agricultural limestone, but only in low pH environments.

Laboratory Aquifer Simulation Model

Figure 5 presents the breakthrough curve for the aquifer simulation model for the first experiment. Chromium was first detected in the effluent approximately 19 days after the influent was spiked with the contaminant. This shows that all Cr(VI) was removed by the barrier material for the first pore volumes of solution. The Cr(VI) concentration of the effluent continued to increase over the next 58 days until the effluent concentration was equal to the influent concentration. The relatively high flow rate used in the experiment caused an unexpected long, drawn-out breakthrough curve. The capacity of the barrier material was exhausted at approximately 66 days. Tests revealed that all chromium in the effluent was in the Cr(VI) state. Small percentages of Cr extracted from the silica sand and limestone give evidence that PAC was the only material directly responsible for removal of Cr from solution. Because almost all Cr removed was Cr(VI), it appears that adsorption of Cr(VI) was the controlling removal mechanism and that little reduction of Cr(VI) to Cr(III) took place.

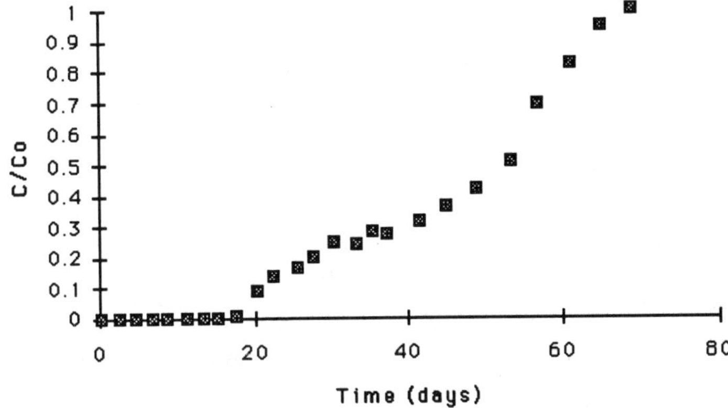

Figure 5. Breakthrough Curve for Cr(VI) on PAC silica sand/limestone

Figure 6 presents the breakthrough curve for the second experiment. Chromium was first detected in the effluent approximately 10 days after the influent was spiked with the contaminant. Although ferric oxide proved to be less effective at adsorbing chromium than PAC, the time of exhaustion was actually longer. Tests revealed that all the chromium in the effluent was Cr(VI). The long, drawn-out breakthrough curve may be due to the relatively high flow rate used in the experiment; it is possible that Cr(VI) may have reduced to Cr(III) in the presence of Fe_2O_3, with the resulting Cr(III) ion forming precipitate $Fe(OH)_3(s)$.

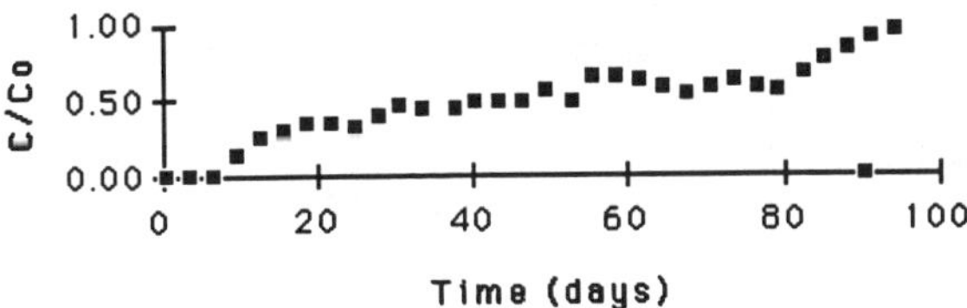

Figure 6. Breakthrough Curve for Cr(VI) on ferric oxide/silica sand/limestone

Computer Modeling

The one-dimensional diffusion-convection equation was used to simulate contaminant transport through porous media (Selim, et al., 1989). This equation is written as:

$$\frac{D\partial^2 C}{\partial x^2} - \frac{V\partial C}{\partial x} = \frac{\partial C}{\partial t}\ \frac{(1+P_d\ K_d)}{\theta}$$

where D is the dispersion coefficient (m2/s), V is average velocity(m/s), C is the equilibrium concentration (mg/l), P_d is the bulk density of the aquifer material, K_d is the distribution coefficient, and θ is the porosity. The aquifer model can be considered a uniform groundwater flow field in the x-direction. The contaminant source is located at the left of the flow field, where a constant concentration leaches into the groundwater. The concentration at this point is Co=constant for time greater than zero. For time less than or equal to zero, the concentration throughout the flow field is zero.

In mathematical terms, the initial and boundary conditions are

Initial Conditions: $C(x,0)=0$ for all x
Boundary Condition 1: $C(0,t)=Co$ for $t > 0$
Boundary Condition 2: $C(0,t)=0$ for $t > 0$

This problem will be solved using a numerical computer model.

Conclusions

1) Because chromium occurs in both anionic and cationic forms in water, it is important to characterize the oxidation state so a suitable geochemical media can be engineered to control their migration.
2) Batch experiments provided conservative estimates for adsorption of Cr(VI) from solution in predicting breakthrough.
3) Based on the aquifer simulation model investigation, either a PAC and limestone barrier or a ferric oxide and limestone barrier would be successful in a trench-based permeable barrier.
4) Increasing the hydraulic retention time will result in better utilization of the barrier media.
5) Adsorption data alone may not adequately predict the movement of chromium in groundwater. Reduction and/or precipitation data may also be required.
6) Barrier mixture maintains permeability and structural integrity without deteriorating over time.

References

Artiola, J. and Fuller, W.H., "Effect of Crushed Limestone Barriers on Chromium Attenuation in Soils," Journal of Environmental Quality, 8(4), pp. 503-510, (1979).

Corapcioglu, M.O. and Huang, C.P., "The Adsorption of Heavy metals on to Hydrous Activated Carbon," Water Research, 21(9), pp.1031-1044, (1987).

Federal Register, 55, 61,11803-11815, (1990).

Griffin, R.A., Au, A.K., and Frost, R.R., "Effect of pH on Adsorption of Chromium from landfill-Leachate by Clay Minerals," Journal of Environmental Science and Health , A12(8), pp.431-449, (1977).

Hingston,F.J., Atkinson,R.J.,Posner,A.M.,and Quirk,J.P.,"Specific Adsorption of Anions,"Nature,215, pp. 1459-1461, (1967).

Hsia, T.H., Lo, S.L., and Lin, C.F., "Interactions of Cr(VI) with Amorphous Iron Oxide: Adsorption Density and Surface Charge," Water Science and Technology 26(1-2), pp.181-188, (1992).

Huang, C.P. and Wu, M.H., "The removal of Chromium from dilute Aqueous Solution By Activated Carbon," Water Research, 21, pp.673-679, (1977).

James, R.J., and Bartlett, R.J., "Behavior of Chromium in Soils: VII. Adsorption and Reduction of Hexavalent Forms," Journal of Evironmantal Quality, 12(2), pp.177-181, (1983).

Jury, W.A., Gardner, W.R., and Gardner, W.H., Soil Physics, John Wiley and Sons, Inc., New York, (1991).

Patterson, J.W., Wastewater Treatment Technology, Ann Arbor Science Publishers, Inc., Ann Arbor, Mich., (1975).

Rai, D. et al.,Geochemical Behavior of Chromium Species, Interim Report, EPRI EA-4544, E.R.P.I., Palo Alto, CA.,(1986).

Rai, D. et al., Chromium Reactions in Geologic Materials, Interim Report, EPRI EA-5741, E.R.P.I., Palo Alto, CA.,(1988).

Richard, F.C. and Bourg, A. C., "Aqueous Geochemistry of Chromium: A Review," Water Resources, 25(7), pp. 807-816, (1991).

Selim, H.M., Amacher, M.C. and Iskandar, I.K., "Modeling the Transport of Chromium (VI) in Soil Columns," Soil Science Society of America, 53, pp. 996-1004, (1989)

Stollenwerk, K.G. and Grove, D.B., "Adsorption and Desorption of Hexavalent Chromium in an Alluvial Aquifer Near Telluride, Colorado," Journal of Evironmantal Quality, 14(1), pp.150-155, (1985).

Stumm, W. and Morgan, J.J., Aquatic Chemistry, John Wiley and Sons, New York, 1981.

Zachara, J.M., Girvin, D.C., Schmidt, R.L., Resch, C.T., "Chromate Adsorption on Amorphous Iron Oxyhydroxide in the presence of Major Groundwater Ions," Environmental Science and Technology, 21(6), pp.589-594, (1987).

Subject Index

Page number refers to first page of paper

Author Index

Page number refers to the first page of paper